Peptide and Peptidomimetic Therapeutics

From Bench to Bedside

Peptide and Peptidomimetic Therapeutics

From Bench to Bedside

Edited by

Nir Qvit
The Azrieli Faculty of Medicine in the Galilee, Bar-Ilan University, Safed, Israel

Samuel J.S. Rubin
Department of Medicine, School of Medicine, Stanford University, Stanford, CA, United States

ELSEVIER

ACADEMIC PRESS
An imprint of Elsevier

Academic Press is an imprint of Elsevier
125 London Wall, London EC2Y 5AS, United Kingdom
525 B Street, Suite 1650, San Diego, CA 92101, United States
50 Hampshire Street, 5th Floor, Cambridge, MA 02139, United States
The Boulevard, Langford Lane, Kidlington, Oxford OX5 1GB, United Kingdom

ISBN: 978-0-12-820141-1

Publisher: Stacy Masucci
Acquisitions Editor: Peter B. Linsley
Editorial Project Manager: Moises Carlo P. Catain
Production Project Manager: Maria Bernard
Cover Designer: Mark Rogers

Typeset by MPS Limited, Chennai, India

Contents

Section 3
Drug discovery

List of contributors

Rina Aharoni Department of Immunology, Weizmann Institute of Science, Rehovot, Israel

Zulfiqar Ahmad Department of Biochemistry, Kirksville College of Osteopathic Medicine, A.T. Still University, Kirksville, MO, United States

Suhail Akhtar Department of Biochemistry, Kirksville College of Osteopathic Medicine, A.T. Still University, Kirksville, MO, United States

N. D. Prasad Atmuri Department of Chemistry, University of British Columbia, Vancouver, British Columbia, Canada

Carmen Avendaño Departamento de Química Orgánica y Farmacéutica, Universidad Complutense, Madrid, Spain

Ivneet Banga Department of Bioengineering, University of Texas, Arlington, TX, United States

Shai Bel Azrieli Faculty of Medicine, Bar Ilan University, Safed, Israel

Łukasz Berlicki Department of Bioorganic Chemistry, Wrocław University of Science and Technology, Wrocław, Poland

Yair Blumberg The Azrieli Faculty of Medicine in the Galilee, Bar-Ilan University, Safed, Israel

Carine Bourguet NMX Research and Solutions Inc., Laval, Québec, Canada

Marlon Henrique Cardoso S-Inova Biotech, Pós Graduação em Biotecnologia, Universidade Católica Dom Bosco, Campo Grande, Mato Grosso do Sul, Brazil; Centro de Analises Proteomicas e Bioquimicas, Programa de Pós-Graduação em Ciências Genômicas e Biotecnologia, Universidade Católica de Brasília, Brasília, Distrito Federal, Brazil; Instituto de Biociências (INBIO), Universidade Federal de Mato Grosso do Sul, Cidade Universitária, Campo Grande, Mato Grosso do Sul, Brazil

Samraggi Choudhury DBT-National Institute of Animal Biotechnology (DBT-NIAB), Hyderabad, Telangana, India

Sitan Diarra Département de Chimie, Université de Montréal, Montréal, Québec, Canada

Marvin N. Driessen Anesthesiology, University Hospital Muenster, Münster, Germany

Michael K. Dunn Ferring Research Institute, Inc., San Diego, CA, United States

Vandon T. Duong Department of Bioengineering, Stanford University School of Medicine, Stanford, CA, United States

Offir Ertracht The Cardiovascular Research Laboratory, Research Institute, Galilee Medical Center, Nahariya, Israel

Emer S. Ferro Pharmacology Department, Biomedical Science Institute, University of São Paulo, São Paulo, SP, Brazil; Department of Biological Regulation, The Weizmann Institute of Science, Rehovot, Israel

Octavio Luiz Franco S-Inova Biotech, Pós Graduação em Biotecnologia, Universidade Católica Dom Bosco, Campo Grande, Mato Grosso do Sul, Brazil; Centro de Analises Proteomicas e Bioquimicas, Programa de Pós-Graduação em Ciências Genômicas e Biotecnologia, Universidade Católica de Brasília, Brasília, Distrito Federal, Brazil

Tatiana Galochkina Université Paris Cité, INSERM, BIGR, Paris, France; Laboratoire d'Excellence GR-Ex, Paris, France

Sonu Gandhi DBT-National Institute of Animal Biotechnology (DBT-NIAB), Hyderabad, Telangana, India

Shikun Ge Department of Biology, Centre of Molecular and Environmental Biology, University of Minho, Braga, Portugal; Chinese-German Joint Laboratory for Natural Product Research, State Key Laboratory of Biological Resources and Ecological Environment of Qinba Areas, College of Biological Science and Engineering, Shaanxi University of Technology, Hanzhong, P.R. China

Jean-Christophe Gelly Université Paris Cité, INSERM, BIGR, Paris, France; Laboratoire d'Excellence GR-Ex, Paris, France

Mayara C.F. Gewehr Pharmacology Department, Biomedical Science Institute, University of São Paulo, São Paulo, SP, Brazil

Rujuta Ghorpade SVKM's NMIMS, Shobhaben Pratapbhai Patel School of Pharmacy and Technology Management, Mumbai, Maharashtra, India

Eric R. Gross Department of Anesthesiology, Perioperative and Pain Medicine, School of Medicine, Stanford University, Stanford, CA, United States

Wessal Hanout The Azrieli Faculty of Medicine in the Galilee, Bar-Ilan University, Safed, Israel

Tamia A. Harris Department of Dermatology, University of Texas Southwestern Medical Center, Dallas, TX, United States; Department of Immunology, University of Texas Southwestern Medical Center, Dallas, TX, United States

Adrienne Joseph Department of Dermatology, University of Texas Southwestern Medical Center, Dallas, TX, United States

Srinivasaraghavan Kannan Bioinformatics Institute, Agency for Science, Technology and Research (A*STAR), Singapore, Singapore

Abidemi Paul Kappo Department of Biochemistry, Molecular Biophysics and Structural Biology Group, Kingsway Campus Auckland Park, University of Johannesburg, Johannesburg, South Africa

Krishna Kumar Department of Chemistry, Tufts University, Medford, MA, United States; Department of Biomedical Engineering, Tufts University, Medford, MA, United States

Maushmi S. Kumar SVKM's NMIMS, Shobhaben Pratapbhai Patel School of Pharmacy and Technology Management, Mumbai, Maharashtra, India

Dilraj Lama Department of Microbiology, Tumor and Cell Biology, Karolinska Institutet, Solna, Stockholm, Sweden

Jolene L. Lau BioMarin Pharmaceutical, Novato, CA, United States

Jianguo Li Bioinformatics Institute, Agency for Science, Technology and Research (A*STAR), Singapore, Singapore; Singapore Eye Research Institute, Singapore, Singapore

Adi Litmanovich The Azrieli Faculty of Medicine in the Galilee, Bar-Ilan University, Safed, Israel; Galilee Medical Center, Nahariya, Israel

Jamie Liu Department of Chemistry, Tufts University, Medford, MA, United States

William D. Lubell Département de Chimie, Université de Montréal, Montréal, Québec, Canada

Lloyd Mabonga Department of Biochemistry and Microbiology, University of Zululand, KwaDlangezwa, South Africa

Subhasis Mahari DBT-National Institute of Animal Biotechnology (DBT-NIAB), Hyderabad, Telangana, India

Priscilla Masamba Department of Biochemistry, Molecular Biophysics and Structural Biology Group, Kingsway Campus Auckland Park, University of Johannesburg, Johannesburg, South Africa

J. Carlos Menéndez Departamento de Química Orgánica y Farmacéutica, Universidad Complutense, Madrid, Spain

Gassan Moady Cardiology Unit, Galilee Medical Center, Nahariya, Israel

Satheesh Kumar Nanjappan Department of Natural Products, National Institute of Pharmaceutical Education and Research, Kolkata, West Bengal, India

Ami Navon Department of Biological Regulation, The Weizmann Institute of Science, Rehovot, Israel

Ines Neundorf Institute for Biochemistry, Department of Chemistry, Faculty of Mathematics and Natural Sciences, University of Cologne, Cologne, Germany

Meital Nuriel-Ohayon Azrieli Faculty of Medicine, Bar Ilan University, Safed, Israel

Dmitriy Orlov Institute of Experimental Medicine, St. Petersburg, Russia

David Paul Department of Pharmaceutical Analysis, St. James College of Pharmaceutical Sciences (SJCOPS), Chalakudy, Kerala, India

Ron Piran The Laboratory for Diabetes and Regenerative Medicine, Azrieli Faculty of Medicine, Bar-Ilan University, Safed, Israel

William Farias Porto S-Inova Biotech, Pós Graduação em Biotecnologia, Universidade Católica Dom Bosco, Campo Grande, Mato Grosso do Sul, Brazil; Porto Reports, Brasília, Distrito Federal, Brazil

Nir Qvit The Azrieli Faculty of Medicine in the Galilee, Bar-Ilan University, Safed, Israel

Samuel J.S. Rubin Department of Medicine, School of Medicine, Stanford University, Stanford, CA, United States

Mussa Saad The Azrieli Faculty of Medicine in the Galilee, Bar-Ilan University, Safed, Israel; Cardiology Department, Ziv Medical Center, Safed, Israel

Deepshikha Shahdeo DBT-National Institute of Animal Biotechnology (DBT-NIAB), Hyderabad, Telangana, India

Olga Shamova Institute of Experimental Medicine, St. Petersburg, Russia

Kathleen M. Sicinski Department of Chemistry, Tufts University, Medford, MA, United States

Abhishek Singh Zoological Survey of India, New Alipore, Kolkata, West Bengal, India; Department of Forensic Biotechnology, National Forensic Sciences University, Gandhinagar, Gujarat, India

Worood Sirhan The Laboratory for Diabetes and Regenerative Medicine, Azrieli Faculty of Medicine, Bar-Ilan University, Safed, Israel

Shruti Surendran Biocon BMS R&D Center, Bengaluru, Karnataka, India

Simon Surprenant Institut de Recherche en Immunologie et Cancérologie, Université de Montréal, Montréal, Québec, Canada

Yaw Sing Tan Bioinformatics Institute, Agency for Science, Technology and Research (A*STAR), Singapore, Singapore

Mukesh Thakur Zoological Survey of India, New Alipore, Kolkata, West Bengal, India

Ekaterina Umnyakova Institute of Experimental Medicine, St. Petersburg, Russia

Özge Ünsal Department of Chemistry, Tufts University, Medford, MA, United States

Chandra S Verma Bioinformatics Institute, Agency for Science, Technology and Research (A*STAR), Singapore, Singapore; Department of Biological Sciences, National University of Singapore, Singapore, Singapore; School of Biological Sciences, Nanyang Technological University, Singapore, Singapore

Nana-Maria Wagner Anesthesiology, University Hospital Muenster, Münster, Germany

Zhipeng A. Wang The First Affiliated Hospital of USTC, Division of Life Sciences and Medicine, University of Science and Technology of China, Hefei, P.R. China; Division of Genetics, Department of Medicine, Brigham and Women's Hospital and Department of Biological Chemistry and Molecular Pharmacology, Harvard Medical School, Boston, MA, United States

Bryan J. Xie Department of Medicine, Stanford University School of Medicine, Stanford, CA, United States

Long Xu Department of Biology, Centre of Molecular and Environmental Biology, University of Minho, Braga, Portugal; Chinese-German Joint Laboratory for Natural Product Research, State Key Laboratory of Biological Resources and Ecological Environment of Qinba Areas, College of Biological Science and Engineering, Shaanxi University of Technology, Hanzhong, P.R. China

Vanessa O. Zambelli Laboratory of Pain and Signaling, Butantan Institute, São Paulo, São Paulo, Brazil; Department of Anesthesiology, Perioperative and Pain Medicine, School of Medicine, Stanford University, Stanford, CA, United States

Xiaoying Zhang Department of Biology, Centre of Molecular and Environmental Biology, University of Minho, Braga, Portugal; Chinese-German Joint Laboratory for Natural Product Research, State Key Laboratory of Biological Resources and Ecological Environment of Qinba Areas, College of Biological Science and Engineering, Shaanxi University of Technology, Hanzhong, P.R. China

Ji-Shen Zheng The First Affiliated Hospital of USTC, Division of Life Sciences and Medicine, University of Science and Technology of China, Hefei, P.R. China

Preface

Samuel J.S. Rubin[1] and Nir Qvit[2]

[1]*Department of Medicine, School of Medicine, Stanford University, Stanford, CA, United States,*
[2]*The Azrieli Faculty of Medicine in the Galilee, Bar-Ilan University, Safed, Israel*

From 1950 to 2008, the pharmaceutical industry produced over 1220 new molecular entities (NMEs) that have contributed dramatically to improving health and extending life expectancy by an average of 2 months each year [1,2]. Despite many improvements in science and technology, the process of drug discovery is extremely challenging, costing an average of 2.6 billion dollars, and requiring 12–15 years from discovery to approval of a drug, if approved. The number of NMEs approved by global regulators such as the US Food and Drug Administration (FDA), the European Medicines Agency, and Japan's Pharmaceutical and Medical Devices Agency has decreased per billion US dollars spent on research and development (R&D) [3]. Since 1950 the number of NMEs approved by the FDA per billion US dollars of R&D spending in the drug industry has halved approximately every 9 years [4]. Based on the observation that the productivity of R&D in the pharmaceutical industry has been falling exponentially, analyst Jack Scannell coined the term "Eroom's law." The name is an ironic play on Moore's law, which states that the number of transistors on an integrated circuit increases exponentially over time. Between 1950 and 2010, the R&D costs of the pharmaceutical industry increased nearly 100-fold, and the number of drugs approved per billion dollars spent has fallen 80-fold [5,6]. As a result, the pharmaceutical industry is exploring various approaches to decrease R&D expenses and increase productivity.

Historically, small-molecule drugs as low molecular weight compounds that can modulate biochemical processes to diagnose, treat, or prevent diseases have been the pillars of medicine and have played an important role in the evolution of drugs. Of the 1222 NMEs approved by the FDA from 1950 to 2008, over 90% (1103) were small-molecule drugs [2]. Small molecules not only dominate the number of approved drugs but have also been some of the biggest blockbusters in the history of the pharmaceutical industry. For example, atorvastatin (Lipitor) generated at least 9 billion dollars per year in revenue for Pfizer during its patent life from 2003 to 2011 [7]. Traditionally, the process of discovery for small-molecule drugs has involved identifying the active ingredient from traditional remedies or serendipitous findings, as with penicillin. Aspirin (acetylsalicylic acid) was identified as the active ingredient of the willow tree, a symbolic medicinal plant. Nearly 4000 years ago, the Babylonians identified the benefit of willow leaves to treat fever, pain, and inflammation [8]; the compound is also mentioned in ancient Egyptian scriptures (Ebers papyrus 1500 BCE) [9], and Hippocrates, "the father of modern medicine," recommended a brew from the willow tree bark to treat various pains (400 BCE) [10]. In 1763 Edmund Stone made an extract from the cortex of a willow tree, dispensed it to 50 patients, and demonstrated its property of fever reduction. Finally, in 1828 Johann Buchner isolated the active ingredient from the willow tree and named it salicin after the willow tree's Latin name.

Past pharmaceutical development was dominated by small molecules, such as penicillin and aspirin, since their pharmacokinetic properties are relatively easy to optimize, their production is uncomplicated, and their development and delivery are relatively straightforward. In addition, high-throughput screening, a technique that utilizes small-molecule chemical compound libraries comprising up to several hundred thousand compounds, can be used to identify and characterize chemical compounds based on their responses in particular assays. However, there are many cases where small-molecule drugs have proved to be ineffective, prone to unacceptable side effects, or ineffective due to a particular disease etiology. In addition, a common feature of small-molecule drugs is that they provide either continuous inhibition or continuous activation of the target, thus engaging compensatory mechanisms leading to drug tolerance, drug resistance, and reduction of therapeutic efficacy over time. Finally, small-molecule drugs often lack specificity, which is thought to contribute to their high failure rate of >90% in clinical trials [11]. To circumvent these problems, the drug discovery field is exploring additional molecular classes with broader capabilities beyond small-molecule drugs.

Biological drugs are a relatively new class of drugs intended to offer safety and efficacy comparable to small molecules. There are several significant differences between biological drugs and small-molecule drugs, though; biological drugs are usually derived from living organisms, their respective molecular size is substantially larger, and manufacturing processes are more complicated. In some contexts, biological drugs include peptides, and the first biologic drug is considered to be humanized insulin (5.8 kDa), which became available in 1982. However, for the purpose of the discussion herein, we will differentiate between biological drugs (proteins and monoclonal antibodies, or mAbs) and peptides. The main difference between biological drugs and peptides is their molecular weight ($\sim$10–150 and $\sim$1–5 kDa, respectively). The difference in molecular weight has led to distinct approaches for the design, engineering, and synthesis of biological drugs and peptides. The Orthoclone OKT3 mAb was the first biological drug approved by the FDA in 1986 for immunosuppression in organ transplantation. From then until 2018, over 80 biological drugs were approved in the United States or Europe for the treatment of various diseases [12]. In 2013, 18 of the approved monoclonal antibodies achieved annual sales of over 1 billion dollars, and 6 of these products had sales of greater than 6 billion dollars [13]. Biological drug development is partially driven by the rapid advancement of our understanding of disease at a molecular level and by the rapid development in related fields, such as genomics, proteomics, and other systems biology approaches that continue to provide important new targets for modulating disease. While small-molecule drugs can interact with the active site of a target that is often conserved, biological drugs can be selected to bind to unique epitopes. This exquisite specificity is the basis for their reduced nonmechanistic (or "off-target") toxicity—a major advantage over small-molecule drugs. However, biological drugs have their own limitations, including high structural complexity, multifaceted manufacturing processes, and risks of immunogenicity. Therefore unique regulatory pathways are required for these molecules. In addition, since biologics are proteins, they all but exclude oral delivery. Finally, biological drugs are relatively large agents, and therefore suitable for targeting circulating antibodies, hormones, and growth factors, or cell surface receptors, integrins, and antigens. Biological drugs are inherently unsuitable for targeting the great majority of disease-associated proteins, which are intracellular [14].

Peptides are short chains of amino acids linked by peptide (amide) bonds and are naturally occurring molecules that regulate many signaling pathways in the body, acting as hormones, neurotransmitters, growth factors, ion channel ligands, and antimicrobials. Over 7000 naturally occurring peptides have been identified, and most of them are involved in the control of cellular homeostasis, from proliferation, differentiation, migration, and cell death, to large-scale angiogenesis and immune responses, which maintain physiological or mediate pathological conditions [15]. Peptides are expected to be cell permeable similarly to small molecules, but also to possess the high selectivity and potency characteristic of biologic drugs. Thus, in various aspects, peptides are the sweet spot between small molecules and biological drugs. Furthermore, peptides can take on different modulatory roles as agonists, antagonists, and allosteric modulators. As a result, peptides have evolved to become a very promising basis for targeting a range of human diseases, starting about 100 years ago with the purification and treatment of a diabetic child with bovine insulin [16]. Since then, many synthetic strategies for developing novel peptides and peptidomimetics (modified peptides), as well as alternative routes for their administration, have emerged [17–19]. Over the past four decades, there has been a rapid growth in the development of therapeutic peptides, with a dramatic increase in the number of peptide-based drugs on the market. Thus there has been an increase in the average number of peptides entering clinical studies, from 1.2 per year (1970s) to over 16.8 per year (2000s) [20,21]. Importantly, peptides have higher success rates in transitioning from phase 1 to phase 2 trials (83%) compared to small molecules and biologics (63% and 77%, respectively), and in transitioning from phase 3 to regulatory review (68%, compared to 61% of small molecules and 63% of biologics) [22]. Taken together, it is not surprising that the number of available therapeutic peptides and peptidomimetics is increasing, with over 80 approved compounds on the market to date, and approximately 150 peptide drugs currently in clinical trials, with another 400–600 peptides in preclinical studies as of 2021 (PMID: 33536635) [15]. In addition, four peptides reached global sales of over 1 billion dollars in 2008, including glatiramer acetate (3.18 billion dollars), leuprolide acetate (2.12 billion dollars), goserelin acetate (1.14 billion dollars), and octreotide acetate (1.12 billion dollars) [20] (for several excellent reviews about peptide drug discovery, see Refs. [15,21,23–26]).

Protein interactions are involved in the control of almost all cellular homeostasis mechanisms. In particular, protein complexes involving protein-protein interactions are critical for cellular signaling networks. Protein complexes are held together by their contact interface, which is usually flat and large, ranging from approximately 300–4800 $Å^2$ [27,28], and typically lack small deep pockets that would bind small molecules (~300–500 $Å^2$) [29–32]. Therefore small molecules cannot effectively target relatively flat protein surfaces and elements lacking a rigid structure, such as protein complexes. Compared to small molecules, peptides demonstrate several advantages for targeting protein complexes [33], such as conformational flexibility (induced fit) [34] and increased selectivity [35–37]. Several studies suggest that 15%–40% of known protein interactions are modulated by endogenous linear peptides [38,39]. In addition, peptides can mimic the properties of one protein in the complex and take on a role as a competitive inhibitor, therefore interrupting the binding of partner protein(s) [30,40–42]. Peptides derived from such intermolecular protein interfaces frequently serve as antagonists; however, if a peptide binds the site of an intramolecular protein interaction, the molecule could potentially serve as an agonist as well [43,44].

Three main approaches have been described to identify bioactive peptides: (1) isolation of naturally occurring peptides, (2) screening of peptide libraries, and (3) rational design. Peptides that play critical roles in signaling pathways and demonstrate potent bioactivity and efficacy as pharmacological compounds have been isolated from prokaryotes to humans, including antimicrobial peptides [45], cyclotides [46], conotoxins [47], knottins [48], and some human hormones [36]. Peptide libraries are divided into three main classes: biological peptide libraries (e.g., phage display [49], yeast two-hybrid assay [50], and split intein cyclization of peptides and proteins (SICLOPPS) [51]), as well as semisynthetic, and fully synthetic peptide libraries. Many approaches have been developed for semi- and fully synthetic libraries, such as SPOT-synthesis or oligonucleotide-encoded expression, and microbial display systems [52,53]. It is important to emphasize that screening of peptide libraries is usually done with large libraries of thousands or millions of peptides, while using rational design approaches (*vide infra*) generally involves small peptide libraries that contain between ten and hundreds of peptides. Several strategies have been used to rationally design peptides, but the most common ones involve bioinformatic and computational methods to integrate data derived from the sequence [44,54–69] and/or structure of a peptide or protein [70–72].

This book provides a comprehensive survey, critique, and in-depth analysis of the important emerging field of peptide and peptidomimetic therapeutics from early research at the laboratory bench to clinical use at the patient bedside. Beginning with a bird's eye review of the field including the past, present, and future trends, chapters then discuss pitfalls as well as advantageous properties and successful approaches to the discovery, design, synthesis, development, production, and use of peptide and peptidomimetic therapeutics for the advancement of human health. The work also delves deep into conceptual and technical methods for identifying, characterizing, and employing these molecules for therapeutic and other applications, including numerous illustrative examples and informative studies. The book emphasizes the latest developments and most likely future directions of subfields while nurturing an understanding of how peptides and peptidomimetics are applied in biomedical sciences for targeting diverse processes. Analysis of preclinical, clinical, and business market data are also provided, in addition to future perspectives. Critical assessments and insights throughout the work are derived from a diverse group of field leaders around the world who each offer unique insights. The chapters and sections are composed with wide accessibility and sufficient context in mind such that each chapter can stand alone for the focused reader while offering advanced insights to the experienced scientist or peptide expert. Taken together, this work provides an extensive and detailed resource to many within and beyond the field. The book is of broad utility to students, academic researchers, industry teams, clinicians, and interested nonscientists.

References

[1] Arias, E., United States life tables, 2004. Natl Vital Stat Rep. 2007;56(9):1–39.

[2] Munos B. Lessons from 60 years of pharmaceutical innovation. Nat Rev Drug Discov 2009;8(12):959–68.

[3] DiMasi JA, et al. Trends in risks associated with new drug development: success rates for investigational drugs. Clin Pharmacol Ther 2010;87(3):272–7.

[4] DiMasi JA, et al. Innovation in the pharmaceutical industry: new estimates of R&D costs. J Health Econ 2016;47:20–33.
[5] Scannell JW, et al. Diagnosing the decline in pharmaceutical R&D efficiency. Nat Rev Drug Discov 2012;11(3):191–200.
[6] Scannell JW, et al. When quality beats quantity: decision theory, drug discovery, and the reproducibility crisis. PLoS One 2016;11(2):e0147215.
[7] World Health Organization. World Health Organization model list of essential medicines. Geneva: World Health Organization; 2019.
[8] Mahdi JG, et al. The historical analysis of aspirin discovery, its relation to the willow tree and antiproliferative and anticancer potential. Cell Prolif 2006;39(2):147–55.
[9] Rainsford KD. Aspirin and related drugs. CRC Press; 2004.
[10] Breitenfeld T, et al. Hippocrates: the forefather of neurology. Neurol Sci 2014;35 (9):1349–52.
[11] Biotechnology Innovation Organization (BIO). Clinical development success rates 2006-2015; 2016.
[12] Kaplon H, et al. Antibodies to watch in 2019. mAbs 2019;11(2):219–38.
[13] Ecker DM, et al. The therapeutic monoclonal antibody market. mAbs 2015;7(1):9–14.
[14] Imming P, et al. Drugs, their targets and the nature and number of drug targets. Nat Rev Drug Discov 2006;5(10):821.
[15] Fosgerau K, et al. Peptide therapeutics: current status and future directions. Drug Discov Today 2015;20(1):122–8.
[16] Banting FG, et al. Pancreatic extracts in the treatment of diabetes mellitus. Can Med Assoc J 1922;12(3):141–6.
[17] Marzinzik AL, et al. Towards intracellular delivery of peptides. CHIMIA Int J Chem 2013;67(12):899–904.
[18] Uhlig T, et al. The emergence of peptides in the pharmaceutical business: From exploration to exploitation. EuPA Open Proteomics 2014;4(0):58–69.
[19] Vorherr T. Modifying peptides to enhance permeability. Future Med Chem 2015;7 (8):1009–21.
[20] Reichert, J. *Development trends for peptide therapeutics*. In: Peptide therapeutics symposium. San Diego, CA; 2010.
[21] Kaspar AA, et al. Future directions for peptide therapeutics development. Drug Discov Today 2013;18(17–18):807–17.
[22] Bhat A, et al. Lead discovery and optimization strategies for peptide macrocycles. Eur J Med Chem 2015;94:471–9.
[23] Qvit N. Development and therapeutic applications of oligonucleotides and peptides. Chim Oggi/Chem Today 2011;29(2):4–7.
[24] Joo SH. Cyclic peptides as therapeutic agents and biochemical tools. Biomol Ther (Seoul) 2012;20(1):19–26.
[25] Tapeinou A, et al. Review cyclic peptides on a merry-go-round; towards drug design. Biopolymers 2015;104(5):453–61.
[26] Qvit N, et al. Peptidomimetic therapeutics: Scientific approaches and opportunities. Drug Discov Today 2017;22(2):454–62.
[27] Pal A, et al. Peptide segments in protein-protein interfaces. J Biosci 2007;32(1):101–11.
[28] Hwang H, et al. Protein-protein docking benchmark version 4.0. Proteins 2010;78 (15):3111–14.
[29] Arkin MR, et al. Small-molecule inhibitors of protein-protein interactions: progressing towards the dream. Nat Rev Drug Discov 2004;3(4):301–17.

[30] Fuller JC, et al. Predicting druggable binding sites at the protein-protein interface. Drug Discov Today 2009;14(3–4):155–61.

[31] Arkin MR, et al. Small-molecule inhibitors of protein-protein interactions: progressing toward the reality. Chem Biol 2014;21(9):1102–14.

[32] Cukuroglu E, et al. Hot spots in protein-protein interfaces: towards drug discovery. Prog Biophys Mol Biol 2014;116(2–3):165–73.

[33] Mochly-Rosen D, et al. Peptide inhibitors of protein-protein interactions. Chim Oggi/ Chem Today 2010;28(1):14–16.

[34] Koshland DE. The key-lock theory and the induced fit theory. Angew Chem Int Ed Engl 1995;33(23–24):2375–8.

[35] Bruno BJ, et al. Basics and recent advances in peptide and protein drug delivery. Ther Deliv 2013;4(11):1443–67.

[36] Rubin S, et al. Cyclic peptides for protein-protein interaction targets: Applications to human disease. Crit Rev Eukaryot Gene Expr 2016;26(3):199–221.

[37] Cunningham AD, et al. Peptides and peptidomimetics as regulators of protein-protein interactions. Curr Opin Struct Biol 2017;44:59–66.

[38] Petsalaki E, et al. Peptide-mediated interactions in biological systems: new discoveries and applications. Curr Opin Biotechnol 2008;19(4):344–50.

[39] London N, et al. Modeling peptide-protein interactions. Methods Mol Biol 2012;857:375–98.

[40] Jones DS, et al. Developing therapeutic proteins by engineering ligand-receptor interactions. Trends Biotechnol 2008;26(9):498–505.

[41] Scott DE, et al. Small molecules, big targets: Drug discovery faces the protein-protein interaction challenge. Nat Rev Drug Discov 2016;15(8):533–50.

[42] Wojcik P, et al. Peptide-based inhibitors of protein-protein interactions. Bioorg Med Chem Lett 2016;26(3):707–13.

[43] Churchill EN, et al. Rationally designed peptide regulators of protein kinase C. Trends Endocrinol. Metab. 2009;20(1):25–33.

[44] Cunningham AD, et al. Specific kinase inhibition using peptides to target kinase-substrate docking. Chim Oggi/Chem Today 2016;34(5):22–5.

[45] Fjell CD, et al. Designing antimicrobial peptides: form follows function. Nat Rev Drug Discov 2012;11(1):37–51.

[46] Jagadish K, et al. Cyclotides, a promising molecular scaffold for peptide-based therapeutics. Biopolymers 2010;94(5):611–16.

[47] Essack M, et al. Conotoxins that confer therapeutic possibilities. Mar Drugs 2012;10 (6):1244.

[48] Moore SJ, et al. Knottins: disulfide-bonded therapeutic and diagnostic peptides. Drug Discov Today Technol 2012;9(1):e1–e70.

[49] Wu CH, et al. Advancement and applications of peptide phage display technology in biomedical science. J Biomed Sci 2016;23:8.

[50] Barreto K, et al. Screening combinatorial libraries of cyclic peptides using the yeast two-hybrid assay. Methods Mol Biol 2014;1163:273–309.

[51] Tavassoli A, et al. Split-intein mediated circular ligation used in the synthesis of cyclic peptide libraries in E. coli. Nat Protoc 2007;2(5):1126–33.

[52] Needels MC, et al. Generation and screening of an oligonucleotide-encoded synthetic peptide library. Proc Natl Acad Sci USA 1993;90(22):10700–4.

[53] Frank R. The SPOT-synthesis technique. J Immunol Methods 2002;267(1):13–26.

[54] Qvit, N., et al. Peptide regulators of protein-protein interactions with selectivity for specific sub-cellular compartments. 21th American peptide symposium. Bloomington, IN: Biopolymers; 2009.

[55] Qvit N, et al. Highly specific modulators of protein kinase C localization: Applications to heart failure. Drug Discov Today: Dis Mech 2010;7(2):e87–93.

[56] Qvit N, et al. A second generation peptide activator derived from the C2 domain of delta protein kinase C and regulates its function in the mitochondria. J Pept Sci 2010.

[57] Qvit N, et al. Rational design of a novel peptide that selectively inhibits one delta PKC function. Biopolymers. Malden, MA: Wiley-Blackwell Commerce Place; 2011.

[58] Qvit N, et al. Selective Inhibition of δPKC-Mediated Pyruvate Dehydrogenase Kinase Phosphorylation Alone is Sufficient to Induce Protection From Ischemia and Reperfusion Injury. American Heart Association; 2012.

[59] Qvit N, et al. Novel selective peptide regulator of protein kinase C based on rational approach design. Biosens Bioelectron 2012;2012:142.

[60] Qi X, et al. A novel Drp1 inhibitor diminishes aberrant mitochondrial fission and neurotoxicity. J Cell Sci 2013;126(Pt 3):789–802.

[61] Qvit, N., et al. Engineering a delta protein kinase C inhibitor that selectively inhibits interaction and phosphorylation of one substrate of the multi-substrate kinase. In: American Society for Biochemistry and Molecular Biology annual meeting. San Diego, CA; 2014.

[62] Qvit N, et al. The many hats of protein kinase C delta: one enzyme with many functions. Biochem Soc Trans 2014;42(6):1529–33.

[63] Qvit N, et al. GAPDH (glyceraldehyde-3-phosphate dehydrogenase) protein-protein interaction inhibitor reveals a non-catalytic role for GAPDH oligomerization in cell death. J Biol Chem 2016;291(26):13608–21.

[64] Hurt CM, et al. Transient receptor potential vanilloid 1 regulates mitochondrial membrane potential and myocardial reperfusion injury. J Am Heart Assoc 2016;5(9):1–13.

[65] Qvit N, et al. Engineered substrate-specific delta PKC antagonists to enhance cardiac therapeutics. Angew Chem Int Ed Engl 2016;55(50):15672–9.

[66] Franco A, et al. Correcting mitochondrial fusion by manipulating mitofusin conformations. Nature 2016;540(7631):74–9.

[67] Heymann HM, et al. Transient receptor potential vanilloid 1 inhibitors block laparotomy- and opioid-induced infarct size reduction in rats. Br J Pharmacol 2017;24:4826–35.

[68] Kornfeld OS, et al. Interaction of mitochondrial fission factor with dynamin related protein 1 governs physiological mitochondrial function in vivo. Scientific Reports 2018;8 (1):14034.

[69] Ferreira JCB, et al. A selective inhibitor of mitofusin 1-betaIIPKC association improves heart failure outcome in rats. Nat Commun 2019;10(1):329.

[70] Tan YS, et al. Stapled peptide design: principles and roles of computation. Drug Discov Today 2016;21(10):1642–53.

[71] Bhardwaj G, et al. Accurate de novo design of hyperstable constrained peptides. Nature 2016;538(7625):329–35.

[72] Fu J, et al. A small shared epitope-mimetic compound potently accelerates osteoclast-mediated bone damage in autoimmune arthritis. J Immunol 2013;191(5):2096–103.

Section 1

History

Chapter 1

Therapeutic peptides: historical perspectives and current development trends

Jolene L. Lau[1] and Michael K. Dunn[2]
[1]*BioMarin Pharmaceutical, Novato, CA, United States,* [2]*Ferring Research Institute, Inc., San Diego, CA, United States*

1.1 Introduction: The evolution of peptide therapeutics

Peptides represent a unique class of pharmaceutical compounds, molecularly poised between small molecules and proteins, yet biochemically and therapeutically distinct from both. As intrinsic signaling molecules for many physiological functions, peptides present an opportunity for therapeutic intervention that closely mimics natural pathways. Indeed, several peptide drugs are essentially "replacement therapies" that add back or supplement peptide hormones in cases where endogenous levels are inadequate or absent. This is exemplified by the isolation of insulin from whole animal tissue and the first therapeutic use of the peptide in the 1920s in diabetics unable to produce sufficient quantities of the hormone [1]. Similar practice continued with the purification of adrenocorticotrophic hormone (ACTH) from livestock pituitary glands to treat a variety of endocrine disorders in patients [2].

The utilization of peptides as therapeutics has evolved over time and continues to evolve with changes in drug development and treatment paradigms (Table 1.1). Peptides isolated from natural sources, such as insulin and adrenocorticotropic peptide (ACTH), provided life-saving medicines in the first half of the 20th century. When sequence elucidation and chemical synthesis of peptides became feasible in the 1950s, synthetic oxytocin and vasopressin also entered clinical use. As venoms of arthropods and cephalopods became recognized as treasure troves of bioactive peptides, isolation of natural products from exotic sources became a popular strategy for identifying new potential therapeutics. The genomic era allowed for the identification and molecular characterization of receptors for many important endogenous

Peptide and Peptidomimetic Therapeutics. DOI: https://doi.org/10.1016/B978-0-12-820141-1.00027-3

TABLE 1.1 Source or chemical nature of early peptides.

Peptide	Source	Introduction to the clinic	Sequence description
Insulin	Isolated from canine and bovine pancreata	1920s	Native
Adrenocorticotropic peptide (ACTH)	Isolated from bovine and porcine pituitary glands	1950s	Native
Calcitonin	Isolated from salmon ultimobranchial gland	1971	Native
Oxytocin	Synthetic	1962	Native
Vasopressin	Synthetic	1962	Native
Octreotide	Synthetic analog of somatostatin	1988	Cyclic octapeptide analog of somatostatin-14
Leuprorelin	Synthetic analog of gonadorelin	1984	Nonapeptide analog of decapeptide gonadorelin

peptide hormones, and industry and academia began to pursue novel peptidic ligands for these receptors.

1.1.1 Moving beyond native peptides

New technologies opened the door for strategies to overcome certain limitations of native peptides, such as short plasma half-life, lack of receptor subtype selectivity, and negligible oral bioavailability. (These limitations of peptides have been described in detail elsewhere [3,4]. We will focus on the characteristics of peptides that have been nominated for human clinical development.) The short half-life of endogenous peptide hormones is explained by the presence of numerous peptidases and excretory mechanisms that inactivate and clear peptides, often in minutes. Lability allows the body to rapidly modulate hormone levels to maintain homeostasis; this serves an important physiological purpose but is inconvenient or unfeasible for dosing in most clinical settings.

Likewise, endogenous peptides may activate multiple receptor subtypes—for example, there are five known receptor subtypes for melanocortin; five for somatostatin; and three for vasopressin. While receptor subtypes bind the same ligand, they are often expressed in different locations and have different physiological functions, allowing for localized release of the ligand to exert a specific local effect, while not affecting distal receptors. Using a single ligand for multiple functions is physiologically economical, as the body needs to produce only one substance to regulate many effects. However, this works best in localized paracrine or endocrine signaling, as activating all receptor subtypes at the same time (*i.e.*, systemic administration) can be disruptive.

For example, the endogenous peptide vasopressin acts on multiple receptors, inducing anti-diuresis via activity at vasopressin receptor 2 (AVPR2) in renal tubular cells and vasoconstriction via vasopressin receptor 1A (AVPR1A) expressed in vascular smooth muscle. This nonselective activity may be useful in homeostasis when the peptide is locally released but may be detrimental in systemic administration in treatment of disease. A selective V2 agonist such as desmopressin may be preferable for treatment of polyuria or polydipsia in central diabetes insipidus, as desmopressin has much lower vasopressor effect than arginine vasopressin [5]. Conversely, a selective vasopressin V1a-receptor agonist was hypothesized to potentially be a better systemic therapeutic for treatment of sepsis, although this approach did not meet primary endpoints in a clinical trial [6].

The development of synthetic peptide chemistry techniques such as solid-phase peptide synthesis allowed for rapid and combinatorial synthesis of analogs incorporating both natural and modified amino acids. This allowed investigators to use medicinal chemistry strategies to make more drug-like peptide molecules by improving the aforementioned liabilities. Peptide analogs of native hormones with improved pharmaceutical properties thus entered the clinic and now represent the majority of peptide drugs.

Most peptides, including endogenous or native peptides with therapeutic potential, cannot be delivered orally due to instability under the low pH and proteolytic conditions found in the GI tract. Furthermore, the high polarity and molecular weight of peptides severely limits intestinal permeability. As oral delivery is often viewed as attractive for supporting patient compliance, the need for injection made peptides a less appealing option for indications that required chronic, outpatient therapy. As a result, most peptide drugs have been developed where parenteral as opposed to oral delivery is acceptable due to either lack of alternatives or administration setting. However, recent successes in formulation technology have transitioned parental peptides such as semaglutide and octreotide to oral delivery, and novel ingested controlled-delivery devices offer the potential for direct injection to the GI tract [4].

1.1.2 Nonpeptide alternatives to peptide therapeutics

Another strategy in overcoming the challenges of targeting peptide receptors lay in a different molecular class: small molecules. In the 1990s, the availability of massive combinatorial chemistry libraries and high-throughput screening (HTS) technologies swung the pendulum in a new direction, towards small molecules that target peptide receptors. Small molecules are generally more suitable for oral delivery and easier to manufacture than peptides; the challenge lies in finding a small molecule that mimics a peptide ligand's receptor binding and selective modulation at an acceptable potency. The number and diversity of scaffolds present in modern screening libraries supported the idea that lead molecules could be identified, optimized, and developed into drugs. Structural biology added another arrow to the quiver by elucidating key molecular interactions at receptor active sites that could be leveraged by any class of molecule.

The small molecule approach has been more successful in some cases than others. At peptidic receptors, small molecules are often less potent than peptides, and small molecules that act as antagonists are easier to identify than agonists. The large ligand-binding site of some peptidic G-protein-coupled receptors (GPCRs) and specific conformational changes required for signal transduction provide significant challenges for small-molecule drug discovery, particularly for Class B GPCRs [7–9]. Nonetheless, orally available small molecules such as losartan and valsartan replaced the peptide saralasin (SARENIN) as angiotensin II receptor blockers for hypertension, and other small-molecule drugs target Class A GPCRs for which no peptide drugs are marketed (Table 1.2).

While overcoming some of the challenges of peptide drugs, these small molecules retain the potential for liabilities that are infrequently associated with peptides, such as cytochrome inhibition leading to drug–drug interactions (DDIs) and side effects caused by off-target binding. Although a significant and important avenue of new discovery, small-molecule ligands for peptidic receptors are not a complete substitute for peptide compounds.

1.1.3 Current state of peptide therapeutics discovery

More recently, a broader and more nuanced appreciation of the potential of peptide therapeutics has emerged. As a result, characteristics of peptides previously viewed as liabilities are no longer as problematic: for example, injection is viewed as an acceptable route of administration for certain indications, partly due to the development of longer-acting peptides or depot formulations that reduce injection frequency.

Despite the availability of multiple classes of oral medications for type 2 diabetes mellitus, the injectable glucagon-like peptide 1 (GLP-1) peptide agonist market has continued to grow since the 2005 approval of exenatide, likely driven by favorable efficacy profiles and rising diabetes rates, and multiple next-generation drug candidates are in development today, as we describe later in this chapter. Teriparatide, a truncated parathyroid hormone prescribed for osteoporosis, provides an anabolic mechanism of action that is distinct from the orally administered bisphosphonates, justifying the use of a daily injectable [12].

TABLE 1.2 Small-molecule drugs that act on peptide receptors.[a]

Drug or drug class	Primary molecular target(s)	Action at receptor	Liabilities of small molecules[b]	Current status of small molecule and peptide drugs
Losartan and other sartans	Angiotensin II receptor 1	Antagonism	DDIs; fetal toxicity	Marketed; displaced the peptide angiotensin-receptor blocker saralasin[c]
Small-molecule opioids (natural, synthetic, and semi-synthetic)	Opioid receptors	Agonism	Multiple effects caused by lack of receptor subtype specificity	Marketed; no peptides are approved in major markets
Tolvaptan and other vaptans	Vasopressin V2 receptor	Antagonism	Hepatotoxicity	Marketed; no peptides are approved
Bosentan and others	Endothelin receptors	Antagonism	Hepatotoxicity; DDIs	Marketed; no peptides are approved
Aprepitant and others	Neurokinin 1 receptor	Antagonism	DDIs	Marketed; no peptides are approved
Elagolix and others	Gonadotrophin-releasing hormone receptors	Antagonism	DDIs (*e.g.*, sufugolix)	Marketed; multiple peptides are currently marketed
Rimegepant and others	CGRP receptor	Antagonism	DDIs (*e.g.*, telcagepant)	Marketed, along with receptor-targeting antibodies; no peptides are approved

[a]*Information for classes of agents with marketed drugs is derived from United States Food and Drug Administration (FDA) labels [10].*
[b]*Not all agents within a given drug class display all of the liabilities outlined in the chart.*
[c]*In addition, saralasin acts as a partial agonist at the angiotensin II type 1 receptor and thus does not reduce blood pressure in patients with low levels of plasma angiotensin II [11].*

In addition, peptide drug candidates are being generated against a range of molecular targets that reach beyond historically-dominant extracellular hormone receptors. Peptides that disrupt protein–protein interactions [13], target cytokine receptors [14], and inhibit intracellular targets [15] have entered the clinic. Phage display has identified new peptides as the launching point for discovery and medicinal chemistry efforts, and novel peptide scaffolds have brought new families of peptide leads into the clinic. This has subsequently culminated in a diverse and robust development landscape for therapeutic peptides.

To date, over 80 peptide drugs have been approved in the United States, Europe, and Japan. Here we review the characteristics of peptide drugs and clinical candidates, therapeutic applications, and prospects for the future.

1.2 The therapeutic peptides dataset

We have compiled a comprehensive set of data related to all peptides that have entered human clinical studies, subject to certain dataset inclusion criteria (Table 1.3). This database includes information on chemical structures, molecular pharmacology, therapeutic indications, clinical trial initiation dates, and global regulatory status. Sources used to populate the database include scientific articles and conference abstracts; patents and patent applications; company websites, press releases, presentations, and financial regulatory filings; public clinical trial and drug information repositories; and proprietary (subscription) databases that collate publicly-available information from the preceding and other sources (*e.g.*, conference posters). Reports derived from an earlier version of this dataset have been published and describe the database curation methodology in greater detail [16,17]. We have since continued to maintain and curate a proprietary version of the data. The analysis presented in this article only includes peptides with development activity geared toward regulatory approval in major pharmaceutical markets (the United States, Europe, and Japan) as of May 2021.

1.2.1 Development status of therapeutic peptides

As of May 2021, this dataset contains information on 584 therapeutic peptides. Of these, 84 have received market authorization in the United States, Europe, and/or Japan[1]; 12 peptides have subsequently been withdrawn, leaving 72 currently approved. The list of approved therapeutic peptides is available in Table 1.4, included at the conclusion of the chapter. 149 peptides are in active clinical development, just under half of which are currently in Phase 2 studies (Fig. 1.1).

1. Two additional peptides, difelikefalin and vosoritide, were approved in 2021 after this analysis was performed in May.

TABLE 1.3 Key inclusion and exclusion criteria for the peptide therapeutics database.

Criteria	Example of an included peptide	Example of an excluded peptide
Each molecular entity was included only once in the database[a]	Exenatide (BYETTA)	Exenatide (BYDUREON)
Lower length limit: two amino acids linked by an amide bond	Carfilzomib (KYPROLIS)	Bortezomib (VELCADE)
Upper length limit: recombinantly-expressed peptides less than 50 amino acids in length, or synthetic peptides of any length[b]	Lixisenatide (LYXUMIA)	Insulin (all products)
Peptides exclusively derived from nonrecombinant bacterial fermentation are excluded; semi-synthetic peptides are included	Voclosporin (LUPKYNIS)	Daptomycin (CUBICIN)
Epitope-specific peptide vaccines are excluded; nonspecific enhancers of immune function are included	Glatiramer (COPAXONE)	Galinpepimut-S
Peptides conjugated to other molecules are included, as long as the drug candidate contains a discrete, functional peptidic domain that otherwise meets the inclusion criteria[b]	Dulaglutide (TRULICITY)	Trebananib
Diagnostics are excluded, but peptides developed as both therapeutics and diagnostics are included	Lutetium DOTATATE	Annexin V-128

[a]*Peptides developed by multiple sponsors or in multiple formulations were included once, with dates and other fields reflecting the earliest program.*
[b]*Only the peptidic portion of a conjugate molecule needs to meet the length limit for inclusion.*

The number of peptides entering clinical development gradually trended upward between 1980 and 2010, with the five-year trailing average peaking at 29 peptides per year in 2010 (Fig. 1.2). The cumulative number of approved peptides exhibited a slight upward trend as well, with 25 peptide approvals occurring from the start of 2010 through this writing.

1.2.2 Physical characteristics of therapeutic peptides

Not surprisingly, the characteristics of peptides entering clinical development have evolved over time. These changes reflect advances in peptide chemistry

TABLE 1.4 List of approved peptides.

Compound name	Current status	First approval	Approved indication(s)	Primary therapeutic area	Approved RoA	Conjugation	Molecular target	Chemical basis
Corticotropin	Approved	1952	Multiple inflammatory diseases; West syndrome	CNS	Subcutaneous	No	MC receptors	Native
Lypressin	Withdrawn	1961	Diabetes insipidus	Urology	Intranasal	No	AVP1 receptor	Native
Angiotensinamide	Withdrawn	1962	Shock and circulatory collapse	Hepatology and Critical Care	Intravenous	No	Angiotensin II receptors	Native
Vasopressin	Approved	1962	Vasodilatory shock; bleeding esophageal varices	Urology	Intravenous	No	AVP1 receptor	Analog
Oxytocin	Approved	1962	Labor induction; postpartum hemorrhage; lactation	Infertility and OB-GYN	Intramuscular; intravenous; intranasal	No	OT receptor	Analog
Eledoisin	Approved	1970s	Dry eye	Ophthalmology	Topical	No	NK1 receptor	Native
Demoxytocin	Withdrawn	1970s	Delivery induction; labor stimulation	Infertility and OB-GYN	Buccal	No	OT receptor	Analog
Somatostatin	Approved	1970s	Bleeding esophageal varices	Hepatology and Critical Care	Intravenous	No	SST receptors	Native
Felypressin	Approved	1970s	Dental anesthesia adjunct	Pain	Subcutaneous	No	AVP1 receptor	Analog
Calcitonin (salmon)	Approved	1971	Hypercalcemia; Paget's disease; osteoporosis	Bones and Connective Tissues	Intranasal; subcutaneous	No	Calcitonin receptor	Native
Ornipressin	Approved	1971	Bleeding esophageal varices	Hematology	Intravenous	No	AVP1 receptor	Analog
Desmopressin	Approved	1972	Diabetes insipidus; nocturia	Urology	Oral; intranasal	No	V2 receptor	Analog

Terlipressin	Approved	1978	Bleeding esophageal varices	Hepatology and Critical Care	Intravenous	No	AVP1 receptor	Analog
Ambamustine	Withdrawn	1980s	Non-Hodgkin's lymphomas	Oncology	Intravenous	No	DNA (cytotoxic agent)	Heterologous
Tetracosactide	Approved	1980	Multiple inflammatory diseases	Endocrinology	Subcutaneous	No	MC receptors	Native
Elcatonin	Approved	1981	Osteoporosis	Bones and Connective Tissues	Intramuscular	No	Calcitonin receptor	Analog
Saralasin	Withdrawn	1981	Renovascular hypertension	Cardiovascular	Intravenous	No	Angiotensin II receptors	Analog
Cargutocin	Withdrawn	1982	Labor induction	Infertility and OB-GYN	Unknown (injectable)	No	OT receptor	Analog
Buserelin	Approved	1984	Prostate cancer; endometriosis; assisted reproduction	Urology	Intranasal; subcutaneous	No	GnRH receptor	Analog
Leuprorelin	Approved	1984	Prostate cancer; endometriosis; uterine fibroids; precocious puberty	Urology	Subcutaneous; intramuscular	No	GnRH receptor	Analog
Thymopentin	Approved	1985	Various autoimmune and infectious diseases	Allergy and Immunology	Subcutaneous; intramuscular	No	HLA-DR	Analog
Enalapril	Approved	1985	Heart failure; hypertension	Cardiovascular	Oral	No	Angiotensin-converting enzyme	Analog
Triptorelin	Approved	1986	Prostate cancer; endometriosis; breast cancer; precocious puberty	Urology	Intramuscular	No	GnRH receptor	Analog

(*Continued*)

TABLE 1.4 (Continued)

Compound name	Current status	First approval	Approved indication(s)	Primary therapeutic area	Approved RoA	Conjugation	Molecular target	Chemical basis
Calcitonin (human)	Approved	1986	Hypercalcemia; Paget's disease; osteoporosis	Bones and Connective Tissues	Intramuscular; intranasal; subcutaneous	No	Calcitonin receptor	Native
Goserelin	Approved	1987	Breast cancer; endometriosis; prostate cancer	Urology	Subcutaneous	No	GnRH receptor	Analog
Lisinopril	Approved	1987	Acute myocardial infarction; heart failure; hypertension	Cardiovascular	Oral	No	Angiotensin-converting enzyme	Analog
Octreotide	Approved	1988	Acromegaly; neuroendocrine tumors	Endocrinology	Subcutaneous; intramuscular; intravenous	No	SST 2 receptor	Analog
Romurtide	Withdrawn	1989	Blood count support in cancer patients	Oncology	Subcutaneous	Yes	NOD2 protein	Analog
Glucagon	Approved	1989	Hypoglycemia	Metabolic	Subcutaneous	No	Glucagon receptor	Native
Gonadorelin	Approved	1989	Assisted reproduction; primary amenorrhea	Infertility and OB-GYN	Subcutaneous	No	GnRH receptor	Native
Nafarelin	Approved	1990	Precocious puberty; endometriosis	Infertility and OB-GYN	Intranasal	No	GnRH receptor	Analog
Histrelin	Approved	1991	Prostate cancer; precocious puberty	Infertility and OB-GYN	Subcutaneous; intramuscular	No	GnRH receptor	Analog

Deslorelin	Withdrawn	1992	Endometriosis; precocious puberty	Infertility and OB-GYN	Subcutaneous	No	GnRH receptor	Analog
Lanreotide	Approved	1994	Acromegaly; neuroendocrine tumors	Endocrinology	Subcutaneous	No	SST receptors	Analog
Carperitide	Approved	1995	Acute decompensated heart failure	Cardiovascular	Intravenous	No	NPR-A	Native
Glatiramer	Approved	1996	Multiple sclerosis	CNS	Subcutaneous	No	MHC II	Heterologous
Sermorelin	Withdrawn	1997	Growth hormone deficiency	Endocrinology	Subcutaneous	No	GHRH receptor	Analog
Eptifibatide	Approved	1998	Acute coronary syndrome managed medically or with percutaneous coronary intervention	Cardiovascular	Intravenous	No	GP IIb IIIa	Analog
Cetrorelix	Approved	1999	Assisted reproduction	Infertility and OB-GYN	Subcutaneous	No	GnRH receptor	Analog
Ganirelix	Approved	1999	Assisted reproduction	Infertility and OB-GYN	Subcutaneous	No	GnRH receptor	Analog
Bivalirudin	Approved	1999	Acute coronary syndrome; heparin-induced thrombocytopenia; thrombosis	Hematology	Intravenous	No	Thrombin	Analog
P-15	Approved (medical device)	1999	Bone grafting	Orthopedics	Implanted	No	$\alpha2\beta1$	Analog
Atosiban	Approved	2000	Preterm labor	Infertility and OB-GYN	Intravenous	No	OT receptor	Analog
Taltirelin	Approved	2000	Spinocerebellar degeneration	CNS	Oral	No	TRH receptor	Analog

(Continued)

TABLE 1.4 (Continued)

Compound name	Current status	First approval	Approved indication(s)	Primary therapeutic area	Approved RoA	Conjugation	Molecular target	Chemical basis
Aviptadil	Approved	2000	Erectile dysfunction	Urology	Intracavernous	No	VIP1 receptor	Analog
Nesiritide	Approved	2001	Acute decompensated heart failure	Cardiovascular	Intravenous	No	NPR-A	Native
Carbetocin	Approved	2001	Postpartum hemorrhage	Infertility and OB-GYN	Intravenous	No	OT receptor	Analog
Teriparatide	Approved	2002	Osteoporosis	Bones and Connective Tissues	Subcutaneous	No	PTH1 receptor	Analog
Abarelix	Withdrawn	2003	Prostate cancer	Urology	Intramuscular	No	GnRH receptor	Analog
Enfuvirtide	Approved	2003	HIV infection	Antimicrobial; Antiviral	Subcutaneous	No	gp41	Analog
Ziconotide	Approved	2004	Pain	Pain	Intrathecal	No	N-type calcium channels	Native
Pramlintide	Approved	2005	Type 1 and type 2 diabetes	Metabolic	Subcutaneous	No	Calcitonin receptor	Analog
Exenatide	Approved	2005	Type 2 diabetes	Metabolic	Subcutaneous	No	GLP-1 receptor	Native
Romiplostim	Approved	2008	Immune thrombocytopenic purpura	Hematology	Subcutaneous	Yes	Thrombopoietin receptor	Heterologous
Degarelix	Approved	2008	Prostate cancer	Urology	Subcutaneous	No	GnRH receptor	Analog
Icatibant	Approved	2008	Hereditary angioedema in C1-esterase-inhibitor deficiency	Cardiovascular	Subcutaneous	No	B2 receptor	Analog

Mifamurtide	Approved	2009	Osteosarcoma	Oncology	Intravenous	Yes	NOD2 protein	Analog
Liraglutide	Approved	2009	Type 2 diabetes	Metabolic	Subcutaneous	Yes	GLP-1 receptor	Analog
Tesamorelin	Approved	2010	HIV lipodystrophy	Metabolic	Subcutaneous	Yes	GHRH receptor	Analog
Peginesatide	Withdrawn	2012	Anemia associated with chronic kidney disease	Hematology	Intravenous; subcutaneous	Yes	EPO receptor	Heterologous
Lucinactant	Approved	2012	Neonatal respiratory distress syndrome	Respiratory	Inhaled	Yes	Surfactant replacement	Heterologous
Linaclotide	Approved	2012	Constipation-dominant irritible bowel syndrome; chronic idiopathic constipation	Gastroenterology	Oral	No	GC-C receptor	Analog
Pasireotide	Approved	2012	Acromegaly; Cushing's disease	Endocrinology	Subcutaneous	No	SST receptors	Analog
Teduglutide	Approved	2012	Short bowel syndrome	Gastroenterology	Subcutaneous	No	GLP-2 receptor	Analog
Carfilzomib	Approved	2012	Multiple myeloma	Oncology	Intravenous	No	Chymotrypsin-like active site in the 20 S proteasome	Analog
Lixisenatide	Approved	2013	Type 2 diabetes	Metabolic	Subcutaneous	Yes	GLP-1 receptor	Analog
Afamelanotide	Approved	2014	Erythropoietic protoporphyria	Dermatology	Subcutaneous	No	MC1 receptor	Analog
Albiglutide	Approved	2014	Type 2 diabetes	Metabolic	Subcutaneous	Yes	GLP-1 receptor	Analog
Dulaglutide	Approved	2014	Type 2 diabetes	Metabolic	Subcutaneous	Yes	GLP-1 receptor	Analog
TDM-621	Approved (Medical device)	2014	Surgical bleeding	Hematology	Injection	No	Extracellular matrix	Heterologous

(*Continued*)

TABLE 1.4 (Continued)

Compound name	Current status	First approval	Approved indication(s)	Primary therapeutic area	Approved RoA	Conjugation	Molecular target	Chemical basis
Etelcalcetide	Approved	2016	Secondary hyperparathyroidism	Endocrinology	Intravenous	No	CaSR	Heterologous
Plecanatide	Approved	2017	Chronic idiopathic constipation	Gastroenterology	Oral	No	GC-C receptor	Analog
Abaloparatide	Approved	2017	Osteoporosis	Bones and Connective Tissues	Subcutaneous	No	PTH1 receptor	Analog
Oxodotreotide	Approved	2017	Neuroendocrine tumors	Oncology	Intravenous	Yes	SST receptors	Analog
Semaglutide	Approved	2017	Diabetes	Metabolic	Subcutaneous, oral	Yes	GLP-1 receptor	Analog
LJPC-501	Approved	2017	Catecholamine-resistant hypotension	Hepatology and Critical Care	Intravenous	No	Angiotensin II receptors	Native
Albuvirtide	Approved	2018	HIV infection	Antimicrobial; Antiviral	Intravenous	Yes	gp41	Analog
Plitidepsin	Approved	2018	Multiple myeloma	Oncology	Intravenous	No	Cytotoxic agent	Analog
Bremelanotide	Approved	2019	Hypoactive sexual desire disorder	Infertility and OB-GYN	Subcutaneous	No	MC receptors	Analog
Bulevirtide	Approved	2020	HDV infection	Antimicrobial; Antiviral	Subcutaneous	Yes	Unknown hepatocyte protein	Analog
Setmelanotide	Approved	2020	Genetic obesity syndromes	Metabolic	Subcutaneous	No	MC4 receptor	Analog

Voclosporin	Approved	2021	Lupus nephritis	Allergy and Immunology	Oral	No	Calcineurin	Analog
Dasiglucagon	Approved	2021	Hypoglycemia	Metabolic	Subcutaneous	No	Glucagon receptor	Analog
Pegcetacoplan	Approved	2021	Paroxysmal nocturnal humoglobinuria	Hematology	Subcutaneous	Yes	Complement C3	Heterologous
Difelikefalin[a]	Approved	2021	Pruritus associated with kidney disease		Intravenous	No		
Vosoritide[b]	Approved	2021	Achondroplasia		Subcutaneous	No		

[a] *Difelikefalin and vosoritide were approved after the original database analysis for this book chapter were performed. Unless otherwise specified, these peptides are not included as approved peptides in the figures and tables.*
[b] *Also link to footnote a*

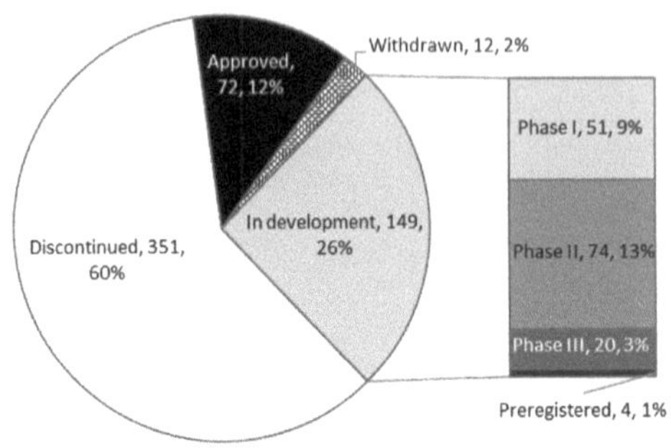

FIGURE 1.1 Current development status of therapeutic peptides. Numbers refer to the number and percentage of all peptides in the given category. "Withdrawn" refers to previously approved products no longer on the market; "Discontinued" refers to programs terminated prior to approval, and the "Active" category encompasses all peptides in active clinical development.

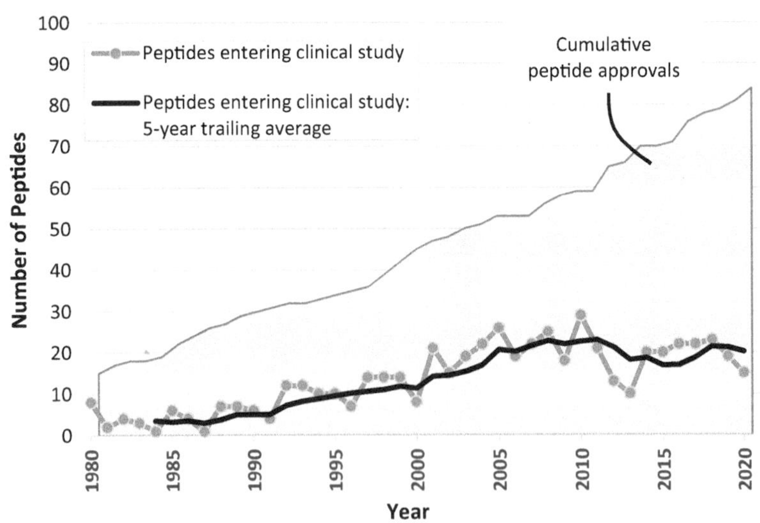

FIGURE 1.2 Cumulative number of peptides approved in major pharmaceutical markets and the number of peptides entering clinical development. Entry into clinical development is defined as the year of the first Phase 1 or pilot human study.

and purification, improved tools for molecular pharmacology, shifting healthcare trends, and the emergence of competing molecular types such as monoclonal antibodies.

1.2.2.1 Peptide length

In the 1980s, nearly all peptides entering clinical development were fewer than 10 amino acids long. Average peptide length has increased in each subsequent decade (Fig. 1.3), largely due to improvements in peptide synthesis and manufacturing technology. A broadened range of high-value molecular targets that are activated by larger peptide ligands, including B class GPCRs, has also played a role. In the 2010s, development candidates were more equally distributed in the various length ranges up to 40 amino acids, suggesting perhaps that innovations in manufacturing efficiency and methodology, including hybrid recombinant-synthetic approaches, have reduced the length limitations on peptide drug development [18].

1.2.2.2 Chemical basis of peptide therapeutics

We characterized the "chemical basis" of peptide drugs with respect to their relationship to endogenous peptide molecules: *native*, *analog*, and *heterologous*. A *native* peptide has the same amino acid sequence as an endogenous peptide product. Although the first native peptides were isolated from mammalian tissues, most native peptides currently on the market are produced synthetically or through recombinant expression. Many biologically active peptides, including venoms and toxins acting on intriguing molecular targets,

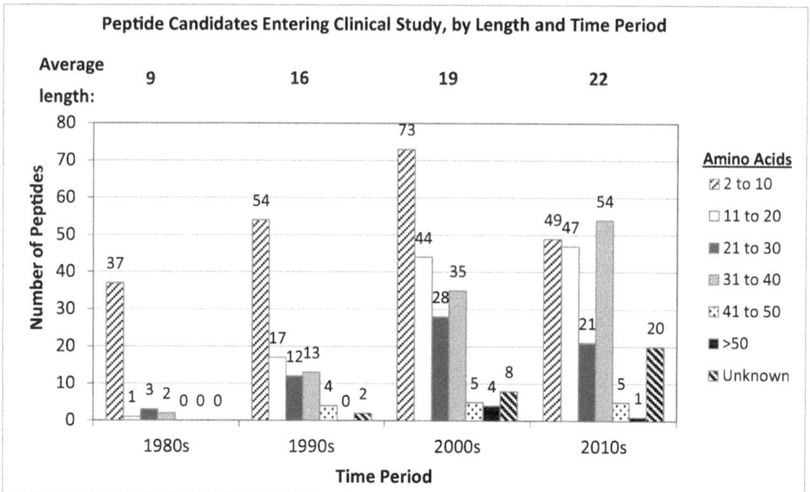

FIGURE 1.3 Length of peptides entering clinical development, by decade. Peptides with unknown length were not included in the average length calculation. If we were unable to determine the exact peptide sequence and/or length from publicly available information but knew that possible sequences were within a specific size range category, the peptide was assigned the average length within that size range (e.g., 35 AA for a peptide known to be between 31 and 40 AA in length).

have subsequently been identified from non-mammalian natural sources and have entered clinical trials, leading to a handful of approvals thus far [19].

The limitations of endogenous peptides provided motivation for researchers to create *analogs*, which are defined here as modified or substituted versions of native peptides with improved drug properties. As such, desmopressin and selepressin are vasopressin analogs with improved selectivity for the AVPR2 and AVPR1A receptors, described earlier in the text; and 8-amino acid octreotide has an increased plasma half-life and increased selectivity for sst_2 and sst_5 receptor subtypes over the native 14-amino acid somatostatin [20,21].

Heterologous peptides were discovered independently of the natural peptide, such as through synthetic library screening, phage display, or other methods. Examples include the peptidic portion of the thrombopoietin receptor agonist romiplostim (AMG531), which was discovered through phage display [22], and the CXCR4 receptor antagonist LY-2510924, discovered via "medium-throughput" screening [23]. In the last decade, approaches for generating libraries of constrained heterologous peptides through chemically-modified phage display have led to multiple peptide and peptide-drug conjugate discovery partnership agreements focused on nontraditional peptide targets [24,25].

The majority of peptide drugs on the market and in development are analogs that build on the intrinsic activity of native hormones with improved pharmaceutical properties (Fig. 1.4). An analog drug discovery program is de-risked with respect to target validation due to the availability of the native peptide as biological precedent. Methods for identifying heterologous peptide drug candidates expand target-space beyond receptors with native peptide or

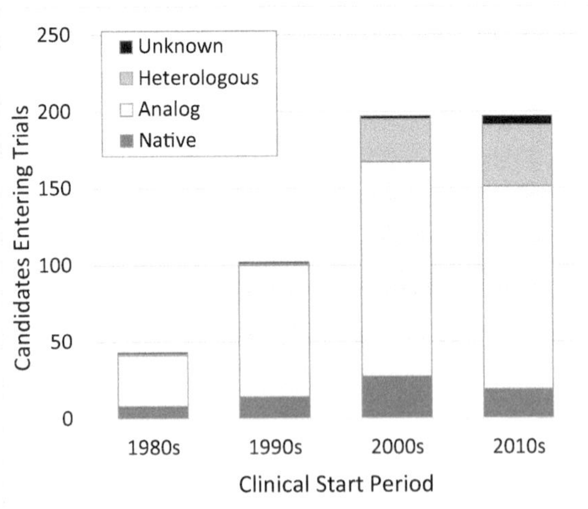

FIGURE 1.4 Chemical basis of peptides entering clinical development.

protein ligands, but the peptide leads may have insufficient potency or selectivity.

1.2.2.3 The rise of peptide conjugates

Conjugation has emerged as a popular mechanism to alter or enhance the properties of peptide and protein drug candidates. The proportion of conjugated peptides and the variety of conjugated moieties has increased over time; over 30% of peptides that entered clinical development in the 2010s were conjugates (Fig. 1.5). Conjugation to polyethylene glycol (PEG), albumin-binding lipids, and proteins such as Fc fragments has been used as a half-life extension strategy, with the first such peptides entering clinical development in the late-1990s. Conjugation can also be used to deliver a cytotoxic payload or imaging agent to specific cell types targeted by the peptide. See Table 1.5 for a description of various conjugation strategies deployed with therapeutic peptide development candidates.

1.2.2.4 Intrinsic peptide stability: primary sequence and higher-order structural protection

In addition to increasing half-life through conjugation and formulation, the intrinsic stability of peptide therapeutics has been improved through two general methods that render peptides less susceptible to protease degradation and/or renal clearance: traditional amino acid modification of primary sequence, and scaffolds based on "knotted" natural products or synthetic tertiary structures, which confer protection by reinforcing or shielding labile bonds. A well-known example of the former is the single amino acid change in teduglutide (Ala2 to Gly2), resulting in a DPP-IV resistant analog of human GLP-2 developed and approved for treatment of short bowel syndrome [26]. Examples of the latter include guanylyl cyclase-C agonists based on the guanylin and uroguanylin structures [27], disulfide-rich peptides originated by Protagonist Therapeutics [28], and bicyclic peptides generated by chemical cross-linking of phage-displayed peptides [24].

1.2.3 Molecular targets of therapeutic peptides

As noted above, many endogenous peptides and analogs thereof have been developed as potential therapeutics; thus, peptide hormone receptors are popular targets for peptide drugs. GPCRs represent the largest class of drug targets for peptides. Although this dominance has shrunk over time, greater than 40% of peptides that entered the clinic in the 2010s targeted GPCRs (Fig. 1.6). Non-GPCR cell surface receptors and other extracellular targets, including antimicrobial agents, make up the majority of the remaining targets. A small number of intracellular targets are still being pursued with the help of cell-penetrating strategies.

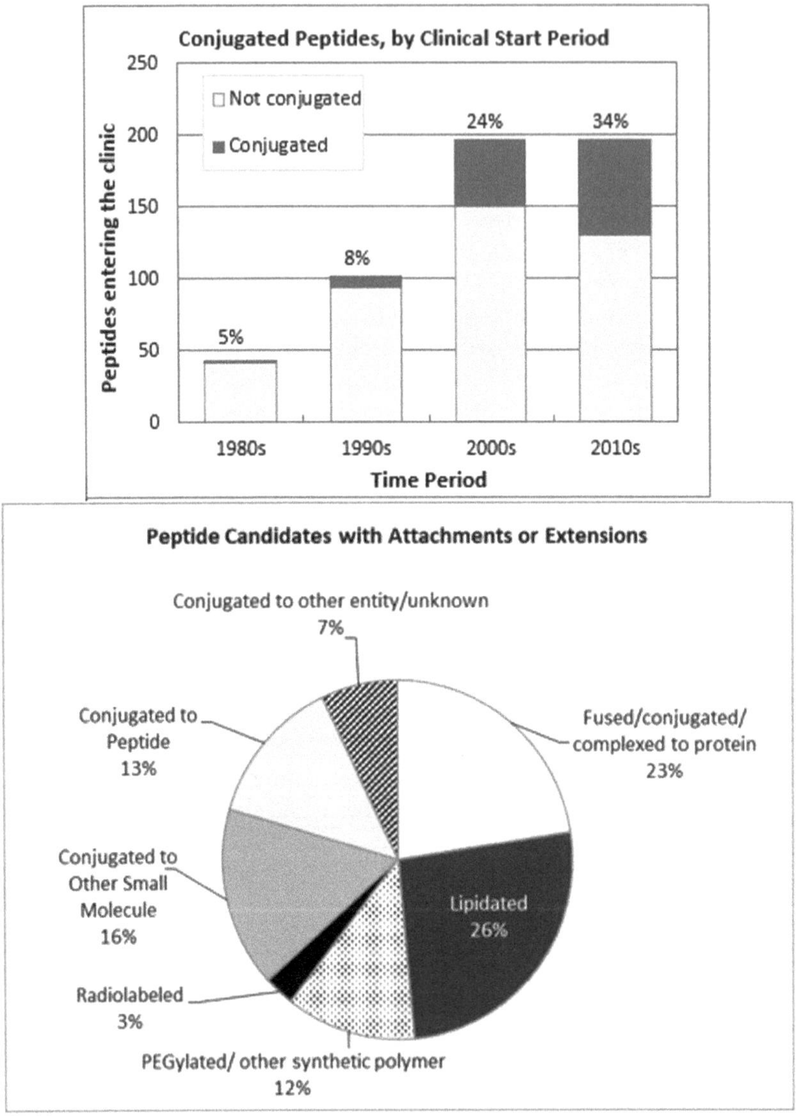

FIGURE 1.5 Number and percentage of conjugated peptides entering clinical development (top) and distribution of all conjugated moieties in the peptide database (bottom). Peptides of unknown conjugation status were assigned to the unconjugated category in the left sub-figure.

Certain targets have rapidly increased in popularity due to the success of a first-in-class peptide, resulting in numerous follow-on research programs pursuing the same mechanism. During the 1980s and 1990s, the

TABLE 1.5 Exemplary peptide conjugates. In this context, "conjugate" may refer to recombinantly produced fusions of a peptide of interest to a protein or peptide motif with a separate function, such as half-life extension or facilitation of cell membrane penetration.

Peptide name	Conjugated moiety	Rationale for conjugation	Current status
Romiplostim	Fc immunoglobulin	Half-life extension	Approved
Liraglutide	Hexadecanoic acid	Half-life extension	Approved
Peginesatide	PEG	Half-life extension	Approved, then withdrawn
NA-1	HIV-TAT peptide	Cell-penetrating peptide	Phase 3
Lutetium-DOTATATE	Lutetium-DOTA	Radiopharmaceutical	Approved
BT1718	Maytansoid toxin	Peptide-drug conjugate	Phase 2

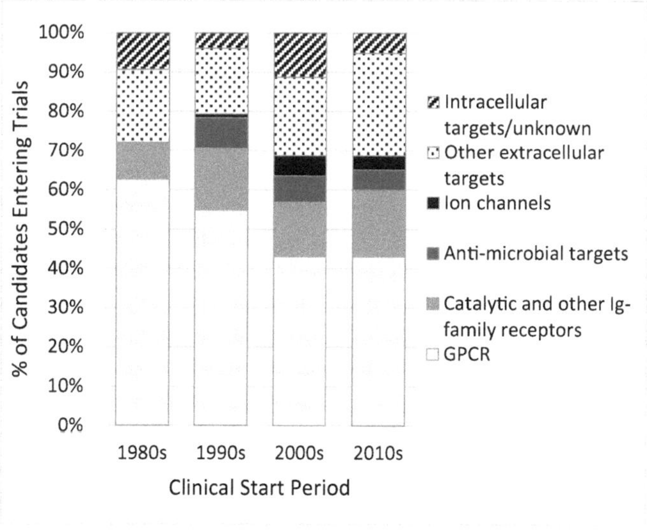

FIGURE 1.6 Molecular targets of peptides entering clinical development.

gonadotropin-releasing hormone (GnRH) receptor was a "hot" target, reflecting the promise of GnRH agonists and antagonists in treating a variety of important conditions regulated by reproductive hormones. GnRH-targeting peptides were formulated into a range of delivery formats and approved for prostate cancer, endometriosis, assisted reproduction, and other indications. Research and development focus shifted more towards small-molecule GnRH antagonists in the 2000s; after decades of effort, elagolix (ORILISSA) became the first small molecule in this class approved by the FDA in 2018, for the treatment of endometriosis.

Another favorite peptide target emerged in the late-1990s when GLP-1 receptor agonists derived from the exendin-4 structures hit the clinic for the treatment of type 2 diabetes and, eventually, obesity. The emergence of metabolic disease as a major worldwide health concern has driven development of agents that target gastrointestinal peptide receptors associated with appetite, insulin secretion, and energy balance [29]. GLP-1 remains a popular target today: by our count, 62 peptides that target GLP-1 receptor have entered the clinic, of which six are approved and 19 are in active clinical development as of May 2021; this count includes GLP-1 receptor antagonists and co-agonists that target other glucagon-family receptors in addition to the GLP-1 receptor.

A current trend is towards polypharmacy, wherein multiple molecular targets are addressed by a single molecule that simultaneously acts on multiple receptors or by a molecule that contains multiple functional domains. Dopastatin (BIM-23A760 or TBR-760) was an early attempt at polypharmacy: a hybrid of a peptidic somatostatin receptor agonist linked to a small-molecule dopamine agonist, designed to address two contributors to neuroendocrine tumor disease pathology [30]. Although dopastatin development was previously discontinued by originator Ipsen, rare endocrine disease specialty company Tiburio launched in 2019 with two chimeric dopamine-somatostatin molecules in its pipeline [31]. Other hybrid molecules that have entered clinical development attempt to target metabolic disease on different fronts, and GLP-1 receptor has an important role in this enthusiasm. "Twincretins" are single peptides that act as agonists of the GLP-1 receptor and the glucagon or glucose-dependent insulinotropic polypeptide (GIP) receptors [32]; for example, tirzepatide is a GLP-1/GIP dual agonist that has recently reported encouraging Phase 3 clinical results [33].

Several peptide-addressable targets for which no drugs are yet approved have shown therapeutic promise in early-stage clinical trials or in preclinical models of disease. For example, putative ligands for the adrenomedullin [34] and adiponectin [35,36] receptors have entered into Phase II clinical trials. The continuing addition of new targets suggests ongoing opportunity for first-in-class peptide therapeutics.

1.2.4 Approvals and uses of therapeutic peptides

Peptides have been investigated across the therapeutic spectrum, reflecting the potential utility across a wide range of indications and perhaps coupled with the cautious optimism that accompanies many development programs. Not surprisingly, the areas of highest concentration for peptide development (at present) are areas of high interest to the pharmaceutical industry: metabolic disease, oncology, and cardiovascular disease (Fig. 1.7).

Interestingly, the therapeutic landscape of approved peptide drugs does not mirror that of peptides in development. For example, many peptides have entered development in oncology indications but few have received approval, which may simply reflect poor success rates in oncology as a whole [37–39]. Since most successful oncology therapeutics either access the cytoplasm to act as cytotoxic agents (small-molecule chemotherapeutics) or bind to cell surface receptors to block pro-proliferative receptor activation and/or target cancer cells for immune system destruction (antibody therapeutics), the role for peptides in oncology may require careful consideration of the molecular target and tumor of interest.

Endogenous ligands for peptide hormone receptors remain the basis for many peptide drugs routinely used in medical practice. Since 2015, eight of the 15 peptides approved by regulatory agencies have been close analogs of endogenous human peptides (Table 1.4). Of the remainder, one approval was for a native peptide (angiotensin II); four are analogs of nonhuman peptides (plecanatide, plitidepsin, bulevirtide, voclosporin); and the others are heterologous ligands (pegcetacoplan and difelikefalin).

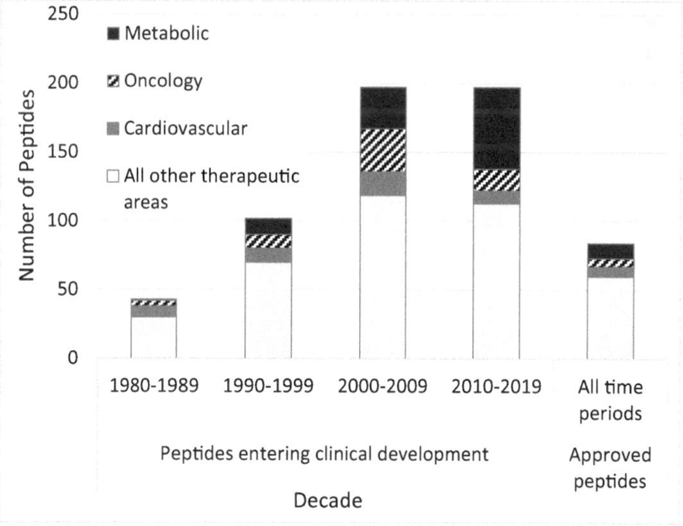

FIGURE 1.7 Primary therapeutic area for peptides by time period of clinical initiation, compared with the primary therapeutic area for approved peptides across all time periods (including pre-1980). This figure highlights peptides in the three most popular therapeutic areas (at present). See the supplementary information for therapeutic area definitions.

1.3 Clinical development timelines and benchmarks for peptides

The median development time for the cohort of peptides approved since 2015 (Fig. 1.8) was 9.7 years and varied widely across programs. Some of the peptides with a short-to-moderate length of clinical development were "me-too" compounds against established targets, in indications with clear clinical and regulatory precedent: the guanylate cyclase-C (GC-C) activator plecanatide for constipation, the liquid-formulated glucagon analog dasiglucagon for hypoglycemia [40], and the GLP-1 agonist semaglutide for type 2 diabetes mellitus.

Since Fig. 1.8 captures only the first approval for a peptide, it does not include development timeline information for label extensions. After the initial approval of semaglutide as a weekly injection, the sponsor obtained an additional approval for semaglutide oral (RYBELSUS), the first oral formulation of a GLP-1 agonist [10]. RYBELSUS uses a combination of strategies including resistance to dipeptidyl peptidase-IV (DPP-IV) proteolysis, half-life extension through lipidation, and permeation enhancement with salcaprozate sodium (SNAC) to enable sufficient oral bioavailability for once-daily dosing [4].

The peptide with the fastest transition to approval was LJPC-501, intravenous angiotensin II for hypotension associated with distributive shock.

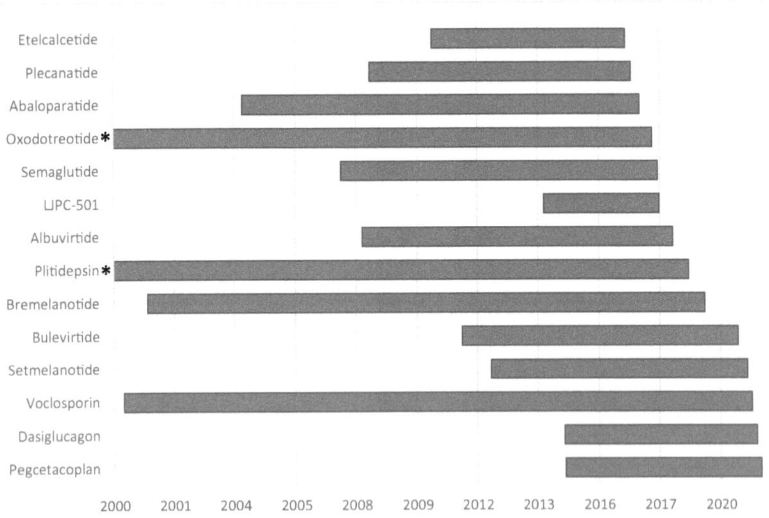

FIGURE 1.8 Duration of clinical development for peptides approved from 2016 through May 2021. The left side of each bar represents initiation of the first clinical trial; the right side of each bar represents the peptide's first approval in a major market. Oxodotreotide and plitidepsin, indicated on the graph with asterisks, entered clinical development prior to 1999.

The physiological effects of this native peptide are well-understood, and the sponsor received approval based on a primary endpoint of mean arterial pressure (MAP) that is directly related to the pharmacological activity of the molecule [10,41].

The pegcetacoplan development program also progressed rapidly through clinical trials in the rare indication of paroxysmal nocturnal hemoglobinuria (PNH) [42]. Pegcetacoplan is a pegylated dimer of cyclic peptides, each copy of which is based on the sequence of the complement C3 inhibitor compstatin, which was discovered via phage display [43]. Although pegcetacoplan is the first approved peptide that targets the complement cascade, others may follow given the commercial success of antibody therapeutics against complement C5 and the potential role for this system across multiple diseases [44].

We also performed a comparison of the probability of success for peptides, as defined by our peptide database, compared to industry-wide benchmarks for new biological entities (NBEs), new molecular entities (NMEs), and peptides. The industry benchmarks we used for comparison come from a recent report by the Biotechnology Innovation Organization (BIO) and collaborators for programs between 2011 and 2020, broken down by drug modality [38]. This success metric looks at the likelihood that a drug candidate will eventually receive regulatory approval at a given phase of development. NBEs comprise antibodies and other proteins, live biologic products, and vaccines, while NMEs primarily comprise small molecules and synthetic peptides.

The BIO methodology counts each disease indication program individually, so a single molecular entity could be represented more than once if it was developed for multiple indications. For example, a somatostatin receptor agonist tested in acromegaly and neuroendocrine tumors would be counted twice, once for each of these indications, and the success and failure of each program would be counted separately. In contrast, the analysis performed herein with our peptide dataset would only account for this program once, based on its highest phase or most successful indication.

Our peptide success rates are higher at each phase of development than peptide success rates derived by BIO, due to the differences in per-program methodology described above (Fig. 1.9). These rates, as well as those provided in other key studies, display similar trends across phases of development: the highest rate of attrition occurs at Phase 2, when programs must show clinical proof of concept.

In the past, industry-wide success rates of biologics have been significantly higher than success rates of NMEs, due to generally higher target specificity and reduced potential for off-target toxicity of antibody therapeutics. This gap has narrowed in recent years, potentially driven by an increased proportion of riskier antibody targets with lower correlation between target engagement and clinical efficacy [45]. Since native and analog peptides have some of the advantages and disadvantages of both

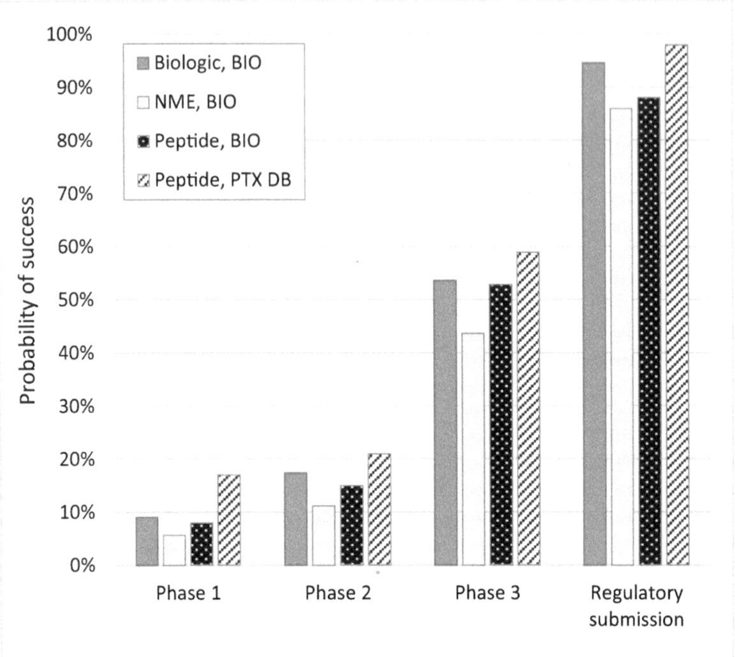

FIGURE 1.9 Probability of success for agents at various stages of clinical development. NME: new molecular entity; "regulatory submission" refers to the approval rate of marketing applications. Success rates for biologic and NME drugs are taken from a 2021 report (see text), and peptide success rates are taken from the authors' peptide therapeutics database (PTX DB).

biologics and small molecules, peptide attrition rates often fall in an intermediate range. We believe that development trends toward heterologous peptides and molecular targets without endogenous peptide ligands will cause peptide success rates to drop, in analogy to biologic success rates.

Furthermore, with the proliferation of "me-too" programs targeting well-clinically-validated receptors, we expect higher rates of strategy-based failures, as drug sponsors working on popular targets trade early-stage clinical attrition for costly and time-consuming late-stage or commercial attrition. One example is the GLP-1 agonist albiglutide, which received marketing approval in the US in 2014 but was discontinued in 2017 due to limited prescribing of the drug in a competitive GLP-1 agonist market [46]. In these mature target landscapes, sponsors of new programs will need to differentiate their programs against current standard-of-care or risk rapid clinical obsolescence.

1.4 The future of peptide therapeutics

From humble beginnings as substances isolated from livestock glands, peptides have established a unique therapeutic niche and will continue to be an important element in the pharmaceutical landscape. We believe that research will continue to identify new peptide opportunities by identifying new endogenous receptor-ligand pairs; uncovering new therapeutic uses for old mechanisms; and deploying new strategies to expand the clinical uses where peptides are a favorable molecular modality.

Improvements in peptide screening and computational biology will continue to support peptide drug discovery. Metabolomic, proteomic, and genomic screening of toxins and other sources of natural products can identify bioactive peptides that may contain unique structural features generated by uncommon post-translational modifications or nonribosomal synthesis [47–49]. An improved understanding of the molecular basis for human genetic disorders can generate new potential therapeutic leads [50,51], and the de-orphanization of poorly-characterized peptide receptors can stimulate research efforts for new receptor-ligand pairs [52,53].

Peptides approved for rare diseases may find increased utility in broader indications. Specific methods to block the complement cascade (pegcetacoplan) and induce satiety via melanocortin receptor 4 (setmelanotide) could provide benefit beyond the rare or ultra-rare indications currently targeted by these recently-approved peptides. In the case of setmelanotide, a wider range of patients with genetically-defined, early-onset obesity syndromes may be successfully treatable; the initial challenge for the peptide's sponsor is to identify these patients, who previously might not have received genetic testing due to the dearth of approved treatment options available.

Cross-pollination between fields of research may help researchers find new therapeutic uses for peptides. Peptide natural products act as potent antimicrobials, and synthetic peptides with membrane-penetrating properties are well established. In one new research angle, screening heterologous peptides for the ability to modulate gut microbiome composition toward a more "healthy" phenotype might provide new treatment opportunities [54].

Finally, new peptide drug delivery, formulation, and half-life extension approaches will further the reach of this unique class of molecules. Efforts are underway to improve the oral availability of peptide therapeutics by increasing drug stability in the gastrointestinal (GI) tract and formulating peptides with permeability enhancers [55], and improving the central nervous system (CNS) availability of peptides through conjugation to carrier molecules or delivery in nanoparticles [56,57].

Acknowledgments

The authors acknowledge Denise Riedl for her thoughtful feedback and comments on the manuscript; Mimika Koletsou and Rob Meadows for developing tools for data mining and visualization; and the many contributors who have carefully curated the peptide dataset since its inception.

References

[1] Banting FG, Best CH, Collip JB, Campbell WR, Fletcher AA. Pancreatic Extracts in the Treatment of Diabetes Mellitus. Can Med Assoc J 1922;12:141–6. Available from: https://www.ncbi.nlm.nih.gov/pubmed/20314060.

[2] Elkinton JR, Hunt Jr AD, Godfrey LL, Mc CW, Rogerson AG, Stokes JJ. Effects of pituitary adrenocorticotropic hormone (acth) therapy. J Am Med Assoc 1949;141:1273–9. Available from: https://doi.org/10.1001/jama.1949.02910180001001.

[3] Di L, Strategic approaches to optimizing peptide ADME properties, AAPS J. 17 (n.d.) 134–143. Available from: https://doi.org/10.1208/s12248-014-9687-3.

[4] Brayden DJ, Hill TA, Fairlie DP, Maher S, Mrsny RJ. Systemic delivery of peptides by the oral route: Formulation and medicinal chemistry approaches. Adv Drug Deliv Rev 2020. Available from: https://doi.org/10.1016/j.addr.2020.05.007.

[5] Kim RJ, Malattia C, Allen M, Moshang T, Maghnie M. Vasopressin and desmopressin in central diabetes insipidus: Adverse effects and clinical considerations. Pediatr Endocrinol Rev 2004;(2 Suppl 1):115–23.

[6] Laterre PF, Berry SM, Blemings A, Carlsen JE, François B, Graves T, et al. Effect of selepressin vs placebo on ventilator- and vasopressor-free days in patients with septic shock: The SEPSIS-ACT Randomized Clinical Trial. JAMA - J Am Med Assoc 2019. Available from: https://doi.org/10.1001/jama.2019.14607.

[7] Hollenstein K, de Graaf C, Bortolato A, Wang MW, Marshall FH, Stevens RC. Insights into the structure of class B GPCRs. Trends Pharmacol Sci 2014;35:12–22. Available from: https://doi.org/10.1016/j.tips.2013.11.001.

[8] Agoulnik AI, Agoulnik IU, Hu X, Marugan J. Synthetic non-peptide low molecular weight agonists of the relaxin receptor 1. Br J Pharmacol 2017;174:977–89. Available from: https://doi.org/10.1111/bph.13656.

[9] Mishra RK, Shum AK, Platanias LC, Miller RJ, Schiltz GE. Discovery and characterization of novel small-molecule CXCR4 receptor agonists and antagonists. Sci Rep 2016;6:30155. Available from: https://doi.org/10.1038/srep30155.

[10] U.S.F.& D. Administration, Drugs@FDA: FDA-Approved Drugs, (n.d.). https://www.accessdata.fda.gov/scripts/cder/daf/ (accessed June 6, 2021).

[11] Hollenberg NK, Williams GH, Burger B, Ishikawa I, Adams DF. Blockade and stimulation of renal, adrenal, and vascular angiotensin II receptors with 1-Sar, 8-Ala angiotensin II in normal man. J Clin Invest 1976;57:39–46. Available from: https://doi.org/10.1172/JCI108266.

[12] Camacho PM, Petak SM, Binkley N, Diab DL, Eldeiry LS, Farooki A, et al. American association of clinical endocrinologists/American college of endocrinology clinical practice guidelines for the diagnosis and treatment of postmenopausal osteoporosis-2020 update. Endocr Pract 2020. Available from: https://doi.org/10.4158/GL-2020-0524SUPPL.

[13] Chang YS, Graves B, Guerlavais V, Tovar C, Packman K, To KH, et al. Stapled α-helical peptide drug development: a potent dual inhibitor of MDM2 and MDMX for p53-dependent cancer therapy. Proc Natl Acad Sci U S A 2013;110:E3445–54. Available from: https://doi.org/10.1073/pnas.1303002110.

[14] Bhandari A, Cheng X, Bourne G, Frederick B, Zhang J, McMahon J, et al. Discovery of novel oral peptide antagonists of IL-23 receptor that are efficacious in a rat model of IBD. Inflamm Bowel Dis 2016;22(Suppl 1):S55. Available from: https://doi.org/10.1097/01.MIB.0000480271.29623.7d.

[15] Birk AV, Liu S, Soong Y, Mills W, Singh P, Warren JD, et al. The mitochondrial-targeted compound SS-31 re-energizes ischemic mitochondria by interacting with cardiolipin. J Am Soc Nephrol 2013;24:1250–61. Available from: https://doi.org/10.1681/ASN.2012121216.

[16] Reichert J, Pechon P, Tartar A, Dunn MK, Development trends for peptide therapeutics: a comprehensive quantitative analysis of peptide therapeutics in clinical development, 2010.

[17] Lau JL, Dunn MK. Therapeutic peptides: historical perspectives, current development trends, and future directions. Bioorganic Med Chem 2018;. Available from: https://doi.org/10.1016/j.bmc.2017.06.052.

[18] Pennington MW, Zell B, Bai CJ. Commercial manufacturing of current good manufacturing practice peptides spanning the gamut from neoantigen to commercial large-scale products. Med Drug Discov 2021. Available from: https://doi.org/10.1016/j.medidd.2020.100071.

[19] Pennington MW, Czerwinski A, Norton RS. Peptide therapeutics from venom: current status and potential, Bioorganic. Med Chem 2018. Available from: https://doi.org/10.1016/j.bmc.2017.09.029.

[20] Hannon JP, Nunn C, Stolz B, Bruns C, Weckbecker G, Lewis I, et al. Drug design at peptide receptors: somatostatin receptor ligands. J Mol Neurosci 2002;18:15–27. Available from: https://doi.org/10.1385/JMN:18:1-2:15.

[21] Fricker G, Bruns C, Munzer J, Briner U, Albert R, Kissel T, et al., Intestinal absorption of the octapeptide SMS 201-995 visualized by fluorescence derivatization, Gastroenterology. 100 (n.d.) 1544–1552. Available from: https://doi.org/10.5555/uri:pii:001650859190651Z.

[22] Bussel JB, Kuter DJ, George JN, McMillan R, Aledort LM, Conklin GT, et al. AMG 531, a Thrombopoiesis-Stimulating Protein, for Chronic ITP. N Engl J Med 2006;355:1672–81. Available from: https://doi.org/10.1056/NEJMoa054626.

[23] Peng SB, Zhang X, Paul D, Kays LM, Gough W, Stewart J, et al. Identification of LY2510924, a novel cyclic peptide CXCR4 antagonist that exhibits antitumor activities in solid tumor and breast cancer metastatic models. Mol Cancer Ther 2015;14:480–90. Available from: https://doi.org/10.1158/1535-7163.MCT-14-0850.

[24] Heinis C, Rutherford T, Freund S, Winter G. Phage-encoded combinatorial chemical libraries based on bicyclic peptides. Nat Chem Biol 2009. Available from: https://doi.org/10.1038/nchembio.184.

[25] Cary DR, Ohuchi M, Reid PC, Masuya K. Constrained peptides in drug discovery and development. Yuki Gosei Kagaku Kyokaishi/Journal Synth Org Chem 2017. Available from: https://doi.org/10.5059/yukigoseikyokaishi.75.1171.

[26] Drucker DJ. The Discovery of GLP-2 and development of teduglutide for short bowel syndrome. ACS Pharmacol Transl Sci 2019. Available from: https://doi.org/10.1021/acsptsci.9b00016.

[27] Waldman SA, Camilleri M. Guanylate cyclase-C as a therapeutic target in gastrointestinal disorders. Gut 2018. Available from: https://doi.org/10.1136/gutjnl-2018-316029.

[28] Modi NB, Cheng X, Mattheakis L, Hwang CC, Nawabi R, Liu D, et al. Single- and multiple-dose pharmacokinetics and pharmacodynamics of PN-943, a Gastrointestinal-restricted oral peptide antagonist of $\alpha 4\beta 7$, in healthy volunteers. Clin Pharmacol Drug Dev 2021. Available from: https://doi.org/10.1002/cpdd.946.

[29] Clemmensen C, Müller TD, Woods SC, Berthoud HR, Seeley RJ, Tschöp MH. Gut-brain cross-talk in metabolic control. Cell 2017;168:758–74. Available from: https://doi.org/10.1016/j.cell.2017.01.025.

[30] Saveaunu A, Datta R, Zhang S, Shen Y, Dong JZ, Graillon T, et al. Novel Somatostatin-Dopamine Chimeric Compound Demonstrates Superior Efficacy in Suppressing Growth Hormone Secretion from Human Acromegalic Tumors Partially Responsive to Current Somatostatin and Dopamine Therapies. Boston, MA: ENDO; 2016. 2016.

[31] Tiburio Therapeutics, Inc., (n.d.). https://www.cydanco.com/cydan-newcos/newco/tiburio-therapeutics-inc/ (accessed June 6, 2021).

[32] Finan B, Ma T, Ottaway N, Müller TD, Habegger KM, Heppner KM, et al. Unimolecular Dual Incretins Maximize Metabolic Benefits in Rodents, Monkeys, and Humans. Sci Transl Med 2013;5. Available from: https://doi.org/10.1126/scitranslmed.3007218 209ra151-209ra151.

[33] Tirzepatide significantly reduced A1C and body weight in people with type 2 diabetes in two phase 3 trials from Lilly's SURPASS program, CISION PR Newswire. (2021). https://www.prnewswire.com/news-releases/tirzepatide-significantly-reduced-a1c-and-body-weight-in-people-with-type-2-diabetes-in-two-phase-3-trials-from-lillys-surpass-program-301229506.html (accessed June 6, 2021).

[34] This Study Collects Information on the safety of inhaled pegylated adrenomedullin (PEG-ADM), How the drug is tolerated and how it affects patients suffering from a type of lung failure that cause fluid to build up in the lungs making breathing difficult (ClinicalTrials. Gov.) (n.d.). https://clinicaltrials.gov/ct2/show/NCT04417036 (accessed June 6, 2021).

[35] ALY688-SR in generally healthy overweight or obese adults, ClinicalTrials.Gov. (n.d.). https://clinicaltrials.gov/ct2/show/NCT04855565 (accessed June 6, 2021).

[36] Phase 2a/3 Safety and Efficacy of ALY688 Ophthalmic Solution in Subjects With Dry Eye Disease (OASIS-1), (n.d.). https://clinicaltrials.gov/ct2/show/NCT04899518 (accessed June 6, 2021).

[37] DiMasi JA, Reichert JM, Feldman L, Malins A. Clinical approval success rates for investigational cancer drugs. Clin Pharmacol Ther 2013;94:329–35. Available from: https://doi.org/10.1038/clpt.2013.117.

[38] Thomas D, Chancellor D, Micklus A, LaFever S, Hay M, Chaudhuri S, et al., Clinical development success rates and contributing factors, 2011-2020, 2021. https://pharmaintelligence.informa.com/~/media/informa-shop-window/pharma/2021/files/reports/2021-clinical-development-success-rates-2011-2020-v17.pdf.

[39] Wong CH, Siah KW, Lo AW. Estimation of clinical trial success rates and related parameters. Biostatistics 2019. Available from: https://doi.org/10.1093/biostatistics/kxx069.

[40] Blair HA. Dasiglucagon: First Approval. Drugs 2021. Available from: https://doi.org/10.1007/s40265-021-01531-z.

[41] Khanna A, English SW, Wang XS, Ham K, Tumlin J, Szerlip H, et al. Angiotensin II for the treatment of vasodilatory shock. N Engl J Med 2017. Available from: https://doi.org/10.1056/nejmoa1704154.

[42] Hillmen P, Szer J, Weitz I, Röth A, Höchsmann B, Panse J, et al. Pegcetacoplan vs eculizumab in paroxysmal nocturnal hemoglobinuria. N Engl J Med 2021. Available from: https://doi.org/10.1056/nejmoa2029073.

[43] Sahu A, Kay BK, Lambris JD. Inhibition of human complement by a C3-binding peptide isolated from a phage-displayed random peptide library. J Immunol 1996;157(2):884–91.

[44] Mastellos DC, Ricklin D, Lambris JD. Clinical promise of next-generation complement therapeutics. Nat Rev Drug Discov 2019. Available from: https://doi.org/10.1038/s41573-019-0031-6.

[45] Galson S, Austin CP, Khandekar E, Hudson LD, DiMasi JA, Califf R, et al. The failure to fail smartly. Nat Rev Drug Discov 2021. Available from: https://doi.org/10.1038/d41573-020-00167-0.

[46] Ernst D. Type 2 diabetes injectable therapy will soon be unavailable, MPR. (2017). https://www.empr.com/home/news/type-2-diabetes-injectable-therapy-will-soon-be-unavailable/ (accessed June 6, 2021).

[47] Ling LL, Schneider T, Peoples AJ, Spoering AL, Engels I, Conlon BP, et al. A new antibiotic kills pathogens without detectable resistance. Nature 2015;517:455–9. Available from: https://doi.org/10.1038/nature14098.

[48] Motley JL, Stamps BW, Mitchell CA, Thompson AT, Cross J, You J, et al. Opportunistic Sampling of Roadkill as an Entry Point to Accessing Natural Products Assembled by Bacteria Associated with Non-anthropoidal Mammalian Microbiomes. J Nat Prod 2017;80:598–608. Available from: https://doi.org/10.1021/acs.jnatprod.6b00772.

[49] Lavergne V, Harliwong I, Jones A, Miller D, Taft RJ, Alewood PF. Optimized deep-targeted proteotranscriptomic profiling reveals unexplored Conus toxin diversity and novel cysteine frameworks. Proc Natl Acad Sci U S A 2015;112:E3782–91. Available from: https://doi.org/10.1073/pnas.1501334112.

[50] Duerrschmid C, He Y, Wang C, Li C, Bournat JC, Romere C, et al. Asprosin is a centrally acting orexigenic hormone. Nat Med 2017. Available from: https://doi.org/10.1038/nm.4432.

[51] Mishra I, Duerrschmid C, Ku Z, He Y, Xie W, Silva ES, et al. Asprosin-neutralizing antibodies as a treatment for metabolic syndrome. Elife 2021. Available from: https://doi.org/10.7554/ELIFE.63784.

[52] Walther A, Riehemann K, Gerke V. A novel ligand of the formyl peptide receptor: Annexin I regulates neutrophil extravasation by interacting with the FPR. Mol Cell 2000;5:831–40. Available from: https://www.ncbi.nlm.nih.gov/pubmed/10882119.

[53] Foster SR, Hauser AS, Vedel L, Strachan RT, Huang XP, Gavin AC, et al. Discovery of human signaling systems: pairing peptides to G protein-coupled receptors. Cell 2019. Available from: https://doi.org/10.1016/j.cell.2019.10.010.

[54] Chen PB, Black AS, Sobel AL, Zhao Y, Mukherjee P, Molparia B, et al. Directed remodeling of the mouse gut microbiome inhibits the development of atherosclerosis. Nat Biotechnol 2020. Available from: https://doi.org/10.1038/s41587-020-0549-5.

[55] Melmed S, Popovic V, Bidlingmaier M, Mercado M, van der Lely AJ, Biermasz N, et al. Safety and efficacy of oral octreotide in acromegaly: results of a multicenter phase III trial. J Clin Endocrinol Metab 2015;100:1699–708. Available from: https://doi.org/10.1210/jc.2014-4113.

[56] Lalatsa A, Schatzlein AG, Uchegbu IF. Strategies to deliver peptide drugs to the brain. Mol Pharm 2014;11:1081–93. Available from: https://doi.org/10.1021/mp400680d.

[57] Hill MD, Martin RH, Mikulis D, Wong JH, Silver FL, Terbrugge KG, et al. E. trial investigators, Safety and efficacy of NA-1 in patients with iatrogenic stroke after endovascular aneurysm repair (ENACT): a phase 2, randomised, double-blind, placebo-controlled trial. Lancet Neurol 2012;11:942–50. Available from: https://doi.org/10.1016/S1474-4422(12)70225-9.

Section 2

Basic science

Chapter 2

Therapeutic peptidomimetics: targeting the undruggable space

Lloyd Mabonga[1], Priscilla Masamba[2] and Abidemi Paul Kappo[2]
[1]Department of Biochemistry and Microbiology, University of Zululand, KwaDlangezwa, South Africa, [2]Department of Biochemistry, Molecular Biophysics and Structural Biology Group, Kingsway Campus Auckland Park, University of Johannesburg, Johannesburg, South Africa

2.1 Introduction

The process of drug discovery is propelled by molecular targets to identify, synthesize, characterizing and screening new therapeutic agents that can selectively target disease-relevant molecular mechanisms or pathways [1]. Advances in smart-drug discovery technology have enabled scientists to design, develop and screen novel therapeutic drug compounds, which can be used effectively in the treatment of various diseases [2]. One emerging approach to target the "undruggable space" is the efficient mimicking of bioactive conformations of peptides yielding molecules that are generally referred to as therapeutic peptidomimetics. This strategy has been a long-standing goal in the design of protein–protein interaction (PPI)-focused smart-drug inhibitors [3,4]. Therapeutic peptidomimetics are analeptic small protein-like chains whose vital elements mimic native peptides or proteins in three-dimensional space and retain the ability to interact with pharmacological targets and yield therapeutically relevant biological effects [5]. They typically arise either from modifications of antecedent peptides or by designing resemblance systems that mimic peptides like peptoids and β-peptides [6]. Therapeutic peptidomimetics offer solutions to the challenges associated with pure peptide-based drugs [7,8]. This chapter focuses on therapeutic peptidomimetics and targeting the undruggable space.

2.2 The "undruggable" space

Over the years, proteomic and genomic revolutions have furnished us with an ever-expanding panorama of mechanistic shrewdness into pathogenesis and disease progression. Mutated genes and pathological protein products

Peptide and Peptidomimetic Therapeutics. DOI: https://doi.org/10.1016/B978-0-12-820141-1.00026-1

have emerged as the basis for modern smart-drug development [9]. The realization of the preeminence of disrupting disease-relevant proteins has increased over the years and investment in this drug-development space continues to skyrocket [10]. Despite the significant increase in investment, the number of new drugs has however plummeted, indicating that a fundamental barrier exists in creating new drugs. Rhetoric challenges have been in existence for these classical protein-based drug modalities for many decades [10]. The challenges can be traced, at least in part, to an insurmountable obstacle that has stymied pharmaceutical scientists—the *undruggable proteins* [9,10]. The unfortunate irony hangs in the balance in finding approved therapies that directly target products of the most frequently activated genes. There are yet no direct-acting inhibitors for mutationally activated disease-relevant proteins found in nearly half of all pathophysiological cues and disease states [7–11].

Validated overwhelming evidence of high-quality biological data for these therapeutic vulnerabilities has been on parade for many centuries. However, many of these attractive like-targets from a biological perspective are still considered prototypically intractable. They are adjudged as "difficult to drug" or "yet to be drugged." Regardless of the tremendous efforts in the past decades to uncover multiple aspects, the proteome space remained beyond the targeting capabilities of the orthodox drug discovery technologies [11–13]. Hence, the concept of the *undruggable space* moniker is strongly colored by history and was fabricated to account for proteins that could not be targeted pharmacologically [10]. It suggests that the "proteome space" lie beyond the reach of human ingenuity in drug discovery parlance. It is an operational definition, based on experiences and hypotheses to date of trying to bind specific kinds of small molecules to specific proteins using specific tests. One obvious challenge with such an empirical definition is that the data supporting the conclusion of undruggability are rarely published as negative results. Much of the discussion of protein druggability is driven by anecdotal accounts of unsuccessful screening efforts or hypothetical arguments. Additionally, this concept has obliterated the potential of the human proteome to yield new drugs and for a very long time, the critical need to target the proteome space remained unfathomable [6]. As a result, the quest to cure incurable diseases today lies in the possibilities of developing drugs that interact with the proteins so-branded "undruggable" [7–13].

To understand the nature of these undruggable proteins and how they might ultimately be targeted to make drugs, it is crucial to understand what makes some protein targets druggable and others undruggable, and the prospects for drugging the undruggable ones [14,15]. According to some studies [16–18], druggable proteins are characterized by the presence of a solvent-accessible surface harboring a deep invagination fringed by hydrophobic amino acid side chains (hydrophobic regions). The hydrophobic pocket helps to engulf small-molecule drugs and engage with the large molecule protein therapeutics (also known as biologics) to maximize contact surface area and produce a stable complex. Surface character analyses of protein structures available in the Protein Data Bank (PDB) suggest that only less than 10% of the proteins encoded in the human proteome possess hydrophobic pockets suitable for tight binding of small drug-like molecules [15–17]. In addition to that, bioinformatics studies predict that more than 85% of all human proteins are located inside the cell making them inaccessible by the large molecule biologics. The combination of hydrophobic pockets and intracellular localization makes such protein targets risky prospects and difficult to the drug [18]. Thus, the majority (as high as 85%) of all existing proteins are in the "undruggable space." The improving technology expertise is ushering in a new dawn in the protein-based drug discovery field and soon the status "undruggable space" may be fleeting [9]. Drugging the "undruggable" seeks to expand and explore the intractable space and address the many therapeutically relevant proteins, which fall outside the "druggable space" [7]. While the phrase is aspirational, it loosely overstates the resilience of the proteome space toward drug discovery. However, progress is being made to 'drug' many of these proteins [13].

The realization that most undruggable targets are poorly suited for engagement by orthodox drug classes has provided a powerful impetus for research endeavors aimed at discovering entirely new structural classes of drugs that could fill the existing gaps in the drug discovery parlance [5–13]. Of interest are therapeutic peptidomimetics. They are synthetic peptide mimicries that synergistically combines the versatile surface-recognition traits of biologics with the synthetic manipulability and cell-permeability of small molecules [9]. Therapeutic peptidomimetics are a fast-growing spectrum of nootropics in modern-day drug discovery. They are a tremendously important diagnostic and therapeutic arsenal that may extrapolate to highly protracted remissions and curative therapies for several incurable diseases [8]. Technological advances point to therapeutic peptidomimetics as precedented or unprecedented tools that can be used to target the "undruggable" proteome space. This positive boost was, nevertheless, weighted against one huge negativity; the nearly catholic consensus among drug discovery professionals of the inability of peptides to be drugs. Peptidomimetics offer solutions to the challenges associated with pure peptide-based drugs.

The majority of the established peptide drugs generally suffer from several delimitations. They are unable to keep constant and well-defined three-dimensional configuration, but instead exist as a huge group of conformational isomers that interconvert on eye brink scale. They have limited stability toward degradation by peptidases. Peptides have poor biologic membrane penetrating properties, for example through intestines and cell membranes. They possess low oral bioavailability, rapid excretion and poor target specificity due to their being flexible in nature [3–6]. Taken together, the listed detriments severely diminish the prospects of peptides becoming drugs, and their combination has effectively debilitated the discovery of what is known as peptide drugs. Consequently, the quest to drug the undruggable space has converged on therapeutic peptidomimetics as one of the potential smart-drug tools to cure incurable diseases.

2.3 Therapeutic peptidomimetics

Therapeutic peptidomimetics proffer conformationally limited structures reducing cross-target interactions. They have better penetration properties through biologic membranes. They though possess improved resistance to proteolytic degradation by peptidases and other enzymes, and as such, are known to harbor immunological resistance to immune responses [3,4]. To achieve these properties, antecedent peptides are therapeutically engineered to incorporate backbone amide bond modifications, unnatural amino acid substitutions, rigid scaffolds and additional hydrophobic residues [5–7]. Though peptidomimetics do not possess a natural peptide backbone structure, they, however, retain the biological aptness to interact with the subject protein of interest by arranging pharmacophores in a required three-dimensional confirmation complementary to protein hot spot regions (binding pockets). Thus, therapeutic peptidomimetics can be rationally designed to achieve desired therapeutic effects on cellular signaling pathways by targeting specific disease-relevant PPI motifs [6,7].

Based on the level of homology between therapeutic peptidomimetics and their native peptide precursors, four prominent classifications (A–D) of peptidomimetics exists. The class A peptidomimetics are the most identical to the antecedent peptide, while class D peptidomimetics show the least similarities (Fig. 2.1) [3,19,20]. Class A and B include peptide-like structures whereas classes C and D encompass small molecular scaffolds. Class A peptidomimetics are short pseudopeptides with minimal modifications relative to the peptide side chains and backbone. They constitute the amino acid sequence of the native peptide incorporating a small number of modified amino acids, which helps to stabilize the bioactive conformation. Their backbone and side chains align closely with the bioactive conformation of the antecedent peptide [20]. Class B mimetics are modified forms of class A mimetics harboring different unnatural amino acids, isolated small-molecule building blocks or major backbone alterations. While still peptidic in nature, Class B mimetics embrace much more sophisticated backbone and side-chain functionalities (e.g., peptoids, β-peptides and α/β-mixed peptides). Foldamers (β- and α/β-peptides as well as peptoids) furnished with side chains aligning topologically squared to the native peptide also form part of this class [3]. The development of class B mimetics using foldamers as validated scaffolds is indispensable for the development of inhibitors [21,22].

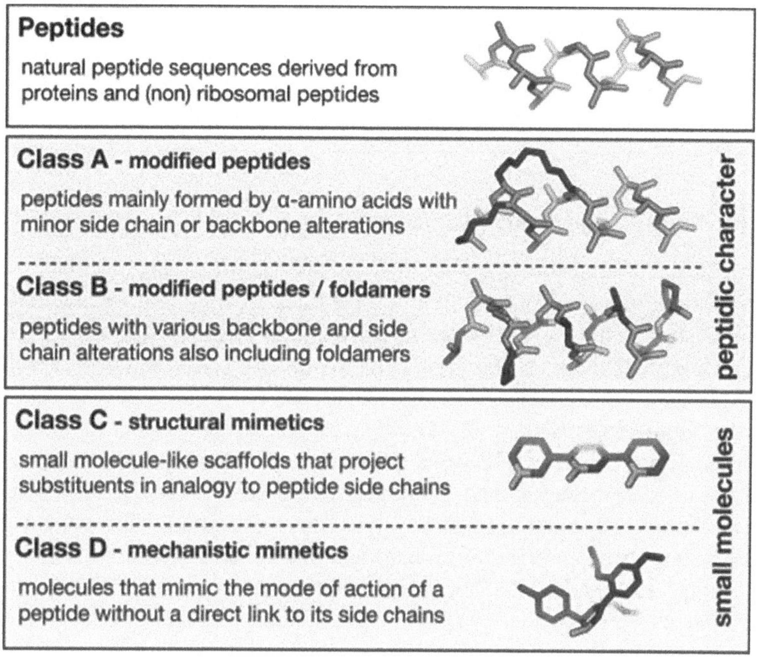

FIGURE 2.1 Classification of peptidomimetics based on structural conformations and degree of similarity to the native peptide precursor. These are divided into four different classes, A–D, where class A peptidomimetics are the most identical to the native peptide and class D peptidomimetics show the least similarities. The dark-colored regions on the different classes of peptides represent inserted modifications. *Adapted from Pelay-Gimeno M, Glas A, Koch O, Grossmann TN. Structure-based design of inhibitors of protein–protein interactions: mimicking peptide binding epitopes. Angew Chem Int Ed 2015;54:8896–927.*

Class C mimetics comprise highly modified structures lacking the entire peptide backbone, which have been replaced by small-molecule characters [3,6]. The display of the central scaffold is such that the conformation of the key residues (e.g., hot spots) in the bioactive orientation of the native peptide is in alignment in comparison to the substituents [3]. Therefore, the replacement of the entire peptide backbone yields bioactive compounds with improved oral bioavailability and pharmacokinetic properties, which conforms to Lipinski's rule of five and hence, may be promising drug candidates [21,22]. Class D mimetics are small-molecule drugs that mimic the mechanism of action of a druggable bioactive native peptide even without the side chain contributing to the functions [3,20]. In this class, the small-molecule drugs elicit their action by binding either into the allosteric pocket or in the active site of a protein. Class D peptidomimetics are usually developed from lead structures obtained from high-throughput screening of synthetic chemical libraries [23].

2.4 Examples of therapeutic peptidomimetics based on applications

This section presents key examples of therapeutic applications of peptidomimetics relative to targeting the undruggable proteome space.

2.4.1 Antimicrobial peptidomimetics

Antimicrobial therapeutic peptidomimetics are an important arsenal to circumvent the limitations of antimicrobial peptides (AMPs) that include failure to demonstrate efficacy over existing treatments, highly uneconomical synthesis, short half-lives in vivo and in vivo toxicity [24–29]. Antimicrobial resistance is a global health-care challenge, and the most worrying bacteria are the multidrug-resistant ESKAPE pathogens: *Enterococcus faecium*, *Staphylococcus aureus*, *Klebsiella pneumoniae*, *Acinetobacter baumannii*, *Pseudomonas aeruginosa*, and *Enterobacter* species [24–29]. AMPs have for some time now been used as possible novel antibiotics; however, a few AMPs have been approved for clinical use due to their side effects [30,31]. The therapeutic potential of antimicrobial peptidomimetics emanates from their high resistance against enzymatic degradation and their ability to target the undruggable space [32]. The most promising antimicrobial peptidomimetics so far optimized include: β-peptides, *N*-alkylated glycines (peptoids), α-peptide/β-peptoids, α/γ N-acylated N-aminoethylpeptides (AApeptides), β-peptoids (*N*-alkylated β-alanine oligomers), and oligoacyllysines (OAKs) (Fig. 2.2) [2].

Peptoids are promising and potential antimicrobial agents with activity reported against *Candida* biofilm [34], tuberculosis [35], fungi [36], cancer cells [37] and the neglected tropical disease leishmaniasis [38]. They have found applications for protein peptoid mutagenesis [39], Huntington's disease [40], positron emission tomography (PET) [41], biomaterials [42], carboxyalkyl peptoid PNAs [43], glycopeptoids [44], DNA-binding peptoids [45], fluorescent analogs [46] and within supra- and macromolecular engineering [47–51]. The use of γ-AApeptides antimicrobial peptidomimetics has been reported against viral infections and the treatment of metastasis in cancer. γ-AApeptides antimicrobial peptidomimetics inhibit viral replication and antitumor activity in various cancer cell lines [52]. They have also been reportedly mentioned in mimicking the Tat peptide of the human immunodeficiency virus (HIV) with a high-affinity binding to the HIV-1 RNA as compared to their antecedent peptides [53,54]. In addition to that, the peptidomimetic POL3026 is another specific novel beta-hairpin mimetic antagonist of CXCR4 that is known to possess potent anti-HIV activity [55].

Other promising antimicrobial peptidomimetics include arylamide and phenylene ethynylene oligomers [56,57], small synthetic antimicrobial peptidomimetics (SMAMPs or SAMPs) [58,59], pyrazole, poly-lysine and

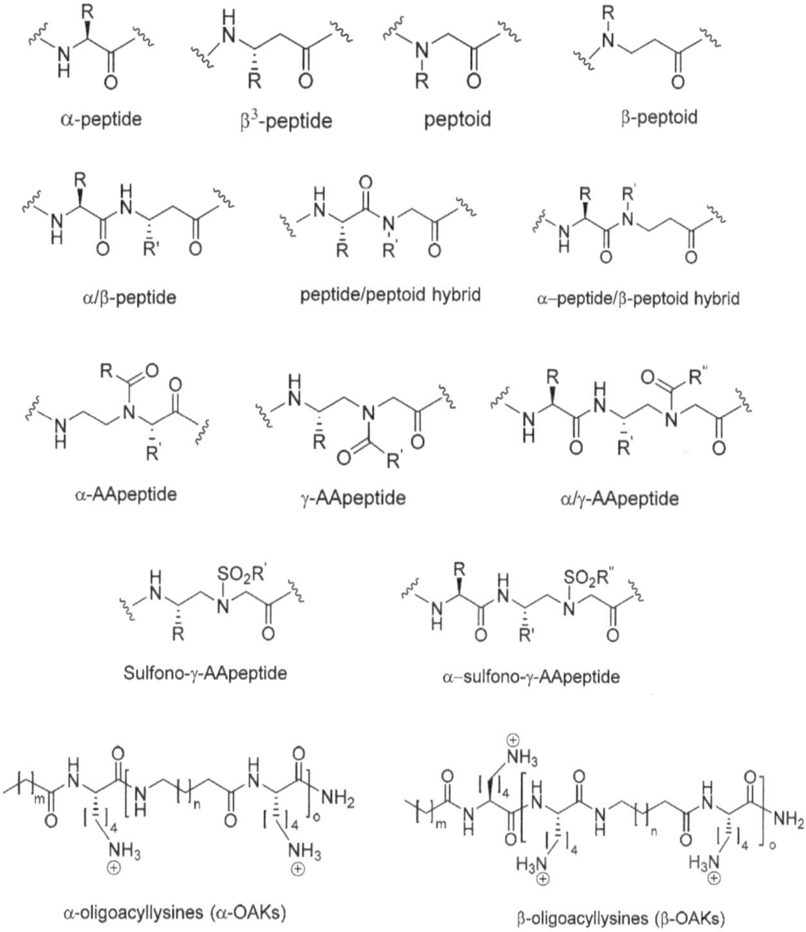

FIGURE 2.2 Basic structures of selected promising antimicrobial peptidomimetics optimized thus far. *Adapted from Molchanova N, Hansen PR, Franzyk H. Advances in development of antimicrobial peptidomimetics as potential drugs. Molecules 2017;22:1430.*

histidine-derived ultrashort antimicrobial peptidomimetics [60], macrocyclic peptidomimetics [61], binaphthyl-based, functionalized oxazoles, biaryl amino acid templates in cyclic β-hairpin cationic antimicrobial peptidomimetics [62], thiazole peptidomimetics [63], 1,4-dihydropyridine cationic peptidomimetics [64], and peptidomimetics containing 3-aminobenzoic acid [65–67]. The advantage of antimicrobial peptidomimetics is that their solid phase synthesis allows easy preparation of large dispositions of analogs by simple substitution or addition of alternative residues using synthetic amino acid analogs. Antimicrobial peptidomimetics in comparison to AMPs, possess enhanced stability in vivo and are often less toxic [33].

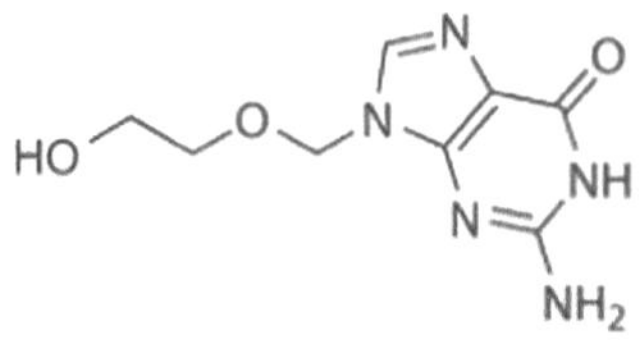

FIGURE 2.3 Molecular structure of acyclovir [also ACV, 9-(2-hydroxyethoxymethyl) guanine, or Zovirax]. ACV is a potent antiviral agent used as a highly specific inhibitor of herpes simplex viruses (HSV) types 1 and 2. *Adapted from Alvarez-Ros MC, Palafox MA. Conformational analysis, molecular structure and solid-state simulation of the antiviral drug acyclovir (zovirax) using density functional theory methods. Pharmaceuticals 2014;7:695–722.*

The quest to find new and effective prodrugs against the herpes simplex virus (HSV) has prompted scientists to investigate antiviral peptidomimetics and investigate a series of acyclovir analogs with thiazole ring containing amino acids such as glycine, alanine, valine, leucine [2,68]. The discovery of the guanine analog of acyclovir (ACV, 9-[(2-hydroxyethoxy) methyl] guanine) (Fig. 2.3) ushered in a new generation of antiviral therapeutic peptidomimetics [69–72]. Acyclovir is a potent antiviral agent, and an inhibitor of the herpes viruses, HSV-1 and HSV-2 with high specificity [73–77]. Most people infected with HIV type 1 (HIV-1) are often infected with HSV-2 which is associated with increased plasma and genital levels of HIV-1. ACV has also been shown to inhibit HIV in human herpes virus (HHV) coinfections [77].

Nevertheless, ACV demonstrates low aqueous solubility and low (14%) bioavailability when orally administered [77,78]. To curb this challenge, therapeutic peptidomimetics of acyclovir esters (Fig. 2.4) have been developed and subsequently tested against HSV-1 and HSV-2 [79–81]. The Val-thiazole-acyclovir peptidomimetic was remarkable, with higher stability than valacyclovir (the first effective prodrug of acyclovir) [2].

The use of synthetic antiprotozoal and antimalarial peptidomimetics is limited by the high price of production and their isolation from natural sources is not a viable solution [2,82–84]. However, Ettari and colleagues synthesized novel antimalarial peptidomimetics decorated with an aspartyl aldehyde warhead [84]. The thioacylal and acylal derivative peptidomimetics possessed increased antiplasmodial activity as compared to their antecedent molecule. Thioacylal derivatives are also promising trypanocidal agents [2,84]. Malaria and leishmaniasis are caused by *Plasmodium* and *Leishmania*, respectively. Peptidomimetic inhibitors of *N*-myristoyltransferase (NMT) from human malaria and leishmaniasis parasites are already developed. NMT is essential in *Leishmania* and subsequently validated as a drug target in *Plasmodium* [84].

2.4.2 Anticancer peptidomimetics

Anticancer therapeutic peptidomimetics designed to target the undruggable proteome space in cancer diagnostics and therapeutics already exist. A few

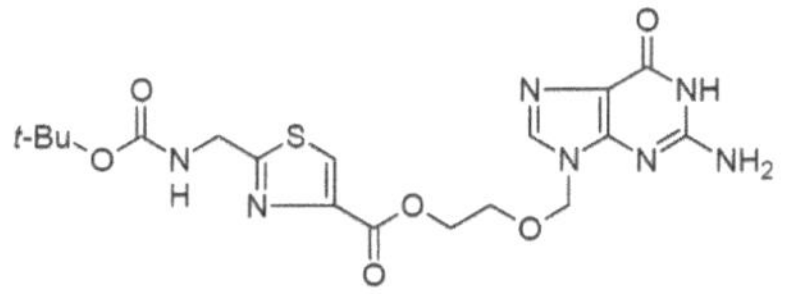

1
Boc-2-aminomethyl-thiazole-acyclovir
(Boc-Gly-(Thz)-ACV)

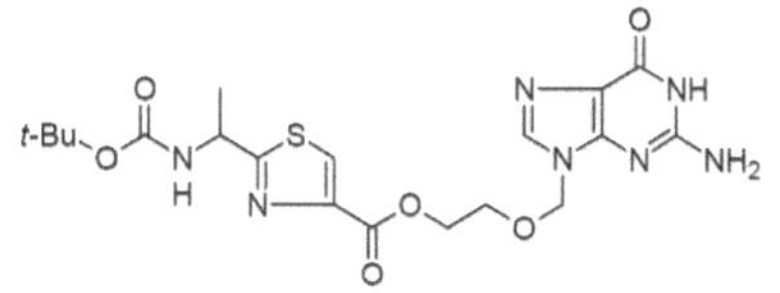

2
Boc-2-Ala-thiazole-acyclovir
(Boc-Ala-(Thz)-ACV)

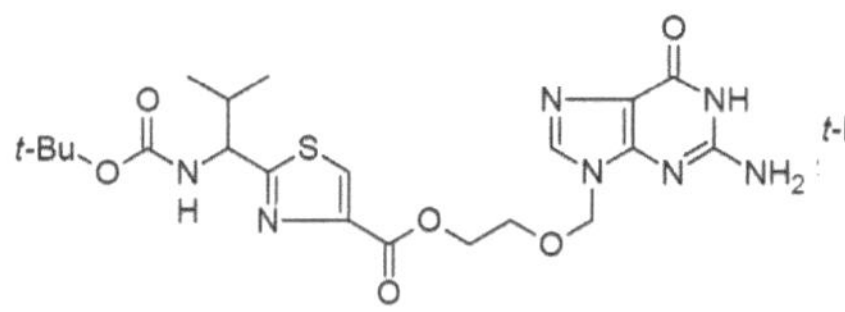

3
Boc-2-Val-thiazole-acyclovir
(Boc-Val-(Thz)-ACV)

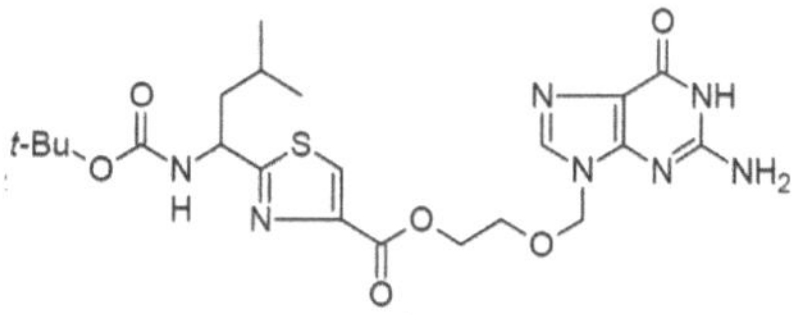

4
Boc-2-Leu-thiazole-acyclovir
(Boc-Leu-(Thz)-ACV)

FIGURE 2.4 Antiviral peptidomimetics developed from acyclovir esters. Valthiazole-acyclovir was the first effective analog of acyclovir. The drug is more stable than the other esters. Ala-thiazole-acyclovir, Gly-thiazole, Leu-thiazole-acyclovir esters are unstable under acidic conditions and undergo rapid hydrolysis to yield their chemical precursor acyclovir. *Adapted from Hristov G, Stankova I. Chemical stability of new acyclovir analogues with peptidomimetics. Sci Pharm 2011;79:259–64.*

drugs are already on the market with some having entered clinical trials. Their application has been tested on different protein models such as small GTPases, transmembrane receptors, apoptosis regulators and transcriptional regulators [68,85]. For many years the kinases were considered undruggable [86]. Nonetheless, researchers pushed forward with the dream of finding drugs that could be selectively used to drug kinases and a small-molecule drug called Imatinib was discovered [10]. Imatinib binds to a small number of kinases inside cells. One of the kinases targeted by Imatinib is BCR-ABL, which is present in patients with a specific type of cancer called chronic myelogenous leukemia (CML) [87]. This relatively rare disease turned out to be a battleground that forged the modern protein-based approach to cancer drug discovery [10]. RAS proteins have been in the limelight of cancer drug discovery in the last millennium [88].

Unfortunately, all previous efforts toward drugging RAS proved unfruitful; hence, the inglorious moniker "undruggable." The recent identification of a new binding pocket interfering with the SOS-*K-RAS* interaction (Fig. 2.5) prompted the interest to re-visit the "previously undruggable" target. Three strategies to address K-RAS directly have been so far established rendering RAS proteins indeed "druggable" [88]. These include targeting the CaaX-Box, the nucleotide-binding domain and a shallow hydrophobic pocket harboring the switches I and II [13]. More so, in recent times, cancer therapy can specifically inhibit both effector pathways downstream of K-RAS via selective MEK (Binimetinib, Cobimetinib, Selumetinib, Trametinib)

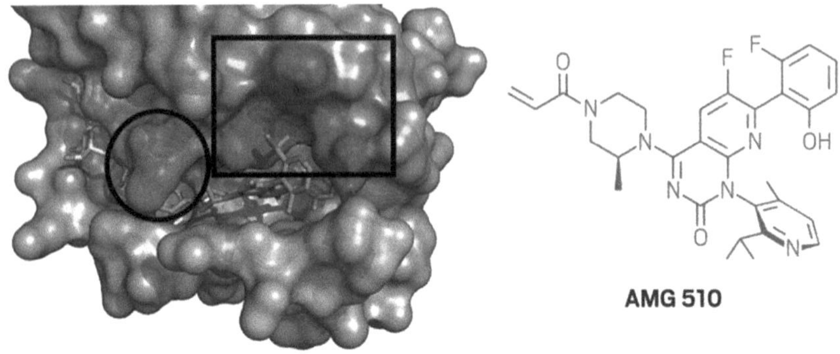

FIGURE 2.5 Surface representation of K-RAS (shown in gray) bound to deoxyguanosine diphosphate (left) and AMG 510 (right). The region surrounded by the circle is the site where AMG 510 covalently attaches to a cysteine residue. The rectangular region shows histidine flipped out by the drug candidate. K-RAS is the most frequently mutated oncogene in solid tumors and was previously regarded as "undruggable" due to lack of "easy-to-find pockets" where therapeutics such as small molecules could bind, thus halting progress in drug development. AMG 510 is the first atropisomer covalent inhibitor of K-RAS to make it into human clinical trials. *Figure from Long GV, Hauschild A, Santinami M, Atkinson V, Mandala M, Chiarion-Sileni V, et al. Adjuvant dabrafenib plus trametinib in stage III BRAF-mutated melanoma. N Engl J Med 2017;377:1813–23.[89]. Used with permission of Amgen Inc.*

[89–93], B-RAF (Dabrafenib, Vemurafenib, Sorafenib, Encorafenib) [94–98], or PI3K (Idealisib, Buparlisib) [98–101] inhibitors.

Myc (avian myelocytomatosis viral oncogene homolog), remains one of the most sought-after oncogene drug targets in cancer, with a family comprising of c-Myc, N-Myc, and L-Myc [87]. c-Myc is a master regulator of gene transcription in the human genome, where it controls the expression of ~30% of genes [102]. Additionally, oncogenic deregulation of c-Myc are predominantly linked to approximately 70% of human cancer malignancies including breast, colon, cervix, lung, bone, brain and blood cancers, accounting for one-seventh of global cancer deaths [103]. c-Myc is overexpressed in approximately 50% of all triple-negative breast cancers (TNBCs); one of the most aggressive breast cancer subtypes [102]. c-Myc is not an easily "druggable" protein and has been negatively branded "undruggable" due to its nuclear localization and with a protein structure, which does not conform to any known secondary elements feature [94]. Moreover, over 30 years of drug development research efforts did not result in any clinically potent Myc inhibitor. The mutationally activated oncoprotein remained obdurate and beyond the targeting capabilities of established technologies. Many inhibitors that have been designed to inhibit the binding of c-Myc to cofactors or downstream effectors have been unsuccessful when applied clinically [97–102]. This is due to insufficient specificity, poor bioavailability, fast metabolism, poor tumor site penetration and consequent inadequate therapeutic efficacy [96]. The development of new technologies has seen the oncoprotein become an indispensable target for drug discovery in cancer [13,14,97]. Few prototype inhibitors have so far entered the early phases of clinical trials [104–109]. However, the direct blocking of inhibition of Myc-dependent tumors downstream gene expression has been unsuccessful despite almost three decades of efforts into its research and development [110].

Novel therapeutic peptidomimetics to inhibit the neuronal differentiation of neuroblastoma cells, as well as interfere with tumorigenesis and neuroblastomas malignancies have been synthesized from the unnatural amino acid-substituted (hydroxyethyl) urea. These peptidomimetics have served as lead compounds for the further development of novel anticancer drugs [19]. HERP5, HERP6 and HERP7 peptidomimetics to inhibit HER2-mediated signaling for cell growth have been developed with two of the compounds (HERP5 and HERP7) having been proven to exhibit antiproliferative activity against breast cancer cell lines overexpressing HER2 protein [20]. In pancreatic tumoral cell lines, bicyclic therapeutic peptidomimetics have been identified. Both CDR-derived peptidomimetics and chemically synthesized cyclic peptidomimetics from gastrin binders proved effective in targeting pancreatic tumor cell lines [3].

2.4.3 Immunodetection peptidomimetics

The widespread application of immunodetection peptidomimetics as a gate away from monoclonal antibodies (mAbs) is a noteworthy example in clinical therapeutics settings [111]. Antibodies are often used to study the molecular basis of physiologic processes, and monoclonal antibodies have been employed as potent diagnostic agents, which can be synergistically coupled to radionuclide, immune-modulatory agents or used in the treatment of cancers [111]. The use of mAbs is limited because their production is cumbersome and creating diverse and therapeutic amounts of useful mAb remain difficult [111]. Hence, the shift into targeting mAb therapeutic mimicries as large-scale immunodetection probes in modern-day drug discovery. Antibody-like-binding immunodetection peptidomimetics (ABiPs) already exist and with some already in clinical trials. One example is the anti-erbB peptidomimetic and the anti-Her2/neu peptidomimetic (AHNP) (Fig. 2.6), used in advanced breast cancer therapy. AHNP is a small *exo*-cyclic peptide derived from the anti-p185 (her2/neu) rhumAb 4D5 (h4D5). It mimics many but not all the antitumor characteristics exhibited by h4D5 [111,112]. AHNP is a mimic of the mAb Herceptin. These Herceptin-like AbiPs are used to detect the Her2/neu antigen at extremely low concentrations using the immunodetection amplification technique (IDAT) [112]. To improve the binding affinity to p185(her2/neu) and the antitumor efficacy, AHNP is alternatively synthesized as an oligomeric fusion protein with a non-immunoglobulin protein scaffold, streptavidin [111,112].

On the other hand, epidermal growth factor receptor (EGFR)-based immunodetection peptidomimetics have also been developed. EGFR is

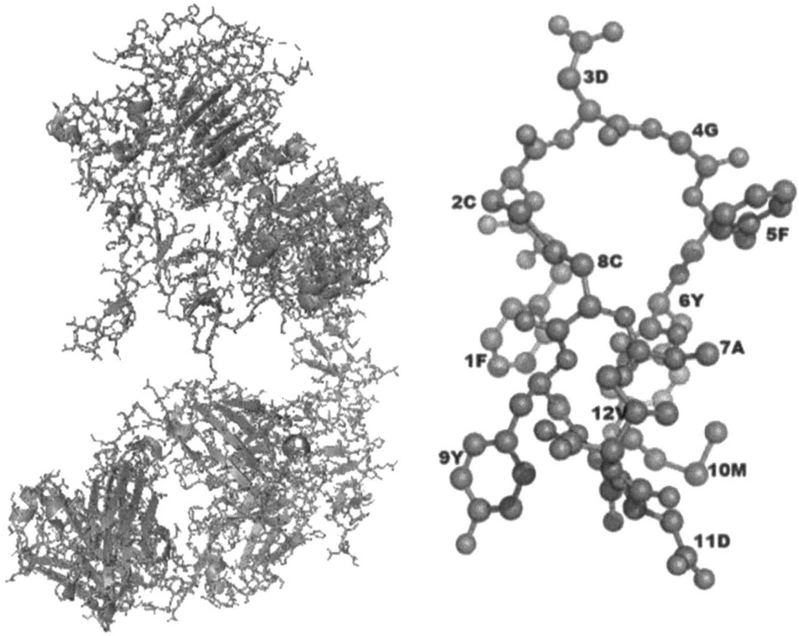

FIGURE 2.6 The native structure of the anti-p185 (her2/neu) rhumAb 4D5 (h4D5) (PDB: 1N8Z) as viewed on Pymol (left) and the ball-and-stick representation of the solution anti-Her2/neu peptidomimetic (AHNP) as determined by NMR (right). Important amino acids in the structure are shown by their one-letter code. *Adapted from Murali R, Greene MI. Structure based antibody-like peptidomimetics. Pharmaceuticals 2012;5:209–35.*

overexpressed in several solid tumors such as breast, prostate, pancreas and lung cancers; however, its overexpression is positively correlated to the metastatic potential of the tumor. Anti-EGFR receptor-binding peptidomimetics (AERP) designed from anti-EGFR antibodies have been utilized as potential small-molecule tumor-specific imaging agents [111–114]. Cardó-Vila and co-workers synthesized an EGFR peptidomimetic drug decoy candidate, $_D$(CARVC)—a synthetic, low molecular weight, retro-inverted, water-soluble peptidomimetic—by in vitro, in cellulo, and in vivo assays. The drug decoy candidate demonstrated antitumor activity [115]. A bioactive anti-EGF receptor peptidomimetic (AERP) was designed using a template made up of the C225 heavy chain CDR3 region, which structurally adopts a β-turn and an antiparallel β-sheet secondary structure (Fig. 2.7). Conformational traits combined with the pharmacophore enhanced the stability of folding, avidity and provided additional functional surfaces for binding [112]. A fully developed highly diverse library of ABiPs consisting of an alternating mAb panel may also be useful for target validation, antigen detection, therapeutics and as a platform for drug development [2,111–115].

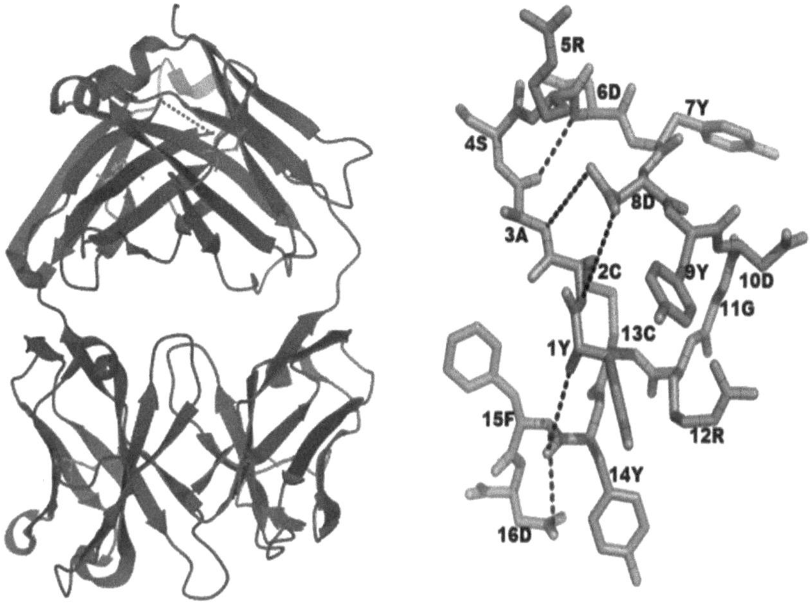

FIGURE 2.7 The cartoon structure of the native C225 heavy chain CDR3 region template used to design AERP (left) and the stick representation of the three-dimensional structure of AERP (right). AERP adopts a β-turn and an antiparallel β-sheet orientation to create a large ring-sized structure. The structure is stabilized by the formation of intramolecular hydrogen bonds, which are depicted by dash lines. Amino acid residues are shown using their one-letter code. *Adapted from Murali R, Greene MI. Structure based antibody-like peptidomimetics. Pharmaceuticals 2012;5:209–35.*

2.5 Conclusion

Accelerating the arrival of next-generation drugs is a grand challenge awaiting scientists. A concerted effort to target the undruggable proteins might reap enormous benefits in terms of yielding powerful new medicines for diverse human diseases. The potential of the proteome space as a drug discovery arsenal may unlock and bring under control all manner of pathologies, from neurodegeneration to cancer to infectious diseases. Researchers are becoming aware of the great benefits that could accrue to society by tackling these undruggable proteins. Indeed, the succeeding chapters describe attempts to solve the problem by designing and developing therapeutic peptidomimetics as smart drugs against these undruggable therapeutic vulnerabilities. A significant number of laboratories are trying to create collections of architecturally complex synthetic mimicries that might be more effective at binding to these challenging protein families. In the course of time, an increasing number of proteins historically branded intractable will likely be considered druggable. The time may even come, far into the future, when all disease-relevant proteins have been targeted successfully with drugs. This will represent the final panorama of human ingenuity as we reach the pinnacle of curing the incurable - when the majority of undruggable proteins will have been drugged.

References

[1] Díaz-Eufracio BI, Jesús Naveja J, Medina-Franco JL. Advances in protein chemistry and structural biology. Chapter Three-Protein-Protein Interact Modulators Epigenetic Therapies 2018;110:65–84.

[2] Kharb R, Rana M, Sharma PC, Yar MS. Therapeutic importance of peptidomimetics in medicinal chemistry. J Chem Pharm Res 2011;3:173–86.

[3] Pelay-Gimeno M, Glas A, Koch O, Grossmann TN. Structure-based design of inhibitors of protein–protein interactions: mimicking peptide binding epitopes. Angew Chem Int Ed 2015;54:8896–927.

[4] Jana T, Ghosh A, Mandal SD, Banerjee R, Saha S. PPIMpred: a web server for high-throughput screening of small molecules targeting protein–protein interaction. R Soc Open Sci 2017;4:160501.

[5] Vagner J, Qu H, Hruby VJ. Peptidomimetics, a synthetic tool of drug discovery. Curr Opin Chem Biol 2008;12:292–6.

[6] Robertson NS, Spring DR. Using peptidomimetics and constrained peptides as valuable tools for inhibiting protein–protein interactions. Molecules 2018;23:959.

[7] Nimmagadda A, Shi Y, Cai J. γ-AApeptides as a new strategy for therapeutic development. Curr Med Chem 2019;26:2313–29.

[8] Stevers LM, Sijbesma E, Botta M, MacKintosh C, Obsil T, Landrieu I, et al. Modulators of 14-3-3 protein-protein interactions. J Med Chem 2018;61:3755–78.

[9] Verdine GL, Walensky LD. The challenge of drugging undruggable targets in cancer: lessons learned from targeting BCL-2 family members. Clin Cancer Res 2007;13:24.

[10] Stockwell BR. The Quest for the Cure: The Science and Stories Behind the Next Generation Of Medicines. New York, USA: Columbia University Press; 2011.

[11] Kaiser J. After decades, progress against an 'undruggable' cancer target. Science 2019;366:561.

[12] Dang CV, Reddy EP, Shokat KM, Soucek L. Drugging the 'undruggable' cancer targets. Nat Rev Cancer 2017;17:502–8.

[13] Mörchen B, Shkura O, Stoll R, Helfrich I. Targeting the "undruggable". RAS - N Strateg - N hope? Cancer Drug Resist 2019;2:813–26.

[14] Liua P, Wanga Y, Lib X. Targeting the untargetable KRAS in cancer therapy. Acta Pharmaceutica Sin B 2019;9:871–9.

[15] Hopkins AL, Groom CR. The druggable genome. Nat Rev Drug Discov 2002;1:727–30.

[16] Hopkins AL, Groom CR. Target analysis: a priority assessment of druggability. Ernst Scher Res Found Workshop 2003;42:11–17.

[17] Russ AP, Lampel S. The druggable genome: an update. Drug Discov Today 2005;10:1607–10.

[18] Arkin MR, Wells JA. Small-molecule inhibitors of protein-protein interactions: progressing towards the dream. Nat Rev Drug Discov 2004;3:301–17.

[19] Ripka AS, Rich DH. Peptidomimetic design. Curr Opin Chem Biol 1998;2:441–52.

[20] Azzarito V, Long K, Murphy NS, Wilson AJ. Inhibition of α-helix-mediated protein–protein interactions using designed molecules. Nat Chem 2013;5:161–73.

[21] Spiegel J, Mas-Moruno C, Kessler H, Lubell WD. Cyclic azapeptide integrin ligand synthesis and biological activity. J Org Chem 2012;77:5271–8.

[22] Kritzer JA, Stephens OM, Guarracino DA, Reznik SK, Schepartz A. β-Peptides as inhibitors of protein–protein interactions. Bioorg Med Chem 2005;13:11–16.

[23] Wade M, Li YC, Wahl GM. MDM2, MDMX and p53 in oncogenesis and cancer therapy. Nat Rev Cancer 2013;13:83–96.

[24] O'Neill J. Tackling drug-resistant infections globally: final report and recommendations. 2016. Available from: https://amr-review.org/sites/default/files/160518_Final%20paper_-with%20cover.pdf (accessed 18.01.20).

[25] Crowther GS, Baines SD, Todhunter SL, Freeman J, Chilton CH, Wilcox MH. Evaluation of NVB302 vs vancomycin activity in an in vitro human gut model of Clostridium difficile infection. J Antimicrob Chemother 2013;68:168–76.

[26] Velden WJ, van Iersel TM, Blijlevens NM, Donnelly JP. Safety and tolerability of the antimicrobial peptide human lactoferrin 1–11 (hLF1–11). BMC Med 2009;7:44.

[27] Andersson DI, Hughes D, Kubicek-Sutherland JZ. Mechanisms and consequences of bacterial resistance to antimicrobial peptides. Drug Resist Updates 2016;26:43–57.

[28] Nizet V. Antimicrobial peptide resistance mechanisms of human bacterial pathogens. Curr Issues Mol Biol 2006;8:11–26.

[29] Bechinger B, Gorr SU. Antimicrobial peptides: mechanisms of action and resistance. J Dent Res 2017;96:254–60.

[30] Pendleton JN, Gorman SP, Gilmore BF. Clinical relevance of the eskape pathogens. Expert Rev Anti-Infect Ther 2013;11:297–308.

[31] Haney EF, Mansour SC, Hancock REW. Antimicrobial peptides: an introduction. In: Hansen PR, editor. In Antimicrobial Peptides: Methods and Protocols. New York, NY, USA: Springer; 2017. p. 3–22.

[32] Avan I, Hall CD, Katritzky AR. Peptidomimetics via modifications of amino acids and peptide bonds. Chem Soc Rev 2014;43:3575–94.

[33] Molchanova N, Hansen PR, Franzyk H. Advances in development of antimicrobial peptidomimetics as potential drugs. Molecules 2017;22:1430.

[34] Luo Y, Bolt HL, Eggimann GA, McAuley DF, McMullan R, Curran T, et al. Peptoid efficacy against polymicrobial biofilms determined by using propidium monoazide-modified quantitative PCR. ChemBioChem 2017;18:111–18.

[35] Kapoor R, Eimerman PR, Hardy JW, Cirillo JD, Contag CH, Barron AE. Efficacy of antimicrobial peptoids against *Mycobacterium tuberculosis*. Antimicrob Agents Chemother 2011;55:3058–62.

[36] Corson AE, Armstrong SA, Wright ME, McClelland EE, Bicker KL. Discovery and characterization of a peptoid with antifungal activity against *Cryptococcus neoformans*. ACS Med Chem Lett 2016;7:1139–44.

[37] Huang W, Seo J, Willingham SB, Czyzewski AM, Gonzalgo ML, Weissman IL, et al. Learning from host-defense peptides: cationic, amphipathic peptoids with potent anticancer activity. PLoS ONE 2014;9:e90397.

[38] Eggimann GA, Bolt HL, Denny PW, Cobb SL. Investigating the anti-leishmanial effects of linear peptoids. ChemMedChem 2015;10:233–7.

[39] Lee BC, Zuckermann RN. Protein side-chain translocation mutagenesis via incorporation of peptoid residues. ACS Chem Biol 2011;6:1367–74.

[40] Chen X, Wu J, Luo Y, Liang X, Supnet C, Kim MW, et al. Expanded polyglutamine-binding peptoid as a novel therapeutic agent for treatment of Huntington's disease. Chem Biol 2011;18:1113–25.
[41] Seo J, Ren G, Liu HG, Miao Z, Park M, Wang YH, et al. In vivo biodistribution and small animal PET of 64Cu-labeled antimicrobial peptoids. Bioconj Chem 2012;23: 1069–79.
[42] Sun J, Zuckermann RN. Peptoid polymers: a highly designable bioinspired material. ACS Nano 2013;7:4715–32.
[43] De Cola C, Manicardi A, Corradini R, Izzo I, De Riccardis F. Carboxyalkyl peptoid PNAs: synthesis and hybridization properties. Tetrahedron 2012;68:499–506.
[44] Szekely T, Roy O, Faure S, Taillefumier C. From glycopeptides to glycopeptoids. Eur J Org Chem 2014;2014:5641–57.
[45] Mao J, Bong D. Synthesis of DNA-binding peptoids. Synlett 2015;26:1581–5.
[46] Unciti-Broceta A, Diezmann F, Ou-Yang CY, Fara MA, Bradley M. Synthesis, penetrability and intracellular targeting of fluorescein-tagged peptoids and peptide-peptoid hybrids. Bioorg Med Chem 2009;17:959–66.
[47] Gangloff N, Ulbricht J, Lorson T, Schlaad H, Luxenhofer R. Peptoids and polypeptoids at the frontier of supra- and macromolecular engineering. Chem Rev 2016;116:1753–802.
[48] Zuckermann RN, Kodadek T. Peptoids as potential therapeutics. Curr Opin Mol Ther 2009;11:299–307.
[49] Trabocchi A, Peptoids GA. In Peptidomimetics in Organic and Medicinal Chemistry: The Art of Transforming Peptides in Drugs. Chichester, UK: John Wiley & Sons, Ltd; 2014.
[50] Fowler SA, Blackwell HE. Structure–function relationships in peptoids: recent advances toward deciphering the structural requirements for biological function. Org Biomol Chem 2009;7:1508–24.
[51] Hansen PR, Munk JK. Synthesis of antimicrobial peptoids. Methods Mol Biol 201;1047:151–59.
[52] Méndez-Samperio P. Peptidomimetics as a new generation of antimicrobial agents: current progress. Infect Drug Resist 2014;7:229–37.
[53] Niu Y, Jones AJ, Wu H, Varani G, Cai J. γ-AApeptides bind to RNA by mimicking RNA-binding proteins. Org Biomol Chem 2011;9:6604–9.
[54] Niu Y, Bai G, Wu H, et al. Cellular translocation of a γ-AApeptide mimetic of Tat peptide. Mol Pharm 2012;9:1529–34.
[55] Moncunill G, Armand-Ugón M, Clotet-Codina I, et al. Anti-HIV activity and resistance profile of the CXC chemokine receptor 4 antagonist POL3026. Mol Pharmacol 2008;73:1264–73.
[56] Tew GN, Scott RW, Klein ML, Degrado WF. De novo design of antimicrobial polymers, foldamers, and small molecules: from discovery to practical applications. Acc Chem Res 2010;43:30–9.
[57] Thaker HD, Cankaya A, Scott RW, Tew GN. Role of amphiphilicity in the design of synthetic mimics of antimicrobial peptides with gram-negative activity. ACS Med Chem Lett 2013;4:481–5.
[58] Haug BE, Stensen W, Kalaaji M, Rekdal Ø, Svendsen JS. Synthetic antimicrobial peptidomimetics with therapeutic potential. J Med Chem 2008;51:4306–14.
[59] Flemming K, Klingenberg C, Cavanagh JP, Sletteng M, Stensen W, Svendsen JS, et al. High in vitro antimicrobial activity of synthetic antimicrobial peptidomimetics against staphylococcal biofilms. J Antimicrob Chemother 2009;63:136–45.

[60] Ahn M, Gunasekaran P, Rajasekaran G, Kim EY, Lee SJ, Bang G, et al. Pyrazole derived ultra-short antimicrobial peptidomimetics with potent anti-biofilm activity. Eur J Med Chem 2017;125:551–64.

[61] Ibrahim MA, Panda SS, Oliferenko AA, Oliferenko PV, Girgis AS, Elagawany M, et al. Macrocyclic peptidomimetics with antimicrobial activity: synthesis, bioassay, and molecular modeling studies. Org Biomol Chem 2015;13:9492–503.

[62] Srinivas N, Moehle K, Abou-Hadeed K, Obrecht D, Robinson JA. Biaryl amino acid templates in place of D-pro-L-pro in cyclic β-hairpin cationic antimicrobial peptidomimetics. Org Biomol Chem 2007;5:3100–5.

[63] Wales SM, Hammer KA, Somphol K, Kemker I, Schroder DC, Tague AJ, et al. Synthesis and antimicrobial activity of binaphthyl-based, functionalized oxazole and thiazole peptidomimetics. Org Biomol Chem 2015;13:10813–24.

[64] Lapidot I, Albeck A, Gellerman G, Shatzmiller S, Grynszpan F. 1,4-dihydropyridine cationic peptidomimetics with antibacterial activity. Int J Pept Res Ther 2015;21:n243–7.

[65] Lohan S, Cameotra SS, Bisht GS. Antibacterial evaluation of structurally amphipathic, membrane active small cationic peptidomimetics: synthesized by incorporating 3-amino benzoic acid as peptidomimetic element. Eur J Med Chem 2014;83:102–15.

[66] Ghosh C, Haldar J. Membrane-active small molecules: designs inspired by antimicrobial peptides. ChemMedChem 2015;10:1606–24.

[67] Liskamp RMJ, Rijkers DTS, Kruijtzer JAW, Kemmink J. Peptides and proteins as a continuing exciting source of inspiration for peptidomimetics. Chem Bio Chem 2011;12:1626–53.

[68] Yana P. Cancer drugs: targeting undruggable proteins. Berkeley Sci J 2018;22.

[69] Alvarez-Ros MC, Palafox MA. Conformational analysis, molecular structure and solid-state simulation of the antiviral drug acyclovir (zovirax) using density functional theory methods. Pharmaceuticals 2014;7:695–722.

[70] Hristov G, Stankova I. Chemical stability of new acyclovir analogues with peptidomimetics. Sci Pharm 2011;79:259–64.

[71] Elion GB, Furman PA, Fyfe JA, de Miranda P, Beauchamp L, Schaeffer HJ. Selectivity of action of an antiherpetic agent, 9-(2-hydroxyethoxymethyl)guanine. Proc Nat Acad Sci USA 1977;74:5716–20.

[72] De Clercq E, Field HJ. Antiviral prodrugs–the development of successful prodrug strategies for antiviral chemotherapy. Br J Pharmacol 2006;147:1–11.

[73] Gao H, Mitra AK. Synthesis of acyclovir, ganciclovir and their prodrugs: a review. Synthesis 2000;3:329–51.

[74] De Clercq E. Antiviral drugs: current state of the art. J Clin Virol 2001;22:73–89.

[75] Emmert DH, Lancaster MD. Treatment of common cutaneous herpes simplex virus infections. Am Fam Physician 2000;61 1697–1607.

[76] Fuchs J, Celum C, Wang J, Hughes J, Sánchez J, Cowan F, et al. Clinical and virologic efficacy of herpes simplex virus type 2 suppression by acyclovir in a multicontinent clinical trial. J Infect Dis 2010;201:1164–8.

[77] Piret J, Boivin G. Resistance of herpes simplex viruses to nucleoside analogues: mechanisms, prevalence, and management. Antimicrob Agents Chemother 2011;55:459–72.

[78] Anand BS, Katragadda S, Nashed YE, Mitra AK. Amino acid prodrugs of acyclovir as possible antiviral agents against ocular HSV-1 infection: interaction with the neutral and cationic amino acid transporter on the cornel epithelium. Curr Eye Res 2004;29:153–66.

[79] Stankova I, Dzimbova T, Shishkov St, Kostova K, Galabov A. Synthesis and biological activity of amino acid ester prodrugs of acyclovir. In: Rolka K, Rekowski P, Silberring J,

editors. Peptides 2006. Proceedings of the 29th European Peptide Symposium. Geneva Kenes Internat; 2007. p. 226–7.

[80] Stankova I, Dzimbova T, Shishkov S. Synthesis and antiviral activity of some amino acid esters of acyclovir. Adv Exp Med Biology. Peptides for Youth: Proceedings of the 20th American Peptide Symposium 2007;169–70.

[81] Stankova I, Dzimbova T, Shishkov ST, Todorov D, Galabov A. Esters of acyclovir with peptidomimetics—synthesis and antiviral activity. in preparation. Sci Pharm 2011;79:259–64.

[82] Júnior EFS, Menezes LFS, Silva de Araújo IF, Schwartz EF. Natural occurrence in venomous arthropods of antimicrobial peptides active against protozoan parasites. Toxins 2019;11:563.

[83] Olaleye TO, Brannigan JA, Roberts SM, Leatherbarrow RJ, Wilkinsonb AJ, Tate EW. Peptidomimetic inhibitors of *N*-myristoyltransferase from human malaria and leishmaniasis parasites. Org Biomol Chem 2014;12:8132–7.

[84] Ettari R, Zappala M, Micale N, Schirmeister T, Gelhaus C, Leippe M, et al. Synthesis of novel peptidomimetics as inhibitors of protozoan cysteine proteases falcipain-2 and rhodesain. Eur J Med Chem 2010;45:3228–33.

[85] Mabonga L, Kappo AP. Peptidomimetics: a synthetic tool for inhibiting protein–protein interactions in cancer. Int J Pept Res Ther 2019;. Available from: https://doi.org/10.1007/s10989-019-09831-5.

[86] Kain K. The future of cancer therapy: an interview with Gerard Evan. Dis Model Mech 2008;1(2– 3):90.

[87] Buchdunger E, Zimmermann J, Mett H, Meyer T, Muller M, Druker BJ, et al. Inhibition of the Abl protein- tyrosine kinase in vitro and in vivo by a 2- phenylaminopyrimidine derivative. Cancer Res 1996;56:100.

[88] Halford B. Amgen unveils its KRas inhibitor in human clinical trials. C&EN 2019;97(14):4.

[89] Long GV, Hauschild A, Santinami M, Atkinson V, Mandala M, Chiarion-Sileni V, et al. Adjuvant dabrafenib plus trametinib in stage III BRAF-mutated melanoma. N Engl J Med 2017;377:1813–23.

[90] Janne PA, van den Heuvel MM, Barlesi F, Cobo M, Mazieres J, Crinò L, et al. Selumetinib plus docetaxel compared with docetaxel alone and progression-free survival in patients with KRAS-mutant advanced non-small cell lung cancer: the SELECT-1 randomized clinical trial. JAMA 2017;317:1844–53.

[91] Pulido JS, Heier JS, Marmorstein AD. Selumetinib in plexiform neurofibromas. N Engl J Med 2017;376:1195.

[92] Signorelli J, Shah Gandhi A. Cobimetinib. Ann Pharmacother 2017;51:146–53.

[93] Dummer R, Schadendorf D, Ascierto PA, Arance A, Dutriaux C, Di Giacomo AM, et al. Binimetinib vs dacarbazine in patients with advanced NRAS-mutant melanoma (NEMO): a multicentre, open-label, randomised, phase 3 trial. Lancet Oncol 2017;18:435–45.

[94] Blundell TL, Burke DF, Chirgadze D, Dhanaraj V, Hyvonen M, Innis CA, et al. Protein-protein interactions in receptor activation and intracellular signalling. Biol Chem 2000;381:955–9.

[95] Brown RV, Danford FL, Gokhale V, Hurley LH, Brooks TA. Demonstration that drug-targeted downregulation of MYC in non-Hodgkins lymphoma is directly mediated through the promoter G-quadruplex. J Biol Chem 2011;286:41018–27.

[96] Sorolla A, Wang E, Golden E, et al. Precision medicine by designer interference peptides: applications in oncology and molecular therapeutics. Oncogene 2019;. Available from: https://doi.org/10.1038/s41388-019-1056-3.

[97] Carabet LA, Rennie PS, Cherkasov A. Therapeutic inhibition of myc in cancer. structural bases and computer-aided drug discovery approaches. Int J Mol Sci 2019;20:120.

[98] Nijenhuis CM, Huitema AD, Blank C, Haanen JB, van Thienen JV, Rosing H, et al. Clinical pharmacokinetics of vemurafenib in BRAF-mutated melanoma patients. J Clin Pharmacol 2017;57:125–8.

[99] Zhang W, Heinzmann D, Grippo JF. Clinical pharmacokinetics of vemurafenib. Clin Pharmacokinet 2017;56:1033–43.

[100] Delord JP, Robert C, Nyakas M, McArthur GA, Kudchakar R, Mahipal A, et al. Phase I dose-escalation and expansion study of the BRAF inhibitor encorafenib (LGX818) in metastatic BRAF-mutant melanoma. Clin Cancer Res 2017;23:5339–48.

[101] Gounder MM, Mahoney MR, Van Tine BA, Ravi V, Attia S, Deshpande HA, et al. Sorafenib for advanced and refractory desmoid tumors. N Engl J Med 2018;379: 2417–28.

[102] Sheiness D, Fanshier L, Bishop JM. Identification of nucleotide sequences which may encode the oncogenic capacity of avian retrovirus MC29. J Virol 1978;28:600–10.

[103] Han G, Wang Y, Bi WZ. c-Myc overexpression promotes osteosarcoma cell invasion via activation of MEK-ERK pathway. Oncol Res 2012;20:149–56.

[104] Ponzielli R, Katz S, Barsyte-Lovejoy D, Penn LZ. Cancer therapeutics: targeting the dark side of Myc. Eur J Cancer 2005;41:2485–501.

[105] McKeown MR, Bradner JE. Therapeutic strategies to inhibit MYC. Cold Spring Harb Perspect Med 2014;4:a014266.

[106] Chen H, Liu H, Qing G. Targeting oncogenic Myc as a strategy for cancer treatment. Signal Transduct Target Ther 2018;3:5.

[107] Whitfield JR, Beaulieu ME, Soucek L. Strategies to inhibit Myc and their clinical applicability. Front Cell Dev Biol 2017;5:10.

[108] Koh CM, Sabo A, Guccione E. Targeting MYC in cancer therapy: RNA processing offers new opportunities. BioEssays N Rev Mol Cell Dev Biol 2016;38:266–75.

[109] Rickman DS, Schulte JH, Eilers M. The expanding world of N-MYC-driven tumors. Cancer Discov 2018;8:150–63.

[110] Blackwood EM, Eisenman RN. Max: a helix-loop-helix zipper protein that forms a sequence-specific DNA-binding complex with Myc. Science 1991;251:1211–17.

[111] Murali R, Liu Q, Cheng X, Berezov A, Richter M, Furuchi K, et al. Antibody like peptidomimetics as large scale immunodetection probes. Cell Mol Biol 2003;49:209–16.

[112] Murali R, Greene MI. Structure based antibody-like peptidomimetics. Pharmaceuticals 2012;5:209–35.

[113] Park BW, Zhang HT, Wu C, Berezov A, Zhang X, Dua R, et al. Rationally designed anti-HER2/neu peptide mimetic disables P185HER2/neu tyrosine kinases in vitro and in vivo. Nat Biotechnol 2000;18:194–8.

[114] Ponde DE, Su Z, Berezov A, Zhang H, Alavi A, Greene MI, et al. Development of anti-EGF receptor peptidomimetics (AERP) as tumor imaging agent. Bioorg Med Chem Lett 2011;21:2550–3.

[115] Cardó-Vilaa M, Giordanoa RJ, Sidmanb RL, Bronka LF, Fanc Z, Mendelsohnc J, et al. From combinatorial peptide selection to drug prototype (II): targeting the epidermal growth factor receptor pathway. PNAS 2010;107:11.

Chapter 3

Tailoring peptides and peptidomimetics for targeting protein–protein interactions

Łukasz Berlicki
Department of Bioorganic Chemistry, Wrocław University of Science and Technology, Wrocław, Poland

3.1 Introduction

Protein–protein interactions (PPIs) are events that are responsible for the control of numerous physiological processes in each living cell. The set of all PPIs in a given organism is called the interactome. The human interactome contains 14000 binary PPIs that are related to nearly all processes in the cell [1]. Among them, there are several interactions that are involved in the development of various human diseases, in particular cancer [2]. Therefore PPIs have become molecular targets for an increasing number of medicinal chemistry projects [3,4], which is stimulated additionally by several drugs introduced to the market, as well as their extremely high potential for treating various cancers. Inhibition of PPIs responsible for immune system control is one of the most remarkable examples, which constitute a basis for the development of cancer immunotherapy [5–8]. This discovery was acknowledged by the Nobel Prize in Medicine in 2018 to Tasuku Honjo and James P. Allison.

However, PPIs are, in general, considered to be challenging molecular targets for small molecules [9]. Usually, the area of contact between proteins is relatively large and these interactions cannot be directly mimicked by low-molecular-weight compounds. Therefore one of the possible approaches is based on finding key interactions between proteins that are covering much smaller areas, so-called hot spots [10]. However, finding hot spots is not always possible, and often small molecules are not effective enough to efficiently inhibit targeted PPIs. The other option considers the application of inhibitors of larger size. Antibodies are one class of molecules that can be applied [11,12]. A large area of possible interactions and well-established

Peptide and Peptidomimetic Therapeutics. DOI: https://doi.org/10.1016/B978-0-12-820141-1.00009-1

methods for their discovery and production are the major advantages of this approach. Unfortunately, antibodies also show several disadvantages, including immunogenicity, poor solid tumor penetration, poor control of pharmacokinetics, and immune-related adverse effects (irAEs) [13]. Finally, medium-sized molecules, in particular peptides and peptidomimetics, can be considered to be highly advantageous for the construction of PPI inhibitors [4]. In this case, the possible large interaction surface of peptides is combined with their synthetic availability, structural diversity and biocompatibility. On the other hand, the major challenges of application for peptides and related compounds as PPI inhibitors is related to control of their three-dimensional structure and enzymatic hydrolysis in the body. The conformational stability of peptides is often low but could be effectively modulated by several well-established methodologies that are based on constraints introduced to the main chain and/or between side chains of an oligomer [14]. In most cases, such modifications also increase metabolic stability of peptide-based compounds as proteases/peptidases show significant specificity toward molecules built from the 20 canonical L-amino acids.

Surfaces of interacting proteins show great structural diversity and many conformational arrangements, in particular, including any known secondary structures, could be involved. Numerous examples of helix mediated PPIs have been extensively studied, in addition to those based on extended structures or loops. MDM2-p53 is the most classical example of an interaction [15,16] involving a helix secondary structure (Fig. 3.1A), while PD-1/PD-L1 [17] and menin-MLL [18] are representative examples of interactions mediated by extended structures and loops, respectively (Fig. 3.1B and C).

It is worth noting that the ability to mimic protein fragments is an intrinsic feature of peptides, which can serve as a starting point for the development of effective PPI inhibitors. Therefore methodologies for construction of peptide-based inhibitors can be divided according to targeted three-dimensional structures, that is, helix, extended and loop [19]. While in the case of extended and loop structures the major methodology is based on the development of macrocyclic peptides (Fig. 3.2A), approaches for helix mimetics are much more diversified (Fig. 3.2B–F). These include stapling of side chains, formation of hydrogen bond surrogates (HBSs) and main chain modifications such as D-peptides, β-peptides and peptoids.

3.2 Helix mimetics

A helix is a compact structure with several intramolecular hydrogen bonds, and therefore constitutes a great scaffold for the construction of peptide-based PPI inhibitors, in particular those mediated by helices [20]. Mimicking helices was widely studied in the field of structural peptide chemistry, and numerous efficient methodologies have been elaborated. Either modifications of side chains or the peptide main chain can be applied for this purpose.

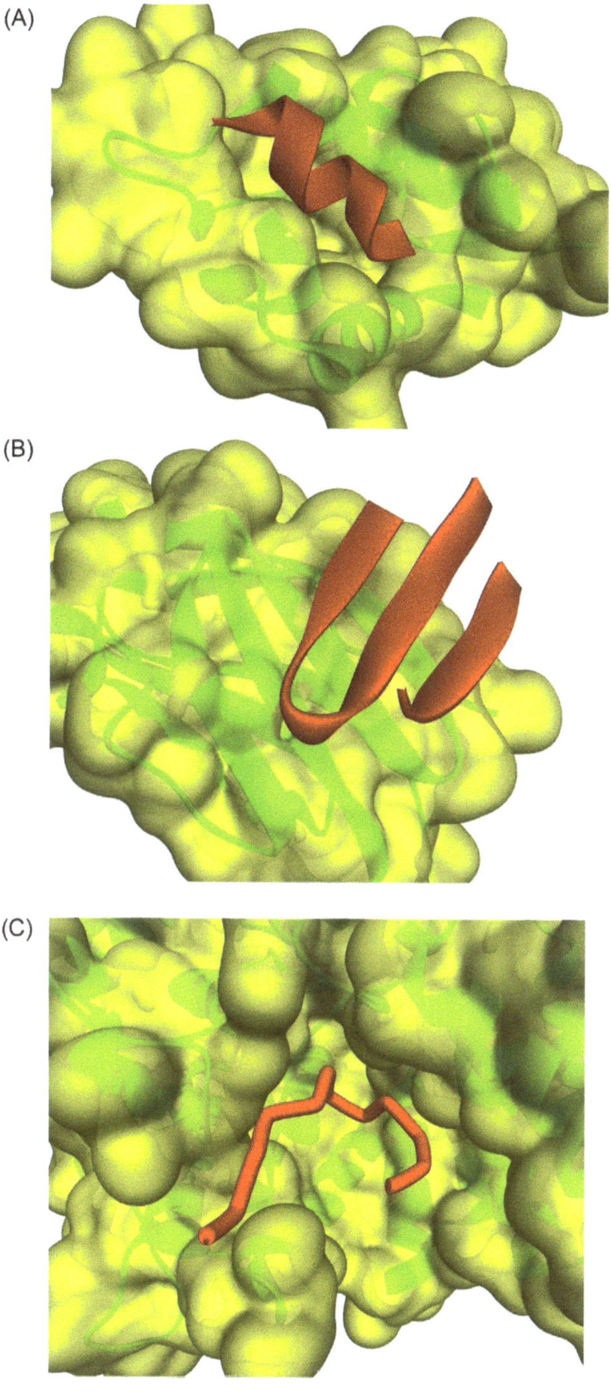

FIGURE 3.1 Examples of protein–protein interactions mediated by various secondary structures: (A) helix—MDM2/p53 (PDB ID 3LNZ), (B) β-sheets—PD-L1/PD-1 (PDB ID 4ZQK) and (C) loop—MLL-menin (PDB ID 4GQ6). The first interaction partner is shown as a ribbon within the transparent surface, while the second one is a dark ribbon. Only the interacting fragment of the second interacting partner is shown. The rest of the second protein is not shown (B) or has been cut off before crystallization (A and C).

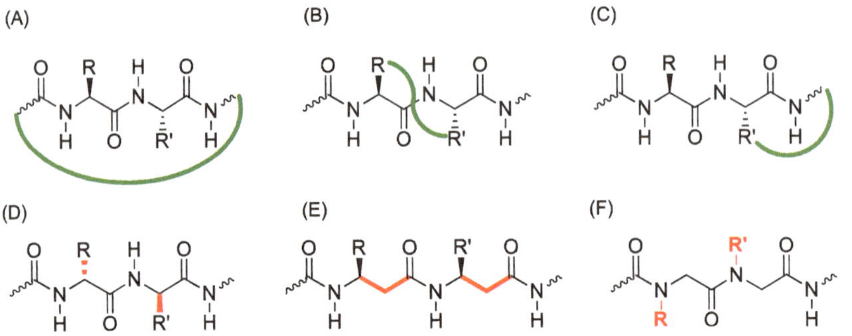

FIGURE 3.2 Major types of peptide modifications that are useful for construction of protein—-protein interaction inhibitors: (A) macrocyclization; (B) side chain to side chain stapling; (C) hydrogen bond surrogates; (D) D-peptides; (E) β-peptides; (F) peptoids. Cyclizations are shown in green, while main chain modifications are in red.

Covalent linking of side chains that are in spatial proximity in helical conformation, so-called stapling, emerged as one of the most widely studied approaches [21,22]. Original hydrocarbon stapling based on metathesis reaction has been further extended by numerous other possibilities, including triazole formation, lactamization, cysteine side chain crosslinking, biaryl linkage formation (by Suzuki reaction or C–H activation), etc. (see Chapter 5 for details) [23]. In the case of PPI inhibition, stapling of peptides in helical conformation shows several advantages. First, the entropic effect of freezing the molecule in the active conformation significantly increases binding affinity. Second, it was demonstrated in some cases that the staple itself can also interact with the target. Third, such modification also enhances the stability of the peptidomimetic against enzymatic degradation, and, fourth, it often increases peptide cellular uptake [24,25]. Numerous advantages of stapled peptides have led to discovery of several PPI inhibitors, including BCL-2 family–BH3 domains [26], MDM2-p53 [27], β-catenin–TCF [28], Rab–GTPase-effector [29], ERα–coactivator protein [30], VDR–coactivator protein [31], Cullin3–BTB [32], and β-arrestin–β-adaptin [33].

Several stapled peptides have been developed as inhibitors of MDM2-p53 interaction [27]. The interaction between p53 and MDM2 proteins occurs via a helical fragment of p53 (Fig. 3.1A). In particular, three residues exposed by this helix, namely Leu, Trp, and Phe, dock to a hydrophobic cavity of MDM2. The helicity of a 16-amino acid linear peptide fragment of p53 is 11%, while analogs that are stapled between residues 7 and 14 showed helicity up to 85%, in the case of peptide SAH–p53-8 (Fig. 3.3). This peptide exhibited inhibitory activity significantly higher than the linear analog (K_d = 55 nM and 410 nM for stapled and linear peptides, respectively). Moreover, in contrast to the linear peptide, the stapled one was cell-permeable and induced cell death.

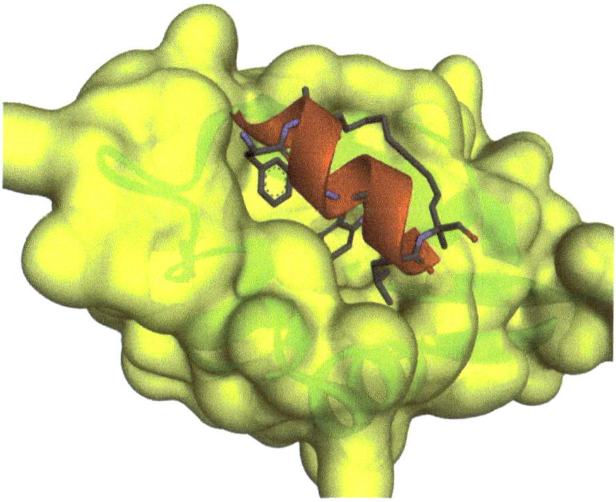

FIGURE 3.3 Crystal structure of the complex of stapled peptide (SAH-p53-8) with MDM2 (PDB ID 3V3B) [34]. MDM2 is shown as a ribbon within a surface. The peptide is shown as a dark ribbon with interacting residues in stick representation.

The crystal structure of stapled peptide SAH-p53−8 in complex with MDM2 (Fig. 3.3) has proven that its mode of binding is analogous to that of linear peptide inhibitors [34]. The conformations of the bound peptides were very similar (close to α-helix), and the Leu, Trp and Phe triad of the stapled peptide was anchored in the MDM2 pocket. However, additional stabilization of binding was achieved by the interaction of the staple with the hydrophobic region of MDM2 (L54, F55, G58, and M62).

HBSs constitute a second group of peptide inhibitors of PPIs, where local cyclization stabilizes helical conformation [35,36]. In this case, the N terminus of the peptide is covalently bound to a third residue's nitrogen atom, which is expected to replace the i-$i+4$ hydrogen bond that initializes the $\alpha-$helix formation. Typically, the synthesis of HBSs is done by ring-closing metathesis reaction of resin-bound bis-olefin modified peptides. *N*-allyl modification is introduced in the third residue, while the N terminus is derivatized with pentenoic acid [37]. Alternatively, a hydrazine bridge could be also applied [38]. The main goal of this approach is the stabilization of helical conformation with minimal changes in structure, which allows for the usage of all amino acid residue side chains for creating interactions with the protein target. This is particularly useful in the case of helical fragments deeply buried in the molecular target.

The HBS methodology was used for developing inhibitors of the following PPIs: HIF-1α/CBP/p300 [39], BCL-xL/Bac BH3 [40] and MTIP/MyoA [41]. The MTIP/MyoA interaction seems to be most challenging among the abovementioned cases, as the helical structure is fully immersed in the target protein (Fig. 3.4). In such a case, the HBS technique appears to be preferred. Comparison of the native peptide with the HBS revealed the same mode of binding, but HBS analogs showed increased binding to MTIP ($IC_{50} = 4.4\ \mu M$ and $IC_{50} = 2.4\ \mu M$ for native and modified peptides, respectively).

The second group of methodologies for the construction of peptide-based PPI inhibitors mimicking helices is related to modifications of the peptide backbone (Fig. 3.2D and E). The most classical approach in this area is based on the application of D-peptides. The major advantage of this technique is related to a significant increase of proteolytic stability of D-peptides in comparison to their L-counterparts. It is worth noting that this strategy can be applied to mimicking any three-dimensional structure. In the case of helix mimicry, the mirror image of the native α-helix (i.e., left-handed helix) is obtained. Therefore interactions between the left-handed helical conformation of the D-peptide with the target protein cannot mimic precisely that of the L-peptide, and the sequence of the D-peptide has to be optimized.

Mirror image phage display is a technique highly effective for the discovery of new D-peptide ligands and was applied to searching for new MDM2 binders [42]. The ^{D}PMI-α peptide was identified with a moderate K_d value (219 nM, Table 3.1). Interestingly, its binding interface to MDM2 was significantly different from that observed in the native interaction, namely Trp/Leu/Leu versus Phe/Trp/Leu. Further optimizations led to the discovery of

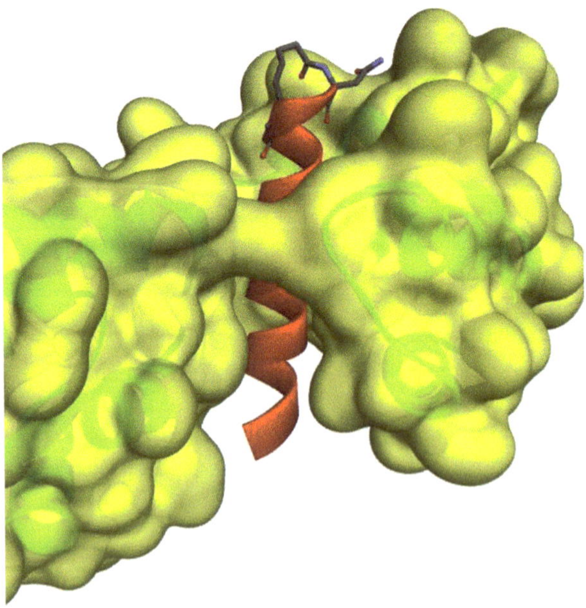

FIGURE 3.4 Crystal structure of hydrogen bond surrogate peptide and myoA tail interacting protein complex (PDB ID 4MZL) [41]. The color scheme is the same as in Fig. 3.3.

TABLE 3.1 Sequence comparison of D-peptide inhibitors of the MDM2-p53 interaction and the native fragment of p53 (L-peptide).

Name	Sequence[a]	K_d (nM)
$^{17-28}$p53	TFSDLWKLLPE	452
^{D}PMI-α	tnwyanlekllr	219
^{D}PMI-β	tawyanfekllr	35
^{D}PMI-γ	hdwwplafeallr	53
p-CF_3 − Phe7 − ^{D}PMI-β	tawyan(*p*-CF_3)fekllr	0.45
^{D}PMI-δ	ta(6-F)wyan(*p*-CF_3)fekllr	0.22

[a]*Residues interacting directly with MDM2 are marked in red; small letters denote D-amino acid residues.*

D-peptides ^{D}PMI-α and ^{D}PMI-γ with enhanced activities (K_d = 35 and 53 nM, respectively) and improved binding mediated by the Trp/Phe/Leu triad [43]. Furthermore, structure-based optimization of D-peptides with non-canonical amino acids produced compounds with subnanomolar activity. An increase of the Phe7 volume in ^{D}PMI-β by *para* substitution with a CF_3 group that filled the space originally occupied by Trp23 of p53 resulted in a K_d of 0.45 nM (Fig. 3.5 and Table 3.1) [44]. Modification of both Trp3 and Phe7 in PMI-β produced the highly active ^{D}PMI-δ. Interestingly, all presented D-peptides share only one amino acid residue with the native p53 fragment (analogous to Leu26 of p53), but the space natively occupied by right-handed α-helix is perfectly filled by the left-handed helix of the D-peptide.

β-Amino acid-containing peptides are one of the most promising groups of peptide-based PPI inhibitors [45,46]. The incorporation of an additional carbon atom in the peptide backbone significantly influences their properties. Surprisingly, a large number of structurally different conformations can be observed for peptides of this class, depending on sequence pattern and relative stereochemistry of neighboring residues. Several types of β-amino acids have already been explored for the construction of biologically active peptides; however, homologated $β^3$-analogs of canonical amino acids and cyclopentane/cyclohexane containing β-amino acids have been applied most frequently [14]. Two general strategies can be applied. First, all amino acid residues can be exchanged with their β-analogs while preserving the same set of side chains or α- and β-residues can be combined in one sequence [47]. In the second option, constrained cyclopentane/cyclohexane β-residues are often chosen to induce appropriate conformational preferences. The great majority of structural studies as well as the biological activity of β-amino

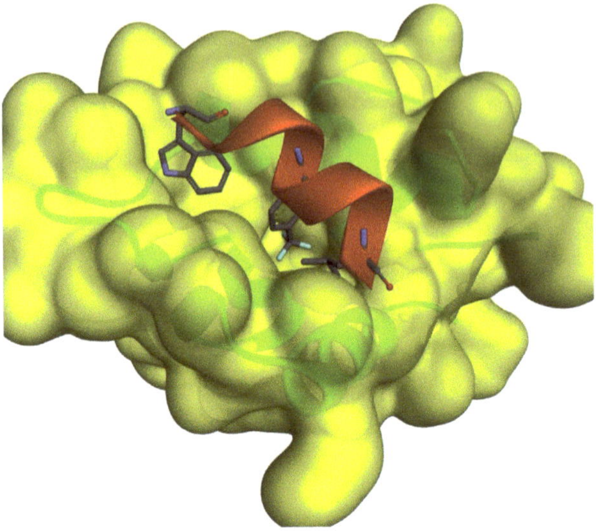

FIGURE 3.5 Crystal structure of D-peptide inhibitor (p-CF_3 − Phe7 − ^{D}PMI-β) bound to MDM2 (PDB ID 3TPX) [44]. The color scheme is the same as in Fig. 3.3.

acid containing peptides are related to helical structures, as precise rules for construction of such peptides have been elaborated [48]. Well-defined conformational preferences of this class of peptides are one of its most important advantages. In particular, such peptides have high tendencies to fold, which is often observed even for short peptides of ten residues [49,50]. Moreover, significant modification of the backbone is responsible for their high resistance to enzymatic hydrolysis. On the other hand, the major challenge of constructing β-amino acid containing biologically active peptides is the appropriate three-dimensional distribution of side chains that are interacting with a molecular target, as conformations adopted by peptides from this class usually do not allow for direct mimicry of native α-peptides.

Helical β-amino acid-containing peptides have been studied as inhibitors of PPIs, and several examples have been published as inhibitors of hDM2-p53 [51], Bcl-2 Family-BH3 Domain [52], and VEGF-VEGFR1 [53]. Bcl-xL ligands that inhibit its interaction with the BH3 domain are considered valuable anticancer drug candidates. β-Amino acid-containing peptide inhibitors of this interaction have been developed by the Gellman group. Initially, backbones with all β-residues or a 1:1 α:β pattern were evaluated, but their inhibitory activity was poor [54]. However, a combination of the α/β pattern applied to the *N*-terminus with an α-peptide tail provided highly active peptides (Fig. 3.6). The incorporation of constrained *trans*-2-aminocyclopentanecarboxylic acid residues was one of the key factors responsible for high affinity. The crystal structure in complex with the BCL-xL protein showed

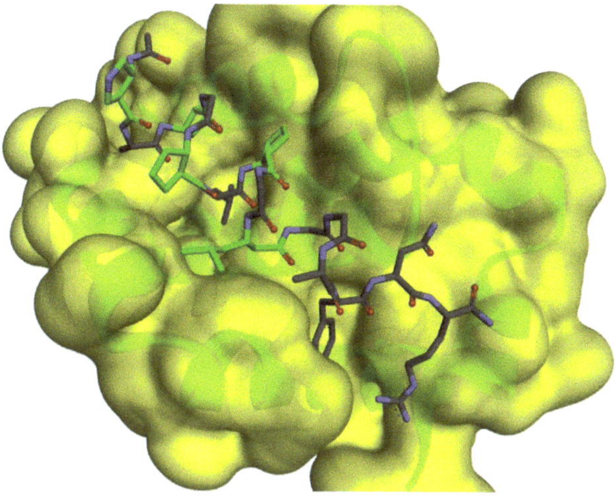

FIGURE 3.6 The crystal structure of the α/β-peptide ($K_i = 2$ nM) in complex with the BCL-xL protein (PDB ID 3FDM) [55]. The protein is shown as a ribbon covered with a surface. The inhibitor is shown in stick representation. Solvent exposed side chains are omitted for clarity.

a similar arrangement of interacting residues for the α/β-peptide as found for the native α-peptide [55]. Further studies have shown that α/β-peptides with an αααβ pattern could achieve very high inhibitory activity combined with high resistance to enzymatic cleavage (with 2–3 orders of magnitude higher half-life times) [56,57].

3.3 Extended structures

Several PPIs occur via interactions of beta-structures [58,59]. In general, these types of targets are considered challenging due to the fact that usually large, flat surfaces are involved. Moreover, mimicking extended structures is much more problematic than helices. In extended structures the number of intramolecular interactions that could stabilize the secondary structure of medium-sized peptides is much smaller than in the case of helices; therefore it is not generally easy to construct scaffolds that can be freely substituted.

The most often applied strategy is based on the synthesis of macrocyclic peptides that can mimic two strands of beta-sheet; however, linear peptides with extended conformations are also able to block some PPIs efficiently.

The PD-1/PD-L1 interaction is one of the most widely studied topics in the area of cancer immunotherapy, as recently approved monoclonal antibody drugs against this target have proven to be highly effective [8]. These two proteins interact via a large surface of β-sheets that are perpendicular to each other, and no significant hot spots on these surfaces have been detected [17]. Large molecules like monoclonal antibodies are excellent inhibitors of

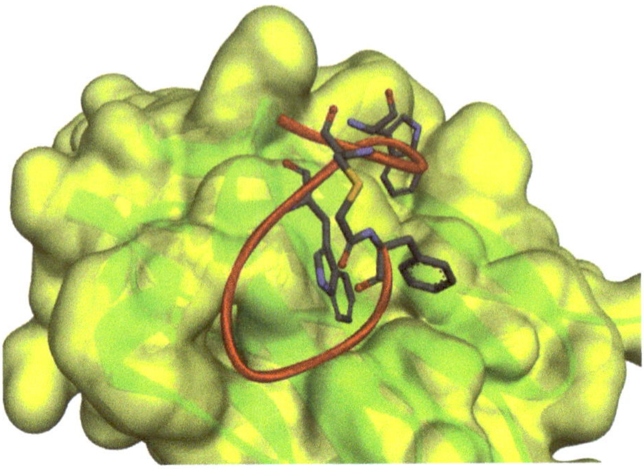

FIGURE 3.7 Crystal structure of the macrocyclic peptide/PD-L1 complex (PDB ID 5O4Y) [62]. The color scheme is the same as in Fig. 3.3.

this PPI [6], while small organic compounds are hardly achievable and only a few examples can be found in the literature [60]. A Bristol-Myers Squibb patent describes a large set of macrocyclic peptides that inhibit the PD-1/PD-L1 interaction at nanomolar levels [61]. The crystal structure of the macrocyclic peptide-PD-L1 complex revealed that the backbone of the inhibitory peptide was perpendicular to the β-sheet structures of PD-L1 [62]. Moreover, major interactions between the protein and the inhibitor are mediated by hydrophobic side chains of tryptophan and phenylalanine residues. Although the interaction area is smaller than that observed for the PD-1/PD-L1 pair, significant nanomolar binding affinity was demonstrated (Fig. 3.7).

3.4 Loops

The interactions between proteins can occur also by other structural elements, including various types of loops. The most commonly used approach in such cases is based on the construction of macrocyclic peptides that could mimic the size and conformation of an interacting loop. Therefore a covalent bond closing the macrocycle should be introduced between molecular fragments that have been found in spatial proximity in the native complex. The introduction of such constraints is aimed at stabilization of the active conformation of the peptide, as well as increasing its proteolytic stability.

The menin-MLL interaction is considered to be a promising target for drugs to treat acute leukemia [63,64]. Interactions between these proteins have been well characterized structurally (Fig. 3.1C) and indicate that the N-terminal fragment of menin docks in a large cleft of MLL in a loop conformation [18]. Interestingly, this crystal structure shows a hydrogen bond between the C-terminal proline (Pro13) and the guanidine side chain of Arg8. Creating a

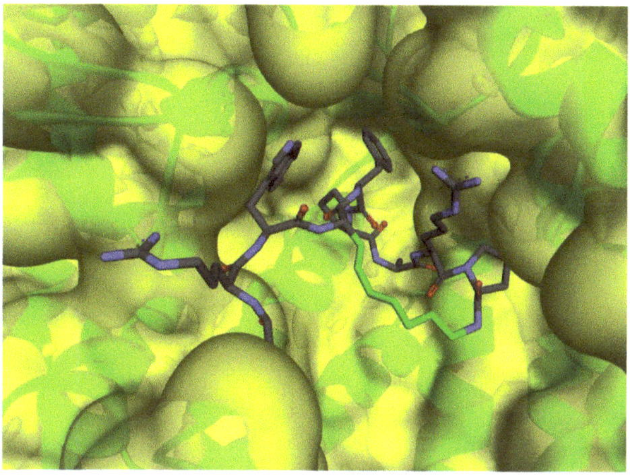

FIGURE 3.8 Crystal structure of the complex of MLL and the macrocyclic inhibitor (PDB ID 4I80, $K_i = 23.8$ nM) [65]. The protein is shown as a ribbon covered with a surface. The inhibitor is shown in stick representation.

macrocycle was possible by the incorporation of a hydrocarbon chain that covalently links the abovementioned residues (Fig. 3.8) [65]. Comparing crystal structures of complexes with the native ligand or the macrocyclic inhibitor revealed that their placement inside the target protein cavity is nearly identical, and all intermolecular interactions are preserved. Further optimization of selected side chains produced a highly active inhibitor of the MLL-menin interaction with a K_i of 4.7 nM.

3.5 Summary

Peptides and peptidomimetics are excellent scaffolds for the construction of PPI inhibitors due to their intermediate size and controllable conformational behavior. Their inherent similarity to protein parts and ability to effectively mimic any required conformation exhibited by native proteins in combination with several methodologies that allow for modulation of metabolic stability and cell permeability provides excellent tools for the development of effective drug candidates.

Acknowledgments

The work was financially supported by the National Science Centre, Poland, Grant No. 2018/31/B/ST5/02631 (to Ł. B.).

References

[1] Rolland T, Taşan M, Charloteaux B, Pevzner SJ, Zhong Q, Sahni N, et al. A proteome-scale map of the human interactome network. Cell 2014;159:1212–26.

[2] Milroy LG, Grossmann TN, Hennig S, Brunsveld L, Ottmann C. Modulators of protein-protein interactions. Chem Rev 2014;114:4695–748.
[3] Nevola L, Giralt E. Modulating protein-protein interactions: the potential of peptides. Chem Commun (Camb) 2015;51:3302–15.
[4] Wójcik P, Berlicki Ł. Peptide-based inhibitors of protein–protein interactions. Bioorg Med Chem Lett 2016;26:707–13.
[5] Okazaki T, Honjo T. PD-1 and PD-1 ligands: from discovery to clinical application. Int Immunol 2007;19:813–24.
[6] Iwai Y, Hamanishi J, Chamoto K, Honjo T. Cancer immunotherapies targeting the PD-1 signaling pathway. J Biomed Sci 2017;24:26 - 0.
[7] Kim JM, Chen DS. Immune escape to PD-L1/PD-1 blockade: seven steps to success (or failure). Ann Oncol 2016;27:1492–504.
[8] Wei SC, Duffy CR, Allison JP. Fundamental mechanisms of immune checkpoint blockade therapy. Cancer Discov 2018;8:1069–86.
[9] Arkin MR, Tang Y, Wells JA. Small-molecule inhibitors of protein-protein interactions: progressing toward the reality. Chem Biol 2014;21:1102–14.
[10] Macalino SJY, Basith S, Clavio NAB, Chang H, Kang S, Choi S. Evolution of in silico strategies for protein-protein interaction drug discovery. Moleculaes 2018;23 E1963 - 0.
[11] Konstantinidou M, Zarganes-Tzitzikas T, Magiera-Mularz K, Holak TA, Dömling A. Immune checkpoint PD-1/PD-L1: is there life beyond antibodies? Angew. Chem. Int. (Ed.) 2018;57:4840–8.
[12] Chiavenna SM, Jaworski JP, Vendrell A. State of the art in anti-cancer mAbs. J Biomed Sci 2017;24:15 - 0.
[13] Nelson AL, Dhimolea E, Reichert JM. Development trends for human monoclonal antibody therapeutics. Nat Rev Drug Discov 2010;9:767–74.
[14] Martinek TA, Fülöp F. Peptidic foldamers: ramping up diversity. Chem Soc Rev 2012;41:687–702.
[15] Murray JK, Gellman SH. Targeting protein-protein interactions: lessons from p53/MDM2. Biopolymers 2007;88:657–86.
[16] Popowicz GM, Dömling A, Holak TA. The structure-based design of Mdm2/Mdmx-p53 inhibitors gets serious. Angew Chem Int Ed Engl 2011;50:2680–8.
[17] Zak KM, Kitel R, Przetocka S, Golik P, Guzik K, Musielak B, et al. Structure of the complex of human programmed death 1, PD-1, and its ligand PD-L1. Structure 2015;23:2341–8.
[18] Shi A, Murai MJ, He S, Lund G, Hartley T, Purohit T, et al. Structural insights into inhibition of the bivalent menin-MLL interaction by small molecules in leukemia. Blood 2012;120:4461–9.
[19] Pelay-Gimeno M, Glas A, Koch O, Grossmann TN. Structure-based design of inhibitors of protein-protein interactions: mimicking peptide binding epitopes. Angew. Chem Int Ed 2015;54:8896–927.
[20] Rezaei Araghi R, Keating AE. Designing helical peptide inhibitors of protein-protein interactions. Curr Opin Struct Biol 2016;39:27–38.
[21] Cromm PM, Spiegel J, Grossmann TN. Hydrocarbon stapled peptides as modulators of biological function. ACS Chem Biol 2015;10:1362–75.
[22] Ali AM, Atmaj J, Van Oosterwijk N, Groves MR, Dömling A. Stapled peptides inhibitors: a new window for target drug discovery. Comput Struct Biotechnol J 2019;17:263–81.
[23] Lau YH, de Andrade P, Wu Y, Spring DR. Peptide stapling techniques based on different macrocyclisation chemistries. Chem Soc Rev 2015;44:91–102.
[24] Kritzer JA. Stapled peptides: how to be quick on the uptake. Nat Chem Biol 2016;12:764–5.

[25] Chu Q, Moellering RE, Hilinski GJ, Kim Y-W, Grossmann TN, Yehab JT-H, et al. Towards understanding cell penetration by stapled peptides. Med Chem Commun 2015;6:111–19.
[26] Walensky LD, Kung AL, Escher I, Malia TJ, Barbuto S, Wright RD, et al. Activation of apoptosis in vivo by a hydrocarbon-stapled BH3 helix. Science 2004;305:1466–70.
[27] Bernal F, Tyler AF, Korsmeyer SJ, Walensky LD, Verdine GL. Reactivation of the p53 tumor suppressor pathway by a stapled p53 peptide. J Am Chem Soc 2007;129:2456–7.
[28] Grossmann TN, Yeh JT, Bowman BR, Chu Q, Moellering RE, Verdine GL. Inhibition of oncogenic Wnt signaling through direct targeting of β-catenin. Proc Natl Acad Sci USA 2012;109:17942–7.
[29] Spiegel J, Cromm PM, Itzen A, Goody RS, Grossmann TN, Waldmann H. Direct targeting of Rab-GTPase-effector interactions. Angew Chem Int Ed Engl 2014;53:2498–503.
[30] Phillips C, Roberts LR, Schade M, Bazin R, Bent A, Davies NL, et al. Design and structure of stapled peptides binding to estrogen receptors. J Am Chem Soc 2011;133:9696–9.
[31] Misawa T, Demizu Y, Kawamura M, Yamagata N, Kurihara M. Structural development of stapled short helical peptides as vitamin D receptor (VDR)-coactivator interaction inhibitors. Bioorg Med Chem 2015;23:1055–61.
[32] de Paola I, Pirone L, Palmieri M, Balasco N, Esposito L, Russo L, et al. Cullin3-BTB interface: a novel target for stapled peptides. PLoS ONE 2015;10:e0121149 e0121149.
[33] Nevola L, Martín-Quirós A, Eckelt K, Camarero N, Tosi S, Llobet A, et al. Light-regulated stapled peptides to inhibit protein-protein interactions involved in clathrin-mediated endocytosis. Angew Chem Int Ed Engl 2013;52:7704–8.
[34] Baek S, Kutchukian PS, Verdine GL, Huber R, Holak TA, Lee KW, et al. Structure of the stapled p53 peptide bound to Mdm2. J Am Chem Soc 2012;134:103–6.
[35] Patgiri A, Jochim AL, Arora PS. A hydrogen bond surrogate approach for stabilization of short peptide sequences in alpha-helical conformation. Acc Chem Res 2008;41:1289–300.
[36] Miller SE, Thomson PF, Arora PS. Synthesis of hydrogen-bond surrogate α-helices as inhibitors of protein-protein interactions. Curr Protoc Chem Biol 2014;6:101–16.
[37] Patgiri A, Menzenski MZ, Mahon AB, Arora PS. Solid-phase synthesis of short α-helices stabilized by the hydrogen bond surrogate approach. Nat Protoc 2010;5:1857–65.
[38] Cabezas E, Satterthwait AC. The hydrogen bond mimic approach: solid-phase synthesis of a peptide stabilized as an alpha-helix with a hydrazone link. J Am Chem Soc 1999;121:3862–75.
[39] Henchey LK, Kushal S, Dubey R, Chapman RN, Olenyuk BZ, Arora PS. Inhibition of hypoxia inducible factor 1—transcription coactivator interaction by a hydrogen bond surrogate α-helix. J Am Chem Soc 2010;132:941–3.
[40] Wang D, Wei Liao W, Arora PS. Enhanced metabolic stability and protein-binding properties of artificial α helices derived from a hydrogen-bond surrogate: application to Bcl-xL. Angew. Chem Int Ed 2005;44:6525–9.
[41] Douse CH, Maas SJ, Thomas JC, Garnett JA, Sun Y, Cota E, et al. Crystal structures of stapled and hydrogen bond surrogate peptides targeting a fully buried protein-helix interaction. ACS Chem Biol 2014;9:2204–9.
[42] Liu M, Pazgier M, Li C, Yuan W, Li C, Lu W. A left-handed solution to peptide inhibition of the p53-MDM2. Interact Angew Chem Int Ed Engl 2010;49:3649–52.
[43] Liu M, Li C, Pazgier M, Li C, Mao Y, Lv Y, et al. D peptide inhibitors of the p53-MDM2 interaction for targeted molecular therapy of malignant neoplasms. Proc Natl Acad Sci USA 2010;107:14321–6.
[44] Zhan C, Zhao L, Wei X, Wu X, Chen X, Yuan W, et al. An ultrahigh affinity d-peptide antagonist Of MDM2. J Med Chem 2012;55:6237–41.

[45] Cabrele C, Martinek TA, Reiser O, Berlicki Ł. Peptides containing beta-amino acid patterns—challenges and successes in medicinal chemistry. J Med Chem 2014;57:9718–39.

[46] Bautista AD;, Craig CJ, Harker EA, Schepartz A. Sophistication of foldamer form and function in vitro and in vivo. Curr Opin Chem Biol 2007;11:685–92.

[47] Johnson LM, Gellman SH. α-Helix mimicry with α/β-peptides. Meth Enzymol 2013; 523:407–29.

[48] Mándity IM, Wéber E, Martinek TA, Olajos G, Tóth GK, Vass E, et al. Design of peptidic foldamer helices: a stereochemical patterning approach. Angew Chem Int Ed Engl 2009;48:2171–5.

[49] De Pol S, Zorn C, Klein CD, Zerbe O, Reiser O. Surprisingly stable helical conformations in alpha/beta-peptides by incorporation of cis-beta-aminocyclopropane carboxylic acids. Angew Chem Int Ed 2004;43:511–14.

[50] Berlicki Ł, Pilsl L, Wéber E, Mándity IM, Cabrele C, Martinek TA, et al. Unique α,β- and α,α,β,β-peptide foldamers based on cis-β-aminocyclopentanecarboxylic acid. Angew Chem Int Ed Engl 2012;51:2208–12.

[51] Kritzer JA, Hodsdon ME, Schepartz A. Solution structure of a beta-peptide ligand for hDM2. J Am Chem Soc 2005;127:4118–19.

[52] Boersma MD, Haase HS, Peterson-Kaufman KJ, Lee EF, Clarke OB, Colman PM, et al. Evaluation of diverse α/β-backbone patterns for functional α-helix mimicry: analogues of the Bim BH3 domain. J Am Chem Soc 2012;134:315–23.

[53] Haase HS, Peterson-Kaufman KJ, Lan Levengood SK, Checco JW, Murphy W, Gellman SH. Extending foldamer design beyond α-helix mimicry: α/β-peptide inhibitors of vascular endothelial growth factor signaling. J Am Chem Soc 2012;134:7652–5.

[54] Sadowsky JD, Fairlie WD, Hadley EB, Lee HS, Umezawa N, Nikolovska-Coleska Z, et al. (α/β + α)-Peptide antagonists of BH3 domain/Bcl-xL recognition: toward general strategies for foldamer-based inhibition of protein – protein interactions. J Am Chem Soc 2007;129:139–54.

[55] Lee EF, Sadowsky JD, Smith BJ, Czabotar PE, Peterson-Kaufman KJ, Colman PM, et al. High-resolution structural characterization of a helical alpha/beta-peptide foldamer bound to the anti-apoptotic protein Bcl-xL. Angew Chem Int Ed 2009;48:4318–22.

[56] Horne WS, Boersma MD, Windsor MA, Gellman SH. Sequence-based design of α/β-peptide foldamers that mimic BH3 domains. Angew Chem Int Ed 2008;47:2853–6.

[57] Peterson-Kaufman KJ, Haase HS, Boersma MD, Lee EF, Fairlie WD, Gellman SH. Residue-based preorganization of BH3-derived α/β-peptides: modulating affinity, selectivity and proteolytic susceptibility in α-helix mimics. ACS Chem Biol 2015;10:1667–75.

[58] Sawyer N, Watkins AM, Arora PS. Protein domain mimics as modulators of protein-protein interactions. Acc Chem Res 2017;50:1313–22.

[59] Wuo MG, Arora PS. Engineered protein scaffolds as leads for synthetic inhibitors of protein-protein interactions. Curr Opin Chem Biol 2018;44:16–22.

[60] Guzik K, Zak KM, Grudnik P, Magiera K, Musielak B, Törner R, et al. Small-molecule inhibitors of the programmed cell death-1/programmed death-ligand 1 (PD-1/PD-L1) interaction via transiently induced protein states and dimerization of PD-L1. J Med Chem 2017;60:5857–67.

[61] Miller MM, Mapelli C, Allen MP, Bowsher MS, Boy KM, Gillis EP, et al. Macrocyclic inhibitors of the PD-1/PD-L1 and CD80(B7-1)/PD-L1 protein-protein interactions. WO Pat 2014;151634 - 0.

[62] Magiera-Mularz K, Skalniak L, Zak KM, Musielak B, Rudzinska-Szostak E, Berlicki Ł, et al. Bioactive macrocyclic inhibitors of the PD-1/PD-L1 immune checkpoint. Angew. Chem Int Ed 2017;56:13732–5.

[63] Grembecka J, He S, Shi A, Purohit T, Muntean AG, Sorenson RJ, et al. Menin-MLL inhibitors reverse oncogenic activity of MLL fusion proteins in leukemia. Nat Chem Biol 2012;8:277–84.

[64] Cierpicki T, Grembecka J. Challenges and opportunities in targeting the menin-MLL interaction. Future Med Chem 2014;6:447–62.

[65] Zhou H, Liu L, Huang J, Bernard D, Karatas H, Navarro A, et al. Structure-based design of high-affinity macrocyclic peptidomimetics to block the menin-mixed lineage leukemia 1 (MLL1) protein-protein interaction. J Med Chem 2013;56:1113–23.

Chapter 4

Advances in peptide synthesis

Zhipeng A. Wang[1,2] and Ji-Shen Zheng[1]
[1]*The First Affiliated Hospital of USTC, Division of Life Sciences and Medicine, University of Science and Technology of China, Hefei, P.R. China,* [2]*Division of Genetics, Department of Medicine, Brigham and Women's Hospital and Department of Biological Chemistry and Molecular Pharmacology, Harvard Medical School, Boston, MA, United States*

4.1 Introduction to peptides

Proteins and peptides play key roles in all living organisms. They are involved in a large number of physiological and biological processes (Fig. 4.1). In the postgenomic era, the complexity of the human proteome has drawn increasing attention due to the ever-growing catalog of documented posttranslational modifications (PTMs) [1]. The functions of PTMs are open questions for structural biology and biophysics. A growing request from cell biology, biochemistry, and molecular biology, to specifically control the structures of target peptides and proteins at the atomic level, is becoming more and more

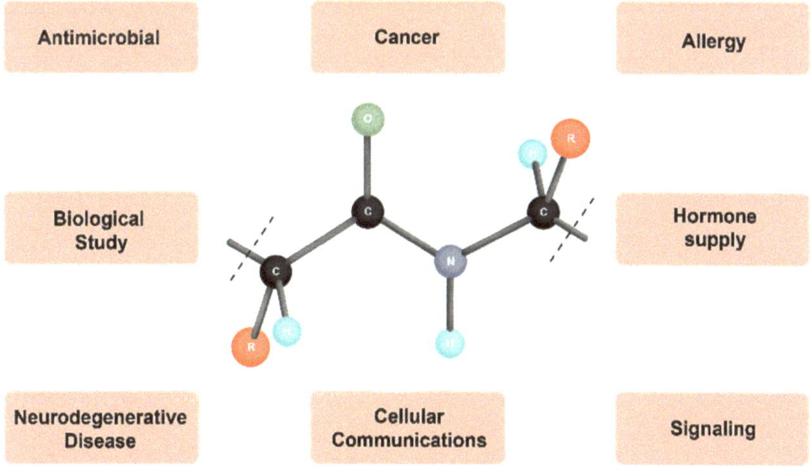

FIGURE 4.1 A wide array of peptide functions and applications.

Peptide and Peptidomimetic Therapeutics. DOI: https://doi.org/10.1016/B978-0-12-820141-1.00004-2

important [2], especially with chemical functional groups not existing in nature [3].

Chemical synthesis is one of the most useful tools to prepare proteins with desired structures [4], and can be traced back to studies by Fischer in the year 1901 [5]. The field received its first Nobel Prize, awarded to Vincent du Vigneaud in 1955, for the first synthesis of the cyclic peptide oxytocin. In this chapter, the development of solid-phase peptide synthesis (SPPS) will be reviewed, including the innovation of solid phase, coupling reagents, and protecting groups, together with their modifications [6]. Furthermore, approaches for accelerated peptide synthesis are also introduced, such as microwave-assisted coupling and microflow-assisted coupling. In addition, several important but challenging synthetic peptides will be summarized regarding further applications [7].

4.2 Fluorenylmethyloxycarbonyl solid-phase peptide synthesis

Achieving the assembly of polypeptides on solid supports is rather straightforward nowadays, and is achieved simply by repeating a cycle of coupling and deprotection, followed by a final cleavage (Fig. 4.2). However, this is the product of a tremendous effort in developing the process of SPPS. Three primary areas of innovation have moved the field forward: the application of organic polymer beads, the development of novel activation reagents, and the use of protecting groups.

4.2.1 Three major developments

There are three main factors in fluorenylmethyloxycarbonyl (Fmoc)-SPPS, namely the composition of the solid-phase resin/beads, the protecting groups, and the activation reagents that drive amide bond formation. The resin is composed of a polymer support or scaffold. Polystyrene (PS) and polyethylene glycol (PEG) are two commonly used polymer materials. The key

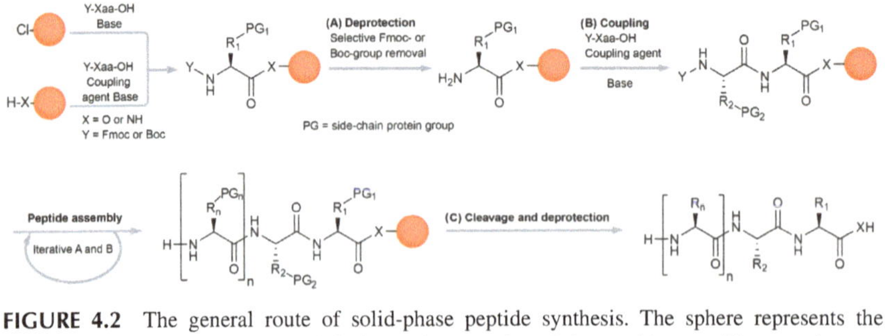

FIGURE 4.2 The general route of solid-phase peptide synthesis. The sphere represents the resin; the white cycle at the lower left shows the repeating of the peptide assembly procedure.

prerequisite is that the resin should have been swelled adequately in dimethylformamide (DMF)/dichloromethane (DCM) [8].

To form an amide bond between two different amino acids, one of the two active functional groups needs to be caged, either the carboxylic group or the amino group. For the amino acids with functional groups, side-chain protection is also needed, such as the thiol of cysteine and the ε-amino group of lysine. Therefore, two main types of protecting group, the temporary protecting group and the permanent protecting group, will be removed only at the end of synthesis. In 1932, Leonidas Zervas first applied the carboxybenzyl (Cbz, Z) protecting group. After careful screening, carbamate was finally chosen as the best candidate to reduce oxazolone formation [9].

Merrifield introduced solid-phase organic synthesis to the peptide field in 1963 [10], for which he was awarded the Nobel Prize in 1984. There are two systems of SPPS: tert-butyloxycarbonyl (Boc)/benzyl (Bzl) chemistry and Fmoc/*t*-butoxycarbonyl (tBu) chemistry. In the Boc/Bzl strategy, trifluoroacetic acid (TFA) is repeatedly used to remove the Boc-protecting group on the α-amino group, and hydrogen fluoride (HF) is used for the last cleavage step. However, HF is extremely toxic, corrosive, and volatile, and it is incompatible with some PTMs, such as phosphorylation and glycosylation [11]. The Fmoc/tBu strategy uses basic conditions for deprotection and has become much more common [12]. Therefore we will focus on Fmoc-SPPS in further discussions.

Compared with Boc-SPPS, Fmoc-SPPS avoids the repeated usage of TFA for deprotection, which makes Fmoc-SPPS applicable as a synthesizer. Furthermore, the final cleavage by TFA instead of HF makes Fmoc-SPPS accessible to most PTMs. However, due to the repeated deprotection with piperidine in Fmoc-SPPS, peptides cannot be cleaved from the resin directly as C-terminal thioesters. Another advantage of Boc-SPPS is that TFA disrupts interactions between peptides on resin, which results in higher coupling efficiency than Fmoc-SPPS.

Biomedical and biophysical applications of synthetic peptides require high yields, which in turn require robust coupling reagents for the sequential formation of backbone amide bonds. For example, the difference between 99% and 95% coupling efficiency in the synthesis of a 10-amino-acid polypeptide reduces the potential yield from 90% to 60%. Several groups of activation reagents have been developed, such as the carbodiimide family, exemplified by diisopropylcarbodiimide (DIC), the 1-hydroxybenzotriazole (HOBt) family, and the related 2-(1H-benzotriazol-1-yl)-1,1,3,3-tetramethyluronium hexafluorophosphate family, exemplified by hexafluorophosphate benzotriazole tetramethyl uronium (HBTU). These have been widely and systematically summarized elsewhere [13].

4.2.2 Extensions of solid-phase peptide synthesis

One challenge of SPPS is that the technique is limited to a size of about 50 amino acids, which is too small for a general protein domain [14]. The Kent group first solved this problem through the development of native chemical ligation (NCL), in which a peptide thioester selectively reacts with an *N*-terminal Cys peptide to give a native peptide bond. Sequential NCL reactions allow the assembly of larger peptides from short sequences prepared by SPPS (Fig. 4.3) [15].

The peptide thioester is one of the critical intermediates for NCL. As stated previously, Fmoc deprotection by base poses a challenge in the solid-phase synthesis of peptide thioesters. Therefore many different strategies have been developed to prepare thioesters, such as the *N*-to-*S* acyl shift strategies [16], together with several different thioester equivalents [17], such as the peptide hydrazide [18]. Most of these strategies require specially modified chemical linkage to the resin. For example, an enamide-assisted *N*-to-*S* acyl transfer strategy was developed, in which the imine product of the acyl transfer undergoes hydrolysis in situ during peptide cleavage, making the process irreversible (Fig. 4.4) [19]. In the peptide hydrazide ligation strategy,

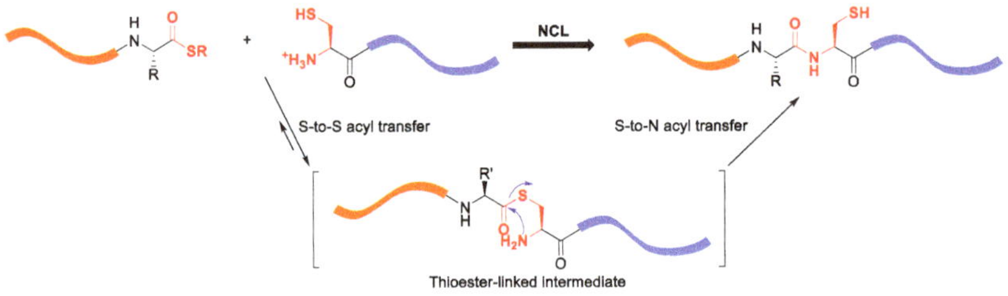

FIGURE 4.3 Native chemical ligation.

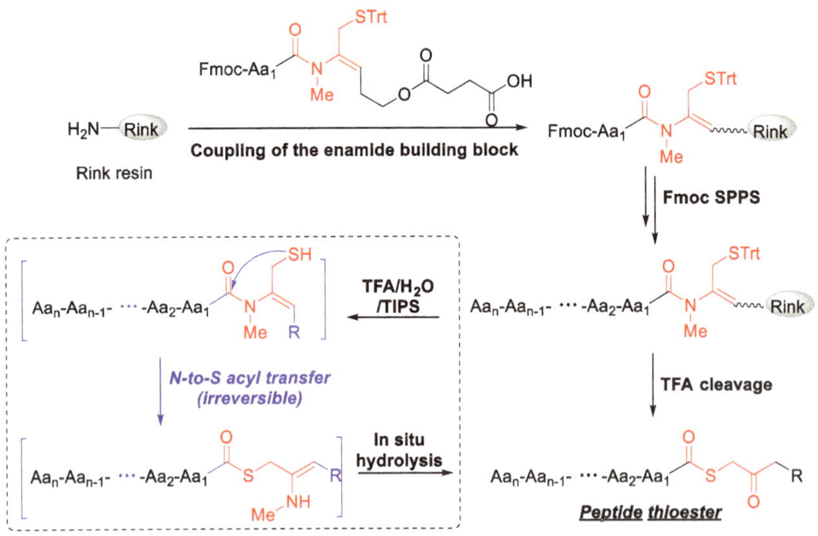

FIGURE 4.4 Fmoc synthesis of peptide thioesters through an irreversible *N*-to-*S* acyl transfer. Rink represents the rink-amide resin.

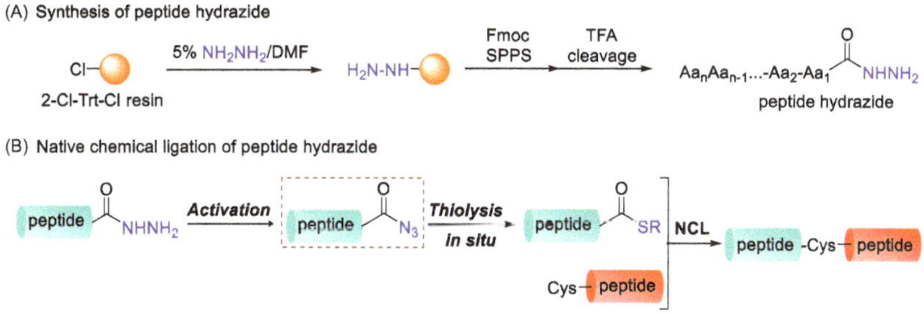

FIGURE 4.5 The native chemical ligation of peptide hydrazides: (A) synthesis of peptide hydrazide from SPPS; (B) peptide hydrazide-based NCL reaction.

a special hydrazine resin is prepared by the treatment of 2-Cl-trityl resin with hydrazine, and the cleaved peptide hydrazide is oxidized first to an acyl azide, before thiolytic formation of the peptide thioester (Fig. 4.5) [20]. The 3,4-diaminobenzoyl linker is another thioester precursor, and resin with this functional group preinstalled is commercially available. Conversion of this linker to a thioester, after finishing the peptide elongation, is accomplished by forming an *N*-acylurea, which is thiolyzed to the thioester (Fig. 4.6B) [21]. The 3,4-diaminobenzoyl linker can also be oxidated into peptidyl benzotriazole after being cleaved from resin, followed by conversion to thioester [22].

Other peptide ligation strategies have also been developed, which rely on functional groups beyond the thiol/thioester pair required for NCL [23]. α-Ketoacid-hydroxylamine (KAHA) ligation, for example, needs special C-terminal peptidyl α-ketoacid and N-terminal hydroxylamine [24]. Traceless Staudinger ligation, on the other hand, requires a specific C-terminal phosphine functional moiety [25]. These versatile ligation strategies are reliant on a wide variety of C-terminal modified peptides, which are uniquely accessible by SPPS. In addition to the variants on the C-terminal, orthogonal protection of the N-terminal, such as thiazole (Thz) or trifluoroacetamidomethyl (Tfacm)-protected Cys, can enable controlled, multistep NCL assembly of whole proteins [26,27]. Enzyme-catalyzed peptide ligation, also known as chemoenzymatic protein synthesis, is another large group of ligation strategies that do not require thiol/thioester; this approach has been applied widely in biochemistry and chemical biology studies [28].

4.3 Approaches for accelerated peptide synthesis

Peptides are important therapeutic molecules. It is critical to rapidly provide the desired new functional peptide variants and quickly optimize these designs for their improved potency, selectivity, and stability. Despite the expanded number of protected amino acid monomers, chemical peptide synthesis is limited by synthesis speeds of minutes to hours per amide bond. To speed up the synthesis procedure, high temperatures can greatly increase the chemical

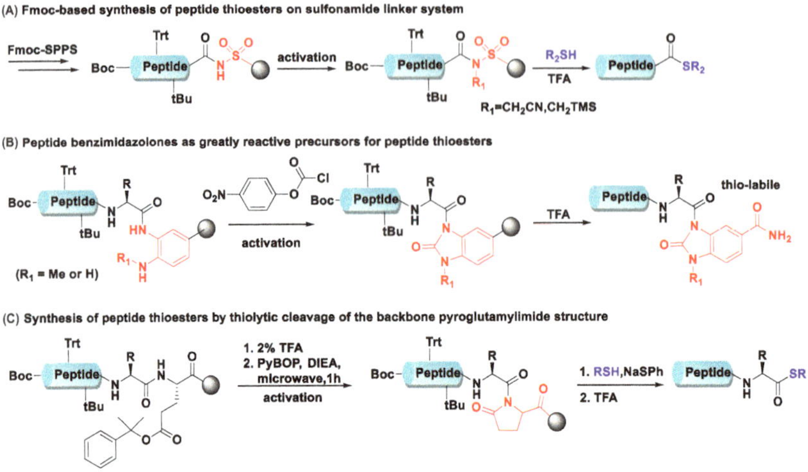

FIGURE 4.6 Thioester synthesis through activation by an external reagent: (A) the preparation of peptide thioester with the sulfonamide linker system; (B) the synthesis of peptide thioester from peptide benzimidazolone; and (C) the conversion of peptide pyroglutamylimide to peptide thioester.

reaction rate. Therefore directly heating the reaction has been developed to accelerate peptide synthesis.

4.3.1 Microwave-assisted peptide synthesis

Among all the heating methods, microwave irradiation stands out for its precise temperature control. With such control, microwave-assisted amino acid coupling and Fmoc deprotection can significantly reduce the reaction time, while also improving product yield and purity [29]. Moreover, mechanistically, microwaves appear to interact with secondary structures, such as the β-strand, and break the formation of β-sheet, which is generally believed to impair coupling [30].

Microwave heating has several main advantages over conductive heating. Microwaves can directly deliver energy to the solution inside the reaction vessel instead of requiring thermal conductivity of the vessel. This helps to heat the solution evenly. Additionally, microwaves, or more essentially, an electromagnetic field, can provide different heating modes, such as monomode or multimode, for variable accessibility. Finally, the most commonly used solvents for SPPS, DMF and *N*-methylpyrrolidone, have favorable microwave heating profiles, making microwave heating faster and more efficient than conductive heating [29].

For microwave-assisted peptide coupling, the most important factor is temperature. Researchers initially started with a temperature of 110°C [31], but more recent reports are based on 50°C–80°C, with lower temperatures producing similarly good yields [29]. It needs to be noted that some amino acids require lower temperature couplings; otherwise, there is a risk of epimerization. To be suitable for coupling under high temperature, different coupling reagents and activation reagents have been carefully screened [32]. Generally speaking, common coupling reagents perform nearly the same with an increased temperature. Some literature has shown that the uronium type of coupling reagent 1-[(1-(cyano-2-ethoxy-2-oxoethylidenaminooxy)-dimethylamino-morpholinomethylene)] methanaminium hexafluorophosphate (COMU) works better than typical HXTU, including HBTU/hexafluorophosphate 1-hydroxy-7-azabenzotriazole tetramethyl uronium (HATU) or benzotriazol-1-yl-oxytripyrrolidinophosphonium hexafluorophosphate (PyBOP). Unlike the uranium and phosphonium reagents, COMU generates aminoacyl oxime-activated ester intermediate (Fig. 4.7) [33]. Microwave approaches have only been developed for Fmoc-SPPS thus far, and typically rely on a 5- to 10-fold excess of amino acids and coupling reagent [34]. Notably, not all the difficult peptides can be easily achieved even with microwave-assisted methods.

Few publications have reported the preparation of very long peptides and mini-proteins by microwave-assisted synthesis, despite being theoretically better suited for this method. However, examples of microwave syntheses do exist, such as the β-amyloid and affibody [30]. The longest microwave-assisted peptide prepared thus far was a 109-mer containing polyglutamine and polyproline (Fig. 4.8A) [35].

A helpful application of microwave-assisted SPPS is the synthesis of phosphopeptides, which is generally done by the coupling of side-chain phosphorylated amino acid building blocks. However, these phosphorylated amino acids, including Ser, Thr, and Tyr, require a double coupling, which can be a huge waste of time and material. With a microwave-assisted coupling step, a shortened time with an improved yield can be easily achieved [29].

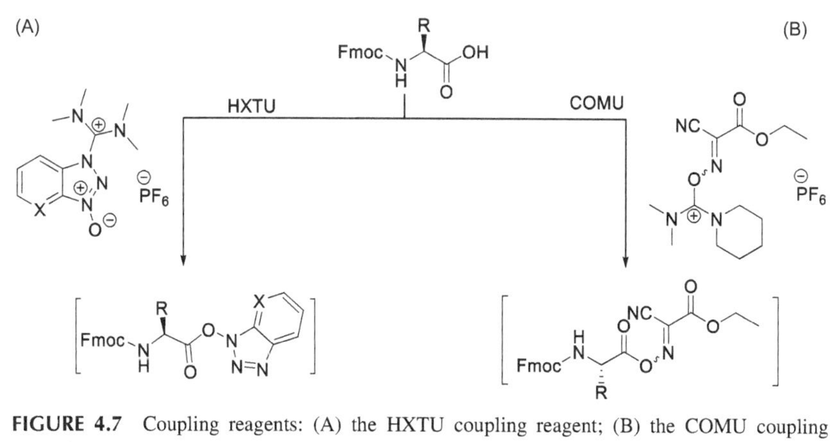

FIGURE 4.7 Coupling reagents: (A) the HXTU coupling reagent; (B) the COMU coupling reagent.

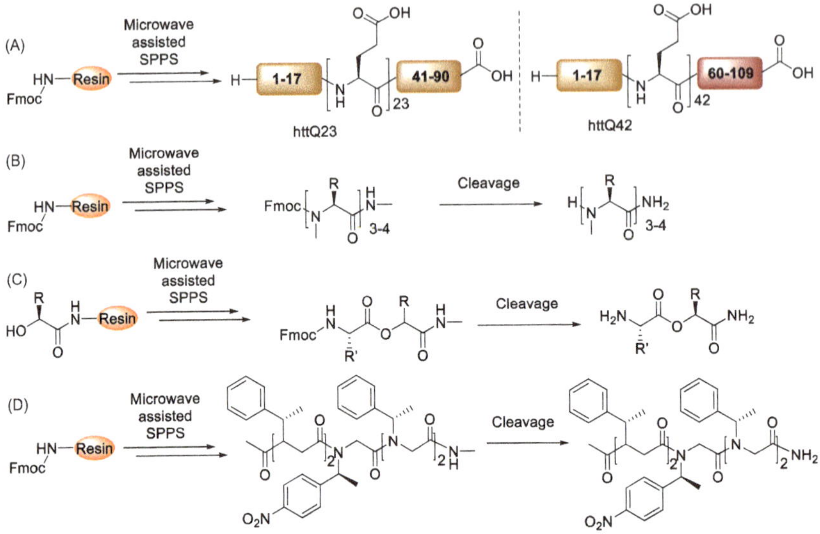

FIGURE 4.8 Microwave-assisted SPPS: (A) the microwave-assisted SPPS of huntingtin exon 1 with a 23/42-glutamine-long poly-Q stretch; (B) the microwave-assisted SPPS of backbone *N*-methylated peptide; (C) the microwave-assisted SPPS of ester-linkage; (D) the microwave-assisted SPPS of peptoid.

Another great example of microwave-assisted SPPS can be found in glycopeptides. As one of the most widely existing PTMs, glycosylation is related to many cellular activities, such as the cell growth and immune response [36]. Glycopeptides can be divided into two main groups, as the *N*-glycoside linking to Asn or the O-glycosides linking to Ser/Thr/Tyr/etc. [37]. Glycosylated amino acid building blocks are generally synthesized and installed through SPPS, where protecting groups are suitable for standard coupling conditions such as HXTU and piperidine [38]. However, due to the large size and complexity of the amino acid building block with glycosylation, the coupling step can be time-consuming and incomplete [39]. In addition, glycosylated amino acids are expensive, which encourages researchers to minimize their usage. Thus improvements in the reaction time and yield will help to advance the field of solid-phase glycopeptide synthesis (SPGS). Microwave-assisted SPGS is compatible with commercial reagents and does not suffer from glycan moiety β-elimination by piperidine [40]. Typical SPGS examples are the synthesis of MUC1-related peptides with glycosylations [41], *O*-pentasaccharides [42], antifreeze glycopeptide derivatives [43], etc. Optimal synthesis temperatures range widely from 40°C to 70°C, but they achieve a high yield and require short coupling times. Although very successful in some syntheses, evidence shows that temperatures above 40°C can lead to the deglycosylation of specific peptide sequences [29].

Microwave-assisted SPPS is particularly powerful in the preparation of backbone-modified peptides and peptide analogs. *N*-methylated peptides, as an example, are a group with multiple bioactivities, which are generally believed to have higher stability and activity. Coupling of an Fmoc and *N*-methyl disubstituted amino group by traditional SPPS generally suffers from low yield and thus requires optimization of coupling conditions. Fortunately, a microwave-assisted coupling condition can greatly improve the situation (Fig. 4.8B) [44]. A recent protocol with (7-azabenzotriazol-1-yloxy)tripyrrolidinophosphonium hexafluorophosphate (PyAOP)/HOAt coupling condition in DMF with 60°C has been developed, which demonstrates that the microwave-assisted reaction is one of the best approaches [45].

Microwave-assisted peptide synthesis has also been applied to peptidomimetics [46], pseudopeptides (Fig. 4.8C) [47], β-peptides [48], and peptoids (Fig. 4.8D) [49]. Peptidomimetics are valuable tools in the development of new protein–protein interaction (PPI) inhibitors. With all the developments so far, special efforts have been applied, such as the new solid-phase material PEG, which compacts with microwaves very well [50].

4.3.2 Flow reactor and peptide reactor

Another development of peptide synthesis is the invention of flow reactor-based SPPS (Fig. 4.9). In 2014, Pentelute and coworkers [51] introduced a rapid procedure for peptide synthesis, which takes about 1–3 min per amino acid. This specially designed system can maintain the maximal concentration of all the synthesis reagents. The flow-based SPPS approach enables a 7-s amide bond formation step, and a total time of 40 s per amino acid. This

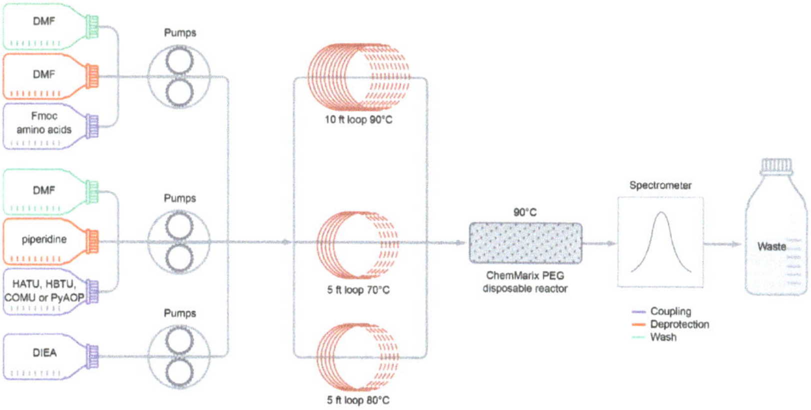

FIGURE 4.9 The general flow diagram of peptide synthesis using a flow reactor. *DMF*: dimethylformamide; *HATU*: hexafluorophosphate 1 hydroxy-7-azabenzotriazole tetramethyl uronium; *HBTU*: hexafluorophosphate benzotriazole tetramethyl uronium; *COMU*: 1-[(1-(cyano-2-ethoxy-2-oxoethylidenaminooxy)-dimethylamino-morpholinomethylene)] methanaminium hexafluorophosphate; *PyAOP*: (7-azabenzotriazol-1-yloxy)tripyrrolidinophosphonium hexafluorophosphate; *DIEA*: *N*,*N*-diisopropylethylamine.

method has been demonstrated even for small proteins, generally with over 100 amino acids. This special flow apparatus was further developed to be fully automated [52]. With the assistance of the rapid flow apparatus, the same laboratory made several protein examples, such as the H2 relaxin [53], *Bacillus amyloliquefaciens* RNase (Barnase) [51], and the designed ankyrin repeat protein (DARPin pE59) [54].

Besides the flow-based peptide coupling, Pentelute et al. [55] also extended the same principle to broader applications. Sortase A, for example, was solidified on a microreactor, which allows the efficient catalysis of peptide ligation in a flow-based manner. In that case, no large excessive glycine oligomer is needed to reduce the side reaction and to drive the equilibrium.

4.4 Advances in chemical synthesis of several important classes of peptides

Peptide synthesis has been widely applied to pharmaceutical and biomedical studies. Some natural peptide products can serve as inhibitors, which has inspired the design of novel peptides that mimic a motif or a secondary structure of a protein. The development of new synthetic methods in peptide synthesis has made scalable preparation of such peptides serving as PPI inhibitors achievable. However, certain important classes of peptide sequences remain difficult to prepare, which demands further efforts in these fields.

4.4.1 Difficult sequences

In the past, peptide-based drugs were mainly inspired by natural peptide sequences with relatively low molecular weight. Recently, relatively longer sequences have been studied by academic and industrial groups. However, many long peptides show a high tendency to aggregate, mostly due to interactions between different polypeptide backbones [56]. The aggregation tendency of a sequence is the basis for labeling it a "difficult peptide" or "difficult sequence" (DS). This is quantitatively defined by a repeatedly low coupling efficiency (lower than 15%) after double coupling [57]. A primary feature of DSs is the formation of intra- or interpeptide β-sheet interactions [58]. The key interaction in the β-sheet formation is noncovalent hydrogen bonding, which heavily depends on the sequence (Fig. 4.10).

A second major contributor to DSs is sequential sterically hindered amino acids. Transmembrane (TM) domains, for instance, are often difficult because they contain many bulky and hydrophobic sequences. Onc pitfall in DS synthesis is the incomplete removal of the Fmoc protecting group [59]. However, no method can yet predict whether a peptide is "difficult" simply based on its sequence, suggesting that the phenomenon of aggregation is determined by multiple factors [60]. Additionally, some difficult peptides only start to

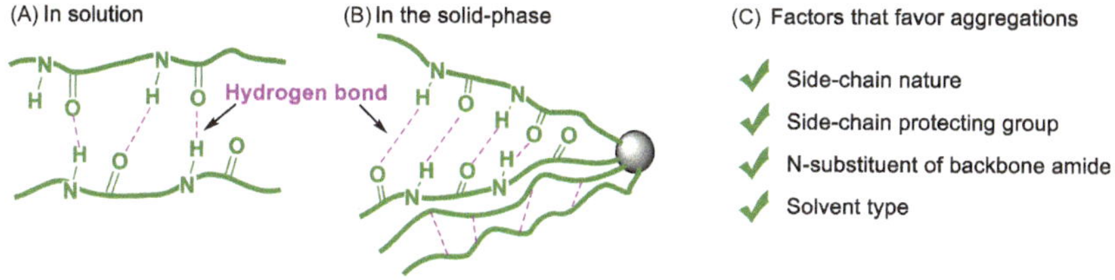

FIGURE 4.10 Hydrogen bond interchain interactions cause aggregations of difficult peptides in solution and in the solid-phase peptide synthesis: (A) interpeptide interactions of DS in solution; (B) interpeptide interactions of DS in the solid phase; (C) factors favoring aggregations.

aggregate after being fully deprotected. In this case, the sequence may appear fine during synthesis but will become a challenge for further purification [61].

One class of natural difficult peptides is those with only one kind of amino acid residue, or homo-oligomers, which have a high chance to aggregate, especially those made of nonpolar residues such as poly(Ala) and poly(Val). These sequences are so hydrophobic that they tend to aggregate on resin and in solution after being deprotected [62]. To deal with these difficult peptides, many different strategies have been gradually developed over the past decades. One simple strategy developed to deal with these sequences is changing the solvent from DMF to dimethyl sulfoxide (DMSO) for swelling the solid phase [63]. Another important group of solvents contains fluorine, such as 2,2,2-trifluoroethanol (TFE) and 1,1,1,3,3,3-hexafluoro-2-propanol (HFIP) [64], which uses the special role of hydrogen bonding of fluorine [65]. The adjustment of pH can enhance the solubility of some ionizable polypeptide chains. The addition of salt, as well as denaturant reagents such as urea, guanidinium chloride (Gn·HCl), or guanidinium thiocyanate (Gn·HSCN), can break up weak interactions [66]. Detergents are another option to help solubilizing difficult sequences.

Many different chemical modification methods have also been developed to solubilize difficult peptides (Fig. 4.11). The earliest strategy is to block the α-amino group chemically, which prevents the amide amine from hydrogen bonding and also suppresses its nucleophilic role in rare side reactions such as diketopiperazine (DKP) [67] or the aspartimide [68]. There are two types of backbone amino protection strategy, "temporary" or "permanent," with temporary strategies preferred for broader biological applicability. Temporary pseudoprolines are a typical example based on Ser, Thr, or Cys, which prevent hydrogen bond formation by mimicking the secondary amide of proline. The *cis*-conformation of these residues also inhibits the formation of β-sheets, which can be easily converted back to native structures under acidic conditions [69]. A permanent backbone-protecting method is methylation of the Nα, which is useful in the synthesis of natural products, and protease-resistant peptides, although it cannot be removed upon peptide cleavage. The incorporation of O-ester depsipeptides into the polyamide chain also disturbed the formation of β-sheets in another trial [70]. This strategy can yield a native amide linkage

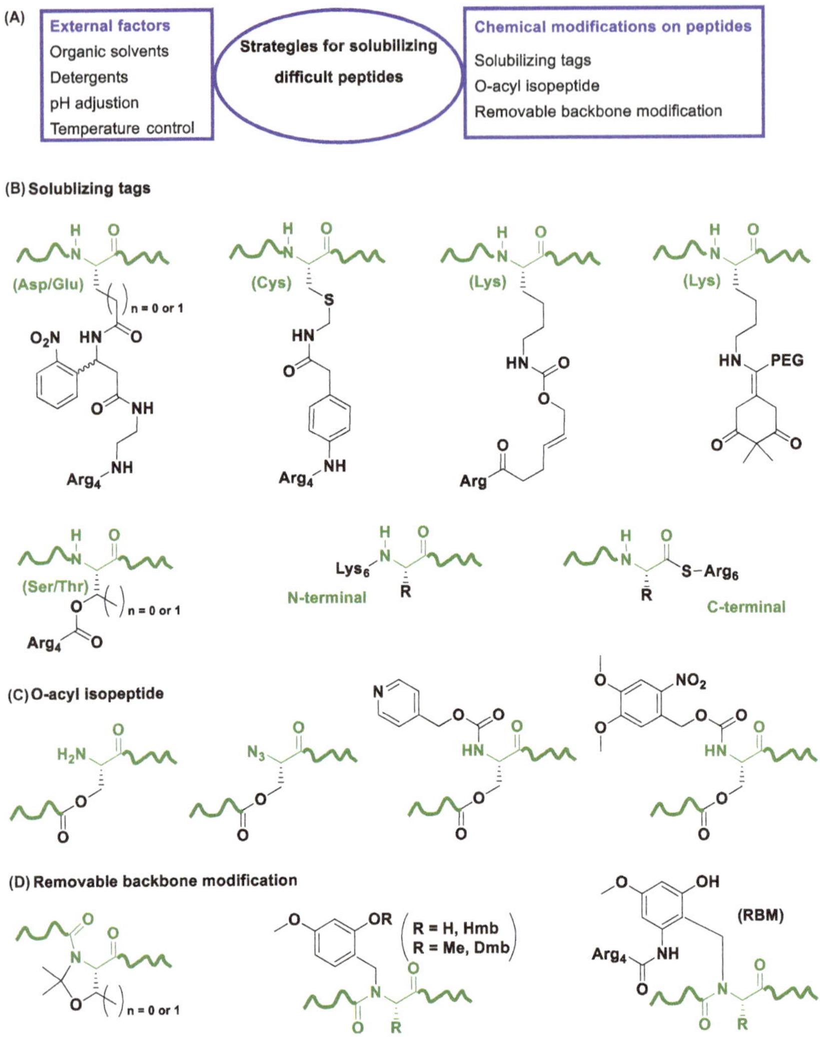

FIGURE 4.11 (A) Solubilizing strategies for difficult peptides. (B) Poly-arginine, poly-lysine, or polyethylene glycol tags incorporated into N-terminal, C-terminal, or amino acid side chains. (C) O-acyl isopeptides. (D) Removable backbone modification.

because of an *O–N* acyl transfer, which occurs after being cleaved from the resin and incubated in basic aqueous conditions [71]. Besides backbone modifications, the amino acid side-chain protecting groups can be tailored to facilitate both SPPS and purification, especially through the use of solubilizing tags. PEG chain is a very popular tag that can be introduced to almost all amino acid side chains [72]. Short peptide chains, typically polar amino acid oligomers, are also a common choice [73]. This is still a very active area of research [74].

4.4.2 Transmembrane peptides

Among the difficult peptides, TM peptides have been attracting significant focus due to the importance of membrane proteins (MPs) [75]. Most drug targets so far are MPs and their derivatives [76]. However, due to their physical chemistry properties, MPs are very challenging for chemical synthesis [77,78].

Improvements to the properties of TMs may be based on physical factors, in which only the external conditions are changed, with no chemical modifications [79]. Another strategy is to disrupt the secondary interaction between peptides by solubilizing tags installed onto the peptides. So far, different solubilizing tags have been applied to the N-terminal, the C-terminal [80], and amino acid side chains [81]. A typical example is the synthesis of the hepatitis C virus (HCV) protein, nonstructural protein 4A (NS4A) (1–54) TM, in which both the N-terminal of the peptidyl thioester and the C-terminal of the cysteine peptide carry poly-lysine tags (Fig. 4.12) [82]. The choice of solubilizing tags can be multiple, such as the PEG derivative [83] or poly-arginine tag [84]. For instance, in the synthesis of virus Type 1 protease, Kent et al. made use of C-terminal polyarginine tag connected by a protease recognizable sequence [80]. After the final ligation, the polyarginine tag can be removed by the activity of the newly ligated full-length protease itself.

A key consideration in designing a synthesis with solubilizing tags is whether they will be removed. In some instances, the ligation products require the tag to remain soluble, while in other cases removal of the tag is necessary for the activity of the ligation product. This has been addressed through the development of conditionally cleavable linkers [85].

To make a better version of the solubilizing peptide to assist both the coupling reaction on resin and the ligation in aqueous condition [86], specially designed structures are needed to fulfill different requirements, such as compatibility with SPPS and NCL, and easy cleavage under orthogonal conditions [87]. Zheng et al. have reported a series of works about three generations of removable backbone modification (RBM) strategies stemming from the 2,4-dimethoxybenzyl (Dmb) unit. The first generation is based on a special modified glycine (Gly) building block with an RBM group to couple the solubilizing peptide (Fig. 4.13) [88].

The same group then refined the RBM strategy for application at other amino acid positions. The acetylated RBM can survive the TFA cleavage and assist the ligation as well. After deacetylation the RBM becomes sensitive to the acidic conditions, which can be cleaved under TFA to release native structures (Fig. 4.14).

FIGURE 4.12 The synthesis of NS4A with poly-lysine tags.

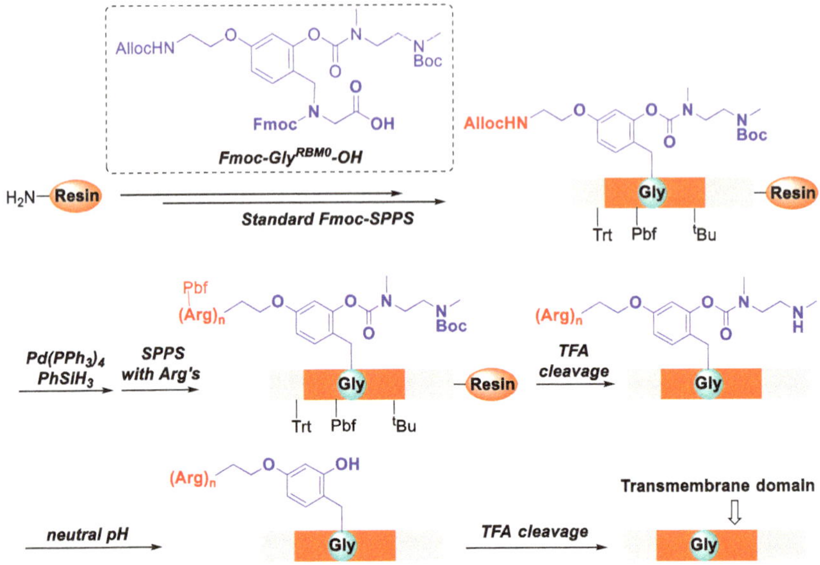

FIGURE 4.13 Fluorenylmethyloxycarbonyl solid-phase peptide synthesis of transmembrane peptides using the first removable backbone modification method.

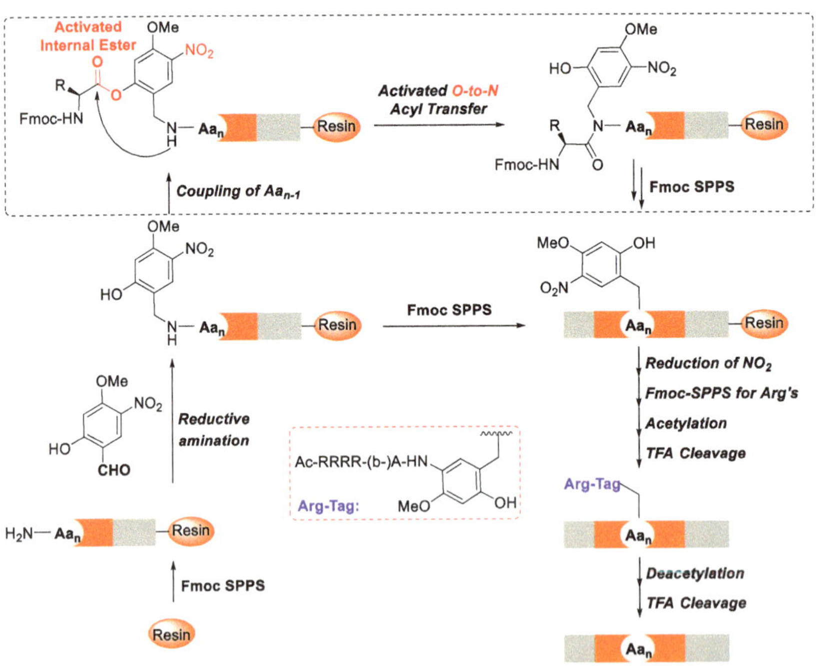

FIGURE 4.14 Fluorenylmethyloxycarbonyl solid-phase peptide synthesis of transmembrane peptides using the second removable backbone modification method.

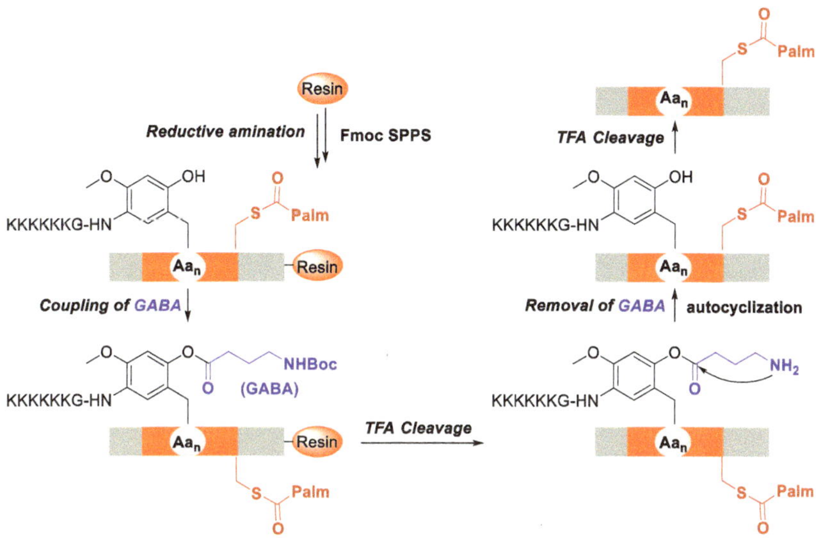

FIGURE 4.15 Fluorenylmethyloxycarbonyl solid-phase peptide synthesis of transmembrane peptides using the third removable backbone modification method.

Recently, the third-generation RBM method was developed [89], which is based on a new autocyclization γ-aminobutyric acid (GABA) group for temporarily protecting the phenol group of RBM (Fig. 4.15) [90]. The third RBM strategy enables the preparation of nucleophilic-sensitive *S*-palmitoylated peptides. The RBM strategies were used to prepare the core TM domain Kir5.1 (64–179), the *S*-palmitoylated sarcolipin, the phosphorylated or *S*-palmitoylated influenza A virus M2 proton channel, the isotopically labeled cation-specific ion channel p7 from the hepatitis C virus, and others.

4.4.3 Cyclic peptides

Cyclic peptides are a group of peptides with covalently linked cyclic structures. The linkage can be generated between the C-terminus and N-terminus, between a terminus and a side chain, or between side chains. Due to the diverse topology and more rigid structures, and thus better resistance to proteases, cyclic peptides are attracting increasing attention to their synthesis, characterization, and applications.

Natural disulfide-containing peptides represent a promising class of bioactive macromolecules, which can bind to target proteins with high affinity and specificity to regulate protein structures and functions. These peptides have received much attention as peptide pharmaceuticals (e.g., FDA-approved Ziconotide) and diagnostic reagents. The methods for the chemical synthesis of disulfide peptides [91] have been thoroughly reviewed elsewhere [92].

However, native disulfide-rich peptides usually suffer from reduction, S–S scrambling, and poor stability, which limits the extent of their application. To overcome these problems, a range of artificial disulfide-bond replacement strategies have been developed to keep the normal structure and function of peptides while improving their stability. The disulfide bond-replaced peptide surrogates have shown attractive prospects in the pharmaceutical industry. Disulfide replacement can yield more rigid cyclic peptides, which are considered as one of the best mimetics for protein secondary structures. After cyclization, peptides are inert to protease, and therefore they are more metabolically stable compared to their corresponding linear peptides. Additionally, the entropy loss during binding to the protein target is reduced due to a more strained structure [91].

A general method for the preparation of disulfide-bond surrogate peptides was chemoselective cyclization of amino acid side chains to give the Se-S bridged, C-S substituted, C-C bridged, or cystathionine analogs (Fig. 4.16A). Nonetheless, many side-chain bridging reactions are not generally compatible with peptides, and only limited types of disulfide surrogates could be formed. Therefore a diaminodiacid (DADA)-based strategy was developed, in which a premade diaminodiacid building block was first assembled into a peptide, and then an intramolecular amide cyclization was conducted to give disulfide-bond surrogates (Fig. 4.16b). However, the synthesis of

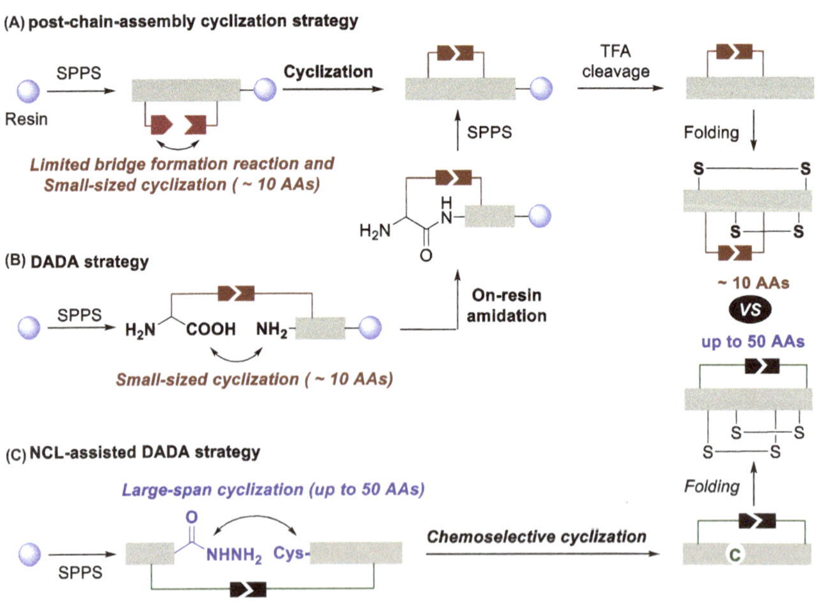

FIGURE 4.16 Strategies for the synthesis of disulfide-bond surrogate peptides: (A) side-chain cyclization strategy, (B) on-resin DADA strategy, and (C) NCL-assisted DADA strategy.

bioactive peptide analogs with relatively large-span disulfide-bond replacement remains challenging using the diaminodiacid method. Thus a new NCL-assisted diaminodiacid (NCL-assisted DADA) method was developed for making challenge disulfide peptide surrogates (Fig. 4.16C) [93].

To inhibit PPIs, cyclic peptide mimics with rigid structures are considered one of the best mimetics for protein secondary structures. After cyclization, peptides are inert to protease, and therefore they are more metabolically stable compared with their corresponding linear peptides. Additionally, the entropy loss during binding to the protein target is also less due to a more strained structure [94].

Reactions used in the synthesis of cyclic peptides can be divided into several main groups based on the linkage patterns: from terminal to terminal, from terminal to the side chain, and from side chain to side chain (Fig. 4.17) [95].

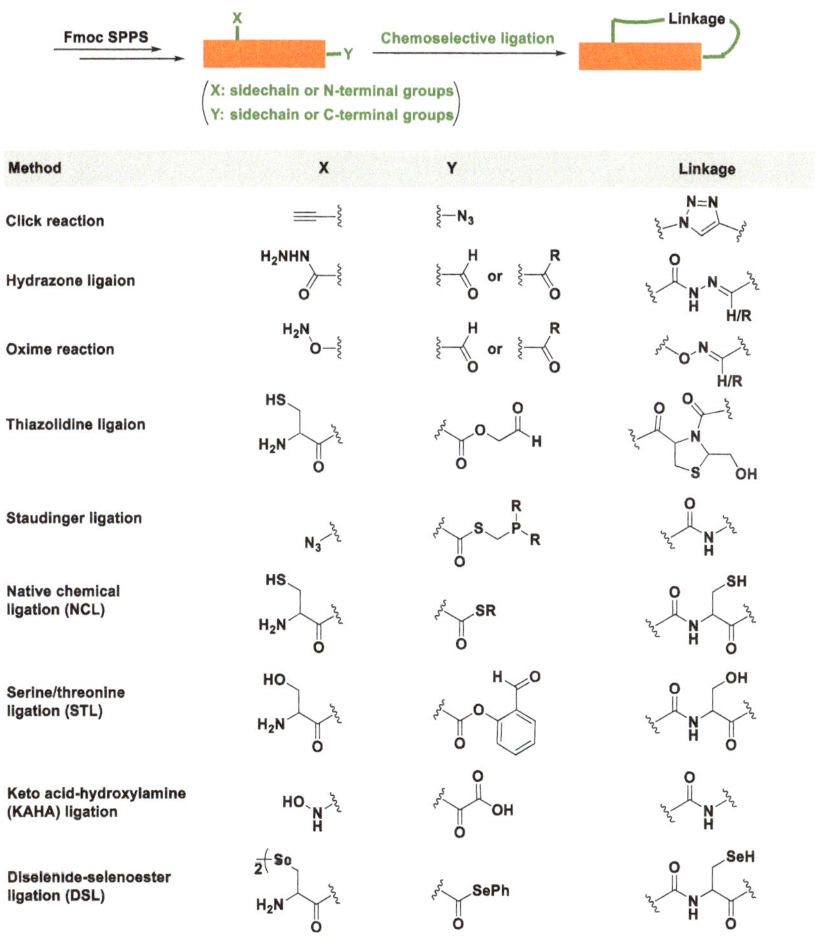

FIGURE 4.17 Chemoselective ligations for cyclic peptide synthesis.

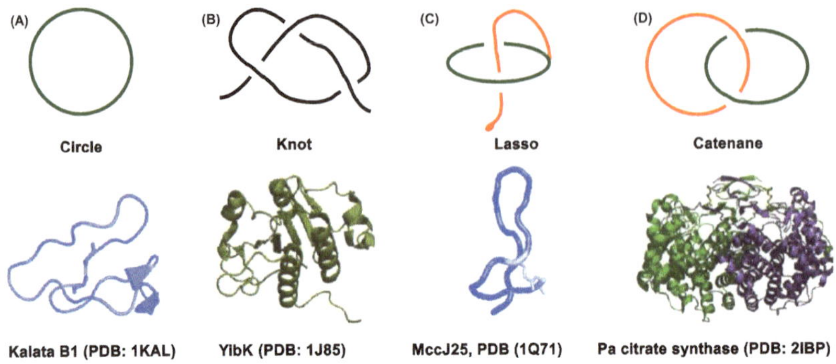

FIGURE 4.18 Topology of natural proteins.

Among them, side-chain ligation was most widely applied for natural peptides, especially the sulfhydryl group on the Cys side chain [96]. On the other hand, for synthetic peptides, more unnatural functional groups can be applied to generate not only cyclic peptides but also more sophisticated structures, such as bicyclic peptides [97]. Xue et al. also developed more complex peptides with three-dimensional cage structures [98]. One potential problem for peptide cyclization is competition between intra- and intermolecular reactions, of which the latter forms peptide oligomers. Cyclization on the solid phase with an excess of solvent has proven viable for favoring the intramolecular reaction [99].

A special group of cyclic peptides known as "stapled peptides" is important for the purpose of α-helix mimicking [100]. Due to the unstable folding of small peptides, a special chemical linkage has been applied to constrain peptide confirmation. Various linkages, such as hydrocarbon linkage [101], the disulfide-bond linkage, dicarboxylic diamine linkage [102], and backbone cyclization [103] have all been explored. These stapled peptides demonstrably mimic native helices in key residues, termed "hot spots," to mediate PPIs [104]. Therefore these peptide secondary structure mimetics may serve medical applications [105].

Besides the typical cyclic peptide structures mentioned earlier, some other structures, such as a knot, lasso, and catenane (Fig. 4.18B–D), have been challenging targets [106]. To date, there is still room for further development of robust and practical methods to synthesize these peptides with complex topologies.

4.5 Conclusions and outlook

Peptide synthesis has been one of the focuses of protein chemical biology research for decades due to its reliability and wide scope. Taking a long view, peptide solid-phase synthesis had become well established before the start of the 21st century. However, there are still difficulties and challenges within the field, such as the so-called difficult sequences. Novel developments, such as microwave-assisted SPPS, have helped address some of these problems. To chemically synthesize diverse peptides with natural as well as nonnatural structures remains one of the main directions for future studies [107]. With further developments of SPPS methods and techniques, we hope to see more widespread biological uses [108] as well as medical applications [109].

Acknowledgments

The authors thank Yi Zheng (Yillustration Studio) for assistance with figure preparation. The authors also thank Samuel D. Whedon from Harvard Medical School for proofreading and suggestions.

References

[1] Wang ZA, Zeng Y, Kurra Y, et al. A genetically encoded allysine for the synthesis of proteins with site-specific lysine dimethylation. Angew Chem Int Ed 2016;56:212–16. Available from: https://doi.org/10.1002/anie.201609452.

[2] Chen X, Tang S, Zheng J-S, et al. Chemical synthesis of a two-photon-activatable chemokine and photon-guided lymphocyte migration in vivo. Nat Commun 2015;6:1–9. Available from: https://doi.org/10.1038/ncomms8220.

[3] Wang ZA, Kurra Y, Wang X, et al. A versatile approach for site-specific lysine acylation in proteins. Angew Chem Int Ed 2017;56:1643–7. Available from: https://doi.org/10.1002/anie.201611415.

[4] El-Faham A, Albericio F. Peptide coupling reagents, more than a letter soup. Chem Rev 2011;111:6557–602. Available from: https://doi.org/10.1021/cr100048w.

[5] Fischer E, Fourneau E. Über einige derivate des glykokolls. Ber dtsch chem Ges 1901;34:2868–79.

[6] Han SY, Kim YA. Recent development of peptide coupling reagents in organic synthesis. Tetrahedron 2004;60:2447–67. Available from: https://doi.org/10.1016/j.tet.2004.01.020.

[7] Carpino LA, Beyermann M, Wenschuh H, Bienert M. Peptide synthesis via amino acid halides. Acc Chem Res 1996;29:268–74. Available from: https://doi.org/10.1021/ar950023w.

[8] Valeur E, Bradley M. Amide bond formation: beyond the myth of coupling reagents. Chem Soc Rev 2009;38:606–31. Available from: https://doi.org/10.1039/B701677H.

[9] Bergmann Max, Zervas Leonidas. "Über ein allgemeines Verfahren der Peptid-Synthese" [On a general method of peptide synthesis]. Berichte der deutschen chemischen Ges 1932;65(7):1192–201. Available from: https://doi.org/10.1002/cber.19320650722.

[10] Merrifield RB. Solid phase peptide synthesis. I. The synthesis of a tetrapeptide. J Am Chem Soc 1963;85:2149–54. Available from: https://doi.org/10.1021/ja00897a025.

[11] Camarero JA, Hackel BJ, De Yoreo JJ, Mitchell AR. Fmoc-based synthesis of peptide α-thioesters using an aryl hydrazine support. J Org Chem 2004;69:4145–51. Available from: https://doi.org/10.1021/jo040140h.

[12] Albericio F, Isidro-llobet A, Mercedes A. Amino acid-protecting groups. Chem Rev 2009;109:2455–504. Available from: https://doi.org/10.1021/cr800323s.

[13] Humphrey JM, Chamberlin AR. Chemical synthesis of natural product peptides: coupling methods for the incorporation of noncoded amino acids into peptides. Chem Rev 1997; 97:2243–66. Available from: https://doi.org/10.1021/cr950005s.

[14] Kent SBH. Novel protein science enabled by total chemical synthesis. Protein Sci 2019;28:313–28. Available from: https://doi.org/10.1002/pro.3533.

[15] Dawson PE, Muir TW, Clark-Lewis I, Kent SBH. Synthesis of proteins by native chemical ligation. Science 1994;266:776. Available from: https://doi.org/10.1126/science.7973629.

[16] Kawakami T, Aimoto S. Peptide ligation using a building block having a cysteinyl prolyl ester (CPE) autoactivating unit at the carboxy terminus. Chem Lett 2007;36:76–7. Available from: https://doi.org/10.1246/cl.2007.76.

[17] Zheng JS, Tang S, Huang YC, Liu L. Development of new thioester equivalents for protein chemical synthesis. Acc Chem Res 2013;46:2475–84. Available from: https://doi.org/10.1021/ar400012w.

[18] Wang ZA. The recent progresses in chemical synthesis of proteins with site-specific lysine post-translational modifications. Curr Org Synth 2019;16:369–84. Available from: https://doi.org/10.2174/1570179416666190328233918.

[19] Zheng JS, Chang HN, Wang FL, Liu L. Fmoc synthesis of peptide thioesters without post-chain-assembly manipulation. J Am Chem Soc 2011;133:11080–3. Available from: https://doi.org/10.1021/ja204088a.

[20] Zheng JS, Tang S, Qi YK, et al. Chemical synthesis of proteins using peptide hydrazides as thioester surrogates. Nat Protoc 2013;8:2483–95. Available from: https://doi.org/10.1038/nprot.2013.152.

[21] Blanco-Canosa JB, Dawson PE. An efficient Fmoc-SPPS approach for the generation of thioester peptide precursors for use in native chemical ligation. Angew Chem Int Ed 2008;120:6957–61. Available from: https://doi.org/10.1002/ange.200705471.

[22] Wang, et al. Peptide o-aminoanilides as crypto-thioesters for protein chemical synthesis. Angewandte Chemie - International Edition 2015. Available from: https://doi.org/10.1002/anie.201408078.

[23] Weller CE, Dhall A, Ding F, et al. Aromatic thiol-mediated cleavage of N-O bonds enables chemical ubiquitylation of folded proteins. Nat Commun 2016;7:1–10. Available from: https://doi.org/10.1038/ncomms12979.

[24] Wucherpfennig TG, Pattabiraman VR, Limberg FRP, et al. Traceless preparation of C-terminal α-ketoacids for chemical protein synthesis by α-ketoacid-hydroxylamine ligation: synthesis of SUMO2/3. Angew Chem Int Ed 2014;53:12248–52. Available from: https://doi.org/10.1002/anie.201407014.

[25] Nilsson BL, Kiessling LL, Raines RT. High-yielding staudinger ligation of a phosphinothioester and azide to form a peptide. Org Lett 2001;3:9–12. Available from: https://doi.org/10.1021/ol006739v.

[26] Durek T, Torbeev VY, Kent SBH. Convergent chemical synthesis and high-resolution X-ray structure of human lysozyme. Proc Natl Acad Sci 2007;104:4846–51. Available from: https://doi.org/10.1073/pnas.0610630104.

[27] Tang S, Si YY, Wang ZP, et al. An efficient one-pot four-segment condensation method for protein chemical synthesis. Angew Chem Int Ed 2015;54:5713–17. Available from: https://doi.org/10.1002/anie.201500051.

[28] Wang, et al. Histone H2B deacylation selectivity: exploring chromatin's dark matter with an engineered sortase. JACS 2022;3360.

[29] Pedersen SL, Tofteng AP, Malik L, Jensen KJ. Microwave heating in solid-phase peptide synthesis. Chem Soc Rev 2012;41:1826–44. Available from: https://doi.org/10.1039/c1cs15214a.

[30] Malik L, Tofteng AP, Pedersen SL, et al. Automated "X-Y" robot for peptide synthesis with microwave heating: application to difficult peptide sequences and protein domains. J Pept Sci 2010;16:506–12. Available from: https://doi.org/10.1002/psc.1269.

[31] Yin JX, Li XL, Wu XL, et al. A rapid, microwave-assisted, one-pot cyclopropanation reaction to synthesize (4-ferrocenyl)benzoylcyclopropane in the solid state. Can J Chem 2005;83:2140–3. Available from: https://doi.org/10.1139/V05-251.

[32] Pedersen SL, Sørensen KK, Jensen KJ. Semi-automated microwave-assisted SPPS: optimization of protocols and synthesis of difficult sequences. Biopolymers 2010;94:206–12. Available from: https://doi.org/10.1002/bip.21347.

[33] Subiros-Funosas R, Acosta GA, El-Faham A, Albericio F. Microwave irradiation and COMU: a potent combination for solid-phase peptide synthesis. Tetrahedron Lett 2009; 50:6200–2. Available from: https://doi.org/10.1016/j.tetlet.2009.08.117.

[34] Shabanpoor F, Hughes RA, Bathgate RAD, et al. Solid-phase synthesis of europium-labeled human INSL3 as a novel probe for the study of ligand-receptor interactions. Bioconjugate Chem 2008;19:1456–63. Available from: https://doi.org/10.1021/bc800127p.

[35] Singer D, Zauner T, Genz M, et al. Synthesis of pathological and nonpathological human exon 1 huntingtin. J Pept Sci 2010;16:358–63. Available from: https://doi.org/10.1002/psc.1252.

[36] McMahon CM, Isabella CR, Windsor IW, et al. Stereoelectronic effects impact glycan recognition. J Am Chem Soc 2020;142:2386–95. Available from: https://doi.org/10.1021/jacs.9b11699.

[37] Buskas T, Ingale S, Boons GJ. Glycopeptides as versatile tools for glycobiology. Glycobiology 2006;16:113R–36R. Available from: https://doi.org/10.1093/glycob/cwj125.

[38] Paulsen H. Syntheses, conformations and X-ray structure analyses of the saccharide chains from the core regions of glycoproteins. Angew Chem Int Ed 1990;29:823–39. Available from: https://doi.org/10.1002/anie.199008233.

[39] Pratt MR, Bertozzi CR. Synthetic glycopeptides and glycoproteins as tools for biology. Chem Soc Rev 2005;34:58–68. Available from: https://doi.org/10.1039/b400593g.

[40] Becker T, Dziadek S, Wittrock S, Kunz H. Synthetic glycopeptides from the mucin family as potential tools in cancer immunotherapy. Curr Cancer Drug Targets 2006;6:491–517. Available from: https://doi.org/10.2174/156800906778194577.

[41] Westerlind U, Schröder H, Hobel A, et al. Tumor-associated MUC1 tandem-repeat glycopeptide microarrays to evaluate serum- and monoclonal-antibody specificity. Angew Chem Int Ed 2009;48:8263–7. Available from: https://doi.org/10.1002/anie.200902963.

[42] Naruchi K, Hamamoto T, Kurogochi M, et al. Construction and structural characterization of versatile lactosaminoglycan-related compound library for the synthesis of complex glycopeptides and glycosphingolipids. J Org Chem 2006;71:9609–21. Available from: https://doi.org/10.1021/jo0617161.

[43] Mahindra A, Sharma KK, Jain R. Rapid microwave-assisted solution-phase peptide synthesis. Tetrahedron Lett 2012;53:6931–5. Available from: https://doi.org/10.1016/j.tetlet.2012.10.028.

[44] Rodríguez H, Suarez M, Albericio F. A convenient microwave-enhanced solid-phase synthesis of short chain N-methyl-rich peptides. J Pept Sci 2010;16:136–40. Available from: https://doi.org/10.1002/psc.1209.

[45] Petersson EJ, Schepartz A. Toward β-amino acid proteins: design, synthesis, and characterization of a fifteen kilodalton β-peptide tetramer. J Am Chem Soc 2008;130:821–3. Available from: https://doi.org/10.1021/ja077245x.

[46] Kritzer JA, Lear JD, Hodsdon ME, Schepartz A. Helical β-peptide inhibitors of the p53-hDM2 interaction. J Am Chem Soc 2004;126:9468–9. Available from: https://doi.org/10.1021/ja031625a.

[47] Joshi BP, Park Jwon, Kim Jmin, et al. Application of microwave method to the solid phase synthesis of pseudopeptides containing ester bond. Tetrahedron Lett 2008;49: 98–101. Available from: https://doi.org/10.1016/j.tetlet.2007.11.004.

[48] Pomerantz WC, Abbott NL, Gellman SH. Lyotropic liquid crystals from designed helical β-peptides. J Am Chem Soc 2006;128:8730–1. Available from: https://doi.org/10.1021/ja062412z.

[49] Gorske BC, Stringer JR, Bastian BL, et al. New strategies for the design of folded peptoids revealed by a survey of noncovalent interactions in model systems. J Am Chem Soc 2009;131:16555–67. Available from: https://doi.org/10.1021/ja907184g.

[50] Camperi SA, Marani MM, Iannucci NB, et al. An efficient strategy for the preparation of one-bead-one-peptide libraries on a new biocompatible solid support. Tetrahedron Lett 2005;46:1561–4. Available from: https://doi.org/10.1016/j.tetlet.2004.12.105.

[51] Simon MD, Heider PL, Adamo A, et al. Rapid flow-based peptide synthesis. ChemBioChem 2014;15:713–20. Available from: https://doi.org/10.1002/cbic.201300796.

[52] Mijalis AJ, Thomas DA, Simon MD, et al. A fully automated flow-based approach for accelerated peptide synthesis. Nat Chem Biol 2017;13:464–6. Available from: https://doi.org/10.1038/nchembio.2318.

[53] Lühmann T, Mong SK, Simon MD, et al. A perfluoroaromatic abiotic analog of H2 relaxin enabled by rapid flow-based peptide synthesis. Org Biomol Chem 2016;14: 3345–9. Available from: https://doi.org/10.1039/c6ob00208k.

[54] Mong SK, Vinogradov AA, Simon MD, Pentelute BL. Rapid total synthesis of DARPin pE59 and barnase. ChemBioChem 2014;15:721–33. Available from: https://doi.org/10.1002/cbic.201300797.

[55] Hartrampf N, Saebi A, Poskus M, et al. Synthesis of proteins by automated flow chemistry. Science 2020;368:980–7. Available from: https://doi.org/10.1126/science.abb2491.

[56] Kent SBH. Chemical synthesis of peptides and proteins. Annu Rev Biochem 1988;57:957–89. Available from: https://doi.org/10.1146/annurev.bi.57.070188.004521.

[57] MiltonSaskia RC, de L, MiltonPaul CF, Adams A. Prediction of difficult sequences in solid-phase peptide synthesis. J Am Chem Soc 1990;112:6039–46.

[58] Miranda LP, Alewood PF. Accelerated chemical synthesis of peptides and small proteins. Proc Natl Acad Sci 1999;96:1181–6. Available from: https://doi.org/10.1073/pnas.96.4.1181.

[59] Atherton E, Woolley V, Sheppard RC. Internal association in solid phase peptide synthesis. Synthesis of cytochrome C residues 66–104 on polyamide supports. Chem Commun 1980;970–1. Available from: https://doi.org/10.1039/C39800000970.

[60] Hyde C, Johnson T, Owen D, et al. Some "difficult sequences" made easy. A study of interchain association in solid-phase peptide synthesis. Int J Pept Protein Res 1994; 43:431–40. Available from: https://doi.org/10.1111/j.1399-3011.1994.tb00541.x.

[61] Jacobsen MT, Petersen ME, Ye X, et al. A helping hand to overcome solubility challenges in chemical protein synthesis. J Am Chem Soc 2016;138:11775–82. Available from: https://doi.org/10.1021/jacs.6b05719.

[62] Adams DJ, Atkins D, Cooper AI, et al. Vesicles from peptidic side-chain polymers synthesized by atom transfer radical polymerization. Biomacromolecules 2008;9:2997–3003. Available from: https://doi.org/10.1021/bm8006693.

[63] Dhalluin C, Boutillon C, Tartar A, Lippens G. Magic angle spinning nuclear magnetic resonance in solid-phase peptide synthesis. J Am Chem Soc 1997;119:10494–500. Available from: https://doi.org/10.1021/ja971795l.

[64] Shen F, Tang S, Liu L. Hexafluoro-2-propanol as a potent cosolvent for chemical ligation of membrane proteins. Sci China Chem 2011;54:110–16. Available from: https://doi.org/10.1007/s11426-010-4188-4.

[65] Berkessel A, Adrio JA, Hüttenhain D, Neudörfl JM. Unveiling the "booster effect" of fluorinated alcohol solvents: aggregation-induced conformational changes and cooperatively enhanced H-bonding. J Am Chem Soc 2006;128:8421–6. Available from: https://doi.org/10.1021/ja0545463.

[66] Madison V, Deber CM, Blout ER. Cyclic peptides. 17. Metal and amino acid complexes of cyclo(pro-gly)4 and analogues studied by nuclear magnetic resonance and circular dichroism. J Am Chem Soc 1977;99:4788–98. Available from: https://doi.org/10.1021/ja00456a042.

[67] Fischer PM. Diketopiperazines in peptide and combinatorial chemistry. J Pept Sci 2003;9:9–35. Available from: https://doi.org/10.1002/psc.446.

[68] Mergler M, Dick F, Sax B, et al. The aspartimide problem in Fmoc-based SPPS. Part I. J Pept Sci 2003;9:36–46. Available from: https://doi.org/10.1002/psc.430.

[69] Huang Y, Jahreis G, Lücke C, Fischer G. Rapid nitrogen inversion pathway in the cis/trans isomerization of selenoxo peptide bonds. Chem Eur J 2013;19:1179–83. Available from: https://doi.org/10.1002/chem.201203721.

[70] Carpino LA, Krause E, Sferdean CD, et al. Synthesis of "difficult" peptide sequences: application of a depsipeptide technique to the Jung-Redemann 10- and 26-mers and the amyloid peptide Aβ(1-42). Tetrahedron Lett 2004;45:7519–23. Available from: https://doi.org/10.1016/j.tetlet.2004.07.162.

[71] Coin I, Dölling R, Krause E, et al. Depsipeptide methodology for solid-phase peptide synthesis: circumventing side reactions and development of an automated technique via depsidipeptide units. J Org Chem 2006;71:6171–7. Available from: https://doi.org/10.1021/jo060914p.

[72] Qin X, Khuong AC, Yu Z, et al. Simplifying alternating peptide synthesis by protease-catalyzed dipeptide oligomerization. Chem Commun 2013;49:385–7. Available from: https://doi.org/10.1039/c2cc36381j.

[73] Terpe K. Overview of tag protein fusions: from molecular and biochemical fundamentals to commercial systems. Appl Microbiol Biotechnol 2003;60:523–33. Available from: https://doi.org/10.1007/s00253-002-1158-6.

[74] Paradís-Bas M, Tulla-Puche J, Albericio F. The road to the synthesis of "difficult peptides. Chem Soc Rev 2016;45:631–54. Available from: https://doi.org/10.1039/c5cs00680e.

[75] Bill RM, Henderson PJF, Iwata S, et al. Overcoming barriers to membrane protein structure determination. Nat Biotechnol 2011;29:335–40. Available from: https://doi.org/10.1038/nbt.1833.

[76] Yildirim MA, Goh KII, Cusick ME, et al. Drug-target network. Nat Biotechnol 2007;25:1119–26. Available from: https://doi.org/10.1038/nbt1338.

[77] Zuo C, Tang S, Zheng JS. Chemical synthesis and biophysical applications of membrane proteins. J Pept Sci 2015;21:540–9. Available from: https://doi.org/10.1002/psc.2721.

[78] Zheng JS, He Y, Zuo C, et al. Robust chemical synthesis of membrane proteins through a general method of removable backbone modification. J Am Chem Soc 2016;138:3553–61. Available from: https://doi.org/10.1021/jacs.6b00515.

[79] Kochendoerfer GG, Salom D, Lear JD, et al. Total chemical synthesis of the integral membrane protein influenza A virus M2: Role of its C-terminal domain in tetramer assembly. Biochemistry 1999;38:11905–13. Available from: https://doi.org/10.1021/bi990720m.

[80] Johnson ECB, Malito E, Shen Y, et al. Modular total chemical synthesis of a human immunodeficiency virus type 1 protease. J Am Chem Soc 2007;129:11480–90. Available from: https://doi.org/10.1021/ja072870n.

[81] Tan Z, Shang S, Danishefsky SJ. Rational development of a strategy for modifying the aggregatibility of proteins. Proc Natl Acad Sci 2011;108:4297–302. Available from: https://doi.org/10.1073/pnas.1100195108.

[82] Bianchi E, Ingenito R, Simon RJ, Pessi A. Engineering and chemical synthesis of a transmembrane protein: the HCV protease cofactor protein NS4A. J Am Chem Soc 1999;121: 7698–9. Available from: https://doi.org/10.1021/ja9901617.

[83] Becker CFW, Oblatt-Montal M, Kochendoerfer GG, Montal M. Chemical synthesis and single channel properties of tetrameric and pentameric TASPs (template-assembled synthetic proteins) derived from the transmembrane domain of HIV virus protein u (Vpu). J Biol Chem 2004;279:17483–9. Available from: https://doi.org/10.1074/jbc.M313212200.

[84] Sato T, Saito Y, Aimoto S. Synthesis of the C-terminal region of opioid receptor like 1 in an SDS micelle by the native chemical ligation: effect of thiol additive and SDS concentration on ligation efficiency. J Pept Sci 2005;11:410–16. Available from: https://doi.org/10.1002/psc.634.

[85] Paradís-Bas M, Tulla-Puche J, Albericio F. Semipermanent C-terminal carboxylic acid protecting group: application to solubilizing peptides and fragment condensation. Org Lett 2015;17:294–7. Available from: https://doi.org/10.1021/ol5033943.

[86] Yoshiya T, Tsuda S, Masuda S. Development of trityl group anchored solubilizing tags for peptide and protein synthesis. ChemBioChem 2019;20:1906–13. Available from: https://doi.org/10.1002/cbic.201900105.

[87] Li JB, Tang S, Zheng JS, et al. Removable backbone modification method for the chemical synthesis of membrane proteins. Acc Chem Res 2017;50:1143–53. Available from: https://doi.org/10.1021/acs.accounts.7b00001.

[88] Zheng JS, Yu M, Qi YK, et al. Expedient total synthesis of small to medium-sized membrane proteins via fmoc chemistry. J Am Chem Soc 2014;136:3695–704. Available from: https://doi.org/10.1021/ja500222u.

[89] Huang DL, Montigny C, Zheng Y, et al. Chemical synthesis of native s-palmitoylated membrane proteins through removable-backbone-modification-assisted Ser/Thr ligation. Angew Chem Int Ed 2020;59:5178–84. Available from: https://doi.org/10.1002/anie.201914836.

[90] Huang DL, Li Y, Liang J, et al. The new salicylaldehyde S,S-propanedithioacetal ester enables N-to-C sequential native chemical ligation and Ser/Thr ligation for chemical protein synthesis. J Am Chem Soc 2020;142:8790–9. Available from: https://doi.org/10.1021/jacs.0c01561.

[91] Góngora-Benítez M, Tulla-Puche J, Albericio F. Multifaceted roles of disulfide bonds. Peptides Therapeutics Chem Rev 2014;114:901–26. Available from: https://doi.org/10.1021/cr400031z.

[92] Jin AH, Muttenthaler M, Dutertre S, et al. Conotoxins: chemistry and biology. Chem Rev 2019;119:11510–49. Available from: https://doi.org/10.1021/acs.chemrev.9b00207.

[93] Qu Q, Gao S, Wu F, et al. Synthesis of disulfide surrogate peptides incorporating large-span surrogate bridges through a native-chemical-ligation-assisted diaminodiacid strategy. Angew Chem Int Ed 2020;59:6037–45. Available from: https://doi.org/10.1002/anie.201915358.

[94] Kim YW, Grossmann TN, Verdine GL. Synthesis of all-hydrocarbon stapled-helical peptides by ring-closing olefin metathesis. Nat Protoc 2011;6:761–71. Available from: https://doi.org/10.1038/nprot.2011.324.

[95] Rubin SJS, Qvit N. Backbone-cyclized peptides: a critical review. Curr Top Med Chem 2018;18:526–55. Available from: https://doi.org/10.2174/1568026618666180518092333.

[96] Chow HY, Zhang Y, Matheson E, Li X. Ligation technologies for the synthesis of cyclic peptides. Chem Rev 2019;119:9971–10001. Available from: https://doi.org/10.1021/acs.chemrev.8b00657.

[97] Wang XS, Chen PHC, Hampton JT, et al. A genetically encoded, phage-displayed cyclic-peptide library. Angew Chem Int Ed 2019;58:15904–9. Available from: https://doi.org/10.1002/anie.201908713.

[98] Li Z, Shao S, Ren X, et al. Construction of a sequenceable protein mimetic peptide library with a true 3D diversifiable chemical space. J Am Chem Soc 2018;140:14552–6. Available from: https://doi.org/10.1021/jacs.8b08338.

[99] Huang YC, Li YM, Chen Y, et al. Synthesis of autophagosomal marker protein LC3-II under detergent-free conditions. Angew Chem Int Ed 2013;52:4858–62. Available from: https://doi.org/10.1002/anie.201209523.

[100] Maity SK, Mann G, Jbara M, et al. Palladium-assisted removal of a solubilizing tag from a Cys side chain to facilitate peptide and protein synthesis. Org Lett 2016;18:3026–9. Available from: https://doi.org/10.1021/acs.orglett.6b01442.

[101] Walensky LD, Bird GH. Hydrocarbon-stapled peptides: principles, practice, and progress. J Med Chem 2014;57:6275–88. Available from: https://doi.org/10.1021/jm4011675.

[102] Cui H-K, Guo Y, He Y, et al. Diaminodiacid-based solid-phase synthesis of peptide disulfide bond mimics. Angew Chem Int Ed 2013;125:9737–41. Available from: https://doi.org/10.1002/ange.201302197.

[103] Gilon C, Halle D, Chorev M, et al. Backbone cyclization: a new method for conferring conformational constraint on peptides. Biopolymers 1991;31:745–50. Available from: https://doi.org/10.1002/bip.360310619.

[104] Wang ZA, Ding XZ, Tian CL, Zheng JS. Protein/peptide secondary structural mimics: design, characterization, and modulation of protein-protein interactions. RSC Adv 2016; 6:61599–609. Available from: https://doi.org/10.1039/c6ra13976k.

[105] Rubin SJS, Tal-Gan Y, Gilon C, Qvit N. Conversion of protein active regions into peptidomimetic therapeutic leads using backbone cyclization and cycloscan—how to do it yourself!. Curr Top Med Chem 2018;18:556–65. Available from: https://doi.org/10.2174/1568026618666180518094322.

[106] Wang XW, Zhang W, Bin. Chemical topology and complexity of protein architectures. Trends Biochem Sci 2018;43:806–17. Available from: https://doi.org/10.1016/j.tibs.2018.07.001.

[107] Qi YK, Si YY, Du SS, et al. Recent advances in the chemical synthesis and semi-synthesis of poly-ubiquitin-based proteins and probes. Sci China Chem 2019; 62:299–312. Available from: https://doi.org/10.1007/s11426-018-9401-8.

[108] Wang ZA, Millard CJ, Lin CL, et al. Diverse nucleosome site-selectivity among histone deacetylase complexes. Elife 2020;9:e57663. Available from: https://doi.org/10.7554/eLife.57663.

[109] Li Y, Cao X, Tian C, Zheng J. Chemical protein synthesis-assisted high-throughput screening strategies for d-peptides in drug discovery. Chin Chem Lett 2020;31:2365–74. Available from: https://doi.org/10.1016/j.cclet.2020.04.015.

Chapter 5

Stapled peptidomimetic therapeutics

Srinivasaraghavan Kannan[1], Dilraj Lama[2], Yaw Sing Tan[1], Jianguo Li[1,3] and Chandra S Verma[1,4,5]

[1]*Bioinformatics Institute, Agency for Science, Technology and Research (A*STAR), Singapore, Singapore,* [2]*Department of Microbiology, Tumor and Cell Biology, Karolinska Institutet, Solna, Stockholm, Sweden,* [3]*Singapore Eye Research Institute, Singapore, Singapore,* [4]*Department of Biological Sciences, National University of Singapore, Singapore, Singapore,* [5]*School of Biological Sciences, Nanyang Technological University, Singapore, Singapore*

5.1 Introduction

Peptides are naturally occurring molecules involved in numerous biological functions, acting as growth factors, neurotransmitters, ligands, etc. Peptides also act as mediators of protein–protein interactions (PPI). PPIs are key mediators of various biological processes and their dysregulation results in diseases [1,2]. They were once deemed undruggable, due to their generally large and flat binding surfaces with no deep cavities that are suitable for small molecule binding [3,4]. A recent study has highlighted that approximately 40% of all PPIs are mediated by short peptides and these could direct the design of peptidic inhibitors of the associated protein interactions [5]. Peptides can disrupt these interactions with high specificity and are characterized by low toxicity and high biocompatibility, thus making them attractive complements as therapeutics to small-molecule inhibitors and large biomolecules (antibodies).

However, when a short peptide sequence of a native interfacial structural element is extracted from the PPI context, it usually is unable to retain its native conformation, since the structural reinforcement provided by the parent protein is missing. These isolated peptides are often unstructured and intrinsically disordered, yielding weak affinities for their respective targets, as the peptides have to undergo disorder to order transition upon binding [6–12]. In addition, their low conformational stabilities also render them susceptible to proteolytic degradation. Finally, they exhibit poor cell permeability [6–12]. These issues are being addressed by strategies such as modifications of

Peptide and Peptidomimetic Therapeutics. DOI: https://doi.org/10.1016/B978-0-12-820141-1.00022-4

the backbones, macrocyclization, the inclusion of nonnatural amino acids, etc., toward the development of molecules with desirable pharmacokinetic and biological properties [6–14].

Macrocyclization involves imposing constraints on peptides by the introduction of chemical cross-links, which are engineered to ensure that the peptide adopts the bound conformation in its apo state; this strategy is widely used to overcome peptide flexibility [10–14]. Constrained peptides have demonstrated increased potency for targets as the entropic cost for adopting a bioactive conformation is reduced. Multiple methods have been used to constrain peptides, including cyclization through disulfide bridges, head-to-tail, head-to-sidechain, sidechain-to-tail, and sidechain to sidechain. Commonly used linkages for peptide macrocyclization are disulfide bonds, lactam bridges, lactones, thioether bonds, biaryl bridges, heterocycles (1,2,3-triazole), hydrogen bonding surrogates and hydrocarbon linkages [10–14]. A popular cyclization technique is stapling via olefin metathesis whereby alpha-methyl alkenyl sidechains are cross-linked [14–19]; this method has been very successful when a helical structure is desired. The appropriate unnatural amino acid precursors are introduced such that there is no clash with the target protein. The linkers that form the staple are diverse in chemistry and lengths, usually spanning the [i, $i+3$], [i, $i+4$], or [i, $i+7$] positions (Fig. 5.1). The main advantage of constraining a peptide with the macrocyclic link into its bound state (or close to) is that the free energy costs associated with reorganizing an unconstrained peptide from its free state to its bound state is minimized. The other advantage brought about is that the peptide spans a large surface of recognition, hence bringing in

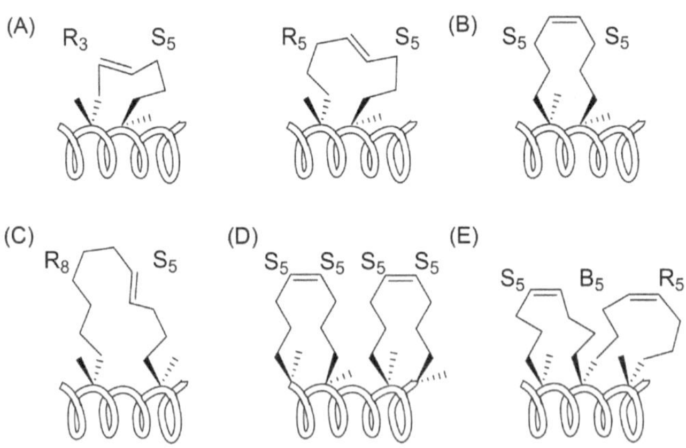

FIGURE 5.1 Stapled peptide architectures. Amino acids required for the formation of the staples are indicated as either R_n or S_n, representing their absolute configuration as well as the number of carbon atoms (n) in their alkenyl sidechain. (A) i, $i+3$ staples with six (left) and eight (right) carbon atoms. (B) i, $i+4$ staple. (C) i, $i+7$ staple. (D) double i, $i+4$ staples. (E) i, $i+4+4$ stitch. *Adapted from Tan YS, Lane DP, Verma CS. Stapled peptide design: principles and roles of computation. Drug Discov Today 2016;21:1642–53. Copyright 2016 Elsevier.*

specificity and hence a reduction in the possibility of off-target effects. The constraints also introduce proteolytic resistance by protecting the backbone from cleavage. Stapling also enhances cell permeability, at least in some classes; while this is assumed to result from stabilizing intramolecular hydrogen bonds and thus reducing the desolvation penalty incurred as the peptide transits from the polar solvent phase to the apolar membrane phase, the exact mechanisms remain unclear. Recent successes in the applications of stapling to nonhelical peptides [20,21] provide for cautious optimism. Stapled peptides have been successfully applied to inhibit several PPIs of the therapeutic potential and have been nicely reviewed elsewhere [14,22,23]. Noteworthy is the case of the inhibition of the interaction between p53 and Mdm2/Mdm4 by ALRN-6924. This is a dual selective stapled peptide inhibitor of Mdm2/Mdm4 that activates p53 and has been successfully advanced to phase II clinical trials [24].

Although the proteolytic stability and cellular uptake of some stapled peptides have improved, these do not necessarily ensure desired bioactivity [25–27]. For example, the desired peptide may be very long, in which case, a single staple is likely insufficient to protect against proteolysis, as was seen for single hydrocarbon and lactam staples of the 30-residue vasoactive intestinal peptide [28]. This can be overcome by introducing multiple staples, as was shown by an approximately threefold improvement in resistance to proteolysis when double [i, $i+4$] hydrocarbon staples were used in a 37-residue peptide from the gp41 envelope glycoprotein, compared to the use of a single staple [26]. Subsequently, the use of a double [i, $i+4$], [i, $i+3$] staple across a different region of the gp41 peptide was also shown to demonstrate stabilization [29]. In both cases, the resultant binding affinities were similar to the affinity of the singly stapled analog. Doubly stapled peptides against Rab8a showed a higher affinity and cellular uptake compared to the single stapled analog and, had a very long half-life in the presence of proteinase K [25]. The use of two [i, $i+7$] staples in a fusion peptide from the respiratory syncytial virus were highly resistant to proteases and very effective in suppressing viral infection in mouse models [27]. If the amino acid *bis*-pentenylglycine (B5) is used, it acts as a junction from where two staples can emerge, and this is referred to as stitching. This allows for multiple combinations of stereochemistry and linker lengths. A rigorous study showed that the i, $i+4+4$ (S5 + B5 + R5) and i, $i+4+7$ (S5 + B5 + S8) result as the most effective combination [30]. A peptide with the latter stitch architecture showed very high helicity and cell permeability compared to an i, $i+7$ stapled analog [31,32]. These peptides are being recognized as holding great potential.

In addition to cyclization, peptides can be further modified by replacing the natural amino acids with nonproteinogenic amino acids to improve their proteolytic stabilities. Nonnatural amino acids such as D-enantiomers, N-methylated, alpha methylated, β-, α,β-didehydro-, α-disubstituted amino acids, azapeptides, retro-inverso peptides, depsipeptides, and peptoids are the most commonly explored to enhance the pharmacological properties of peptide-based therapeutics [33]. The incorporation of alternative modifications toward imparting more drug-like properties to the peptides often results in losses in target affinity. In summary, a general strategy that will ensure protease resistance, cellular uptake and high target affinity of constrained peptides is still lacking.

In this chapter, we discuss our recent efforts in the design and development of stapled peptide-based inhibitors against several targets that are associated with various cancers. In the first part, we discuss the development of L-amino acid stapled peptide inhibitors of eukaryotic initiation factor 4E (eiF4E). Next, we discuss the design of all D-amino acid staples and stitched peptide inhibitors of Mdm2. In the third part, we discuss the design of non-helical stapled peptide inhibitors against intracellular PPIs that are mediated by extended peptide fragments. Finally, we discuss the use of computational methods to determine membrane permeability and aggregation propensities of the stapled peptides.

5.2 Development of stapled peptide inhibitors of eukaryotic initiation factor 4E

5.2.1 Eukaryotic initiation factor 4E in translation initiation, regulation and oncogenesis

The eukaryotic initiation factor protein assembly "eIF4F" is involved in the translation of capped mRNAs and is a heterotrimeric complex comprising of the cap-binding protein eIF4E, the helicase enzyme eIF4A and the scaffolding protein eIF4G [34,35]. eIF4E is involved in specifically recognizing the 7-methylguanosine (m^7G) cap structure present at the 5′ end of capped mRNAs [36,37]. This protein-mRNA system subsequently interacts with the 43S preinitiation ribosomal subunit to form the 48S initiation complex which scans the 5′-untranslated region (5′-UTR) of mRNAs for the AUG start codon [38]. Once the start codon is detected, the 60S ribosomal subunit is attached which leads to the formation of a higher order macromolecular machinery that initiates the process of protein synthesis [38]. The 5′-UTR of most of the cellular mRNAs are short and unstructured which makes it relatively easier for the 48S machinery to efficiently scan and identify the initiation codon [37,39]. This class of mRNAs is classified as "strong mRNAs." The 5′-UTRs of proto-oncogenes on the other hand are lengthy with multiple stable secondary structures which hinder the scanning and recognition of the start codon, resulting in a slower and more regulated rate of their translation; these are classified as "weak mRNAs" [37,39].

Capped cellular mRNAs are proposed to compete with one another for binding to the eIF4F initiation complex for translation. eIF4E is the rate-limiting component of the ternary complex and the levels of its expression and/or regulation are thought to be critical for mRNA translation [35].

Weak mRNAs are found to be more sensitive to the elevated levels of eIF4E expression and this contributes to malignancy by disproportionately increasing the rate of translation of oncogenic proteins. The biological activity of eIF4E is also regulated by the translation repressor 4E-binding proteins (4E-BPs) [40]. They keep it in an inactive state through PPIs and prevent its association with the scaffold eIF4G protein, thus impeding the formation of the eIF4F complex [40]. Phosphorylation of 4E-BPs through the mTOR pathway can disrupt/prevent its interaction with eIF4E which is then free to proceed for translation initiation [41]. eIF4E itself undergoes phosphorylation at the residue S209 by Mnk kinases [42]. A complete regulatory role of eIF4E phosphorylation in translation is not entirely clear though it seems to correlate with an increase in the translation efficiency of weak mRNAs involved in tumor progression [42,43]. A molecular model of this regulation through simulation of the phosphorylated states of eIF4E and its impact on mRNA-cap recognition has been postulated [44]. Phosphorylation of eIF4E after translation initiation reduces the protein's affinity for the cap and this facilitates a faster rate of eIF4E dissociation from the mRNA bound state as compared to its unphosphorylated derivative. This will increase the cellular concentration of free eIF4E which can be hijacked by competing proto-oncogenic mRNAs for their specific translation.

Overexpression/deregulation of eIF4E is associated with the development of multiple cancer-related malignancies and targeting of eIF4E for therapeutics is an attractive clinical opportunity [45,46]. Hence, multiple approaches are being undertaken to target different molecular aspects of eIF4E biology. This includes inhibition of eIF4E phosphorylation and/or phosphorylated eIF4E [47,48], hindering mRNA recognition by eIF4E [49], blocking eIF4E PPIs [50] or preventing the translation of eIF4E mRNA itself [51]. All these different strategies have their own challenges and consequently have achieved different levels of success. A water-mediated recognition module in the cap-binding pocket of eIF4E that facilitates an efficient eIF4E:mRNA-cap recognition has been recently identified [52]. This molecular feature has the potential to be efficiently exploited in the design/optimization of new inhibitors targeted specifically against mRNA-cap binding to eIF4E.

Inhibiting PPIs between eIF4E and eIF4G is an attractive avenue for the development of broad-spectrum anticancer therapeutics. However, the protein interacting interface of eIF4E is a largely planar surface devoid of deep and well-formed binding grooves [53]. The affinities of currently reported small-molecule PPI inhibitors of eIF4E are in the micromolar range [50,54]; the lower potency of these molecules probably results from targeting such a challenging molecular landscape. The presence of a transiently exposed cryptic pocket in the region has however been reported [53]. This cavity can accommodate small molecules and provide a favorable environment for their binding which emphasizes its potential druggability. Hence, it is a promising avenue to be considered in the rational design and/or optimization of new classes of potent eIF4E inhibitors directed toward the PPI interface on eIF4E.

5.2.2 Design of Stapled peptide inhibitors of eukaryotic initiation factor 4E

Peptide-based inhibitors offer alternative options for the inhibition of PPIs involving eIF4E. In this regard, the experimentally determined structure of eIF4E in complex with a 12-mer peptide derived from eIF4G1 (eIF4G^{D5S}; Fig. 5.2A) [55] served as an excellent starting point for the development of such inhibitors against eIF4E [56–59]. In the complex structure, eIF4G^{D5S} peptide is bound with a disordered N-terminal region and an almost orthogonally oriented helical C-terminal. The peptide contains the specificity

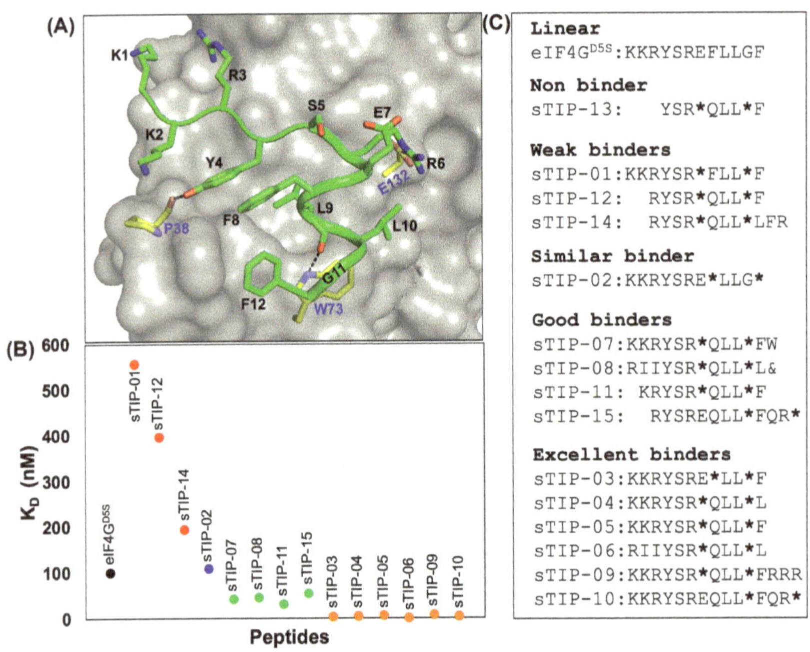

FIGURE 5.2 (A) Crystal structure of a 12mer eIF4G^{D5S} peptide in complex with eIF4E (PDB ID: 4AZA). The protein is represented on the surface, while the backbone of the peptide is shown in cartoon and side chain in stick representations. Hydrogen bonds and salt-bridge interactions formed by the peptide with eIF4E residues are explicitly shown in sticks and indicated with dashed lines. This figure is partly reproduced from Lama et al. [56]. (B) eIF4E binding affinities of the sTIP family of stapled-peptides. The K_D values of the respective peptides are obtained from Lama et. al. [56,57]. They are clustered into different subgroups based on the K_D values relative to the linear eIF4G^{D5S} peptide. (C) The sequences of the 15 sTIP peptides categorized on the basis of their subgroups. sTIP-13 did not show any significant binding and hence is classified as a nonbinder. "*" represents the sites for hydrocarbon linker; & = Lys(ButPhI).

determining "^{4}YXXXXLΦ^{10}" motif, (where Φ signifies any hydrophobic residue) which is critical for eIF4E recognition. Y4 forms a hydrogen-bond with P38, L9 docks into a preformed shallow hydrophobic pocket and its backbone is involved in hydrogen-bond interaction with the side chain of W73. The conserved hydrophobic residue "Φ" in eIF4G1 is L10 which packs against a hydrophobic surface on eIF4E. In addition, salt-bridge interactions between R6 and E132 are also important for potent binding. Using this structure as the basis for a rational design and optimization scheme, several stapled-peptide inhibitors of eIF4E were developed (Fig. 5.2B and C). These are collectively referred to as the "sTIP family of stapled-peptides" [56–58] (Fig. 5.3). These sTIP stapled peptides exhibited a range of binding affinities from uM to nM. The following sections describe the design of the eiF4E

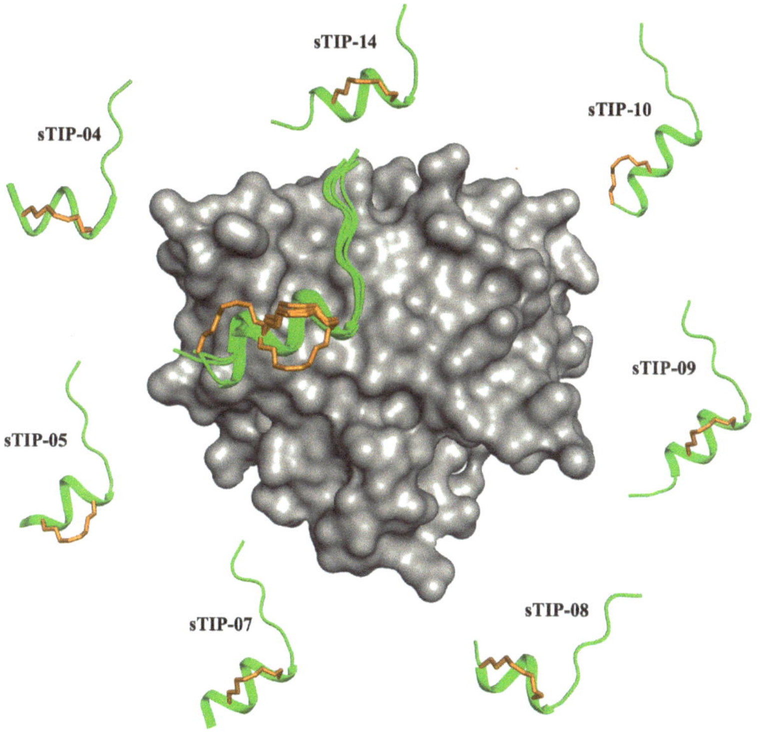

FIGURE 5.3 Superimposition of the crystal structures of sTIP family members whose structures have been resolved in complex with eIF4E. sTIP-04 (PDB ID: 4BEA), sTIP-05 (PDB ID: 5ZJY), sTIP-07 (PDB ID: 5ZJZ), sTIP-08 (PDB ID: 5ZK9), sTIP-09 (PDB ID: 5ZML), sTIP-10 (PDB ID: 5ZK5) and sTIP-14 (PDB ID: 5ZK7). The protein is shown in surface, peptide in cartoon and the hydrocarbon linker in stick representations. *Reproduced from Lama D, Liberatore AM, Frosi Y, Nakhle J, et al. Structural insights reveal a recognition feature for tailoring hydrocarbon stapled-peptides against the eukaryotic translation initiation factor 4E protein. Chem Sci. 2019;10: 2489–500.*

inhibiting stapled peptides, the molecular determinants that influence their potency to interact with eIF4E and the general understandings derived from their design and development.

sTIP-01 was derived/designed by substituting the solvent-exposed residues E7 and G11 separated by a helical turn in the parent eIF4G^{D5S} peptide with an *i*, *i* + 4 hydrocarbon linker (Fig. 5.2A). From a design perspective, it was considered to be a viable substitution; however, the binding affinity of the stapled peptide was found to be reduced significantly compared to eIF4G^{D5S} (Fig. 5.2B and C). The covalent linkage created a physical strain on the peptide structure both in its bound and unbound states, thus significantly impeding the optimum conformation required for binding, resulting in the reduced affinity [56]. This peptide is a classic example of how stapling and binding may not necessarily have a linear relationship.

The N-terminal region of the eIF4G^{D5S} peptide is positively charged (^{1}KKR^{3}), adopts a disordered conformation when bound to eIF4E and does not form any specific interactions with eIF4E (Fig. 5.2A). However, the positive charge at the N-terminal region of the peptide appears to be important for protein recognition, as a stapled-peptide with no basic terminal residues (sTIP-13) showed negligible eIF4E binding, while deletion of the two lysine residues (sTIP-12) significantly weakened the binding (Fig. 5.2B and C). The loss in affinity due to the N-terminal truncation can be partially regained by introducing an F10L substitution together with "^{11}FR^{12}" C-terminal extension (sTIP-14) that occupies the hydrophobic surface on the eIF4E more efficiently. These alterations enhanced the affinity of the stapled peptide sTIP-14 but it was still a weak binder as compared to the linear parent peptide (Fig. 5.2B). Interestingly, deletion of only the first lysine residue (K1) resulted in a potent binder (sTIP11) (Fig. 5.2B) suggesting that the intrinsically disordered region at the N-terminal can be truncated by a residue without any detrimental effects on its electrostatic recognition of eIF4E.

The sTIP-02 derivative obtained by replacement of eIf4E interacting residues F8 and F12 from eIF4G^{D5S} with an *i*, *i* + 4 linker (Fig. 5.2A) exhibited binding that was similar to the parent linear peptide. Although the staple linker was expected to interact favorably with the eIF4E protein, no significant gain in binding affinity (Fig. 5.2B) was observed [56]. Investigation of the unbound form of the peptide suggested that the constraint induces the unbound peptide into metastable conformations which were very distinct from the bound state structure [56], and hence the peptide would have to undergo significant conformational changes to adopt the bound conformation. The entropic penalty for conformational reorganization undermined the enthalpic gain from the interactions of the linker with the protein that resulted in net negligible improvement in the peptide's binding affinity.

Peptide derivatives (sTIP-07 and sTIP-08) with an extension (W13 and &13 ["Lys(ButPhI)"], respectively) at the C-terminal exhibited significantly improved kinetic and thermodynamic properties (Fig. 5.2B and C).

Further stapling the extended C-terminal with an i, $i+4$ staple and simultaneous deletion of N-terminal residues resulted in the sTIP-15 peptide that exhibited affinity which was better than the linear counterpart which carries the charged residues at the N-terminus. Moreover, this sTIP-15 exhibited a lower degree of disorder and higher helical content—a favorable structural framework for lead optimization (Fig. 5.2B) [57]. Stapled peptide sTIP-03 is an i, $i+3$ variant obtained by inserting the hydrocarbon linker at residues F8 and G11 on the 12mer eIF4G^{D5S} linear peptide (Fig. 5.2A). This substitution positioned the linker in a configuration that facilitated its partial interaction with the protein surface [56]. The constraint imposed by the staple in this case preorganized the solution ensemble of the peptide toward a conformation that was similar to the bound state structure [56]. The lower degree of reorganization required for binding along with the energetic contribution from the linker produced a potent stapled-peptide derivative with low nanomolar affinity (Fig. 5.2B). Additional derivatives of i, $i+4$ stapled peptides, with rationally designed/incorporated single or double substitutions and C-terminal extension resulted in more potent binders of eIF4E (Fig. 5.2B and C). The 12mer sTIP-04, 05 and 06 along with 15mer sTIP-09 and 10 i, $i+4$ stapled peptides are the most potent peptides known/reported to date for the inhibition of eIF4E. The substitutions and C-terminal extensions introduced in these peptides released the conformational strain created by the linker and facilitated the formation of novel hot spot interactions, which were critical in their development as excellent binders of eIF4E (Fig. 5.2B) [56,57]. These are important case studies that emphasize the scenarios where stapling has to be supplemented with appropriate sequence optimization for desired outcomes.

The sTIP peptides are currently the only known inhibitors of eIF4E that specifically interact at its PPI interface with nanomolar affinity (Fig. 5.3). They are therefore a source of excellent lead compounds for drug development targeting eIF4E for potential applications in a broad range of cancer-related therapeutics. The future development of these peptides should involve further optimization of their physicochemical properties through tailoring of the terminal segments and exploration of alternate linker variations in the formation of the staple. A major area of focus going forward will also be to systematically examine and rationally improve the cellular uptake efficiency of these peptides which currently remains one of the major roadblocks in their progress and development toward biomedical applications.

5.3 Development of all D-amino acid stapled peptides

Stapled peptides have successfully been used to inhibit several PPIs of therapeutic potential; yet the route to the clinic is fraught with challenges. A major rate-limiting step has been the engineering of proteolytic stability that can ensure sustained target binding and cellular activity. While the staple can protect the region of the peptide that it spans from protease-mediated degradation, often the residues that are located outside of the macrocycle are susceptible to degradation [25,26]. An alternative is the use of all D-peptides and these are hyperstable against proteolysis and have displayed strong binding affinity against diverse targets, including VEGF−VEGF-receptor [60,61], PD-1−PD-L1 [62], human immunodeficiency virus type 1 (HIV-1) entry [63], GLP1R [64] PTH1R [64] and p53−Mdm2 [65,66]. Unfortunately, these all-D-peptides also suffer from a lack of membrane permeability and hence are not biologically active.

An option that was explored was to combine stapling and D-amino acid peptides by Kannan et al. [32], using the well-characterized p53−Mdm2/Mdm4 interaction as a test case. The transcription factor p53 is crucial for guarding the cell in response to various stress signals through activating cell cycle arrest, apoptosis or senescence [67]. p53 can be inactivated in cancers through mutations or overexpression of its negative regulators Mdm2 and Mdm4. This has led to the search and development of molecules that can target mutant p53 or that can be potent inhibitors of the interactions between p53 and Mdm2 or Mdm4 [68,69].

The latter has been extensively studied and exploits the observation that an α-helical segment in the N-terminal transactivation domain of p53 binds into a hydrophobic groove on the surface of the N-terminal domain of both Mdm2 and Mdm4. Three hydrophobic residues from p53, Phe19, Trp23, and Leu26 embed deep into a hydrophobic pocket in Mdm2 (or Mdm4) and are key determinants of the interaction (Fig. 5.4A). The p53 peptide in solution is largely disordered and adopts the helical conformation upon binding. This interaction has been successfully mimicked by small molecules, peptides and biologics which can easily compete out p53 from the surface of Mdm2/4 [70]. However, most small molecules developed against Mdm2 are ineffective at inhibiting Mdm4 largely because of distinct structural differences in the p53 peptide-binding pocket between Mdm2 and Mdm4. In addition, small-molecule inhibitors of Mdm2 that have entered clinical trials have

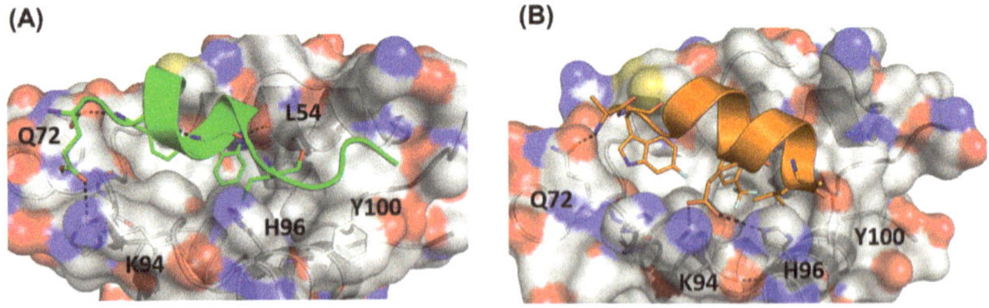

FIGURE 5.4 Structural representation of crystal structures of (A) p53−Mdm2 (PDB: 1YCR) and (B) ^{d}PMI-δ−Mdm2 (PDB: 3TPX) complex. Mdm2 is shown as surface and bound peptide is shown as a cartoon with interacting residues highlighted in sticks. Hydrogen-bond interactions are shown as dotted lines (black).

been largely unsuccessful owing to toxic effects in patients [70]. Resistance to the effect of these molecules also resulted from overexpression of Mdm4 in tumors. Dual inhibition of Mdm2 and Mdm4 has only been successfully achieved by ALRN-6924, which is an L-amino acid stapled peptide that has also shown promise in clinical trials [24].

In parallel, a linear peptide made up of all D-amino acids using a mirror image phage display screening technology was developed by Liu et al. [65,66] and this was modified to yield a 220 pM inhibitor of Mdm2 (referred to as ^{d}PMI). The peptide however was unable to enter cells by itself and needed a nanoparticle to transport it and elicit biological functionality. Crystallographic studies [65,66] showed that the peptide bound Mdm2 in a conformation similar to that adopted by the all L-amino acid wild-type peptide. As expected, the helix was left-handed and projected the sidechains of 6-F-^{d}Trp3, *p*-CF_3-^{d}Phe7 and ^{d}Leu11 into the hydrophobic pocket of Mdm2 (Fig. 5.4B) as seen in the complex of wild-type p53 peptide and Mdm2 (Fig. 5.4A).

Using this structure of the complex between ^{d}PMI and Mdm2, Kannan et al. [32] developed a variety of single stapled, double stapled and stitched peptide inhibitors of Mdm2 with varying degrees of inhibition of Mdm2/Mdm4, cell permeability and biological activity. In the following section, we discuss the computational design, development and validation of these peptides.

5.3.1 Design of all D-stapled peptide inhibitors of Mdm2

In-silico alanine scanning was used for the identification of peptide residues that are critical for binding and positions that can be replaced by staple linkers without perturbing the binding to Mdm2. Several *i*, $i + 4$ and *i*, $i + 7$ stapled all D-peptides were designed with appropriate stereocenters that mirror image the standard strategies used for stapling right-handed alpha-helices made up of L-amino acids. The resultant stapled all-D-peptides were chemically synthesized and tested in biophysical, biochemical and cell-based assays.

Stapling is often reported to increase the helicity of the free peptide in solution. Stapling all-D-peptides resulted in increased helicity, ranging from approximately 25% to 40% as compared to the linear peptide (~20% helicity). Next, models of the complexes between these stapled ^{d}PMI-δ peptides and Mdm2 were constructed by introducing the staples into the crystal structure of the complex between the linear ^{d}PMI-δ peptide and Mdm2 and each complex was subjected to molecular dynamics (MD) simulations. The stapled peptides remained stable and largely helical (~95%) during the MD simulations, with the three critical residues 6-F-^{d}Trp3, p-CF3-^{d}Phe7, and ^{d}Leu11 buried in the hydrophobic pocket of Mdm2. Interestingly, the hydrocarbon linkers in the different stapled ^{d}PMI peptides adopted diverse conformations. Some staples engaged in interactions with the surface of Mdm2 and some staples remained largely solvent-exposed, without engaging the Mdm2 surface (Fig. 5.5). Using biophysical and biochemical assays, it was found that the peptides exhibited binding affinities ranging from nM to uM, with three peptides exhibiting affinities better than the linear counterpart, one a nonbinder and the rest exhibiting reduced binding affinities compared to the linear counterpart (Table 5.1). Isothermal titration calorimetry (ITC) experiments confirmed the 1:1 stoichiometric binding and gave further insights into the enthalpy-entropy compensation in the binding of these peptides. The linear ^{d}PMI-δ peptide and ^{d}PMI-δ(1−5) stapled peptide have reduced helicity in solution, and yet are characterized by a favorable enthalpy of binding. It is clear that the entropic penalty incurred as the peptides adopt a more ordered state upon binding Mdm2 is compensated for by the favorable enthalpic changes. In contrast, ^{d}PMI-δ(5−12) and ^{d}PMI-δ(6−10) had higher helicities (32% and 40%, respectively) and pay smaller entropic costs upon binding which compensates for the reduced enthalpies. As expected, both the linear and stapled peptides were found to be resistant to proteolytic degradation in whole-cell homogenate and human plasma.

Next, the ability of these peptides to enter cells was tested against HCT116 cells and their ability to activate p53 was examined by appropriate reporters. The linear all-D-peptide remained inactive even at high concentrations (50 μM), three of the six stapled ^{d}PMI-δ peptides were also inactive even at high concentrations, while three peptides showed dose responsive increases in p53 activity (Table 5.1).

Cell permeable macrocyclic peptides are often hydrophobic in nature and can disrupt the outer membranes of cells, resulting in cellular leakage and the associated stress can activate p53. To explore the source of the p53 activation by these stapled peptides, membrane integrity assays [measured through lactate dehydrogenase (LDH) release] were carried out. In addition, the peptides were also subjected to counter screen assays to explore whether the activation of p53 originated in disruption of the p53-Mdm2 complex or

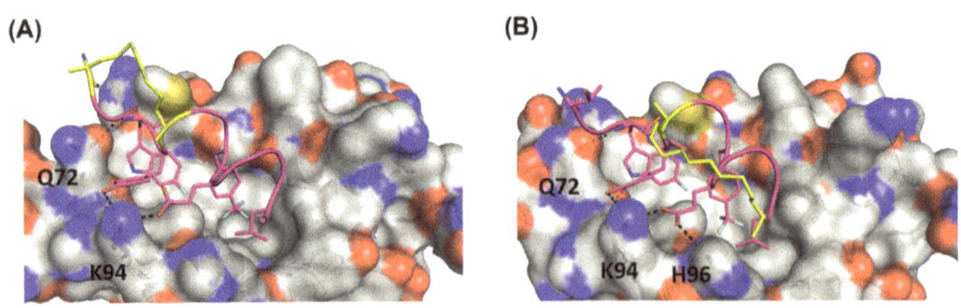

FIGURE 5.5 Structural representations of MD snapshots of (A) ^{d}PMI-δ(1−5)−Mdm2 and (B) ^{d}PMI-δ(5−12)−Mdm2 complexes. Mdm2 is shown as surface and bound peptide is shown as a cartoon with interacting residues highlighted in sticks. The hydrocarbon linker is highlighted in sticks. Hydrogen bond interactions are shown as dotted lines.

TABLE 5.1 Secondary structure, binding and cellular activity of linear stapled, double stapled, and stitched ^{d}PMI−δ peptides determined through various biophysical and biochemical methods.

Peptide	CD helicity (%)	FP Kd (nM)	SPR Kd (nM)	ITC (kcal/mol)			P53 activity EC_{50} (μM)	LDH release EC_{50} (μM)	Counter screen EC_{50} (μM)
				ΔH	-TΔS	ΔG			
^{d}***PMI-δ:*** T A (6FW) Y A N (p-CF3-F) E K L L R	20.4	41.4 ± 1.2	<1	− 13.8	3.04	− 10.8	> 50	> 50	> 50
^{d}***PMI-δ(1−5):*** ***R_5*** A (6FW) Y ***R_5*** N (p-CF3-F) E K L L R	24.7	16.8 ± 1.3	<1	− 15.8	4.2	− 11.6	13.7 ± 1.2	24 ± 3	> 50
^{d}***PMI-δ(2−6):*** T ***R_5*** (6FW) Y A ***R_5*** (p-CF3-F) E K L L R	46.7	7450 ± 1.2	> 500	n.d	n.d	n.d	> 50	17.7 ± 1.9	13.9 ± 4.2
^{d}***PMI-δ(2−9):*** T ***S_8*** (6FW) Y A N (p-CF3-F) E ***R_5*** L L R	26.5	232 ± 1.4	7.4	n.d	n.d	n.d	> 50	> 50	> 50
^{d}***PMI-δ(5−9):*** T A (6FW) Y ***R_5*** N (p-CF3-F) E ***R_5*** L L R	38.0	142.3 ± 1.1	29	n.d	n.d	n.d	> 50	46.5 ± 1.2	47 ± 3.0
^{d}***PMI-δ(5−12):*** T A (6FW) Y ***R_5*** N (p-CF3-F) E K L L ***R_5***	31.8	33.9 ± 1.4	<1	− 8.25	− 1.83	− 10.1	21.3 ± 2.6	> 50	> 50
^{d}***PMI-δ(6−10):*** T A (6FW) Y A ***R_5*** (p-CF3-F) E K ***R_5*** L R	39.9	118 ± 1.2	59	− 9.35	− 1.86	− 11.2	30.3 ± 3.1	18.5 ± 0.7	39 ± 4.0
^{d}***PMI-δ(1−5,9−12):*** ***R_5*** A (6FW) Y ***R_5*** N (p-CF3-F) E ***S_5*** L L ***R_5***	52.1	62.9 ± 1.1	<1	n.d	n.d	n.d	4 ± 0.4	> 50	> 50

(*Continued*)

TABLE 5.1 (Continued)

Peptide	CD helicity (%)	FP Kd (nM)	SPR Kd (nM)	ITC (kcal/mol)			P53 activity EC_{50} (μM)	LDH release EC_{50} (μM)	Counter screen EC_{50} (μM)
				ΔH	-TΔS	ΔG			
^{d}PMI-δ(1–5–12): ***R_5*** A (6FW) Y ***B_5*** N (p-CF3-F) E K L L ***R_8***	20.0	1.5 ± 1.5	<1	n.d	n.d	n.d	34.6 ± 7.6	>50	>50
ATSP-7041: L T F ***S_8*** E Y W A Q (Cba) ***R_5*** S A A	49.6	33.8 ± 1.6	<1	n.d	n.d	n.d	4.2 ± 2.1	>50	>50

Source: Data from Kannan S, et al. Macrocyclization of an all-d linear α-helical peptide imparts cellular permeability. Chem Sci. 2020;11:5577–91.

whether alternate targets in the cell could also be engaged, resulting in p53 activation. The linear peptide and two of the staple peptides did not exhibit any activity in LDH release and counter screen assays probably due to being cell impermeable. In contrast stapled peptides ^{d}PMI-δ(1−5) and ^{d}PMI-δ(6−10) exhibited p53 activity but also resulted in considerable LDH leakage and were active in counter screen assays, suggesting that the activation of p53 could result from membrane disruption and/or nonspecific binding (Table 5.1). The most active stapled peptide ^{d}PMI-δ(5−12) was not active in LDH leakage assay or counter screen activity, suggesting that it enters cells and disrupts the p53-Mdm2 complex, resulting in p53 activation.

5.3.2 Development of double stapled and stitched peptides

The next modification to ^{d}PMI-δ(5−12) was examined with the introduction of a second staple. The introduction of two staples in a peptide required sufficient spacing between the two pairs of nonnatural amino acids to avoid any cross-reactivity during olefin metathesis and this often required a longer peptide sequence. Studies had shown that additional staples can often impart increased helicity, cell permeability, improve proteolytic stability and increase cellular activity [26,27,71]; examples include pathways in Rab8a GTP-ase [25], HIV-1 [30], respiratory syncytial virus entry [27], Ral GTP-ase [71], estrogen receptor-α [31] and BCL9 [72]. Double-stapled peptides can also be designed contiguously with a common attachment/anchoring point and are referred to as "stitched" peptides. Synthesis of these peptides by using RCM reactions was demonstrated recently by Hilinski et al. [30].

Variations of the two stapled peptides ^{d}PMI-δ(1−5) and ^{d}PMI-δ(5−12) were constructed to yield a double stapled peptide stitched peptide ^{d}PMI-δ(1−5, 9−12) consisting of i, $i+4$ and i, $i+3$ staples and a stitched peptide ^{d}PMI-δ(1−5−12) which had residue 5 as the common attachment point for the two staples (Fig. 5.6). The double stapled peptide did not show any increase in helicity (20%, Table 5.1) while the stitched peptide showed

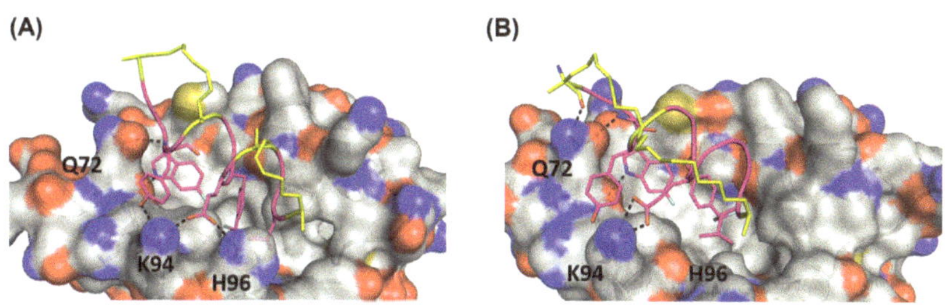

FIGURE 5.6 Structural representations of MD snapshots of (A) double staple ^{d}PMI-δ(1−5, 9−12)−Mdm2 and (B) stitched ^{d}PMI-δ(1−5−12)−Mdm2 complex. Mdm2 is shown as surface and bound peptide is shown as a cartoon with interacting residues highlighted in sticks. The hydrocarbon linker is highlighted in sticks. Hydrogen bond interactions are shown as dotted lines.

increased helicity (52%, Table 5.1). The increased helicity of the stitched peptide was in agreement with other reports, whereas the reduced helicity of the double stapled peptide was surprising. Nevertheless, both the double stapled and stitched peptides displayed tight binding in biophysical and biochemical experiments, underscoring the important balance between enthalpy and entropy in their binding. The stitched peptide displayed enhanced cellular activity compared to that displayed by the singly stapled peptides (Table 5.1). The lack of LDH leakage (Table 5.1) and negligible counter screen activity suggests that ^{d}PMI-δ(1−5−12) enters cells without membrane disruption and results in the activation of p53 by inhibiting the Mdm2-p53 complex (Table 5.1); in contrast, ^{d}PMI-δ(1−5, 9−12) did not have significant cellular activity (Table 5.1). While studies suggest similarities in trends in single and double stapled peptides [25−32], lack of cell activity in ^{d}PMI-δ(1−5, 9−12) demonstrates that the structure−function relationships that modulated cellular activity are complex and enigmatic. The mechanism underlying the increased permeability of the stitched peptide can partly be attributed perhaps to increased helicity and hydrophobicity. However, the double stapled peptide has a longer hydrophobic linker and yet its lack of permeability suggests that the mechanisms of cell permeability are more complex and more systematic studies are needed to tease out the mechanistic details. Nevertheless, it is clear that macrocyclization of an all-D-linear peptide can enhance both binding and cellular activities and holds the potential to be developed as a potential therapeutic.

5.4 Development of nonhelical stapled peptides

The peptide stapling technique was originally developed to stabilize α-helices. However, a significant portion of residues found at protein−protein interfaces are nonhelical [73]. Restricting the stapling technique to helical peptides severely limits the spectrum of PPIs that can be targeted. There have been recent attempts to extend the stapling technique to the stabilization of nonhelical peptide conformations, but these remain few and far between. Unlike α-helical peptides, only one type of macrocyclization reaction has been used to staple nonhelical peptides thus far. Azido and alkynyl sidechains can react in a copper (I)-catalyzed azide-alkyne cycloaddition (CuAAC) "click" reaction, resulting in the formation of a triazole linkage (Fig. 5.7A). This reaction was used to stabilize a model β-hairpin peptide called trpzip [74]. The β-hairpin content was evaluated using nuclear magnetic resonance, which showed that side chain length affects the extent of stabilization. The optimal residues for β-hairpin stabilization were found to be an N-terminal alkynyl residue with two methylene groups and a C-terminal azido residue with one methylene group.

In subsequent related work, Park and Waters used propargylglycine and azidolysine residues to staple an adenosine triphosphate (ATP)-binding β-hairpin peptide called WKWK [75]. They varied the position of the

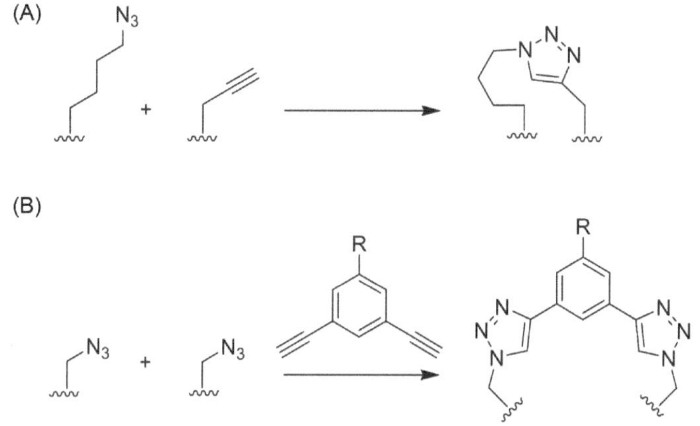

FIGURE 5.7 CuAAC reactions for forming stapled β-hairpin peptides. (A) One-component cycloaddition. (B) Two-component double-click cycloaddition.

triazole staple in the β-hairpin, from hydrogen-bonded positions, to nonhydrogen bonded positions, and terminal positions. Regardless of the staple position, the stapled peptides exhibited enhanced thermal stability and proteolytic resistance as compared to the corresponding unstapled peptides. Based on circular dichroism results, the optimal placement of the staple for stabilizing the β-hairpin structure is at the hydrogen-bonded positions. Binding to ATP was unaffected by stapling.

So far, the early work on stapling nonhelical peptides was performed on model peptides that have no known protein targets. The first bioactive stapled nonhelical peptides were reported in 2017 [21]. Both studies reported the use of the two-component double-click CuAAC reaction, which involves two amino acid sidechains with azido groups and a dialkynyl linker (Fig. 5.7B), to stabilize the conformations of peptides with nonregular conformations.

In the first study, stapled peptides were designed to bind to importin α to block the entry of the transcription factor HNF1β into the nucleus. The overexpression of HNF1β is associated with ovarian cancer and inhibition of its function could be achieved by blocking its nuclear import. Based on the crystal structure of an HNF1β-derived peptide bound to importin α, Wiedmann et al. [21] designed stapled peptides to compete for binding with the nuclear localization signal (NLS) of HNF1β at the NLS binding site of importin α. In the crystal structure of the HNF1β NLS peptide bound to importin α, the peptide was shown to bind in an extended and unstructured conformation. MD simulations of the complex followed by binding free energy decomposition and computational alanine scanning analysis were used to identify the HNF1β residues that are important for binding to importin α and suitable residues that could be replaced by staple linkers. Molecular modeling helped to determine the optimal lengths of the azido amino acids and dialkynyl urea linker needed to stabilize the extended conformation of the HNF1β NLS peptide. A set of four stapled HNF1β peptides with different linker lengths and types were synthesized and evaluated in biophysical and cell-based assays. The designed lead stapled peptide exhibited an enhanced binding affinity for importin α and cell permeability compared to the linear wild-type peptide (Fig. 5.8).

In another study, constrained peptides were designed to block the interaction of the tankyrase proteins with their substrate protein partners [76]. Tankyrase 1 and tankyrase 2 (TNKS) are poly(ADP-ribose) polymerases that catalyze the transfer of ADP-ribose from NAD^+ to other proteins. Increased expression of TNKS is associated with cancer development. The inhibition of TNKS could be useful for the treatment of several cancers; however, small-molecule inhibitors that target the active site of TNKS have met with limited success due to toxicity associated with off-target effects. The ankyrin domain of TNKS, which comprises a series of ankyrin-repeat clusters (ARCs), is an alternative target for drug development as it is involved in the recognition of substrate proteins. Therefore targeting the ARC domain offers the possibility for the development of a new class of highly specific inhibitors of TNKS.

In crystal structures of the fourth ARC domain of TNKS2 (TNKS2 ARC4) in complex with substrate-derived peptides, the peptides are shown to bind in an extended conformation. Based on these structures, Xu et al. [76] designed stapled peptides that mimic the extended conformation of an 8-residue consensus peptide bound at the tankyrase-substrate interface. MD simulations coupled with computational alanine scanning were used to identify residues that are critical for binding and potential sites for the incorporation of staple linkers. Several stapled peptides with different azido side chain lengths, dialkynyl linker types, and staple positions were synthesized and tested for binding to tankyrase 2. Two of these stapled peptides were found to bind better than the linear consensus peptide and were able to disrupt the TNKS interaction with the Axin protein. Despite adopting an extended conformation, these constrained extended peptides displayed resistance to proteolytic degradation. In contrast, the unconstrained linear peptide

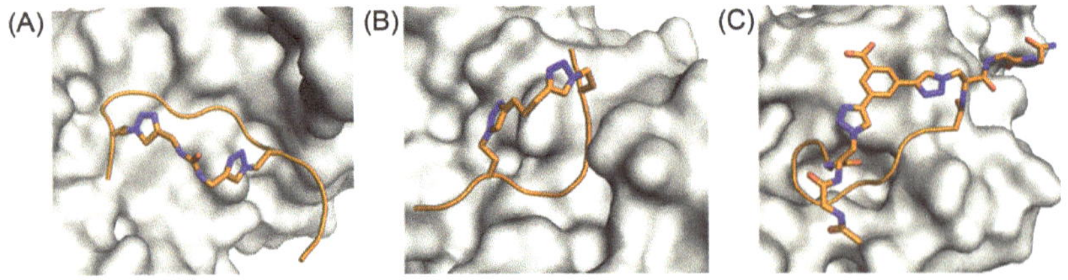

FIGURE 5.8 Structures of stapled nonhelical peptides. (A) A model of a stapled HNF1β peptide bound to importin α1. (B) Crystal structure of a stapled consensus peptide bound to tankyrase 2 (PDB 5BXO). (C) Crystal structure of a stapled CK2β peptide bound to CK2α (PDB ID: 6Q4Q).

was proteolytically unstable. This confirms that macrocyclization greatly enhances the stability of the peptide. Cell permeability was conferred by attaching a cell-penetrating peptide sequence to the linker. This ability to introduce different functionalities via the linker is a major advantage of employing two-component CuAAC stapling. The permeabilized stapled peptide exhibited cell activity in a dose-dependent manner.

Most recently, the two-component CuAAC stapling technique was applied to the stabilization of a β-hairpin peptide that binds to CK2α [77]. CK2 is a protein kinase overexpressed in various cancers. Inhibition of CK2 function is a promising strategy for cancer therapy. CK2 exists as a tetramer made up of two types of subunits: CK2α and CK2β. The catalytic function resides on the CK2α subunit while the CK2β subunit performs regulatory functions such as control of substrate specificity, cellular localization and enzymatic activity. Inhibitors have been developed to bind at the ATP binding site in CK2α. However, these inhibitors come with specificity issues as the ATP binding pocket is well conserved across all kinases. There have been increasing efforts to develop non-ATP-competitive inhibitors of CK2 to avoid off-target effects. The CK2α/β interface is an alternative target for the selective inhibition of CK2.

Given that the CK2α/β interface is mostly shallow, peptides are an ideal class of molecule to target this PPI. A β-hairpin disulfide-linked cyclic peptide called *Pc* had been designed to mimic the CK2α-interacting region of CK2β and was shown to inhibit the CK2α/β interaction. However, *Pc* is unsuitable for use as a drug as it lacks cell permeability and is constrained by a disulfide bridge that is unstable in the reducing environment of the cytosol. Using the crystal structure of CK2β bound to *Pc*, MD simulations and binding free energy decomposition were used by Ierge et al. [77] to identify critical residues for binding and potential sites for incorporation of a chemically stable two-component triazole staple to replace the disulfide bridge (Fig. 5.8). Several stapled peptides with different linker types were designed, synthesized and tested experimentally. One of the stapled peptides displayed improved affinity as compared to *Pc*. Ligand-mapping MD simulations that used chlorobenzene as the probe molecule were then used to suggest mutations to further improve the potency of the stapled peptide (Fig. 5.8) [78–80]. Cell permeability was conferred by the functionalization of the staple with arginine residues. The lead peptide exhibits a fivefold improvement in binding affinity over *Pc*, is stable under physiological conditions, inhibits cancer cell proliferation and induces apoptosis.

Although the stapling technique was originally developed to constrain α-helical peptides, applications of the method to constrain the conformations of nonhelical peptides are emerging. As nonhelical peptides mediate a significant portion of PPI, it is anticipated that more instances of stapled nonhelical peptides will be reported in the near future, as new drug targets are explored and peptide stapling techniques based on new chemistries are developed. Although only the CuAAC click reaction has been used to staple nonhelical peptides so far, other staple architectures such as hydrocarbon, lactam and thioether may be suitable as well and are just awaiting validation.

5.5 Stapling modulates membrane permeability and aggregation propensity of peptide therapeutics

A major bottleneck in the use of peptides and peptidomimetics is their limited cell permeability, and several strategies are being explored to overcome this. Although several all-hydrocarbon stapled α-helical peptides are membrane-penetrating [81–83], the associated mechanism remains enigmatic and it appears that there may not be a unique mechanism associated with the process. There appears to be some correlation between the resultant helicity and hydrophobicity of a peptide upon stapling and its membrane permeability [84–86]. Some insights into the mechanism are discussed next.

5.5.1 Stapling modulates peptide membrane permeability

From a thermodynamic point of view, when a peptide is helical, its polar backbone atoms are engaged in intramolecular hydrogen bonds. The resultant free energy cost associated with dehydration as the peptide enters the apolar membrane is thus minimized. In contrast, the unstapled peptides are largely unstructured, hence their backbone atoms are exposed to water molecules, and this imposes a large desolvation-free energy barrier to membrane entry. The hydrophobicity of the staple chain will further assist in solvating the peptides by the membrane lipids. Thus the high membrane permeability of the stapled peptides likely arises from two effects: (1) the helical constraint that results in reduced exposure of the polar surface area and (2) the staple associated increase in the hydrophobic surface.

Of course, there is the need to optimize the length of the hydrocarbon chain and the two stapling positions, which requires molecular insights into the mode of action of the hydrocarbon stapled helical peptide with model lipid membranes.

5.5.2 Stapling affects peptide aggregation propensity

Therapeutic peptides usually contain a large content of hydrophobic residues and are prone to aggregate because (1) most of the protein–protein binding interface that the peptide is designed to target is characterized by large hydrophobic areas and (2) high hydrophobicity is needed for favorable membrane permeability. Therefore solubility can be a big issue in the development of peptide therapeutics. A variety of peptide stapling strategies have been shown to modulate the solubility of the peptide [83,87–90]. By fine-tuning the chemistry of the staple chain, Smeenk et al. successfully designed a stapled peptide scaffold with high water solubility [83]. Another widely used stapling strategy is triazole stapling [87]. Jacobsen et al. have shown that stapling with a chain containing the triazole group can significantly enhance the solubility of a 3_{10} helix peptide [90]. Kawamoto et al. also designed a series of triazole-containing hydrocarbon stapled α-helical peptides targeting beta-catenin BCL9 interactions and found that the position of the triazole along the hydrocarbon chain can significantly affect the peptide solubility by modulating the peptide structure [78].

Hydrocarbon stapling usually favors aggregation because of the increased hydrophobicity. Aggregation can be further enhanced if the polar residues are replaced by the incorporation of the staple carrying residues [91]. For example, a hydrocarbon stapled peptide NYAD-1 targeting HIV displayed low solubility than its linear analog CAI because of self-aggregation mediated by hydrophobic interactions between the hydrocarbon chain and the aromatic residues of the peptide [88]. However, in some cases, the hydrocarbon chain can also result in enhanced solubility. Luong et al. used hydrocarbon stapling to design antimicrobial peptides and found that hydrocarbon stapling can either increase or decrease peptide aggregation. For a peptide with a short linker sequence, hydrocarbon stapling favors aggregation because of enhanced overall hydrophobicity. However, in the presence of a longer linker sequence, hydrocarbon stapling was found to inhibit aggregation, which according to the authors occurs because a peptide with a longer linker sequence can fold into more compact structures with hydrophobic residues buried inside, resulting in enhanced solubility [92]. Therefore peptide stapling can enhance or reduce peptide aggregation depending on the peptide sequence, the chemical nature of the staple chain and the staple position. Mechanistic understanding of the effects of peptide stapling on peptide solubility can help the design of better stapled peptides.

5.6 Concluding remarks

Peptide-based inhibitors demonstrate high target specificity, high biocompatibility and low toxicity and hence are being explored as next-generation therapeutics. Despite limitations such as conformational and proteolytic instability, aggregation propensity and lack of cell permeability, there has been significant progress during the last two decades, including some progress in their development as therapeutics. Modifications to bypass some of the liabilities by the introduction of constraints such as stapling or by manipulations of the amino acid backbone has resulted in successful peptidomi metics. Stapling has been used to constrain peptides in their active/bound conformations and also to protect the peptides against proteolysis; a surprising benefit has been the emergence of cell permeability in some peptides. Further modifications such as the incorporation of D-amino acids have resulted in complete resistance to proteolysis, much increased stability and bioavailability (paving the way for exploring oral administration).

The peptide stapling technique was originally developed and widely used to stabilize α-helices. The α-helix is the most common motif found at protein–protein interfaces. However, a significant portion of residues found at protein–protein interfaces are nonhelical. Most successful stapling efforts have targeted helical peptides, but some progress has been made in the successful stabilization of nonhelical peptide conformations. It is anticipated that more instances of stapled nonhelical peptides will be reported in the near future as new drug targets are explored and peptide stapling techniques based on new chemistries are developed.

The hydrocarbon stapled peptides were originally developed to constrain the unbound peptides in conformations they adopt upon binding their target receptors, thus gaining affinity; however, a beneficial effect of this turned out to be enhanced metabolic stability and surprisingly, cell permeability. The mechanism behind cell permeability of these hydrocarbon staple peptides still remains unclear, although an intuitively appealing hypothesis is that hydrocarbon stapling enhances the overall hydrophobicity of the peptide, resulting in enhanced membrane permeability. However, not all stapled peptides can enter cells, so clearly there are other mechanisms at play. Undoubtedly, there will be a greater understanding of this in the near future and this will catalyze the developments of molecules that will be therapeutically interesting and expand the druggable space of PPI targets.

Acknowledgments

The authors thank A*STAR (grant IDs: H17/01/a0/010, IAF111213C) for support. DL is supported by a grant from the Swedish Research Council (2013_08807).

References

[1] Petta I, Lievens S, Libert C, Tavernier J, De Bosscher K. Modulation of protein–protein interactions for the development of novel therapeutics. Mol Ther 2015;24:707–18.

[2] Macalino SJY, et al. Evolution of in silico strategies for protein-protein interaction drug discovery. Molecules. 2018;23:1963.

[3] Scott DE, Bayly AR, Abell C, Skidmore J. Small molecules, big targets: drug discovery faces the protein–protein interaction challenge. Nat Rev Drug Discov 2016;15:533–50.

[4] Ran X, Gestwicki JE. Inhibitors of protein-protein interactions (PPIs): an analysis of scaffold choices and buried surface area. Curr Opin Chem Biol 2018;44:75–86.

[5] Nevola L, Giralt E. Modulating protein–protein interactions: the potential of peptides. Chem Commun 2015;51:3302–15.

[6] Lau JL, Dunn MK. Therapeutic peptides: historical perspectives, current development trends, and future directions Bioorg. Med Chem 2017;26:2700–7.

[7] Fosgerau K, Hoffmann T. Peptide therapeutics: current status and future directions. Drug Discov Today 2015;20:122–8.

[8] Bakail M, Ochsenbein F. Targeting protein–protein interactions, a wide open field for drug design. Comptes Rendus Chimie 2016;19:19–27.
[9] Henninot A, Collins JC, Nuss JM. The current state of peptide drug discovery: back to the future. J Med Chem 2018;61:1382–414.
[10] Morrison C. Constrained peptides' time to shine? Nat Rev Drug Discov 2018;17:531–3.
[11] Valeur E, et al. New modalities for challenging targets in drug discovery. Angew Chem Int Ed 2017;56:10294–323.
[12] Cary DR, Ohuchi M, Reid PC, Masuya K. Constrained peptides in drug discovery and development. J Synth Org Chem, Jpn 2017;75:1171–8.
[13] Vinogradov A, et al. Macrocyclic peptides as drug candidates: recent progress and remaining challenges. J Am Chem Soc 2019;141:4167–81.
[14] Sawyer TK, et al. Macrocyclic α helical peptide therapeutic modality: a perspective of learnings and challenges. Bioorg Med Chem 2018;26:2807–15.
[15] Walensky LD, Bird GH. Hydrocarbon-stapled peptides: principles, practice, and progress. J Med Chem 2014;57:6275–88.
[16] Tan YS, Lane DP, Verma CS. Stapled peptide design: principles and roles of computation. Drug Discov Today 2016;21:1642–53.
[17] Ali AM, et al. Stapled peptides inhibitors: a new window for target drug discovery. Computational Struct Biotechnol J 2019;17:263–81.
[18] Klein M. Stabilized helical peptides: overview of the technologies and its impact on drug discovery. Expert Opin Drug Discov 2017;12:1117–25.
[19] Lerge J, et al. Stapled peptides as a new technology to investigate protein–protein interactions in human platelets. Chem Sci 2018;9:4638–43.
[20] Xu W, et al. Macrocyclized extended peptides: inhibiting the substrate-recognition domain of tankyrase. J Am Chem Soc 2017;139:2245–56.
[21] Wiedmann MM, et al. Development of cell-permeable, non-helical constrained peptides to target a key protein–protein interaction in ovarian cancer. Angew Chem IntEd 2017;56:524–9.
[22] Iyer VV. A review of stapled peptides and small molecules to inhibit protein-protein interactions in cancer. Curr Med Chem 2016;23:3025–43.
[23] Xie X, Gao L, Shull AY, Teng Y. Stapled peptides: providing the best of both worlds in drug development. Future Med Chem 2016;16:1969–80.
[24] Funda Meric-Bernstam F, et al. Phase I trial of a novel stapled peptide ALRN-6924 disrupting MDM4- and MDM2- mediated inhibition of WT p53 in patients with solid tumors and lymphomas. J Clin Oncol 2017;35 2505–2505.
[25] Cromm PM, et al. Protease-resistant and cell-permeable double-stapled peptides targeting the Rab8a GTPase. ACS Chem Biol 2016;11:2375–82.
[26] Bird GH, et al. Hydrocarbon double-stapling remedies the proteolytic instability of a lengthy peptide therapeutic. PNAS 2010;107:14093–8.
[27] Gaillard V, et al. A short double-stapled peptide inhibits respiratory syncytial virus entry and spreading. Antimicrob Agents Chemother 2017;61 e02241-16.
[28] Giordanetto F, et al. Stapled vasoactive intestinal peptide (VIP) derivatives improve VPAC2 agonism and glucose-dependent insulin secretion. ACS Med Chem Lett 2013;4:1163–8.
[29] Bird G, Irimia A, Ofek G, et al. Stapled HIV-1 peptides recapitulate antigenic structures and engage broadly neutralizing antibodies. Nat Struct Mol Biol 2014,21:1058–67.
[30] Hilinski GJ, et al. Stitched α-helical peptides via bis ring-closing metathesis. J Am Chem Soc 2014;136:12314–22.
[31] Speltz TE, et al. A "cross-stitched" peptide with improved helicity and proteolytic stability. Org Biomol Chem 2018;16:3702–6.

[32] Kannan S, et al. Macrocyclization of an all-d linear α-helical peptide imparts cellular permeability. Chem Sci 2020;11:5577–91.

[33] Qvit N, Rubin SJ, Urban TJ, Mochly-Rosen D, Gross ER. Peptidomimetic therapeutics: scientific approaches and opportunities. Drug Discov Today 2017;22:454–62.

[34] Merrick WC. eIF4F: a retrospective. J Biol Chem 2015;290:24091–9.

[35] Pelletier J, Graff J, Ruggero D, Sonenberg N. Targeting the eIF4F translation initiation complex: a critical nexus for cancer development. Cancer Res 2015;75:250–63.

[36] Rhoads RE. eIF4E: new family members, new binding partners, new roles. J Biol Chem 2009;284:16711–15.

[37] Jia Y, Polunovsky V, Bitterman PB, Wagner CR. Cap-dependent translation initiation factor eIF4E: an emerging anticancer drug target. Med Res Rev 2012;32:786–814.

[38] Jackson R, Hellen CUT, Pestova TV. The mechanism of eukaryotic translation initiation and principles of its regulation. Nat Rev Mol Cell Biol 2010;11:113–27.

[39] Mignone F, Gissi C, Liuni S, Pesole G. Untranslated regions of mRNAs. Genome Biol 2002;3:004.

[40] Peter D, Igreja C, Weber R, Wohlbold L, et al. Molecular architecture of 4E-BP translational inhibitors bound to eIF4E. Mol Cell 2015;57:1074–87.

[41] Qin X, Jiang B, Zhang Y. 4E-BP1, a multifactor regulated multifunctional protein. Cell Cycle 2016;15:781–6.

[42] Proud CG. Mnks, eIF4E phosphorylation and cancer. Biochim Biophys Acta 2015;1849: 766–73.

[43] Scheper GC, Proud CG. Does phosphorylation of the cap-binding protein eIF4E play a role in translation initiation? Eur J Biochem 2002;269:5350–9.

[44] Lama D, Verma C. Deciphering the mechanistic effects of eIF4E phosphorylation on mRNA-cap recognition. Protein Sci 2020;6:1373–86.

[45] Borden KL. Targeting the oncogene eIF4E in cancer: from the bench to clinical trials. Clin Invest Med 2011;34:E315.

[46] Hsieh AC, Ruggero D. Targeting eukaryotic translation initiation factor 4E (eIF4E) in cancer. Clin Cancer Res 2010;16:4914–20.

[47] Diab S, et al. MAP kinase-interacting kinases—emerging targets against cancer. Chem Biol 2014;21:441–52.

[48] Gu Y, et al. Small-molecule induction of phospho-eIF4E sumoylation and degradation via targeting its phosphorylated serine 209 residue. Oncotarget. 2015;6:15111–21.

[49] Chen X, et al. Structure-guided design, synthesis, and evaluation of guanine-derived inhibitors of the eIF4E mRNA-cap interaction. J Med Chem 2012;55:3837–51.

[50] Moerke NJ, et al. Small-molecule inhibition of the interaction between the translation initiation factors eIF4E and eIF4G. Cell 2007;128:257–67.

[51] Bhattacharyya D, Nguyen K, Basu S. Rationally induced RNA:DNA G-quadruplex structures elicit an anticancer effect by inhibiting endogenous eIF-4E expression. Biochemistry 2014;53:5461–70.

[52] Lama D, Pradhan M, Brown C, Eapen R, Joseph T, Kwoh CK, et al. Water-bridge mediates recognition of mRNA Cap in eIF4E. Structure 2017;25:188–94.

[53] Lama D, Brown C, Lane D, Verma C. Gating by tryptophan 73 exposes a cryptic pocket at the protein-binding interface of the oncogenic eIF4E protein. Biochemistry 2015;54:6535–44.

[54] Cencic R, Hall DR, Robert F, du Y, et al. Reversing chemoresistance by small molecule inhibition of the translation initiation complex eIF4F. Proc Natl Acad Sci USA 2011;108:1046–51.

[55] Brown CJ, Verma CS, Walkinshaw MD, Lane DP. Crystallization of eIF4E complexed with eIF4GI peptide and glycerol reveals distinct structural differences around the cap-binding site. Cell Cycle 2009;8:1905–11.

[56] Lama D, Quah S, Verma C, Lakshminarayan R, et al. Rational optimization of conformational effects induced by hydrocarbon staples in peptides and their binding interfaces. Sci Rep 2013;3:3451.

[57] Lama D, Liberatore AM, Frosi Y, Nakhle J, et al. Structural insights reveal a recognition feature for tailoring hydrocarbon stapled-peptides against the eukaryotic translation initiation factor 4E protein. Chem Sci 2019;10:2489–500.

[58] Lama D, Quah S, Brown C, Lane D, et al. Stapled-peptides targeting the protein-binding interface of eukaryotic translation initiation factor 4E (eIF4E) protein. J Biomol Struct Dyn 2015;33:102–3.

[59] Zhou W, Quah ST, Verma CS, Liu Y, et al. Improved eIF4E binding peptides by phage display guided design: plasticity of interacting surfaces yield collective effects. PLoS One 2012;7:e47235.

[60] Mandal K, et al. Chemical synthesis and X-ray structure of a heterochiral {D-protein antagonist plus vascular endothelial growth factor} protein complex by racemic crystallography. Proc Natl Acad Sci USA 2012;109:14779–84.

[61] Uppalapati M, et al. A potent D-protein antagonist of VEGF-A is nonimmunogenic, metabolically stable, and longer-circulating in vivo. ACS Chem Biol 2016;11:1058–65.

[62] Chang HN, et al. Blocking of the PD-1/PD-L1 interaction by a D-peptide antagonist for cancer. Immunother Angew Chem, Int Ed 2015;54:11760–4.

[63] Welch BD, et al. Design of a potent D-peptide HIV-1 entry inhibitor with a strong barrier to resistance. J Virol 2010;84:11235–44.

[64] Garton M, et al. Method to generate highly stable d-amino acid analogs of bioactive helical peptides using a mirror image of the entire PDB. PNAS 2018;115:1505–10.

[65] Liu M, et al. A left handed solution to peptide inhibition of the p53-MDM2 interaction. Angew Chem, Int Ed 2010;49:3649–52.

[66] Liu M, et al. D-peptide inhibitors of the p53-MDM2 interaction for targeted molecular therapy of malignant neoplasms. Proc Natl Acad Sci USA 2010;107:14321–6.

[67] Lane DP. p53, guardian of the genome. Nature 1992;358:15–16.

[68] Karni-Schmidt O, Lokshin M, Prives C. The roles of MDM2 and MDM4 in cancer. Annu Rev Pathol 2016;11:617–44.

[69] Pei D, Zhang Y, Zheng J. Regulation of p53: a collaboration between Mdm2 and Mdm4. Oncotarget 2012;3:228–35.

[70] Tisato V, Voltan R, Gonelli A, Secchiero P, Zauli G. MDM2/X inhibitors under clinical evaluation: perspectives for the management of hematological malignancies and pediatric cancer. J Hematol Oncol 2017;10:133.

[71] Thomas JC, et al. Inhibition of Ral GTPases using a stapled peptide approach. J Biol Chem 2016;291:18310–25.

[72] Kawamoto SA, et al. Design of triazole-stapled BCL9 α-helical peptides to target the β-catenin/B-cell CLL/lymphoma 9 (BCL9) protein-protein interaction. J Med Chem 2012;55:1137–46.

[73] Guharoy M, Chakrabarti P. Secondary structure based analysis and classification of biological interfaces: identification of binding motifs in protein-protein interactions. Bioinformatics 2007;23:1909–18.

[74] Celentano V, Diana D, De Rosa L, Romanelli A, Fattorusso R, D'Andrea LD. β-Hairpin stabilization through an interstrand triazole bridge. Chem Commun 2012;48:762–4.

[75] Park JH, Waters ML. Positional effects of click cyclization on β-hairpin structure, stability, and function. Org & Biomol Chem 2013;11:69–77.

[76] Xu W, Lau YH, Fischer G, Tan YS, Chattopadhyay A, de la Roche M, et al. Macrocyclized extended peptides: inhibiting the substrate-recognition domain of tankyrase. J Am Chem Soc 2017;139:2245–56.

[77] Iegre J, Brear P, Baker DJ, Tan YS, Atkinson EL, Sore HF, et al. Efficient development of stable and highly functionalised peptides targeting the CK2α/CK2β protein–protein interaction. Chem Sci 2019;10:5056–63.

[78] Tan YS, Sledz P, Lang S, Stubbs CJ, Spring DR, Abell C, et al. Using ligand-mapping simulations to design a ligand selectively targeting a cryptic surface pocket of polo-like kinase 1. Angew Chem Int (Ed.) 2012;51:10078–81.

[79] Tan YS, Spring DR, Abell C, Verma C. The use of chlorobenzene as a probe molecule in molecular dynamics simulations. J Chem Inf Model 2014;54:1821–7.

[80] Tan YS, Spring DR, Abell C, Verma CS. The application of ligand-mapping molecular dynamics simulations to the rational design of peptidic modulators of protein–protein interactions. J Chem Theory Comput 2015;11:3199–210.

[81] Barducci A, Bonomi M, Parrinello M. Metadynamics. WIREs Computat Mol Sci 2011;1: 826–43.

[82] Chee SMQ, Wongsantichon J, Soo Tng Q, Robinson R, Joseph TL, Verma C, et al. Structure of a stapled peptide antagonist bound to nutlin-resistant Mdm2. PLoS One 2014;9:e104914.

[83] Smeenk LEJ, Dailly N, Hiemstra H, van Maarseveen JH, Timmerman P. Synthesis of water-soluble scaffolds for peptide cyclization, labeling, and ligation. Org Lett 2012;14:1194–7.

[84] Eiríksdóttir E, Konate K, Langel Ü, Divita G, Deshayes S. Secondary structure of cell-penetrating peptides controls membrane interaction and insertion. Biochimica et Biophysica Acta (BBA) - Biomembranes 2010;1798:1119–28.

[85] Chu Q, Moellering RE, Hilinski GJ, Kim YW, Grossmann TN, Yeh JTH, et al. Towards understanding cell penetration by stapled peptides. MedChemComm 2015;6:111–19.

[86] Sun TL, Sun Y, Lee CC, Huang HW. Membrane permeability of hydrocarbon-cross-linked peptides. Biophys J 2013;104:1923–32.

[87] Lau YH, de Andrade P, Wu Y, Spring DR. Peptide stapling techniques based on different macrocyclisation chemistries. Chem Soc Rev 2015;44:91–102.

[88] Bhattacharya S, Zhang H, Cowburn D, Debnath AK. Novel structures of self-associating stapled peptides. Biopolymers 2012;97:253–64.

[89] Sharma K, Strizhak AV, Fowler E, Wang X, Xu W, Hatt Jensen C, et al. Water-soluble, stable and azide-reactive strained dialkynes for biocompatible double strain-promoted click chemistry. Org & Biomol Chem 2019;17:8014–18.

[90] Jacobsen Ø, Maekawa H, Ge NH, Görbitz CH, Rongved P, Ottersen OP, et al. Stapling of a 310-helix with click chemistry. J Org Chem 2011;76:1228–38.

[91] Wu Y, Li YH, Li X, Zou Y, Liao HL, Liu L, et al. A novel peptide stapling strategy enables the retention of ring-closing amino acid side chains for the Wnt/β-catenin signalling pathway. Chem Sci 2017;8:7368–73.

[92] Luong HX, Kim DH, Lee BJ, Kim YW. Antimicrobial and hemolytic activity of stapled heptapeptide dimers. Bull Korean Chem Soc 2016;8:1199–203.

Chapter 6

Ring-closing metathesis/ transannular cyclization to azabicyclo[X.Y.0]alkanone dipeptide turn mimics for biomedical applications

N. D. Prasad Atmuri[1], Simon Surprenant[2], Sitan Diarra[3], Carine Bourguet[4] and William D. Lubell[3]

[1]Department of Chemistry, University of British Columbia, Vancouver, British Columbia, Canada, [2]Institut de Recherche en Immunologie et Cancérologie, Université de Montréal, Montréal, Québec, Canada, [3]Département de Chimie, Université de Montréal, Montréal, Québec, Canada, [4]NMX Research and Solutions Inc., Laval, Québec, Canada

6.1 Introduction

Peptide structures exist typically in a dynamic fluctuation between multiple conformers [1]. Their conformational equilibrium is responsible for engaging various receptors in different molecular recognition events and accounts in part for the paradigm of high activity and limited stability exhibited by peptide structures. In peptide-based drug discovery, the conformational constraint is often used to offset the natural balance in favor of greater activity without off-target effects, such as susceptibility to peptidase degradation [2,3], which limit efficacy and bioavailability [4]. Ideally through topological restriction [5], the ultimate so-called peptidomimetic, whether a completely synthetic nonpeptide molecule or pseudo-peptide, will exhibit enhanced biological activity, physicochemical properties and a pharmacokinetic profile to serve as a therapeutic prototype [6,7].

The development of a peptidomimetic may follow a hierarchical design approach [8]. In the exploration of the structure–activity relationships leading from peptide to therapeutic prototype, tools are essential for dissecting ligand–receptor interactions, mechanisms of action and signaling pathways responsible for the activity. Molecular scaffolds are particularly desired for

Peptide and Peptidomimetic Therapeutics. DOI: https://doi.org/10.1016/B978-0-12-820141-1.00029-7

restricting the geometry of the peptide backbone [ϕ (phi), ψ (psi) and ω (omega) dihedral angles] and side-chains [χ (chi) torsion angles] to mimic those in natural secondary structures, such as in β- and γ-turns, α-helices and β-strands, which are predisposed for molecular recognition [9–11].

β-Turns are the third most abundant natural secondary structure after α-helices and β-strands. Often called reverse turns, β-turns serve two primary roles. In peptide folding, they can change the orientation of the peptide by 180 degrees and place the α-carbons of the first and fourth residues at a distance $\leq$ 7 Å to accommodate a hydrogen bond between the carbonyl oxygen and amide nitrogen of the i and $i+3$ residues (Fig. 6.1). Moreover, the protuberance exposes side chains to the external environment for receptor engagement. β-Turn geometry is particularly relevant for G protein-coupled receptor and antibody recognition [12,13], as well as posttranslational modifications, such as glycosylation [14–16]. Strategies for stabilizing β-turn conformations in peptides offer the potential to diminish entropic costs for folding and thereby enhance receptor binding affinity [17].

The application of covalent constraint to stabilize β-turn conformers has garnered significant interest since an α-amino γ-lactam (so-called Freidinger-Veber lactam, Agl) residue was used to replace Gly^6 in luteinizing hormone-releasing hormone, and the resulting pGlu-His-Trp-Ser-Tyr-Agl-Leu-Arg-Pro-Gly-NH_2 analog exhibited an 8.9-fold improvement in potency relative to the native peptide [18]. Inspired by this precedent and natural azabicyclo[X.Y.0]alkanone amino acid structures, such as those

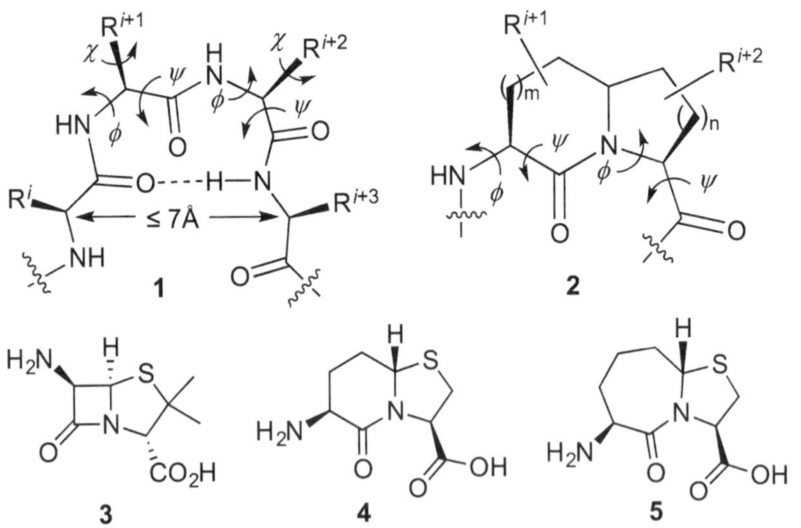

FIGURE 6.1 General structures of the β-turn (**1**) and azabicycloalkanone amino acid mimic (**2**), and representative penicillin, thiaindolizidinone, and thiapyrroloazepinone dipeptide mimics (**3–5**).

found in penicillin antibiotics, thiaindolizidinone (e.g., **4**) and thiapyrroloazepinone (e.g., **5**) dipeptides were created and employed to restrain peptide folding and orient pharmacophores in enzyme inhibitors (Fig. 6.1) [19–21]. Subsequently, a variety of azabicycloalkanone amino acid structures have been synthesized possessing different ring sizes, stereochemistry and substitution patterns for probing structure–activity relationships of various biologically relevant peptides [11,22–36].

The application of azabicyclo[X.Y.0]alkanone sets having different ring sizes and substitution patterns have demonstrated utility for studying peptide structure–activity relationships in part because such modifications have subtle yet significant consequences on conformation (Fig. 6.2) [34]. For example, bicyclic analogs **6** were used to study the conformation of the central residues in the second mitochondria-derived activator of caspase (Smac) Ala-Val-Pro-Ile peptide sequence to identify potent antagonists which bound with high affinity the inhibitor of apoptosis proteins (IAPs) and blocked cancer cell line growth [23,24]. Replacement of the thiaindolizidinone moiety within a nonselective ligand using quinolizidinone and indolizidine-2- and 9-one systems (e.g., **7**) provided a selective opioid receptor-like (ORL1) receptor antagonist [25,26]. Potent inhibitors of angiotensin-converting enzyme (ACE) and neutral endopeptidase (NEP) with activity both in vitro and

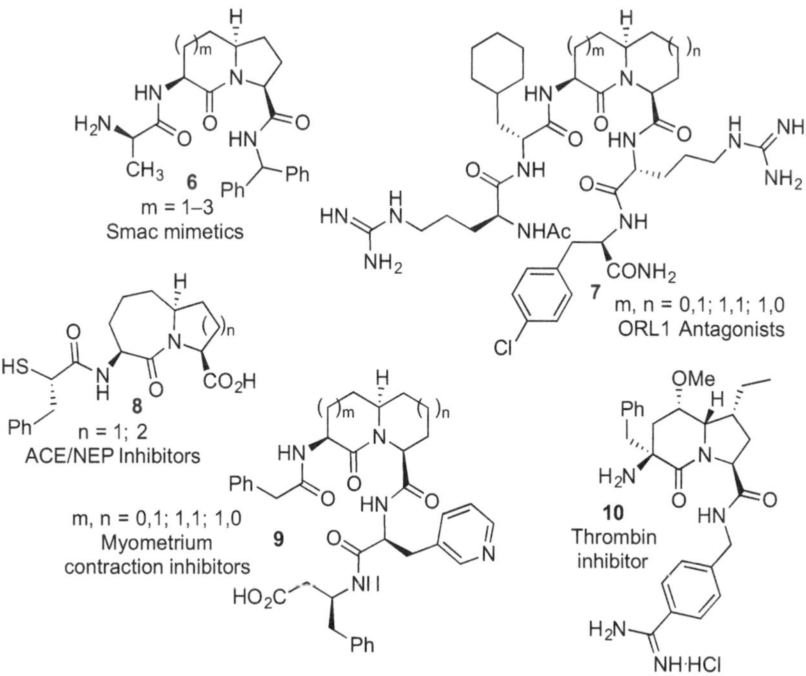

FIGURE 6.2 Representative sets of azabicyclo[X.Y.0]alkanone amino acid derivatives (**6–10**).

in vivo resulted from the application of a series of 7,6- and 7,5-fused bicyclic systems (e.g., **8**) [27,28]. Moreover, the application of a series of bicyclic ring systems (e.g., **9**) demonstrated the importance of the (3*S*,6*S*,9*S*)-indolizidin-2-one amino acid residue for orienting pharmacophores to modulate the prostaglandin F2α receptor by a biased allosteric mechanism that ultimately delays labor in animal models [29–31]. In addition, the significance of ring substituents for selectivity and potency has been illustrated in indolizidin-2-one inhibitors of thrombin (e.g., **10**) [32].

The utility of sets of azabicyclo[X.Y.0]alkanone amino acids as illustrated above promoted interest to develop a general strategy for synthesizing these bicyclic ring systems with stereocontrol, and supporting the potential for adding side-chains onto ring carbons to mimic effectively both peptide backbone and side-chain conformation and function [33–36]. A ring-closing metathesis/transannular cyclization (RCM-TAC) approach was considered a promising strategy for preparing a variety of azabicyclo[X.Y.0]alkanone amino acid systems from a common route. Stereoselective construction of bicyclic tertiary amines had previously been reported using transannular reactions of secondary amine nucleophiles onto olefins and epoxides [37]. As detailed in this chapter, the coupling of various ω-unsaturated amino acid components (e.g., **11** and **12**) was employed to prepare diverse dipeptides **13**, which were cyclized by ring-closing metathesis (RCM) to furnish unsaturated lactam precursors **14** (Scheme 6.1) [38,39]. The application of the 2,4-dimethoxybenzyl (Dmb) group to offset the amide equilibrium in favor of the dipeptide *E-cis* isomer promoted cyclization to form unsaturated lactams having smaller ring sizes and proved fortuitous for obtaining specific diastereomers by transannular cyclization. Iodoamidation of various 8–10-membered unsaturated lactams gave unambiguously 5,5-, 5,6-, 6,4-, 6,5-, 6,6-, and 7,5-fused bicycles by way of a shared synthetic route [39–42].

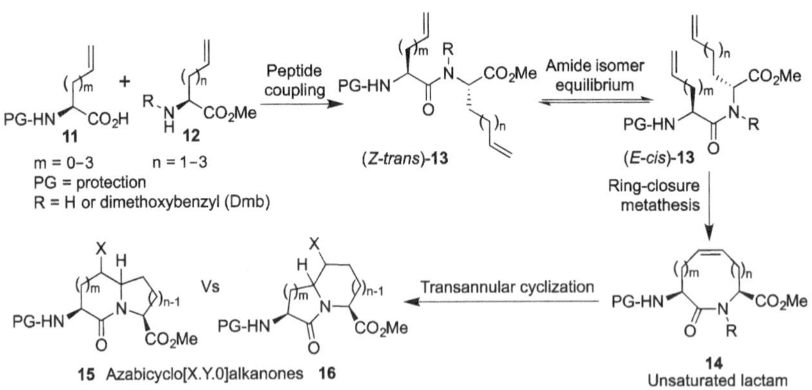

SCHEME 6.1 General plan of a ring-closing metathesis/transannular cyclization approach for azabicyclo[X.Y.0]alkanone amino acid synthesis.

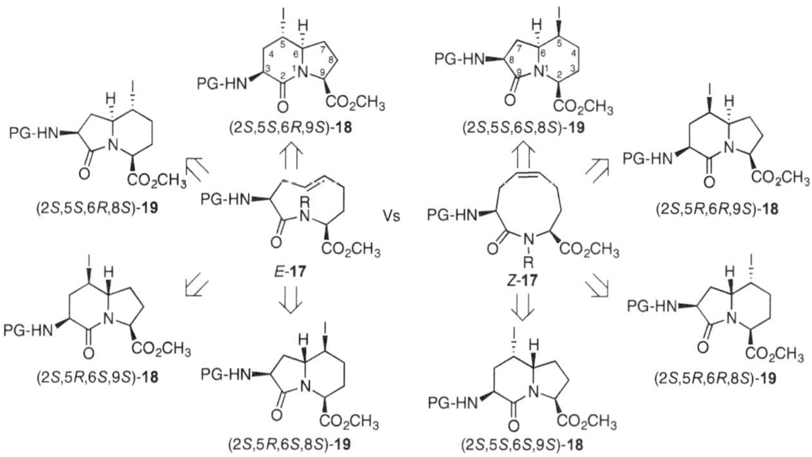

FIGURE 6.3 "One ring to rule them all" contingent on olefin stereochemistry and facial attack.

Iodolactamization of a single ring system can in principle give rise to two different bicycles (e.g., **15** and **16**), each having four possible diastereomers contingent on the facial attack on the starting olefin geometry. For example, *E*- and *Z*-hexahydro-1H-azonines **17** could respectively react to give two stereoisomers of both indolizidine-2- and 9-one analogs **18** and **19**. Governance of the transannular cyclization offers the reward of preparing selectively various bicycles with condition control in a diversity-oriented approach from a common precursor (Fig. 6.3) [42,43]. Considering that the resulting iodide (e.g., **18** and **19**) can serve as a handle for appending various side chain groups onto the ring system, one unsaturated lactam could offer a wealth of regiochemical, stereochemical and substituent patterns for exploring various backbone and side-chain presentations.

6.2 Ring-closing metathesis/transannular cyclization approach for making bicycles with different ring size

A modular method for the synthesis of azabicyclo[X.Y.0]alkanone dipeptide mimics was conceived to provide a set of bicyclic ring systems from a common strategy. The conceptual process featured (1) the coupling of two ω-olefin amino acids to form a dipeptide, (2) RCM to provide an unsaturated lactam, and (3) electrophilic transannular cyclization (TAC) to furnish the bicyclic mimic (Scheme 6.1). For the implementation of this method, effective syntheses were needed to prepare enantiomerically pure ω-olefin amino acids of different chain lengths. Conditions were also required to cyclize the dipeptide into the unsaturated lactam with control of the double bond geometry and to perform stereoselectively the electrophilic TAC. The stereochemistry of the final bicyclic dipeptide mimics was designed to result from the implementation of inexpensive readily available natural amino acids as enantiomerically pure sources of chirality.

6.2.1 ω-Olefin amino acid synthesis

ω-Amino acids of different chain lengths were critical components for the RCM-TAC strategy to produce a variety of azabicyclo[X.Y.0]alkanone dipeptide mimics (Scheme 6.1). Enantiomerically pure ω-amino acids of four to seven carbon atom chain-lengths were desired to prepare a range of bicyclic structures composed of ring sizes typically between five and seven members. Effective processes were developed to synthesize the ω-amino acids from inexpensive methionine and serine as chiral starting materials.

2-Aminobut-3-enoic acid (vinylglycine) has been synthesized from a variety of starting materials using different methods [44,45]. A reliable and scalable approach for obtaining this building block in enantiomerically pure form proved to be the dehydrosulfenylation of a methionine-derived sulfoxide as described by Rapoport [44,45]. The method enabled the synthesis of enantiomerically pure vinylglycinates with different protecting groups on multiple gram scales, without racemization (Scheme 6.2).

2-Aminopent-4-enoic acid (allylglycine) has been prepared using a variety of synthetic processes [38,46,47]. The Negishi cross-coupling of vinyl bromide with the zincate derived from β-iodoalaninate **27** was developed to provide effectively multiple grams of enantiomerically pure allylglycine **28a** from serine as an inexpensive starting material (Scheme 6.3) [39,47]. (2*R*)- and (2*S*)-Methyl *N*-Boc-β-iodoalaninates (**27**) were respectively synthesized on multiple gram scales in three steps from inexpensive D- and L-serine by routes featuring the Appel reaction [48]. The mild conditions of the Appel and Negishi reactions avoided significant β-elimination to dehydroalanine and maintained stereochemical integrity during manipulations of the iodoalanine building block, which has been used as a key intermediate for the preparation of other ω-olefin amino acids. Upon activation with 1,2-dibromoethane and chlorotrimethylsilane (TMSCl), zinc inserts readily into the C-I bond. The resulting organometallic species was cross-coupled to vinyl bromide by the Negishi reaction in good yield.

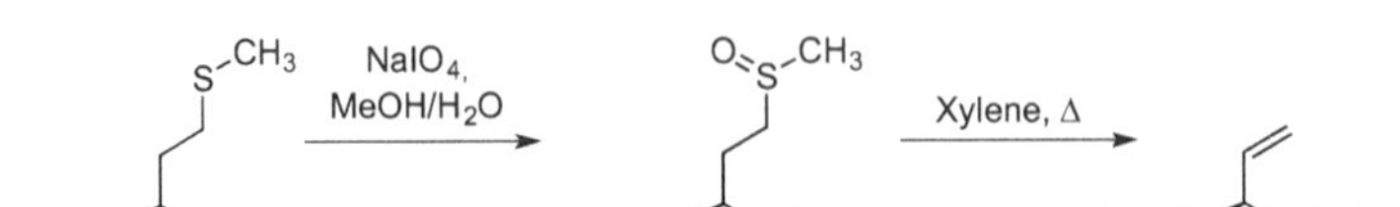

20: R = *t*-Bu
21: R = CH_3

22a: R = *t*-Bu (92%)
23c: R= Me (95%)

24a: R = *t*-Bu (52%)
25b: R = Me (71% from **21b**)
25c: R = Me (>99%)

PG = **a**, Boc, **b**, Fmoc, **c**, Cbz

SCHEME 6.2 Synthesis of vinylglycine building blocks.

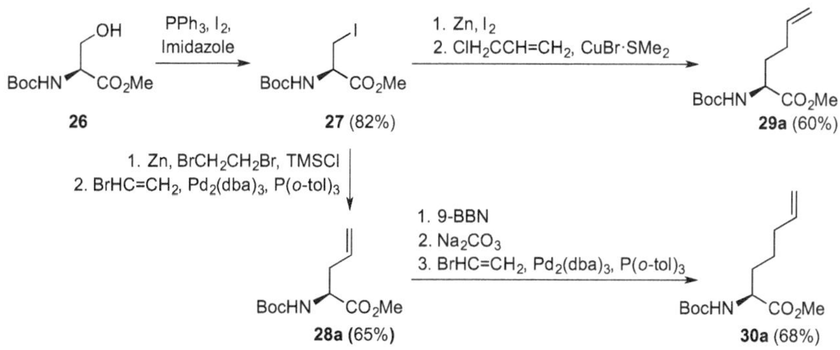

SCHEME 6.3 Synthesis of ω-unsaturated α-amino acid building blocks of 5−7 carbon chain lengths.

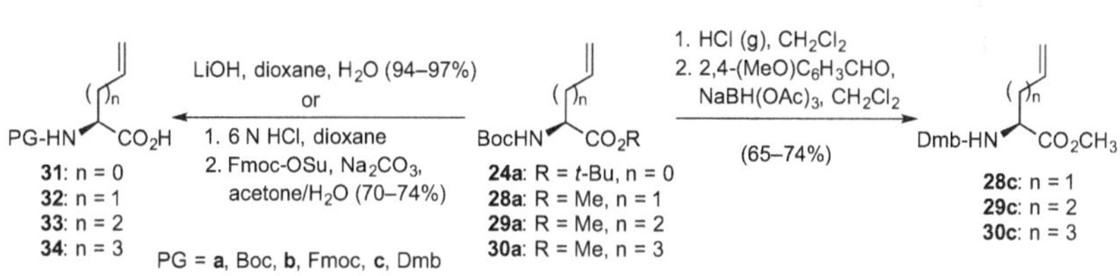

SCHEME 6.4 Protecting group shuffle on ω-unsaturated α-amino acid building blocks.

2-Aminohex-5-enoic acid (homoallylglycine) has been made by several routes, among which the method developed by Jackson was particularly effective on multiple gram scales [48]. Employing β-iodoalaninate **27** as starting material, *N*-Boc-homoallylglycine methyl ester (**29a**) was prepared by insertion of zinc followed by copper-catalyzed cross-coupling to allyl chloride.

2-Aminohept-6-enoic acid (homohomoallylglycine) has been synthesized using a few methods [49]. Considering the capacity to prepare protected allylglycine on multiple gram scales as described above, a method was developed to employ this enantiomerically pure building block in a two-carbon homologation [39]. Hydroboration of allylglycine **28a** with 9-borabicyclo [3.3.1]nonane (9-BBN) and cross-coupling of the resulting borane intermediate to vinyl bromide under Suzuki conditions afforded *N*-Boc-homohomoallylglycine methyl ester (**30a**) on gram scale without racemization.

With the various ω-olefin amino esters in hand, protection group exchanges were performed to provide suitable building blocks for the subsequent peptide coupling—RCM-TAC sequence to access azabicyclo[X.Y.0] alkanone dipeptide mimics (Scheme 6.4). For example, Boc-protected ω-olefin amino acid methyl esters **28a**−**30a** were respectively saponified using LiOH to give the corresponding acids **32a**−**34a**. Conversion to Fmoc-protected amino acid counterparts **32b**−**34b** was achieved by a complete deprotection of *N*-(Boc)amino esters **28a–30a** using 6N HCl followed by reprotection with Fmoc-OSu. The Boc protecting group was selectively cleaved from **28a–30a** with HCl gas in dichloromethane and after free-basing, the amine was reacted with 2,4-dimethoxybenzaldehyde and subjected to reductive amination using $NaBH(OAc)_3$ to provide secondary amino esters **28c–30c**. The acid labile 2,4-dimethoxybenzyl (Dmb) group was used to lower the barrier for tertiary amide *cis–trans* isomerization to favor cyclization to smaller lactam rings (*vide infra*, Scheme 6.1).

6.2.2 Dipeptide formation and ring-closing metathesis

Formation of the dipeptides (e.g., **35** and **36**) required for RCM was achieved from the appropriate protected ω-olefin amino acid (e.g., **31–34**, PG = Boc and Fmoc) and ester (e.g., **28–30**, PG = H and Dmb) building blocks in good yields (55%–87%) using the coupling reagent 1-[*bis*(dimethylamino) methylene]-1H-1,2,3-triazolo[4,5-b]pyridinium 3-oxide hexafluoro-phosphate (HATU) and *N*-ethylmorpholine in DCM (Scheme 6.5).

With the linear dipeptide in hand, RCM was performed using the first-generation of Grubbs catalyst (Schemes 6.6 and 6.7) [50]. In the formation of the 10-membered unsaturated lactam, cyclization of the secondary and tertiary amides **37** (**38**) and **41h** (**42h**) gave respectively *E*- and *Z*-olefins **39** (**40**) and **43h** (**44h**) [38,51]. Smaller ring lactams could not be formed from secondary amides, but tertiary amides reacted to provide *Z*-unsaturated lactams in good yields, due in part to the Dmb group lowering the barrier for *cis–trans* amide isomerization to bring the olefinic ends in closer proximity. The olefin geometry in 7- to 10-membered lactams **39**, **40**, **43d–i** and **44a–h** was ascertained using X-ray crystallography and NMR spectroscopy [38,39].

The impact of the amine protecting group was studied in the transannular cyclization of the 9-membered lactam. The corresponding hydrochloride salt **47** was obtained upon treatment of *N*-(Boc)amino lactam **43e** with gaseous HCl in DCM, and protected subsequently with a series of different groups (Scheme 6.8). Phthalimide (Phth) **48** and 2-nitrobenzenesulfonylamide (*o*-NBS)

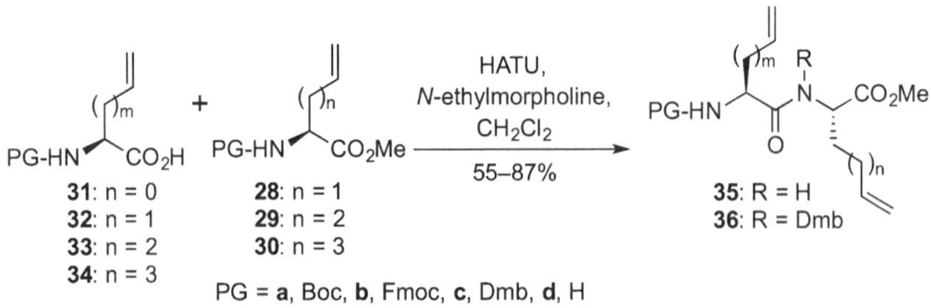

SCHEME 6.5 General approach for dipeptide synthesis.

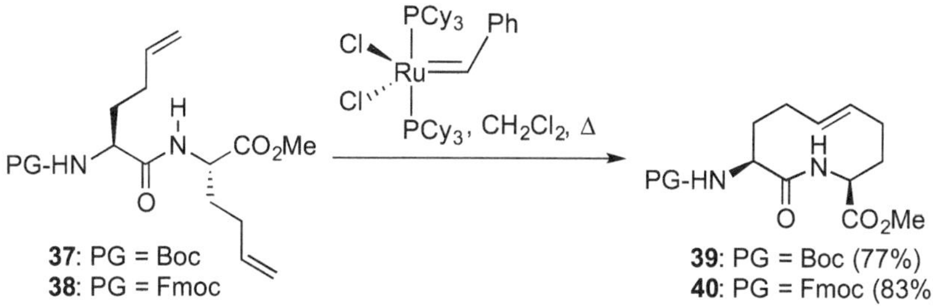

SCHEME 6.6 Ten-membered *E*-unsaturated lactam synthesis.

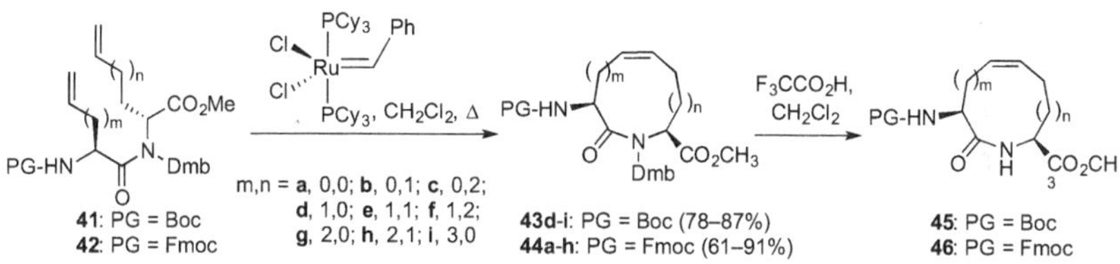

SCHEME 6.7 *Z*-Unsaturated lactam synthesis by ring-closing metathesis of Dmb-dipeptides.

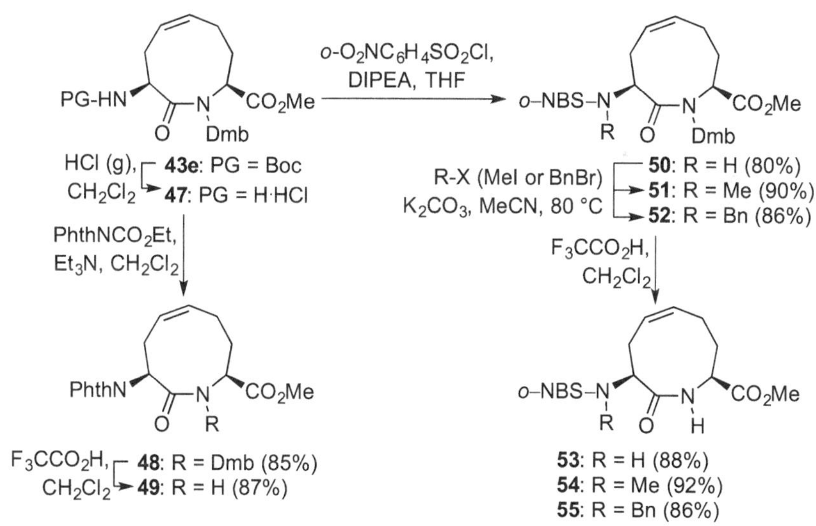

SCHEME 6.8 Exocyclic amine modification of lactam **43e**.

50 were respectively obtained in 85% and 80% yields on treatment of hydrochloride **47** with *N*-carbethoxyphthalimide and 2-nitrobenzenesulfonyl chloride and a tertiary amine in MeCN and THF. The 9-fluorenylmethyl carbamate (Fmoc) **44e** was also obtained upon treatment with Fmoc-OSu and Na_2CO_3 in acetone/water in a 75% yield. Alkylation of sulfonamide **50** with methyl iodide and benzyl bromide with K_2CO_3 in MeCN at 80°C gave respectively *N*-methyl and *N*-benzyl sulfonamides **51** and **52** in 90% and 86% yields. The Dmb group was removed from macrocycles **48** and **50–52** using 50% TFA in DCM to afford macrocycles **49** and **53–55** in 86 – 92% yields.

6.2.3 Transannular cyclization

With a robust method in hand, unsaturated lactams were assembled possessing 8- to 10-membered rings, an olefin located at different positions, and a series of amine and amide protection strategies. Electrophilic TAC was next examined to prepare different azabicyclo[X.Y.0]alkanone dipeptides. The ring size, site of unsaturation, amine and amide protection, and reaction conditions, all played influential roles in the transannular cyclization, which could typically be used to provide selectively specific bicyclic systems.

Two 10-membered lactams were studied in the transannular cyclization (Fig. 6.4). For example, *N*-(Boc)amino *E*-Δ^6–10-membered lactam **39** was converted to *N*-(Boc)amino pyrroloazepinone (3*S*,6*R*,7*S*,10*S*)-**56** in 48% yield using iodine in THF at reflux followed by reprotection with Boc_2O and triethylamine in dichloromethane [40]. X-ray crystallography confirmed the stereochemistry of the ring fusion and iodine-bearing carbons of the azabicyclo[5.3.0]alkanone ring system (3*S*,6*R*,7*S*,10*S*)-**56**. To avoid reprotection, $NaHCO_3$ was added to the cyclization conditions, which afforded stereo- and regio-selectively (3*S*,6*R*,7*S*,10*S*)-**56** in 62% yield [52]. The same stereochemical outcome was obtained upon replacement of the Boc-group with Fmoc-protection, which provided respectively bicycle (3*S*,6*R*,7*S*,10*S*)-**57** in 45% and 67% yields using iodine in THF at reflux in the absence and presence of $NaHCO_3$. Alternatively, azabicyclo[4.4.0]alkanone bicycle (3*S*,5*R*,6*R*,10*S*)-**58** was obtained from *Z*-Δ^5–10-membered lactam **44f**. Treatment of Dmb protected lactam **44f** with iodine in THF at reflux provided stereo- and regio-selectively quinolizidinone (3*S*, 5*R*,6*R*,10*S*)-**58** in 52% yield.

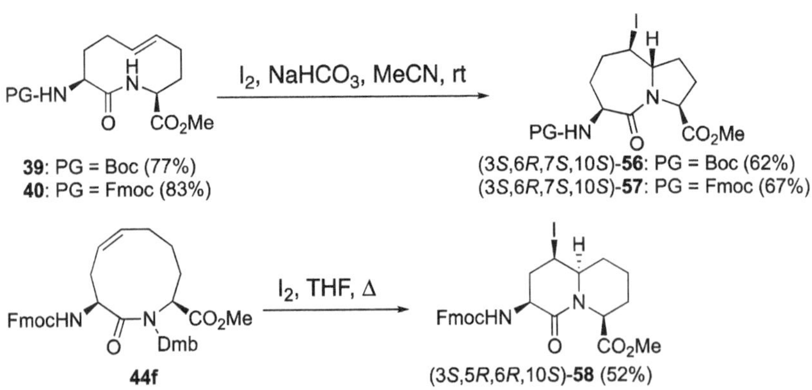

FIGURE 6.4 Transannular cyclization of 10-membered lactams **39**, **40**, and **44f**.

Two different unsaturated 9-membered lactams reacted to effectively furnish azabicyclo[X.Y.0]alkanone ring systems by way of transannular cyclization (Fig. 6.5). Indolizidin-2- and 9-one amino esters were synthesized with various degrees of regio- and stereo-control contingent on olefin location and amine and amide protection. For example, *Z*-Δ^5–9-membered tertiary lactam **44e** reacted with iodine in THF at reflux to provide 5-iodo indolizidin-2-one (3*S*,5*R*,6*R*,9*S*)-**59a** as a single isomer in 86% yield [40]. Moreover, Dmb analog **44e** reacted with phenylselenyl bromide, instead of iodine, as electrophile for initiating cyclization to provide selectively phenylselenyl indolizidine-2-one (3*S*,5*R*,6*R*,9*S*)-**60** in 79% yield. On the other hand without Dmb, *Z*-Δ^5–9-membered secondary lactam **46e** reacted under similar conditions with iodine in THF at reflux to yield two diastereoisomeric iodides (3*S*,5*R*,6*R*,9*S*)- and (3*S*,5*S*,6*S*,9*S*)-**59a** in 47% and 26% yields, respectively [40]. Secondary lactam **46e** reacted, however, with iodine in the presence of (diacetoxy)iodobenzene (DIB) in MeCN to yield selectively 5-iodo indolizidin-2-one diastereoisomer (3*S*,5*S*,6*S*,9*S*)-**59a** in 89% yield. Changing the exocyclic amine protection from a Fmoc to a phthalimide group offered entry to the 5-iodo indolizidin-9-one regio-isomer (3*S*,5*R*,6*R*,9S)-**61** [42]. For example, treatment of phthalimide *Z*-Δ^5–9-membered lactam **48** with iodine and DIB in MeCN gave 5-iodo indolizidin-2-one (3*S*,5*S*,6*S*,9*S*)-**59b**

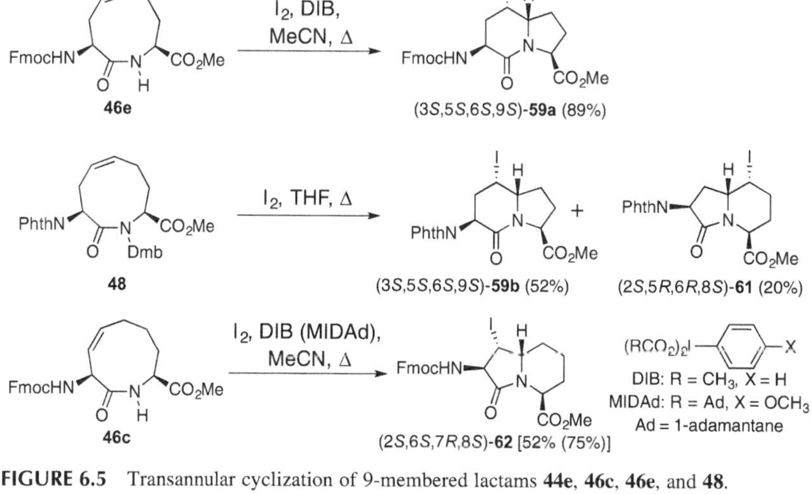

FIGURE 6.5 Transannular cyclization of 9-membered lactams **44e**, **46c**, **46e**, and **48**.

and 5-iodo indolizidin-9-one (2*S*,5*R*,6*R*,8S)-**61** in 52% and 20% yields, respectively.

7-Iodo indolizidin-9-one (2*S*,6*S*,7*R*,8S)-**62** was synthesized from *Z*-Δ^4–9-membered secondary lactam **46c** by cyclization using iodine and DIB in MeCN in 52% yield. Without DIB, secondary lactam **46c** did not cyclize using iodine alone [41]. Moreover, the yield of 7-iodo indolizidin-9-one (2*S*,6*S*,7*R*,8S)-**62** was raised to 75% by switching from DIB to the more electron-rich and sterically bulky 4-methoxy-iodosobenzene di(adamantane-1-carboxylate) (MIDAd) as an additive in the iodolactamization [41].

Two different unsaturated 8-membered lactams underwent transannular cyclization to yield azabicyclo[X.Y.0]alkanone ring systems (Fig. 6.6). In both cases, hypervalent iodine species were essential additives for inducing cyclization. For example, *Z*-Δ^4–8-membered lactam **46b** reacted with DIB and MIDAd as additives with iodine in MeCN to provide pyrrolizidinone (3*R*,4*S*,5*S*,8S)-**62** in 62% and 78% yields, respectively. On the other hand, *Z*-Δ^5–8-membered lactam **46d** reacted with iodine and MIDAd in MeCN at reflux to provide three bicycles: pyrrolizidinones (3*S*,5*R*,6*R*,8S)- and (3*S*,5*S*,6*S*,8S)-**64** as well as fused 6,4-bicycle (3*S*,5*S*,6*S*,8S)-**65** in 21%, 19%, and 8% yields, respectively.

6.3 Installation of ring substituents to mimic side chains

Sidechains play important roles in the recognition of peptide structures. In peptide mimicry with azabicyclo[X.Y.0]alkanone amino acids, methods that enable the introduction of ring substituents offer enhanced utility by providing means to replicate both backbone conformation and side-chain function. Ideally, multiple ring substituents may be introduced onto the heterocycle skeleton by way of a common intermediate. In the RCM-TAC strategy for

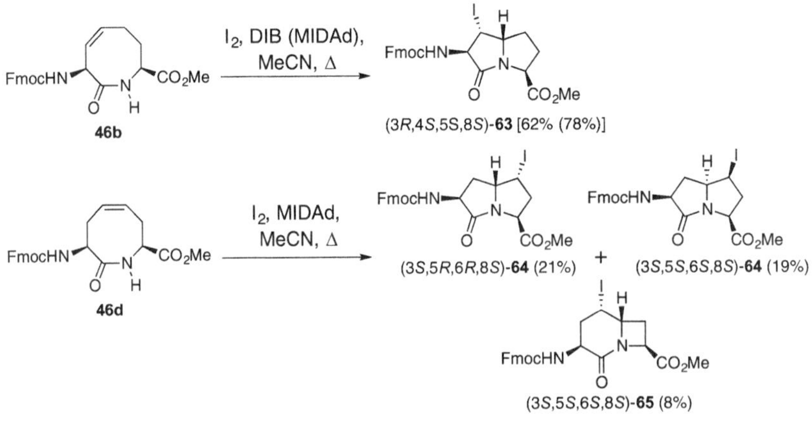

FIGURE 6.6 Transannular cyclization of 8-membered lactams **46b** and **46d**.

azabicyclo[X.Y.0]alkanone dipeptide mimic synthesis, side chains have been added to the resulting iodo bicycles by way of routes featuring nucleophilic substitution and elimination of the iodide followed by functionalization of the resulting olefin. In addition, syntheses of substituted ω-olefin amino acids offer potential building blocks for introducing side chains onto bicycles.

6.3.1 Nucleophilic substitution

A general method for ring substituent introduction was developed by way of diastereoselective iodide displacement from 6-iodo-pyrroloazepin-2-ones (e.g., **56** and **57**) using various silver salts in S_N1 reactions [52]. Silver salts avoided elimination and favored ionization for nucleophilic attack. For example, silver trifluoromethanesulfonate served as a nonnucleophilic means for iodide ionization. Various alcohols and water reacted with iodides **56** and **57** in acetonitrile by way of a carbocation intermediate to provide the corresponding 6-alkoxy and hydroxy azabicyclo[5.3.0]alkanones (e.g., **66a–e**) in 77%–91% yields (Scheme 6.9). The reaction proceeded with net retention of the iodide stereochemistry due to attack from the least hindered face of the bicyclic carbocation intermediate. The less nucleophilic phenol reacted as a solvent with iodide and AgOTf at 50°C to give a 53% yield of phenyl ether **66f**. Silver azide and silver nitrate reacted with iodide in acetonitrile at

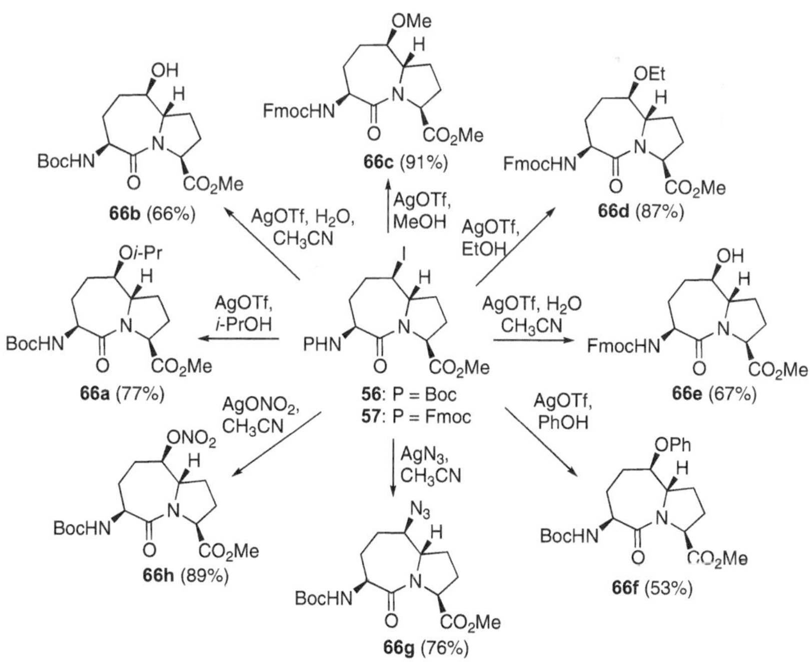

SCHEME 6.9 Iodide displacement by S_N1 reactions using silver salts.

60°C and room temperature to provide azide **66g** and nitrate **66h** in 76% and 89% yields, respectively. On the other hand, attempts were unsuccessful using silver cyanide and silver thiocyanate under various conditions including sonication, high temperature and microwave irradiation, likely due to their low dissociation constants [52].

6.3.2 Adding side-chain functionality by elimination

β-Elimination of iodide to provide enamines (e.g., **68** and **71**) followed by olefin functionalization has given entry to various ring-substituted bicycles from 5-iodo indolizidin-2-ones (e.g., **59a** and **67**) [31]. For example, indolizidin-2-ones with different ring fusion configurations were synthesized by olefin reduction. Diverse aryl substituents were appended to the olefin using palladium-catalyzed oxidative Heck chemistry. In addition, allylic oxidation has been used to obtain 5- and 7-hydroxy Δ^5-indolizidine-2-ones.

β-Elimination of *N*-Fmoc protected iodide (3*S*,5*R*,6*R*,9*S*)-**59a** with triethylamine in acetonitrile at 80°C caused concomitant removal of the amine protection and double bond formation (Scheme 6.10) [31]. Enamine **68** was obtained after *N*-protection with Boc_2O. Hydrogenation of enamine **68** with Pd/C in EtOH gave concave isomer (6*S*)-**69**. Under similar elimination conditions, *N*-phenylacetyl protected iodide **67** was converted to enamine **71**. Hydrogenolysis of benzyl ester **71** afforded enamine acid **72** contaminated with concave indolizidine-2-one (6*S*)-**74**. Hydride reduction of enamine **71** using sodium cyanoborohydride in EtOH in presence of catalytic acetic acid gave a 3:1 mixture of indolizidine-2-one diastereomers (6*R*)- and (6*S*)-**73**. Benzyl ester hydrogenolysis provided the corresponding acid (6*R*)-**74** in 91% yield.

Olefin **71** acted as a common precursor in a series of C-C bond forming reactions featuring oxidative Heck reactions using palladium acetate as a

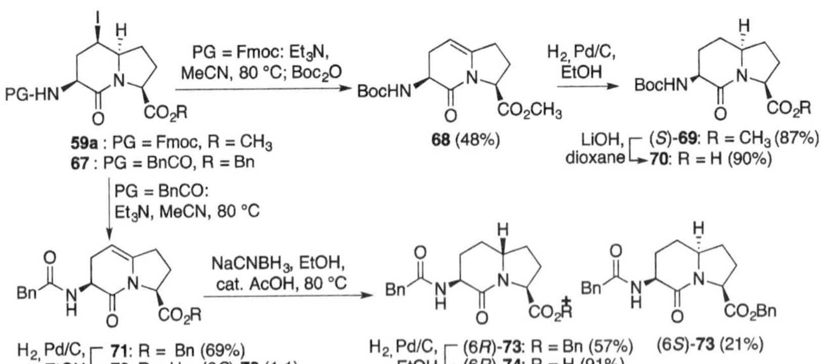

SCHEME 6.10 Access to unsaturated and saturated indolizidin-2-ones by β-elimination and reduction.

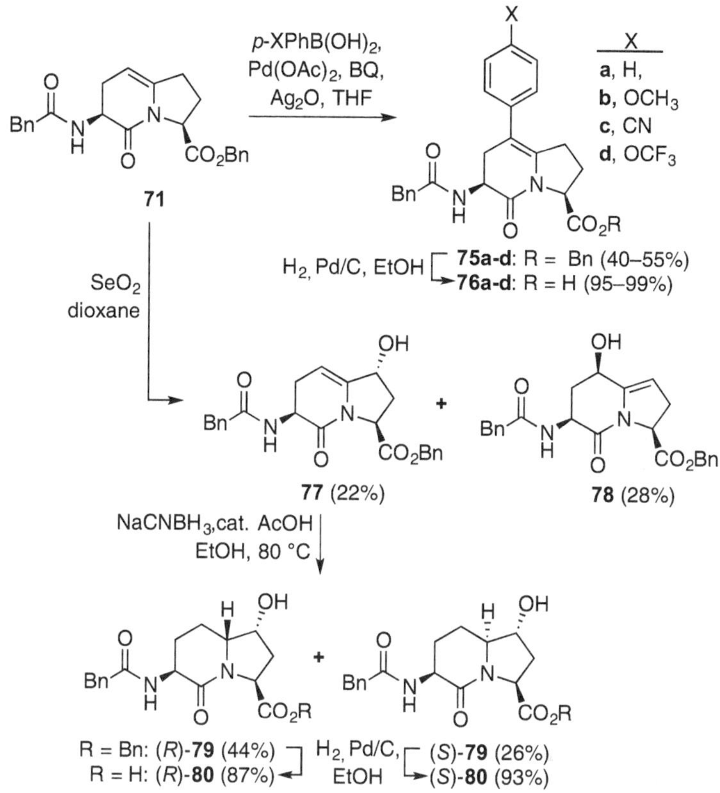

SCHEME 6.11 Introduction of side-chain diversity by olefin functionalization.

catalyst and a set of aromatic boronic acids (Scheme 6.11). 5-Aryl-Δ^5-indolizidin-2-ones **75a–d** were obtained in 40%–55% yields from enamine **71** using electron-rich and poor aryl boronates, $Pd(OAc)_2$, silver oxide and 1,4-benzoquinone (BQ) in THF at 80°C [31,53]. Allylic oxidation of olefin **71** using selenium dioxide and *tert*-butyl hydroperoxide (TBHP) in dioxane produced a separable mixture of 5-hydroxy-Δ^6- and 7-hydroxy-Δ^5-indolizidin-2-ones **78** and **77** in 28% and 22% yields, respectively [31,54]. Reduction of 7-hydroxy-Δ^5-indolizidin-2-one **77** using sodium cyanoborohydride and catalytic acetic acid in EtOH at 80°C gave separable concave and convex diastereoisomers (6*R*)- and (6*S*)-**79** in 44% and 26% yields, respectively. Carboxylic acids **76a–d** and **80** were prepared by benzyl ester hydrogenolysis in 87%–99% yields.

6.3.3 Synthesis and application of substituted ω-olefin amino acids

ω-Olefin amino acids are the key building blocks in the RCM-TAC strategy for making azabicyclo[X.Y.0]alkanone amino acids. In principle, their substituted variants may be employed to provide bicycles possessing ring substituents. Substituted ω-unsaturated amino acids have been synthesized by a variety of approaches [48–62]. For example, β-allyl aspartates (2*S*,3*R*)-**83** and (2*S*,3*S*)-**84** have been respectively prepared from diastereoselective alkylation of *N*-(Cbz)aspartate diesters **81** and **82** (Scheme 6.12) [58,59]. A series of β-allyl aspartates (2*S*,3*R*)-**83a–e** were produced with high diastereoselectivity by employing a variety of allyl halides in alkylation of a dianion intermediate generated with two equivalents of LiHMDS in THF using HMPA (or DMPU) as determinant additive [58]. Alternatively, (2*S*,3*S*)-β-allyl aspartate **84** was prepared effectively by way of a phenylselenyl intermediate and diastereoselective radical chemistry using allyltributylstannane and azobisisobutyronitrile (AIBN) [59].

4-Cloromethyl-2-aminohexenoates (2*S*,4*R*)- and (2*S*,4*S*)-**85** were synthesized in four steps from L-serine by a route featuring copper-catalyzed S_N2' allylation of 1,4-dichlorobut-2-ene using the zincate derived from *N*-(Boc)-β-iodoalanine **27** (Scheme 6.13) [55,60]. Intramolecular chloride displacement gave efficient access to enantiopure vinylprolines (2*S*,4*R*)- and (2*S*,4*S*)-**86** [60]. Alternatively, vinylproline was synthesized in eight steps from (2*S*,4*R*)-*trans*-4-hydroxproline [61].

Substituted ω-olefin amino acids have been examined in a few syntheses of azabicyclo[X.Y.0]alkanone amino acids. For example, β-allyl aspartates

SCHEME 6.12 Aspartate β-allylation with 1,2 asymmetric induction.

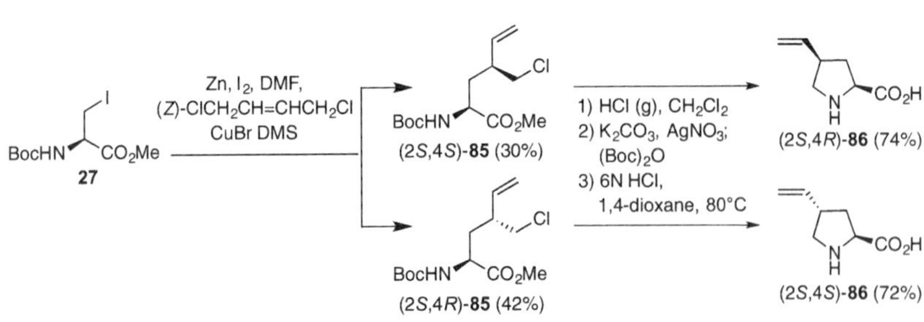

SCHEME 6.13 Synthesis of chloromethyl-aminohexenoates and vinylprolines from β-iodoalanine.

(2*S*,3*R*)- and (2*S*,3*S*)-**87** were respectively employed in the syntheses of thiaindolizidinones (2*S*,5*RS*,7*R*,8*S*)- and (2*S*,5*R*,7*S*,8*S*)-**89**, which may serve as constrained Asp-Gly dipeptide mimics [62]. Olefin oxidation using OsO_4 and $NaIO_4$ followed by condensation of the resulting aldehydes (2*S*,3*R*)- and (2*S*,3*S*)-**88** with cysteine followed by esterification and lactam formation provided thiaindolizidinones by way of thiazolidine intermediates (Scheme 6.14). Ring-fusion stereochemistry was dictated by β-allyl aspartate-derived semi-aldehyde **88**: (2*S*,3*R*)-**88** gave a 2:1 mixture of (2*S*,5*R*,7*R*,8*S*)- and (2*S*,5*S*,7*R*,8*S*)-**89**; (2*S*,3*S*)-**88** yielded selectively (2*S*,5*R*,7*S*,8*S*)-**89**. In addition, α-helix inducer tricycle **93** has been synthesized from 4- and 5-vinylprolines **90** and **91** by way of a route featuring peptide coupling and RCM (Scheme 6.15) [63].

6.4 Conformational analysis of unsaturated lactam and azabicyclo[X.Y.0]alkanone amino esters by X-ray crystallography illustrates relationships with β-turn geometry

X-ray crystallographic and spectroscopic analyses of the unsaturated lactam and bicyclic amino acid analogs have demonstrated the potential for these

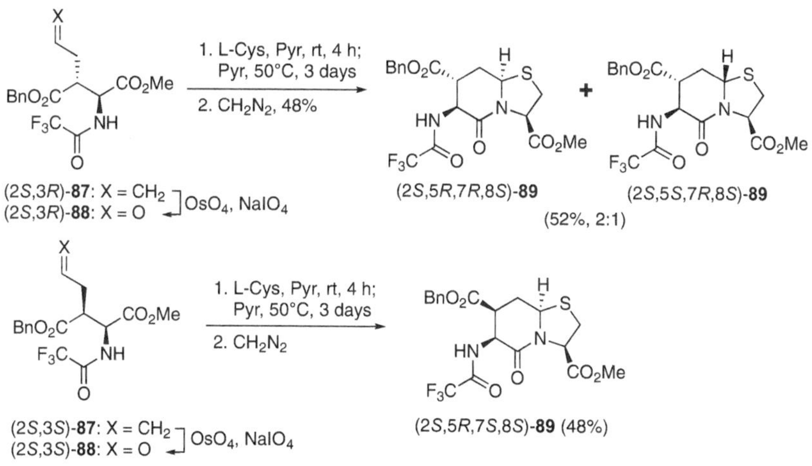

SCHEME 6.14 Constrained Asp-Gly mimic syntheses from β-allyl aspartates.

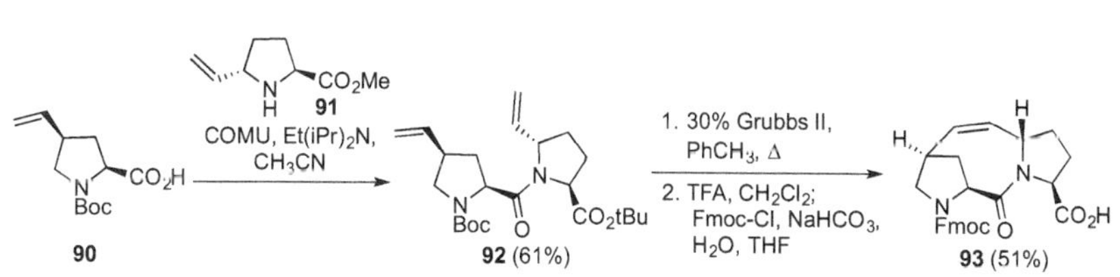

SCHEME 6.15 Synthesis of α-helix inducer tricycle from 5-vinylproline.

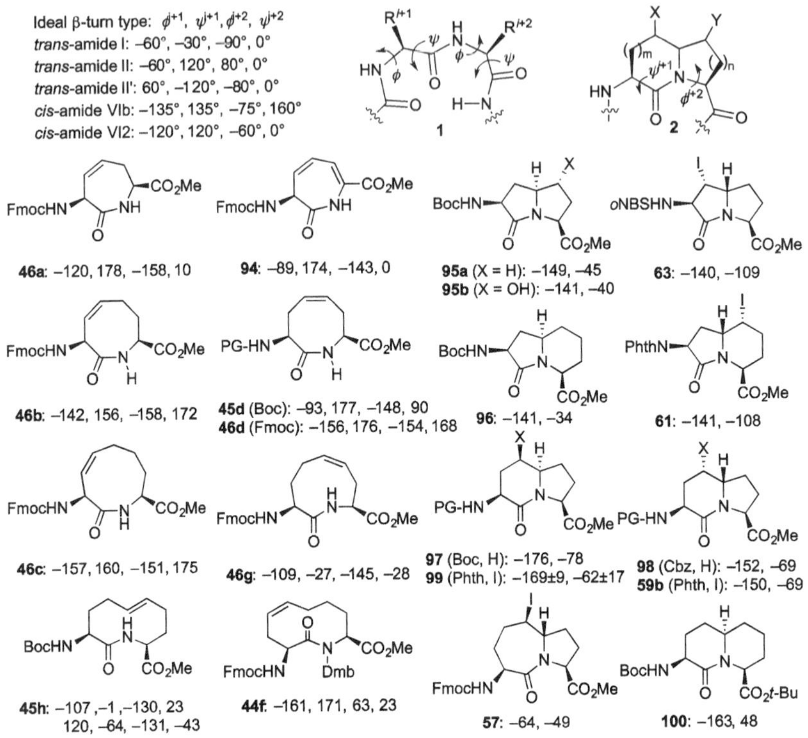

FIGURE 6.7 Comparison of dihedral angle values of ideal β-turns, lactams and bicycles [39–42,64].

constrained dipeptides to mimic the central residues of ideal types of β-turn secondary structures (Fig. 6.7). The unsaturated lactam and bicycle constraints, both can serve to rigidify the backbone for studying peptide conformation-activity relationships. Moreover, substituents on the azabicyclo[X.Y.0]alkanone may mimic the orientation and function of peptide side chains.

6.4.1 Crystal structures of 7–10-member unsaturated lactams

In most cases, 7–10-member lactams adopt *E-cis*-amide geometry (Fig. 6.7). On the other hand, 9- and 10-member lactams **46g** and **44f** exhibited *Z-trans*-amide bonds in the crystalline state. In the *E-cis*-amide analogs, the dihedral angles of 8- (e.g., **46d, 46b**) and 9-membered (**46c**) lactams compared well with those of the central residues of ideal type VI β-turns (except for ϕ^{i+2}). The torsion angles of 7-member lactam **46a** reflected values of an ideal type VI2 β-turn. A geometry similar to an ideal type I β-turn featuring a *Z-trans*-amide was adopted by 9-member lactam **46g**. Moreover, 10-member lactams **45h** and **44f**, which possess respectively *trans-* and *cis*-olefin geometry exhibited backbone dihedral angles similar to those of ideal type I and II β-turns, but the dihedral angles of the *N*-terminal residue of the latter resembled an extended parallel β-pleated sheet.

6.4.2 Crystal structures of azabicyclo[X.Y.0]alkanones

Intra-cyclic dihedral angle values [ψ^{i+1} and ϕ^{i+2}] from X-ray analyses were used to compare different azabicyclo[X.Y.0]alkanones having consistent L,L (*S*,*S*)- or D,D (*R*,*R*)-dipeptide backbone configuration to minimize ambiguities from influences of protection groups and crystal packing effects on backbone geometry. Backbone geometry differed subtly contingent upon heterocycle size, configuration, and ring substituent (Fig. 6.7). In the cases of indolizidin-2- and -9-one systems the backbone geometry exhibited torsion angle values similar to the central residue of an ideal type II′ β-turn ($-120 \pm 55°$ and -80 ± 46) with significant variations contingent on ring fusion and substituents [42]. Other bicycles hovered similarly around type II′ β-turn dihedral angle geometry; however, iodo pyrroloazepinone **57** adopted values more closely in line with the central residues of a type I β-turn.

6.5 Biomedical application of azabicyclo[X.Y.0]alkanone dipeptide mimics as prostaglandin F2α receptor modulators to delay preterm birth

Preterm birth (PTB, birth before 37 weeks of gestation) is associated with high mortality and morbidity rates [65]. An unmet healthcare urgency with the highest per patient expense, PTB is estimated to cost employer-sponsored health plans $6 billion annually [66]. Each year there are 15 million babies born prematurely globally, resulting in 1.1 million deaths, most occurring in low and middle-income countries [65]. Investment in research is needed to both understand the factors implicated in premature labor and to develop effective treatments for prolonging gestation.

Pregnant women in premature labor are routinely treated with labor delaying drugs called tocolytic agents, which have generally short term effects with potential side effects to mother and infant: beta-mimetics, calcium channel blockers, oxytocin receptor blockers, magnesium sulfate, and prostaglandin inhibitors [67]. Contemporary tocolytic agents are typically used to stop premature contractions for a maximum of 48 h, enough time to place mothers into special care units and begin treatment with antenatal corticosteroids to reduce risks of infant mortality and morbidity [67,68]. Offering a temporary respite in contractions, such tocolytic drugs do not prolong gestation and may incur adverse effects to mother and infant [68].

6.5.1 Targeting the prostaglandin F2α receptor

In the search for an efficient specific tocolytic agent to arrest premature uterine contractions and prolong gestation without side effects, the prostaglandin

F2α (PGF2α) receptor (FP) was targeted because of relevant roles in labor and childbirth [30,69,70]. Notably, FP expression increases at the initial phases of parturition [69]. Moreover, FP knock-out mice do not go into labor and must be delivered by the caesarian section [70]. In addition, FP is a member of the G-protein-coupled receptor (GPCR) family which are usually regulated by allosterism [71,72].

Conceptualizing that portions of the receptor could serve to modulate the protein by an allosteric mechanism of action, a set of peptide fragments derived from the second extracellular loop of FP were synthesized and tested [73]. Previously, extracellular loops of GPCRs were shown to be important for ligand recognition and allosteric modulation [74]. Short peptides derived from the second extracellular loop were hypothesized to engage the receptor and interfere with dynamic conformational changes in the presence of the native endogenous "orthosteric" ligand, PGF2α, causing biased signaling that would negatively allosterically modulate pathways to contractions [30].

Among the peptides tested, an L-peptide (L-**101**, ILGHRDYK, Fig. 6.8) exhibited activity with a short half-life [75]. The enantiomeric sequence was examined to enhance metabolic stability, and the D-peptide PDC113 (**101**, ilGhrdyk, small letters represent D-amino acids) exhibited inhibitory activity in a porcine eye vessel microvascular contraction assay and reduced more than 80% of the vasomotor response induced by PGF2α with an IC_{50} of 340 nmol/L [73]. Based on the teachings of Rudinger [76], investigation of the side chain pharmacophores responsible for activity led to the replacement of D-arginine with D-citrulline in PDC31 (**102**), which inhibited >85% of PGF2α-induced micro-vessel contractility with an IC_{50} of 13 nmol/L as well as exhibited efficacy in inhibiting uterine contractions and delaying labor in induced mice, ovine and bovine models [30,77]. In a Phase 1b clinical trial, peptide **102** was shown to reduce safely intra-uterine pressure in healthy women treated for primary dysmenorrhea featuring prostaglandin-dependant excessive uterine contractility leading to severe pain during menstruation [78].

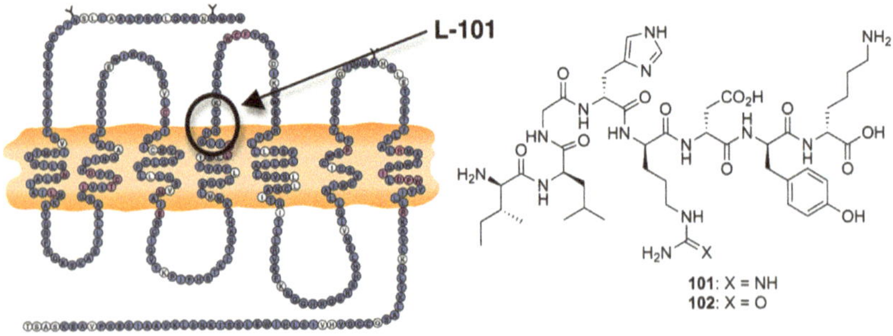

FIGURE 6.8 Lead peptides **101** and **102** were derived from the extracellular loop of FP. *Adapted from Bourguet CB, Claing A, Laporte SA, Hébert TE, Chemtob S, Lubell WD. Synthesis of azabicycloalkanone amino acid and azapeptide mimics and their application as modulators of the prostaglandin F2α receptor for delaying preterm birth. Can J Chem 2014;92:1031–40.*

6.5.2 From peptide to azabicyclo[X.Y.0]alkanone mimetics that delay labor in mice

Study of structure–activity relationships in the interests of gaining insight into the biologically-active backbone conformation and the mechanism of action of peptide **102**, as well as enhancing pharmacokinetic properties, led to the application of azabicyclo[X.Y.0]alkanone amino acids in pursuit of a peptide mimic. Substitution of bicyclic amino acids for the Gly-D-His region was pursued in part due to earlier reports of this dipeptide situating at the central positions of a biologically active β-turn conformation in bombesin [79]. Among a set of fused ring analogs which were shown by X-ray crystallography to adopt dihedral angle geometries similar to the central residues of ideal β-turns (Fig. 6.7) [29–31,80], the most active possessed the (2*S*,6*S*,9*S*)-indolizidin-2-one amino acid residue [30]. Subsequent medicinal chemistry led to replacement of the lipophilic D-Ile-D-Leu dipeptide by a phenylacetyl group, and conversion of the D-Arg(D-Cit)-D-Asp-D-Tyr-D-Lys sequence into a pyridylalaninyl-β-homophenylalanine surrogate to deliver PDC113.824 (**103**, Fig. 6.8), which reduced smooth muscle contraction (98%) with high potency ($IC_{50} = 1.1$ nM) ex vivo. Moreover, indolizidin-2-one **103** delayed labor up to ≈20 h in mice induced with lipopolysaccharide (LPS) as a mimic of bacterial pathogen-associated molecular patterns during infection [31,81], which are attributed to 40%–70% of all PTBs [82,83]. Indolizidin-2-one **103** is a promising small-molecule tocolytic agent and a useful tool for studying the allosteric mechanism of action for modulation of FP signaling pathways implicated in PTB [81]. Although inactive in the absence of the endogenous orthosteric ligand (PGF2α), indolizidin-2-one **103** influenced the rate of exchange of labeled and unlabeled PGF2α, and in the presence of PGF2α, acted respectively as a positive allosteric modulator (PAM) of the $G_{\alpha q}$-extracellular signal-regulated kinase-1/2 (ERK-1/2, mitogen-activated protein, MAP) cascade and as a negative allosteric modulator (NAM) of the $G_{\alpha 12/13}$-Rho-associated protein kinase (ROCK) signaling pathway leading to cytoskeletal reorganization [81].

6.6 Conclusion

Application of azabicyclo[X.Y.0]alkanone amino acids has accelerated translational research to convert peptide leads into drug-like prototypes. Proven effective for exploring structure–activity relationships, these constrained dipeptide surrogates can dissect out active conformers and orient pharmacophores to improve selectivity and enhance receptor engagement. Modular assembly of these tools for peptide mimicry has been achieved by a peptide coupling/RCM-TAC strategy. Commencing with effective syntheses of ω-unsaturated amino acid building blocks of four to seven carbon chain lengths using methionine and serine as inexpensive chiral precursors, a versatile synthetic strategy has proven effective for preparing novel iodo

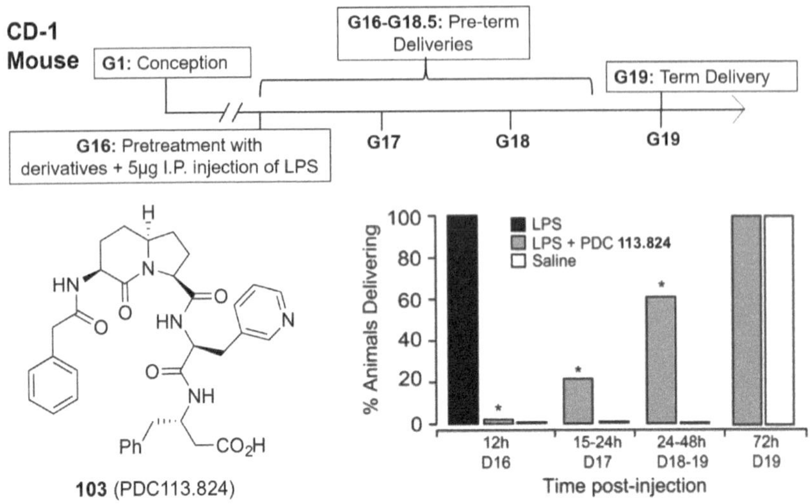

FIGURE 6.9 Indolizidin-2-one **103** (PDC113.824) delayed labor up to ≈20 h in LPS-induced mice. At 16 days of gestation, timed-pregnant (CD-1) mice were treated with either saline or **103** using osmotic pumps. Within 15 min after placement of the pumps, animals were injected intraperitoneally with lipopolysaccharide (LPS) *Escherichia coli* endotoxin to mimic the inflammatory component of human preterm labor. Animals were inspected every hour for the first 18 h and every 2 h thereafter to document the timing of birth. Data represent $N=5$ or 6 animals per treated group and $N=18$ for the saline-injected controls. $^*p<0.05$ compared to saline treated by Fisher's exact test. *Adapted from Goupil E, Tassy D, Bourguet C, Quiniou C, Wisehart V, Pétrin D, et al. A novel biased allosteric compound inhibitor of parturition selectively impedes the prostaglandin F2α-mediated Rho/ROCK signaling pathway. J Biol Chem 2010;285: 25624–36.*

substituted pyrrolizidinone, indolizidin-2-one, indolizidin-9-one, quinolizidinone, and pyrroloazepinone amino acid analogs by way of transannular cyclization on isomeric unsaturated 8-, 9-, and 10-membered lactams (Figs. 6.9 and 6.10). The use of hypervalent(III)iodine additives improved yields, enabled cyclization of smaller-ring unsaturated lactams, and influenced stereochemical outcomes in the transannular cyclization, which was achieved by amine protection and conditions which could be tuned to give stereoselectively specific iodo bicycles with regio-control.

Harnessing the iodide group as a handle for installing ring substituents, a diverse selection of side-chain functional groups has been used for creating bicyclic mimics of various peptides. To address the unmet need of delaying PTB, novel peptide prototypes were designed using constrained mimicry of backbone geometry and side-chain function, and these molecules demonstrated in vivo activity. The modular RCM-TAC method offers broad utility for synthesizing a wide array of bicyclic dipeptide mimics to investigate peptide structures for various biomedical applications.

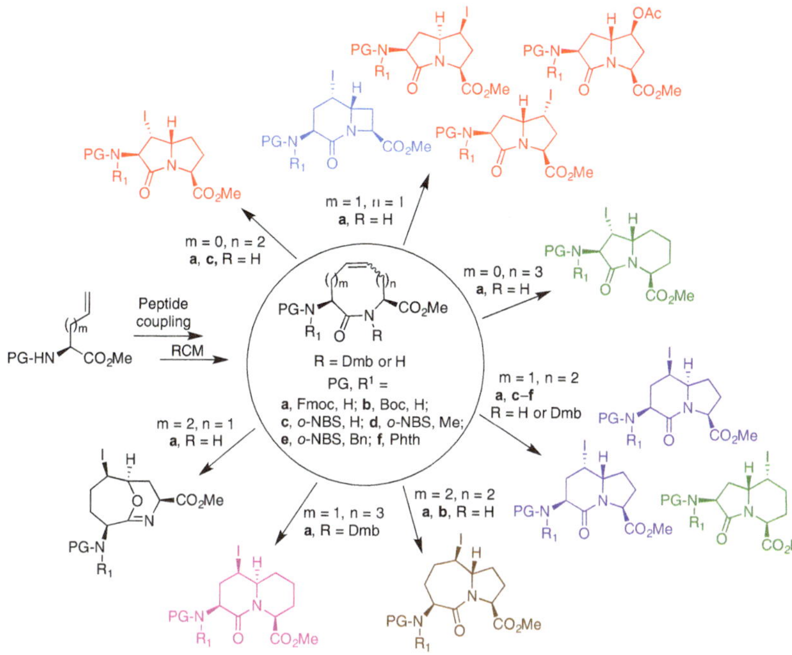

FIGURE 6.10 Graphical representation of various azabicycloalkanone amino acid analogs synthesized using a common ring-closing metathesis/transannular sequence.

Acknowledgments

We thank the Natural Sciences and Engineering Research Council of Canada (NSERC) for funding (Discovery Research Project #04079), the Fonds de recherche nature et technologie Quebec for the Centre in Green Chemistry and Catalysis (FRQNT-2020-RS4–265155-CCVC), and the Canadian Institutes of Health Research (CIHR) and NSERC the for the Collaborative Health Research Project "Treatment of Preterm Birth with Prostaglandin-F2alpha Receptor Modulators" No. 337381. We thank Dr. Xin Hou for assistance in preparing Fig. 6.9.

References

[1] Veber DF, Freidinger RM. The design of metabolically-stable peptide analogs. Trends Neurosci 1985;8:392–6.

[2] Smith III AB, Hirschmann R, Pasternak A, Akaishi R, Guzman MC, Jones DR, et al. Design and synthesis of peptidomimetic inhibitors of HIV-1 protease and renin. Evidence for improved transport. J Med Chem 1994;37:215–18.

[3] Lee VH, Yamamoto A. Penetration and enzymatic barriers to peptide and protein absorption. Adv Drug Deliv Rev 1989;4:171–207.

[4] Pichereau C, Allary C. Therapeutic peptides under the spotlight. Eur Biopharm Rev 2005;5:88–91.

[5] Hruby VJ. Conformational and topographical considerations in the design of biologically active peptides. Biopolymers 1993;33:1073–82.

[6] Vlieghe P, Lisowski V, Martinez J, Khrestchatisky M. Synthetic therapeutic peptides: science and market. Drug Discov Today 2010;15:40–56.

[7] Sawyer TK. Ch. 1: Renaissance in peptide drug discovery: the third wave. In: Peptide-based Drug Discovery: Challenges and New Therapeutics; 2017. p. 1–34. doi:10.1039/9781788011532-00001e; ISBN: 978-1-78801-153-2.

[8] Marshall GR. A hierarchical approach to peptidomimetic design. Tetrahedron 1993;49:3547–58.

[9] Whitby LR, Ando Y, Setola V, Vogt PK, Roth BL, Boger DL. Design, synthesis, and validation of a β-turn mimetic library targeting protein–protein and peptide–receptor interactions. J Am Chem Soc 2011;133:10184–94.

[10] Loughlin WA, Tyndall JD, Glenn MP, Fairlie DP. Beta-strand mimetics. Chem Rev 2004;104:6085–118.

[11] (a) Merritt HI, Sawyer N, Arora PS. Bent into shape: folded peptides to mimic protein structure and modulate protein function. Peptide Science 2020;e24145.
(b) Malde AK, Hill TA, Iyer A, Fairlie DP. Crystal structures of protein-bound cyclic peptides. Chem Rev 2019;119:9861–e29914.

[12] Tyndall JD, Pfeiffer B, Abbenante G, Fairlie DP. Over one hundred peptide-activated G protein-coupled receptors recognize ligands with turn structure. Chem Rev 2005;105:793–826.

[13] Hinds MG, Welsh JH, Brennand DM, Fisher J, Glennie MJ, Richards NG, et al. Synthesis, conformational properties, and antibody recognition of peptides containing beta-turn mimetics based on α-alkylproline derivatives. J Med Chem 1991;34:1777–89.

[14] O'Connor SE, Imperiali B. Conformational switching by asparagine-linked glycosylation. J Am Chem Soc 1997;119:2295–6.

[15] Price JL, Powers DL, Powers ET, Kelly JW. Glycosylation of the enhanced aromatic sequon is similarly stabilizing in three distinct reverse turn contexts. Proc Natl Acad Sci. 108:14127–32.

[16] Otvos Jr. L, Thurin J, Kollat E, Urge L, Mantsch HH, Hollosi M. Glycosylation of synthetic peptides breaks helices: phosphorylation results in distorted structure. Int J Pept Protein Res 1991;38:476–82.

[17] Staneva I, Wallin S. Binding free energy landscape of domain-peptide interactions. PLoS Comput Biol 2011;7:e1002131.

[18] Freidinger RM, Veber DF, Perlow DS, Saperstein R. Bioactive conformation of luteinizing hormone-releasing hormone: evidence from a conformationally constrained analog. Science 1980;210:656–8.

[19] (a) Wyvratt M, Tischler M, Ikeler T, Springer J, Tristram E, Patchett A. In bicyclic inhibitors of angiotensin-converting enzyme, peptides: structure and function. In: Proceedings 8th American Peptide Symposium. Rockford, III, Pierce Chemical Co; 1983;551–4.
(b) Wyvratt MJ, et al. Recent developments in the design of angiotensin-converting enzyme inhibitors. Med Res Rev 1985;5:483–531.

[20] Nagai U, Sato K. Synthesis of a bicyclic dipeptide with the shape of β-turn central part. Tetrahedron Lett 1985;26:647–50.

[21] Nagai U, Sato K, Nakamura R, Kato R. Bicyclic turned dipeptide (BTD) as a β-turn mimetic; its design, synthesis and incorporation into bioactive peptides. Tetrahedron 1993;49:3577–92.

[22] Hanessian S, McNaughton-Smith G, Lombart H-G, Lubell WD. Design and synthesis of conformationally constrained amino acids as versatile scaffolds and peptide mimetics. Tetrahedron 1997;53:12789–854.

[23] Sun H, Lu J, Liu L, Yi H, Qiu S, Yang C-Y, et al. Nonpeptidic and potent small-molecule inhibitors of cIAP-1/2 and XIAP proteins. J Med Chem 2010;53:6361–7.

[24] Cong H, Xu L, Wu Y, Qu Z, Bian T, Zhang W, et al. Inhibitor of apoptosis protein (IAP) antagonists in anticancer agent discovery: current status and perspectives. J Med Chem 2019;62:5750–72.

[25] Halab L, Becker JA, Darula Z, Tourwé D, Kieffer BL, Simonin F, et al. Probing opioid receptor interactions with azacycloalkane amino acids. Synthesis of a potent and selective ORL1 antagonist. J Med Chem 2002;45:5353–7.

[26] Van Cauwenberghe S, Simonin F, Cluzeau J, Becker JA, Lubell WD, Tourwé D. Structure – activity study of the ORL1 antagonist Ac-Arg-D-Cha-Qaa-D-Arg-D-p-ClPhe-NH_2. J Med Chem 2004;47:1864–7.

[27] Robl JA, Cimarusti MP, Simpkins LM, Brown B, Ryono DE, Bird JE, et al. Dual metalloprotease inhibitors. 6. Incorporation of bicyclic and substituted monocyclic azepinones as dipeptide surrogates in angiotensin-converting enzyme/neutral endopeptidase inhibitors. J Med Chem 1996;39:494–502.

[28] Bralet J, Schwartz J-C. Vasopeptidase inhibitors: an emerging class of cardiovascular drugs. Trends Pharmacol Sci 2001;22:106–9.

[29] Bourguet CB, Goupil E, Tassy D, Hou X, Thouin E, Polyak F, et al. Targeting the prostaglandin F2α receptor for preventing preterm labor with azapeptide tocolytics. J Med Chem 2011;54:6085–97.

[30] Bourguet CB, Claing A, Laporte SA, Hébert TE, Chemtob S, Lubell WD. Synthesis of azabicycloalkanone amino acid and azapeptide mimics and their application as modulators of the prostaglandin F2α receptor for delaying preterm birth. Can J Chem 2014;92: 1031–40.

[31] Mir M, Atmuri F, Bourguet NDP, Fores CB, Hou JR, Chemtob X, et al. Utility of aza-amino acyl proline and indolizidinone amino acid residues for peptide mimicry: conception of prostaglandin F2α receptor allosteric modulators that delay preterm birth. J Med Chem 2019;62:4500–25.

[32] (a) Hanessian S, Therrien E, Granberg K, Nilsson I. Targeting thrombin and factor VIIa: design, synthesis, and inhibitory activity of functionally relevant indolizidinones. Bioorg Med Chem Lett 2002;12:2907–11.
(b) Hanessian S, Balaux E, Musil D, Olsson L-L, Nilsson I. Exploring the chiral space within the active site of α-thrombin with a constrained mimic of D-Phe-Pro-Arg—design, synthesis, inhibitory activity, and X-ray structure of an enzyme–inhibitor complex. Bioorg Med Chem Lett 2000;10:243–7.

[33] Cluzeau J, Lubell WD. Design, synthesis, and application of azabicyclo [XY 0] alkanone amino acids as constrained dipeptide surrogates and peptide mimics. Peptide Sci 2005; 80:98–150.

[34] Halab L, Gosselin F, Lubell WD. Design, synthesis, and conformational analysis of azacycloalkane amino acids as conformationally constrained probes for mimicry of peptide secondary structures. Peptide Sci 2000;55:101–22.

[35] (a) Cordero FM, Valenza S, Machetti F, Brandi A. Design and synthesis of a new bicyclic dipeptide isostere. Chem Commun 2001;1590–1.
(b) Fustero S, Mateu N, Albert L, Aceña JL. Straightforward stereoselective access to cyclic peptidomimetics. J Org Chem 2009;74:4429–32.

(c) Vartak AP, Skoblenick K, Thomas N, Mishra RK, Johnson RL. Allosteric modulation of the dopamine receptor by conformationally constrained type VI β-turn peptidomimetics of Pro-Leu-Gly-NH_2. J Med Chem 2007;50:6725–9.

[36] Khashper A, Lubell WD. Design, synthesis, conformational analysis and application of indolizidin-2-one dipeptide mimics. Org Biomol Chem 2014;12:5052–70.

[37] (a) Sudau A, Münch W, Bats JW, Nubbemeyer U. Planar chirality: cycloaddition and transannular reactions of optically active azoninones that contain (E)-olefins. Chem Eur J 2001;7:611–21.

(b) White JD, Hrnciar P. Synthesis of polyhydroxylated pyrrolizidine alkaloids of the alexine family by tandem ring-closing metathesis-transannular cyclization.(+)-Australine. J Org Chem 2000;65:9129–42.

(c) White JD, Hrnciar P, Yokochi AF. Tandem ring-closing metathesis transannular cyclization as a route to hydroxylated pyrrolizidines. Asymmetric synthesis of (+)-Australine. J Am Chem Soc 1998;120:7359–60.

(d) Jensen T, Mikkelsen M, Lauritsen A, Andresen TL, Gotfredsen CH, Madsen R. A concise synthesis of castanospermine by the use of a transannular cyclization. J Org Chem 2009;74:8886–9.

(e) Brock EA, Davies SG, Lee JA, Roberts PM, Thomson JE. Asymmetric synthesis of the tropane alkaloid (+)-pseudococaine via ring-closing iodoamination. Org Lett 2012;14:4278–81.

(f) Brock EA, Davies SG, Lee JA, Roberts PM, Thomson JE. Asymmetric synthesis of polyhydroxylated pyrrolizidines via transannular iodoamination with concomitant N-debenzylation. Org Lett 2011;13:1594–7.

[38] Kaul R, Surprenant S, Lubell WD. Systematic study of the synthesis of macrocyclic dipeptide β-turn mimics possessing 8-, 9-, and 10-membered rings by ring-closing metathesis. J Org Chem 2005;70:3838–44.

[39] Atmuri NDP, Lubell WD. Insight into transannular cyclization reactions to synthesize azabicyclo [XY Z] alkanone amino acid derivatives from 8-, 9-, and 10-membered macrocyclic dipeptide lactams. J Org Chem 2015;80:4904–18.

[40] Surprenant S, Lubell WD. From macrocycle dipeptide lactams to azabicyclo[X.Y. 0]alkanone amino acids: a transannular cyclization route for peptide mimic synthesis. Org Lett 2006;8:2851–4.

[41] Atmuri NDP, Reilley DJ, Lubell WD. Peptidomimetic synthesis by way of diastereoselective iodoacetoxylation and transannular amidation of 7–9-membered lactams. Org Lett 2017;19:5066–9.

[42] Atmuri NDP, Lubell WD. Stereo-and regiochemical transannular cyclization of a common hexahydro-1 H-azonine to afford three different indolizidinone dipeptide mimetics. J Org Chem 2020;85:1340–51.

[43] Gosselin F, Lubell WD. Rigid dipeptide surrogates: syntheses of enantiopure quinolizidinone and pyrroloazepinone amino acids from a common diaminodicarboxylate precursor. J Org Chem 2000;65:2163–71.

[44] (a) Hallinan KO, Crout DHG, Errington W. Simple synthesis of L- and D-vinylglycine (2-aminobut-3-enoic acid) and related amino acids. J Chem Soc Perkin Trans 1994;1:3537–43.

(b) Organ MG, Xu J, N'Zemba B. A modular, general and enantiospecific strategy for the synthesis of CVS 1778 analogs: inhibitors of factor Xa. Tetrahedron Lett 2002;43:8177–80.

(c) Rose NGW, Blaskovich MA, Wong A, Lajoie GA. Synthesis of enantiomerically enriched β,γ-unsaturated-α-amino acids. Tetrahedron 2001;57:1497–507.

(d) Sicherl F, Cupido T, Albericio F. A novel dipeptidomimetic containing a cyclic threonine. Chem Commun 2010;46:1266–8.
(e) Dondoni A, Giovannini PP, Perrone D. Cross-metathesis of C-allyl iminosugars with alkenyl oxazolidines as a key step in the synthesis of C-iminoglycosyl α-amino acids. 1 A route to iminosugar containing C-glycopeptides. J Org Chem 2005;70:5508–18.
(f) Berkowitz DB, Maiti G. Following an ISES lead: the first examples of asymmetric Ni (0)-mediated allylic amination. Org Lett 2004;6:2661–4.
(g) Itaya T, Shimizu S, Nakagawa S, Morisue M. Syntheses of optically active, protected and unprotected vinylglycines. Chem Pharm Bull 1994;42:1927–30.

[45] (a) Afzali-Ardakani A, Rapoport H. L-Vinylglycine. J Org Chem 1980;45:4817–20.
(b) Masuda Y, Maruyama C, Kawabata K, Hamano Y, Doi T. Synthesis of (2S,3R,4R)-3,4-dihydroxyarginine and its inhibitory activity against nitric oxide synthase. Tetrahedron 2016;72:5602–11.

[46] (a) Myers AG, Gleason JL, Yoon T, Kung DW. Highly practical methodology for the synthesis of D-and L-α-amino acids, N-protected α-amino acids, and N-methyl-α-amino acids. J Am Chem Soc 1997;119:656–73.
(b) Rémond E, Bayardon JRM, Ondel-Eymin M-JL, Jugé S. Stereoselective synthesis of unsaturated and functionalized L-NHBoc amino acids, using Wittig reaction under mild phase-transfer conditions. J Org Chem 2012;77:7579–87.
(c) Workman JA, Garrido NP, Sançon J, Roberts E, Wessel HP, Sweeney JB. Asymmetric [2, 3]-rearrangement of glycine-derived allyl ammonium ylids. J Am Chem Soc 2005;127:1066–7.
(d) Xu P-F, Lu T-J. Selective synthesis of either enantiomer of α-amino acids by switching the regiochemistry of the tricyclic iminolactones prepared from a single chiral source. J Org Chem 2003;68:658–61.

[47] Atmuri NDP, Lubell WD. Preparation of N-(Boc)-allylglycine methyl ester using a zinc-mediated, palladium-catalyzed cross-coupling reaction. Org Synth 2015;92:103–16.

[48] (a) Rodríguez A, Miller DD, Jackson RF. Combined application of organozinc chemistry and one-pot hydroboration–Suzuki coupling to the synthesis of amino acids. Org Biomol Chem 2003;1:973–7.
(b) Trost BM, Rudd MT. Chemoselectivity of the ruthenium-catalyzed hydrative diyne cyclization: total synthesis of (+)-cylindricine C, D, and E. Org Lett 2003;5:4599–602.

[49] (a) Chattopadhyay SK, Sil S, Mukherjee JP. Synthesis of 2-aminosuberic acid derivatives as components of some histone deacetylase inhibiting cyclic tetrapeptides. Beilstein J Org Chem 2017;13:2153–6.
(b) Fanelli R, Jeanne-Julien L, René A, Martinez J, Cavelier F. Stereoselective synthesis of unsaturated α-amino acids. Amino Acids 2015;47:1107–15.
(c) Shin J-A, Kim J, Lee H, Ha S, Lee H-Y. $Cu(OTf)_2$-promoted 1,4-addition of alkyl bromides to dehydroalanine. J Org Chem 2019;84:4558–65.
(d) Suzuki N, Suzuki T, Ota Y, Nakano T, Kurihara M, Okuda H, et al. Design, synthesis, and biological activity of boronic acid-based histone deacetylase inhibitors. J Med Chem 2009;52:2909–22.
(e) Traoré M, Mietton F, Maubon D, Peuchmaur M, Francisco Hilário F, Pereira de Freitas R, et al. Flexible synthesis and evaluation of diverse anti-apicomplexa cyclic peptides. J Org Chem 2013;78:3655–75.

[50] Trnka TM, Grubbs RH. The development of $L_2X_2Ru = CHR$ olefin metathesis catalysts: an organometallic success story. Acc Chem Res 2001;34:18–29.

[51] Fink BE, Kym PR, Katzenellenbogen JA. Design, synthesis, and conformational analysis of a proposed type I β-turn mimic. J Am Chem Soc 1998;120:4334–44.

[52] Godina TA, Lubell WD. Mimics of peptide turn backbone and side-chain geometry by a general approach for modifying azabicyclo[5.3.0]alkanone amino acids. J Org Chem 2011;76:5846–9.

[53] (a) Crowley JD, Hänni KD, Lee AL, Leigh DA. [2]Rotaxanes through palladium active-template oxidative heck cross-couplings. J Am Chem Soc 2007;129:12092–3.
(b) Koley M, Dastbaravardeh N, Schnürch M, Mihovilovic MD. Palladium(II)-catalyzed regioselective ortho arylation of sp^2 C–H bonds of N-aryl-2-amino pyridine derivatives. ChemCatChem 2012;4:1345–52.
(c) Lee A-L. Enantioselective oxidative boron Heck reactions. Org Biomol Chem 2016;14:5357–66.

[54] (a) Shen W, Coburn CA, Bornmann WG, Danishefsky SJ. Concise total syntheses of DL-camptothecin and related anticancer drugs. J Org Chem 1993;58:611–17.
(b) Anderson RJ, Raolji GB, Kanazawa A, Greene AE. A novel, expeditious synthesis of racemic camptothecin. Org Lett 2005;7:2989–91.

[55] Dunn MJ, Jackson RFW, Pietruszka J, Turner D. Synthesis of enantiomerically pure unsaturated α-amino acids using serine-derived zinc/copper reagents. J Org Chem 1995; 60:2210–15.

[56] (a) Humphrey JM, Bridges RJ, Hart JA, Chamberlin R. 2,3-Pyrrolidinedicarboxylates as neurotransmitter conformer mimics: enantioselective synthesis via chelation-controlled enolate alkylation. J Org Chem 1994;59:2467–72.
(b) Pedregal C, Collado I, Escribano A, Ezquerra J, Domínguez C, Mateo AI, et al. 4-Alkyl- and 4-cinnamylglutamic acid analogues are potent GluR5 kainate receptor agonists. J Med Chem 2000;43:1958–68.

[57] Feng Z, Lubell WD. Synthesis of enantiopure 7-[3-azidopyl]indolizidin-2-one amino acid. A constrained mimic of the peptide backbone geometry and heteroatomic side-chain functionality of the Ala-Lys dipeptide. J Org Chem 2001;66:1181–5.

[58] Hanessian S, Margarita R, Hall A, Luo X. 1,2-Asymmetric induction in dianionic allylation reactions of amino acid derivatives-synthesis of functionally useful, enantiopure aspartates and constrained scaffolds. Tetrahedron Lett 1998;39:5883–6.

[59] Hanessian S, Yang H, Schaum R. Hydrogen-bonding as a stereocontrolling element in free-radical C-allylation reactions: vicinal, proximal, and remote asymmetric induction in the amino acid series. J Am Chem Soc 1996;118:2507–8.

[60] Mulamreddy R, Atmuri NDP, Lubell WD. 4-Vinylproline. J Org Chem 2018;83:13580–6.

[61] Beyermann M, Rothemund S, Heinrich N, Fechner K, Furkert J, Dathe M, et al. A role for a helical connector between two receptor binding sites of a long-chain peptide hormone. J Biol Chem 2000;275:5702–9.

[62] Ndungu JM, Gu X, Gross DE, Ying J, Hruby VJ. A simple and efficient synthesis of an Asp-Gly dipeptide mimetic. Tetrahedron Lett 2004;45:3245–7.

[63] Hack V, Reuter C, Opitz R, Schmieder P, Beyermann M, Neudörfl JM, et al. Efficient α-helix induction in a linear peptide chain by N-capping with a bridged-tricyclic diproline analogue. Angew Chem Int (Ed.) 2013;52:9539–43.

[64] Rao MHR, Pinyol E, Lubell WD. Rigid dipeptide mimics: synthesis of enantiopure C6-functionalized pyrrolizidinone amino acids. J Org Chem 2007;72:736–43.

[65] Chang HH, Larson J, Blencowe H, Spong CY, Howson CP, Cairns-Smith S, et al. Preventing preterm births: analysis of trends and potential reductions with interventions in 39 countries with very high human development index. Lancet 2013;381:223–34.

[66] Grosse SD, Waitzman NJ, Yang N, Abe K, Barfield WD. Employer-sponsored plan expenditures for infants born preterm. Pediatrics 2017;140:e20171078.
[67] Haas DM, Imperiale TF, Kirkpatrick PR, Klein RW, Zollinger TW, Golichowski AM. Tocolytic therapy: a meta-analysis and decision analysis. Obstet Gynecol 2009;113:585–94.
[68] Haas DM, Benjamin T, Sawyer R, Quinney SK. Short-term tocolytics for preterm delivery—current perspectives. Int J Women's Health 2014;6:343.
[69] Brodt-Eppley J, Myatt L. Prostaglandin receptors in lower segment myometrium during gestation and labor. Obstet Gynecol 1999;93:89–93.
[70] Sugimoto Y, Yamasaki A, Segi E, Tsuboi K, Aze Y, Nishimura T, et al. Failure of parturition in mice lacking the prostaglandin F receptor. Science 1997;277:681–3.
[71] Presland J. Identifying novel modulators of G protein-coupled receptors via interaction at allosteric sites. Curr Opin Drug Discov Devel 2005;8:567–76.
[72] Kenakin TP. Ligand detection in the allosteric world. J Biomol Screen 2010;15:119–30.
[73] Peri KG, Quiniou C, Hou X, Abran D, Varma DR, Lubell WD, et al. THG113: a novel selective FP antagonist that delays preterm labor. Semin Perinatol 2002;389–97.
[74] Pathe-Neuschäfer-Rube A, Neuschäfer-Rube F, Püschel GP. Role of the ERC motif in the proximal part of the second intracellular loop and the C-terminal domain of the human prostaglandin F2α receptor (hFP-R) in G-protein coupling control. Biochem J 2005;388:317–24.
[75] Poupart J, Hou X, Chemtob S, Lubell WD. Application of N-dodecyl L-peptide to enhance serum stability while maintaining inhibitory effects on myometrial contractions ex vivo. Molecules 2019;24:4141.
[76] Rudinger J. The design of peptide hormone analogs. Drug Design 1971;20:319–419.
[77] Hirst JJ, Parkington HC, Young IR, Palliser HK, Peri KG, Olson DM. Delay of preterm birth in sheep by THG113.31, a prostaglandin F2α receptor antagonist. Am J Obstet Gynecol 2005;193:256–66.
[78] Böttcher B, Laterza R, Wildt L, Seufert R, Buhling K, Singer C, et al. A first-in-human study of PDC31 (prostaglandin F2α receptor inhibitor) in primary dysmenorrhea. Hum Reprod 2014;29:2465–73.
[79] Coy DH, Heinz-Erian P, Jiang N-Y, Sasaki Y, Taylor J, Moreau J, et al. Probing peptide backbone function in bombesin. A reduced peptide bond analogue with potent and specific receptor antagonist activity. J Biol Chem 1988;263:5056–60.
[80] Boukanoun MK, Hou X, Nikolajev L, Ratni S, Olson D, Claing A, et al. Investigation of the active turn geometry for the labour delaying activity of indolizidinone and azapeptide modulators of the prostaglandin F2α receptor. Org Biomol Chem 2015;13:7750–61.
[81] Goupil E, Tassy D, Bourguet C, Quiniou C, Wisehart V, Pétrin D, et al. A novel biased allosteric compound inhibitor of parturition selectively impedes the prostaglandin F2α-mediated Rho/ROCK signaling pathway. J Biol Chem 2010;285:25624–36.
[82] Kemp MW. Preterm birth, intrauterine infection, and fetal inflammation. Front Immunol 2014;5:574.
[83] Nadeau-Vallée M, Obari D, Palacios J, Brien M-È, Duval C, Chemtob S, et al. Sterile inflammation and pregnancy complications: a review. Reproduction 2016;152:R277–92.

Section 3

Drug discovery

Chapter 7

The current state of backbone cyclic peptidomimetics and their application to drug discovery*

Samuel J.S. Rubin[1] **and Nir Qvit**[2]
[1]*Department of Medicine, School of Medicine, Stanford University, Stanford, CA, United States,* [2]*The Azrieli Faculty of Medicine in the Galilee, Bar-Ilan University, Safed, Israel*

7.1 Introduction

Naturally occurring peptides identified either by isolation of natural products or by screening libraries are often not directly suitable for use as convenient therapeutics because they have intrinsic limitations, including poor chemical and physical stability and a short circulating plasma half-life. Therefore, modifications can be employed to overcome these limitations, such as N-terminal and C-terminal modifications [1], peptide bond modifications [2], incorporation of nonproteinogenic amino acids [3], and cyclization [4–6]. All of these alterations are used to create peptidomimetics from a parent peptide. Peptidomimetics (modified peptides) usually demonstrate high potency as well as improved stability and increased selectivity compared to their linear analogs, making them ideal candidates for biological research and therapeutic lead compounds [7–12].

Historically, several models have been proposed to describe the conformation of molecules upon their binding (e.g., protein and protein/peptide). The initial "Lock and Key" model was introduced by Emil Fisher in 1894. In this model, it is assumed that both molecules have a specific, fixed geometric shape and orientation that is complementary to the other molecule, allowing them to fit perfectly into each other like a lock and a key [13]. Daniel E. Koshland Jr. introduced a more flexible "Induced Fit" model in 1958, which takes into account the flexibility of molecules and the influence

* Portions of this chapter were adapted from Rubin SJS, Qvit N. Backbone-cyclized peptides: a critical review. Curr Top Med Chem. 2018;18, (7): 526–555. Copyright Bentham Science Publishers, 2018.

Peptide and Peptidomimetic Therapeutics. DOI: https://doi.org/10.1016/B978-0-12-820141-1.00021-2

of each molecule on the other molecule [14,15]. Cyclic peptides are restricted in their conformation and thus their induced fit upon binding to a target is limited—either preventing a cyclic peptide from achieving the bioactive conformation, or in an ideal scenario, maintaining a bioactive conformation. Consequently, to identify an active cyclic peptide(s) with desirable biological activity, a library of cyclic peptides should be developed to screen the conformational space [16].

Henceforth, we focus on cyclic peptides, which generally demonstrate improved properties compared to their linear counterparts. However, in some cases, cyclization fails to improve a linear molecule, usually due to (1) the use of important bioactive residues in cyclization, which blocks critical interaction sites and/or (2) restriction of the peptide in an inactive conformation. "Traditional" cyclization methods (e.g., head-to-tail, side chain-to-tail, side chain-to-side chain, and side chain-to-head, Fig. 7.1) frequently and inevitably interfere with the bioactive epitopes of a linear compound, resulting in inactive molecules due to obstruction of side chains necessary for peptide bioactivity, which is particularly important for shorter peptides [17]. Furthermore, "traditional" cyclization is limited by the number of cyclic peptides that can be generated while keeping the natural/primary sequence of the peptide. Therefore, frequently cyclic peptides generated by "traditional" cyclization will result in a rigid and inactive constrained conformation.

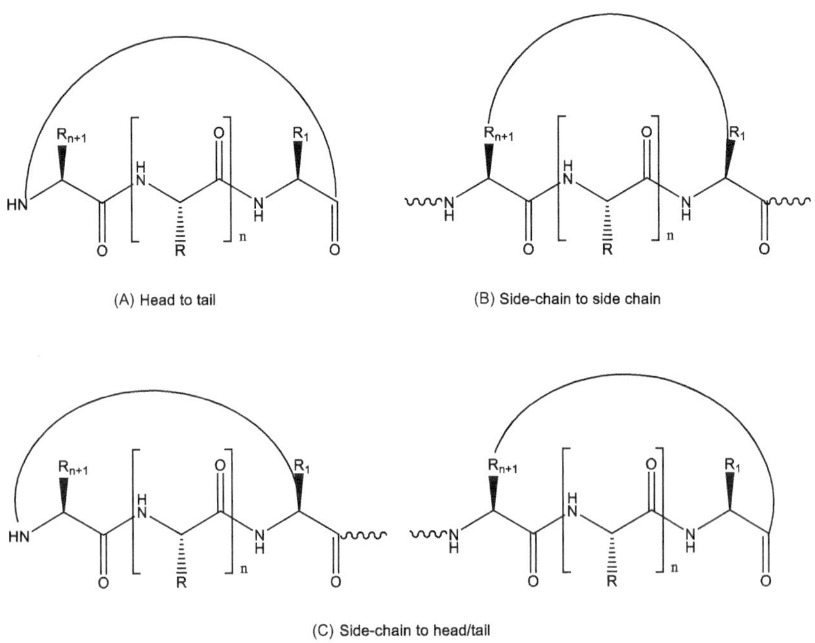

FIGURE 7.1 "Classical" modes of peptide cyclization.

Several methods have been introduced to overcome limitations associated with traditional peptide cyclization, such as click chemistry [18,19] (for a recent review see [20]), ring-closing metathesis (RCM) [21,22] (for a recent review see [23]), stapled peptides [24–26] (for a recent review see [27]), native chemical ligation [28–30] (for a recent review see [31]), and backbone cyclization pioneered by Chaim Gilon [32] (for recent reviews see [33,34]). These methods apply various synthetic approaches for peptide cyclization, including but not limited to cyclization via lactam bridges [35,36], lactone bridges [7,37], and disulfide bridges [38]. Herein, our primary focus is on the rational design of backbone-cyclized peptides using lactam and disulfide bridges.

Backbone cyclization is a method to develop cyclic peptides without utilizing the residues that are part of the natural linear peptide, which may be essential for the bioactivity of the peptide. This approach is particularly valuable for short peptides and/or peptides in which all the amino acids are important for the bioactivity [39]. In backbone cyclization, functionalized alkyl spacers anchored to the peptide backbone are connected to another functionalized spacer, to a functional amino acid side chain on the peptide, or the N- or C-terminus. The cyclization is accomplished using a building unit composed of a functional atom (e.g., sulfur, nitrogen, or carbon) with an orthogonal protecting group, covalently bonded via a spacer to the backbone. The main advantage of this method is that the cyclization linkage is formed between backbone atoms and not the atoms of the side chain functional groups, which are typically critical for binding and/or biological function. Backbone cyclization offers additional modes of cyclization, in addition to the "traditional" modes of cyclization, such as backbone-to-backbone, backbone-to-N terminus, backbone-to-C terminus, and backbone-to-side chain (Fig. 7.2).

Bridge chemistry is also important for the bioactivity of the cyclic peptides. This feature is defined by the type(s) of functional groups used to form the bridge. Several different approaches have been introduced for backbone cyclization, including backbone disulfide bridges [4,40], backbone amide bridges [41], backbone metal bridges [42,43], backbone urea bridges [44], and backbone azo bridges [45]. For instance, the replacement of lactam bridges with backbone disulfide bridges was used to support on-resin backbone cyclization and enhance peptide selectivity [40]. Furthermore, backbone metal cyclization via two hemi-chelating groups on the peptide was used to implement simultaneous cyclization and metal labeling for targeted imaging [42,43]. This approach could also be utilized for further development of peptide receptor radionuclide therapy [46]. All these formats have significant effects on the biological activity and functional capacity of peptide libraries.

Currently, there are various types of building blocks that can be used for backbone cyclization, offering almost unlimited opportunities to explore the conformational space of cyclic peptides using the cycloscan approach [47]. Cycloscan is a method in which a library of cyclic peptides is prepared using an identical primary sequence; however, the peptides in the library differ

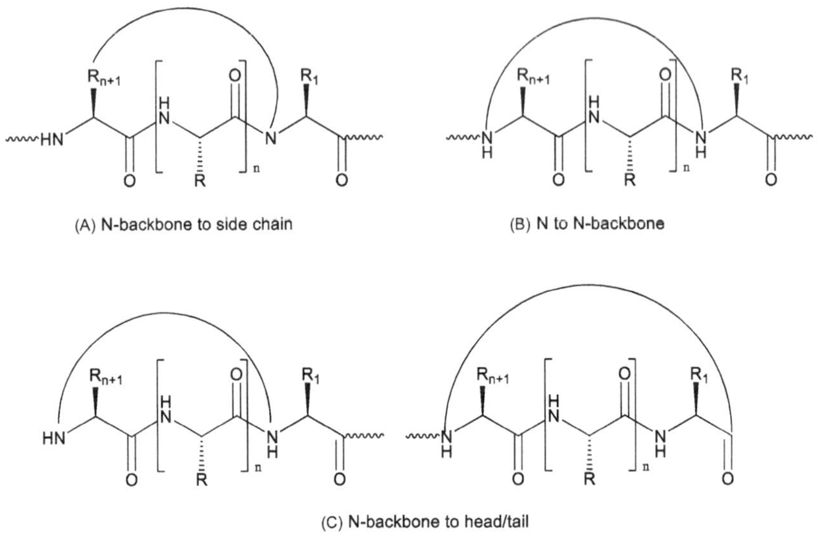

FIGURE 7.2 Modes of backbone cyclization demonstrated by examples of N-backbone cyclization.

from each other in their conformation, as a result of the cyclization approach and bridge features. A cycloscan library of backbone-cyclized peptides that all have the same primary sequence (the natural epitope or the pharmacophore) but different building blocks and bridge elements can be used to efficiently screen the conformational space of a parent peptide.

Backbone cyclization and cycloscan have been used to modify a myriad of bioactive natural peptides, such as substance P [48,49], somatostatin [40,43,50–52], pheromone biosynthesis activating neuropeptide (PBAN) [53], nuclear factor-kappa B (NF-κB) [54,55], insulin-like growth factor-I receptor (IGF-IR) [4,56], human immunodeficiency virus (HIV)-1 integrase [57], bovine pancreatic trypsin inhibitor (BPTI) [58], protein kinase B (PKB) [59], protein kinase C (PKC) and its substrates [60], as well as LACK (Leishmania's receptor for activated C-kinase) and TRACK (Trypanosoma receptor for activated C-kinase) proteins [61]. In many cases, backbone-cyclized peptides have been shown to improve the pharmacological properties compared to linear analogs, such as selectivity [49,51], stability [61,62] and paracellular permeability [63,64].

In summary, backbone cyclization is a method to develop focused, cyclic peptide libraries based on linear peptides and uses mainly unnatural building blocks with an extra linker of customizable length covalently bonded to a backbone functional group for the peptide cyclization. The library is screened for bioactivity and the most active peptides are selected as lead compounds. For each lead compound, a focused library of cyclic peptides is synthesized. Most (or all) of the peptides in the library are derived from the same natural/primary sequence, while factors such as size, location and chemistry of the cyclization diversify the library.

In this chapter, we discuss the applications of backbone-cyclized peptides in basic research and drug discovery. While not the focus of this section, we briefly review solid-phase peptide synthesis and discuss various approaches for the synthesis and development of backbone-cyclized peptides using alpha-amino acid derivative building units.

7.2 Peptide synthesis and backbone cyclization

The chemical synthesis of peptides and peptidomimetics is most frequently performed using the solid-phase synthesis (SPS) approach pioneered by Merrifield [65], which became a common procedure that can be completely automated and/or used by nonexperts [66,67]. In SPS, amino acids/building blocks are bonded to an insoluble material (e.g., a bead) and synthesized via a stepwise procedure in a reactant solution (Fig. 7.3). The advantages of this method are: (1) reagents can be used in excess to drive reactions to completion, and later can be removed easily, (2) additional molecules/functions can easily be added to the molecule, such as glucose groups [68], photoaffinity labels, radioactive tracers, or crosslinkers [69,70], and (3) incorporation of extended sets of molecules, such as D-amino acids, N-methyl amino acids, and nonproteinogenic amino acids, which often improve peptide stability since these units are generally more resistant to proteolysis and cannot always be added by recombinant expression. Moreover, nonproteinogenic amino acids can maintain desirable features of natural alpha-amino acids, and often enhance bioactivity of the natural compounds. As a result, SPS is widely used for the synthesis of peptides, peptidomimetics, short proteins, nucleic acids, and oligosaccharides [71,72].

Building blocks can be highly diverse, allowing for the development of a variety of backbone-cyclized peptides. There are four main components in each building block that determine its chemical properties: (1) The amino acid from which the building block is derived, (2) the functional group, (3) the protecting group, and (4) the alkyl chain. Here, we will focus on N^{α}-(ω-functionalized alkylated) building unit derivatives of natural amino acids and their application to the synthesis of cyclic peptides [32] and peptoids [64] (Fig. 7.4). These building blocks have also been used for the synthesis of pseudo peptides [73] and peptide nucleic acids (PNAs) [74]. Moreover, many other building blocks are available, including alpha-amino acid N-substitutions (e.g., N-Me and N-alkyl) [75], alpha-carbon modifications [2,76], and beta- and gamma-amino acids derivatives [77–79]. In addition, a plethora of nonproteinogenic building blocks are currently available, such as aldol building units [80], beta-lactam building blocks [81,82], and sugar-conjugated amino acids [83,84].

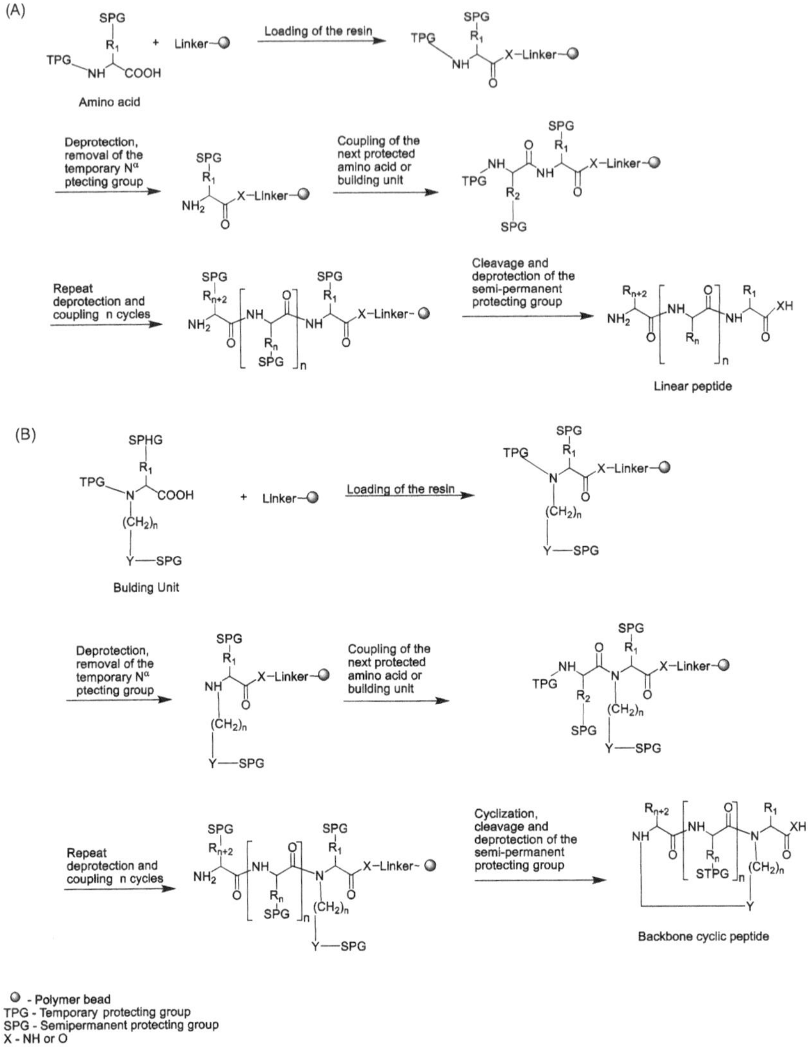

FIGURE 7.3 The chemical synthesis of peptides and peptidomimetics using solid phase synthesis.

7.3 Backbone cyclic peptides mimicking natural peptides

7.3.1 Neuropeptide family

Neuropeptides are encoded by the human genome and allow neurons to communicate with each other. While some neuropeptides function as neurotransmitters, more broadly neuropeptides support widespread neuronal communication networks. The molecules can be classified by their site of origin, their target site, their mechanistic/functional features, and/or their chemical homology.

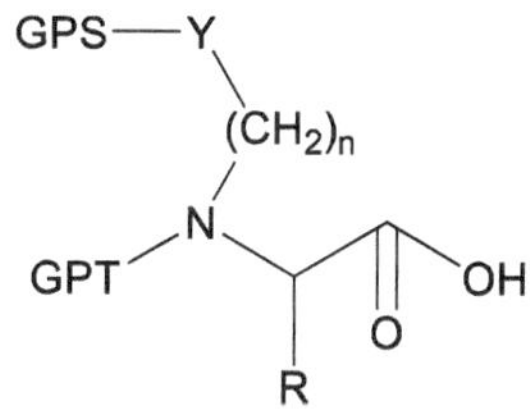

R = Pharmacophore of the unit
Y = Functional group that is used to form the bridge
SPG = Protecting group of the functional group (Semipermanent protecting group)
n = Length of the alkyl chain
TPG = Temporary protecting group

FIGURE 7.4 General structure of Nα-(ω-Y-alkyl) building units.

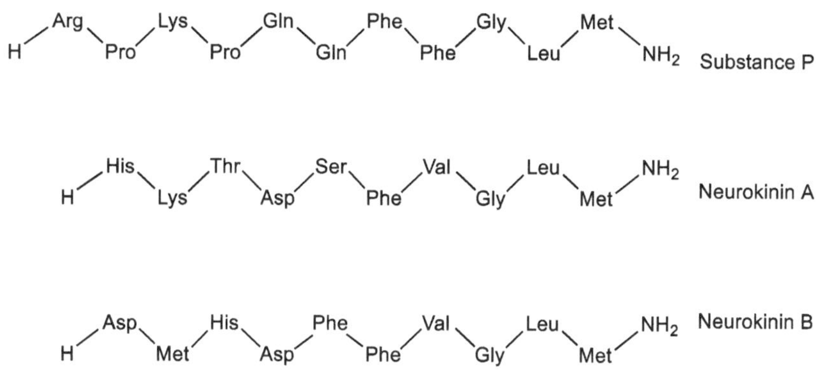

FIGURE 7.5 Linear structures of Substance P and two other tachykinin peptides.

7.3.1.1 Substance P

Substance P is a member of the neurokinin family, a subfamily of the tachykinin peptides that also includes neurokinin A and neurokinin B (Fig. 7.5). Substance P is an eleven-amino acid peptide with an amidation at the C-terminus. This neuropeptide is produced from a polyprotein precursor after the splicing of the preprotachykinin A gene. It is ubiquitous in the human body, throughout the central and peripheral nervous systems, as well as in the liver, lung, and placenta. Substance P has been associated with transmission of pain stimuli, related to inflammation, hepatitis and cancer, as well as bacterial and viral infection. Each neurokinin receptor maintains specificity via selective peptides (neurokinin 1 by substance P, neurokinin 2 by neurokinin A and neurokinin 3 by neurokinin B), although the selectivity of each receptor for the corresponding peptide is not high [85,86]. Despite studies of substance P dating to the 1930s, only small molecule NK1 antagonists (e.g., aprepitant) with limited efficacy have been approved for the treatment of nausea [87]. There have been many efforts to identify alternative modes of drugging substance P and/or its receptor, NK1, which have yet to reconcile in vitro activity with clinical efficacy [88].

Bitan et al. developed a biased library containing backbone bicyclic peptide analogs of Substance P. Several backbone-cyclized analogs with the N^{α}-(ω-thioalkylene) building unit demonstrated relevant bioactivity (EC_{50} 20 nM toward neurokinin 1) and specificity (they were not active for neurokinin 2 and neurokinin 3 receptors) (Fig. 7.6) [89]. In a follow-up study, Bitan et al. prepared a biased library of bicyclic peptides in which the backbone peptides were cyclized on resin using N^{α}-(ω-thioalkylene) building units to produce lactam and disulfide rings. The activities of the peptides were about three orders of magnitude lower than the native agonist (EC_{50} in the low micromolar range), probably due to overrigidification from bicyclization (Fig. 7.7) [90].

In an additional study, Bitan et al. developed a library of backbone-cyclized peptides and performed structure–activity relationship (SAR) studies. In early systematic studies where each peptide bond at the C-terminus was replaced by an N-methylated peptide bond, the authors identified methylations important for activity and selectivity for a single receptor [91–93]. Next, they developed metabolically stable neurokinin-1-specific activators (e.g., SP6, EC_{50} 20 nM), which completely lacked activity toward off-target receptors (neurokinin-2 and neurokinin-3) (Fig. 7.8) [48]. To gain insight into the relationship between conformation and bioactivity the authors solved the nuclear magnetic resonance (NMR) structures of two backbone-cyclized analogs, which were similar to other bioactive substance P analogs [94].

Based on the same scaffold, Byk et al. developed and characterized a backbone hexapeptide library derived from the C-terminal hexapeptide of substance P. The library was designed using NMR data and molecular modeling of a short, selective peptide analog of neurokinin-1, WS-septide, which was identified previously (Fig. 7.9) [92,93]. The authors determined the bioactivity and specificity of the backbone-cyclized peptides for neurokinin-1 receptor, and also identified a correlation between bioactivity and peptide ring size. An

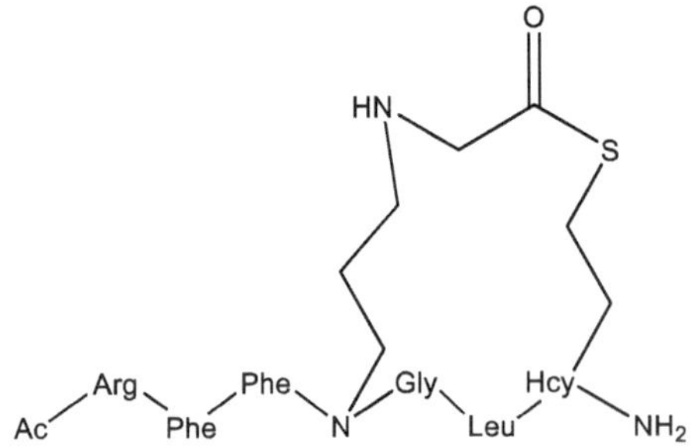

Hcy - Homocysteine

FIGURE 7.6 Structure of the backbone cyclic analog I of Substance P [89].

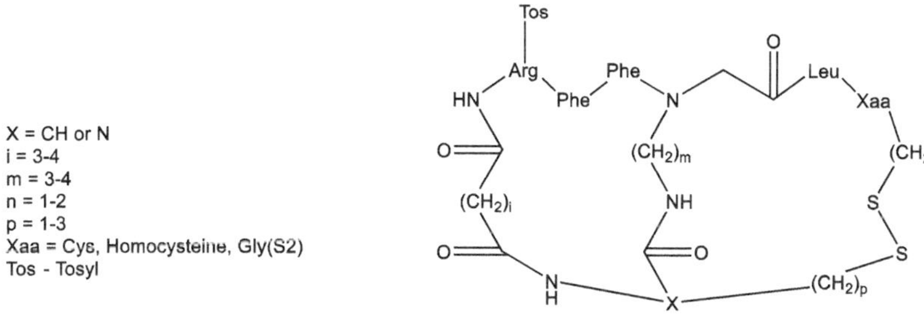

FIGURE 7.7 Structure of the backbone bicyclic analog of Substance P [90].

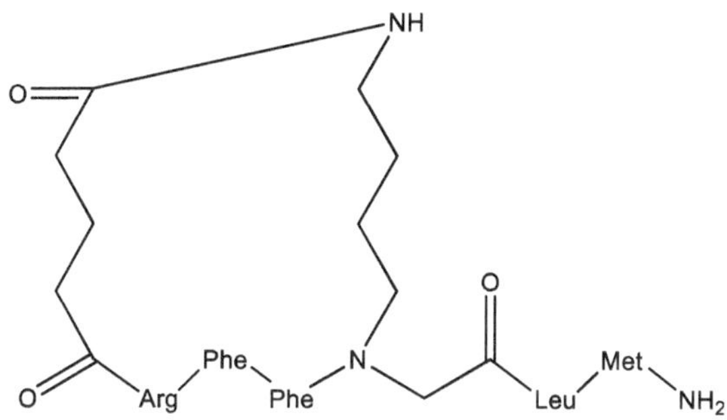

FIGURE 7.8 Structure of the backbone cyclic analog SP6 of Substance P [48].

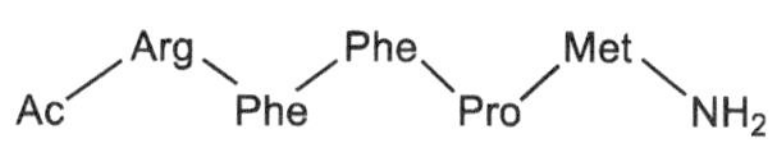

FIGURE 7.9 Linear structure of the WS-septide [92,93].

analog with a ring size of 20 atoms was the most active and selective backbone-cyclized peptide. Importantly, this analog was also more metabolically stable compared to the linear peptide (Fig. 7.10) [49].

7.3.1.2 Pheromone biosynthesis activating neuropeptide

PBAN is a 33-amino acid neuropeptide member of the pyrokinin/PBAN family, which activates the biosynthesis of various pheromones in moths. Members of this family play a key role in a variety of physiological and behavioral functions, including contraction of the locust hindgut, egg diapause in the larvae and stimulation of sex pheromone biosynthesis in female moths. Female moths release PBAN into their hemolymph during the scotophase to stimulate the biosynthesis of neuropeptides for attracting conspecific males [95].

SAR studies revealed that the C-terminal of PBAN is conserved in numerous members of the pyrokinin/PBAN family and may be important for its bioactivity. Two backbone-cyclized peptide libraries using N^{α}-(ω-aminoalkylene) building units were developed to identify antagonists capable of inhibiting sex

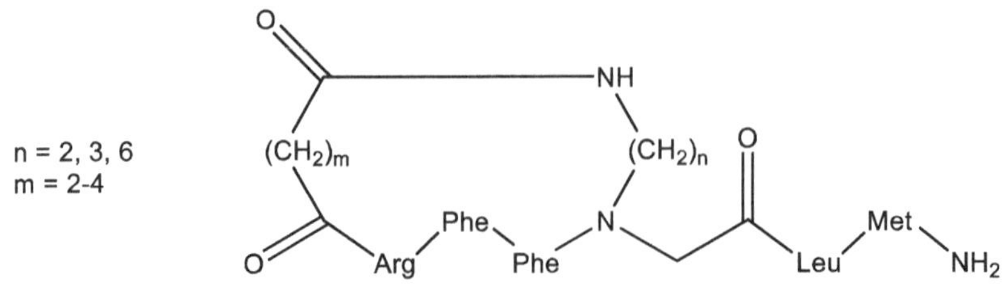

FIGURE 7.10 Structure of the backbone cyclic analog of Substance P [49].

pheromone biosynthesis in female *Heliothis peltigera* moths. The first library with Ser amino acids was designed based on the C-terminal hexapeptide sequence Tyr–Phe–Ser–Pro–Arg–Leu–NH_2, while the second library was based on the sequence of the lead Arg–Tyr–Phe–(D)Phe–Pro–Arg–Leu–NH_2 antagonist identified previously [96]. Out of the two libraries, four backbone-cyclized peptides (all from the second library with (D)Phe) exhibited antagonist activity and completely inhibited the sex pheromone biosynthesis at 1 nmol. Interestingly, the substitution of (D)Phe with Ser in these analogs completely abolished their bioactivity (Fig. 7.11) [41,53]. Next, Altstein et al. optimized the cyclization bridge size while maintaining building unit location. The authors evaluated different bridge sizes derived from the length of functionalized spacers, using various linkers. Backbone-cyclized peptides in which the bridge amide bond was after two or four carbons (peptides 21 and 25, respectively) were highly active compared to an inactive backbone-cyclized peptide with the same sequence and ring size but an amide bond bridge after three carbons (peptide 23) [41].

7.3.1.3 Bradykinin

Bradykinin is a nonapeptide thought to rely on a turn structure for its biological activity. The peptide induces vasodilation, mediating inflammation through the vasculature and likely pain as well. Bradykinin binds B1 receptors (constitutively expressed) and B2 receptors (expressed as a result of tissue injury). Icatibant is a cyclic peptidomimetic bradykinin B2 receptor antagonist, which was approved in 2008 by the European Commission for the treatment of hereditary angioedema (HAE) in adults with C1-esterase-inhibitor deficiency [97].

Schumann et al. synthesized backbone-cyclized peptide analogs with N-carboxy alkylated, N-amino alkylated amino acids and dipeptide building blocks synthesized in solution beforehand. The authors used these compounds to study the proposed bioactive turn conformation. Indeed, only compound 3, which was backbone-cyclized between positions 2 and 3, exhibited significant agonist activity in a rat uterus assay. This validated the hypothesis that the bradykinin turn structure is important for its biological activity (Fig. 7.12) [98].

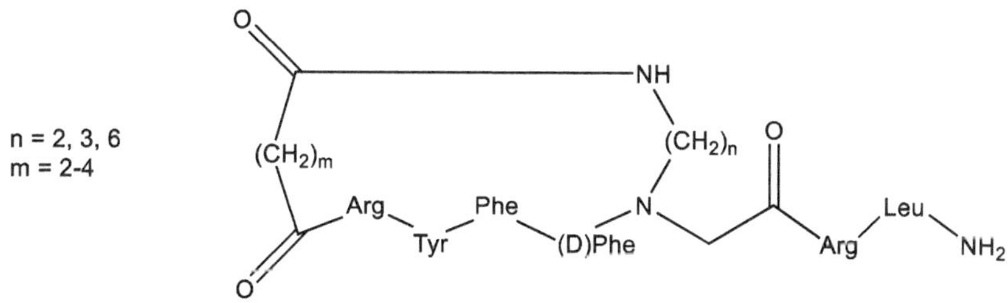

FIGURE 7.11 Structure of the backbone cyclic analog of PBAN [41,53].

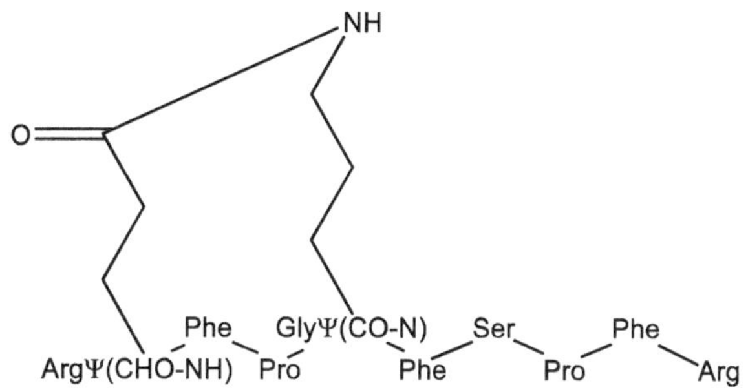

FIGURE 7.12 Structure of the backbone cyclic analog of bradykinin, compound 3 [98].

7.3.2 Peptide hormones

Peptide hormones regulate endocrine processes in the body. Most are hydrophilic molecules that bind cell surface and/or soluble molecules, although some may be transported to the cytoplasm where they bind intracellular receptors. Peptide hormones engage signaling pathways via second messenger molecules and exhibit high level and long-range control and integration of homeostatic processes.

7.3.2.1 Somatostatin

Somatostatin, also growth hormone–inhibiting hormone (GHIH) or somatotropin release-inhibiting factor (SRIF), is an endogenous cyclic tetradecapeptide hormone (Fig. 7.13), which also has an alternative active form of 28 amino acids (the short form with another 14 amino acids at one end). The molecule is produced by many tissues, and it regulates a wide variety of physiological processes by inhibiting the secretion of other hormones, such as growth hormone, thyroid-stimulating hormone, cholecystokinin, and insulin. Somatostatin regulates the endocrine system via interactions with five different ubiquitously distributed G-protein-coupled somatostatin seven-transmembrane receptors (sst1-sst5). Although somatostatin exerts multiple biological activities and has broad inhibitory effects (antisecretory, antiproliferative and antiangiogenic), its clinical use is limited by a broad spectrum

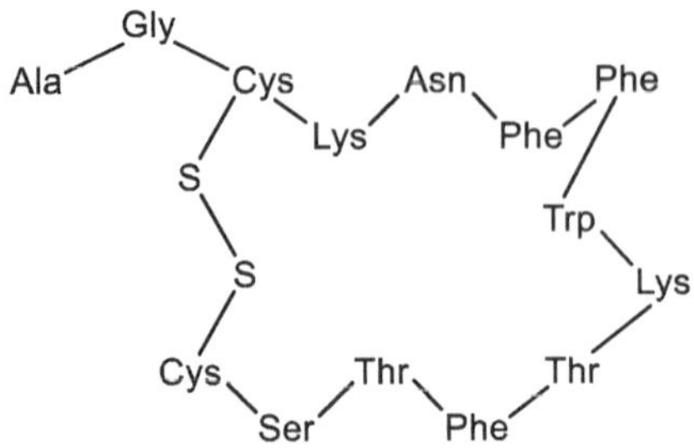

FIGURE 7.13 Structure of somatostatin.

of off-target biological responses and a short half-life (1–3 min). In the 1980s, octreotide (a cyclic octapeptide somatostatin analog, brand name Sandostatin) was introduced to the clinic. Several additional cyclic peptides soon followed, including pasireotide (Signifor), lanreotide (Lanreotide) and vapreotide (Sanvar), all of which demonstrated longer half-lives and increased pharmacologic efficacy (Fig. 7.14). The majority of these analogs bind strongly to two out of the five human somatostatin receptors (hsst2 and hsst5) and have a moderate or low affinity for the other receptors (hsst3 and hsst1, respectively). The affinity for hsst4 varies between the different analogs, and the compounds are used to treat patients with neuroendocrine tumors and pituitary adenomas [52,99–101].

Gilon et al. synthesized a backbone-cyclized analog of somatostatin, PTR 3046, using N^{α}-(ω-carboxyalkylene) building units. The compound binds selectively to the seven-transmembrane domain somatostatin receptor 5 (sst5) with an IC_{50} of 67 nM for sst5 and an IC_{50} greater than 1000 nM for sst1–4 and is highly resistant to enzymatic degradation. PTR 3046 inhibited bombesin- and caerulein-induced amylase and lipase release from the pancreas, but not growth hormone or glucagon release, in animal studies (Fig. 7.15) [50].

Gazal et al. synthesized a library of somatostatin analogs that all share a common core sequence. Diversity in the library was derived from the mode of cyclization, either from the N- or C-terminus to an N^{α}-(ω-thioalkylene) building unit or from one building unit to another building unit. Some of the analogs had a high binding affinity to hsst2 and hsst5 in the low nanomolar range and demonstrated significant metabolic stability compared to somatostatin (Fig. 7.16) [40,102].

Afargan et al. also developed a stable backbone-cyclized analog of somatostatin, PTR 3173, using N^{α}-(ω-carboxyalkylene) building units. This molecule is bound with high affinity to three somatostatin seven-transmembrane domain receptors (3 nM for hsst2, 7 nM for hsst4 and 6 nM for hsst5) and inhibited accumulation of forskolin-stimulated cAMP (cyclic adenosine monophosphate). In animal studies, PTR 3173 selectively and potently inhibited growth hormone release, by 1000-fold and over 10,000-fold more than glucagon and insulin, respectively (Fig. 7.17) [51].

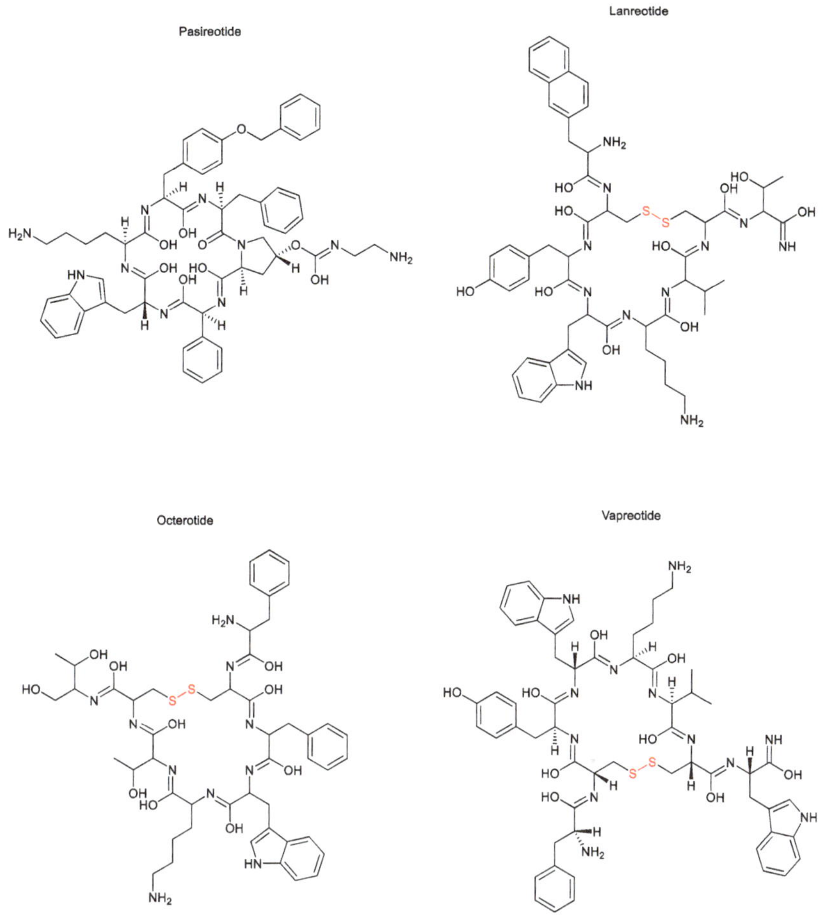

FIGURE 7.14 Structures of octreotide, pasireotide, vapreotide, and lanreotide. Disulfide bonds are marked in red for lanreotide, octreotide, and vapreotide.

Falb et al. developed a bicyclic backbone peptide analog of somatostatin, PTR 3205, with N^{α}-(ω-carboxyalkylene) building units. PTR 3205 was specific and selective for hsst2 (IC_{50} 3.7 nM) and at least two orders of magnitude lower for other somatostatin receptors. The 3D structure of PTR 3205 revealed a similar region to other known somatostatin analogs, such as compound L-363,301 (Merck) and Sandostatin, located in the pharmacophoric region containing the Phe-(D)Trp-Lys-Thr motif (Fig. 7.18) [103].

Fridkin et al. synthesized a library of backbone metal cyclic peptides using a unique approach, which can simultaneously support radiolabeling of peptides for imaging and/or therapeutic purposes [104,105]. Combinatorial libraries were developed using two hemi-chelating arms and N^{α}-(ω-carboxyalkylene) building units to bind the metal and perform the backbone peptide

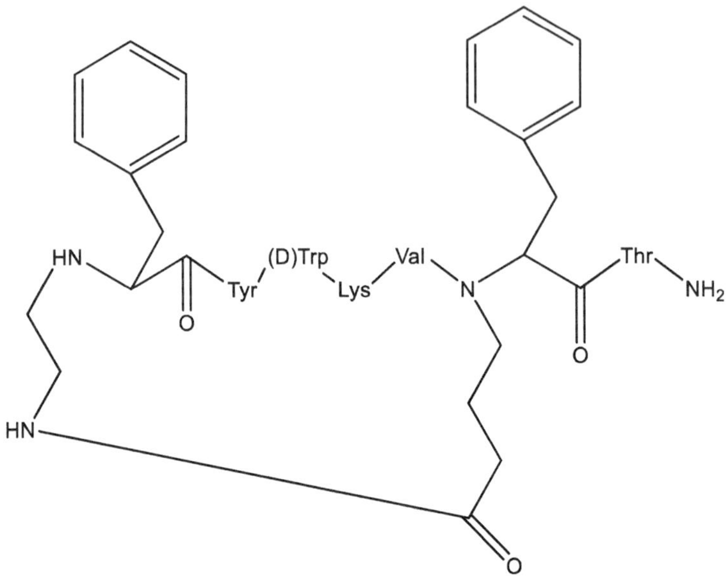

FIGURE 7.15 Structure of the backbone cyclic analog of somatostatin, PTR 3046 [50].

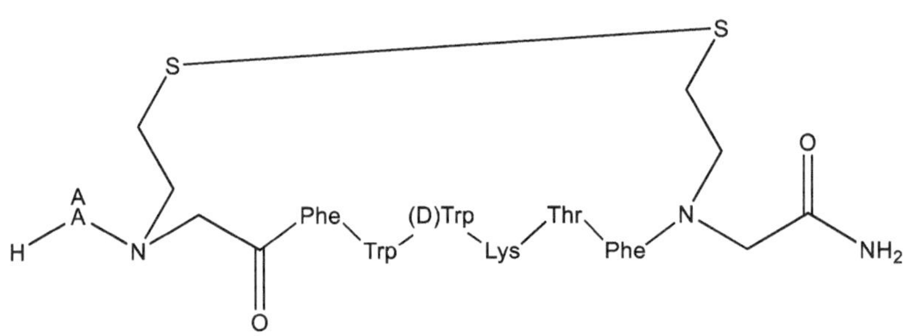

FIGURE 7.16 Structures of the backbone cyclic analogs of somatostatin [40,102].

cyclization. Frothy eight rhenium cyclic somatostatin analogs were developed, in which five demonstrated high bioactivity via in vitro binding to hsst2 with IC_{50} values between 1 and 3 nM (Fig. 7.19) [43].

7.3.2.2 Gonadotropin-releasing hormone

Gonadotropin-releasing hormone (GnRH) is a 10-amino acid peptide hormone produced in the hypothalamus, which stimulates the release of follicle-stimulating hormone (FSH) and luteinizing hormone (LH) from the anterior pituitary. Receptors for GnRH are overexpressed in breast, ovarian, and prostate cancers, which suggests its utility as a therapeutic target. Originally,

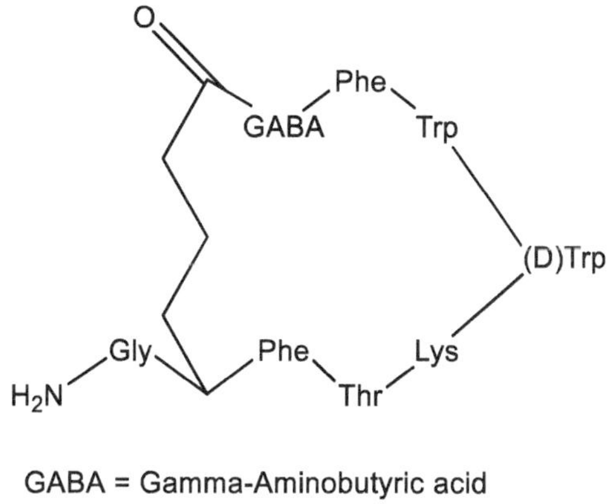

FIGURE 7.17 Structure of another backbone cyclic analog of somatostatin, PTR 3173 [51].

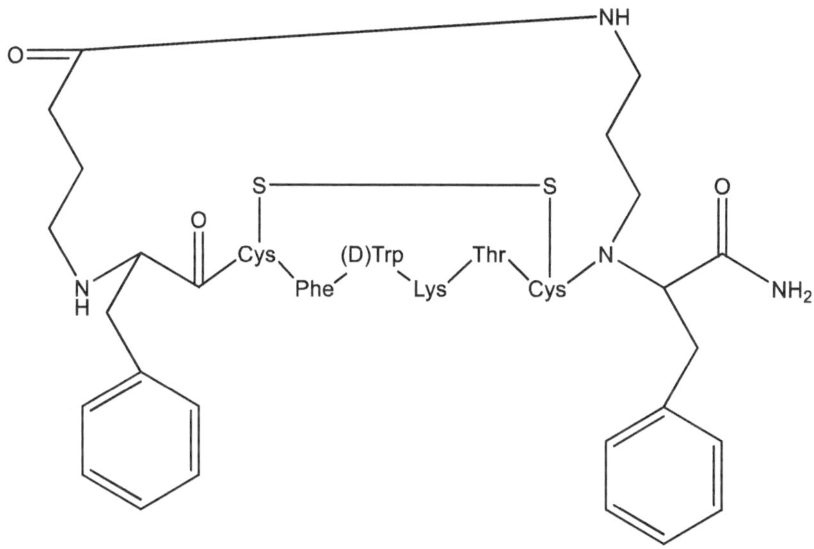

FIGURE 7.18 Structure of another backbone cyclic analog of somatostatin, PTR 3205 [103].

natural GnRH was used for therapeutic purposes, but analogs have since been developed [106]; leuprolide is used to treat cancer, endometriosis, and precocious puberty, but often requires continuous infusion due to its short half-life [107].

Barda et al. designed a backbone-metal cyclized ^{99m}Tc-radiolabeled analog of GnRH to target receptors on malignant cells. They synthesized a GnRH analog with two hemi-chelator groups, the compound bound with high affinity to rat pituitary membranes, which express GnRH receptors, with an IC_{50} in the nanomolar range (similar to native GnRH) (Fig. 7.20) [42].

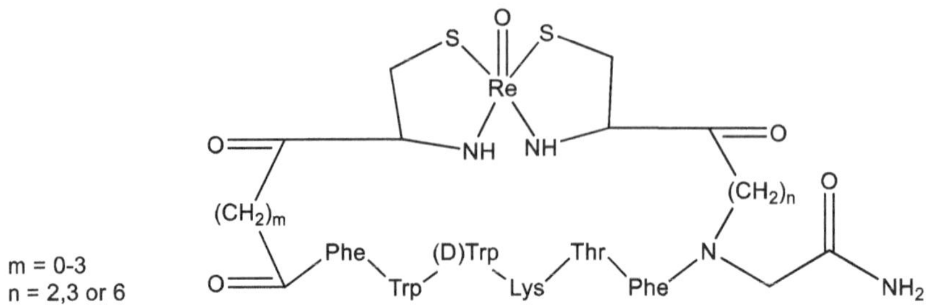

FIGURE 7.19 Structure of an additional backbone cyclic analog of somatostatin [43].

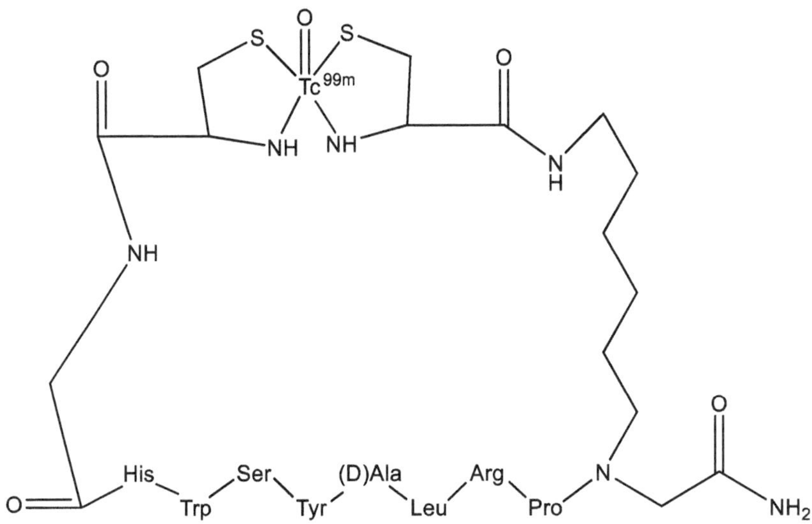

FIGURE 7.20 Structure of the backbone cyclic analog of GnRH [42].

7.3.2.3 Alpha-melanocyte-stimulating hormone

Alpha-melanocyte-stimulating hormone (α-MSH, or melanotropin) is an endogenous tridecapeptide hormone and member of the melanocortin family (Fig. 7.21). α-MSH is involved in feeding behavior and melanogenesis, which is responsible for pigmentation of the hair and skin through a process mediated by melanocortin receptor MC_1. The synthetic linear α-MSH analog afamelanotide was recently approved in Europe and is currently under review by the FDA for use in preventing skin damage from sun exposure [108].

Hess et al. developed a library of backbone-cyclized peptides using N^{α}-(ω-aminoalkylene) building units, derived from the tetrapeptide sequence His-Phe-Arg-Trp shown to be critical for α-MSH bioactivity. Peptide 1 was selective for the melanocortin-4 receptor (MC4R), highly stable and demonstrated transcellular penetration in enterocytes. The peptide was found in the brain following oral administration to rats, and reduced food consumption by

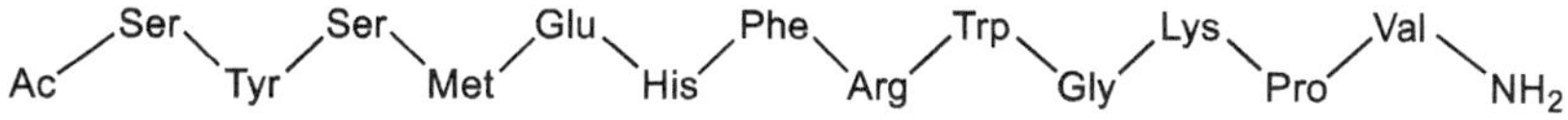

FIGURE 7.21 Linear structure of α-MSH.

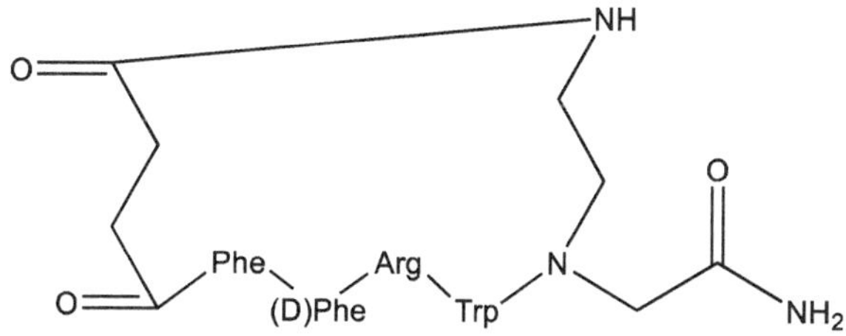

FIGURE 7.22 Structure of the backbone cyclic analog of α-MSH [109].

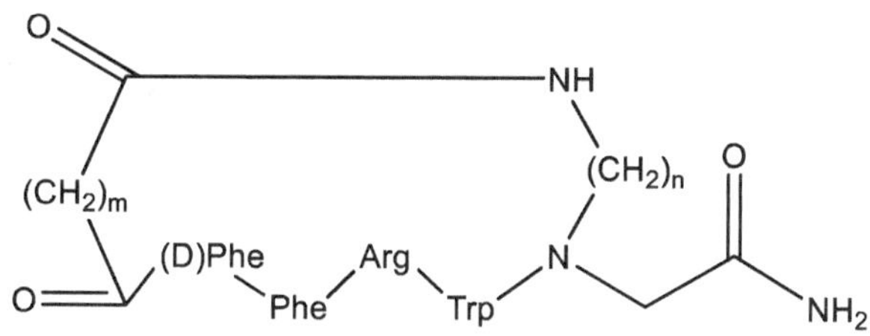

FIGURE 7.23 Structure of another backbone cyclic analog of α-MSH [62].

approximately 50% when administered to mice (0.5 mg/kg) (Fig. 7.22) [109]. Next, they developed a second backbone-cyclized peptide library and an N-methylated library using N^{α}-(ω-aminoalkylene) building units derived from the pentapeptide sequence Phe-(D)Phe-Arg-Trp-Gly. However, all backbone-cyclized peptides had similar or lower bioactivity compared to the lead compound [109]. Though, all the N-methylated backbone-cyclized peptides in this study had significantly reduced biological activity compared to the most active backbone-cyclized peptide. The authors suggest that the reduced bioactivity was due to departure from the active conformation and an increase in rigidity of the backbone-cyclized peptides (Fig. 7.23) [62].

A follow-up study by Ovadia et al. illustrated an interesting approach with hybrid compounds composed of peptides and peptoids, the latter of which are constructed from amino acids where the side chain of the alpha-carbon is moved to the nitrogen. Based on the same Phe–(D)Phe–Arg–Trp–Gly pentapeptide sequence derived from α-MSH, the authors developed a backbone-cyclized peptide-peptoid library using N^{α}-(ω-aminoalkylene) building units to target the MC4R. Two backbone-cyclized peptide-peptoid analogs demonstrated high stability and increased intestinal permeability compared to their linear counterparts, and one backbone cyclic peptide-peptoid analog was highly bioactive (nanomolar range) for MC4R (Fig. 7.24) [64].

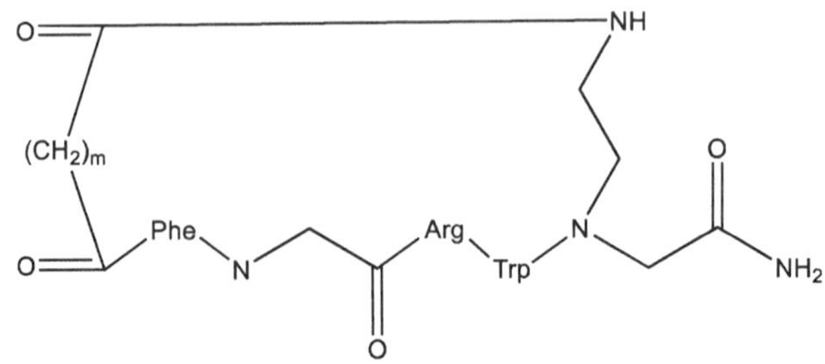

FIGURE 7.24 Structure of an additional backbone cyclic analog of α-MSH [64].

7.3.3 Protein mimetics

Cyclic peptides are particularly useful for the mimicry of protein domains, as their conformation can be fixed in the desired structure. As described earlier, backbone cyclization uniquely preserves functional side chain moieties often critical for protein mimetics. For these key advantages and other druglike properties, backbone cyclic peptides are especially suited to the mimicry of proteins and protein domains.

7.3.3.1 Insulin-like growth factor I receptor

IGF-IR is a cell surface receptor tyrosine kinase activated by IGF-I and IGF-II endocrine hormones. IGF-IR is key for the regulation of cell growth and metabolism, and its activation is critical in various cancer-related signaling pathways. Moreover, overexpression of IGF-IR is related to the development of resistance to anticancer therapeutics, and in some cases, it can be used as a biomarker for cancer. IGF-I, also somatomedin C, is a 70-amino acid single-chain polypeptide with three disulfide bridges, which is produced mainly by the liver and also in some other target tissues. IGF-I plays a significant role in human physiology and development [110]. Mecasermin (Increlex) is a synthetic analog of IGF-I, which is used for the treatment of growth failure in children with IGF-I deficiency [111].

Qvit et al. developed a 34-member library of macrocyclic peptides derived from the sequence of the IGF-IR activation loop. Synthesis was performed either using N^{α}-(ω-thioalkylene) building units and commercial amino acids or using functionalized aldehydes. The library was based on tyrosine residues of the IGF-IR activation loop. The best compound inhibited the activity of IGF-IR (IC_{50} 6 μM), while no inhibitory effect was detected on the closely related tyrosine kinase epidermal growth factor receptor (EGFR) (Fig. 7.25). These macrocyclic molecules have properties of peptides, such as modularity, flexibility, and stability, and also possess drug-like properties of small molecules that comply with Veber [112] and Lipinski [113] rules. The novel approach presented in this study for the conversion of active protein sequences into small macrocyclic

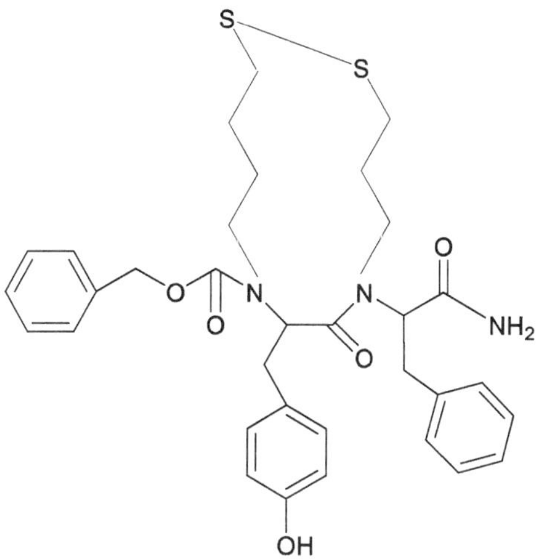

FIGURE 7.25 Structure of the backbone cyclic analog of IGF-IR [4,56].

peptides was another important step toward the incorporation of peptides and peptidomimetics into the drug discovery toolkit [4,56].

7.3.3.2 Bovine pancreatic trypsin inhibitor

BPTI is a small protein (58 residues) that was approved as an antifibrinolytic drug under the trade name Trasylol. However, the drug was withdrawn from the market in 2007 due to the risk of complications or death [114]; BPTI has since become available for limited use. As a result, there was an opportunity to develop smaller peptide compounds with drug activity and increased safety. The BPTI protein inhibits serine proteases, such as trypsin, kallikrein, chymotrypsin, and plasmin, and is often sourced from bovines [115]. A short domain creates a loop-like structure stabilized by a disulfide bond, which is critical for binding of BPTI to trypsin [116]. Equipped with this information, Kasher et al. developed a backbone bicyclic nonapeptide antagonist that acts as an inhibitor of trypsin (Ki ~76 μM) (Fig. 7.26). Compared to the natural protein, BPTI (Ki $\sim 10^{-13}$ M) [117], inhibition by this peptidomimetic is modest. Differences in the biological activity may be due to the replacement and omission of several residues, as well as conformational restriction [58]. In a follow-up study, Kasher et al. prepared a backbone-cyclized peptide library based on the same short domain that creates a loop-like structure. Although the bioactivity of the backbone-cyclized peptides in this library did not improve significantly, the authors suggested using the structure of the peptides that they solved to further develop more bioactive compounds (Fig. 7.27) [118].

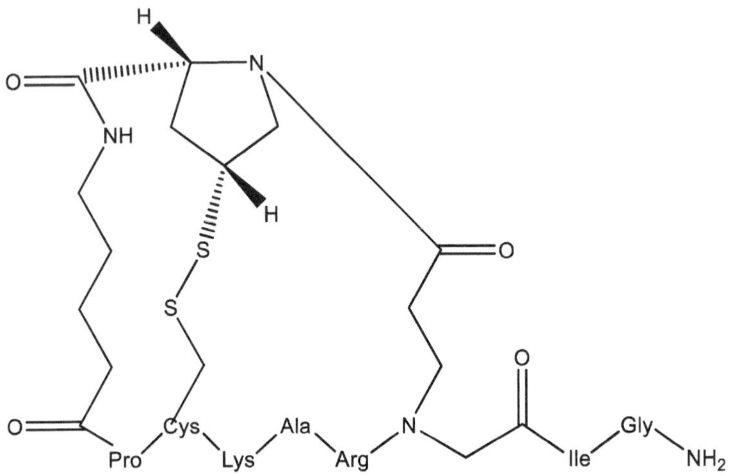

FIGURE 7.26 Structure of the backbone cyclic analog of BPTI [58].

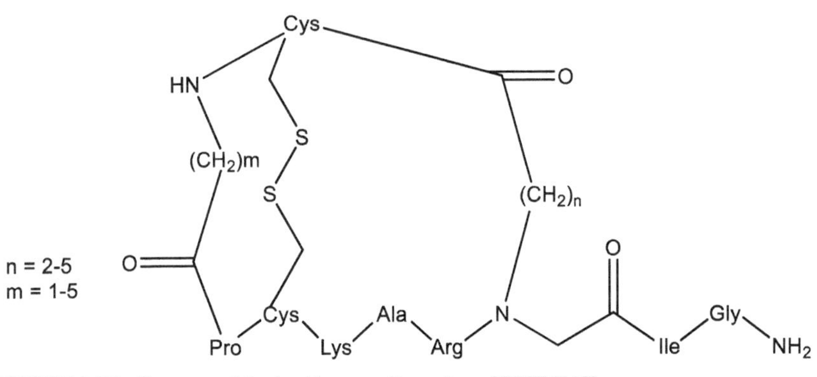

FIGURE 7.27 Structure of the backbone cyclic analog of BPTI [118].

7.3.3.3 Human immunodeficiency virus

HIV has been the target of numerous therapeutic programs making use of backbone-cyclized peptides, and its study also led to the identification of cell-penetrating peptides (CPPs). The HIV trans-activating transcriptional activator (TAT) can be taken up from media into cells by its protein transduction domain (PTD), which led to the development of many CPPs. The linear sequence has been used as a CPP to deliver small molecules, peptides and proteins into cells. For example, the fusion of the TAT PTD to the 120 kDa beta-galactosidase protein resulted in the delivery of the functional protein to most tissues in mice, including the brain [119]. Although it is clear that TAT allows these molecules to cross the cell membrane, there is significant debate in the field over the mechanism [120]. Nevertheless, CPPs have demonstrated efficacy and safety in clinical trials for diagnostic (e.g., contrast agent delivery) and therapeutic (e.g., drug delivery) indications [121]. These molecules are effective vehicles for the transport of diverse cargos into cells, yet their use is limited by the inherent lack of target cell specificity. Thus, it is important to consider the potential toxicity of any CPP-conjugated payload; several enhanced CPP systems are being developed to address this issue [122], and there are also other novel approaches for delivering cargo into cells [123,124].

In addition to facilitating cellular import through the PTD, the TAT protein contains a nuclear localization signal (NLS) with an arginine-rich motif (ARM) that mediates nuclear localization within the cell. Peptide mimicry of this sequence can either facilitate or inhibit nuclear import depending on the context, and several groups have applied backbone cyclization to develop enhanced analogs. Friedler et al. developed a backbone-cyclized peptide library using the N^{α}-(ω-aminoalkylene) building unit, which mimics the ARM of the HIV-1 TAT protein, based on the NMR structure of TAT; the molecules were screened for their ability to mediate nuclear import of corresponding bovine serum albumin conjugates in cells. The authors identified one backbone-cyclized peptide, TAT11, with active NLS properties. They also studied the mechanism of nuclear import, concluding that the peptide used the same pathway as the native TAT NLS. The backbone-cyclized peptide inhibited nuclear import mediated by the TAT NLS in a selective manner, and also inhibited the binding of the HIV-1 Rev ARM to its RNA binding element (Rev response element) (K_i 5 nM) (Fig. 7.28) [125].

In another study based on the NMR structure of HIV-1 matrix protein (MA) (Fig. 7.29), Friedler et al. developed a library of backbone-cyclized peptides using the N^{α}-(ω-aminoalkylene) building unit for cyclization. The authors screened for bioactivity by inhibition of nuclear import by an NLS on bovine serum albumin in cells. In the initial screen, they identified a lead compound ($IC_{50} = 3\ \mu M$). Based on the lead compound, they developed a second library that led to the discovery of a very potent backbone-cyclized

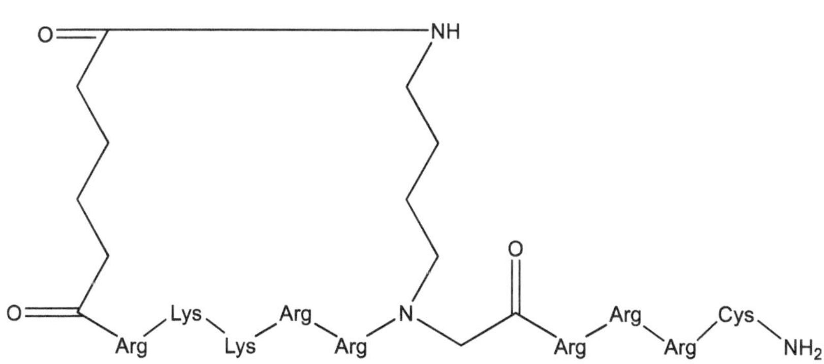

FIGURE 7.28 Structure of the backbone cyclic analog of the ARM in the HIV-1 TAT [125].

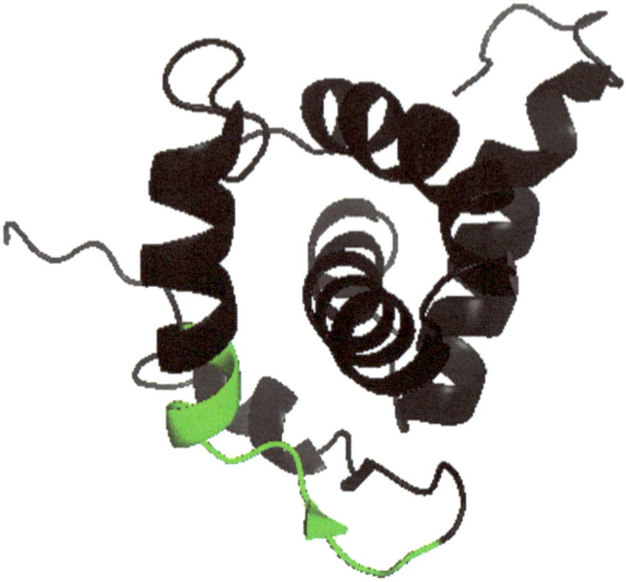

FIGURE 7.29 NMR structure of the HIV-1 matrix protein (MA). Peptides were derived from the domain shown in green.

peptide, BCvir (IC_{50} 35 nM, compared to the linear parent HIV-1 MA NLS peptide IC_{50} 12 μM). In addition, the novel peptide reduced HIV-1 production by 75% in cultured human cells and was highly stable (Fig. 7.30) [126].

Hariton-Gazal et al. constructed a library of backbone-cyclized peptides with N^{α}-(ω-aminoalkylene) building units, based on the previously identified peptide BCvir, derived from the NLS of HIV-1 MA [126]. In this library, the scaffold was constant, but the amino acid residue at the C-terminus was different in each backbone-cyclized peptide. The authors found that while the C-terminus was occupied with bulky hydrophobic amino acids, such as Phe and Nal (naphthyl Ala), the backbone-cyclized peptides were highly bioactive, inhibiting nuclear import in vitro (inhibition of 70% and 90%, respectively); polar amino acid residues, such as Asn and Cys, did not exhibit an inhibitory effect (Fig. 7.31) [127].

Rev is an attractive drug target because it provides an essential signal for HIV-1 mRNAs to be exported from the nucleus for translation. Chaloin et al. developed a library of backbone-cyclized peptides using N^{α}-(ω-aminoalkylene) building units and simultaneous multipeptide SPS. The authors evaluated in vitro inhibition of HIV-1 replication in chronically infected T lymphocytes. The backbone-cyclized peptide Rev-BCP 14 bound Rev responsive element and distorted the major groove of viral RNA in molecular dynamics simulations (IC_{50} 6 μM). Further, the compound interfered with the synthesis of gag and env proteins, likely by preventing the intracellular accumulation of unspliced viral RNA (Fig. 7.32) [128].

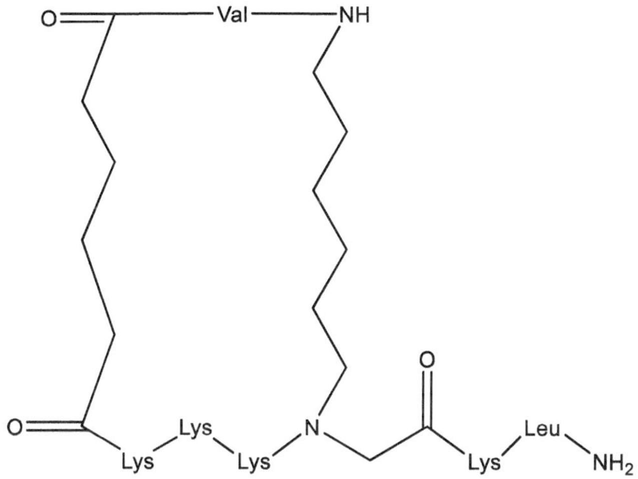

FIGURE 7.30 Structure of the backbone cyclic analog of the HIV-1 MA, BCvir [126].

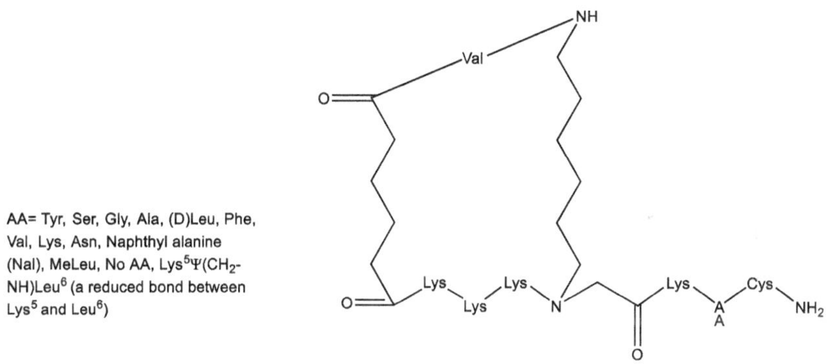

FIGURE 7.31 Structure of additional backbone cyclic analogs of the HIV-1 MA [127].

HIV-1 integrase (IN) is another drug target because it mediates the integration of retroviral DNA into the host genome. Hayouka et al. developed a backbone-cyclized library using N^{α}-(ω-aminoalkylene) building units that mimic the HIV-1 IN binding loop which interacts with its cellular cofactor, lens epithelium-derived growth factor (LEDGF/p75). Biological screening identified one backbone-cyclized peptide, c(MZ 4−1), as a potent and stable inhibitor of HIV-1 IN activity. c(MZ 4−1) binds (low micromolar range) and inhibits integrase catalytic activity by 90% (at 100 nM) in vitro, and also inhibits HIV-1 replication in cells (Fig. 7.33) [57]. In a follow up study, Hayouka and colleagues evaluated the effect of backbone cyclization combining N-alkylation with side-chain cyclization, for which the building units are commercially available and therefore easier to implement in synthesis. Using a similar biological assay, they identified inhibitors of integrase

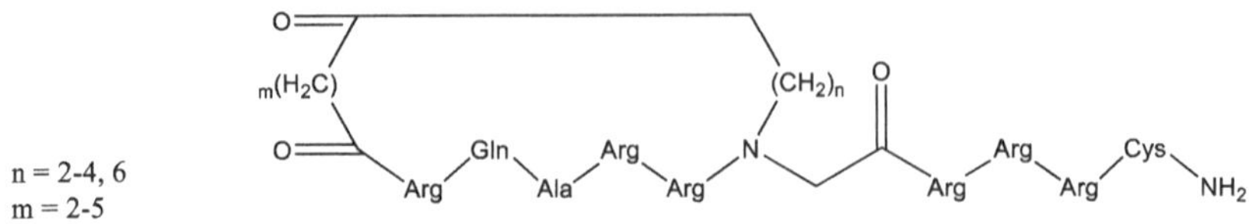

FIGURE 7.32 Structure of the backbone cyclic analog of Rev, Rev-BCP 14 [128].

FIGURE 7.33 Structure of the backbone cyclic analog of the HIV-1 IN, c(MZ 4−1) [57].

and found that backbone-cyclized peptides were generally more stable due to the effects of N-alkylation. However, the most potent side-chain cyclic peptide was almost as potent as some backbone-cyclized peptides and could serve as an alternative to backbone cyclization when synthesis is prohibitive (Fig. 7.34) [129].

7.3.3.4 Leishmania's receptor for activated C-kinase

Leishmaniasis is one of the most neglected tropical diseases in terms of drug discovery and development. In over 88 countries, Leishmania threatens millions of the poorest people, especially those living in rural areas. Up to two million people are infected with Leishmania each year, and 70% of those who die from leishmaniasis are children. There are two main forms of leishmaniases, visceral leishmaniasis (VL), the most severe form of the disease in humans, and cutaneous leishmaniasis (CL), the most common form of leishmaniasis in the New World. CL is endemic in 82 countries, with an incidence of approximately 1.5 million cases per year. Leishmania's receptor for activated C-kinase (LACK) is a scaffold protein required for the viability of the parasite and for its establishment in the host [130]. LACK plays an important role in the early phase of Leishmania infection [131]; LACK-deficient parasites are not viable, and parasites expressing lower levels of LACK fail to parasitize even immune-compromised mice [132].

Qvit et al. developed a small library of backbone-cyclized peptides derived from the *Leishmania sp.* parasite scaffold protein LACK. One of these peptides, p4d (Fig. 7.35), exhibited leishmanicidal activity in culture and was also effective at reducing parasitemia and increasing survival in

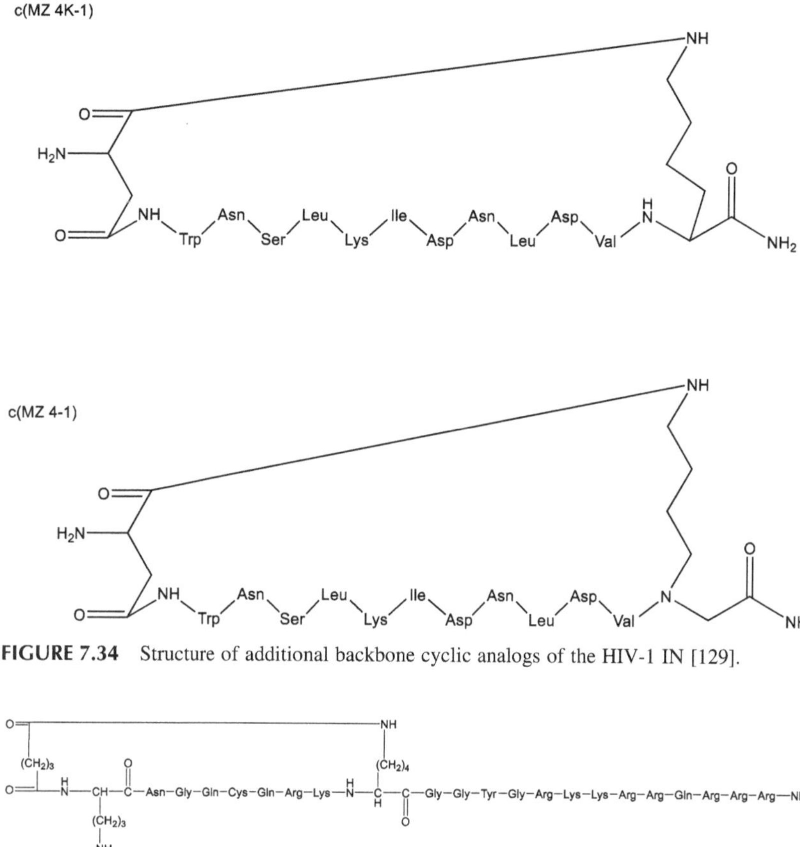

FIGURE 7.34 Structure of additional backbone cyclic analogs of the HIV-1 IN [129].

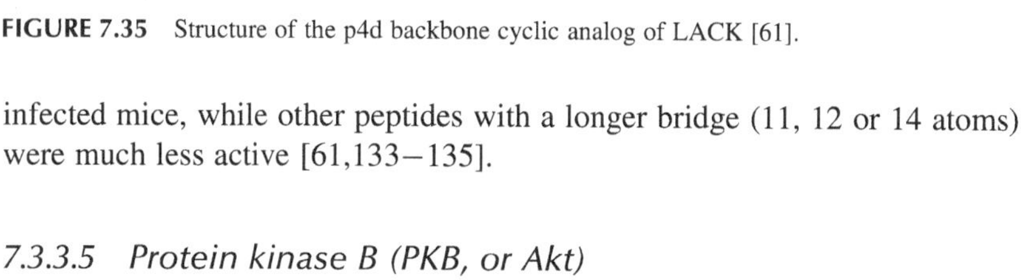

FIGURE 7.35 Structure of the p4d backbone cyclic analog of LACK [61].

infected mice, while other peptides with a longer bridge (11, 12 or 14 atoms) were much less active [61,133–135].

7.3.3.5 Protein kinase B (PKB, or Akt)

Protein kinase B (PKB, or Akt) is a serine/threonine kinase member of the AGC subfamily that has three mammalian isoforms, PKBα, PKBβ and PKBγ, which all have a conserved structure. PKB regulates cellular metabolism, survival, motility, transcription, and cell cycle progression by binding and phosphorylating various substrates. Moreover, PKB signaling is dysregulated in several human diseases, such as cancer, diabetes, and schizophrenia [136]. Phospholipid and small molecule inhibitors of PKB have been developed for the treatment of cancer but demonstrated no success in clinical trials thus far; only small molecule inhibitors of the broader phosphoinositide 3-kinase (PI3K)/Akt pathway have been approved [136].

Hurevich et al. synthesized a library of urea backbone-cyclized peptides using N^{α}-(ω-aminoalkylene) building units, creating analogs of glycogen synthase kinase 3 to target PKB as lead compounds for treating cancer. Several peptides inhibited PKB activity, one of which was particularly active (c(YM3–6) IC_{50} 0.16 μM) (Fig. 7.36) [44].

Tal-Gan et al. synthesized two backbone-cyclized peptide libraries using N^{α}-(ω-aminoalkylene) building units to study the effect of the cyclization approach, bridge chemistry, and ring size on the bioactivity of backbone-cyclized peptides. Interestingly, backbone-cyclized peptides in which the cyclization was done by backbone-to-backbone urea bridge were more potent than N-terminus-to-backbone amide peptides. Two analogs, compound 43 and compound 46, were 10-fold more active than the parent linear peptide (IC_{50} 0.16 and 0.17 μM, respectively) (Fig. 7.37) [59].

7.3.3.6 Nuclear factor-kappa B (NF-κB)

Nuclear factor-kappa B (NF-κB) is a protein complex transcription factor that controls cytokine production and cell survival through the regulation of DNA transcription and is present in almost all cell types. Over-activation of NF-κB is related to inflammatory diseases, infection, and cancer. The protein is involved in many signal transduction cascades and plays a pivotal role in the ubiquitin-proteasome system. NF-κB forms a complex with inhibitor kappa B (IκB), which binds beta-transduction repeat-containing protein (β-TrCP), enabling NF-κB to enter the cell nucleus [137]. Several approved small molecule and monoclonal antibody therapeutics have been shown to modulate the NF-κB signaling cascade, but it is believed that none of these molecules directly modulate NF-κB [137].

Qvit et al. developed an approach for the synthesis of backbone-cyclized phosphorylated peptides derived from the conserved phosphorylated sequence of inhibitor kappa B, $DS(PO_3)GXXS(PO_3)$ where X is any amino acid, which binds β-TrCP. These novel backbone-cyclized peptides prevented the release of NF-κB and its translocation to the nucleus, which could be used to reduce its negative effects in disease. Several backbone-cyclized

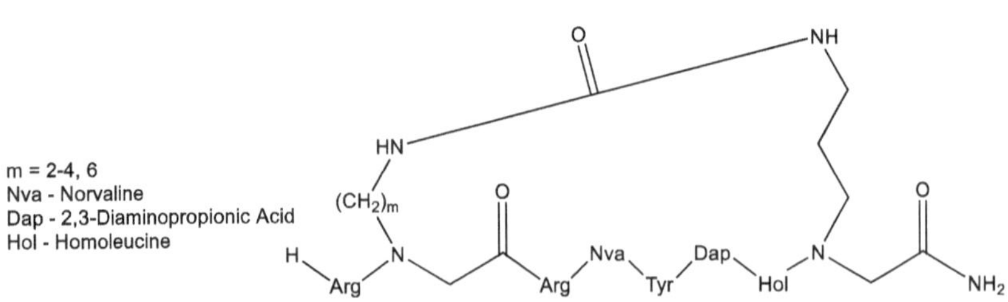

FIGURE 7.36 Structure of the backbone cyclic analog of PKB, c(YM3–6) [44].

Nva - Norvaline
Dap - 2,3-Diaminopropionic Acid
Hol - Homoleucine
Compound 43 c(YM3-6) m=6, n=3
Compound 46 c(YM4-4) m=4, n=4

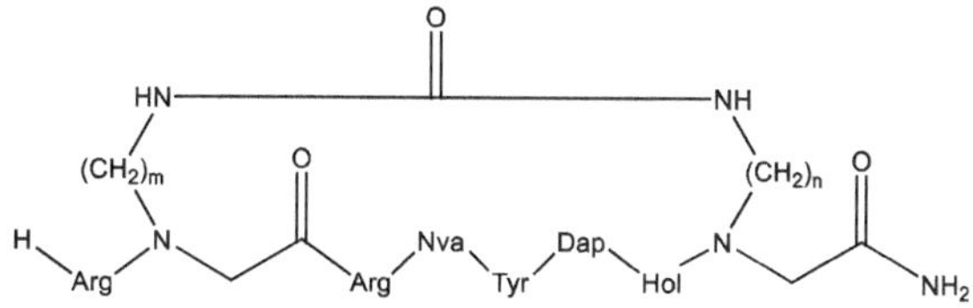

FIGURE 7.37 Structure of an additional backbone cyclic analog of PKB [59].

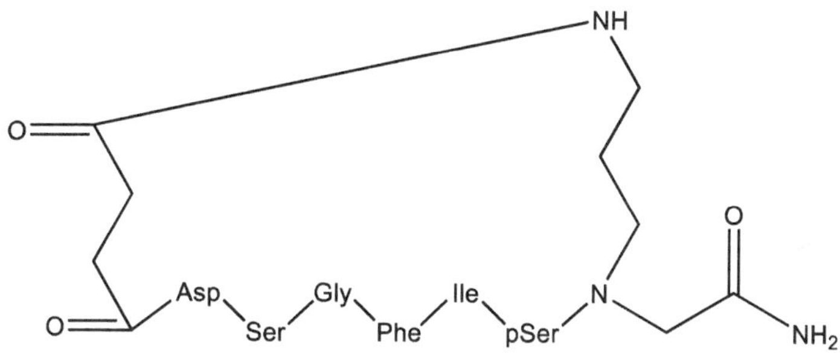

FIGURE 7.38 Structure of the backbone cyclic analog of NF-κB, IκB31−37(pBC-2,3) [54].

phosphopeptides with N^{α}-(ω-aminoalkylene) building units demonstrated high bioactivity in vitro, with the best compound, $I\kappa B_{31-37}$ (pBC-2,3), achieving approximately 90% inhibition of IκB ubiquitylation at a concentration of 3 μM (Fig. 7.38) [54,55,138].

7.3.3.7 Human leukocyte antigen class II histocompatibility, D related beta chain (HLA-DRB1)

Human leukocyte antigen class II histocompatibility, D related beta chain (HLA-DRB1) is expressed on antigen-presenting cells (APCs) and plays a critical role in the immune system by presenting antigenic peptides processed from extracellular proteins to T helper cells. HLA-DRB1 shared epitope alleles have been associated with an increased incidence of rheumatoid arthritis (RA), which is an autoimmune inflammatory disease that primarily affects joints, causing pain, swelling, stiffness, and loss of function. The cause of RA is not known, but likely involves a combination of genetic and environmental factors. Treatment generally includes lifestyle changes, pharmacological immunosuppression, and/or surgery to slow joint damage and reduce pain, but there is no curative regimen [139].

Many RA patients have a five-amino acid shared epitope motif in the DRB1-chain of the HLA protein, which may activate nitric oxide production. Naveh et al. developed a library of backbone-cyclized peptides to stabilize the alpha-helix structure of the DRB1 domain. They synthesized a small library of backbone-cyclized peptides to mimic the shared epitope motif. Although the study identified a stable backbone-cyclized peptide, c(HS4−4), which activated nitric oxide (NO) production in the low nanomolar range, the authors suggest that this lead compound could be useful for informing the development of potent inhibitory antagonists as well (Fig. 7.39) [140,141].

7.3.3.8 Chemokine (C-C motif) receptor 2 (CCR2)

Chemokine (C-C motif) receptor 2 (CCR2) mediates monocyte chemotaxis, in cases of both beneficial (antitumor) and pathological (autoimmune) inflammatory infiltration. Moreover, dysregulation of CCR2 has been linked to multiple sclerosis and RA, amongst other diseases. CCR2 binds via a helical dimerization site to its ligand, C-C motif chemokine ligand 2 (CCL2, also monocyte chemoattractant protein 1 or MCP1), by helix-helix interactions. An approach to inhibit these potentially pathogenic interactions involved the development of backbone-cyclized peptides, which are an optimal scaffold for helical peptidomimetics [142]. Despite the limited historical efficacy of other CCR2 antagonists [143], several small molecules targeting CCR2, such as PF-6309 and CCX872 for pancreatic cancer, have shown promise in recent clinical trials [144].

Hurevich et al. synthesized peptidomimetics using N^{α}-(ω-aminoalkylene) building units and microwave-assisted urea backbone cyclization to construct a "helix-walk" library based on the CCR2 dimerization region sequence. The authors identified potent dimerization sites from the CCR2 G-protein-coupled receptor (GPCR), which they used to synthesize a short linear peptide. Since linear peptides do not form helical structures in solution, they performed backbone cyclization to assemble a cyclic helix mimetic from the short linear peptide. The "helix-walk" refers to the construction of the cyclic library, whereby the dimerization site mimetics were cyclized between ring

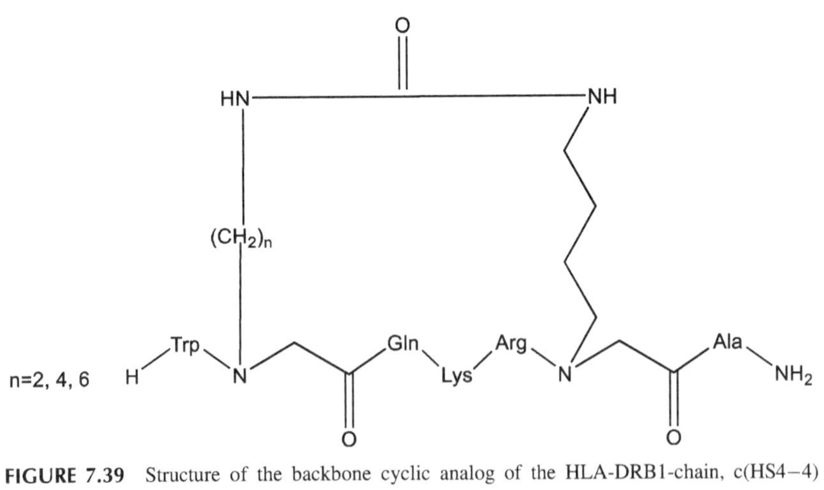

FIGURE 7.39 Structure of the backbone cyclic analog of the HLA-DRB1-chain, c(HS4–4) [140].

positions *i* and *i* + 4, i and *i* + 7, or *i* and *i* + 3, which allowed for frame shift backbone cyclization scanning. Hurevich et al. used this cycloscan approach to identify the optimal characteristics for the backbone-cyclized peptidomimetic to inhibit the binding of CCR2 to its ligand by testing the effect of bridge position on bioactivity. They developed a library of backbone-cyclized peptides to mimic the helical conformation region of the CCR2 dimerization site, which is important for immune responses and has been implicated in the pathogenesis of immune-mediated diseases such as multiple sclerosis. The stable helix structure of backbone-cyclized peptidomimetics produced library hits such as peptides hCCR2(61−67), which specifically inhibited CCR2-mediated chemotactic migration in cultured monocytes. In a bridge position scan, systematic replacement of the building units that form the bridge and adjustment of their position revealed that one backbone-cyclized peptide, M3D-1 (analog of hCCR2(61−67)), specifically inhibited CCR2-mediated chemotactic migration in cells (Fig. 7.40) [145].

7.3.3.9 Src homology region 2 domain-containing phosphatase-1 (SHP-1)

SHP-1 is a protein tyrosine phosphatase expressed by hematopoietic cells and is involved in signaling pathways that regulate cellular activation, growth, proliferation, differentiation, cell cycle, and oncogenesis. Intracellular adaptor proteins help recruit SHP-1, and its tandem Src homology (SH2) domains mediate binding to phospho-tyrosines on target proteins; SHP-1 then dephosphorylates transmembrane receptor cytoplasmic domains and removes phosphates from other intracellular kinases. Several small-molecule SHP-1 antagonists have been suggested for development as anti-cancer agents [146,147].

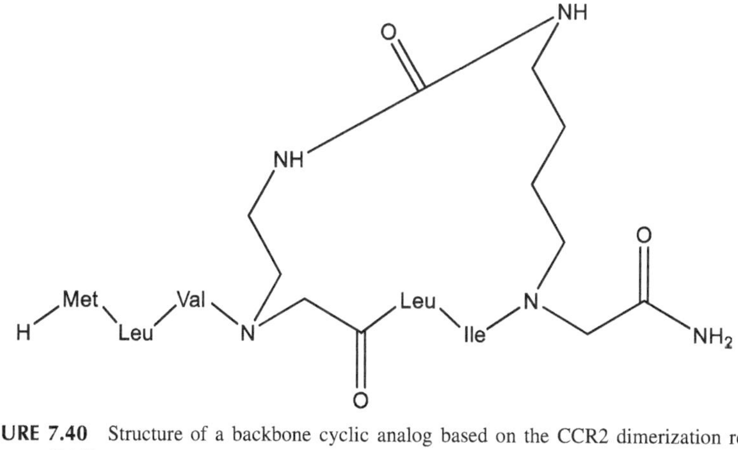

FIGURE 7.40 Structure of a backbone cyclic analog based on the CCR2 dimerization region sequence [145].

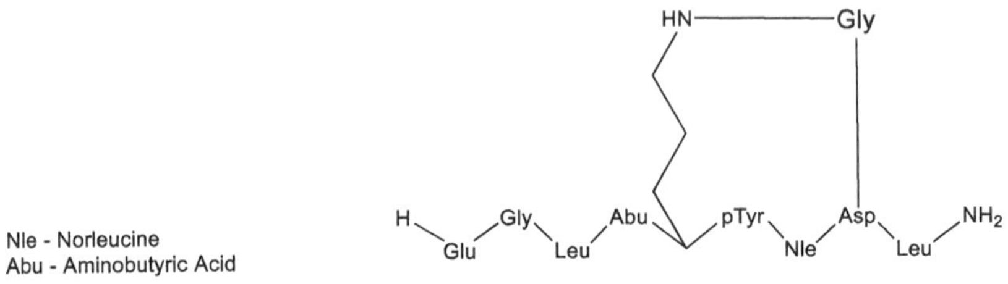

FIGURE 7.41 Structure of the backbone cyclic analog of SHP-1 [148].

Zoda et al. synthesized backbone-to-side chain cyclized octapeptides based on the ligand consensus sequence, which binds the N-terminal SH2 domain of SHP-1 and activates the protein. The authors used dipeptide building blocks and N-functionalized tyrosine derivatives to overcome challenges associated with the condensation of N-protected amino acids. Using these building blocks, they performed SAR studies, which incorporated ring size, flexibility, and hydrophobicity of the bridge. The authors found that backbone-cyclized peptide activity increased (up to 14-fold greater than basal activity) with bridge chain length due to greater flexibility and presumably more favorable binding conformations (Fig. 7.41) [148].

7.4 The future of backbone cyclic peptidomimetics

Over the past century, scientists have used peptides and peptidomimetics to address a variety of problems in biology, ranging in application from research tools to therapeutic leads, and from broad conceptual to focused mechanistic studies. We predict that fields hindered by obstacles to pharmaceutical innovation, such as the development of new antibiotics, will be revived in part by the adoption of peptide or peptidomimetic formats – particularly through the integration of backbone-cyclized peptides, which possess important therapeutic qualities. Further applications of rationally designed or otherwise synthetic peptides may be demonstrated in related fields such as agricultural sciences and environmental microbiology, where numerous examples of natural backbone-cyclized peptides can already be seen. We anticipate that the diverse chemistries, which can be integrated into backbone-cyclized peptides, will present opportunities for the incorporation of more complex features, such as photo- or enzyme-cleavable units or small molecule drug conjugates.

References

[1] Brinckerhoff LH, et al. Terminal modifications inhibit proteolytic degradation of an immunogenic MART-1(27–35) peptide: implications for peptide vaccines. Int J Cancer 1999;83(3):326–34.

[2] Avan I, et al. Peptidomimetics via modifications of amino acids and peptide bonds. Chem Soc Rev 2014;43(10):3575–94.

[3] Yin H. Constrained peptides as miniature protein structures. ISRN Biochem 2012;2012:1–15.
[4] Qvit N, et al. Synthesis of a novel macrocyclic library: discovery of an IGF-1R inhibitor. J Comb Chem 2008;10(2):256–66.
[5] Qvit N. Microwave-assisted synthesis of cyclic phosphopeptide on solid support. Chem Biol Drug Des 2014;85(3):300–5.
[6] Zorzi A, et al. Cyclic peptide therapeutics: past, present and future. Curr Opin Chem Biol 2017;38:24–9.
[7] White CJ, et al. Contemporary strategies for peptide macrocyclization. Nat Chem 2011;3 (7):509–24.
[8] Roxin A, et al. Flexible or fixed: a comparative review of linear and cyclic cancer-targeting peptides. Future Med Chem 2012;4(12):1601–18.
[9] Tsomaia N. Peptide therapeutics: targeting the undruggable space. Eur J Med Chem 2015;94:459–70.
[10] Fosgerau K, et al. Peptide therapeutics: current status and future directions. Drug Discov Today 2015;20(1):122–8.
[11] Russo A, et al. Targeting “undruggable” proteins: design of synthetic cyclopeptides. Curr Med Chem 2016;23(8):748–62.
[12] Qvit N, et al. Cyclic peptides for protein-protein interaction targets. Curr Top Med Chem 2018;18(7):525 -525.
[13] Fischer E. Einfluss der configuration auf die wirkung der enzyme. Ber Dtsch Chem Ges 1894;27:2985–93.
[14] Koshland DE. Application of a theory of enzyme specificity to protein synthesis. Proc Natl Acad Sci USA 1958;44(2):98–104.
[15] Koshland DE. The key-lock theory and the induced fit theory. Angewandte Chemie International Edition in English 1995;33(23–24):2375–8.
[16] Powell GB, et al. Combinatorial peptide libraries: mining for cell-binding peptides. Chem Rev 2014;114(2):1020–81.
[17] Rubin S, et al. Cyclic peptides for protein-protein interaction targets: applications to human disease. Crit Rev Eukaryot Gene Expr 2016;26(3):199–221.
[18] Kolb HC, et al. Click chemistry: diverse chemical function from a few good reactions. Angewandte Chemie International Edition in English 2001;40(11):2004–21.
[19] Tornoe CW, et al. Peptidotriazoles on solid phase: [1,2,3]-triazoles by regiospecific copper(i)-catalyzed 1,3-dipolar cycloadditions of terminal alkynes to azides. J Org Chem 2002;67(9):3057–64.
[20] Castro V, et al. CuAAC: An efficient click chemistry reaction on solid phase. ACS Comb Sci 2016;18(1):1–14.
[21] Kazmaier U, et al. A straightforward approach towards cyclic peptides via ring-closing metathesis-scope and limitations. Org Biomol Chem 2005;3(1):136–45.
[22] Hill TA, et al. Constraining cyclic peptides to mimic protein structure motifs. Angew Chem Int Ed Engl 2014;53(48):13020–41.
[23] Kotha S, et al. Design and synthesis of polycycles, heterocycles, and macrocycles via strategic utilization of ring-closing metathesis. Chem Rec 2018;18(11):1613–32.
[24] Blackwell HE, et al. Highly efficient synthesis of covalently cross-linked peptide helices by ring-closing metathesis. Angewandte Chemie International Edition in English 1998;37 (23):3281–4.
[25] Tan YS, et al. Stapled peptide design: principles and roles of computation. Drug Discov Today 2016;21(10):1642–53.

[26] Bird GH, et al. Chemical synthesis of hydrocarbon-stapled peptides for protein interaction research and therapeutic targeting. Curr Protoc Chem Biol 2011;3(3):99−117.

[27] Ali AM, et al. Stapled peptides inhibitors: a new window for target drug discovery. Comput Struct Biotechnol J 2019;17:263−81.

[28] Dawson PE, et al. Synthesis of proteins by native chemical ligation. Science 1994;266 (5186):776−9.

[29] Tulla-Puche J, et al. On-resin native chemical ligation for cyclic peptide synthesis. J Org Chem 2004;69(12):4101−7.

[30] Clark RJ, et al. Native chemical ligation applied to the synthesis and bioengineering of circular peptides and proteins. Biopolymers 2010;94(4):414−22.

[31] Conibear AC, et al. Native chemical ligation in protein synthesis and semi-synthesis. Chem Soc Rev 2018;47(24):9046−68.

[32] Gilon C, et al. Backbone cyclization—a new method for conferring conformational constraint on peptides. Biopolymers 1991;31(6):745−50.

[33] Rubin SJS, et al. Backbone-cyclized peptides: a critical review. Curr Top Med Chem 2018;18(7):526−55.

[34] Rubin SJ, et al. Conversion of protein active regions into peptidomimetic therapeutic leads using backbone cyclization and cycloscan−how to do it yourself!. Curr Top Med Chem 2018;18(7):556−65.

[35] Montalbetti CA, et al. Amide bond formation and peptide coupling. Tetrahedron 2005;61 (46):10827−52.

[36] Pattabiraman VR, et al. Rethinking amide bond synthesis. Nature 2011;480(7378):471−9.

[37] Parenty A, et al. Macrolactonizations in the total synthesis of natural products. Chem Rev 2006;106(3):911−39.

[38] Gongora-Benitez M, et al. Multifaceted roles of disulfide bonds. Peptides as therapeutics. Chem Rev 2014;114(2):901−26.

[39] Demmer O, et al. Design of cyclic peptides. In Peptide and Protein Design for Biopharmaceutical Applications. John Wiley & Sons, Ltd.; 2009. p. 133−76.

[40] Gazal S, et al. Human somatostatin receptor specificity of backbone-cyclic analogues containing novel sulfur building units. J Med Chem 2002;45(8):1665−71.

[41] Altstein M, et al. Backbone cyclic peptide antagonists, derived from the insect pheromone biosynthesis activating neuropeptide, inhibit sex pheromone biosynthesis in moths. J Biol Chem 1999;274(25):17573−9.

[42] Barda Y, et al. Backbone metal cyclization: novel 99mTc labeled GnRH analog as potential SPECT molecular imaging agent in cancer. Nucl Med Biol 2004;31(7):921−33.

[43] Fridkin G, et al. Backbone metal-cyclization: a novel approach for simultaneous peptide cyclization and radiolabeling. Application to the combinatorial synthesis of rhenium-cyclic somatostatin analogs. Nucl Med Biol 2005;32(1):39−50.

[44] Hurevich M, et al. Novel method for the synthesis of urea backbone cyclic peptides using new Alloc-protected glycine building units. J Peptide Sci 2010;16(4):178−85.

[45] Fridkin G, et al. Azo cyclization: peptide cyclization via azo bridge formation. J Pept Res 2002;60(2):104−11.

[46] Dash A, et al. Peptide receptor radionuclide therapy: an overview. Cancer Biother Radiopharm 2015;30(2):47−71.

[47] Gilon C, et al. Cycloscan: backbone cyclic conformational libraries of peptides. In: 24th Conference of the European, Peptide Society. Edinbrough, Scotland; 1998.

[48] Bitan G, et al. New backbone cyclic substance P analogs. Lett Pept Sci 1995;2(3−4):121−4.

[49] Byk G, et al. Synthesis and biological activity of NK-1 selective, N-backbone cyclic analogs of the C-terminal hexapeptide of substance P. J Med Chem 1996;39(16):3174–8.
[50] Gilon C, et al. A backbone-cyclic, receptor 5-selective somatostatin analogue: synthesis, bioactivity, and nuclear magnetic resonance conformational analysis. J Med Chem 1998;41(6):919–29.
[51] Afargan M, et al. Novel long-acting somatostatin analog with endocrine selectivity: potent suppression of growth hormone but not of insulin. Endocrinology 2001;142(1):477–86.
[52] Ovadia O, et al. Improvement of drug-like properties of peptides: The somatostatin paradigm. Expert Opin Drug Discov 2010;5(7):655–71.
[53] Zeltser I, et al. Insect neuropeptide antagonist. Part II. Synthesis and biological activity of backbone cyclic and precyclic PBAN antagonists. J Pept Res 2001;58(4):275–84.
[54] Qvit N, et al. Design and synthesis of backbone cyclic phosphorylated peptides: the IκB model. Biopolymers 2009;91(2):157–68.
[55] Qvit N, et al. Design and synthesis of backbone cyclic phosphopeptides: the IκB model. In 20th American Peptide Symposium. Montreal, Canada; 2009.
[56] Qvit N, et al. Novel macrocyclic libraries with spatial diversity (Sibs): application for the discovery of IGF-1R inhibitor. In 28th European Peptide Symposium & 3rd International. Prague, Czech Republic; 2004.
[57] Hayouka Z, et al. Cyclic peptide inhibitors of HIV-1 integrase derived from the LEDGF/p75 protein. Bioorg Med Chem 2010;18(23):8388–95.
[58] Kasher R, et al. Synthesis of a bicyclic BPTI mimetic containing 4-thioproline replacing Cys(38). Lett Peptide Sci 1998;5(2–3):101–3.
[59] Tal-Gan Y, et al. Backbone cyclic peptide inhibitors of protein kinase B (PKB/Akt). J Med Chem 2011;54(14):5154–64.
[60] Qvit N, et al. Selective phosphorylation inhibitor of delta protein kinase C-pyruvate dehydrogenase kinase protein-protein interactions: application for myocardial injury in vivo. J Am Chem Soc 2016;138(24):7626–35.
[61] Qvit N, et al. Scaffold proteins LACK and TRACK as potential drug targets in kinetoplastid parasites: development of inhibitors. Int J Parasitol: Drugs Drug Resist 2016;6:74–84.
[62] Linde Y, et al. Structure-activity relationship and metabolic stability studies of backbone cyclization and N-methylation of melanocortin peptides. Biopolymers 2008;90(5):671–82.
[63] Hess S, et al. Effect of structural and conformation modifications, including backbone cyclization, of hydrophilic hexapeptides on their intestinal permeability and enzymatic stability. J Med Chem 2007;50(24):6201–11.
[64] Ovadia O, et al. The effect of backbone cyclization on PK/PD properties of bioactive peptide-peptoid hybrids: the melanocortin agonist paradigm. Bioorg Med Chem 2010;18(2):580–9.
[65] Merrifield RB. Solid phase peptide synthesis I. The synthesis of a tetrapeptide. J Am Chem Soc 1963;85:2149–54.
[66] Qvit N, et al. A laboratory preparation of aspartame analogs using simultaneous multiple parallel synthesis methodology. J Chem Educ 2007;84(12):1988–91.
[67] Qvit N. Laboratory for undergraduate students: preparation of aspartame analogs using simultaneous multiple parallel synthesis methodology. Abstracts of papers of the American Chemical Society.
[68] Nuti F, et al. Microwave-assisted solid phase synthesis of backbone cyclic glycopeptide libraries. In: 30 European Peptide Symposium. Helsinki, Finland; 2008.

[69] Qvit N, et al. Using novel benzophenons to study microbial sugar recognition and translocation by photo-affinity labeling/photolabeling. In: 29th European Peptide Symposium. Gdansk, Poland; 2006.

[70] Qvit N, et al. Development of bifunctional photoactivatable benzophenone probes and their application to glycoside substrates. Biopolymers 2008;90(4):526–36.

[71] Wang Q, et al. Expanding the genetic code for biological studies. Chem Biol 2009;16 (3):323–36.

[72] GroB A, et al. Synthetic peptides as protein mimics. Front Bioeng Biotechnol 2016;3 (211).

[73] Ngyen H, et al. Transition metal complexes of a cyclic pseudo hexapeptide: synthesis, complex formation and catalytic activities. J Pept Sci 2008;14(9):1010–21.

[74] Gambari R. Peptide nucleic acids: a review on recent patents and technology transfer. Expert Opin Ther Pat 2014;24(3):267–94.

[75] Fernandez-Llamazares AI, et al. Backbone N-modified peptides: how to meet the challenge of secondary amine acylation. Biopolymers 2015;104(5):435–52.

[76] Wu H. Gamma-AApeptides as a new class of peptidomimetics: synthesis, structures, and functions. Chemistry 2015;21(6):2501–7.

[77] Cheng RP, et al. Beta-peptides: from structure to function. Chem Rev 2001;101 (10):3219–32.

[78] Seebach D, et al. The world of beta- and gamma-peptides comprised of homologated proteinogenic amino acids and other components. Chem Biodivers 2004;1(8):1111–239.

[79] Seebach D, et al. Beta-peptidic peptidomimetics. Acc Chem Res 2008;41(10):1366–75.

[80] Sasmal S, et al. Synthesis of cyclic peptidomimetics from aldol building blocks. Journal of Organic Chemistry 2002;67(17):6260–3.

[81] Palomo C, et al. Asymmetric synthesis of beta-lactams through the Staudinger reaction and their use as building blocks of natural and nonnatural products. Curr Med Chem 2004;11(14):1837–72.

[82] Alcaide B, et al. Beta-lactams: versatile building blocks for the stereoselective synthesis of non-beta-lactam products. Chem Rev 2007;107(11):4437–92.

[83] Gruner SAW, et al. Carbohydrate-based mimetics in drug design: sugar amino acids and carbohydrate scaffolds. Chem Rev 2002;102(2):491–514.

[84] Paolini A, et al. A convenient microwave-assisted synthesis of N-glycosyl amino acids. Tetrahedron Lett 2007;48(16):2901–4.

[85] Munoz M, et al. Involvement of substance P and the NK-1 receptor in human pathology. Amino Acids 2014;46(7):1727–50.

[86] Mander K, et al. Advancing drug therapy for brain tumours: a current review of the pro-inflammatory peptide Substance P and its antagonists as anti-cancer agents. Recent Pat CNS Drug Discov 2014;9(2):110–21.

[87] Aziz F. Neurokinin-1 receptor antagonists for chemotherapy-induced nausea and vomiting. Ann Palliative Med 2012;1(2):130–6.

[88] Keller M, et al. Lack of efficacy of the substance P (neurokinin1 receptor) antagonist aprepitant in the treatment of major depressive disorder. Biol Psychiatr 2006;59 (3):216–23.

[89] Bitan G, et al. Backbone cyclization of the C-terminal part of substance P. Part 1: the important role of the sulphur in position 11. J Pept Sci 1996;2(4):261–9.

[90] Bitan G, et al. Synthesis and biological activity of novel backbone-bicyclic substance-P analogs containing lactam and disulfide bridges. J Pept Res 1997;49(5):421–6.

[91] Wormser U, et al. Highly selective agonists for substance P receptor subtypes. EMBO J 1986;5(11):2805–8.
[92] Laufer R, et al. Characterization of a neurokinin B receptor site in rat brain using a highly selective radioligand. J Biol Chem 1986;261(22):10257–63.
[93] Laufer R, et al. Desensitization with a selective agonist discriminates between multiple tachykinin receptors. J Pharmacol Exp Therap 1988;245(2):639–43.
[94] Behrens S, et al. Structure-activity relationship of the ring portion in backbone-cyclic C-terminal hexapeptide analogs of substance P. NMR and molecular dynamics. Int J Pept Protein Res 1996;48(6):569–79.
[95] Rafaeli A. Pheromone biosynthesis activating neuropeptide (PBAN): regulatory role and mode of action. Gen Comp Endocrinol 2009;162(1):69–78.
[96] Gilon C, et al. Rationally designed neuropeptide antagonists: anovel approch for generation of environmentally friendly insecticides. invertebrate. Neuroscience 1997;3:245–50.
[97] Bork K, et al. Icatibant. Nat Rev Drug Discov 2008;7(10):801–2.
[98] Schumann C, et al. Synthesis and biological activities of new side chain and backbone cyclic bradykinin analogues. J Pept Res 2002;60(2):128–40.
[99] Theodoropoulou M, et al. Somatostatin receptors: from signaling to clinical practice. Front Neuroendocrinol 2013;34(3):228–52.
[100] Rai U, et al. Therapeutic uses of somatostatin and its analogues: current view and potential applications. Pharmacol Ther 2015;152:98–110.
[101] Narayanan S, et al. Role of somatostatin analogues in the treatment of neuroendocrine tumors. Hematol Oncol Clin North Am 2016;30(1):163–77.
[102] Gazal S, et al. Synthesis of novel protected N-alpha(omega-thioalkyl) amino acid building units and their incorporation in backbone cyclic disulfide and thioetheric bridged peptides. J Pept Res 2001;58(6):527–39.
[103] Falb E, et al. A bicyclic and hsst2 selective somatostatin analogue: design, synthesis, conformational analysis and binding. Bioorg Med Chem 2001;9(12):3255–64.
[104] Giblin MF, et al. Design and characterization of alpha-melanotropin peptide analogs cyclized through rhenium and technetium metal coordination. Proc Natl Acad Sci USA 1998;95(22):12814–18.
[105] Melendez-Alafort L, et al. Labeling peptides with rhenium-188. Int J Pharm 1999;182 (2):165–72.
[106] Fujino M, et al. Synthetic analogs of luteinizing hormone releasing hormone (LH-RH) substituted in position 6 and 10. Biochem Biophys Res Commun 1974;60(1):406–13.
[107] Crowley Jr. WF, et al. The biologic activity of a potent analogue of gonadotropin-releasing hormone in normal and hypogonadotropic men. N Engl J Med 1980;302 (19):1052–7.
[108] Minder EI, et al. Pharmacokinetics and pharmacodynamics of afamelanotide and its clinical use in treating dermatologic disorders. Clin Pharmacokinet 2017;56(8):815–23.
[109] Hess S, et al. Backbone cyclic peptidomimetic melanocortin-4 receptor agonist as a novel orally administrated drug lead for treating obesity. J Med Chem 2008;51 (4):1026–34.
[110] Farabaugh SM, et al. Role of IGF1R in breast cancer subtypes, stemness, and lineage differentiation. Front Endocrinol (Lausanne) 2015;6:59.
[111] Khwaja OS, et al. Safety, pharmacokinetics, and preliminary assessment of efficacy of mecasermin (recombinant human IGF-1) for the treatment of Rett syndrome. Proc Natl Acad Sci USA 2014;111(12):4596–601.

[112] Veber DF, et al. Molecular properties that influence the oral bioavailability of drug candidates. J Med Chem 2002;45(12):2615–23.

[113] Lipinski CA, et al. Experimental and computational approaches to estimate solubility and permeability in drug discovery and development settings. Adv Drug Deliv Rev 1997;23(1–3):3–25.

[114] Mangus RS, et al. Aminocaproic acid (Amicar) as an alternative to Aprotinin (Trasylol) in liver transplantation. Transplant Proc 2014;46(5):1393–9.

[115] Chang JY. Distinct folding pathways of two homologous disulfide proteins: bovine pancreatic trypsin inhibitor and tick anticoagulant peptide. Antioxid Redox Signal 2011;14 (1):127–35.

[116] Huber R, et al. Structure of the complex formed by bovine trypsin and bovine pancreatic trypsin inhibitor. II. Crystallographic refinement at 1.9A resolution. J Mol Biol 1974;89 (1):73–101.

[117] Vincent JP, et al. Trypsin-pancreatic trypsin inhibitor association. Dynamics of the interaction and role of disulfide bridges. Biochemistry 1972;11(16):2967–77.

[118] Kasher R, et al. Miniaturized proteins: the backbone cyclic proteinomimetic approach. J Mol Biol 1999;292(2):421–9.

[119] Schwarze SR, et al. In vivo protein transduction: delivery of a biologically active protein into the mouse. Science 1999;285(5433):1569–72.

[120] Bechara C, et al. Cell-penetrating peptides: 20 years later, where do we stand? FEBS Lett 2013;587(12):1693–702.

[121] Guidotti G, et al. Cell-penetrating peptides: from basic research to clinics. Trends Pharmacol Sci 2017;38(4):406–24.

[122] Huang Y, et al. Curb challenges of the "Trojan Horse" approach: smart strategies in achieving effective yet safe cell-penetrating peptide-based drug delivery. Adv Drug Deliv Rev 2013;65(10):1299–315.

[123] Blum AP, et al. Activating peptides for cellular uptake via polymerization into high density brushes. Chemical Science 2016;7(2):989–94.

[124] Sharei A, et al. A vector-free microfluidic platform for intracellular delivery. Proc Natl Acad Sci U S A 2013;110(6):2082–7.

[125] Friedler A, et al. Development of a functional backbone cyclic mimetic of the HIV-1 Tat arginine-rich motif. J Biol Chem 2000;275(31):23783–9.

[126] Friedler A, et al. Backbone cyclic peptide, which mimics the nuclear localization signal of human immunodeficiency virus type 1 matrix protein, inhibits nuclear import and virus production in nondividing cells. Biochemistry 1998;37(16):5616–22.

[127] Hariton-Gazal E, et al. Inhibition of nuclear import by backbone cyclic peptidomimetics derived from the HIV-1 MA NLS sequence. Biochim Biophys Acta 2002;1594 (2):234–42.

[128] Chaloin L, et al. Potent inhibition of HIV-1 replication by backbone cyclic peptides bearing the Rev arginine rich motif. J Biomed Sci 2007;14(5):565–84.

[129] Hayouka Z, et al. A comparative study of backbone vs side chain peptide cyclization: application for HIV-1 integrase inhibitors. Bioorg Med Chem 2012;20(10):3317–22.

[130] Gomez-Arreaza A, et al. Leishmania mexicana: LACK (Leishmania homolog of receptors for activated C-kinase) is a plasminogen binding protein. Exp Parasitol 2011;127 (4):752–61.

[131] Mougneau E, et al. Expression cloning of a protective Leishmania antigen. Science 1995;268(5210):563–6.

[132] Kelly BL, et al. Leishmania major LACK antigen is required for efficient vertebrate parasitization. J Exp Med 2003;198(11):1689–98.

[133] Qvit N, et al. Targeting neglected diseases based on rational approach design; proof of concept: novel peptides for inhibiting leishmaniasis. In: 61st Annual Meeting of the American Society of Tropical Medicine and Hygiene. Atlanta, GA, USA; 2012.

[134] Qvit N, et al. Peptides that target protein-protein interactions as an anti-parasite strategy. Chimica Oggi/Chem Today 2014;32(6):62–6.

[135] Qvit N, et al. Development of a backbone cyclic peptide library as potential antiparasitic therapeutics using microwave irradiation. J Visual Exp 2016;107:e53589.

[136] Schwartz PA, et al. Protein kinase biochemistry and drug discovery. Bioorg Chem 2011;39(5–6):192–210.

[137] Zhang Q, et al. 30 years of NF-kappaB: a blossoming of relevance to human pathobiology. Cell 2017;168(1–2):37–57.

[138] Qvit N, et al. Inhibiting NF-kappa B by mimicking the I kappa B conserved sequence using backbone cyclic peptides. Biopolymers. Hoboken, NJ: John Wiley & Sons Inc.; 2007.

[139] Sardar S, et al. Old and new therapeutics for Rheumatoid Arthritis: in vivo models and drug development. Immunopharmacol Immunotoxicol 2016;38(1):2–13.

[140] Naveh S, et al. Developing potent backbone cyclic peptides bearing the shared epitope sequence as rheumatoid arthritis drug-leads. Bioorg Med Chem Lett 2012;22(1):493–6.

[141] Fu J, et al. A small shared epitope-mimetic compound potently accelerates osteoclast-mediated bone damage in autoimmune arthritis. J Immunol 2013;191(5):2096–103.

[142] Hurevich M, et al. Allosteric inhibition of G-protein coupled receptor oligomerization: strategies and challenges for drug development. Curr Top Med Chem 2014;14(15):1842–63.

[143] Horuk R. Chemokine receptor antagonists: overcoming developmental hurdles. Nat Rev Drug Discov 2009;8(1):23–33.

[144] Nywening TM, et al. Targeting tumour-associated macrophages with CCR2 inhibition in combination with FOLFIRINOX in patients with borderline resectable and locally advanced pancreatic cancer: a single-centre, open-label, dose-finding, non-randomised, phase 1b trial. Lancet Oncol 2016;17(5):651–62.

[145] Hurevich M, et al. Backbone cyclic helix mimetic of chemokine (C-C motif) receptor 2: A rational approach for inhibiting dimerization of G protein-coupled receptors. Bioorg Med Chem 2013;3958–66.

[146] Kundu S, et al. Novel SHP-1 inhibitors tyrosine phosphatase inhibitor-1 and analogs with preclinical anti-tumor activities as tolerated oral agents. J Immunol 2010;184(11):6529–36.

[147] Liu CY, et al. Novel sorafenib analogues induce apoptosis through SHP-1 dependent STAT3 inactivation in human breast cancer cells. Breast Cancer Res 2013;15(4):R63.

[148] Zoda MS, et al. Syntheses and activities of backbone-side chain cyclic octapeptide ligands with N-functionalized phosphotyrosine for the N-terminal SH2-domain of the protein tyrosine phosphatase SHP-1. J Pept Sci 2010;16(8):403–13.

Chapter 8

Pharmacokinetics and pharmacodynamics of peptidomimetics

Satheesh Kumar Nanjappan[1], Shruti Surendran[2] and David Paul[3]
[1]Department of Natural Products, National Institute of Pharmaceutical Education and Research, Kolkata, West Bengal, India, [2]Biocon BMS R&D Center, Bengaluru, Karnataka, India, [3]Department of Pharmaceutical Analysis, St. James College of Pharmaceutical Sciences (SJCOPS), Chalakudy, Kerala, India

8.1 Introduction

Biologics and therapeutic proteins are proteins or peptides expressed in cellular systems that serve as drug molecules. The absorption, distribution, metabolism, and excretion (ADME) of biologics is governed by processes distinct from small molecules. The discovery of naturally occurring bioactive peptides created an abundant source of pharmacophores from which drug discovery and development led to the class of peptidomimetics. Natural endogenously occurring peptides exhibit desirable therapeutic characteristics but are often restricted in application by rapid proteolysis and inadequate membrane permeability. However, modification of naturally occurring peptides to develop peptidomimetic analogs has created a promising class of therapeutics with great utility [1].

Peptides possess some of the properties of both antibodies and small molecules and, hence, they bridge the gap between antibody drugs and small molecules. Scaffolds can be made into peptidomimetics with small molecule properties while retaining some amino acids and binding specificity to maintain characteristics of proteins [2]. Peptides have limitations in terms of delivery, as most peptides are not orally available and have a short half-life in vivo due to proteolytic degradation. Also, most peptides are cleared from the kidneys after intravenous (IV) administration [3]. However, these limitations can be overcome using different design strategies [4,5] or modifications to ensure stability against enzymatic and thermal degradation [6,7].

Peptide and Peptidomimetic Therapeutics. DOI: https://doi.org/10.1016/B978-0-12-820141-1.00014-5

Most peptide drugs do not follow the "rule of five" proposed by Lipinski et al. [8] and modified by Veber et al. [9]. Despite several limitations of peptide-based drugs, many have made it to the market and, hence, there has been increased interest in the pharmaceutical industry to bring more peptide-based drugs to the market. Examples of drugs that have reached the $1 billion market include glatiramer acetate (an immunomodulatory drug used to treat multiple sclerosis), leuprolide acetate, goserelin acetate, and octreotide. There are more than 50 FDA-approved peptide-based drugs on the market and more than 140 in the pipeline [10,11]. However, the number of available peptide drugs relative to all approved drugs is very small, at around 1.5% of all drugs on the market. Growth in the number of peptide-based drugs will be seen when we consider that the number of peptides entering clinical trials has been around 20 per year recently [12].

Peptidomimetic drugs bridge the gap between simple peptides and non-peptide synthetic structures. Peptidomimetics have numerous advantages, such as stability, bioavailability, and often potency as surrogates of naturally-occurring peptides. Peptidomimetics form the basis of important enzyme inhibitor families that act as receptor agonists and antagonists. In their broadest sense, peptidomimetics sometimes refer to organic molecules mimicking some properties of peptide ligands. A great deal of the drugs created were conceived as mimics of substances known to interact with the target biological system. Peptidomimetic molecules are otherwise regular organic molecules conceived to mimic the structure of an endogenous peptide. However, peptidomimetics offer a wide array of advantages compared to peptides. The intrinsic drawbacks of peptides, ranging from susceptibility to enzymatic degradation, toxicity, as well as high production cost, can largely be addressed by design of peptidomimetics [13]. This chapter focuses on the pharmacokinetics and pharmacodynamics (PK-PD) of peptidomimetic therapeutics.

8.2 Pharmacokinetics of peptidomimetics

The *in vivo* ADME properties of peptides and proteins are different from that of small molecule therapeutics. An important characteristic of peptides is their short half-life in the bloodstream because of the proteolytic action of proteases and peptidases. In the case of endogenous peptides, the short half-life is desirable for the regulation of their concertation and function. For example, glucagon-like peptide-1 (GLP-1) is responsible for lowering blood glucose in patients suffering from diabetes. Nevertheless, it has a short half-life of fewer than two minutes *in vivo*, thereby requiring frequent and high doses for therapeutic applications. This has encouraged the development of long-acting peptidase-resistant GLP-1 analogs, such as exenatide and liraglutide [14]. On the contrary, proteins are responsible for transport such as immunoglobins and plasma proteins like albumin have half-lives of several days that ensure the maintenance of physiologically necessary concentration in the bloodstream [15].

8.3 Absorption

Presently, most of the peptide and protein drugs are given as parenteral administration as they have poor bioavailability by other routes of administration. IV, subcutaneous (SC), and intramuscular (IM) administration are the major route of administration. The highest concentration of proteins and peptide are attained by IV administration as it surpasses the gastrointestinal tract (GIT), which is responsible for the presystemic metabolism. On the other hand, IV administration may not always give the required concentration-time profile depending on the biological activity of the product. In such cases, IM or SC administrations are good substitutes. SC injections have more patient compliance and lower administration cost as they allow self-administration, particularly with the introduction of microneedles and pen devices. In comparison to IV, SC and IM have low systemic bioavailability because of the presystemic degradation [16]. SC and IM absorption kinetics depends on the molecule.

After SC and IM administration, protein and peptide therapeutics enter into the systemic circulation by either blood capillaries or lymphatic vessels. The higher the molecular weight higher the absorption through the lymphatic system for SC and IM routes of administration. In general, macromolecules having a molecular weight of more than 16 kDa are mainly absorbed into the lymphatics, whereas those under 1 kDa are generally absorbed into the blood circulation [17–19]. SC absorption is also dependent upon the charge of the macromolecule: the charges in the extracellular matrix seem to be of the same magnitude, that is, the collagen is positively charged and the hyaluronan is negatively charged, moreover, the proteoglycans are also negatively charged that makes the interstitial charge [19]. The net negative charge and the related ionic interactions with SC-administered proteins result in slower transport for more positively rather than negatively charged proteins [20]. Several approved peptides and proteins such as insulin, enfuvirtide, and pramlintide's are administered as SC injections. After SC and IM administration might lead to flip-flop pharmacokinetics, where the absorption process is slower than the elimination process [21]. In flip-flop pharmacokinetics, the terminal half-life is prolonged following SC and IM administration compared to IV as the rate-limiting step is the absorption instead of elimination. Examples of peptides that demonstrate flip-flop pharmacokinetics include cetrorelix, degarelix, and lixisenatide [22,23].

Since parenteral administration is invasive and inconvenient, there have been various attempts to deliver therapeutic proteins and peptides needle-free through noninvasive routes such as oral, buccal, pulmonary, intranasal, and transdermal administration. Poor permeation through biological membranes of proteins and peptides is the foremost difficulty with this route of administration. As a result, bioavailability decreases especially when the molecular mass is beyond 700 Da. For permeation through the epithelium, a minimum level of lipophilicity is required for facilitating the absorption process, without which there is no passive transcellular permeation and the alternative paracellular pathway is restricted to proteins and peptides with molecular weight less than 200 Da [24,25].

In comparison with the parenteral route and another noninvasive route of administration oral route of administration has the advantage of convenience and greater patient compliance. However, the GIT has enzymes, which degrade proteins and peptides, in turn, poses physiological and biochemical barriers [26]. Poor solubility and permeability due to their large size and hydrophilicity are the physiological barriers. Proteins and peptides undergo acidic hydrolysis and enzymatic degradation due to the acidic environment of the gastric fluid and secreted enzymes in the endothelial cells constitute the biochemical barriers. These barriers account for the poor bioavailability of orally administered peptides. Different strategies have been explored to improve bioavailability which includes different formulation approaches, modification in amino acid structure, conjugation with hydrophobic ligands, use of permeation enhancers, enzyme inhibitors [26,27]. However, at present, there are only a few proteins and peptides that can be administered orally in the market. For example, Desmopressin is a synthetic analog of the natural hormone vasopressin its cyclic structure repels intestinal degradation [24]. Cyclosporine is another small cyclic peptide, which can be orally administered. The cyclic structure prevents its degradation from the proteolytic endopeptidases and its lipophilic properties favor its permeability across the intestinal mucosa [28].

Though the buccal and the sublingual region has a small surface area, it is rich in blood capillaries and lymphatic vessels. Direct absorption and circumventing first-pass hepatic metabolism and GIT are the advantages of this route of administration [29]. The major drawback includes trouble in retaining the drug in the mouth for a long time and tasting. Irrespective of the drawback, this route of administration is preferred for drugs, which have no oral bioavailability [30].

The intranasal route of administration offers similar advantages as that of inhalation such as rapid absorption, bypassing first-pass metabolism and a high vascularity surface area of the nasopharynx [29]. Besides, the nasal epithelium has high permeability allowing a higher molecular weight threshold. Bioavailability through this route is low because of the presence of the nasal mucosal cavity and epithelial cell lining the mucosal cavity metabolic enzymes. Moreover, drug doses through this route are restricted to small amounts as the nasal cavity has a small surface area. This route of administration has shown success in the case of small peptides that are inefficient by the oral route. Peptides deliver by the intranasal route in the market are salmon calcitonin, desmopressin synthetic analog and nafarelin [24].

The pulmonary inhalation route of administration offers a large lung surface area for absorption, high vascularity at the site of administration and bypassing the first pass hepatic metabolism [30]. As the drugs are quickly absorbed and reach the peak concentration faster than the oral route through this route of administration it is preferred for drugs that require rapid onset of action. On the contrary, there are absorption barriers that include lining fluid of alveoli, macrophages of alveoli, and epithelium of alveoli that hinder the drug permeation into the systemic circulation or start degradation [24]. Presently, there is no therapeutic protein or peptide marketed formulation given by this route of administration.

The major benefits of the transdermal route of administration are reduced GIT metabolism and first-pass metabolism, patient convenience and medication adherence, and constant release of the drug into the systemic circulation. Different methods to facilitate transdermal delivery are iontophoresis (i.e., applying low-level electric current) and sonoporation (i.e., applying low-frequency ultrasound), which increases skin permeability. Therapeutic doses of insulin, interferon-γ, and epoetin-α have all been successfully delivered transdermally via sonophoresis. Transdermal iontophoresis has also been applied for the delivery of a host of proteins and peptides, including leuprolide, insulin, growth hormone-releasing factor, calcitonin, and parathyroid hormone [29].

8.4 Distribution

The distribution of drugs is determined by the rate of extravasation, distribution, and excretion from the tissues. For a particular therapeutics, the rate of distribution is dependent on its physicochemical properties, protein binding, and carrier-mediated transport. As most of the protein therapeutics have a higher molecular weight is contrary to the small molecules, their apparent volume of distribution is usually limited to the volume of the extracellular space due to their limited mobility secondary to impaired passage through membranes. Unlike small molecules where diffusion is the transport mechanism, convective transport is the primary mechanism for transport of large proteins to the interstitial space, distribution in the interstitial space, and transport out of the tissue into the lymphatic capillaries [17]. Since the flow rate of the lymph is very low it leads to the extended lymphatic transit time of therapeutic proteins which results in overdue uptake of therapeutic proteins into the systemic circulation after IM or SC administration [31]. Transcellular migration via endocytosis is a minor route of the distribution of therapeutic proteins [32]. Therapeutic peptides and proteins generally follow a biexponential plasma concentration-time profile that can be described by a two-compartment pharmacokinetic model after an IV administration [15]. For example in the case of clenoliximab, secukinumab, somatostatin [33–35]. In this model, the central compartment represents mainly the vascular space and the interstitial space of well-perfused organs and the peripheral compartment represents the interstitial space of slow equilibrating tissues. The distribution of proteins in the central compartment after an IV administration has the volume of distribution slightly higher than the plasma volume (i.e., 3–8 L). The total volume of distribution frequently comprises 14–20 L not more than two to three times the initial volume of distribution [36,37]. Binding to endogenous protein structure is another factor, which is responsible for the distribution of therapeutic peptides and proteins. Specific binding proteins that are involved in the transport and regulation of physiologically active endogenous peptides and proteins frequently interact with each other. For example, growth hormone, recombinant human DNases [38]. Protein binding not only affects the pharmacokinetic activity but also affects the pharmacodynamics activity. Therapeutic proteins and peptides may also bind nonspecifically to plasma proteins (e.g., metkephamid binding to albumin) [39]. Site-specific receptor-mediated uptake can also influence the distribution of therapeutic proteins and peptides. Low volume of distribution should not be considered as low tissue penetration as receptor-mediated specific uptake can result in effective tissue concentration despite the small volume of distribution [40].

8.5 Metabolism

Peptidomimetics can address limitations of the parent peptide by providing higher metabolic stability, bioavailability and enhanced receptor affinity and selectivity [41]. The end products of protein metabolism are amino acids, which are reused in the endogenous amino acid pool for the de novo biosynthesis of structural or functional proteins [42]. The metabolic rate of therapeutic protein and peptide degradation depends on the molecular weight, size, lipophilicity, charge, functional groups, glycosylation pattern, and secondary or tertiary structure [17,43]. Metabolism of peptidomimetics by gut microbiota and the intestine has a considerable impact on their pharmacokinetics as well as therapeutic efficiency will be a key parameter in their development as leads for gut microbiota-targeted small molecules [44]. Farhadi et al. reported in silico the design of peptide inhibitors against pregnane X receptor candidates to control drug metabolism. Farhadi et al. also reported in silico tools for designing peptidomimetics enhancing endoribonucleolytic activities of acinetobacter MazF toxin as the novel antibacterial candidates metabolism [45,46]. Lopinavir is a peptidomimetic molecule and it contains a hydroxyl ethylene scaffold that mimics the peptide linkage targeted by the HIV-1 protease enzyme but which itself cannot be cleaved, thus preventing the activity of the HIV-1 protease. It undergoes extensive oxidative metabolism through CYP3A isozymes. Around 12 metabolites have been identified in vitro, with the C-4 oxidation products M1, M3, and M4 being the predominant metabolites found in plasma [47,48]. Hepatic metabolism may also contribute considerably to protein and peptide metabolism. Proteolysis generally starts with endopeptidases that attack the middle part of the protein, with the subsequent oligopeptides, which are additionally degraded by exopeptidases. The ultimate metabolites of proteins, amino acids and dipeptides are finally reutilized in the endogenous amino acid pool. The amino acid sequence in the protein determines the rate of hepatic metabolism [42]. The uptake of proteins and peptides into the hepatocytes is an important step in the hepatic metabolism of proteins and peptides. Small peptides with sufficient hydrophobicity cross the hepatocyte membrane by passive diffusion but in the case of large proteins uptake is facilitated by receptor-mediated transport processes [49,50].

8.6 Excretion

Generally, the elimination of therapeutic peptides and proteins are the same as endogenous proteins or peptides. Metabolic elimination pathways such as renal or biliary excretion are insignificant for most proteins [51]. If biliary excretion happens, it is usually accompanied by the following metabolic degradation in the GIT. In contrast to small molecule drugs, metabolic degradation of therapeutic peptides and protein can occur anywhere in the body by proteolysis as proteases and peptidases are ubiquitous throughout the body. Blood, kidneys, liver and GIT have an ample amount of proteases and peptidases which makes it an important site for proteolytic degradations. Owing to the nonspecific proteolysis of few proteins in the blood as well as the active cellular uptake, the clearance of protein drugs can exceed cardiac output [42]. Typically, a molecular weight less than 10 kDa of protein or peptide can easily be filtered through the glomeruli in the kidney, which is followed by hydrolytic degradation in the brush border membrane of proximal tubule cells. The filtration rate falls for the protein of molecular weight of more than 30 kDa [17,43]. In addition to molecular weight, charge selectivity has also been reported for glomerular filtration, that is, anionic charge macromolecules have better selectivity than neutral which has better selective than cationic macromolecules. There are three mechanisms for glomerular filtration (GF) of peptides and proteins:

1. GF of bulky peptides and proteins, followed by reabsorption into endocytic vesicles in the proximal tubule and subsequent hydrolysis into small peptide fragments and amino acids [52].
2. GF is followed by intraluminal metabolism, predominantly by exopeptidases in the luminal brush border membrane of the proximal tubule. The resulting peptide fragments and amino acids are reabsorbed into the systemic circulation [53,54].
3. Peritubular extraction of peptides and proteins from postglomerular capillaries with subsequent intracellular metabolism [55,56].

For mechanisms (1) and (2), glomerular filtration is the dominant, rate-limiting step, as subsequent degradation processes are not saturable under physiologic conditions [36]. For proteins that are resistant to proteolysis, renal elimination is the predominant clearance mechanism.

In the case of peptides and proteins, a considerable fraction of peptide and protein dose can be attached to the receptor, which leads to elimination via receptor-mediated uptake followed by intracellular metabolism. This process of endocytosis is not limited to hepatocytes and can occur in other cells, which includes the therapeutic target as well. Drug binding to the receptors is saturable within therapeutic concentration or at a low concentration between the drug and the receptors, as the number of receptors is limited. Hence, receptor-mediated elimination leads to nonlinear pharmacokinetics [57].

8.7 Pharmacodynamics of peptidomimetics

Peptidomimetics offer conformationally restricted structures, potentially minimizing cross-target interactions, better transport properties through biologic membranes, and improved resistance to degradation by peptidases and other enzymes [58]. To achieve these properties, peptides have been chemically modified to include unnatural amino acid substitutions, backbone amide bond modifications, rigid scaffolds or the addition of hydrophobic residues [59]. Peptidomimetics retain the ability to interact with the same target protein of interest by modeling the essential functional groups in a required three-dimensional pattern harmonizing to the binding pocket in a protein [60].

The process of peptidomimetic design is initiated by forging structure–activity relationships which inform a minimal active sequence or the chief pharmacophore elements; also they recognize those essential residues responsible for a biological effect. After this, the three-dimensional arrangements of these features are studied by applying structural constraints. By doing so the peptide complexity reduces and the native pharmacophore model is expressed in three-dimensional space. This way of modeling keeps up with the re-assembly of critical elements and the nonpeptide entities on a new modified platform that has the improved pharmacophore attached to the receptor [61,62].

Each day the new classes and therapeutic applications of peptidomimetics are emerging into the scientific domain. The most well-known "natural" peptidomimetic is the plant-derived alkaloid morphine, which acts by mimicking the N-terminal Tyr–Gly–Gly–Phe of the endorphin neuropeptides [63]. In addition, the γ-AA peptides bind to HIV RNA with high specificity and affinity, suggesting their potential application as anti-HIV agents. These molecules also mimic host defense peptides and display potent and broad-spectrum activity toward a range of drug-resistant bacterial pathogens with additional potential as anticancer agents. For instance, γ-AA peptides have shown great promise in targeted imaging of tumors in the mouse model, and they are also capable of disrupting p53/DNA interactions, and thus antagonize the STAT3 signaling pathway. Recently, from combinatorial screening, γ-AA peptides are identified to inhibit Aβ peptide aggregation, and thus they can be developed into potential anti-Alzheimer's disease agents [64,65].

Most FDA-approved antiviral agents are peptidomimetics and macrocyclic compounds that interact with the active site of a targeted protease [66]. Recently a review article by Pillaiyar et al. provided deep insight into the strategies using peptidomimetics to combat human Coronaviruses. MG11 is an analog of minigastrin, a peptide with high affinity and specificity toward the cholecystokinin-2 receptor (CCK2R), which is overexpressed by different tumors. Triazolo-peptidomimetics based novel radiolabeled MG11 analogs were reported to have the potential for the diagnosis of cancer and therapy. Grob et al. reported that "triazole scan" yielded peptidomimetics with improved resistance to enzymatic degradation and enhanced affinity toward the CCK2R [67]. The selected pharmacological activities of peptidomimetics are displayed in Table 8.1.

TABLE 8.1 Pharmacological activities of selected peptidomimetics.

Pharmacological activity	Mechanism of action	References
Anti-HIV	Vetrivel et al. *stated,* HIV initiates its infection by the interaction of its capsid envelope glycoprotein gp120 and human CD4. The present approach for clinical anti-HIV drugs inhibit viral replication after infection and failed due to the high mutation activity of HIV. A peptidomimetic has been developed from the CD4 binding natural peptide NMWQKVGTPL. The resulting mimetic showed 170 times higher binding affinity to CD4 and 5 times higher proteolytic stability compared to the original peptide. The study also pointed out that an additional advantage of possessing a lower molecular weight also added to its better pharmacological properties.	[68]
Antimicrobial	Molchanova et al. performed work on antimicrobial α-peptide/β-peptoid lysine-based peptidomimetics that exhibit high potency against clinical drug-resistant *Pseudomonas aeruginosa* strain obtained from cystic fibrosis patients. These clinical strains possess *phoQ* and/or *pmrB* mutations that confer high resistance to colistin, the last-resort antibiotic for the treatment of infections caused by *P. aeruginosa.* The lead peptidomimetic LBP-2 demonstrated a 12-fold improved antipseudomonal activity as compared to colistin sulfate as well as favorable killing kinetics, similar antibiofilm activity.	[69]

(Continued)

TABLE 8.1 (Continued)

Pharmacological activity	Mechanism of action	References
Antiviral	Hristov et al. in the search for new and effective prodrugs against the herpes simplex virus, at pH 7.4 Val-thiazole-acyclovir peptidomimetic showed stability higher than that of valacyclovir.	[70]
Analgesic	Matalińska et al. reported design, synthesis and evaluation of receptor affinity, analgesic activity and cytotoxicity of a hybrid peptide, AWL3020. The peptide includes two pharmacophores, one of δ-opioid receptor (δOR) agonists and one of neurokinin-1 receptor (NK1R) antagonists. The substance shows good analgesic action which is however weaker than that of morphine.	[71]
Dental hygiene	Wang et al. worked on potential biomarkers from salivary proteins and peptides for dental caries. Among these markers named Statherin, a cysteine-labeled peptide (DpSpSEEKC) and biomimetic peptide DE-11 extended by a mineralization hydrophilic tail composed of consecutive acidic amino acids capable of absorbing calcium and phosphate ions were synthesized and exhibited beneficial effects on the remineralization of demineralized tooth enamel in vitro.	[72]
Cardiovascular	Recio et al. Apolipoprotein A-I (apoA-I) and apoE mimetic peptides and their role in cholesterol transport during atherosclerosis, suppressors of cytokine signaling (SOCS)1-derived peptides and annexin1 as potent inhibitors of inflammation, incretin mimetics and their function in glucose-insulin tolerance, among others. Incretin mimetics possess the advantages of having low hypoglycemia risk, the ability to address postprandial hyperglycemia and the potential for weight reduction. GLP-1 mimetic peptides have also been shown to exert cardioprotective effects in cardiovascular-related death, nonfatal MI, and nonfatal stroke.	[73–75]

(Continued)

TABLE 8.1 (Continued)

Pharmacological activity	Mechanism of action	References
Anticancer	Ehexige et al. had demonstrated RNAi mediated silencing of STAT3 expression in the tumor tissue in B16F10 mouse model of melanoma. A peptidomimetic-based lipid nanoparticle (LNPs) for the delivery of siRNA in a mouse model of melanoma was designed systematically administered. Remarkably, sequential administration of siRNA against STAT3 induced profound silencing of STAT3 expression in tumor tissue, which resulted in significant downregulation of PD-L1, leading to significant inhibition of tumor growth through inhibition of tumor immune checkpoint.	[76]
Metabolic	Barnhart et al. evaluated the ligand-directed peptidomimetic CKGGRAKDC-GG-$_D$(KLAKLAK)$_2$ (henceforth termed adipotide) in obese old-world monkeys. It reduced the body weight of the treated animals. They found a rapid reduction in obesity without any adverse effects; in the second study a remarkable reduction in white adipose tissue was observed by magnetic resonance imaging and X-ray absorptiometry.	[77]
Antiobesity	Kolonin et al. reported an antiobesity therapy based on targeted induction of apoptosis in the vasculature of adipose tissue. They used in vivo phage display to isolate a peptide motif (sequence CKGGRAKDC) that homes to white fat vasculature. They also showed that the CKGGRAKDC peptide associates with prohibitin, a multifunctional membrane protein, and establish prohibitin as a vascular marker of adipose tissue.	[78]
Neurological	Robert et al. reported that a pharmaceutical composition of the β-amyloid peptide analogs in a pharmaceutically acceptable carrier can be administered to a subject for neuromodulation. The β-amyloid peptide analogs, while lacking neurotoxicity, effectively provides for protective activity against β-amyloid toxicity.	[79]

8.8 Current perspectives

The rapid outbreak of the novel Coronavirus disease 2019 (COVID-19) caused by severe acute respiratory syndrome Coronavirus 2 (SARS-CoV-2) in China followed by its spread around the world poses a serious global concern for public health. To conquer viral infections, the inhibition of proteases essential for proteolytic processing of viral polyproteins is a conventional therapeutic strategy. Fischer et al. computationally screened a compound library of over 687 million compounds for binding at the recently solved crystal structure of the main protease of SARS-CoV-2. Among 687 compounds, 11 drug-like compounds had improved binding free energy to the target protease, in contrast to the cocrystallized peptidomimetic lead compound that suffered from poor pharmacokinetic properties [80]. As an integral part of the novel Coronavirus, the viral main protease (M^{pro}, also called $3CL^{pro}$) represents an attractive target for therapy. Proteases involved in both viral entry (host TMPRSS) and its replication/transcription (3C-like protease M or M^{pro}) were studied as a potential target for SARS-CoV-2 therapeutics [81,82]. Zhang et al. determined the crystal structure of the unliganded M^{pro} at 1.75 Å resolution and used this structure to guide the optimization of a series of alpha-ketoamide inhibitors. The main goal of these optimization efforts was to improve the pharmacokinetic properties of the compounds [83]. Thus peptidomimetics can rapidly be adapted to address emerging biomedical problems. VanPatten et al. discussed the advantages of S protein- and ACE2-targeted peptidomimetics and other potential targets that have been studied using peptide-based therapeutics for COVID-19 [82].

8.9 Conclusions

In recent years, peptidomimetics have gained significant importance in various therapeutic activities areas, such as immunology, endocrinology, urology, and oncology. As described in this chapter, considerable research and development have been done to enhance and improve peptide-based therapeutics using peptidomimetics. Conformationally restrained structures can be used to enhance activity at the desired target site. Peptidomimetics have improved ADME properties when compared to peptides and small molecules. The addition of hydrophobic residues or replacement of amide bonds can provide better transport properties across cellular membranes, as well as a reduced rate of degradation by peptidases and other enzymes. Several examples cited in this chapter support the fact that peptidomimetics are highly advantageous compared to conventional drugs. This has made way for the advanced understanding of peptidomimetic PK-PD. Further and deeper study is required to validate the PK-PD of peptidomimetics, with specificity remaining the primary challenge.

Acknowledgments

The authors thank Dr. V. Ravichandiran, Director of the National Institute of Pharmaceutical Education & Research in Kolkata for his encouragement and support.

References

[1] Qvit N, Rubin SJ, Urban TJ, Mochly-Rosen D, Gross ER. Peptidomimetic therapeutics: scientific approaches and opportunities Drug Discov. Today 2017;454–62.

[2] Stevenson CL. Advances in peptide pharmaceuticals. Curr Pharm Biotechnol 2009;10(1):122–37.

[3] Banga AK. Therapeutic peptides and proteins: formulation, processing, and delivery systems. 3rd ed. Boca Raton, FL: CRC Press, Taylor & Francis Group; 2015. p. 400–10.

[4] Hess S, Ovadia O, Shalev DE, Senderovich H, Qadri B, Yehezkel T, et al. Effect of structural and conformation modifications, including backbone cyclization, of hydrophilic hexapeptides on their intestinal permeability and enzymatic stability. J Med Chem 2007;50(24):6201–11.

[5] Pollaro L, Heinis C. Strategies to prolong the plasma residence time of peptide drugs. MedChemComm 2010;1:319–24.

[6] Craik DJ, Swedberg JE, Mylne JS, Cemazar M. Cyclotides as a basis for drug design. Expert Opin Drug Discov 2012;7(3):179–94.

[7] Poth AG, Chan LY, Craik DJ. Cyclotides as grafting frameworks for protein engineering and drug design applications. Biopolymers 2013;100(5):480–91.

[8] Lipinski CA, Lombardo F, Dominy BW, Feeney PJ. Experimental and computational approaches to estimate solubility and permeability in drug discovery and development settings. Adv Drug Deliv Rev 1997;23(1–3):3–25.

[9] Veber DF, Johnson SR, Cheng HY, Smith BR, Ward KW, Kopple KD. Molecular properties that influence the oral bioavailability of drug candidates. J Med Chem 2002;45(12):2615–23.

[10] Otvos Jr L, Wade JD. Current challenges in peptide-based drug discovery. Front Chem 2014;2:62.

[11] Vlieghe P, Lisowski V, Martinez J, Khrestchatisky M. Synthetic therapeutic peptides: Science and market. Drug Discov Today 2010;15(1–2):40–56.

[12] Lax R. The Future of Peptide Development in the Pharmaceutical Industry. PharManuf Int Pept Rev 2010;2:10–15.

[13] Méndez-Samperio P. Peptidomimetics as a new generation of antimicrobial agents: current progress. Infect Drug Resist 2014;7:229.

[14] Deacon CF. Therapeutic strategies based on glucagon-like peptide 1. Diabetes 2004;53(9):2181–9.

[15] Meibohm B, Derendorf H. Basic concepts of pharmacokinetic/pharmacodynamic (PK/PD) modelling. Int J Clin Pharmacol Therapeut 1997;35(10):401–13.

[16] Zhao L, Ji P, Li Z, Roy P, Sahajwalla CG. The antibody drug absorption following subcutaneous or intramuscular administration and its mathematical description by coupling physiologically based absorption process with the conventional compartment pharmacokinetic model. J Clin Pharmacol 2013;53(3):314–25.

[17] Lin JH. Pharmacokinetics of biotech drugs: peptides, proteins and monoclonal antibodies. Curr Drug Metab 2009;10(7):661–91.

[18] Porter C, Edwards G, Charman SA. Lymphatic transport of proteins after sc injection: implications of animal model selection. Adv Drug Deliv Rev 2001;50(1–2):157–71.

[19] Richter WF, Bhansali SG, Morris ME. Mechanistic determinants of biotherapeutics absorption following SC administration. AAPS J 2012;14(3):559–70.

[20] Mach H, Gregory SM, Mackiewicz A, Mittal S, Lalloo A, Kirchmeier M, et al. Electrostatic interactions of monoclonal antibodies with subcutaneous tissue. Therap Deliv 2011;2(6):727–36.

[21] Yáñez JA, Remsberg CM, Sayre CL, Forrest ML, Davies NM. Flip-flop pharmacokinetics–delivering a reversal of disposition: challenges and opportunities during drug development. Therap Deliv 2011;2(5):643–72.

[22] Jadhav PR, Agersø H, Tornøe CW, Gobburu JV. Semi-mechanistic pharmacodynamic modeling for degarelix, a novel gonadotropin releasing hormone (GnRH) blocker. J Pharmacokinet Pharmacodyn 2006;33(5):609–34.

[23] Nagaraja NV, Pechstein B, Erb K, Klipping C, Hermann R, Niebch G, et al. Pharmacokinetic and pharmacodynamic modeling of cetrorelix, an LHRH antagonist, after subcutaneous administration in healthy premenopausal women. Clin Pharmacol Therap 2000;68(6):617–25.

[24] Brown LR. Commercial challenges of protein drug delivery. Expert Opin drug Deliv 2005;2(1):29–42.

[25] Singh R, Singh S, Lillard JW. Past, present, and future technologies for oral delivery of therapeutic proteins. J Pharm Sci 2008;97(7):2497–523.

[26] Mahato RI, Narang AS, Thoma L, Miller DD. Emerging trends in oral delivery of peptide and protein drugs. Crit Rev Therap Drug Carr Syst 2003;20(2&3).

[27] Karsdal M, Henriksen K, Bay-Jensen A, Molloy B, Arnold M, John M, et al. Lessons learned from the development of oral calcitonin: the first tablet formulation of a protein in phase III clinical trials. J Clin Pharmacol 2011;51(4):460–71.

[28] Holt, D., Mueller, E., Kovarik, J., Richard, F., & Kutz, K. Sandimmun neoral pharmacokinetics: impact of the new oral formulation. In: Transplant Proceedings, 1995.

[29] Antosova Z, Mackova M, Kral V, Macek T. Therapeutic application of peptides and proteins: parenteral forever? Trends Biotechnol 2009;27(11):628–35.

[30] Senel S, Kremer M, Katalin N, Squier C. Delivery of bioactive peptides and proteins across oral (buccal) mucosa. Curr Pharm Biotechnol 2001;2(2):175–86.

[31] Ezan E. Pharmacokinetic studies of protein drugs: past, present and future. Adv Drug Deliv Rev 2013;65(8):1065–73.

[32] Baxter LT, Zhu H, Mackensen DG, Jain RK. Physiologically based pharmacokinetic model for specific and nonspecific monoclonal antibodies and fragments in normal tissues and human tumor xenografts in nude mice. Cancer Res 1994;54(6):1517–28.

[33] Mould DR, Davis CB, Minthorn EA, Kwok DC, Elliott MJ, Luggen ME, et al. A population pharmacokinetic-pharmacodynamic analysis of single doses of clenoliximab in patients with rheumatoid arthritis. Clin Pharmacol Therap 1999;66(3):246–57.

[34] Tang L, Meibohm B. Pharmacokinetics of peptides and proteins. Pharmacokin Pharmacodynam Biotech Drugs 2006;17–44.

[35] Bruin G, Loesche C, Nyirady J, Sander O. Population pharmacokinetic modeling of secukinumab in patients with moderate to severe psoriasis. J Clin Pharmacol 2017;57(7):876–85.

[36] Colburn W. Peptide, peptoid, and protein pharmacokinetics/pharmacodynamics. Petides Peptoids Proteins 1991;3:94–115.

[37] Meibohm B, Derendorf H. Pharmacokinetic/pharmacodynamic studies in drug product development. J Pharm Sci 2002;91(1):18–31.

[38] Mohler M, Cook J, Lewis D, Moore J, Sinicropi D, Championsmith A, et al. Altered pharmacokinetics of recombinant human deoxyribonuclease in rats due to the presence of a binding protein. Drug Metab Disposition 1993;21(1):71–5.

[39] Taki Y, Sakane T, Nadai T, Sezaki H, Amidon GL, Langguth P, et al. First-pass metabolism of peptide drugs in rat perfused liver. J Pharm Pharmacol 1998;50(9):1013–18.
[40] Kuwabara T, Uchimura T, Kobayashi H, Kobayashi S, Sugiyama Y. Receptor-mediated clearance of G-CSF derivative nartograstim in bone marrow of rats. Am J Physiol-Endocrinol Metab 1995;269(1):E1–9.
[41] Lenci E, Trabocchi A. Peptidomimetic toolbox for drug discovery. Chem Soc Rev 2020;49(11):3262–77.
[42] Meibohm B. Pharmacokinetics of protein-and nucleotide-based drugs. Biomaterials for delivery and targeting of proteins and nucleic acids. Boca Raton: CRC Press; 2004. p. 275–94.
[43] Meibohm B, Zhou H. Characterizing the impact of renal impairment on the clinical pharmacology of biologics. J Clin Pharmacol 2012;52(S1):54S–62S.
[44] Gabr, M.T., Deganutti, G. & Reynolds, C.A. Peptidomimetic-based approach toward inhibitors of microbial trimethylamine lyases. Chem Biol Drug Des, 2020.
[45] Farhadi T. In silico designing of peptide inhibitors against pregnane X receptor: the novel candidates to control drug metabolism. Int J Peptide Res Therap 2018;24(3):409–20.
[46] Farhadi T, Hashemian SM, Farhadi Z. In silico designing of peptidomimetics enhancing endoribonucleolytic activities of acinetobacter MazF toxin as the novel anti-bacterial candidates. Int J Peptide Res Therap 2019;26:1061–71.
[47] Kumar GN, Jayanti V, Lee RD, Whittern DN, Uchic J, Thomas S, et al. In vitro metabolism of the HIV-1 protease inhibitor ABT-378: species comparison and metabolite identification. Drug Metab Disposition 1999;27(1):86–91.
[48] Lopinavir:Drug Databank, <https://go.drugbank.com/drugs/DB01601> [assessed 12.20].
[49] Hosseinimehr SJ, Tolmachev V, Orlova A. Liver uptake of radiolabeled targeting proteins and peptides: considerations for targeting peptide conjugate design. Drug Discov Today 2012;17(21–22):1224–32.
[50] Vickers A, Fischer V, Connors S, Fisher RL, Baldeck J-P, Maurer G, et al. Cyclosporin A metabolism in human liver, kidney, and intestine slices. Comparison to rat and dog slices and human cell lines. Drug Metab Disposition 1992;20(6):802–9.
[51] Chanson P, Timsit J, Harris AG. Clinical pharmacokinetics of octreotide. Clin Pharmacokin 1993;25(5):375–91.
[52] Rabkin R, Ryan M, Duckworth WC. The renal metabolism of insulin. Diabetologia 1984;27(3):351–7.
[53] Carone FA, Peterson DR, Flouret G. Renal tubular processing of small peptide hormones. J Lab Clin Med 1982;100(1):1–14.
[54] Carone F, Peterson D. Hydrolysis and transport of small peptides by the proximal tubule. Am J Physiol-Renal Physiol 1980;238(3):F151–8.
[55] Groves CE, Morales M, Wright SH. Peritubular transport of ochratoxin A in rabbit renal proximal tubules. J Pharmacol Exp Therap 1998;284(3):943–8.
[56] Nielsen S, Nielsen J, Christensen EI. Luminal and basolateral uptake of insulin in isolated, perfused, proximal tubules. Am J Physiol-Renal Physiol 1987;253(5):F857–67.
[57] Tang L, Persky AM, Hochhaus G, Meibohm B. Pharmacokinetic aspects of biotechnology products. J Pharm Sci 2004;93(9):2184–204.
[58] Moradi S, Soltani S, Ansari AM, Sardari S. Peptidomimetics and their applications in antifungal drug design. Anti-Infective Agents Med Chem (Former Curr Med Chem-Anti-Infective Agents) 2009;8(4):327–44.
[59] Ahn JM, Boyle NA, MacDonald MT, Janda KD. Peptidomimetics and peptide backbone modifications. Mini Rev Med Chem 2002;2(5):463–73.

[60] Vagner J, Qu H, Hruby VJ. Peptidomimetics, a synthetic tool of drug discovery. Curr OpChem Biol 2008;12(3):292–6.
[61] Hruby VJ, Cai M. Design of peptide and peptidomimetic ligands with novel pharmacological activity profiles. Annu Rev Pharmacol Toxicol 2013;53:557–80.
[62] Zhang X, Betzi S, Morelli X, Roche P. Focused chemical libraries–design and enrichment: an example of protein–protein interaction chemical space. Future Med Chem 2014;6(11):1291–307.
[63] Wiley RA, Rich DH. Peptidomimetics derived from natural products. Med Res Rev 1993;13(3):327–84.
[64] Cunningham AD, Qvit N, Mochly-Rosen D. Peptides and peptidomimetics as regulators of protein–protein interactions. Curr OpStruct Biol 2017;59–66.
[65] Nimmagadda A, Shi Y, Cai J. γ-AApeptides as a new strategy for therapeutic development. Curr Med Chem 2019;26(13):2313–29.
[66] Chang KO, Kim Y, Lovell S, Rathnayake AD, Groutas WC. Antiviral drug discovery: norovirus proteases and development of inhibitors. Viruses 2019;11(2):197.
[67] Grob NM, Häussinger D, Deupi X, Schibli R, Behe M, Mindt TL. Triazolo-peptidomimetics: novel radiolabeled minigastrin analogs for improved tumor targeting. J Med Chem 2020;63(9):4484–95.
[68] Vetrivel U, Sankar P, Nagarajan NK, Subramanian G. Peptidomimetics based inhibitor design for HIV–1 gp120 attachment protein. J Proteom Bioinform 2009;2:481–4.
[69] Molchanova N, Wang H, Hansen PR, Høiby N, Nielsen HM, Franzyk H. Antimicrobial activity of α-peptide/β-peptoid lysine-based peptidomimetics against colistin-resistant *Pseudomonas aeruginosa* isolated from cystic fibrosis patients. Front Microbiol 2019;10:275.
[70] Hristov G, Stankova I. Chemical stability of new acyclovir analogues with peptidomimetics. Sci Pharmaceut 2011;79(2):259–64.
[71] Matalińska J, Lipiński PF, Kotlarz A, Kosson P, Muchowska A, Dyniewicz J. Evaluation of receptor affinity, analgesic activity and cytotoxicity of a hybrid peptide, AWL3020. Int J Peptide Res Therap 2020;1–15.
[72] Wang K, Zhou X, Li W, Zhang L. Human salivary proteins and their peptidomimetics: values of function, early diagnosis, and therapeutic potential in combating dental caries. Arch Oral Biol 2019;99:31–42.
[73] Recio C, Maione F, Iqbal AJ, Mascolo N, De Feo V. The potential therapeutic application of peptides and peptidomimetics in cardiovascular disease. Front Pharmacol 2017;7:526.
[74] Lovshin JA, Drucker DJ. Incretin-based therapies for type 2 diabetes mellitus. Nat Rev Endocrinol 2009;5(5):262.
[75] Wroge J, Williams NT. Glucagon-like peptide-1 (GLP-1) receptor agonists in cardiac disorders. Ann Pharmacother 2016;50(12):1041–50.
[76] Ehexige E, Bao M, Bazarjav P, Yu X, Xiao H, Han S, et al. Silencing of STAT3 via peptidomimetic LNP-mediated systemic delivery of RNAi Downregulates PD-L1 and inhibits melanoma growth. Biomolecules 2020;10(2):285.
[77] Barnhart KF, Christianson DR, Hanley PW, Driessen WH, Bernacky BJ, Baze WB, et al. A peptidomimetic targeting white fat causes weight loss and improved insulin resistance in obese monkeys. Sci Transl Med 2011;3(108) 108ra112-108ra112.
[78] Kolonin MG, Saha PK, Chan L, Pasqualini R, Arap W. Reversal of obesity by targeted ablation of adipose tissue. Nat Med 2004;10(6):625–32.
[79] Robert A, Nichols KF, Victor JH. Neuroprotective beta-amyloid core peptides and peptidomimetic derivatives. <https://patents.google.com/patent/US20190233487A1/en>; 2017 [assessed 11.20].

[80] Fischer A, Sellner M, Neranjan S, Lill MA, Smieško M. Inhibitors for Novel Coronavirus protease identified by virtual screening of 687 million compounds. 2020.

[81] Jin Z, Du X, Xu Y, Deng Y, Liu M, Zhao Y, et al. Structure of M pro from SARS-CoV-2 and discovery of its inhibitors. Nature 2020;582(7811):1–5.

[82] VanPatten S, He M, Altiti AF, Cheng K, Ghanem MH, Al-Abed Y. Evidence supporting the use of peptides and peptidomimetics as potential SARS-CoV-2 (COVID-19) therapeutics. Future Med Chem 2020;12(18):1647–56.

[83] Zhang L, Lin D, Sun X, Rox K, Hilgenfeld R. X-ray structure of main protease of the novel Coronavirus SARS-CoV-2 enables design of α-ketoamide inhibitors. BioRxiv; 2020.

Chapter 9

Formulation of peptides and peptidomimetics

Rujuta Ghorpade and Maushmi S. Kumar
SVKM's NMIMS, Shobhaben Pratapbhai Patel School of Pharmacy and Technology Management, Mumbai, Maharashtra, India

9.1 Introduction

With over 7000 peptides playing a critical role in human physiology, they have been seen as a great area upon which to look for the discovery of novel drug-delivery systems, especially because of their excellent pharmacological and intrinsic properties [1]. The advent of peptide therapeutics was marked in the 1920s when the first synthetic peptide, insulin, came into use. Since then, synthesizing, designing, and delivering peptides and their analogs for a multitude of indications, including oncology, infectious diseases, endocrinology, and immunology have been key drivers of the biopharmaceutical industry [2].

As the future of healthcare is gearing toward personalized medicine, which offers enhanced target specificity and reduced side effects and toxicity, peptides indeed appear as lucrative candidates for researchers. The ability of amino acids to have targeted interactions with a receptor or ligand proves superior to many chemical moieties that often produce off-target toxicity. However, the physical and chemical instability of peptides has now encouraged the approach of designing peptidomimetics in suitable drug-delivery systems to combat these disadvantages of native peptides [3].

At the grassroot level, we know that peptides are macromolecules that play a vital role in countless biological processes, from catalyzing biochemical reactions to providing structural integrity to a cell to DNA transcription. The specificity of these processes can be greatly attributed to several degrees of a peptide structure. These include a primary structure that elucidates on the covalent binding between amino acids, a secondary structure that highlights the turns and loops (i.e., α-helices and β-sheets) of a polypeptide chain, a tertiary structure that results from the interactions between the side chains of various amino acids, and a quaternary structure that refers to the

Peptide and Peptidomimetic Therapeutics. DOI: https://doi.org/10.1016/B978-0-12-820141-1.00017-0

stereochemical arrangement of protein subunits with respect to one another. It is the secondary structure that is of particular interest to researchers as it plays a key role in mediating protein–protein interactions [4]. The inception of peptidomimetics results from the disadvantageous high conformational flexibility and low in vivo stability of natural peptides despite their excellent bioactivity.

Peptidomimetics are molecules that mimic natural peptides or proteins and thus produce the same biological effects as their natural role models. They pose higher proteolytic stability and higher bioavailability and provide improved selectivity or potency. The development of peptidomimetics arises from the fact that conventional peptides bind to their target receptor site via side-chain interactions only, thereby implying that the amino acid backbone can be modified or even replaced, as long as the apt 3D conformation required to bind is supported [5]. Here, we aim at giving the reader a complete overview of the various challenges posed while formulating peptides and how approaches devised over years have successfully circumvented such challenges, along with the latest research in the progress of this approach. Furthermore, we highlight peptidomimetics as a novel formulation approach in itself, along with the design strategies of a peptidomimetic and its real-life applications in the current market.

9.2 Challenges in formulating peptides

Formulating peptides cannot be compared with formulating a molecule synthesized in a chemical laboratory. The major set of challenges that peptides pose to any formulation scientist is their high degree of chemical instability (and even physical instability in certain cases), pronounced levels of stereoisomerism, and the tendency to aggregate.

9.2.1 Chemical instability

Oxidation, deamidation, and disulfide scrambling are the primary pathways of degradation of a peptide via chemical means. The triggers for these pathways vary and are influenced by factors like the type of amino acid residues present, its three-dimensional conformation, and the degree of peptide chain flexibility. Oxidation is seen to be predominant in the case of peptide chains having a large number of His, Tyr, Trp, and Met residues [6], while deamidation is seen in either highly flexible protein sequences or where Asn and Gln residues are majorly present [4]. Disulfide scrambling is followed by hydrolytic damage that eventually leads to degradation or aggregation of the peptide. Optimizing the pH of the buffer system and creating a lyophilized formulation are two strategies to avoid problems with peptide degradation or aggregation brought on by disulphide scrambling [7,8]. Alternatively, the attachment of stabilizing proteins to the primary peptide molecule can also be used to reduce the redox potential of peptides such as by blocking thiol–disulfide exchange done in tryptic peptides derived from human growth hormone [9,10].

9.2.2 Stereoisomerism

Peptides do not possess a higher order of structural characterization the way proteins do, and often exist in a dynamic equilibrium with various structural forms. Hence, the conformer of a peptide present in a formulation may not have so much resemblance with the actual "active" conformer that binds at the receptor site. Either the receptor induces a conformational fit or a peptide binds to the membrane and diffuses in two dimensions as a function of its partition coefficient. For this reason, spectroscopic methods are rather used to identify the less reactive conformer so that under formulation development, the equilibrium can be shifted toward this conformer to reduce its proneness to degradation [11]. The points of stereoisomerism exhibited by a peptide in vivo can be detected by using stereoselective antibodies that act as chiral selectors [12] and assaying the degree of shift of bond angle by using spectroscopic techniques or by newer, more sensitive methodologies such as using these antibody conjugates in a protein conjugation assay [13]. It can be further protected by tagging isomerase enzyme inhibitors with the peptide drug [14].

9.2.3 Self-association

Since peptides are amphiphilic in nature, at higher concentrations of the peptide active pharmaceutical ingredient (API), they have a tendency to aggregate at their more hydrophobic ends and form an amphipathic alpha-helical dimer, tetramers, and even micelles. Hence, it must be realized by the formulator that the aggregating nature of the peptide adversely affects its absorption through various physiological membranes and syringe ability in the case of a parenteral formulation. The complete study of factors causing aggregation of peptides can be elucidated via a combination of analytical techniques such as Fourier transform infrared spectroscopy, size exclusion chromatography (SEC), and design of experiments, which can effectively bypass the need to conduct accelerated studies (except in the case of pH-associated studies) to yield stress test results [15].

9.2.4 Physical instability

The aggregating nature of peptides further paves the way into the overall physical instability of a formulation because of its influence on the zeta potential. Zeta potential is a measure of the magnitude of electrostatic interactions between charged surfaces [16]. Either the zeta potential could induce a net positive force of attraction between peptide monomers and eventually lead to fibril formation (ordered layering into beta sheets) or gelation, or conversely, it could lead to the formation of a stable system containing the desired therapeutic peptide in its right form. The zeta potential can therefore be accurately influenced by restricting pH and temperature fluctuations, decreasing the surface tension, and choosing the right excipients (e.g., cyclodextrins are seen to delay the rate of aggregation by shielding the hydrophobic amino acid side chains) [11]. Stability issues may also arise because of the impurities present in the formulation, which could be product-related, such as by-products formed by degradation of the bulk protein, or process-related, such as remnants of host cell proteins, culture media components, lipids, polysaccharides, and viruses [7]. Such challenges associated with peptides need to be identified and looked after during the preformulation stage itself along with the other standardized tests mentioned in the compendia. Currently, the techniques used to analyze a peptide include SEC, sodium dodecyl sulfate–polyacrylamide gel electrophoresis (SDS-PAGE), reversed-phase high-performance liquid chromatography (RP-HPLC), capillary zone electrophoresis, high-pressure ion-exchange chromatography, tryptic digest RP-HPLC, and Raman spectroscopy, although each of these comes with its own advantages and disadvantages. While performing the bioassay of peptides, it is also crucial to take care of the nature of the mobile phase used, especially its pH, as peaks may not resolve well if a peptide fragment gets ionized [17]. The parameters aforementioned briefly indicate the challenges faced at a lab-scale. A complete review on the extrapolation of the data obtained from these bioanalytical tools along with plausible deviations on a manufacturing scale has been highlighted through the work of Patro et al. [18]. Concurrently, newer analytical methods are also in the process of being developed. One such test used for measuring the degree of proteolysis is called the "pulse proteolysis" technique, which involves urea-induced unfolding followed by brief enzymatic digestion that results in the development of relative resistance to proteolysis that is quantified via microfluidics-based capillary electrophoresis [19] (Box 9.1).

BOX 9.1 Key objectives when choosing a formulation approach for peptides.

1. Enhances protein/peptide solubility
2. Can provide controlled release of protein/peptide molecules
3. Can provide sustained release instead of burst release to avoid undesirable side effects
4. Improves biodistribution of protein/peptides
5. High target specificity for disease tissue in vivo

9.3 Formulation approaches for peptides

The choice of approach for every peptide depends upon a number of factors. This includes the degree of folding of the peptide chain and its effects on the structure–activity relationship of the target site, degradative pathways that could hinder the optimum delivery of the peptide molecule to the target site, the release rate of the drug from the delivery system, and the storage conditions needed for the peptide to remain pharmacologically active. Hence, the approach used in formulating a peptide cannot be "one-size-fits-all." Only a thorough understanding of the peptide of interest via extensive preformulation studies will give clarity on approaches to customize or combine to yield the best possible formulation [20].

The classical approaches for the formulation of peptides have been addressed here in detail.

9.3.1 Chemical modifications

The covalent attachment of nontoxic, nonimmunogenic polymers such as polyethylene glycol (PEG) and glucose or mannose to a peptide is referred to as PEGylation, glycosylation, and mannosylation, respectively [21]. These polymers or monosaccharides are either branched or linear and can be engineered to a choice of molecular weights. The advantages offered by such chemical modifications include enhanced bioavailability, decreased dosing frequency due to prolonged residence in the body, decreased degradation by metabolic enzymes, optimized pharmacokinetics, improved safety and efficacy profile, a reduction or elimination of protein immunogenicity, improved drug solubility, and stability to hydrophobic drugs and endogenous proteins [22].

9.3.2 Polyethylene glycosylation

PEG conjugation is the most widely used form of chemical modification in biotechnology. It works by masking the protein's surface at its N-terminus or unpaired cysteines with varying sizes of PEG polymers and causing an increase in the molecular size of the polypeptide, thereby reducing its renal ultrafiltration, preventing the approach of antibodies or antigen processing cells and reducing the degradation by proteolytic enzymes. Linear PEGylation can be carried out either by alkylation (which does not modify the charge on the peptide) or via acylation. Branched PEGylation can take place by having the amino acid or the peptide arm in between PEG and the attached macromolecule, thus forming an umbrella-like structure. The latter offers higher protection of the protein surface compared to the linear form, thus resulting in higher retention in blood, lower immunogenicity, and decreased enzyme inactivation during the conjugation [23]. A few marketed examples of PEGylated peptides include PEG-Intron (peginterferon alpha-2b) and PEGasys (peginterferon alpha-2a) for hepatitis, Neulasta (pegfilgrastim) for leukopenia, and Oncaspar (pegaspargase) for cancer [24]. Studies have shown that PEGylated antimicrobial peptides like KYE28, when given in the form of a microgel, have high peptide load, membrane disruption capacities, and greater peptide release as opposed to their non-PEGylated counterparts [25]. A 2.5 times enhancement of the half-life of immune-modulating agent, thymosin alpha 1 (Tα1), was seen using site-specific PEGylation by condensing the N-terminal amino group of Tα1 with 3-maleimidopropionic acid [26]. PEGylated-HVGGSSV protein was seen to target the TIP1 protein surface in cancerous cells, thereby making it a potential tool in anticancer targeted therapy [27].

9.3.3 Glycosylation

Glycosylation under protein engineering is a means of exploiting a post-translationally glycosylated peptide to produce an entity with a greater negative charge and enhanced hydrodynamic size [28]. As a consequence, the synthetically glycosylated peptide is seen to have reduced renal clearance, modulated receptor-mediated endocytosis, and is shielded from proteolytic degradation sites [29]. The glycan moiety can be introduced onto the peptide chain at highly specific glycosylation sites, such as Asn−Xxx−Thr/Ser (Xxx being any amino acid except proline) via N-linkage or O-linkage. However, the introduction of glycans requires complete sialylation of the galactose termini of the glycan, without which asialoglycoprotein receptor in the liver or a host of other glycan-specific receptors on endothelial or Küpffer cells metabolize the glycosylated peptide, thus reducing its half-life [30]. The most notable example of the marketed glycosylated therapeutic peptide is Aranesp (darbepoetin alfa), which contains hyperglycosylated erythropoietin analog via two N-linkages [24]. Glycosylation can be also used in tuning the conformation of a peptide resulting in α-to-β transition. This also results in the peptide gaining added stability in its conformation, as seen in the case of Peptide T (H−Ala−Ser−Thr−Thr−Thr−Asn−Tyr−Thr−OH) on N-linked glycosylation which stabilized the Type I (III) β-turn [31,32].

9.3.4 Mannosylation

Mannosylation is a novel type of glycosylation that involves covalent attachment of an alpha-mannopyranosyl residue to the indole C2 carbon atom of tryptophan via a C−C link [33]. The main difference lies in the fact that sugar is linked to carbon rather than a reactive atom such as nitrogen or oxygen. Some naturally occurring *C*-mannosylated proteins include ribonuclease 2, interleukin-12, properdin, and mindin. However, the significance of C-mannosylation in them remains unclear [34].

9.3.5 Enzyme inhibitors

It is challenging for peptides to retain their structural integrity before reaching the target site because of the presence of a large variety of proteases in the body. Enzyme inhibition is an approach often devised in the case of dosage forms administered from nonparenteral routes such as oral, buccal, pulmonary, and nasal [35]. The ways to circumvent enzymatic degradation include coadministration of protease, designing formulations such that they minimize contact of the peptide or protein with proteases, or modifying peptide or protein structure [36]. Protease inhibitors can be classified as (1) amino acid-based (N-acetyl cysteine), (2) peptide-based (bacitracin), (3) polypeptide-based (aprotinin), and (4) nonamino acid-based (diisopropyl fluorophosphate). The choice of inhibitor depends upon the location of the target enzyme and its cellular distribution. However, a protease inhibitor may or may not work for enhancing the bioavailability of a given protein. It also largely depends upon the degree of folding in a peptide and the concentration of the inhibitor used [37]. The focus since the early 2010s has been greatly shifted toward exploiting the disadvantages of protease activity to the benefits of the formulator. There has been extensive research on developing peptide formulations having enzyme-regulated degradation at the target site for more effective drug delivery [38].

9.3.6 Penetration enhancers

A penetration or absorption enhancer is a compound that provides rapid permeation of the drug through biological membranes. These excipients are used especially in the case of designing oral and transdermal formulations. Chitosan and medium-chain fatty acids (such as caprylate, caprate, and laurate) are the most commonly used penetration enhancers. They enhance absorption, increasing the hydrophilicity and bulk of the peptide, thus increasing its paracellular permeability. Penetration enhancers also act by other mechanisms, including fluidization of the plasma membrane, loosening of the tight junctions between cells, inhibition of proteases, and increasing the thermodynamic activity of proteins prone to self-association. However, regulatory concerns come into play as penetration enhancers also lead to mucosal tissue irritation, the extent of mucosal damage, and the rate of tissue damage repair [36]. Alternatively, the use of microneedles (in the case of dermal delivery) of peptides has been seen as advantageous, as shown by Zhao et al. [39] in an attempt to deliver insulin transdermally via auto-antigen-coated microneedles. Furthermore, even cell-penetrating peptides (CPPs), when coadministered with therapeutic peptides, have been seen to increase the latter's intracellular absorption [35].

9.3.7 Colloidal systems

9.3.7.1 Liposomes

Liposomes are microscopic vesicles composed of lipid membranes surrounding discrete aqueous compartments. Liposomes are easy to prepare, highly biocompatible, have controllable in vivo properties, and hence reduce dosing frequency and can be loaded with a broad array of drugs, including peptides, DNA, and diagnostic agents [40]. They are especially seen as advantageous drug-delivery systems for peptides as the aqueous interior of the liposome would uphold the structure and conformation of the protein, while the lipid exterior may help to improve absorption across physiological barriers. The use of a liposomal formulation may also alter the biodistribution of the drug, prolonging the circulation time or helping in site-specific targeting. By and large, a peptide or protein is expected to reside in the aqueous compartment of the liposome. However, the hydrophobic portion of the protein may interact with the external lipid membranes of the liposome, thus causing hindrance in its bioactivity. Conversion of the peptide to an intermediate conformation called the "molten globular state" may help in the prevention of such unforeseen interactions. Since liposomes are not stable over long-term use, and they are often lyophilized [41]. However, extreme temperatures may inactivate the peptide or may even cause leaching of the carrier system. Such issues can be curbed by using cryoprotectants and membrane stabilizers. To further enhance the stability of liposomes, surface PEGylation is carried out to prevent metabolism via the RES mechanism [42].

Encapsulation of an endostatin peptide in liposomes of around 100 nm with a -4 to -36 mV zeta potential resulted in stabilization of the peptide into the liposomal core, thus achieving an optimal drug release rate [43]. Encapsulation of collagen mimetic peptide-tethered vancomycin liposomes in collagen-based scaffold also demonstrated the potential expansion of therapeutic peptides in topical formulations [44]. Even the liposomal preparation containing nanoparticles of Antimycin A (an anticancerous agent used in the treatment of lung cancer), a cholesterol core, and a mitochondria-penetrating peptide having a phenylalanine-arginine-phenylalanine-lysine peptide sequence, showed a greater degree of cellular uptake and mitochondria targeting in A549 cells [45].

9.3.7.2 Micro- and nanoparticles

Microparticles consist of spherical polymeric particles with sizes ranging from 1 to 250 μm (ideally $<$125 μm), while nanoparticles are spherical particles with a size of under 1 μm. These formulations can be further derivatized as micro/nanocapsules and micro/nanospheres. The biggest advantage offered by micro- and nano-dispersed polymeric systems is their ease of administration by a single injection, as opposed to other larger-sized controlled-release dosage forms that require a surgical incision. They are prepared via techniques such as solvent evaporation, organic phase separation, interfacial polymerization, emulsion polymerization, and spray drying, although only a select few techniques are used from those mentioned above due to the restrictions imposed by stability-sensitive peptides. Both biodegradable polyesters [such as poly(D, L-lactide), poly(D, L-lactide-*co*-glycolide), poly(caprolactone), polyaminoacids, etc.], and nonbiodegradable polyesters (polymethyl methacrylate, polystyrene, and polyamide) are used in the preparation of the microparticle matrix. Poly(D,L-lactide-*co*-glycolide) is the most commonly used polymer for peptide formulations involving microparticles, while the most commonly used polymer for nanoparticles is polyalkyl cyanoacrylate. The proposed mechanisms of drug release in the case of micro/nanoparticles are (1) liberation due to polymer erosion or degradation, (2) self-diffusion through pores, (3) release from the surface of the polymer, and (4) pulsed delivery initiated by the application of an oscillating magnetic or sonic field. As in the case of liposomes, precautions must be taken that the polymeric carrier system does not interact with a peptide of interest, causing an escalation into degradation pathways. Furthermore, due to low loading capacities and efficiencies of such particulate formulations, modified methods such as encapsulation of peptides into polylactico-glycolic acid (PLAGA) microspheres via modified solvent evaporation method using a double emulsion have been devised [46].

9.3.7.3 Reverse micelles

Reverse micelles are amphiphilic polymers consisting of a hydrophilic core shielded by a hydrophobic shell. This formulation approach provides enhanced solubility and stability of the peptide drug that is encapsulated into the core along with a greater degree of permeability through biological membranes. The standard method of preparation of reversed micelles is using sodium bis(2-ethylhexyl) sulfosuccinate and D_2O in isooctane [47]. The delivery of 20-nm insulin particles through hydrophilic peptide-containing oily formulations involving the freeze-drying of water-in-oil emulsions (i.e., reverse micelles) resulted in a significant reduction in the plasma glucose level of fasting diabetic rats after oral administration [48]. Micelle encapsulation of cyclic RGD peptide-conjugated poly(ethylene glycol)-*co*-poly(lactic acid) polymeric micelle dramatically increased its solubility, thus alleviating its serum sequestration [49].

9.3.7.4 Emulsions

Peptide emulsions are biphasic systems having a particle size of the droplets usually between 100 and 500 nm. The association of peptides with oil/water-based carriers improves the delivery of these molecules by offering protection against intestinal proteases, improving stability in the presence of gut surfactants, and altering cell permeability and lymphatic secretion. Hence, emulsions serve as a method of drug delivery via both enteral and parenteral routes. Care must be taken that the shearing force is minimum while preparing an emulsion as it may denature the peptide present. One could also opt for the use of a triphasic emulsion system (such water-in-oil-in-water) depending on the physicochemical properties of the peptide of interest. Better yet, one could consider preparing peptide formulations in a microemulsion carrier system, as these carrier systems do not involve the use of any shearing force and the presence of a surfactant only provides additional thermodynamic stability to the peptide. Albeit there is an exciting opportunity posed by emulsions in the formulation of peptides, the sterilization techniques available are limited due to the stability issues that come with a peptide [50].

9.3.7.5 *Mucoadhesive polymeric system*

Mucus-lined organ systems such as the nose and lungs have been increasingly used routes of peptide drug delivery due to the large surface area for absorption they offer along with boosted patient compliance. The progress of mucoadhesive sprays and patches is still under clinical development, although a few have made their way into the market. Mucoadhesive polymers such as hydroxymethyl propyl cellulose, chitosan, and carbopol provide an increased residence time for the peptide formulation onto the mucosal membrane to enhance drug concentration at the target site. Furthermore, these polymers, owing to their high molecular weight and hydrophilic nature, provide resistance against GIT enzymes [51], protect against undesirable pH environments, and increase the rate of absorption through physiological tight junctions [52,53]. Mucoadhesive polymers, if thiolated, perform better than conventional polymers as they form strong disulfide bonds with the cysteine residues of mucous glycoproteins [38].

9.3.8 General excipients

9.3.8.1 *Buffer systems*

Buffers are a vital component in any protein formulation. They not only regulate the pH of the system but also stabilize peptides by a variety of mechanisms in liquid formulations, frozen solutions, or the solid state. Additionally, buffers can also provide increased conformational stability of proteins, whether by ligand binding or by an excluded solute mechanism, and can alter the colloidal stability of proteins and reduce interfacial damage [54]. The choice of the buffer used is based on the pH that provides maximum peptide stability. A few commonly used buffers are listed in Table 9.1.

9.3.8.2 *Solvent system*

Glycerol erythritol and sorbitol are the most preferred solvents because they not only stabilize the protein by making it thermodynamically unfavorable for it to aggregate but further ensure that they increase the transition temperature of the protein. However, they act as peptide stabilizers at low concentrations. At higher

TABLE 9.1 Buffer systems used in formulation of parenteral peptide solutions [7].

Buffering agent	Buffering range
Citrate	pH 2.1–6.2; pK of 3.15, 4.8, and 6.4
Phosphate	pH 6.2–8.2
Succinate	pH 3.2–6.6; pK of 4.2 and 5.6
Acetate	pH 3.8–5.8
Histidine	pK of 1.8, 6.0, and 9.0
Glycine	pK of 2.35 and 9.8
Arginine	pK of 2.18 and 9.1
Triethanolamine	pH 7.0–9.0
Tris-hydroxymethylaminomethane	pK 8.1

concentrations of the solvent, they may facilitate denaturation of the peptide API. Other biomolecules used as carriers include albumin [Epogen (erythropoietin), Alferon N (interferon α-n3), and Abbokinase (urokinase)], small neutral amino acids like glycine [such as Kogenate FS (Factor VIII), Humatrope (hGH)] or amino acids having charged side chains like Asp, Glu, Arg, Lys, and His [Activase (tPA)], and carbohydrates [55].

9.3.8.3 *Preservatives*

As peptides are biological products, they are naturally susceptible to microbial degradation. Although regulatory agencies generally recommend single-dose injectables to avoid the use of preservatives, in an unavoidable situation, sodium bisulfite and thiomersal are seen as safe and effective preservatives. Certain other preservatives that have also been reported to induce protein unfolding and aggregation in the increasing order of their effectiveness are m-cresol > phenol > benzyl alcohol > phenoxyethanol > chlorobutanol [56].

9.3.8.4 *Stabilizers*

These include β-cyclodextrins and their derivatives, PEGs, low concentration of salts like NaCl, KF, phosphate, and Na_2SO_4, surfactants like polysorbate 80 and polysorbate 20, urea, polyvinylpyrrolidone, and ammonium tartarate [7].

9.4 Formulation strategies for peptidomimetics

Incorporating large biomolecules such as peptides into formulations and the issues that arise when combating all the challenges imposed by them have paved the way for the development of a smarter means to imitate the advantages offered by peptides. One such approach is the development of peptidomimetics. Peptidomimetics can be broadly defined as compounds that "mimic" a certain pharmacophore of a natural protein, thus retaining the ability to interact with a given biological target and produce the same biological effect. Peptidomimetics are designed to surmount the challenges posed by naturally occurring peptides such as premature proteolysis, low availability, and less receptor selectivity. In Table 9.2, all the important peptidomimetics explored in various therapeutic areas are listed. Briefly, this designing process involves identifying the vital elements of peptides needed for producing a biological effect and remodeling these elements with a nonpeptide entity on a new platform for improved pharmacophore–receptor affinity [57]. There are several approaches particularly used in the formulation of peptidomimetics, which are described below.

9.4.1 Prodrug approach

Prodrug strategies comprise a transient alteration of the physicochemical properties of a given peptidomimetic through reversible chemical modification. The resulting prodrugs are designed to enhance chemical stability, alter aqueous solubility, or improve bioavailability while maintaining the inherent pharmacological properties of the parent drug [67]. The prodrug approach is only successful when there is assured conversion of the prodrug to the parent drug, either through enzymatic or nonenzymatic catalyzed reactions at the target site. Often, the lipophilic nature of a peptidomimetic renders it either poorly soluble in vivo at physiological pH or highly unstable and hence, hinders its absorption. The introduction of an ionizable moiety linked to the functional derivatives of the peptidomimetic molecule can greatly improve solubility. Commonly used ionizable moieties are phosphate monoester derivative, *N*,*O*-acyl group, *N*-alkylation of the peptide bond, and thiol prodrugs prepared via thioesterification [68].

It was seen that the cyclic prodrug prepared from RGD peptidomimetic by linking the amino and carboxylic acid groups to an acyloxyalkoxy promoiety resulted in an improved degree of cell permeation and enhanced antithrombotic activity compared to the parent mimetic [69]. Similarly, a prodrug of cyclic cidofovir (cHPMPC) synthesized by the esterification of the phosphonic group via the serine side-chain hydroxyl group of Val-Ser-CO_2Me dipeptide was seen to have increased stability and enhanced transport compared with the parent drug in a rat model [70]. McMurray et al. [71] also showed the increased selectivity of a phosphate-based prodrug of PM-73G derived from pTyr-Leu-Pro-Gln toward the SH2 domain of Stat3, which is often overly activated during a multitude of cancers. In recent

TABLE 9.2 Therapeutic areas explored for the use of peptidomimetics.

Therapeutic area	Peptide	Peptidomimetic	References
Diabetes	Glycine and alanine based DPP-IV enzyme substrates that inhibit DPP-IV's action on breakdown of GLP-1	Vidagliptin, Teneligliptin, Anagliptin, and Saxagliptin	[58]
	Incretins including GLP-1 and GIP-1 that augment the secretion of insulin	Liraglutide, Lixisenatide, Albiglutide, Dulaglutide, and Semaglutide	[59]
Cancer	Bcl-2 homology domain 3 (BH3), which leads to apoptosis	Venetoclax	[60]
	Farnesyltransferase, whose inhibition leads to the prevention of the proper functioning of the Ras (protein), which is commonly abnormally active in cancer	BMS-214662, Tipifarnib (R-115777), L778123, Lonafarnib (SCH-66336), and SCH-226374	
	Proteasome substrates that inhibit the process of protein "recycling" in the cell, thus allowing proteins to build up until destroying the cell	Bortezomib, MG-132, and Epoxomycin	
	Matrix metallopreptidase (MMP) substrates, which on inhibition of MMP decrease tumor invasion and slow the rate of cancer progression	Batimastat, Marimastat, Prinomastat, and Tanomastae	
Antimicrobial	Defensin analogs that boost host immunity	mPE and PMX519	[61]
	Cyclic lipopeptide from *Streptomyces roseoporus* that exerts bactericidal effects by disrupting multiple aspects of bacterial cell membrane function	Daptomycin	[62]

(*Continued*)

TABLE 9.2 (Continued)

Therapeutic area	Peptide	Peptidomimetic	References
	β-hairpin antibiotic protegrin I (PG-I), which induces strong cytotoxic effects on bacterial cell membranes	POL7001 and POL7080	[63]
Cardiac	GP IIb/IIIa inhibitors, which are potent inhibitors of platelet aggregation that provide marked protection from ischemic events in patients undergoing percutaneous coronary intervention	Eptifibatide and Tirofiban	[64]
	Vasopressin analogs maintain serum osmolality and therefore extracellular fluid volume, thus maintaining blood volume and blood pressure	Desmopressin acetate and Terlipressin acetate	
	Bradykinin B2 receptor inhibitors, which are used to treat acute episodes of swelling and inflammation associated with hereditary angioedema	Icatibant acetate	
	Angiotensin II receptor antagonists, which modulate renal function and blood pressure	Saralasin	
	SOCS1 inhibition of gene transfer decelerates the transition to heart failure via inhibition of the gp130/JAK/STAT pathway	JAK2 Tkip and SOCS1-KIR	[59]
	apoE inhibitors act as ligands for low-density lipoprotein (LDL)	ATI-5261	

(Continued)

TABLE 9.2 (Continued)

Therapeutic area	Peptide	Peptidomimetic	References
	receptors that play a role in cholesterol metabolism and cardiovascular disease by causing an increase in atherogenic lipoprotein levels		
	Annexin-A1 analogs that reduce levels of cardiac necrosis, inflammation, and fibrosis following myocardial infarction	CGEN-855A	
Obesity	Mimetics of apoA-I, which is the major protein component of HDL particles in plasma that reduce cardiovascular disease	18 A, 4 F, ETC-642	[65]
	Melanocyte-stimulating hormone analog involved in regulation of feeding behavior, mood, sleep–wake cycle, and energy balance	BL3020–1, ALB-127158(a)	
	Neuropeptide Y2R inhibitors; Y2R acts as a neurotransmitter in the brain and autonomic nervous system to increase food intake and storage of energy as fat	Neuromedin-U, Catestatin, and PYY-1119	
Obstetrics	Oxytocin analog that plays a role in social bonding, sexual reproduction, childbirth, and the period after childbirth	Carbetocin	[66]

times, the development of a boronic ester prodrug of serine proteases, prolyl oligopeptidase, has paved its made into the research into a "drug-like" treatment strategy for neurodegenerative diseases [72].

9.4.2 Addition of unnatural amino acids

The incorporation of unnatural amino acids into peptidomimetics can improve metabolic stability while retaining selectivity and therapeutic effectivity. Unnatural amino acids support resistance to proteolysis and hence increase the oral bioavailability of the formulation. Furthermore, unnatural amino acids prevent the trigger of an immune response that would otherwise be generated in the case that amino acid sequences are identified by the body. Synthetic unnatural amino acids or modified noncanonical amino acids from other organisms can be substituted for canonical amino acids (the 20 proteinogenic amino acids encoded in the human body) to augment such peptide sequences [73].

The substitution of the lipophilic appendages of 5,5-; 6,5-; 7,5-; and 8,5-fused 1-aza-2-oxobicyclo[X.3.0]alkane with a proline-based unnatural amino acid *N*-Boc-L-octahydroindole-2-carboxylic acid resulted in the successful mapping of the unknown topology of peptide receptors and in developing potent receptor antagonists like bradykinin B_2 receptor antagonist [74]. Also, an unnatural amino acid-substituted gamma-secretase inhibitor peptidomimetic P3–250 showed enhanced stability while alleviating symptoms of Alzheimer disease [75]. Modifications of the P4, P3, and P2 positions of matriptase-2 inhibitor peptidomimetic (sequence Y-Y-V-R) with unnatural amino acids also increased its selectivity for matriptase-2 [76].

9.4.3 Inversion of chirality

Most amino acids in nature exist in the *L*-amino acid configuration. The D-amino acid configuration is excreted as is by the human body because of the inability of the body's protease enzymes to hydrolyze and cleave it. Hence, the incorporation of D-amino acids into peptidomimetics improves the metabolic stability compared with L-amino acids. The conformational preference of a peptide can be understood by conducting ellipsometry studies.

The replacement of *L*-Leu residue in LHSpTA by *D*-Leu, followed by acylation via *L*-Pro, paved the way for their posing as potential inhibitors of polo-like kinase 1 polo-box domain in HeLa cells [77]. Even the attachment of α-chiral/β-peptoid residues onto a cell-penetrating peptidomimetic demonstrated its higher adsorption through anionic lipid bilayers and an increased overall mean residue ellipticity [78].

9.4.4 Backbone cyclization

Cyclic peptidomimetics are polypeptoids having a cyclic ring structure. The fashion of cyclization can be head-to-tail, side chain-to-side chain, head-to-side chain, and side chain-to-tail cyclization. Several approaches have been used to generate cyclic mimetics, such as backbone cyclization, peptide stapling, and native chemical ligation. These structures can be formed with chemically

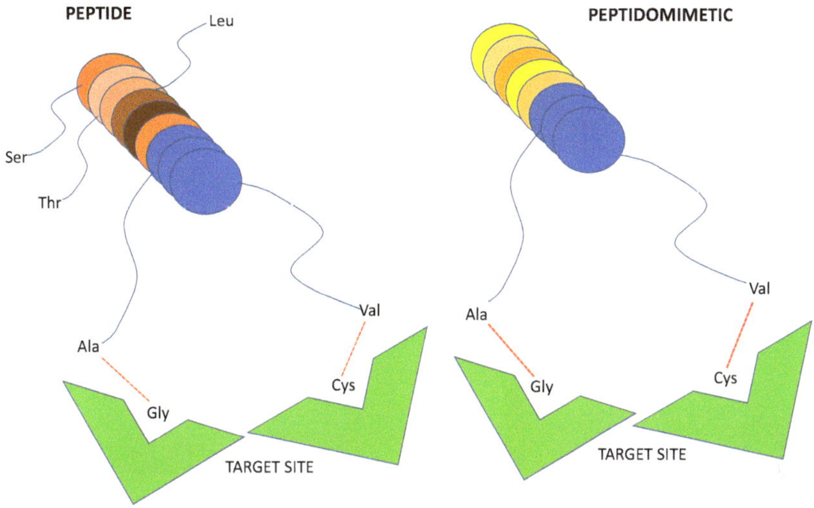

FIGURE 9.1 Evolution of a peptidomimetic from a peptide.

stable bonds such as amide, lactone, ether, thioether, or disulfide bonds. Cyclization enhances conformational stability and improves biological activity compared to linear analogs. The resultant molecules have increased receptor selectivity and affinity and are resistant to hydrolysis via enzymes because of their conformational constraint and rigidity [64]. Fig. 9.1 shows the evolution of a peptidomimetic from a peptide, where only the vital elements of a peptide are conformationally retained while being attached to a different amino acid/nonamino acid-based backbone.

9.5 Future directions and conclusions

The formulation and design of peptides and peptidomimetics have been among the most extensively researched fields under proteomics due to their attractive therapeutic profile. Most peptide formulations and delivery platforms have been founded with the goal of extending the circulating half-life of the peptide along with increased patient compliance. However, each approach comes with its own advantages and disadvantages, due to which they often remain stuck in the nascent phases of clinical trials. Fig. 9.2 depicts the advantages of peptidomimetics over peptides, such as conformational constraint leading to optimal bioavailability at the target site, greater site specificity, and metabolic stability due to no off-target binding.

Also, while efforts are underway in switching from conventional parenteral formulations to oral, nasal, and pulmonary formulations, its reproduction in large-scale production remains a challenge. We hope to see a greater number of peptide and peptidomimetic drugs (especially peptidomimetic drugs) reach clinical trials, with the development of novel strategies of drug delivery such as the

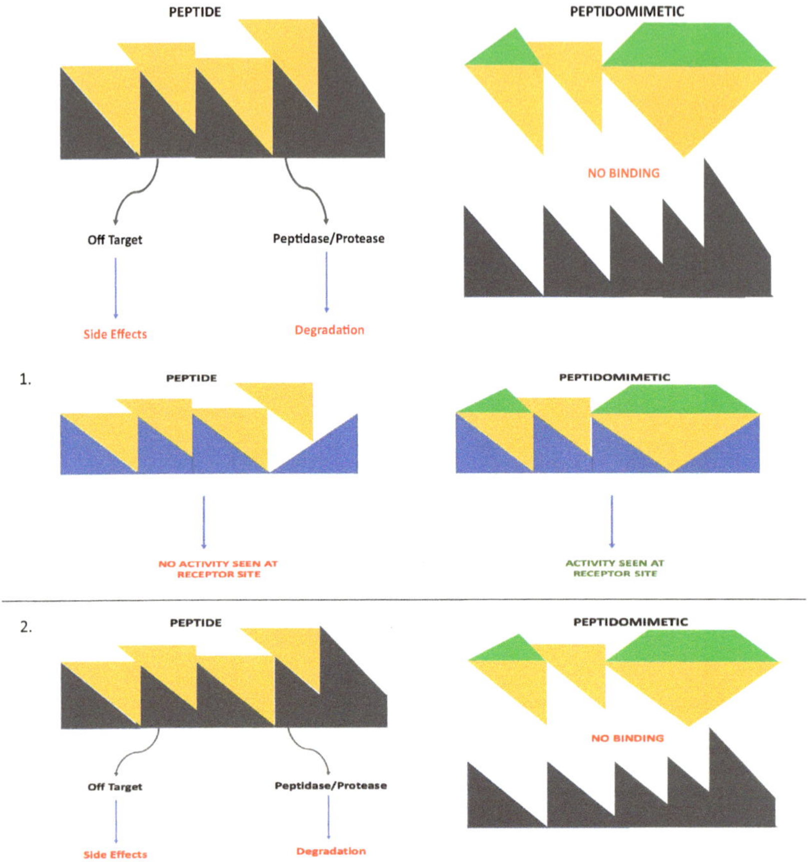

FIGURE 9.2 Advantages of peptidomimetics compared to peptides.

use of endogenous peptides as carriers for peptide-based drugs or formulations that could control the molecular recognition mechanism of a peptide and exploit its binding affinity for protein–protein interactions, and the evolution of newer delivery systems such as nanomedicine in the near future.

Acknowledgments

The authors thank SVKM's NMIMS for their support.

References

[1] Fosgerau K, Hoffmann T. Peptide therapeutics: current status and future directions. Drug Discov Today 2014;20.

[2] Preet P. Peptides: a new therapeutic approach. Int J Curr Pharm Res 2018;10(2):29.

[3] Patel A, Cholkar K, Mitra AK. Recent developments in protein and peptide parenteral delivery approaches. Therap Deliv 2014;5(3):337–65.

[4] Manning RWP, M. C. Peptide formulation: challenges and strategies. Innov Pharm Technol 2009;28:64–8.

[5] Grauer A, König B. Peptidomimetics—a versatile route to biologically active compounds. Eur J Org Chem 2009;30:5099–111.

[6] Underberg WJM, Hoitink MA, Reubsaet JLE, Waterval JCM. Separation and detection techniques for peptides and proteins in stability research and bioanalysis. J Chromatogr B: Biomed Sci Appl 2000;742(2):401–9.

[7] Banga A. Stability of therapeutic peptides and proteins. Therapeutic peptides and proteins, 3. Taylor & Francis LLC; 2005. p. 68–87.

[8] Evers A, Pfeiffer-Marek S, Bossart M, Heubel C, Stock U, Tiwari G, et al. Peptide optimization at the drug discovery-development interface: tailoring of physicochemical properties toward specific formulation requirements. J Pharm Sci 2019;108(4): 1404–14.

[9] Strader MB, Kassa T, Meng F, Wood FB, Hirsch RE, Friedman JM, et al. Oxidative instability of hemoglobin E (β26 Glu→Lys) is increased in the presence of free α subunits and reversed by α-hemoglobin stabilizing protein (AHSP): relevance to HbE/β-thalassemia. Redox Biol 2016;8:363–74.

[10] Chandrasekhar S, Epling DE, Sophocleous AM, Topp EM. Thiol–disulfide exchange in peptides derived from human growth hormone. J Pharm Sci 2014;103(4):1032–42.

[11] Payne RW, Manning MC. Peptide formulation: challenges and strategies. Innov Pharm Technol Ingred 2009;28:64–8.

[12] Hofstetter H, Hofstetter O. Antibodies as tailor-made chiral selectors for detection and separation of stereoisomers. TrAC Trends Anal Chem 2005;24(10):869–79.

[13] Fu C, Zhang Z, Zhou S, Pritts WA, Zhang Q. Assessing localized conformational stability of antibody-drug conjugate by protein conformation assay. J Pharm Biomed Anal 2020;179:113020.

[14] Fujimoto DF, Pinilla C, Segall AM. New peptide inhibitors of type IB topoisomerases: similarities and differences vis-a-vis inhibitors of tyrosine recombinases. J Mol Biol 2006;363(5):891–907.

[15] Feng YW, Ooishi A, Honda S. Aggregation factor analysis for protein formulation by a systematic approach using FTIR, SEC and design of experiments techniques. J Pharm Biomed Anal 2012;57:143–52.

[16] Salgin S, Salgin U, Bahadir S. Zeta potentials and isoelectric points of biomolecules: The effects of ion types and ionic strengths. Int J Electrochem Sci 2012;7(12): 12404–14.

[17] Cholewinski M, Lückel B, Horn H. Degradation pathways, analytical characterization and formulation strategies of a peptide and a protein calcitonine and human growth hormone in comparison. Pharmaceutica Acta Helvetiae 1996;71(6):405–19.

[18] Patro SY, Freund E, Chang B. Protein formulation and fill-finish operations. Biotechnol Annu Rev 2002;8:55–84.

[19] Iyer LK, Phanse R, Xu M, Lan W, Krause ME, Bolgar M, et al. Pulse proteolysis: an orthogonal tool for protein formulation screening. J Pharm Sci 2019;108(2):842–50.

[20] Jorgensen L, Nielson HM. Delivery technologies for biopharmaceuticals: peptides, proteins, nucleic acids and vaccines. Wiley; 2009.

[21] Gupta H, Sharma A. Recent trends in protein and peptide drug delivery systems. Asian J Pharmaceut 2009;3(2):69–75.

[22] Joseph M, Trinh HM, Mitra AK. Chapter 7 - Peptide and Protein-Based Therapeutic Agents. In: A. K. Mitra, K. Cholkar, & A. B. T.-E. N. for D. Mandal Drug Delivery and Medical Devices (eds.), Micro and Nano Technologies. 2017; p. 145–67.

[23] Veronese FM. Peptide and protein PEGylation: a review of problems and solutions. Biomaterials 2001;22(5):405–17.

[24] Marshall SA, Lazar GA, Chirino AJ, Desjarlais JR. Rational design and engineering of therapeutic proteins. Drug Discov Today 2003;8(5):212–21.

[25] Nordström R, Nyström L, Ilyas H, Atreya HS, Borro BC, Bhunia A, et al. Microgels as carriers of antimicrobial peptides—effects of peptide PEGylation. Colloids Surf A Physicochem Eng Asp 2019;565:8–15.

[26] Peng G, Pan X, Hu H, Xu Y, Wu C. N-terminal site-specific PEGylation enhances the circulation half-life of Thymosin alpha 1. J Drug Deliv Sci Technol 2019;49:405–12.

[27] Kapoor V, Singh AK, Rogers BE, Thotala D, Hallahan DE. PEGylated peptide to TIP1 is a novel targeting agent that binds specifically to various cancers in vivo. J Controlled Rel 2019;298:194–201.

[28] Zhang Y, Knapp S. Simplified beta-glycosylation of peptides. Tetrahedron 2018;74 (23):2891–903.

[29] Sinclair AM, Elliott S. Glycoengineering: the effect of glycosylation on the properties of therapeutic proteins. J Pharm Sci 2005;94(8):1626–35.

[30] Beals JM, Shanafelt AB. Enhancing exposure of protein therapeutics. Drug Discov Today Technol 2006;3(1):87–94.

[31] Liang F-C, Chen RP-Y, Lin C-C, Huang K-T, Chan SI. Tuning the conformation properties of a peptide by glycosylation and phosphorylation. Biochem Biophys Res Commun 2006;342(2):482–8.

[32] Urge L, Gorbics L, Otvos L. Chemical glycosylation of peptide T at natural and artificial glycosylation sites stabilizes or rearranges the dominant reverse turn structure. Biochem Biophys Res Commun 1992;184(2):1125–32.

[33] Furmanek A, Hofsteenge J. Protein C-mannosylation: facts and questions. Acta Biochim Polon 2000;47(3):781–9.

[34] Morishita S, Suzuki T, Niwa Y, Dohmae N, Simizu S. Dpy-19 like 3-mediated C-mannosylation and expression levels of RPE-spondin in human tumor cell lines. Oncol Lett 2017;14(2):2537–44.

[35] Pandita D, Shah S, Pathak R. Review on novel approaches for oral protein and peptide drug delivery. World J Pharm Res 2018;7(2):317–44.

[36] Lee VHL. Protease inhibitors and penetration enhancers as approaches to modify peptide absorption. J Controlled Rel 1990;13(2):213–23.

[37] Amancha KP, Hussain A. Effect of protease inhibitors on pulmonary bioavailability of therapeutic proteins and peptides in the rat. Eur J Pharm Sci 2015;68:1–10.

[38] Dan, N. An update pharm strateg oral delivery therapeutic peptides proteins, 2019, 7.

[39] Zhao X, Coulman SA, Hanna SJ, Wong FS, Dayan CM, Birchall JC. Formulation of hydrophobic peptides for skin delivery via coated microneedles. J Controlled Rel 2017;265:2–13.

[40] Cleland JL, Langer R. Formulation and delivery of proteins and peptides design and development strategies. Pharm Res Develop, Genentech Inc, South San Francisco, CA 94080 Dep Chem Eng 1994;1:1–19.

[41] Tan ML, Choong PFM, Dass CR. Recent developments in liposomes, microparticles and nanoparticles for protein and peptide drug delivery. Peptides 2010;31(1):184–93.

[42] Torchilin VP. Recent advances with liposomes as pharmaceutical carriers. Nat Rev Drug Discov 2005;4(2):145–60.

[43] Rezaei N, Mehrnejad F, Vaezi Z, Sedghi M, Asghari SM, Naderi-Manesh H. Encapsulation of an endostatin peptide in liposomes: stability, release, and cytotoxicity study. Colloids Surf B Biointerfaces 2020;185:110552.

[44] Thapa RK, Kiick KL, Sullivan MO. Encapsulation of collagen mimetic peptide-tethered vancomycin liposomes in collagen-based scaffolds for infection control in wounds. Acta Biomater 2020;103:115–28.

[45] Mallick S, Thuy LT, Lee S, Park J-I, Choi JS. Liposomes containing cholesterol and mitochondria-penetrating peptide (MPP) for targeted delivery of antimycin A to A549 cells. Colloids Surf B Biointerfaces 2018;161:356–64.

[46] Couvreur P, Puisieux F. Nano- and microparticles for the delivery of polypeptides and proteins. Adv Drug Deliv Rev 1993;10(2–3):141–62.

[47] Banerjee A, Onyuksel H. Peptide delivery using phospholipid micelles. WIREs Nanomed Nanobiotechnol 2012;4(5):562–74.

[48] Wang T, Wang N, Hao A, He X, Li T, Deng Y. Lyophilization of water-in-oil emulsions to prepare phospholipid-based anhydrous reverse micelles for oral peptide delivery. Eur J Pharm Sci 2010;39(5):373–9.

[49] Chen X, Tai L, Gao J, Qian J, Zhang M, Li B, et al. A stapled peptide antagonist of MDM2 carried by polymeric micelles sensitizes glioblastoma to temozolomide treatment through p53 activation. J Controlled Rel 2015;218:29–35.

[50] New RRC, Kirby CJ. Solubilisation of hydrophilic drugs in oily formulations. Adv Drug Deliv Rev 1997;25(1):59–69.

[51] Lueβen HL, Bohner V, Pérard D, Langguth P, Verhoef JC, de Boer AG, et al. Mucoadhesive polymers in peroral peptide drug delivery. V. Effect of poly(acrylates) on the enzymatic degradation of peptide drugs by intestinal brush border membrane vesicles. Int J Pharmaceut 1996;141(1):39–52.

[52] Lueßen HL, Rentel C-O, Kotzé AF, Lehr C-M, de Boer AG, Verhoef JC, et al. Mucoadhesive polymers in peroral peptide drug delivery. IV. Polycarbophil and chitosan are potent enhancers of peptide transport across intestinal mucosae in vitro. J Controlled Rel 1997;45(1):15–23.

[53] Borchard G, Lueβen HL, de Boer AG, Verhoef JC, Lehr C-M, Junginger HE. The potential of mucoadhesive polymers in enhancing intestinal peptide drug absorption. III: effects of chitosan-glutamate and carbomer on epithelial tight junctions in vitro. J Controlled Rel 1996;39(2):131–8.

[54] Zbacnik TJ, Holcomb RE, Katayama DS, Murphy BM, Payne RW, Coccaro RC, et al. Role of Buffers in Protein Formulations. J Pharm Sci 2017;106(3):713–33.

[55] Muralidhara BK, Wong M. Critical considerations in the formulation development of parenteral biologic drugs. Drug Discov Today, 2020.

[56] Lai MC, Topp EM. Solid-state chemical stability of proteins and peptides. J Pharm Sci 1999;88(5):489–500.

[57] Shah A, Barathi B, Nair A, Rajam C. Peptidomimetics as a cutting edge tool for advanced healthcare. Int J Pharm Sci Res 2017;8(12):4992–5000.

[58] Gupta V, Kalra S. Choosing a gliptin. Indian J Endocrinol Metab 2011;15(4):298–308.

[59] Recio C, Maione F, Iqbal AJ, Mascolo N, De Feo V. The potential therapeutic application of peptides and peptidomimetics in cardiovascular disease. Front Pharmacol 2017;7:526.

[60] Avendaño López C, Menendez JC. Peptidomimetics in cancer chemotherapy. Clin Transl Oncol Publ Federation Span Oncol Societies Natl Cancer Inst Mexico 2007;9:563–70.

[61] Ryan LK, Freeman KB, Masso-Silva JA, Falkovsky K, Aloyouny A, Markowitz K, et al. Activity of potent and selective host defense peptide mimetics in mouse models of oral candidiasis. Antimicrob Agents Chemother 2014;58(7):3820–7.

[62] Beiras-Fernandez A, Vogt F, Sodian R, Weis F. Daptomycin: a novel lipopeptide antibiotic against Gram-positive pathogens. Infect Drug Resistance 2010;3:95–101.

[63] Srinivas N, Jetter P, Ueberbacher BJ, Werneburg M, Zerbe K, Steinmann J, et al. Peptidomimetic antibiotics target outer-membrane biogenesis in *Pseudomonas aeruginosa*. Science 2010;327(5968):1010–13.

[64] Qvit N, Rubin SJS, Urban TJ, Mochly-Rosen D, Gross ER. Peptidomimetic therapeutics: scientific approaches and opportunities. Drug Discov Today 2017;22(2):454–62.

[65] Kumar MS. Peptides and peptidomimetics as potential antiobesity agents: overview of current status. Front Nutr 2019;6:11.

[66] Sawyer TK. Discovery: the third wave. The Royal Society of Chemistry; 2017.

[67] Pauletti GM, Gangwar S, Siahaan TJ, Jeffrey AT, Borchardt R. Improvement of oral peptide bioavailability: peptidomimetics and prodrug strategies. Adv Drug Deliv Rev 1997;27 (2):235–56.

[68] Oliyai R. Prodrugs of peptides and peptidomimetics for improved formulation and delivery. Adv Drug Deliv Rev 1996;19:275–286.

[69] Song X, Xu CR, He HT, Siahaan TJ. Synthesis of a novel cyclic prodrug of RGD peptidomimetic to improve its cell membrane permeation. Bioorg Chem 2002;30(4):285–301.

[70] Peterson LW, Kashemirov BA, Eriksson U, Kim J-S, Mitchell S, Kijek P, et al. Serine side-chain-linked peptidomimetic prodrugs of cidofovir and cyclic cidofovir: C-ester effects on transport and activation. Antivir Res 2008;78(2):A46.

[71] McMurray J, Klostergaard J, Auzenne EJ, Liao WSL, Lu Z, Mandal PK, et al. 422 Targeting the SH2 domain of Stat3 with phosphopeptide mimetic prodrugs leads to tumor growth inhibition and down-regulation of phosphoTyr705 Stat3 and angiogenic pathways. Eur J Cancer Suppl 2010;8(7):134.

[72] Plescia J, Dufresne C, Janmamode N, Wahba AS, Mittermaier AK, Moitessier N. Discovery of covalent prolyl oligopeptidase boronic ester inhibitors. Eur J Medicinal Chem 2020;185:111783.

[73] Stevenazzi A, Marchini M, Sandrone G, Vergani B, Lattanzio M. Amino acidic scaffolds bearing unnatural side chains: An old idea generates new and versatile tools for the life sciences. Bioorg Med Chem Lett 2014;24(23):5349–56.

[74] Belvisi L, Colombo L, Colombo M, Di Giacomo M, Manzoni L, Vodopivec B, et al. Practical stereoselective synthesis of conformationally constrained unnatural proline-based amino acids and peptidomimetics. Tetrahedron 2001;57(30):6463–73.

[75] Liao Y-F, Wang B-J, Liao C-Y, Wu S-Y, Hu M-K. P3–250: Unnatural amino acid-substituted (hydroxyethyl)urea peptidomimetics inhibit gamma-secretase with enhanced stability. Alzheimer's Dement 2006;2(3, Supplement):S448–9.

[76] St-Georges C, Désilets A, Béliveau F, Ghinet M, Dion SP, Colombo É, et al. Modulating the selectivity of matriptase-2 inhibitors with unnatural amino acids. Eur J Med Chem 2017;129:110–23.

[77] Li Z, Zhang Z, Chen Y, Tang S, Lin T, Huang J, et al. Design, synthesis and evaluation of d-amino acid-containing peptidomimetics targeting the polo-box domain of polo-like kinase 1. Bioorg Chem 2019;85:534–40.

[78] Jing X, Yang M, Kasimova MR, Malmsten M, Franzyk H, Jorgensen L, et al. Membrane adsorption and binding, cellular uptake and cytotoxicity of cell-penetrating peptidomimetics with α-peptide/β-peptoid backbone: effects of hydrogen bonding and α-chirality in the β-peptoid residues. Biochim Biophys Acta (BBA) - Biomembr 2012;1818(11):2660–8.

Chapter 10

Medical use of cell-penetrating peptides: how far have they come?

Ines Neundorf
Institute for Biochemistry, Department of Chemistry, Faculty of Mathematics and Natural Sciences, University of Cologne, Cologne, Germany

10.1 Introduction

More than 30 years ago cell-penetrating peptides (CPPs) entered the area of drug delivery. The plasma membrane is composed of a lipid bilayer giving the cell its shape and, more importantly, restricting the in- and out-flow of biomolecules. CPPs, usually short peptide sequences, can overcome this plasma membrane and carry various cargos into cells. Over the last decades, this property has made them valuable tools in basic research as well as various medical applications. In particular, CPPs give hope for molecules such as negatively charged nucleic acids or therapeutic proteins, where further clinical development is hampered due to limited or restricted cellular uptake and thus low target site accumulation. Since their discovery in 1988, a meanwhile unclear number of CPPs has been reported encompassing sequences that come from different sources, for example, naturally (protein) derived, chimeric or synthetic. For instance, CPPsite 2.0 (http://www.crdd.osdd.net) counts 1700 CPPs and offers data about the secondary structure, activity, and uptake mechanism of CPPs. Additionally, researchers can find informative reports about the successful application of CPPs as transport vectors both in vitro and in vivo [1].

Unfortunately, until now, no FDA-approved CPP-including therapeutic has made it to the market. The reasons are multifold and include issues of in vivo stability, target selectivity, tolerability, and efficiency. Although this fact is disappointing and hard to accept despite the long time CPPs have been investigated, actually there is a real promise that CPP-based therapeutics will soon be launched, since many of them are currently enrolled in clinical trials.

Peptide and Peptidomimetic Therapeutics. DOI: https://doi.org/10.1016/B978-0-12-820141-1.00001-7

The present chapter will briefly introduce CPPs, how their activity is improved and in which current medical fields they may be applied. Furthermore, this work presents a brief update on the progress made since some of the last reports on CPPs investigated within preclinical and clinical studies [2–6].

10.1.1 Cell-penetrating peptide structure and mechanism at a glance

To get a more comprehensive view of all the facets of CPPs, including history, structures, uptake mechanisms, and all the different fields of application, it is highly recommended that interested readers refer to recent literature on this topic [7,8]. This work presents only a brief introduction to the topic and provides insights into the classification and mechanism of cellular entry of CPPs.

10.1.1.1 Cell-penetrating peptide classification

CPPs are a highly diverse group of molecules making it difficult to find the right criteria for classification. Very often, they are categorized either based on their physicochemical characteristics or simply on their origin. Using the latter way, CPPs are ranged by being derived from natural protein sequences, which would include the first reported CPPs Tat (Tat_{48-60}: GRKKRRQRRRPPQ) [9,10] and penetratin ($pAntp_{43-48}$: RQIKIWFQNRRMKWKK) [11] or being synthetic peptides, such as the poly-arginine peptides Arg_5 to Arg_9 [12], or chimeric sequences, for instance with Transportan (GWTLNSAGYLLGKINLKALAALAKKIL) as a representative [13]. However, due to several reasons, nowadays CPPs are more likely ranked according to their physicochemical nature into cationic, amphipathic, or hydrophobic CPPs [14]. The group of hydrophobic CPPs is the smallest one, whereas many examples for amphipathic and cationic CPPs exist. For instance, in our group, we developed the amphipathic CPP sC18 (GLRKRLRKFRNKIKEK) and derivatives thereof [15–17] and used them for various drug delivery purposes in vitro and in vivo [18–21].

10.1.1.2 Mechanisms of cellular entry

In spite of the long-time of CPP application, it is astonishing that there is still no absolute agreement about how those peptides enter cells. CPPs can use two main uptake pathways, which are direct translocation and processes based on endocytosis. Direct translocation often occurs at higher concentrations, while endocytosis is in many cases seen at lower doses and for CPP-cargo constructs [22]. Different key factors are responsible to induce the one or other mechanism, whereas CPP sequence and structure, concentration, nature of the attached cargo, as well as target lipid membrane composition give reason to be the most crucial ones. For both mechanisms, it is commonly agreed that CPPs first interact with membrane components at the outer surface of the plasma membrane, most probably via electrostatic interaction between the residues of the cationic amino acids and negatively charged groups of glycan chains. The resulting direct contact to membrane phospholipids may then lead to changes in conformation and aggregation, finally inducing membrane translocation. Moreover, several receptor types, particularly membrane proteoglycans, are nowadays discussed to be involved in the internalization of CPPs, which prefer endocytosis [22].

Interestingly, it has been recently shown that for direct penetration of CPPs cholesterol re-organization and lipid-de-packing are decisive factors, which are influenced by the nature (amphipathic vs purely cationic) of the CPP applied to the target membrane [23]. Thus the more amphipathic the CPPs become, the more they can modulate membrane fluidity facilitating change in peptide orientation with insertion of the hydrophobic residues into the membrane. In fact, cationic CPPs mainly interact with the membrane by the arginine residues and are driven by electrochemical forces. This leads to loosening of the lipid packing and entry of the peptides inside the cells [24]. Although also amphipathic CPPs usually contain a certain amount of cationic amino acids, their interaction with the membrane is driven more by their ability to form secondary structures, in many cases α-helices. Those amphipathic helices then modify lipid organization and insert, according to different models [22], into the hydrophobic core of the lipid bilayer, forming in most cases small membrane pores. Another factor for the direct translocation is the membrane potential that is produced by the ion concentration imbalance across the membrane. It is thought that positively charged CPPs may further enhance the membrane potential, opening membrane pores through which CPPs penetrate the bilayer [25].

Otherwise, most CPPs and CPP-cargo constructs enter cells by energy-dependent endocytosis, which involves different pathways, namely macropinocytosis, clathrin/caveolin-mediated, as well as clathrin/caveolin-independent endocytosis. When carrying cargos of increased size, endocytosis is the main uptake pathway observed for CPPs. Endocytosis comes with the disadvantage that CPPs and CPP-cargo constructs are entrapped within endosomes. Thus different strategies have usually to be applied to guarantee efficient delivery of the cargo to its target site (see below).

The uptake mechanism of CPPs significantly varies and is dependent on various factors, which, in addition, may trigger their ability to switch between different uptake mechanisms. As a result, it is often not predictable by which uptake pathway cell entry occurs and it has, therefore to be defined for every CPP or CPP-cargo construct within the specific experimental setting. This fact should be considered when starting to work with CPP-related applications. Indeed, it should be seriously thought of to elucidate the exact entry mechanism, particularly when in vivo studies are planned. The results would be important not only to avoid possible side reactions, but also to provide a chance for improving and enhancing the internalization activity of the CPP studied, and to obtain the expected biological response. This clearer knowledge about all required conditions will further promote the development of a safe and efficient drug delivery system.

10.2 Application of cell-penetrating peptides in biomedicine

10.2.1 Methods of cell-penetrating peptide-cargo construct formation

CPPs have been widely used to deliver therapeutic molecules inside the cell, to the cytosol or nucleus, or other intracellular destinations. For this, cargos are either attached by covalent or noncovalent methods [7]. Interestingly, in many preclinical and clinical studies the CPPs penetratin and Tat are used, probably because these belong to the first discovered CPPs, and have been extensively studied concerning cellular uptake mechanism, secondary structure and in vivo properties.

10.2.1.1 Covalent cell-penetrating peptide-cargo conjugates

Covalent coupling is indeed the strategy that has first been utilized for the delivery of small molecule drugs and proteins [2]. Small chemotherapeutics are often covalently linked to CPPs using in many cases linkers that allow for selective cleavage when the CPP-cargo construct has reached the cell interior. For example, overexpression of specific proteases is related to many diseases including cancer [26]. Therefore proteolytic cleavage sites for the lysosomal protease cathepsin B have been introduced in many studies to promote cargo release after lysosomal accumulation. Also, we have already demonstrated the benefit of this strategy [27,28]. External cleavage is in many applications realized by targeting the Zinc dependent matrix metalloproteases, for which an overexpression in different tumors has been demonstrated [26]. Moreover, cargos are frequently attached via disulfide bonds to gain the intracellular release of the drug within the reductive milieu of the endosomes. Many other linkages are possible, which are commonly used for the generation of peptide-drug conjugates, such as the formation of oxime bonds, triazoles, maleimide thioethers, etc. [29].

Regarding therapeutically relevant peptide or protein sequences, these are typically fused in line with the CPP sequence (at the *N*- or *C*-terminal, dependent on activity), either by using recombinant techniques or with the help of solid-phase peptide synthesis.

Peptide nucleic acids (PNAs) are synthetic DNA mimetics that have proven their high applicability in molecular biology and diagnostic assays as hybridization probes [30]. Also here, various strategies for safe intracellular delivery of covalent PNA-CPP constructs in mammalian as well as in bacterial cells have been already reported [31–34]. Other synthetic antisense oligonucleotides that are covalently fused to CPPs comprise phosphorodiamidate morpholino oligomers (PMOs) in which the five-membered ribofuranosyl ring in DNA and RNA is replaced by a six-membered morpholino ring [33,35]. Each morpholino ring is linked through an uncharged phosphorodiamidate group compared to the negatively charged phosphodiester found in natural DNA or RNA. Owing to their exon skipping activity, PMOs have been widely used in biomedical applications [35].

10.2.1.2 Noncovalent cell-penetrating peptide-cargo complexes

The noncovalent strategies have the advantage of creating the cargo-CPP complex by simple coincubation. In this way, not only proteins and nucleic acids can be intracellularly delivered, also the uptake of small molecule drugs might be facilitated. For nucleic acids and proteins/peptides, such coincubation techniques principally lead to the formation of electrostatic complexes. Compared to this, when CPPs are coadministered with small molecule drugs, the idea is more that CPPs help to form small membrane pores through which the drugs successfully enter the cells.

In the case of using pDNA, the cargo has to reach the cell nucleus for being processed, and therefore nuclear localization sequences are often additionally used to get there. Compared to this, siRNA has to be effectively delivered to the cytosol, increasing the interest in developing CPPs that are not entrapped within endosomes, or readily escape out of them after the endocytotic uptake. Recently, a new family of peptides, namely the WRAP peptides, suitable for siRNA delivery was presented [36]. WRAP peptides are arginine and tryptophan-rich amphipathic peptides able to form stable nanoparticles and to enroll siRNA molecules into cells.

Another concept that has more and more evolved through the past years is to embed CPPs within a multimodal component system [37]. This means that CPPs are one part of a more sophisticated drug delivery system, which is often built on inorganic or organic nanoparticles that additionally involve bioactive drugs, targeting sequences and also polyethylene glycol (PEG) chains. This strategy is intended to increase the targeting and metabolic stability of the drug as well as the CPP [38]. Within those multimodal structures, CPPs can be coated on surfaces by biophysical forces or also covalently attached to the particles.

10.2.2 Medical fields with cell-penetrating peptide applications

According to the World Health Organization, heart diseases account for the highest mortality rates, while the second leading cause of death worldwide is cancer (with about 1 in 6 deaths). Fittingly, CPP-based therapies are principally found in both fields under clinical development, although the focus is more toward CPP-based therapeutics with anticancer activities. Nevertheless, owing to their high potential, various other areas of clinical interest emerged spanning neurological disorders, inflammation, or pain.

But how do we know if a given CPP-based strategy leads to a biological response?

From all studies reported thus far involving CPPs and CPP-drug systems, one might propose that they can indiscriminately target all cells. Reality shows that this is not the case and that, apparently, CPPs do not reach all cells when applied in in vivo models. That is why Brock et al. previously proposed a paradigm change toward a more opportunistic CPP research in which CPP-based therapeutics should be developed after the relevant target cells for the CPP studied were identified in in vivo studies [39]. This top-down approach would certainly improve the clinical translation of therapeutics including CPPs. For instance, it was demonstrated that only some CPPs can cross the blood–brain-barrier (BBB). Although this observation is a clear prerequisite in terms of protecting the central nervous system from toxicity, it gives evidence that in reality only a few sequences can be developed for brain-related conditions. Another aspect that is important and demands study in more detail is the surrounding environment of the target tissue. Consequently, when studying cancer-related drugs, it is particularly important to know the tumor physiology. Therefore predictive tumor models should be utilized to get a clearer idea of the pathophysiology of CPPs [40]. Moreover, the translation from in vitro to in vivo studies often fails, since the preclinical studies are commonly performed in absence of physiological conditions. Thus unexpected in vivo performance arises and as a consequence, the development of the CPP-based therapeutic often arrests at the preclinical stage. It is, therefore of utmost importance to find the right test systems and in vitro/in vivo models that incorporate also specific physiological conditions such as pH, salt content, presence of proteases and other harmful molecules etc. to increase the probability to find positive hits for an efficient CPP-related medical treatment.

10.3 How to improve cell-penetrating peptide-based applications

There are several challenges that have to be addressed in an effective clinical application of CPPs (Fig. 10.1). Besides the already mentioned unspecific uptake of CPPs, they are prone to proteolytic degradation, and their encapsulation in endosomes limits the efficacy of the cargo. However, during the last years, several strategies have been explored to tackle these challenges.

10.3.1 Improving proteolytic resistance of cell-penetrating peptides

The low oral bioavailability of CPPs is a limitation that comes with most peptide therapeutics and is mainly a result of rapid degradation when inside the gastrointestinal tract. In fact, this is one of the greatest challenges toward

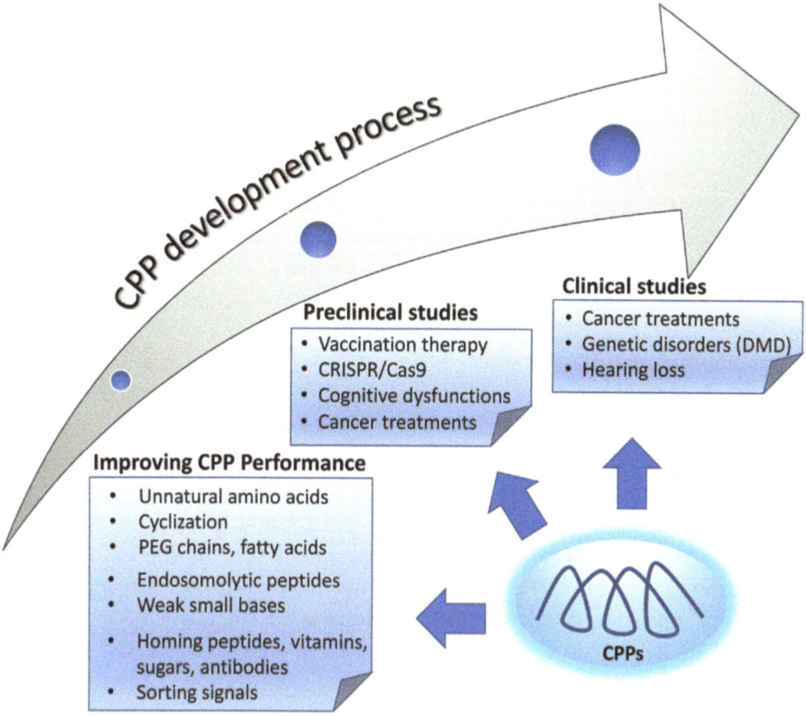

FIGURE 10.1 CPP development process: CPPs have to be improved before being tested in (pre)clincinical studies.

clinical application of most peptide therapeutics [41]. The reasons are the harsh gastric pH and digestion by the present proteolytic enzymes. On the other side, when administered intravenously, CPPs are prone to fast clearance out of the body which competes with the slow release of the free drug. As a result, in many preclinical studies, CPP-drug conjugates are topically administered or, if possible, directly injected into the target site.

To improve their bioavailability, CPPs are modified by different means, such as the introduction of PEG chains or fatty acids to increase molecular weight and lipophilicity. The latter is also a critical factor that might support permeability. In addition, cyclization has been proven quite effective, not only for increasing proteolytic stability but also for guiding CPPs directly into the cytosol. In fact, it has been previously demonstrated that cyclized CPPs have a higher propensity to accumulate in the cytosol and somehow escape encapsulation in vesicles [42]. Further, the replacement of natural amino acids by noncanonical residues (mainly D-versions) is a reliable strategy to increase their overall stability.

As mentioned, CPPs are often part of peptide/protein nanoparticle conjugates. The use of nanoparticles offers several benefits and they are particularly suited to enhance retention and circulation of drugs and to shield them against attacks from proteolytic enzymes [43].

10.3.2 Improving cytosolic accumulation of cell-penetrating peptides

As soon as CPP-cargo conjugates are applied to cells, they usually internalize by endocytosis mechanisms. Consequently, the cargo has to be released from the endosomes to reach its target site. Selective endosomolysis, or permeation/disruption of the endosome membrane, is required to ensure that the plasma membrane will not be affected. Usually, this is realized by fusing endosomolytic peptide sequences into the CPP. Those peptides may specifically interact with the inner endosome membrane featuring more negative charges compared to the zwitterionic cell membrane. For instance, this concept was recently demonstrated for modified variants of the hemolytic peptide Mastoparan X [44]. Otherwise, endosomolytic peptides can address the pH decrease of maturing endosomes. By the present increasing acidic pH environment, they undergo structural changes and specifically interact with the endosomal membrane resulting in the formation of pores that release the CPP-cargo construct. Those peptide sequences are in many cases derived from viral fusion proteins. For instance, the HA2 peptide is derived from the hemagglutinin protein of the influenza virus and represents a quite often and successfully used sequence [22].

Other endosomolytic agents are small molecule weak bases, with chloroquine being the most famous one. These molecules accumulate within endosomes after protonation triggered by the acidic pH and cause endosome swelling and disruption. Often, they are additionally applied when CPP-based nucleic acid transfection is studied to ensure effective localization of the cargo at the target destination (e.g., nuclei in the case of pDNA). In this context, a recent report highlighted a novel compound with endosomolytic activity, namely the small molecule UN7938, and showed its versatility to increase the cytosolic delivery of DNA-Tat complexes [45]. Notably, although chloroquine shares similar features, it showed only poor activity compared to UN7938, which, therefore represents a promising alternative.

10.3.3 Improving cell selectivity and intracellular targeting of cell-penetrating peptides

One of the major limitations in the use of CPPs is their intrinsic ability to target nearly all cells, making them unselective. Therefore homing peptide sequences or other targeting ligands, such as carbohydrates, vitamins, antibodies or aptamers that specifically recognize membrane receptors overexpressed on the target cells are frequently attached to the CPP-cargo conjugate. This is usually pursued by covalent coupling of the targeting unit and the CPP, while different linkages are possible. In one of our previous studies, we have made use of the integrins, cell adhesion mediators, which attract great interest in cancer research. We have generated dual-targeting peptide-drug conjugates composed of an integrin-targeting peptidomimetic and a CPP. Interestingly, besides successful active targeting mediated by the targeting unit, we found that the conjugates are mainly internalized supported by the CPP [28]. In fact, the CPP still outcompeted the targeting unit making it challenging to guarantee safe tumor targeting in an in vivo setting. However, since the uptake took place after very short contact times, this approach may be worth studying in more detail.

Remarkably, various CPPs have an intrinsic preference to target cancer cells. This phenomenon is based on the specific cancer environment, as well as on several specific characteristics of cancer cells. For instance, they display more negatively charged phospholipids at the outer plasma membrane compared to healthy cells, and cationic CPPs are attracted more to cancer cell membranes. Those and other specific physiological conditions can be used to modulate, or to modify CPPs with specific probes (e.g., pH sensitive) and to selectively direct them to cancer cells [46,47].

For several treatments, it is important that the cargo reaches defined intracellular cell compartments or organelles to exert its therapeutic effect. For this, CPPs are differently modified to facilitate drug delivery to distinct destinations. Of note is that some CPPs have the intrinsic ability to target the cell nuclei or nucleoli, while others do not. Therefore every CPP sequence should be studied carefully before evaluating therapeutic treatment strategies. However, several approaches exist to direct CPPs to the nucleus, mitochondria, or also other subcellular locations [48]. In many cases, sorting sequences (signal peptides) from proteins, which help to intracellularly localize them to their target destinations, are utilized and fused to CPPs. In this way, we have previously developed CPP-drug conjugates that actively accumulate inside the matrix of mitochondria [49].

Overall, recent years have endowed promising new strategies to improve the biological performance of CPPs. In particular, cyclic CPPs have proven to be effective to circumvent problems with enzymatic cleavage and to increase cytosolic delivery. Moreover, adding targeting moieties and equipping CPPs with more selectivity potentially prevents side effects and leads to high levels of drugs at the target site.

10.4 Cell-penetrating peptides in preclinical studies

Despite the above-mentioned pitfalls that come with CPPs, for many researchers, they still remain a vital element in the development of therapeutics for various areas of clinical interest. It would be beyond the scope of this chapter to summarize all possible treatments and ongoing interest in the design of CPP-related therapeutics, so several selected examples are discussed herein (Table 10.1) [3,4,8,50]. Above all, it may be of most interest that several pharmaceutical companies have pushed forward the development of CPP-based drugs, which, after successful preclinical studies, are also enrolled in phase I/II clinical studies, as will be reported below.

TABLE 10.1 Selected pharmaceutical companies with examples of cell-penetrating peptide-based therapeutics currently evaluated in preclinical and clinical studies.

Preclinical studies			
Company	Field of interest		References
Feldan	Delivery of CRISPR-associated nuclease (Cas) RNP for rapid entry into mouse airway epithelia		[51,52]
Phion	RALA peptides for noncovalent delivery to the central nervous system		[53,54]
Peptomyc	Omomyc peptides with anti-Myc activity		[55]
PYC	Phylomers fused to antisense oligonucleotide therapeutics targeting genetic diseases		[56]
Clinical studies			
Company	Compound	Application	clinicaltrials.gov ID
Sarepta	SRP-5051	Duchenne's muscular dystrophy	NCT04004065
Aileron	ALRN-6924	Advanced solid tumors, lymphomas with wild-type p53	NCT02264613
Amal	ATP128	Colorectal cancer	NCT04046445
Pep-Therapy	PEP-010	Breast and ovarian cancer	Final stages of preparation for phase I/II trials
APIM	ATX-101	Myeloma and bladder cancer	n.a.
Bicycletherapeutics	BT1718	Advanced solid tumors, breast cancer, nonsmall lung cancer	NCT03486730
Auris Medical	AM-111	Hearing loss	Completed

Abbreviation: n.a., not available.

10.4.1 Cell-penetrating peptides in vaccination

There is a noteworthy interest in using CPPs in vaccination to improve antigen uptake by antigen-presenting cells [57]. In this respect, CPPs are applied as a safe alternative or additive to classical adjuvant formulations. Various CPPs, including, for example, Tat, Penetratin, MPG, and VP22, have been successfully investigated as transporters for protein- or DNA-based vaccines producing promising vaccine delivery systems [8]. Up to now, no clinical data is available demonstrating the efficacy of CPP-based vaccines in humans. However, since the first clinical trials for other CPP-based therapeutics showed predicted efficacy, there is reasonable hope that for CPP-based vaccines clinical trials will be soon initiated. Very active in this field is the group of A. Bolhassani, who recently demonstrated in mice an improved antigen-specific immune response by simultaneous use of adjuvants and CPPs [58].

10.4.2 Cell-penetrating peptide-based strategies in cognitive dysfunction

Remarkably, coadministration of CPPs seems to be a promising way of improving the effort of peptide therapeutics. One advantage that is offered by this administration strategy is surely the ease of drug formulation. Notably, by intranasal coadministration of drug and CPP, the CPP assists in the localization of the drug to the brain. Recently, Alzheimer's disease (AD) has been suggested as a "type 3 diabetes" occurring in the brain, and, therefore antidiabetic drugs have been considered as potential new drugs against dementia [59]. In this respect, the peptide hormone leptin, which is a potential therapeutic for obesity, was successfully shuttled to the brain via intranasal coadministration with penetratin [60]. Indeed, it was proven that leptin was efficiently absorbed into the systemic circulation and that it accumulated in the anterior part of the brain, where it could execute its function. In another study, again penetratin was investigated to effectively deliver exendin-4 and insulin from nose to brain [61]. Exendin-4 was additionally applied, since, in a former study, the direct administration of insulin turned out to be attenuated when administered in aged mice with severe cognitive dysfunction. Indeed, intranasal coadministration of penetratin to exendin-4 with supplemental insulin gave hope for the treatment of progressive cognitive dysfunction. The group of Takeda-Morishita is pushing efforts in this direction and based on these successful preclinical studies, probably also clinical trials will be soon initiated.

10.4.3 Cell-penetrating peptides in CRISPR/Cas9 development

With the discovery of the CRISPR/Cas9 system, gene editing has become a viable biomedical tool [62]. Also here, CPPs have recently entered the field to support the efficient delivery of the CRISPR/Cas9 components [63].

An interesting study was recently reported by Krishnamurthy et al., who created shuttle-peptides, which they noncovalently combined with CRISPR-associated nuclease (Cas) RNP to allow rapid entry into mouse airway epithelia [51]. The shuttles comprised an improved endosomolytic peptide sequence able to bind and transiently destabilize endosomal membranes, combined with an optimized CPP [52]. No short-term toxicity was observed when the system was applied intranasally in mice, and heterogeneous delivery to the airways of mice was appreciated. Indeed, the presented strategy may provide a real breakthrough for cells difficult to transduce. Since this technology was developed together with Feldan Therapeutics, the near future may see clinical studies.

10.4.4 Other companies developing cell-penetrating peptide-based drugs in preclinical trials

Phion therapeutics is offering a drug delivery system using RALA peptide-based nanoparticles. RALA was first invented by Helen McCarthy's lab and is a 30 amino acid long cationic and pH-responsive peptide forming a secondary amphipathic alpha-helical structure [53]. It is mainly used for a non-covalent nucleic acid delivery and has proven immense potential for cargo delivery to the central nervous system [54].

Another CPP-based therapy is used by Peptomyc to inhibit the Myc transcription factor which shows great promise in a variety of cancer types. Omomyc is a Myc-mutant that sequesters Myc in complexes with low DNA binding efficiency and inhibits Myc transcriptional activator function [55]. Peptomyc is now developing cell-permeable Omomyc peptides that exert strong anti-Myc activity resulting in the death of cancer cells.

Recently, a platform for the discovery of CPPs derived from bacterial and viral genomes has been presented by Hoffmann et al. [56]. The identified so-called Phylomers were shown to have great potential in the direct intracellular delivery of peptides, proteins and nucleic acid analog cargos. In vivo applicability was validated by the transport of an oligonucleotide therapeutic fused to a Phylomer CPP in a disease model of Duchenne's muscular dystrophy (DMD). DMD is a genetic disorder based on a genetic defect of the DMD gene encoding for the protein dystrophin. Mutations in this gene cause muscle weakness and wasting. This CPP identification technology is now further established by PYC Therapeutics and combined with antisense oligonucleotide (ASO) therapeutics creating precise and highly effective drugs for targeting different genetic diseases. Also, a CPP-ASO targeting retinitis pigmentosa is currently in preclinical studies.

10.5 Cell-penetrating peptides in clinical studies

To approve that potential drug treatment is safe and efficient for the general population, it is tested in clinical trials in human volunteers. Despite the high number of CPP-related therapeutics reported thus far, only a very few numbers have made it into this significant stage of drug development. One recent review found 25 CPPs being in clinical phase I-III trials applied to inflammation, cancer, pain, heart disease, and aging indications [5]. Meanwhile, all these recent studies are completed. Notably, the vast majority of the currently studied CPP-based therapeutics undergo clinical trials for use in cancer medicine. Following some actual examples are presented (Table 10.1), while some more detailed information is also found in the recent literature [3,6].

10.5.1 Duchenne's muscular dystrophy

SRP-5051 is an experimental RNA-targeted treatment developed by Sarepta Therapeutics to treat DMD patients amenable to exon 51 skipping. This company established next-generation PMO-based chemistries for advanced RNA-targeted therapies, and SRP-5051 includes the so-called PPMO strategy that attaches a CPP for enhanced delivery of the PMO. The used PMO (Eteplirsen/AVI-4658 [64]) was recently enrolled in clinical studies and approved under an accelerated approval provision by the FDA in 2016 (on the market as Exondys 51). Since the clinical benefit was highly debated the approval came along with a lot of criticism by the clinical community, and indeed, more studies including higher patient numbers were requested. However, SRP-5051 is now being studied in clinical phase II trials demonstrating promising results.

10.5.2 Cancer indications

Many CPP-based therapeutics currently being studied in clinical trials address cancer. For instance, hydrocarbon-stapled peptides developed initially by the Verdine lab [65] demonstrate great promise for p53-dependent cancer therapy [66,67]. Fog Pharma is developing so-called Helicons, which are hydrocarbon-stapled peptides suitable for the treatment of various cancer types. The compound ALRN-6924, which acts through activation of p53 through dual inhibition of the oncogenes MDMX and MDM2 is meanwhile involved in several clinical Phase I/II studies. Given its broad spectrum of activity, Aileron Therapeutics is also investigating the effect of ALRN-6924 in the treatment of leukemia and lymphoma.

Cancer cells can activate different immune checkpoint pathways that harbor immunosuppressive functions, and the idea to use immune checkpoint inhibitors to target cancer was a revolutionary milestone in immuno-oncology [68]. In 2018 this concept was awarded jointly to James P. Allison and Tasuku Honjo with the Nobel prize in Physiology and Medicine. AMAL therapeutics offers an immunization platform that consists of a CPP for antigen delivery (so-called CPP-cargo-TLR peptides). One of their compounds,

ATP128, is a therapeutic vaccine for the treatment of colorectal cancer that entered phase I clinical trials in 2019.

Another interesting compound, PEP-010, is in the final stages of preparation before phase I/II clinical studies in 2020. PEP-010was invented by PEP-therapy and specifically dissociates the complex of caspase-9 and PP2A, restoring normal apoptosis in cancer cells. The compound comprises a CPP sequence coupled to an interfering active peptide that blocks the caspase-9/PP2A interaction [69].

Protein-protein interaction blocking peptides were also invented by APIM therapeutics for the treatment of several cancers. Compound ATX-101 contains a novel PCNA-interacting motif called APIM and a CPP and is currently being evaluated in clinical phase I studies. Interestingly, PEP-010 was highly effective when used in combination with conventional anticancer agents [70].

Bicycle therapeutics presents another promising strategy creating constrained bicyclic cell-permeable peptide conjugates that include both toxin conjugates and targeted innate immune activators designed for a rapid systemic drug delivery. With their compound BT1718, which is an antitubulin toxin suitable for the treatment of advanced solid tumors [71], they are currently conducting clinical phase I/II trials.

10.5.3 Neurological indications

Auris Medical is dedicated to developing therapeutics that address medical needs in neurotology and mental health. Amongst others, the company focuses on compounds to treat inner hear loss, for example, Sonsuvi (AM-111), which is a CPP that inhibits the JNK stress kinase. By blocking JNK, AM-111 protects stress-injured cochlear cells and helps to prevent or reduce chronic hearing loss [72]. Although the phase III clinical study of AM-111 did not meet endpoints, a clinically and statistically significant treatment effect in the profound hearing loss subpopulation was seen. Moreover, AM-111 was recently granted fast track status by the FDA.

10.6 Concluding remarks

CPPs are seen as promising modalities to shuttle membrane-impermeable drugs to their intracellular target sites. So, how far have they come?

Since their discovery in 1988, CPPs have been used with great success in many applications for the delivery of molecules of various origins. It became more and more clear that translation into clinical settings is highly dependent on the careful choice of preclinical models and adequate physiological conditions. Moreover, the CPP depicted also frequently exhibits different biological responses depending on the biological system studied, and thus must be carefully selected. As a result, researchers have started screening large peptide libraries for cell-penetrating and cargo-delivery abilities in very specific settings [73].

As already mentioned, many studies have demonstrated that CPPs can be modified according to specific demands, resulting in more proteolytically stable and more selective carriers. Additionally, endosomal release is highly enhanced by using different strategies. All these improvements led to a promising number of CPP-related therapeutics in clinical trials. Notably, during the past years, interest in the development of CPP-based therapeutics for cancer treatments has greatly increased. Although examples are less numerous, ongoing preclinical studies show that also other therapeutic fields will soon present new hits worth studying in clinical settings. For instance, uptake into nasal epithelium was demonstrated for penetratin covalently coupled to resveratrol [74], and the conjugate proved to be efficient for the treatment of nasal conditions. Also in the field of neurological disorders, such as cerebral ischemia and myocardial injuries, CPPs have demonstrated potential in preclinical trials [4]. Since several pharmaceutical companies are currently evaluating CPP-related treatments already in late clinical studies, there is much hope that soon the first CPP-therapeutic will be approved.

References

[1] Agrawal P, Bhalla S, Usmani SS, Singh S, Chaudhary K, Raghava GPS, et al. CPPsite 2.0: a repository of experimentally validated cell-penetrating peptides. Nucleic Acids Res 2016;44:D1098–103. Available from: https://doi.org/10.1093/nar/gkv1266.

[2] Kurrikoff K, Gestin M, Langel Ü. Recent in vivo advances in cell-penetrating peptide-assisted drug delivery. Expert Opin Drug Deliv 2016;13:373–87. Available from: https://doi.org/10.1517/17425247.2016.1125879.

[3] Jauset T, Beaulieu ME. Bioactive cell penetrating peptides and proteins in cancer: a bright future ahead. Curr Opin Pharmacol 2019;47:133–40. Available from: https://doi.org/10.1016/j.coph.2019.03.014.

[4] Guidotti G, Brambilla L, Rossi D. Cell-penetrating peptides: from basic research to clinics. Trends Pharmacol Sci 2017;38:406–24. Available from: https://doi.org/10.1016/j.tips.2017.01.003.

[5] Habault J, Poyet JL. Recent advances in cell penetrating peptide-based anticancer therapies. Molecules 2019;24. Available from: https://doi.org/10.3390/molecules24050927.

[6] Kurrikoff K, Langel Ü. Recent CPP-based applications in medicine. Expert Opin Drug Deliv 2019;16:1183–91. Available from: https://doi.org/10.1080/17425247.2019.1665021.

[7] Kalafatovic D, Giralt E. Cell-penetrating peptides: design strategies beyond primary structure and amphipathicity. Molecules 2017;22. Available from: https://doi.org/10.3390/molecules22111929.

[8] Kardani K, Milani A, Shabani SH, Bolhassani A. Cell penetrating peptides: the potent multi cargo intracellular carriers. Expert Opin Drug Deliv 2019;16:1227–58. Available from: https://doi.org/10.1080/17425247.2019.1676720.

[9] Green M, Loewenstein PM. Autonomous functional domains of chemically synthesized human immunodeficiency virus tat trans-activator protein. Cell 1988;55:1179–88. Available from: https://doi.org/10.1016/0092-8674(88)90262-0.

[10] Frankel AD, Pabo CO. Cellular uptake of the tat protein from human immunodeficiency virus. Cell 1988;55:1189–93. Available from: https://doi.org/10.1016/0092-8674(88)90263-2.

[11] Derossi D, Joliot AH, Chassaing G, Prochiantz A. The third helix of the Antennapedia homeodomain translocates through biological membranes. J Biol Chem 1994;269:10444–50.

[12] Futaki S, Suzuki T, Ohashi W, Yagami T, Tanaka S, Ueda K, et al. Arginine-rich peptides. An abundant source of membrane-permeable peptides having potential as carriers for intracellular protein delivery. J Biol Chem 2001;276:5836–40. Available from: https://doi.org/10.1074/jbc.M007540200.

[13] Pooga M, Soomets U, Hällbrink M, Valkna A, Saar K, Rezaei K, et al. Cell penetrating PNA constructs regulate galanin receptor levels and modify pain transmission in vivo. Nat Biotechnol 1998;16:857–61. Available from: https://doi.org/10.1038/nbt0998-857.

[14] Milletti F. Cell-penetrating peptides: classes, origin, and current landscape. Drug Discov Today 2012;17:850–60. Available from: https://doi.org/10.1016/j.drudis.2012.03.002.

[15] Neundorf I, Rennert R, Hoyer J, Schramm F, Löbner K, Kitanovic I, et al. Fusion of a short HA2-derived peptide sequence to cell-penetrating peptides improves cytosolic uptake, but enhances cytotoxic activity. Pharmaceuticals 2009;2:49–65. Available from: https://doi.org/10.3390/ph2020049.

[16] Horn M, Reichart F, Natividad-Tietz S, Diaz D, Neundorf I. Tuning the properties of a novel short cell-penetrating peptide by intramolecular cyclization with a triazole bridge. Chem Commun 2016;52:2261–4. Available from: https://doi.org/10.1039/c5cc08938g.

[17] Hoyer J, Schatzschneider U, Schulz-Siegmund M, Neundorf I. Dimerization of a cell-penetrating peptide leads to enhanced cellular uptake and drug delivery. Beilstein J Org Chem 2012;8:1788–97. Available from: https://doi.org/10.3762/bjoc.8.204.

[18] Kuhlmann N, Chollet C, Baldus L, Neundorf I, Lammers M. Development of substrate-derived sirtuin inhibitors with potential anticancer activity. ChemMedChem 2017;12:1703–14. Available from: https://doi.org/10.1002/cmdc.201700414.

[19] Splith K, Bergmann R, Pietzsch J, Neundorf I. Specific targeting of hypoxic tumor tissue with nitroimidazole-peptide conjugates. ChemMedChem 2012;7:57–61. Available from: https://doi.org/10.1002/cmdc.201100401.

[20] Geldmacher Y, Splith K, Kitanovic I, Alborzinia H, Can S, Rubbiani R, et al. Cellular impact and selectivity of half-sandwich organorhodium(III) anticancer complexes and their organoiridium(III) and trichloridorhodium(III) counterparts. J Biol Inorg Chem 2012;17:631–46. Available from: https://doi.org/10.1007/s00775-012-0883-2.

[21] Horn M, Neundorf I. Design of a novel cell-permeable chimeric peptide to promote wound healing. Sci Rep 2018;8:16279–91. Available from: https://doi.org/10.1038/s41598-018-34684-1.

[22] Ruseska I, Zimmer A. Internalization mechanisms of cell-penetrating peptides. Beilstein J Nanotechnol 2020;11:101–23. Available from: https://doi.org/10.3762/bjnano.11.10.

[23] Almeida C, Maniti O, Di Pisa M, Swiecicki J-M, Ayala-Sanmartin J. Cholesterol re-organisation and lipid de-packing by arginine-rich cell penetrating peptides: Role in membrane translocation. PLoS One 2019;14:e0210985. Available from: https://doi.org/10.1371/journal.pone.0210985.

[24] Murayama T, Masuda T, Afonin S, Kawano K, Takatani-Nakase T, Ida H, et al. Loosening of lipid packing promotes oligoarginine entry into cells. Angew Chemie - Int Ed 2017;56:7644–7. Available from: https://doi.org/10.1002/anie.201703578.

[25] Gao X, Hong S, Liu Z, Yue T, Dobnikar J, Zhang X. Membrane potential drives direct translocation of cell-penetrating peptides. Nanoscale 2019;11:1949–58. Available from: https://doi.org/10.1039/c8nr10447f.

[26] Tripathi PP, Arami H, Banga I, Gupta J, Gandhi S. Cell penetrating peptides in preclinical and clinical cancer diagnosis and therapy. Oncotarget 2018;9:37252–67. Available from: https://doi.org/10.18632/oncotarget.26442.

[27] Splith K, Hu W, Schatzschneider U, Gust R, Ott I, Onambele LA, et al. Protease-activatable organometal-peptide bioconjugates with enhanced cytotoxicity on cancer cells. Bioconjug Chem 2010;21:1288–96. Available from: https://doi.org/10.1021/bc100089z.

[28] Feni L, Parente S, Robert C, Gazzola S, Arosio D, Piarulli U, et al. Kiss and run: promoting effective and targeted cellular uptake of a drug delivery vehicle composed of an integrin-targeting diketopiperazine peptidomimetic and a cell-penetrating peptide. Bioconjug Chem 2019;30:2011–22. Available from: https://doi.org/10.1021/acs.bioconjchem.9b00292.

[29] Feni L, Neundorf I. The current role of cell-penetrating peptides in cancer therapy. Vol. 1030, 2017. <https://doi.org/10.1007/978-3-319-66095-0_13>.

[30] Pellestor F, Paulasova P, Hamamah S. Peptide nucleic acids (PNAs) as diagnostic devices for genetic and cytogenetic analysis. Curr Pharm Des 2008;14:2439–44. Available from: https://doi.org/10.2174/138161208785777405.

[31] Turner Y, Wallukat G, Säälik P, Wiesner B, Pritza S, Oehlke J. Cellular uptake and biological activity of peptide nucleic acids conjugated with peptides with and without cell-penetrating ability. J Pept Sci 2010;16:71–80. Available from: https://doi.org/10.1002/psc.1198.

[32] El-Andaloussi S, Johansson HJ, Holm T, Langel Ü. A novel cell-penetrating peptide, M918, for efficient delivery of proteins and peptide nucleic acids. Mol Ther 2007;15:1820–6. Available from: https://doi.org/10.1038/sj.mt.6300255.

[33] Lebleu B, Moulton HM, Abes R, Ivanova GD, Abes S, Stein DA, et al. Cell penetrating peptide conjugates of steric block oligonucleotides. Adv Drug Deliv Rev 2008;60:517–29. Available from: https://doi.org/10.1016/j.addr.2007.09.002.

[34] Barkowsky G, Lemster AL, Pappesch R, Jacob A, Krüger S, Schröder A, et al. Influence of different cell-penetrating peptides on the antimicrobial efficiency of PNAs in *Streptococcus pyogenes*. Mol Ther Nucleic Acids 2019;18:444–54. Available from: https://doi.org/10.1016/j.omtn.2019.09.010.

[35] Moulton HM. Cell-penetrating peptides enhance systemic delivery of antisense morpholino oligomers. Meth Mol Biol 2012;867:407–14. Available from: https://doi.org/10.1007/978-1-61779-767-5_26.

[36] Konate K, Dussot M, Aldrian G, Vaissière A, Viguier V, Neira IF, et al. Peptide-based nanoparticles to rapidly and efficiently "wrap 'n roll" siRNA into cells. Bioconjug Chem 2019;30:592–603. Available from: https://doi.org/10.1021/acs.bioconjchem.8b00776.

[37] Bolhassani A, Jafarzade BS, Mardani G. In vitro and in vivo delivery of therapeutic proteins using cell penetrating peptides. Peptides 2017;87:50–63. Available from: https://doi.org/10.1016/j.peptides.2016.11.011.

[38] Silva S, Almeida AJ, Vale N. Combination of cell-penetrating peptides with nanoparticles for therapeutic application: a review. Biomolecules 2019;9. Available from: https://doi.org/10.3390/biom9010022 pii: E22.

[39] Collado Camps E, Brock R. An opportunistic route to success: towards a change of paradigm to fully exploit the potential of cell-penetrating peptides. Bioorg Med Chem 2018;26:2780–7. Available from: https://doi.org/10.1016/j.bmc.2017.11.004.

[40] Van Den Brand D, Massuger LF, Brock R, Verdurmen WPR. Mimicking tumors: toward more predictive in vitro models for peptide- and protein-conjugated drugs. Bioconjug Chem 2017;28:846–56. Available from: https://doi.org/10.1021/acs.bioconjchem.6b00699.

[41] Rastogi S, Shukla S, Kalaivani M, Singh GN. Peptide-based therapeutics: quality specifications, regulatory considerations, and prospects. Drug Discov Today 2019;24:148–62. Available from: https://doi.org/10.1016/j.drudis.2018.10.002.

[42] Lattig-Tunnemann G, Prinz M, Hoffmann D, Behlke J, Palm-Apergi C, Morano I, et al. Backbone rigidity and static presentation of guanidinium groups increases cellular uptake of arginine-rich cell-penetrating peptides. Nat Commun 2011;2:453. Available from: https://doi.org/10.1038/ncomms1459.

[43] Spicer CD, Jumeaux C, Gupta B, Stevens MM. Peptide and protein nanoparticle conjugates: versatile platforms for biomedical applications. Chem Soc Rev 2018;47:3574–620. Available from: https://doi.org/10.1039/c7cs00877e.

[44] Azuma Y, Imai H, Kawaguchi Y, Nakase I, Kimura H, Futaki S. Modular redesign of a cationic lytic peptide to promote the endosomal escape of biomacromolecules. Angew Chemie - Int Ed 2018;57:12771–4. Available from: https://doi.org/10.1002/anie.201807534.

[45] Allen J, Najjar K, Erazo-Oliveras A, Kondow-Mcconaghy HM, Brock DJ, Graham K, et al. Cytosolic delivery of macromolecules in live human cells using the combined endosomal escape activities of a small molecule and cell penetrating peptides. ACS Chem Biol 2019;14:2641–51. Available from: https://doi.org/10.1021/acschembio.9b00585.

[46] Gronewold A, Horn M, Ranđelović I, Tóvári J, Muñoz Vázquez S, Schomäcker K, et al. Characterization of a cell-penetrating peptide with potential anticancer activity. Chem Med Chem 2017;12:42–9. Available from: https://doi.org/10.1002/cmdc.201600498.

[47] Xiang B, Jia XL, Qi JL, Yang LP, Sun WH, Yan X, et al. Enhancing siRNA-based cancer therapy using a new pH-responsive activatable cell-penetrating peptide-modified liposomal system. Int J Nanomed 2017;12:2385–405. Available from: https://doi.org/10.2147/IJN.S129574.

[48] Cerrato CP, Künnapuu K, Langel Ü. Cell-penetrating peptides with intracellular organelle targeting. Expert Opin Drug Deliv 2017;14:245–55. Available from: https://doi.org/10.1080/17425247.2016.1213237.

[49] Klimpel A, Neundorf I. Bifunctional peptide hybrids targeting the matrix of mitochondria. J Control Rel 2018;291:147–56. Available from: https://doi.org/10.1016/j.jconrel.2018.10.029.

[50] Kurrikoff K, Aphkhazava D, Langel Ü. The future of peptides in cancer treatment. Curr Opin Pharmacol 2019;47:27–32. Available from: https://doi.org/10.1016/j.coph.2019.01.008.

[51] Krishnamurthy S, Wohlford-Lenane C, Kandimalla S, Sartre G, Meyerholz DK, Théberge V, et al. Engineered amphiphilic peptides enable delivery of proteins and CRISPR-associated nucleases to airway epithelia. Nat Commun 2019;10:1–12. Available from: https://doi.org/10.1038/s41467-019-12922-y.

[52] Del'Guidice T, Lepetit-Stoffaes JP, Bordeleau LJ, Roberge J, Théberge V, Lauvaux C, et al. Membrane permeabilizing amphiphilic peptide delivers recombinant transcription factor and CRISPR-Cas9/Cpf1 ribonucleoproteins in hard-to-modify cells. PLoS One 2018;13:e0195558. Available from: https://doi.org/10.1371/journal.pone.0195558.

[53] Mccarthy HO, McCaffrey J, Mccrudden CM, Zholobenko A, Ali AA, McBride JW, et al. Development and characterization of self-assembling nanoparticles using a bio-inspired amphipathic peptide for gene delivery. J Control Rel 2014;189:141–9. Available from: https://doi.org/10.1016/j.jconrel.2014.06.048.

[54] Jena L, McErlean E, McCarthy H. Delivery across the blood-brain barrier: nanomedicine for glioblastoma multiforme. Drug Deliv Transl Res 2019;1–15. Available from: https://doi.org/10.1007/s13346-019-00679-2.

[55] Soucek L, Helmer-Citterich M, Sacco A, Jucker R, Cesareni G, Nasi S. Design and properties of a Myc derivative that efficiently homodimerizes. Oncogene 1998;17:2463–72. Available from: https://doi.org/10.1038/sj.onc.1202199.

[56] Hoffmann K, Milech N, Juraja SM, Cunningham PT, Stone SR, Francis RW, et al. A platform for discovery of functional cell-penetrating peptides for efficient multi-cargo intracellular delivery. Sci Rep 2018;8:12538–54. Available from: https://doi.org/10.1038/s41598-018-30790-2.

[57] Skwarczynski M, Toth I. Cell-penetrating peptides in vaccine delivery: facts, challenges and perspectives. Ther Deliv 2019;10:465–7. Available from: https://doi.org/10.4155/tde-2019-0042.

[58] Alizadeh S, Irani S, Bolhassani A, Sadat SM. Simultaneous use of natural adjuvants and cell penetrating peptides improves HCV NS3 antigen-specific immune responses. Immunol Lett 2019;212:70–80. Available from: https://doi.org/10.1016/j.imlet.2019.06.011.

[59] De Felice FG, Lourenco MV, Ferreira ST. How does brain insulin resistance develop in Alzheimer's disease? Alzheimer's Dement 2014;10:S26–32. Available from: https://doi.org/10.1016/j.jalz.2013.12.004.

[60] Khafagy ES, Kamei N, Fujiwara Y, Okumura H, Yuasa T, Kato M, et al. Systemic and brain delivery of leptin via intranasal coadministration with cell-penetrating peptides and its therapeutic potential for obesity. J Control Rel 2020;319:397–406. Available from: https://doi.org/10.1016/j.jconrel.2020.01.016.

[61] Kamei N, Okada N, Ikeda T, Choi H, Fujiwara Y, Okumura H, et al. Effective nose-to-brain delivery of exendin-4 via coadministration with cell-penetrating peptides for improving progressive cognitive dysfunction. Sci Rep 2018;8. Available from: https://doi.org/10.1038/s41598-018-36210-9.

[62] Sander JD, Joung JK. CRISPR-Cas systems for editing, regulating and targeting genomes. Nat Biotechnol 2014;32:347–50. Available from: https://doi.org/10.1038/nbt.2842.

[63] Luther DC, Lee YW, Nagaraj H, Scaletti F, Rotello VM. Delivery approaches for CRISPR/Cas9 therapeutics in vivo: advances and challenges. Expert Opin Drug Deliv 2018;15:905–13. Available from: https://doi.org/10.1080/17425247.2018.1517746.

[64] Lim KRQ, Maruyama R, Yokota T. Eteplirsen in the treatment of Duchenne muscular dystrophy. Drug Des Devel Ther 2017;11:533–45. Available from: https://doi.org/10.2147/DDDT.S97635.

[65] Schafmeister CE, Po J, Verdine GL. An all-hydrocarbon cross-linking system for enhancing the helicity and metabolic stability of peptides. J Am Chem Soc 2000;122:5891–2. Available from: https://doi.org/10.1021/ja000563a.

[66] Walensky LD, Bird GH. Hydrocarbon-stapled peptides: principles, practice, and progress. J Med Chem 2014;57:6275–88. Available from: https://doi.org/10.1021/jm4011675.

[67] Mitra S, Montgomery JE, Kolar MJ, Li G, Jeong KJ, Peng B, et al. Stapled peptide inhibitors of RAB25 target context-specific phenotypes in cancer. Nat Commun 2017;8. Available from: https://doi.org/10.1038/s41467-017-00888-8.

[68] Darvin P, Toor SM, Sasidharan Nair V, Elkord E. Immune checkpoint inhibitors: recent progress and potential biomarkers. Exp Mol Med 2018;50:165–76. Available from: https://doi.org/10.1038/s12276-018-0191-1.

[69] Arrouss I, Nemati F, Roncal F, Wislez M, Dorgham K, Vallerand D, et al. Specific targeting of Caspase-9/PP2A interaction as potential new anti-cancer therapy. PLoS One 2013;8:e60816. Available from: https://doi.org/10.1371/journal.pone.0060816.

[70] Søgaard CK, Blindheim A, Røst LM, Petrović V, Nepal A, Bachke S, et al. "Two hits - one stone" increased efficacy of cisplatin-based therapies by targeting PCNA's role in both DNA repair and cellular signaling. Oncotarget 2018;9:32448–65. Available from: https://doi.org/10.18632/oncotarget.25963.

[71] Bennett G, Rigby M, Lutz B, Park P, Keen N. Abstract B135: the mechanism of action of BT1718, a novel small-molecule drug conjugate for the treatment of solid tumors expressing MT1-MMP, 2018, p. B135. <https://doi.org/10.1158/1535-7163.targ-17-b135>.

[72] Suckfuell M, Lisowska G, Domka W, Kabacinska A, Morawski K, Bodlaj R, et al. Efficacy and safety of AM-111 in the treatment of acute sensorineural hearing loss: a double-blind, randomized, placebo-controlled phase II study. Otol Neurotol 2014;35:1317–26. Available from: https://doi.org/10.1097/MAO.0000000000000466.

[73] Kauffman WB, Guha S, Wimley WC. Synthetic molecular evolution of hybrid cell penetrating peptides. Nat Commun 2018;9:2568–78. Available from: https://doi.org/10.1038/s41467-018-04874-6.

[74] Kim Y, Hwang S, Khalmuratova R, Kang S, Lee M, Song Y, et al. α-Helical cell-penetrating peptide-mediated nasal delivery of resveratrol for inhibition of epithelial-to-mesenchymal transition. J Control Rel 2020;317:181–94. Available from: https://doi.org/10.1016/j.jconrel.2019.11.034.

Chapter 11

Intracellular peptides as drug prototypes

Mayara C.F. Gewehr[1], Ami Navon[2] and Emer S. Ferro[1,2]
[1]Pharmacology Department, Biomedical Science Institute, University of São Paulo, São Paulo, SP, Brazil, [2]Department of Biological Regulation, The Weizmann Institute of Science, Rehovot, Israel

11.1 Introduction

Proteasomes, which support critical cellular functions [1], do not convert proteins directly to amino acids; rather, the proteasome cleaves proteins into smaller peptide fragments, which on average have 10 amino acids [2]. Peptides processed by the proteasome are well known to play critical immunological functions associated with major histocompatibility complex class I (MHC-I) [3]. Peptides produced by the proteasome that are not MHC-I antigens were for a long time considered to be rapidly digested into amino acids to prevent the accumulation of potentially toxic protein fragments [3]. However, additional evolutionary functions for peptides processed by proteasomes were evidenced, as MHC-I appears on/after cartilaginous fish during evolution, while proteasome-structural related proteases are present in evolution on prokaryotes [4–6].

In addition to well-known extracellular functions (i.e., binding to receptors or associating to MHC-I or MHC-II), peptides can also function within cells to modulate protein-protein, protein-RNA and protein-DNA interactions [7–10]. Endogenous peptides with possible antibiotic functions were described early in the 1950s, in Gram-negative *Pseudomonas hydrophila* [11] and yeast *Torula utilis* (*Candida utilis*) [12]. A rationally designed peptide that perturbs FOXO4-p53 nuclear interaction, and selectively causes p53 nuclear exclusion in senescent cells, restores tissue homeostasis protecting animals from aging and cytotoxicity of chemotherapy [13,14]. Synthetic designed peptides were also shown to disrupt the docking of protein kinases to specific anchoring proteins, which have been successfully used to treat cardiac ischemia [15–17]. Designed peptides are also well known to selectively bind microRNAs regulating gene expression [18–21]. A very selective

Peptide and Peptidomimetic Therapeutics. DOI: https://doi.org/10.1016/B978-0-12-820141-1.00002-9

peptide binds and downregulates pre-miR-21, increasing apoptosis, suppressing cell proliferation, invasion and migration [20,21].

Proteasome-processed functional intracellular peptides (InPeps) are considered endogenous, protein-like, components of the cells (Fig. 11.1). This putative novel concept of functional mammalian InPeps, was proposed after the development of a substrate-capture assay, which applied catalytically inactive forms of either thimet oligopeptidase (EC 3.4.24.15; THOP1) or neurolysin (EC 3.4.24.16; Nln) to identify InPeps on the murine brain, kidney, spleen and adipose tissues [22–25]. Unlike standard bottom-up proteomics experiments, these experiments to identify InPeps did not digest the proteins and characterize the resulting peptides, but instead, worked with the peptides naturally present within the samples. InPeps were further sequenced directly from the low molecular weight fraction of cells and/or tissues homogenates using liquid-chromatography in tandem with mass spectrometry (LC/MS/MS), without the need of the substrate capture assay [26–32]. To date, multiple research groups have identified InPeps in human cell lines [26,33], human tissues [34,35], rodents [36–41], zebrafish [42], yeast [29], and plants [30,31]. Following identification of their amino acid sequences, InPeps have been chemically synthesized and their biological and/or pharmacological functions evaluated in different situations [43].

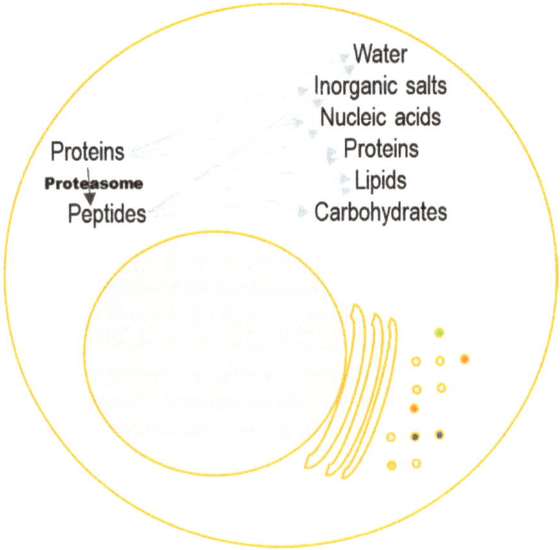

FIGURE 11.1 Proteins interact actively and dynamically with different components to regulate all cell functions. Peptides have great intrinsic biochemical similarities to proteins. Therefore it is quite predictable that peptides have cellular functionality similar to proteins as far as they exist.

Protein-protein, protein-RNA and protein-DNA interactions may be dynamically regulated by endogenous InPeps, which are constantly generated within cells after proteasome digestion of proteins [44]. Genes and gene products function not in isolation but as components of intricate networks of macromolecules (i.e., DNA, RNAs, proteins) and metabolites linked through biochemical or physical interactions [45,46]. Distinct genetic variants, the genotype, lead to distinct phenotypic outcomes, the phenotype, through multiple perturbations of these macromolecular networks [47]. The biological significance of proteasome and oligopeptidases activities products are possibly fine-tuning this intricate macromolecular network, the interactome [48,49]. Here, we intend to present the concept of InPeps as novel functional peptides derived from proteasome and oligopeptidases activities. InPeps can increase the complexity of building cellular macromolecular complexes and, perhaps, how we currently understand cells functioning.

Only recently, a group of InPeps produced during proteasomal degradation of proteins, have been proposed to scape instantaneously and complete degradation to amino acids. Recently, InPeps are functional both inside and outside the cells, which shed light on new possibilities to modulate cell function. Most of the cellular functions already described for InPeps within cells are modulatory, suggesting perhaps an ability to fine-tune cellular activity. The proteasome is at the center of this novel concept as it is responsible for the initial generation of InPeps from proteins.

11.2 Evidence that intracellular peptides come from proteasomal protein degradation

Several evidences were accumulated along the past years to suggest that proteasome process InPeps on mammals, yeast and plants [26,29,32,33,50–56]. Treatment of human embryonic kidney 293T (HEK293) cells or SH-SY5Y with epoxomicin, carfilzomib, MG132, MG262, MLN2238, AM114, or lactacystin β-lactone, mainly caused a decrease in levels of most InPeps; although, carfilzomib and MLN2238 elevated the levels of some peptides [54,56]. Bortezomib (a first-in-class proteasome inhibitor approved for the treatment of myeloma in the relapsed setting posttransplant) and MG662 caused an unexpected increase in the levels of most InPeps in HEK293 and SH-SY5Y cells [54]. Cells treated with a combination of epoxomicin and bortezomib showed large increases in the levels of most InPeps, relative to the levels in cells treated with either compound alone [51]. The paradoxical increase in peptide levels upon treatment with bortezomib and other proteasome inhibitors may be the result of allosteric effects of these compounds on the proteasome; it could also be the result of a complex set of peptides that, at the same time they are being generated inside the proteasome chamber, interfere with proteasome activity either as inhibitors [57] or activators [58]. Downstream of the proteasome, it is possible that InPeps are stabilized by interactions with cellular proteins or other molecules [59]. Evidence for stabilization of cytosolic peptides has been provided from studies examining antigenic peptides, and appears to involve binding to heat-shock proteins [60]. The therapeutic effect of bortezomib and carfilzomib as anticancer drugs is generally considered to be through alteration of protein turnover. However, these drugs produce a rapid and dramatic change in the cellular peptidome, increasing the levels of some peptides and decreasing the levels of other peptides. Therefore it is possible that the therapeutic and/or side effects of bortezomib and carfilzomib are mediated in part through the changes in the cellular peptidome [61].

Possible off-target effects of proteasome inhibitors showed that bortezomib did not inhibit tripeptidyl peptidase 2 and only weakly inhibited cellular aminopeptidase activity, as did some of the other proteasome inhibitors [54]. Potent inhibitors of tripeptidyl peptidase 2 (butabindide) and cellular aminopeptidases (bestatin) did not substantially alter the cell peptidome, indicating that the increase in peptide levels due to proteasome inhibitors is not a result of off-target peptidase inhibition [54]. The calpains were also considered to be likely candidates to initially generate InPeps, based on their broad distribution in many cell types, their preference for hydrophobic residues (which fits the observed cleavage sites that proteasome generates InPeps), and their known preference for limited cleavages of proteins rather than a degradative role. However, elevating intracellular Ca^{++} with the calcium ionophore A23187 does not substantially alter the levels of the majority of the observed peptides, suggesting that the calpains are not involved with the production of InPeps [33]. Together, these results suggest a mechanism in which the proteasome cleaves proteins into InPeps that are readily detected, and then the proteasome further degrades many of these InPeps into smaller fragments [51].

In HEK293 cells mild heat shock (42°C for 1 h) or treatment with the deubiquitinase inhibitor b-AP15 led to higher levels of ubiquitinated proteins but did not significantly increase the levels of InPeps [51]. These results may suggest that substrate ubiquitination may not be necessary for InPeps production by the proteasome, whereas further deubiquitinase inhibitors need to be tested to certify this possibility. Treatment with cycloheximide, an inhibitor of protein translation, did not substantially alter the levels of InPeps, suggesting that InPeps are produced from proteins that were already synthesized; these data also suggest that InPeps are not coming directly from mRNA translation [51].

Proteasome has three distinctive catalytic sites, including a chymotrypsin-like (the β5 subunit) that cleaves proteins after hydrophobic residues. Another subunit (the β2 subunit) has trypsin-like activity and cleaves proteins after basic residues. A third proteasome subunit (the β1 subunit) cleaves proteins after acidic amino acids residues [4]. InPeps C-terminal residues are highly represented by either hydrophobic or basic amino acids, which correspond to the most active proteasome subunits. The most frequently seen amino acids at the P1 location, Schechter and Berger nomenclature [62], are leucine, phenylalanine, and alanine. In addition, some preference for other hydrophobic residues (such as methionine, tyrosine, tryptophan, and valine) and basic (lysine and arginine) were observed [33,50]. A similar analysis of the residues in the P1' site of cleavage, shows an abundance of alanine, glutamate, lysine, and serine. When adjusted to the amino acid abundance in human proteins, these four amino acids are still among the most frequent in the P1′ position, in addition to cysteine and arginine [33]. Altogether, these data corroborate that InPeps are generated by the proteasome [2,63,64].

Furthermore, the size of InPeps range from approximately 578 to 3256 Da, with an average of approximately 1374 Da; these masses are within the range of proteasome products [2,63]. About 50% of the InPeps identified correspond to internal protein fragments, while the other 50% correspond to either N- or C-terminal fragments [29,50,52,54–56,61,65]. It was previously shown that mammalian 26S proteasomes degraded certain proteins exclusively from either the N- or C-terminus, showing no directional preference for other proteins. This property results from interactions of the substrate's termini with the regulatory ATPase and could be predicted based on the calculated relative stabilities of the N- and C-termini. Surprisingly, the direction of entry into the proteasome affected markedly the spectrum of peptides released and consequently influenced the efficiency of MHC-I antigen presentation [66]. Approximately 30% of the yeast peptides correspond to the N- or C-terminus of the protein [29]. Proteasomal degradation would be expected to produce a larger number of internal fragments than N- or C-terminal fragments. It is possible that InPeps, similarly to the generation of the p50 subunit of NF-kappaB [67], are additional cases in which the proteasome processes a longer precursor into shorter active functional peptides [43].

Corroborating the suggestion that InPeps are proteasome activity products distinctive from antigenic MHC-I-associated peptides, comparative analysis using isotope labeling plus semiquantitative mass spectrometry evaluated InPeps profile on transporter-associated with antigen-processing 1/ß2 microglobulin double-knockout mice (TAP1/β2m$^{-/-}$) and C57BL/6 wild-type (WT) animals. Overall, 92 distinctive peptides were identified, and most were shown to have a similar concentration in both TAP1/β2m$^{-/-}$ and WT mice. On TAP1/β2m$^{-/-}$ mice, some peptides showed a modest increase or decrease (~twofold), whereas a glycine-rich peptide derived from the C-terminal of neurogranin (KGPGPGGPGGAGGARGGAGGGPSGD) showed a substantial increase (≥sixfold). Thus TAP1/β2m$^{-/-}$ have only a small influence on the general InPep profile, suggesting that the presence of InPeps at least in neuronal tissue were not associated with MHC-I antigens. Moreover, Milner and colleagues examined the effect of epoxomicin and bortezomib on the rate of synthesis of HLA-bound peptides and cellular proteins in MCF-7 cells [68]. Although the rate of synthesis of many HLA-bound peptides was decreased when cells were treated with the proteasome inhibitors for 4 h, other peptides showed no effect or even an increase in their rates of synthesis in response to the proteasome inhibitors [68,69]. Similarly, the rate of cellular protein synthesis was generally decreased for most proteins, but some were not affected or had elevated rates of synthesis. A comparison of the proteins listed on [69] with the proteins found in [54] revealed ten proteins in common for which data were available for both epoxomicin and bortezomib. Two of these proteins showed a decrease in levels of InPeps and also a decrease in protein synthesis (gene names PARK7 and RBM3). Another protein (gene name PEBP1) showed a decrease in InPeps and protein synthesis with epoxomicin and no substantial change with bortezomib (within 10% of the DMSO control). However, none of the other seven proteins showed a correlation between the rate of protein synthesis and the levels of InPeps after treatment with bortezomib or epoxomicin (gene names of these proteins are: PPIA, TMSB10, EIF5A, ERH, MIF, UBA52, and RPLP2) [54].

The peptidome of the modified yeast *Saccharomyces cerevisiae* [70] was compared to the peptidomes of mammalian cell lines and tissues. Altogether, 297 yeast peptides derived from 75 proteins were identified [29]. The yeast peptides were similar to those of the human peptidome in an average size and amino acid composition. Inhibition of proteasome activity with either bortezomib or epoxomicin led to decreased levels of some yeast peptides, suggesting that these yeast peptides were also generated by the proteasome. Most yeast and human peptides are derived from a subset of abundant proteins, many with functions involving cellular metabolism or protein synthesis and folding. Of the 75 yeast proteins that give rise to peptides, 24 have orthologs that give rise to human and/or mouse peptides and for some, the same region of the proteins were found in the human, mouse, and yeast peptidomes [29]. The catalytic core of the eukaryotic proteasome is structurally related to the archaebacterial proteasome, which contains seven identical catalytically active subunits [4]. Evolution added complexity to the comparatively simple prokaryotic proteasome—in eukaryotes, there are seven distinct beta subunits, only three of which are catalytically active [4]. This feature is conserved from yeast to humans, suggesting that the proteasome may have evolved from a purely degradative function in prokaryotes to a peptide-processing function in eukaryotes. The proteasome is commonly known to function in the degradation of unneeded or damaged proteins. However, it is equally valid to change our current view to consider additional functions for the eukaryotic proteasome as processing InPeps [53,61].

The findings that InPeps are present in yeast, and that some of the peptides arise from orthologs of proteins found to produce InPeps in human cells, supports the idea that the production of InPeps is an evolutionary conserved biological process. Indeed, hundreds of InPeps originating from small protein precursors with unknown functions were identified in moss [32]. InPeps profiles in moss were rapidly changed upon salicylic acid (SA) treatment, which also increased proteasome activity. Dozens of unique peptides contained proteasome subunit β1c cleavage sites were identified, both in control and SA treated moss, further suggesting a potential role of the ubiquitin-proteasome system in the generation of InPeps in plants [32]. Plant cells contain a mixture of 26S and 20 proteasomes that perform ubiquitin-dependent and ubiquitin-independent proteolysis, while several other proteases also participate in endogenous protein degradation in plants [71,72]. InPeps potentially generated by proteasomes originated from protein precursors involved in the metabolism of carbohydrates, proteolysis, and the oxidation/reduction processes [32]. InPeps secreted by plants are further subjected to further proteolysis, supporting the idea that InPeps were not just artifacts due to the extraction procedure, but rather were really native constituents of plants [32]. These results shed more light on the functional role of InPeps generated by proteasome activity.

Although previous studies have not considered the fact that free natural InPeps processed by proteasome could be functional, peptide domains within protein structures have been shown to exist throughout entire protein structures. London and colleagues have created a nonredundant database of high-resolution structures of peptide-protein complexes (termed peptiDB) to investigate the structural basis of these types of interactions [73]. An analysis of this database shows that peptides bind to proteins in a fashion that minimizes the conformational changes of the protein partner while maximizing the enthalpy that is gained by hydrogen bonds and packing. A binding shape analysis shows that peptides tend to bind in a more planar fashion than structured and unstructured proteins do, which may allow those peptides to overcome their own configurational entropy loss upon binding. In some cases, the free protein structure is known to be unstable and is stabilized only upon binding to the peptide. Similar to protein interactions, peptide binding is mediated by "hot spot" hydrophobic residues that have a high frequency of leucine residues [74]. The number of these "hot spots" is roughly correlated with the length of the peptide (e.g., for every three residues, another anchoring "hot spot" is added to provide sufficient binding energy); although a longer peptide could be designed, it is difficult to improve the overall binding affinity [75]. Coincidently, the most frequently identified InPeps in peptidomic studies were mainly generated by the proteasome and had a hydrophobic amino acid, which was frequently leucine, at their C-terminus [29,54–56,61]. These data suggest that the proteasome generates InPeps with an increased structural ability to bind proteins due to both their size and the presence of such hydrophobic residues. Altogether, these evidences suggest that proteasome activity evolved processing functional InPeps within cells.

11.3 Experimental data on the biological significance of intracellular peptides

The seminal work suggests InPeps biological significance was produced on angiotensin-converting enzyme (ACE) transgenic mice [23]. ACE three-copy mice fed with a high-fat diet had lower body and peri-epididymal adipose tissue weights than did 1- and 2-copy mice, whereas on a regular diet ACE 3-copy mice have to eat more to maintain the same body weight [23]. On a high-fat diet, ACE 3-copy mice ate the same but weighed less than 1- and 2-copy mice, indicating a higher metabolic rate of ACE 3-copy mice. These metabolic differences of ACE transgenic mice were not affected by treatment with losartan, an antagonist of angiotensin II (ang II) type 1 receptor (AT1R) [23]. THOP1 substrate-capture assay was used to reveal that the adipose tissue of the animals containing different copy numbers of ACE gene, confined each a different profile of InPeps [23]. InPeps isolated from ACE transgenic animals' adipose tissues were identified by *de novo* mass spectrometry sequencing (a total of 39 InPeps derived from distinctive intracellular protein precursors were manually sequenced). These InPeps contained posttranslational modification (PTM) motifs for different protein kinases; 33% contained protein kinase C (PKC), 23% for protein kinase A and 10% for casein kinase II PTM motifs [24]. Synthetic peptides phosphorylated at these PTM sites have their kinetics of interaction and degradation by THOP1, Nln, and ACE largely modified [24]. These data corroborated previous reports that phosphorylation affects peptide degradation [76,77], and suggested that PTM could be of relevance for InPeps regulatory function, as well as to increase their half-life within cells. Furthermore, two of these InPeps containing a PKC phosphorylation motif (*TQR*) (LVVYPWTQRY, 0.8–4 μM; VVYPWTQRY, 8 μM), were shown to inhibit in vitro the phosphorylation of PepTag C1 fluorescent peptide (PLSRTLSVAAK, 0.08 μM; Promega, Madison, Wisconsin, United States) [23]. Functioning as natural competitors of PKC targets within cells, InPeps could be of biological significance as natural regulators of protein phosphorylation [23], that could affect signaling networks within adipose cells, changing the energetic metabolism observed for ACE transgenic mice [23]. These were seminal reports suggesting that InPeps were putative natural regulators of protein interactions within cells [44]. Therefore InPeps could have important implications in human diseases, including obesity and type 2 diabetes. Indeed, previous studies were conducted to deliver rationally-designed synthetic peptides into the cells, and low nanomolar concentrations of rationally-designed synthetic peptides are often sufficient to affect biological processes [13,16,17,78,79].

Proteasome activity generates InPeps that eventually can become processed by intracellular *endo*-oligopeptidases such as THOP1 [26,80,81] and/or Nln [82–84]. THOP1 and Nln only hydrolyzes a restricted group of peptides in the optimal range of 8–12 amino acids in length [84–86], and were never shown to degrade proteins possibly due to their catalytic site deeply located in the bottom of a narrow channel [82,87,88]. Therefore experimental manipulation of THOP1 and/or Nln activities is helpful to investigate InPeps (without interfering with protein degradation) biological function downstream of the proteasome [25,89–93]. Nln was first identified in rat brain synaptic membranes and shown to ubiquitously participate in the catabolism of bioactive peptides such as neurotensin and bradykinin [94–97]. Nln null ($Nln^{-/-}$) mice have increased glucose tolerance, insulin sensitivity, and gluconeogenesis, and increased liver mRNA for several genes related to gluconeogenesis [92]. Isotopic labeling semiquantitative peptidomic analysis suggests an increase in specific InPeps in gastrocnemius and epididymal adipose tissue, which likely is involved with the increased glucose tolerance and insulin sensitivity in the $Nln^{-/-}$ mice. The possible function of THOP1 in the development of diet-induced obesity (DIO) and insulin resistance was recently investigated utilizing a murine model of hyperlipidic DIO with both C57BL6 WT and THOP1 null ($THOP1^{-/-}$) mice. After 24 weeks of being fed a hyperlipidic diet (HD), $THOP1^{-/-}$ and WT mice ingested similar chow and calories; however, the $THOP1^{-/-}$ mice gained 75% less body weight and showed neither insulin resistance nor nonalcoholic fatty liver steatosis when compared to WT mice. $THOP1^{--/-}$ mice had increased adrenergic-stimulated adipose tissue lipolysis as well as a balanced level of expression of genes and microRNAs associated with energy metabolism, adipogenesis, or inflammation. These differences converge to a healthy phenotype of $THOP1^{-/-}$ fed an HD, compared to WT mice fed the same HD. The molecular mechanism that links THOP1 to energy metabolism was suggested to involve InPeps, in which the relative levels were identified to change in the adipose tissue of WT and $THOP1^{-/-}$ mice. InPeps were observed by molecular modeling to interact with both pre-miR-143 and pre-miR-222 [89]. These data suggest a possible novel regulatory mechanism for regulating gene expression, where InPeps could directly bind and regulate microRNA function that is well known to affect gene expression. Taken altogether, both THOP1 and Nln were shown to play essential functions on energy metabolism. Therefore InPeps could be the link between THOP1/Nln and energy metabolism, which further suggest their possible biological significance (Fig. 11.2).

Additional evidence that InPeps play a role in adipose cells energy metabolism and differentiation were followed. Using isotope-labeling semiquantitative peptidomics and liquid-chromatography in tandem with mass spectrometry (LC/MS/MS) analyses [98], several InPeps were identified in the epididymal adipose tissue of the Wistar rats that were fed either a high-caloric Western diet (WD) or a control diet (CD). The levels of three InPeps were increased by the WD compared to CD. The results of affinity chromatography using these latter InPeps as baits suggested that in the cytoplasm of epididymal adipose tissue from either WD or CD rats, distinctive proteins bind to these peptides. Reintroduced into 3T3-L1 adipocytes, InPeps increased insulin-stimulated glucose uptake in 3T3-L1 adipocytes [99]. Together, these data suggested that within the adipose tissue InPeps could be binding to specific proteins and as a consequence facilitating insulin-induced glucose uptake in 3T3-L1 adipocytes [99].

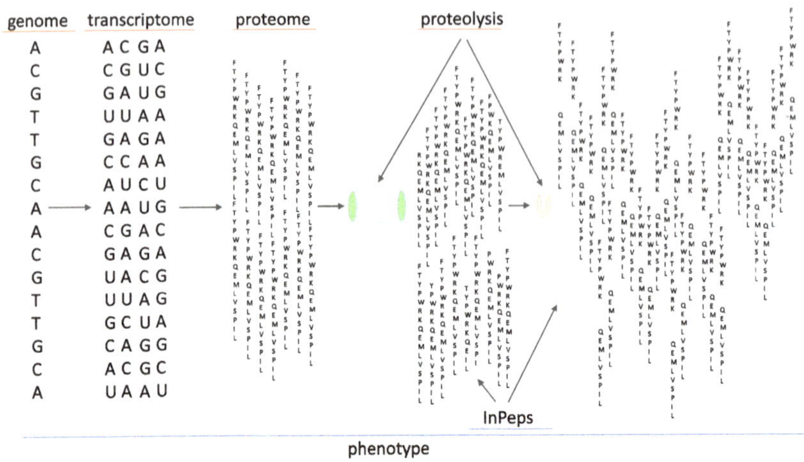

FIGURE 11.2 The "dogma of biology" establishes the flow of molecular information in cells from DNA (genome) to RNA (transcriptome) to proteins (proteome). Neither genome, transcriptome nor proteome working alone defines the phenotype of a cell or organism (i.e., gene expression depends on chromatin unpacking through histone posttranslational modifications). Proteolysis increases the complexity of the proteome, processing it to functional InPeps as new components of the intricate network of macromolecules and their metabolites within cells. InPeps have an average size of 10 amino acids, and all structural requirements to modulate protein interactions. One big challenge is to understand the mechanism of action of InPeps, and how these microproteins modulate the phenotype of a cell or individual. This will allow the development of novel molecules for the treatment of human diseases.

InPeps profile was observed to change from preadipocytes to adipocytes [100]. Using LC/MS/MS, a total of 181 downregulated InPeps and 89 upregulated InPeps were identified after preadipocytes differentiated into adipocytes (fold change >1.5 and p-value <0.05). The possible biological functions of the identified InPeps were evaluated by gene ontology (GO) and pathway analysis of their precursors. The GO and pathway analysis indicated that at least 10 of these InPeps, containing from 6 to 26 amino acids in length, might have bioactive effects in adipogenesis and metabolic diseases [100]. A recent study about the biological significance of InPeps was conducted evaluating the mechanisms of brown adipose tissue (BAT) activation [34]. InPeps were evaluated by LC/MS/MS in BAT adipocytes treated with forskolin (FSK). A total of 4370 peptides were identified in BAT adipocytes, of which 951 were upregulated and 379 were downregulated after FSK treatment. The differentially expressed peptides were evaluated via GO enrichment, KEGG pathway, and protein-protein interaction (PPI) network analysis. These bioinformatic analyses demonstrated that the ECM-receptor interaction GO term (these interactions lead to direct or indirect control of cellular activities such as adhesion, migration, differentiation, proliferation, and apoptosis) was the most enriched, and that collagen alpha-related proteins exhibited the highest degree of PPI [34]. Selected candidate InPeps were further validated via assessing the expression levels of uncoupling protein 1 and peroxisome proliferator-activated receptor-gamma coactivator in BAT adipocytes exposed to InPeps. Four peptides separately derived from TSC22 domain family protein 1 (T22D1), bromodomain and WD repeat-containing protein 1 (BRWD1), protein piccolo (PCLO), and collagen alpha-1 (III) chain (CO3A1) increased the expression levels of uncoupling protein1 (UCP-1) and peroxisome proliferator-activated receptor-gamma coactivator (PGC-1α). ECM-receptor interaction may play an important role in the process of FSK-stimulated BAT activation, and the InPeps derived from pT22D1, pBRWD1, pPCLO and pCO3A1, were suggested to be promoting BAT thermogenesis [34]. These results further corroborate that InPeps have biological significant functions.

Not only in mammals but also in plants, the levels of InPeps have been suggested to change during differentiation [30,31]. Native InPeps pools of *Physcomitrella patens* moss were investigated in two developmental stages as well as in protoplasts [30,31]. Over 20,000 unique endogenous peptides, ranging in size from 5 to 78 amino acid residues, were identified. Both the protonema and protoplast state, plastid proteins served as the main source of peptides and that their major fraction formed outside of chloroplasts. In general, the composition of InPeps pools was very different between cell types. In gametophores, stress-related proteins, for example, late embryogenesis abundant proteins, were among the most productive precursors. The Driselase-mediated protonema conversion to protoplasts led to a peptide generation "burst," with a several-fold increase in the number of components in the latter. Degradation of plastid proteins in protoplasts was accompanied by suppression of photosynthetic activity. Peptide pools in plant cells are not merely a product of waste protein degradation but may serve as important functional components for plant metabolism. Peptide "burst" is suggested as a form of biotic stress response that might produce peptides with antimicrobial activity from originally functional proteins [30,31]. Therefore similar to mammals, InPeps could constitute a large pool of putative functional molecules in plants.

Furthermore, InPeps isolated using the inactive THOP1 substrate-capture assay [26] were chemically synthesized covalently bound through a reductive Cys-Cys bond to a cell-penetrating peptide (CPP; CRKKRRQRRR) [25,101]. These InPeps were then reintroduced into Chinese hamster ovarian-S (CHO-S) or in human embryonic kidney 293 (HEK293) cells and analyzed their effect on G protein-coupled receptor (GPCR) signal transduction. A mixture containing four of these peptides (20−80 μM) significantly inhibited the increase in the extracellular acidification response triggered by ang II in CHO-S cells transfected with the AT1R-CHO-S. Subsequently, either alone or in a mixture, these peptides increased luciferase gene-reporter transcription in AT1R CHO-S cells stimulated with ang II and in HEK293 cells treated with isoproterenol [25]. These peptides without the CPP covalently bound failed to affect GPCR-mediated cellular responses. All four functional peptides were shown in vitro to competitively inhibit the degradation of a synthetic substrate by THOP1. Using individual peptides as baits in affinity columns, several proteins involved in GPCR signaling were identified, including α-adaptin A and dynamin 1; these results suggested that InPeps could be interacting with specific proteins, and that was altering their activities and as a consequence modulating GPCR signal transduction [25]. Surface plasma resonance experiments showed that distinct InPeps can either inhibit or enhance PPIs, depending on the peptide/protein pair that was analyzed [102]. Together, these data support the suggestions that InPeps could be regulating protein interactions to interfere with GPCR signal transduction [44]. Moreover, overexpression of THOP1 in both CHO-S and HEK293 cells was sufficient to reduce luciferase activation triggered either GPCR agonists ang II or isoproterenol [25]. The enzymatic activity of THOP1 was also inhibited in HEK293 cells using a specific siRNA sequence [93]. In these cells, a total of 113 distinct peptides could be reproducibly detected in three independent experiments, while the amino acid sequence of 53 peptides was identified after LC/MS/MS sequencing analysis using the MASCOT search program followed by manual interpretation of the respective spectra [93]. The peptides identified ranged in size from 7 to 22 amino acids and were derived from 21 different proteins. About half of the peptides identified had altered levels in cells with THOP1 suppressed expression, compared to control cells. Peptides that increased were 20 (ratio ≥ 1.8), possibly representing substrates of THOP1, and peptides that decreased (ratio ≤ 0.7) were 6, possibly representing products of THOP1 [93]. Peptides showed no changes in their relative concentration were 87, which are unlikely to serve as substrates or products of THOP1 [93]. siRNA-induced decreased expression of THOP1 was sufficient to potentiate GPCR signal transduction activation by isoproterenol (1–10 μM), evaluated by luciferase gene-reporter assays. The protein kinase A inhibitor KT5720 (1 μM), reduced the positive effects of the THOP1 siRNA on isoproterenol signal transduction. Thus conversely to what was observed with overexpression [25], THOP1 inhibition by siRNA inhibited GPCR signal transduction of HEK293 cells stimulated by isoproterenol [93]. Altogether, these data suggested InPeps that were altered by THOP1 inhibition, were affecting GPCR signal transduction, and that THOP1 was controlling the levels of functional InPeps that participate in GPCR signal transduction [25,93]. Therefore these results provided the first evidence that InPeps (either being reintroduced within cells or having their levels modulated by THOP1 absence, overexpression, or inhibition), could alter the complex macromolecular interactions within cells, altering signal transduction; these results also suggested that InPeps could be used as drug-prototypes for pharmacological treatment of human diseases [43].

A recent study suggested the importance of Nln for the treatment of patients with acute myeloid leukemia (AML) [103]. Nln chemical inhibition with the compound 3-[(2 S)-1-[(3 R)-3-(2-Chlorophenyl)-2-(2-fluorophenyl) pyrazolidin-1-yl]-1-oxopropan-2-yl]-1-(adamantan-2-yl)urea (R2) or genetic knockdown using shRNA, impaired oxidative metabolism and disrupted the formation of respiratory chain supercomplexes (RCS). Although R2 may inhibit both Nln and THOP1 [104], Nln knockdown does not affect the expression of THOP1 [103]. Inhibiting Nln on primary AML and normal hematopoietic cells in vivo using the R2, reduced the leukemic burden in mice without toxicity. Moreover, inhibiting Nln targeted the AML stem cells, as evidenced by decreased engraftment in secondary experiments. In contrast, inhibiting Nln did not reduce the engraftment of normal hematopoietic cells. These results demonstrate that pharmacological inhibition of Nln impairs leukemic cell growth in vitro and in vivo. The mechanism of action of Nln inhibition improving AML remains elusive, and could involve InPeps. Indeed, previous peptidomic analysis of peripherical tissues of $Nln^{-/-}$ mice, using isotopic labeling semiquantitative peptidomic analyses and LC/MS/MS, identified 736 peptide spectra, all of them fragments of intracellular proteins (i.e., InPeps) [92]. In the gastrocnemius two peptides derived from troponin I, fast skeletal muscle, SADAMLKALLGSKHK and DMEVKVQKSSKELEDMNQKL, increased more than twofold in $Nln^{-/-}$ mice. In the adipose tissue, two different peptides, one derived from the acyl-CoA-binding protein (VEKVDELKKKYGI) and one derived from the hemoglobin subunit α (LASVSTVLTSKYR), increased more than twofold in $Nln^{-/-}$ mice. In the liver, several endocannabinoid peptides containing the mouse hemopressin (PVNFKLLSH) endocannabinoid signature (SDLHAHKLRVDPVNFK, RVDPVNFKLLS, VDPVNFKLL, VDPVNFKLLSH, and RVDPVNF), or the previously identified antinociceptive AGH peptide (AGHLDDLPGALSA) decreased more than twofold in $Nln^{-/-}$ mice. In the soleus, only the peptide FASFPTTK derived from the hemoglobin α subunit decreased more than twofold $Nln^{-/-}$ mice. These data suggest a great specificity of Nln for metabolizing InPeps and/or a compensatory mechanism occurring in the $Nln^{-/-}$ mice. Both chemical and genetic inhibition of Nln produced similar results in mediating RCS formation, which impairs leukemic cell growth in vitro and in vivo. Therefore the therapeutic effects of Nln inhibition on AML could be mediated through specific InPeps. Additional investigations should be conducted to identify these specific InPeps mediating RCS formation that impairs leukemic cell growth in vitro and in vivo upon Nln inhibition [103].

A direct correlation of the biological significance of InPeps is lacking, mainly due to the absence of methods to determine InPeps absolute concentrations within cells and tissues. Without knowing the absolute concentrations of InPeps within cells and tissues, it is virtually impossible to deliver appropriate concentrations of InPeps into the cells to obtain effects at their endogenous concentrations. Moreover, the lack of a feasible quantitative method to determine the concentration of InPeps in cells and tissues limits the studies to determine their steady-state concentrations in health and diseases states, or during regular challenges of the cells. The use of proteasome inhibitors (to modulate the levels of InPeps) also limits the conclusions ascertain the biological significance of InPeps, because proteasome inhibition itself causes multiple effects on the cells [105]. On the other hand, manipulation of THOP1 and Nln have been very helpful to investigate InPeps biological function [25,89–93], mainly because these *endo*-oligopeptidases only hydrolyze a restricted group of peptides and cannot degrade proteins [82,84–88].

11.4 Evidence for biological significance of intracellular peptides in the central nervous system

THOP1 or Nln are peptidase of broad tissue distribution in mammals, with high expression in the central nervous system (CNS) [106,107]. C57BL6 null mice for either THOP1 (THOP1$^{-/-}$) or Nln (Nln$^{-/-}$) have alterations of CNS InPeps profile [90,91]. In Nln$^{-/-}$ mice brain, 31 peptides (only one being an opioid neuropeptide) were increased that could represent substrates of Nln, and a slight decrease of six other InPeps may represent products of Nln [91]. Some of the InPeps founded altered in Nln$^{-/--}$ mice brain, were also present in THOP1$^{-/-}$ mice brain (ASKGLGSDLDSSLASL, KGLGSDLDSSLASL, NDFASAVRILEV, and ADKVPKTAENF). Among these latter four InPeps, a slight reduction was observed in the prefrontal cortex, but not in the striatum nor hippocampus, of THOP1$^{-/-}$, for the peptides ASKGLGSDLDSSLASL and KGLGSDLDSSLASL (derived from clathrin coat assembly protein AP180), and NDFASAVRILEV (derived from cytochrome c oxidase subunit 5A) [90]. These results suggest that Nln and THOP1 have both similar and distinctive substrates/products in the mouse brain, which may be related to their distinctive subcellular localization in neurons; THOP1 is mainly localized in the nucleus and Nln is mainly localized in dendrites [108,109]. THOP1$^{-/-}$ have alterations in specific behaviors, immune-stimulated neurodegeneration, and infection-induced inflammation. InPeps are possible mediators of THOP1 biological function, suggesting that InPeps can play diverse physiological functions. The reduced expression of proteasome ß5 subunit founded on THOP1$^{-/-}$ mouse brain could be a biological response of these animals to compensate for the lack of THOP1 and other peptidases. However, further studies are yet necessary to clarify the possible molecular mechanisms linking behavior changes in proteasome ß5 subunit and InPeps [90].

Previous studies using LC/MS/MS, identified 271 different neuropeptides in WT mouse brain extracts, which arose from 44 distinct secretory pathway proteins [27]. In parallel, a total of 427 different peptides were identified that arose from 112 distinct intracellular proteins: approximately 2/3 of these peptides arise from cytosolic proteins, 1/4 from mitochondrial proteins and the remaining peptides arise from proteins located in the nucleus, lysosomes, or membrane compartments [27]. Therefore LC/MS/MS analysis of WT mouse brain extracts suggested that InPeps are present in great numbers and diversity in the mouse brain; because LC/MS/MS is not directly quantitative, it is not possible to conclude that the number of neuropeptides was smaller due to their lower concentrations in mouse brain compared to InPeps [27]. Label-free quantitative peptidomic using LC/MS/MS, analyzed daytime and nighttime variations in peptides from rat suprachiasmatic nucleus (SCN). Together with neuropeptides, InPeps were subjected to daytime and nighttime variations and could be playing a physiological role to control circadian rhythm [41,110,111].

A total of 344 peptides were identified by LC/MS/MS sequence analysis of peptides in cultured brain slices [55]; 75% of these peptides identified in cultured slices matched those previously detected in extracts of mice brains [27]. Of the 344 peptides identified in extracts of cultured brain slices, the vast majority (294 peptides) were not detected in the media. The failure to detect a peptide in the culture medium could mean it was really not present, or it may be present at low levels below the detection limit of the assay or otherwise difficult to detect due to the inherent properties of the peptide. To address this, the number of times each peptide was observed in the various analyses was investigated. Approximately one-half of the peptides found in slices but not in the media (156 out of 294) were seen in a single LC/MS/MS run of the slice extracts, indicating that they are low abundance or otherwise difficult to detect by LC/MS/MS. However, 114 peptides of the 294 peptides were detected in 2–4 of the LC/MS/MS runs of slices, and 24 peptides were detected in five or more of the LC/MS/MS runs of slices, but never in the LC/MS/MS runs of media. These peptides are likely to represent those present in slices and not secreted into the media at appreciable levels. Similarly, of the 28 peptides seen only in the media and not in slices, 14 of these were seen only a single time and thus are likely to represent peptides that are more difficult to detect by LC/MS/MS. However, the other half of the "media-only" peptides were detected multiple times in the media; three of these peptides were detected in five LC/MS/MS runs of media but never in the LC/MS/MS runs of extracts of brain slices [55]. Moreover, the dorsal root ganglion (DRG) and its anatomically and functionally associated spinal nerve and ventral and dorsal roots are important components of the peripheral sensory-motor system in mammals. The cells within these structures are well known to use several peptides as intercellular signaling molecules. Additional mass spectrometry-based characterizations of peptides contained within and secreted from DRG were performed [112]. In summary, 1216 peptides were identified by both PEAKS and ProsightPC softwares; 1074 peptides were exclusively identified by PEAKS and 434 by ProsightPC. From these, at least 60% of peptides identified in DRG tissue extracts were derived from intracellular proteins, while in the remaining 40% several peptides were nonneuropeptides with many from neuronal structural proteins. InPeps that were found been released by DRG cells arose from proteins such as hemoglobin (subunits alpha and beta), vimentin, cytochrome b-c1 complex, histone H2A type 2-A, myelin basic protein, peptidyl-prolyl *cis-trans* isomerase A, protein S100-A4, serine/arginine-rich splicing factor 2, neurofilament 3, neurofilament light polypeptide, and microtubule-associated protein 4, among others [112]. Together, these results suggested that InPeps could be both inside and secreted from neuronal cells. Because proteins that give rise to InPeps lack a signal-peptide sequence, the secretory mechanism responsible to release InPeps from cells remains yet uncharacterized. One possibility is that InPeps, similar to gas and cannabinoid neurotransmitters, are produced and released on demand and are not accumulated in secretory vesicles before being released [113].

Moreover, repeated exposure to amphetamine (AMPH) is known to induce long-lasting behavioral changes, referred to as sensitization, that are accompanied by various neuroadaptations in the brain [39,114,115]. Matrix-assisted laser desorption/ionization time-of-flight mass spectrometry followed by LC/MS/MS was used to show variations on the signal intensities of a subset of InPeps, in addition to neuropeptides present in the dorsal striatum, nucleus accumbens and medial prefrontal cortex of AMPH-sensitized male Sprague-Dawley rats [39]. These data suggest that InPeps may be playing, together with neuropeptides, a regulatory role in neuronal adaptation during AMPH sensitization [39].

11.5 Intracellular peptides profiles dramatically change in neurodegenerative diseases

THOP1 gene suppression that was shown to affect InPep profiles in mouse brain also affected the clinical performance of these mice in neurodegenerative experimental autoimmune encephalomyelitis (EAE), an animal model of multiple sclerosis, a chronic neuroinflammatory demyelinating disorder of the CNS with a marked neurodegenerative component involving $CD4^+$ T cells [116]. THOP1 knockout mice have worse clinical scores in EAE from 17th to 26th day postimmunization (dpi) compared to WT mice, which was associated to increased TNF-α levels in the dorsal hippocampus and spinal cord at the 26th dpi [90]. These data suggest that THOP1 is protecting mice against neuroimmune-induced degeneration. It is possible that THOP1 processing-activity of both antigens and InPeps is contributing to neurodegenerative/neuroinflammatory disorders.

Huntington's disease is the result of a long polyglutamine tract in the gene encoding huntingtin protein, which in turn causes a large number of cellular changes and ultimately results in neurodegeneration of striatal neurons [117]. Although many theories have been proposed, the precise mechanism by which the polyglutamine expansion causes cellular changes is not certain [118]. Some evidence supports the hypothesis that the long polyglutamine tract inhibits the proteasome, leading to general inhibition of protein turnover and eventual cell death [119,120]. Mouse striatal cell lines expressing either 7 glutamines

(ST*Hdh*$^{Q7/Q7}$) or 111 glutamines in the huntingtin protein, either heterozygous (ST*Hdh*$^{Q7/Q111}$) or homozygous (ST*Hdh*$^{Q111/Q111}$) constitute an in vitro model for investigating the cell biology of Huntington's disease [121]. Proteasome extracted from ST*Hdh* $^{Q111/Q111}$ cells have reduced β5 and β1 activity, and elevated β2 activity, using small synthetic substrates [29]. The heterozygous ST*Hdh* $^{Q7/Q111}$ cells shown a similar reduction in proteasome β5 activity as found in the homozygous ST*Hdh* $^{Q111/Q111}$ cells. These ST*Hdh* $^{Q111/Q111}$ or ST*Hdh* $^{Q7/Q111}$ cell lines expressing huntingtin showed a large reduction in nearly all of the InPeps detected. Treatment of ST*Hdh*$^{Q7/Q7}$ cells with proteasome inhibitors epoxomicin or bortezomib also caused a large reduction in most InPeps, suggesting that they are products of proteasome-mediated cleavage of cellular proteins. These data support the hypothesis that InPeps could be playing critical functions in neurodegenerative diseases, yet of unknown mechanisms of action [52].

Because of a spontaneous mutation that arose in 1976, *Purkinje cell degeneration* (*pcd*) mice lose their Purkinje cells beginning around 3 weeks postnatal, and also undergo loss of retinal photoreceptor cells, olfactory bulb mitral cells, and some thalamic neurons [122,123]. In addition to the original *pcd* mutation, there are several additional spontaneous mutations, chemically induced mutations, and targeted gene disruptions of the same genetic *loci*, all of which produce ataxia due to Purkinje cell death, regardless of mutation or strain of mouse [124]. In 2002, the mutation was mapped to a gene encoding a protein with homology to metallocarboxypeptidases, named Nna1/CCP1 [125]. Unlike other metallocarboxypeptidases that are secreted or function within the secretory pathway, Nna1/CCP1 lacks a signal peptide, contains a putative ATP/GTP binding domain, and is cytosolic [126,127]. Hundreds of peptides derived from cytosolic and mitochondrial proteins were founded greatly elevated in pcd mouse hypothalamus, amygdala, cortex, prefrontal cortex, and striatum. These data suggested that brain neurodegeneration affects the levels of InPeps, although the precise function of InPeps in neurodegeneration remains elusive [50]. The major proteins detected on two-dimensional gel electrophoresis of *pcd* mice were present in mutant and WT mouse cortex and hypothalamus at comparable levels. Proteasome activity was normal in these brain regions of pcd mice. These data suggested that the increase in cellular InPeps levels in the *pcd* mice is due to reduced degradation of the peptides downstream of the proteasome. Quantitative peptidomics technique compared InPeps levels in WT and *pcd* mice, examining adult heart, spleen, and brain, and presymptomatic 3 weeks-old amygdala and cerebellum. Contrary to adult mouse brain, young *pcd* brain and adult heart and spleen did not show a large increase in InPeps levels. The pattern of changes for the InPeps and secretory pathway peptides in *pcd* mice was generally similar to the pattern observed in mice lacking primary cilia. Collectively, these results suggest that InPeps accumulation in adult *pcd* mouse brain is a secondary effect and is not due to the role of Nna1/CCP1 in peptide turnover

[28]. Therefore future studies should investigate possible enzymes involved on postproteasome InPeps degradation on pcd mice; THOP1 and Nln are putative candidate enzymes. Both nondegenerating and degenerating regions of pcd mouse brain, but not the WT mouse brain, show elevated autophagy, which can be triggered by a decrease in amino acid levels. It is possible that InPeps could be connected by some means to autophagy. Additional investigations are needed to address this possibility.

11.6 Intracellular peptides disruption in anterior temporal lobe and corpus callosum of postmortem schizophrenic brains

Schizophrenia is a complex disorder hypothesized to develop from a combination of genetic, neurodevelopmental, and environmental factors [128]. Molecules that are directly involved in the pathogenesis of schizophrenia and may serve as biomarker candidates can be identified with "omics" approaches such as proteomics and peptidomics [128]. A peptidomic study in postmortem schizophrenic brain anterior temporal lobe (ATL) and corpus callosum (CC) areas was performed using LC/MS/MS and a label-free ion quantification technique based on data-dependent acquisition (DDA). Results of these latter studies indicated alterations in a specific InPep derived from neurogranin (GPGGPGGAGVARGGAGGG) in both the ATL and CC and, a decrease of another InPep (GTKAVTKYTSSK; PepH), a fragment of histone H2B type 1-H, in the ATL [35]. PepH is highly conversed in the C-terminus (amino acid fragment 113–126) of the histone H2B (isoforms type 1-C/E/F/G/I, 1A-O, type 2-E/F and type 3B). PepH sequence could be released from histone H2B after a single cleavage at the V-S peptide bond by the action of the nuclear enzyme with chymotrypsin-like enzyme activity. Indeed, we have previously shown that most of the InPeps previously identified were generated by the action of proteasomal β5 subunit of chymotrypsin-like activity [51,54–56]. The proteasome is essential for the integrity of the genome and proper gene expression, and dysregulation of proteasome function disrupts many basic cellular processes that are important in health and disease [1]. Recently, proteasome inhibition was shown to affect chromatin modifications and RNA Pol II transcription in MCF-7 breast cancer cells treated with the proteasome inhibitor MG132, causing time-dependent and pathway-specific, targeted changes in gene expression. The changes in gene expression are concomitant with a distinct reorganization of the chromatin architecture that involves spreading of the H3K4me3 into the gene bodies, together with dynamic recruitment and processivity of the RNA Pol II at differentially expressed genes [105]. Proteasome inhibition with distinctive proteasome inhibitors was also shown to largely change InPeps profile in human cell lines [29,54,61]. Histone H2B InPep VNDIFERI was shown to increase (>fivefold) in human cells treated with bortezomib (50–500 nM, 30–90 min), while decreasing (<0.5-fold) in cells treated with epoxomicin

(2 μM, 1 h). Additionally, H2B InPeps such as SFVNDIFER, SFVNDIFERI, STITSREIQTAVR, LLLPGELAKHAVSE, SFVNDIFERIAGE and ADTLWDIQKDLKDL, were increased (> fivefold) by bortezomib (500 nM, 30–90 min) treatment [54]. These results strongly suggest that cells treated with distinctive proteasome inhibitors could have different phenotypes due to different InPeps processed.

PepH covalently coupled at its C-terminus to a CPP (YGRKKRRQRRR; pepH-CPP) was tested in serum-deprived Neuro2A cells, shown a protective effect against cell death [35]. Cells were also challenged with lipopolysaccharide (LPS), and pepH-CPP was able to prevent the endotoxic effects of LPS. Alone, pepH without the CPP was not effective to protect neuro 2A cells from LPS cytotoxicity, suggesting that pepH needs to be internalized to affect cell survival. These findings that InPeps such as pepH are altered in specific regions of schizophrenic patient brains, and when reintroduced in neuronal cells protect from cell death, shed new light on the biological effects of InPeps in psychiatric diseases [35].

The level of compaction and gene repression within heterochromatic regions can be modulated by histone modifications, such as acetylation of H3 K56 and H4 K16, as well as monoubiquitylation of H2B K123. Investigating the role of the carboxy-terminus of histone H2B in yeast heterochromatin formation. Fierz et al. [129] identified that the disorderly compaction of chromatin induced by a mutation at H2B T122 specifically hinders telomeric heterochromatin formation. These data suggest that an appropriately arrayed chromatin mediated by H2B C-terminus is required for silent information regulator proteins binding, identifying an intrinsic property of the nucleosome that is required for the establishment of telomeric heterochromatin. This requirement is also likely to exist in higher eukaryotes, as the AVTKY motif of H2B is evolutionarily conserved. H2B T122 is positioned within the highly conserved AVTKY motif of the αC helix of H2B present in the pepH. The histone H2B type 1-H protein that gives rise to pepH is located in the xMHC region (extended major histocompatibility region) along with other histones (HIST1H2BD, HIST1H2BC, HIST1H2BG and HIST1H4K). The xMHC is a region of around 8-Mb on chromosome 6 containing over 250 protein-coding genes. Chromosome 6 constitutes about 6% of the human genome, and the xMHC contains the most significant association for schizophrenia. There are previous reports of histone dysregulation in schizophrenia, Huntington's disease, and autism [130]. Therefore it is plausible to suggest that pepH, found both in control and schizophrenic patients' brains, could be functional in participating in chromatin dynamic structure organization as it contains the highly functional AVTKY motif of the histone 2B C-terminus [129] (Fig. 11.3). Additional histone derived InPeps, including the VNDIFERI (Fig. 11.3), could also play a function on nucleosome formation and dynamics of chromatin structure important for controlling gene expression.

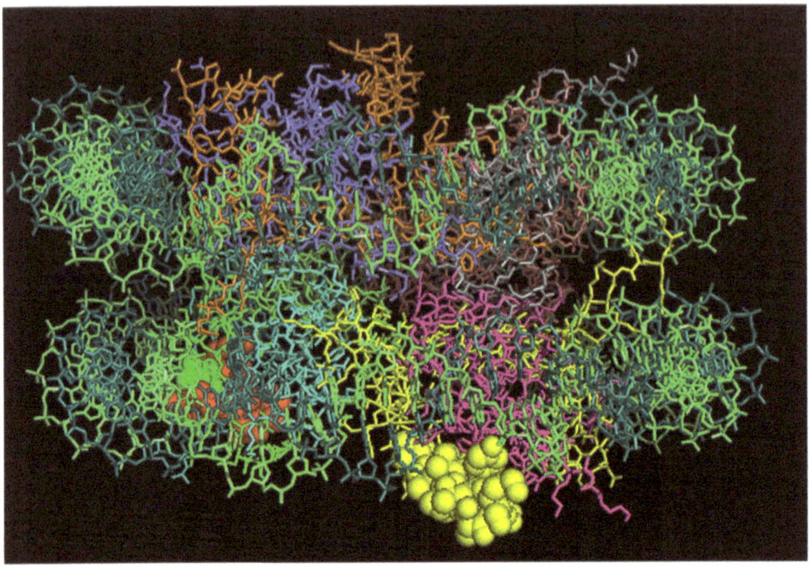

FIGURE 11.3 Simulation of pepH (*right, balls - at the front of the nucleosome structure*) on H2B within nucleosome based on data from the Protein Data Bank (PDB) [131]. Another previously identified human histone H2B InPep (VNDIFERI) [54,56] is shown (*left, balls - behind the nucleosome structure*).

The finding that InPeps derived from histones are present within structural regions directly involved in nucleosome formation add new insights on the possible function of InPeps in the modulation of gene expression. These concerns should also apply to studies using proteasome inhibitors [61,105], which will change not only the levels of proteins but also of InPeps within the cells. Because psychiatric disorders are considerably more difficult to diagnose in their early stages, usually by the time the diagnosis is clear and clinical treatment can be started, the disorder is already established and thus of greater severity. Consequently, the scientific community has been searching for biomarker candidates that can aid the early detection of such disorders and for novel therapeutics to improve treatment or at least delay disease progression. Moreover, key molecules involved in the establishment of psychiatric diseases may help the understanding of their pathogenesis and thus drive the development of more effective treatments.

11.7 Intracellular peptides as tools for therapeutic development

InPeps were suggested to induce apoptosis in cancer cells [57,132], to affect the immune response activating the expression of proteasome ß5 subunit [58], and to protect neurons from undergoing cell death [35]. The rationale used to identify these novel functional InPeps was similar: biologically significant cellular components will change their relative concentration upon cell challenge. One universal biological process with specific and well-defined phases is the cell cycle. HeLa cells were synchronized using thymidine double blockage [133]. HeLa cells were arrested in S phase and at different times (4, 10, and 16 h, at which ~84% of cells were in S, G_1, and G_2/M phases, respectively; control cells were asynchronous), harvested, analyzed by flow cytometry, and used for InPep isolation, isotope labeling, LC/MS/MS, and semiquantitative analysis [57]. Pep5 (WELVVLGKL), a fragment of G_1/S-specific cyclin D2, was the only InPep over 17 identified shown a twofold increase specifically during the S phase of HeLa cell cycle [57]. Pep5 covalently bound at its C-terminus to a CPP (YGRKKRRQRRR; pep5-CPP, 5−100 μM) induced cell death in several tumor cell lines and reduced the volume of rat C6 glioblastoma tumors in vivo [57,132]. In human MDA-MB-231 breast cancer cells, pep5-CPP caused a major disruption to stress fiber integrity [132]. Pep5-CPP induced permanent phosphorylation of extracellular signal-regulated kinase (ERK1/2) in MDA-MB-231 cells, synchronized in G1/S or S phase [132]. Affinity chromatography followed by LC/MS/MS identified CLIC1 and plectin as the only two proteins that interacted with pep5 in both asynchronous and synchronized MDA-MB-231 cells. These interactions could explain the long-lasting ERK1/2 phosphorylation and the cytoskeleton perturbations in the MDA-MB-231 cells, in which the stress fibers' integrity is affected by pep5-CPP treatments [132]. These data suggest that pep5 has potential therapeutic properties for treating specific types of cancers, such as breast cancer cells [132]. The sequence of pep5 is present only in cyclin D2 and is not found in the other cyclins. There is total homology of the pep5 sequence within cyclin D2 from a wide range of organisms, including humans, rats, bovines, mice, and chickens, suggesting its biological significance. Cyclin D2 must be degraded by the proteasome during G_1/S transition to allow cell cycle progression, which suggests that pep5 can be endogenously generated by the proteasome. At relatively high concentrations (100 μM) pep5 can almost completely inhibit the proteasome proteolytic activity [57]. Therefore it is possible that pep5 could play a function regulating the activity of proteasome during cell cycle.

When investigating the pharmacological activity of pep5-CPP, the substitution of any leucine for alanine drastically reduced the peptide cell death property [57]. A recent study made a single amino acid substitution on pep5 and found 6 derived peptides with higher activity than the original pep5 [134]. These pep5-derived sequences were synthesized using a method for synthesizing cell-penetrating functional peptide hetero-dimers, containing a disulfide bond that is cleaved by the intracellular reducing environment upon delivery into the cytosol [134]. The system used Fmoc-Lys(ivDde)-OH to synthesize CPP and functional peptides in one molecule to selectively form a disulfide bond of heterodimeric peptides [134]. These exciting results seem to corroborate the current perception that covalent conjugation of peptides to CPP disrupt at least partially their biological activity. These results also suggest that InPep generated inside cells may have biological activity at lower concentrations than currently used to deliver using regular CPP.

EL28 (VGSELIQKY) is another InPep with pharmacological activity [58]. EL28 corresponds to amino acids 251–259 of the human 19 S ATPase regulatory subunit 4, and was initially identified because of its increase in the extracts of HeLa cells treated with interferon-gamma [58]. In vitro, EL28-CPP (VGSELIQKYYGRKKRRQRRR; 50 μM) increased proteasome chymotrypsin, trypsin and acidic activities, further suggesting that InPeps can modulate proteasome activity during protein degradation [58]. In vivo, EL28-CPP (100 μM) potentiated the ability of interferon-gamma to stimulate the expression of the immunoproteasome β5i subunit and to increase the proliferation of $CD8^+$ T-cells. The EL28-CPP but not its free form improved and positively modulated the secondary IgG antibovine serum albumin immune responsiveness elicited in high antibody-responder mice. These results suggested that EL28 is a functional InPep produced by proteasome that can function to potentiate interferon-gamma activity [58].

InPeps such as VFDVELLKLE (VFD10) and VFDVELL (VFD7) were shown to modulate PPIs between CaM or 14–3-3 epsilon with unidentified proteins from the soluble fraction of the rat brain [102]. VFD10 and VFD7 are fragments of the FK506-binding protein, recently shown to be processed by the proteasome in HEK293 cells [54–56]. The steady-state intracellular concentration of the VFD7 peptide in HEK293 cells was determined using LC/MS/MS. A representative Ms spectrum of the natural VFD7 that was extracted from 1.5×10^7 HEK293 cells was compared to 1 nmol of the synthetic VFD7 peptide, each of the samples labeled with distinctive isotopes. Under these conditions, the absolute concentration of the VFD7 peptide in HEK293 cells was established to be approximately 16 ± 3 μM; according to a previous report, the individual volume of a HEK293 cell was determined to be 1.5 pL [135]. VFD7 (10–50 μM) was introduced into HEK293 cells using transient cell-permeabilization with CHAPS 0.1%, raising the intracellular calcium concentration [102]. A possible interpretation of these results is that intracellular delivery of relatively high concentrations of VFD7 (10–50 μM) disrupted its steady-state equilibrium within cells causing an increase of its free (nonassociated to high molecular weight components of the cells) form. It is well established that an increase in intracellular calcium concentration is a common signaling mechanism because it exerts allosteric regulatory affects on many enzymes and proteins. Calcium controls processes as diverse as fertilization, proliferation, development, learning and memory, contraction and secretion, and must be accomplished within the context of calcium being highly toxic [136]. It is known that disruption of FK506-binding protein (FKBP) binding to ryanodine receptors results in channel sensitization, dominant subconductance states, and increased cytoplasmic Ca^{2+} [137]. Therefore it is possible that the delivery of VFD7 into HEK293 cells have disrupted its steady-state equilibrium, causing an increase of FKBP unassociated to ryanodine receptors, and consequently the increase in intracellular calcium concentrations. One question frequently asked is, "how does the overexpression of a protein affect intracellular InPep levels?" If we can extrapolate the results obtained from VFD7 assays, one answer to this question could be: considering the overexpressed protein to be a substrate of the proteasome, and its products stable InPeps, its overexpression would have functional consequences. Because allowing changes in the homeostatic equilibrium of InPeps will change the macromolecular organization within cells, a similar concept may apply when using proteasome inhibitors.

Recently, the intracellular effects of VFD7 and related peptides on cognate PPIs were evaluated *in vivo* using HEK293 cells [138]. Internalization of peptides was achieved using a C-terminus covalently bound CPP (YGRKKRRQRRR). The NanoBiT® luciferase structural complementation reporter system was used to evaluate the effects of VFD7 and related peptides on PPI. Interacting proteins were encoded by a pair of plasmids vectors each encoding, simultaneously, either FK506-binding protein (FKBP) or FKBP-binding domain (FRB) of mechanistic target of rapamycin complex 1 (mTORC1). Interaction of FKBP-FRB within HEK293 cells were induced by addition of rapamycin. These experiments suggested that rapamycin-induced interaction between FKBP-FRB within HEK293 cells was inhibited at nano molar concentrations by VFD7-cpp (10-500 nM) and by FDVELLYG RKKRRQRRR (VFD6-CPP; 1-500 nM); other VFD7-CPP related peptides were either no effective or much less effective on inhibiting rapamycin induced FKBP-FRB interaction within HEK293 cells [138]. Computational molecular dynamics simulations indicated that VFD7-CPP, VFD6-CPP, VFAVELLYGRKKKRRQRRR (VFA7-CPP) and VFEVELLYGRKKKR RQRRR (VFA7-CPP) can bind both to FKBP and FRB protein surfaces. However, only VFD7-CPP and VFD6-CPP could induce changes on FKBP structure. These data suggesting that InPeps binding to protein surfaces could not be sufficient to alter PPI, which cannot exclude additional functional consequences. Indeed, InPeps that were extracted from HEK293 cells were mainly associated to macromolecular components (i.e., proteins and/or nucleic acids). These data suggested that InPeps intracellular proteolytic stability and mechanism of action inhibiting PPI within cells, could involve binding to macromolecular components of the cells. In a model of cell death induced by hypoxia-reoxygenation, VFD6-CPP (1 μM) increased viability of mouse embryonic fibroblasts cells (MEF) expressing mTORC1 regulated autophagy-related gene 5 (Atg5), but not in autophagy-deficient MEF cells lacking the expression of Atg5. Thus, VFD6-CPP could have therapeutic application reducing undesired side effects of rapamycin long-term treatments [138]. These data corroborate previous evidence that InPeps have biological significance binding to macromolecular components of the cells. Thus, InPeps could be valuable tools for the rational design of therapeutic molecules targeting intracellular PPI.

Peptide DIIADDEPLT (Pep19) is such a compound with CB_1R inverse agonist activity that completely lacks CNS activity, as evidenced by lack of c-Fos expression induction in the brain, cell toxicity, induction of the cannabinoid tetrad, depression, and anxiety in Wistar rats [139]. Pep19 (0.1–0.6 mg/kg) administrated via oral or intraperitoneal routes reduced plasma glucose, triacylglycerol and cholesterol concentrations, and blood pressure and adiposity indices, without altering heart rate in glucose-induced obese Wistar rats compared to nonobese controls [139]. Pep19 treatment activated pERK1/2 and protein kinase B (AKT) signaling pathways, with concomitant induction of uncoupling protein 1 (UCP1) expression in 3T3-L1 differentiated adipocytes [139]. Pep19 (0.6 mg/kg) oral treatment increased the number and reduced the size of inguinal adipocytes. Together, these data suggest that Pep19 has a peripherical mechanism of action that enhances adipose tissue metabolism. Therefore Pep19 represents a cannabinoid compound capable of improving fat metabolism and activating energy metabolism without CNS adverse effects.

The study of InPeps is a rapidly developing field. An additional review of InPep cell biology and pharmacological applications was recently published [43]. Hemopressins, the prototype of secreted InPeps, and pep19 were also recently reviewed elsewhere [43,128,140–144].

11.8 Concluding remarks

Structural proteins are in a dynamic steady state of synthesis and degradation [145,146]. Protein turnover and the role of both lysosomes [147] and proteasome [1,148] have been extensively investigated over the past six to seven decades. InPeps, in contrast, represent a new concept with many open questions and their biological significance seems evident from studies conducted with both mammals and plants. Supported by basic cell biology concepts, InPeps were hypothesized as proteasome products that naturally regulate protein interactions from within cells [23,44].

Accumulated evidence confirmed that the proteasome processes InPeps. The mechanism of action of InPeps, for example, as regulators of protein interactions, and most recently of gene expression, still requires further study. Considering that proteins are involved in virtually all cellular processes, it would be not surprising that InPeps could also play multiple functions. Direct reintroduction of InPeps covalently fused to CPP allow their cellular internalization and seems to be a useful tool so far. There are many possible ways to perform these challenges but a protein interaction system that used small luciferases as reporters seems to be very useful for such a proposal [149]. Reintroducing InPeps within cells also allow investigating the effect on microRNA and/or gene expression. Together, these future experiments should provide further evidences about InPeps biological significance and mechanism of action (Fig. 11.4). Still, a long way should be paved in order to understand InPeps function downstream the proteasome.

Oligopeptidases such as THOP1 and Nln were shown to play a role in processing InPeps downstream of the proteasome. Among others, null THOP1 mice presented with higher fat metabolism and greater resistance to DIO, while Nln null mice have improved blood glucose uptake. InPeps were suggested to play a role in the phenotype differences presented by THOP1 and Nln null mice, considering that these two oligopeptidases were extensively shown not to hydrolyze proteins [89–92]. Moreover, the mRNA expression profiles of additional peptidases, such as dipeptidyl peptidase 4 (DPP4), neprilysin (NEP), insulin-degrading enzyme, ACE1, prolyl-oligopeptidase, and that of proteasome beta5-subunit (Protβ5), were shown to be modulated in THOP1 null mice adipose tissue [89]; NEP, ACE1, Protβ5 were also seen modulated in specific brain regions of THOP1 null mice [90]. Differential expression of proteasome and several other peptidases could also interfere with InPeps profile, both inside and outside the cells. All these intricate regulatory possibilities support the biological relevance of InPeps.

InPeps can be chemically synthesized and converted into drugs. Indeed, InPeps have pharmacological activity both outside and inside cells. Hemopressins and their derivatives are amongst the most well-characterized InPeps acting outside cells on type 1 cannabinoid receptors [22,37,140,143,150–158]. InPeps such as pep5 [57], EL28 [58], pepH [35], VFD10 and VFD7 [102] were shown to have pharmacological activity only inside the cells. Therefore behind academic

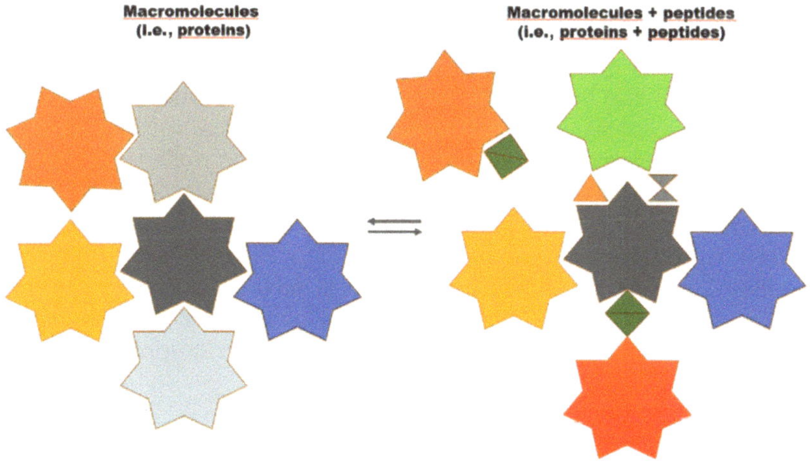

FIGURE 11.4 This simplified example illustrates the possible effect that InPeps could have on building a macromolecular organization of cells. InPeps could interfere with and/or join macromolecular building blocks.

purposes, InPeps are also prototypes for drug development, which brings economic value to further scientific investigations.

InPeps already identified are generated only from a relatively small subset of proteins, and within most of these proteins, a subset of the potential fragments implies that these peptides are not merely protein degradation products but are selectively produced and/or selectively retained from plants to mammals [29,32]. For the abundant cellular proteins that give rise to many internal peptides detected in the peptidome, it is possible that the proteasome is involved in their production and the observed fragments, therefore represent normal protein turnover intermediates. However, because many InPeps found in the cellular peptidomes represent only a limited region of the protein, typically the N- or C-termini, it is likely that these peptides are preferentially cleaved. Some evidence for this hypothesis has been recently obtained in a variety of systems including *C.elegans* [159], *Drosophila* [160], and mammalian cells [25,102,132].

Considering that on average 10,000 proteins of 500 amino acids in size are expressed in a given cell and that the proteasome generates peptides of 10 amino acids on average [2,63], hypothetically, 500,000 distinctive peptides can be produced by the proteasome from the entire cell proteome. One peptide from each protein is frequently used in self-recognition by the immune system [161], constituting 2% of the above 500,000 estimated number of proteasome generated peptides. The number of InPeps identified in mammalian cells and tissues greatly vary, from approximately 100 to 4000 peptides [26,27,34,41,93]. Therefore from 0.02% to 0.8% (less than half) of the 500,000 maximum estimated number of peptides produced by the proteasome could be identified as InPeps. In the moss *Physcomitrella patens*, over 20,000 unique endogenous peptides, many of which could be associated with proteasome activity [31,32], ranging in size from 5 to 78 amino acid residues, were identified in gametophore, protonema, and protoplast cells [30]. If the diversity of InPeps is really greater in plants than in mammals, InPeps could have a yet uninvestigated evolutionary biological significance. Although speculative, these numbers corroborate previous assumptions that most proteasome activity products are rapidly digested within cells [3,59]. On the other hand, several recent studies produced evidences that even in small numbers, InPeps can be functional and alterations in their intracellular/extracellular profile induce measurable changes in cell function. Therefore rather than the last step in protein function, proteasome activity processing InPeps is a further step in protein function and another source of lead compounds for therapeutic modulation of cellular processes.

Funding

Emer S. Ferro, Ph.D., is supported by the Erna and Jakob Michael "Visiting Professorship," Department of Biological Regulation, Weizmann Institute of

Science, Rehovot, Israel. This work was also supported by grants 2016/04000-3 and 2019/27149-0 from the São Paulo Research Foundation (FAPESP) and 302809/2016-3 from the Brazilian National Council of Research and Technological Development (CNPq).

References

[1] Hershko A, Ciechanover A. The ubiquitin system. Annu Rev Biochem 1998;67(1):425–79.

[2] Kisselev AF, Akopian TN, Woo KM, Goldberg AL. The sizes of peptides generated from protein by mammalian 26 and 20 S proteasomes. Implications for understanding the degradative mechanism and antigen presentation. J Biol Chem 1999;274(6):3363–71.

[3] Goldberg AL, Cascio P, Saric T, Rock KL. The importance of the proteasome and subsequent proteolytic steps in the generation of antigenic peptides. Mol Immunol 2002;39 (3–4):147–64.

[4] Bochtler M, Ditzel L, Groll M, Hartmann C, Huber R. The proteasome. Annu Rev Biophys Biomol Struct 1999;28:295–317.

[5] Hughes AL, Nei M. Evolutionary relationships of the classes of major histocompatibility complex genes. Immunogenetics 1993;37(5):337–46.

[6] Kasahara M, McKinney EC, Flajnik MF, Ishibashi T. The evolutionary origin of the major histocompatibility complex: polymorphism of class II alpha chain genes in the cartilaginous fish. Eur J Immunol 1993;23(9):2160–5.

[7] Pawson T, Nash P. Protein-protein interactions define specificity in signal transduction. Genes Dev 2000;14(9):1027–47.

[8] Pawson T, Scott JD. Signaling through scaffold, anchoring, and adaptor proteins. Science (N York, NY) 1997;278(5346):2075–80.

[9] Mochly-Rosen D. Localization of protein kinases by anchoring proteins: a theme in signal transduction. Science 1995;268(5208):247–51.

[10] Dagliyan O, Proctor EA, D'Auria KM, Ding F, Dokholyan NV. Structural and dynamic determinants of protein-peptide recognition. Structure 2011;19(12):1837–45.

[11] Connell GE, Watson RW. Intracellular peptides of *Pseudomonas hydrophila*. Biochim Biophys Acta 1957;24(1):226–7.

[12] Mc MI. Synthesis of intracellular peptides in *Torula utilis*. J Biol Chem 1958;231(2):777–85.

[13] Baar MP, Brandt RMC, Putavet DA, Klein JDD, Derks KWJ, Bourgeois BRM, et al. Targeted apoptosis of senescent cells restores tissue homeostasis in response to chemotoxicity and aging. Cell 2017;169(1):132–47 e16.

[14] Serrano M. Ageing: tools to eliminate senescent cells. Nature 2017;545(7654):294–6.

[15] Ferreira JCB, Campos JC, Qvit N, Qi X, Bozi LHM, Bechara LRG, et al. A selective inhibitor of mitofusin 1-betaIIPKC association improves heart failure outcome in rats. Nat Commun 2019;10(1):329.

[16] Kiyuna LA, Albuquerque RPE, Chen CH, Mochly-Rosen D, Ferreira JCB. Targeting mitochondrial dysfunction and oxidative stress in heart failure: challenges and opportunities. Free Radic Biol Med 2018;129:155–68.

[17] Cunningham AD, Qvit N, Mochly-Rosen D. Peptides and peptidomimetics as regulators of protein-protein interactions. Curr Opin Struct Biol 2017;44:59–66.

[18] Pai J, Hyun S, Hyun JY, Park SH, Kim WJ, Bae SH, et al. Screening of pre-miRNA-155 binding peptides for apoptosis inducing activity using peptide microarrays. J Am Chem Soc 2016;138(3):857–67.

[19] Heaphy S, Dingwall C, Ernberg I, Gait MJ, Green SM, Karn J, et al. HIV-1 regulator of virion expression (Rev) protein binds to an RNA stem-loop structure located within the Rev response element region. Cell 1990;60(4):685–93.

[20] Bose D, Nahar S, Rai MK, Ray A, Chakraborty K, Maiti S. Selective inhibition of miR-21 by phage display screened peptide. Nucleic Acids Res 2015;43(8):4342–52.

[21] Sakamoto K, Otake K, Umemoto T. Discovery of peptidic miR-21 processing inhibitor by mirror image phage display: a novel method to generate RNA binding D-peptides. Bioorg Med Chem Lett 2017;27(4):826–8.

[22] Rioli V, Gozzo FC, Heimann AS, Linardi A, Krieger JE, Shida CS, et al. Novel natural peptide substrates for endopeptidase 24.15, neurolysin, and angiotensin-converting enzyme. J Biol Chem 2003;278(10):8547–55.

[23] Heimann AS, Favarato MH, Gozzo FC, Rioli V, Carreno FR, Eberlin MN, et al. ACE gene titration in mice uncovers a new mechanism for ACE on the control of body weight. Physiol Genomics 2005;20(2):173–82.

[24] Machado MFM, Cunha FM, Berti DA, Heimann AS, Klitzke CF, Rioli V, et al. Substrate phosphorylation affects degradation and interaction to endopeptidase 24.15, neurolysin, and angiotensin-converting enzyme. Biocheml Biophys Res Commun 2006;339(2).

[25] Cunha FM, Berti DA, Ferreira ZS, Klitzke CF, Markus RP, Ferro ES. Intracellular peptides as natural regulators of cell signaling. J Biol Chem 2008;283(36):24448–59.

[26] Berti DA, Morano C, Russo LC, Castro LM, Cunha FM, Zhang X, et al. Analysis of intracellular substrates and products of thimet oligopeptidase in human embryonic kidney 293 cells. J Biol Chem 2009;284(21):14105–16.

[27] Fricker LD. Analysis of mouse brain peptides using mass spectrometry-based peptidomics: implications for novel functions ranging from non-classical neuropeptides to microproteins. Mol Biosyst 2010;6(8):1355–65.

[28] Berezniuk I, Sironi JJ, Wardman J, Pasek RC, Berbari NF, Yoder BK, et al. Quantitative peptidomics of Purkinje cell degeneration mice. PLoS One 2013;8(4) e60981-e.

[29] Dasgupta S, Yang C, Castro LM, Tashima AK, Ferro ES, Moir RD, et al. Analysis of the yeast peptidome and comparison with the human peptidome. PLoS One 2016;11(9): e0163312.

[30] Fesenko IA, Arapidi GP, Skripnikov AY, Alexeev DG, Kostryukova ES, Manolov AI, et al. Specific pools of endogenous peptides are present in gametophore, protonema, and protoplast cells of the moss *Physcomitrella patens*. BMC Plant Biol 2015;15:87 -.

[31] Fesenko I, Khazigaleeva R, Govorun V, Ivanov V. Analysis of endogenous peptide pools of *Physcomitrella patens* Moss. Methods Mol Biol 2018;1719:395–405.

[32] Filippova A, Lyapina I, Kirov I, Zgoda V, Belogurov A, Kudriaeva A, et al. Salicylic acid influences the protease activity and posttranslation modifications of the secreted peptides in the moss *Physcomitrella patens*. J Pept Sci 2019;25(2):e3138.

[33] Gelman JS, Sironi J, Castro LM, Ferro ES, Fricker LD. Peptidomic analysis of human cell lines. J Proteome Res 2011;10(4):1583–92.

[34] Li Y, Wang X, Wang F, You L, Xu P, Cao Y, et al. Identification of intracellular peptides associated with thermogenesis in human brown adipocytes. J Cell Physiol 2019;234(5):7104–14.

[35] Cafe-Mendes CC, Ferro ES, Torrao AS, Crunfli F, Rioli V, Schmitt A, et al. Peptidomic analysis of the anterior temporal lobe and corpus callosum from schizophrenia patients. J Proteom 2017;151:97–105.

[36] Gelman JS, Dasgupta S, Berezniuk I, Fricker LD. Analysis of peptides secreted from cultured mouse brain tissue. Biochim Biophys Acta 2013;1834(11):2408–17.

[37] Gelman JS, Sironi J, Castro LM, Ferro ES, Fricker LD. Hemopressins and other hemoglobin-derived peptides in mouse brain: comparison between brain, blood, and heart peptidome and regulation in Cpefat/fat mice. J Neurochem 2010;113(4):871–80.
[38] Romanova EV, Lee JE, Kelleher NL, Sweedler JV, Gulley JM. Mass spectrometry screening reveals peptides modulated differentially in the medial prefrontal cortex of rats with disparate initial sensitivity to cocaine. AAPS J 2010;12(3):443–54.
[39] Romanova EV, Lee JE, Kelleher NL, Sweedler JV, Gulley JM. Comparative peptidomics analysis of neural adaptations in rats repeatedly exposed to amphetamine. J Neurochem 2012;123(2):276–87.
[40] Lee JE, Zamdborg L, Southey BR, Atkins Jr. N, Mitchell JW, Li M, et al. Quantitative peptidomics for discovery of circadian-related peptides from the rat suprachiasmatic nucleus. J Proteome Res 2013;12(2):585–93.
[41] Southey BR, Lee JE, Zamdborg L, Atkins Jr. N, Mitchell JW, Li M, et al. Comparing label-free quantitative peptidomics approaches to characterize diurnal variation of peptides in the rat suprachiasmatic nucleus. Anal Chem 2014;86(1):443–52.
[42] Teixeira CMM, Correa CN, Iwai LK, Ferro ES, Castro LM. Characterization of intracellular peptides from zebrafish (*Danio rerio*) brain. Zebrafish 2019;16(3):240–51.
[43] de Araujo CB, Heimann AS, Remer RA, Russo LC, Colquhoun A, Forti FL, et al. Intracellular peptides in cell biology and pharmacology. Biomolecules 2019;9(4).
[44] Ferro ES, Hyslop S, Camargo AC. Intracellullar peptides as putative natural regulators of protein interactions. J Neurochem 2004;91(4):769–77.
[45] Sahni N, Yi S, Zhong Q, Jailkhani N, Charloteaux B, Cusick ME, et al. Edgotype: a fundamental link between genotype and phenotype. Curr Opin Genet Dev 2013;23 (6):649–57.
[46] Luck K, Kim DK, Lambourne L, Spirohn K, Begg BE, Bian W, et al. A reference map of the human binary protein interactome. Nature 2020;580(7803):402–8.
[47] Vidal M, Cusick ME, Barabasi AL. Interactome networks and human disease. Cell 2011;144(6):986–98.
[48] Kovacs IA, Luck K, Spirohn K, Wang Y, Pollis C, Schlabach S, et al. Network-based prediction of protein interactions. Nat Commun 2019;10(1):1240.
[49] Menche J, Sharma A, Kitsak M, Ghiassian SD, Vidal M, Loscalzo J, et al. Disease networks. Uncovering disease-disease relationships through the incomplete interactome. Science 2015;347(6224):1257601.
[50] Berezniuk I, Sironi J, Callaway MB, Castro LM, Hirata IY, Ferro ES, et al. CCP1/Nna1 functions in protein turnover in mouse brain: Implications for cell death in Purkinje cell degeneration mice. FASEB J 2010;24(6):1813–23.
[51] Dasgupta S, Fishman MA, Castro LM, Tashima AK, Ferro ES, Fricker LD. Effect of protein denaturation and enzyme inhibitors on proteasomal-mediated production of peptides in human embryonic kidney cells. Biomolecules 2019;9(6):207 -.
[52] Dasgupta S, Fishman MA, Mahallati H, Castro LM, Tashima AK, Ferro ES, et al. Reduced levels of proteasome products in a mouse striatal cell model of huntington's disease. PLoS One 2015;10(12):e0145333.
[53] Ferro ES, Rioli V, Castro LM, Fricker LD. Intracellular peptides: from discovery to function. EuPA Open Proteom 2014;3:143–51.
[54] Dasgupta S, Castro LM, Dulman R, Yang C, Schmidt M, Ferro ES, et al. Proteasome inhibitors alter levels of intracellular peptides in HEK293T and SH-SY5Y cells. PLoS One 2014;9(7):e103604.

[55] Gelman JS, Sironi J, Berezniuk I, Dasgupta S, Castro LM, Gozzo FC, et al. Alterations of the intracellular peptidome in response to the proteasome inhibitor bortezomib. PLoS One 2013;8(1):e53263.

[56] Fricker LD, Gelman JS, Castro LM, Gozzo FC, Ferro ES. Peptidomic analysis of HEK293T cells: effect of the proteasome inhibitor epoxomicin on intracellular peptides. J Proteome Res 2012;11(3):1981–90.

[57] de Araujo CB, Russo LC, Castro LM, Forti FL, do Monte ER, Rioli V, et al. A novel intracellular peptide derived from g1/s cyclin d2 induces cell death. J Biol Chem 2014;289(24):16711–26.

[58] Monte ER, Rossato C, Llanos RP, Russo LC, de Castro LM, Gozzo FC, et al. Interferon-gamma activity is potentiated by an intracellular peptide derived from the human 19S ATPase regulatory subunit 4 of the proteasome. J Proteom 2017;151:74–82.

[59] Reits E, Griekspoor A, Neijssen J, Groothuis T, Jalink K, van Veelen P, et al. Peptide diffusion, protection, and degradation in nuclear and cytoplasmic compartments before antigen presentation by MHC class I. Immunity 2003;18(1):97–108.

[60] Lev A, Takeda K, Zanker D, Maynard JC, Dimberu P, Waffarn E, et al. The exception that reinforces the rule: crosspriming by cytosolic peptides that escape degradation. Immunity 2008;28(6):787–98.

[61] Fricker LD. Proteasome inhibitor drugs. Annu Rev Pharmacol Toxicol 2020;60:457–76.

[62] Schechter I, Berger A. On the size of the active site in proteases. I. Papain. Biochem Biophys Res Commun 1967;27(2):157–62.

[63] Kisselev AF, Akopian TN, Woo KM, Goldberg AL. The sizes of peptides generated from protein by mammalian 26 and 20S proteasomes. J Biol Chem 1999;274(6):3363–71.

[64] Bard JAM, Bashore C, Dong KC, Martin A. The 26S proteasome utilizes a kinetic gateway to prioritize substrate degradation. Cell 2019;177(2):286–98 e15.

[65] Dasgupta S, Fishman MA, Castro LM, Tashima AK, Ferro ES, Fricker LD. Effect of protein denaturation and enzyme inhibitors on proteasomal-mediated production of peptides in human embryonic kidney cells. Biomolecules 2019;9(6).

[66] Berko D, Tabachnick-Cherny S, Shental-Bechor D, Cascio P, Mioletti S, Levy Y, et al. The direction of protein entry into the proteasome determines the variety of products and depends on the force needed to unfold its two termini. Mol Cell 2012;48(4):601–11.

[67] Cohen S, Lahav-Baratz S, Ciechanover A. Two distinct ubiquitin-dependent mechanisms are involved in NF-kappaB p105 proteolysis. Biochem Biophys Res Commun 2006;345 (1):7–13.

[68] García-Medel N, Sanz-Bravo A, Barnea E, Admon A, López de Castro JA. The origin of proteasome-inhibitor resistant HLA class I peptidomes: a study with HLA-A*68:01. Mol Cell Proteom 2012;11(1) M111.011486.

[69] Milner E, Gutter-Kapon L, Bassani-Strenberg M, Barnea E, Beer I, Admon A. The effect of proteasome inhibition on the generation of the human leukocyte antigen (HLA) peptidome. Mol Cell Proteom 2013;12(7):1853–64.

[70] Fleming JA, Lightcap ES, Sadis S, Thoroddsen V, Bulawa CE, Blackman RK. Complementary whole-genome technologies reveal the cellular response to proteasome inhibition by PS-341. Proc Natl Acad Sci U S A 2002;99(3):1461–6.

[71] Dametto A, Buffon G, Dos Reis Blasi EA, Sperotto RA. Ubiquitination pathway as a target to develop abiotic stress tolerance in rice. Plant Signal Behav 2015;10(9): e1057369.

[72] Kurepa J, Wang S, Li Y, Smalle J. Proteasome regulation, plant growth and stress tolerance. Plant Signal Behav 2009;4(10):924–7.

[73] London N, Movshovitz-Attias D, Schueler-Furman O. The structural basis of peptide-protein binding strategies. Structure 2010;18(2):188–99.

[74] Stein A, Aloy P. Contextual specificity in peptide-mediated protein interactions. PLoS one 2008;3(7) e2524-e.

[75] Sood VD, Baker D. Recapitulation and design of protein binding peptide structures and sequences. J Mol Biol 2006;357(3):917–27.

[76] Benore-Parsons M, Seidah NG, Wennogle LP. Substrate phosphorylation can inhibit proteolysis by trypsin-like enzymes. Arch Biochem Biophys 1989;272(2):274–80.

[77] Kaspari A, Diefenthal T, Grosche G, Schierhorn A, Demuth HU. Substrates containing phosphorylated residues adjacent to proline decrease the cleavage by proline-specific peptidases. Biochim Biophys Acta 1996;1293(1):147–53.

[78] Chen CH, Ferreira JCB, Mochly-Rosen D. ALDH2 and cardiovascular disease. Adv Exp Med Biol 2019;1193:53–67.

[79] Qvit N, Rubin SJS, Urban TJ, Mochly-Rosen D, Gross ER. Peptidomimetic therapeutics: scientific approaches and opportunities. Drug Discov Today 2017;22(2):454–62.

[80] Kessler JH, Khan S, Seifert U, Le Gall S, Chow KM, Paschen A, et al. Antigen processing by nardilysin and thimet oligopeptidase generates cytotoxic T cell epitopes. Nat Immunol 2011;12(1):45–53.

[81] Silva CL, Portaro FC, Bonato VL, de Camargo AC, Ferro ES. Thimet oligopeptidase (EC 3.4.24.15), a novel protein on the route of MHC class I antigen presentation. Biochem Biophys Res Commun 1999;255(3):591–5.

[82] Ray K, Hines CS, Rodgers DW. Mapping sequence differences between thimet oligopeptidase and neurolysin implicates key residues in substrate recognition. Protein Sci 2002;11 (9):2237–46.

[83] Oliveira V, Campos M, Hemerly JP, Ferro ES, Camargo AC, Juliano MA, et al. Selective neurotensin-derived internally quenched fluorogenic substrates for neurolysin (EC 3.4.24.16): comparison with thimet oligopeptidase (EC 3.4.24.15) and neprilysin (EC 3.4.24.11). Anal Biochem 2001;292(2):257–65.

[84] Oliveira V, Campos M, Melo RL, Ferro ES, Camargo AC, Juliano MA, et al. Substrate specificity characterization of recombinant metallo oligo-peptidases thimet oligopeptidase and neurolysin. Biochemistry 2001;40(14):4417–25.

[85] Dando PM, Brown MA, Barrett AJ. Human thimet oligopeptidase. Biochem J 1993;294 (Pt 2):451–7.

[86] Camargo AC, Gomes MD, Reichl AP, Ferro ES, Jacchieri S, Hirata IY, et al. Structural features that make oligopeptides susceptible substrates for hydrolysis by recombinant thimet oligopeptidase. Biochem J 1997;324(Pt 2):517–22.

[87] Ray K, Hines CS, Coll-Rodriguez J, Rodgers DW. Crystal structure of human thimet oligopeptidase provides insight into substrate recognition, regulation, and localization. J Biol Chem 2004;279(19):20480–9.

[88] Teixeira PF, Masuyer G, Pinho CM, Branca RMM, Kmiec B, Wallin C, et al. Mechanism of peptide binding and cleavage by the human mitochondrial peptidase neurolysin. J Mol Biol 2018;430(3):348–62.

[89] Gewehr MCF, Teixeira AAS, Santos BAC, Biondo LA, Gozzo FC, Cordibello AM, et al. The relevance of thimet oligopeptidase in the regulation of energy metabolism and diet-induced obesity. Biomolecules 2020;10(2):E321.

[90] Santos NBD, Franco RD, Camarini R, Munhoz CD, Eichler RAS, Gewehr MCF, et al. Thimet oligopeptidase (EC 3.4.24.15) key functions suggested by knockout mice phenotype characterization. Biomolecules 2019;9(8):382.

[91] Castro LM, Cavalcanti DM, Araujo CB, Rioli V, Icimoto MY, Gozzo FC, et al. Peptidomic analysis of the neurolysin-knockout mouse brain. J Proteom 2014;111:238–48.

[92] Cavalcanti DM, Castro LM, Rosa Neto JC, Seelaender M, Neves RX, Oliveira V, et al. Neurolysin knockout mice generation and initial phenotype characterization. J Biol Chem 2014;289(22):15426–40.

[93] Russo LC, Castro LM, Gozzo FC, Ferro ES. Inhibition of thimet oligopeptidase by siRNA alters specific intracellular peptides and potentiates isoproterenol signal transduction. FEBS Lett 2012;586(19):3287–92.

[94] Checler F, Vincent JP, Kitabgi P. Degradation of neurotensin by rat brain synaptic membranes: involvement of a thermolysin-like metalloendopeptidase (enkephalinase), angiotensin-converting enzyme, and other unidentified peptidases. J Neurochem 1983;41(2):375–84.

[95] Checler F, Vincent JP, Kitabgi P. Purification and characterization of a novel neurotensin-degrading peptidase from rat brain synaptic membranes. J Biol Chem 1986;261(24):11274–81.

[96] Checler F, Dauch P, Masuo Y, Vincent JP. Neurotensin and neuromedin N are differently metabolized in ventral tegmental area and nucleus accumbens. J Neurochem 1991;56(4):1320–8.

[97] Rioli V, Kato A, Portaro FC, Cury GK, te Kaat K, Vincent B, et al. Neuropeptide specificity and inhibition of recombinant isoforms of the endopeptidase 3.4.24.16 family: comparison with the related recombinant endopeptidase 3.4.24.15. Biochem Biophys Res Commun 1998;250(1):5–11.

[98] Che FY, Yan L, Li H, Mzhavia N, Devi LA, Fricker LD. Identification of peptides from brain and pituitary of Cpe(fat)/Cpe(fat) mice. Proc Natl Acad Sci USA 2001;98(17):9971–6.

[99] Berti DA, Russo LC, Castro LM, Cruz L, Gozzo FC, Heimann JC, et al. Identification of intracellular peptides in rat adipose tissue: insights into insulin resistance. Proteomics 2012;12(17):2668–81.

[100] Wang X, Xu S, Chen L, Shen D, Cao Y, Tang R, et al. Profiling analysis reveals the potential contribution of peptides to human adipocyte differentiation. Proteom Clin Appl 2018;12(6):e1700172.

[101] Soughayer JS, Wang Y, Li H, Cheung SH, Rossi FM, Stanbridge EJ, et al. Characterization of TAT-mediated transport of detachable kinase substrates. Biochemistry 2004;43(26):8528–40.

[102] Russo LC, Asega AF, Castro LM, Negraes PD, Cruz L, Gozzo FC, et al. Natural intracellular peptides can modulate the interactions of mouse brain proteins and thimet oligopeptidase with 14-3-3epsilon and calmodulin. Proteomics 2012;12(17):2641–55.

[103] Mirali S, Botham A, Voisin V, Xu C, St-Germain J, Sharon D, et al. The mitochondrial peptidase, neurolysin, regulates respiratory chain supercomplex formation and is necessary for AML viability. Sci Transl Med 2020;12(538).

[104] Hines CS, Ray K, Schmidt JJ, Xiong F, Feenstra RW, Pras-Raves M, et al. Allosteric inhibition of the neuropeptidase neurolysin. J Biol Chem 2014;289(51):35605–19.

[105] Kinyamu HK, Bennett BD, Bushel PR, Archer TK. Proteasome inhibition creates a chromatin landscape favorable to RNA Pol II processivity. J Biol Chem 2020;295(5):1271–87.

[106] Thul PJ, Åkesson L, Wiking M, Mahdessian D, Geladaki A, Ait Blal H, et al. A subcellular map of the human proteome. Sci (N York, NY) 2017;356(6340).

[107] Uhlen M, Oksvold P, Fagerberg L, Lundberg E, Jonasson K, Forsberg M, et al. Towards a knowledge-based human protein atlas. Nat Biotechnol 2010;28(12):1248–50.

[108] Fontenele-Neto JD, Massarelli EE, Gurgel Garrido PA, Beaudet A, Ferro ES. Comparative fine structural distribution of endopeptidase 24.15 (EC3.4.24.15) and 24.16 (EC3.4.24.16) in rat brain. J Comp Neurol 2001;438(4):399–410.

[109] Massarelli EE, Casatti CA, Kato A, Camargo AC, Bauer JA, Glucksman MJ, et al. Differential subcellular distribution of neurolysin (EC 3.4.24.16) and thimet oligopeptidase (EC 3.4.24.15) in the rat brain. Brain Res 1999;851(1-2):261–5.

[110] Golan K, Kollet O, Markus RP, Lapidot T. Daily light and darkness onset and circadian rhythms metabolically synchronize hematopoietic stem cell differentiation and maintenance: the role of bone marrow norepinephrine, tumor necrosis factor, and melatonin cycles. Exp Hematol 2019;78:1–10.

[111] Markus RP, Fernandes PA, Kinker GS, da Silveira Cruz-Machado S, Marcola M. Immune-pineal axis-acute inflammatory responses coordinate melatonin synthesis by pinealocytes and phagocytes. Br J Pharmacol 2018;175(16):3239–50.

[112] Tillmaand EG, Yang N, Kindt CA, Romanova EV, Rubakhin SS, Sweedler JV. Peptidomics and secretomics of the mammalian peripheral sensory-motor system. J Am Soc Mass Spectrom 2015;26(12):2051–61.

[113] Gelman JS, Fricker LD. Hemopressin and other bioactive peptides from cytosolic proteins: are these non-classical neuropeptides? AAPS J 2010;12(3):279–89.

[114] Robinson TE, Berridge KC. The neural basis of drug craving: an incentive-sensitization theory of addiction. Brain Res Brain Res Rev 1993;18(3):247–91.

[115] Bartlett E, Hallin A, Chapman B, Angrist B. Selective sensitization to the psychosis-inducing effects of cocaine: a possible marker for addiction relapse vulnerability? Neuropsychopharmacology 1997;16(1):77–82.

[116] Koudriavtseva T, Mainero C. Neuroinflammation, neurodegeneration and regeneration in multiple sclerosis: intercorrelated manifestations of the immune response. Neural Regen Res 2016;11(11):1727–30.

[117] Cepeda C, Tong XP. Huntington's disease: from basic science to therapeutics. CNS Neurosci Ther 2018;24(4):247–9.

[118] Labbadia J, Morimoto RI. Huntington's disease: underlying molecular mechanisms and emerging concepts. Trends Biochem Sci 2013;38(8):378–85.

[119] Venkatraman P, Wetzel R, Tanaka M, Nukina N, Goldberg AL. Eukaryotic proteasomes cannot digest polyglutamine sequences and release them during degradation of polyglutamine-containing proteins. Mol Cell 2004;14(1):95–104.

[120] Hunter JM, Lesort M, Johnson GV. Ubiquitin-proteasome system alterations in a striatal cell model of Huntington's disease. J Neurosci Res 2007;85(8):1774–88.

[121] Trettel F, Rigamonti D, Hilditch-Maguire P, Wheeler VC, Sharp AH, Persichetti F, et al. Dominant phenotypes produced by the HD mutation in STHdh(Q111) striatal cells. Hum Mol Genet 2000;9(19):2799–809.

[122] Mullen RJ, Eicher EM, Sidman RL. Purkinje cell degeneration, a new neurological mutation in the mouse. Proc Natl Acad Sci U S A 1976;73(1):208–12.

[123] O'Gorman S, Sidman RL. Degeneration of thalamic neurons in "Purkinje cell degeneration" mutant mice. I. Distribution of neuron loss. J Comp Neurol 1985;234(3):277–97.

[124] Wang T, Morgan JI. The Purkinje cell degeneration (pcd) mouse: an unexpected molecular link between neuronal degeneration and regeneration. Brain Res 2007;1140:26–40.

[125] Fernandez-Gonzalez A, La Spada AR, Treadaway J, Higdon JC, Harris BS, Sidman RL, et al. Purkinje cell degeneration (pcd) phenotypes caused by mutations in the axotomy-induced gene, Nna1. Science 2002;295(5561):1904–6.

[126] Harris A, Morgan JI, Pecot M, Soumare A, Osborne A, Soares HD. Regenerating motor neurons express Nna1, a novel ATP/GTP-binding protein related to zinc carboxypeptidases. Mol Cell Neurosci 2000;16(5):578–96.

[127] Kalinina E, Biswas R, Berezniuk I, Hermoso A, Aviles FX, Fricker LD. A novel subfamily of mouse cytosolic carboxypeptidases. FASEB J 2007;21(3):836–50.

[128] Cafe-Mendes CC, Ferro ES, Britto LR, Martins-de-Souza D. Using mass spectrometry-based peptidomics to understand the brain and disorders such as Parkinson's disease and schizophrenia. Curr Top Med Chem 2014;14(3):369–81.

[129] Fierz B, Chatterjee C, McGinty RK, Bar-Dagan M, Raleigh DP, Muir TW. Histone H2B ubiquitylation disrupts local and higher-order chromatin compaction. Nat Chem Biol 2011;7(2):113–19.

[130] Sanders AR, Göring HH, Duan J, Drigalenko EI, Moy W, Freda J, et al. Transcriptome study of differential expression in schizophrenia. Hum Mol Genet 2013;22(24):5001–14.

[131] Fujii Y, Wakamori M, Umehara T, Yokoyama S. Crystal structure of human nucleosome core particle containing enzymatically introduced CpG methylation. FEBS Open Bio 2016;6(6):498–514.

[132] Russo LC, Araujo CB, Iwai LK, Ferro ES, Forti FL. A cyclin D2-derived peptide acts on specific cell cycle phases by activating ERK1/2 to cause the death of breast cancer cells. J Proteom 2017;151.

[133] Gallo CJ, Koza RA, Herbst EJ. Polyamines and HeLa-cell DNA replication. Biochem J 1986;238(1):37–42.

[134] Kozaki I, Shimizu K, Honda H. Disulfide linked hetero dimeric peptide arrays for screening functional peptides inside cells. J Biosci Bioeng 2020;129(5):613–18.

[135] Gillen CM, Forbush 3rd. B. Functional interaction of the K-Cl cotransporter (KCC1) with the Na-K-Cl cotransporter in HEK-293 cells. Am J Physiol 1999;276(2):C328–36.

[136] Berridge MJ, Lipp P, Bootman MD. The versatility and universality of calcium signalling. Nat Rev Mol Cell Biol 2000;1(1):11–21.

[137] Steele TWE, Samso M. The FKBP12 subunit modifies the long-range allosterism of the ryanodine receptor. J Struct Biol 2019;205(2):180–8.

[138] Robinson CV, Sali A, Baumeister W. The molecular sociology of the cell. Nature 2007;450(7172):973–82.

[139] Reckziegel P, Festuccia WT, Britto LRG, Jang KLL, Romao CM, Heimann JC, et al. A novel peptide that improves metabolic parameters without adverse central nervous system effects. Sci Rep 2017;7(1):14781.

[140] Bomar MG, Galande AK. Modulation of the cannabinoid receptors by hemopressin peptides. Life Sci 2013;92(8-9):520–4.

[141] Gewehr MCF, Silverio R, Rosa-Neto JC, Lira FS, Reckziegel P, Ferro ES. Peptides from natural or rationally designed sources can be used in overweight, obesity, and type 2 diabetes therapies. Molecules 2020;25(5).

[142] Macedonio G, Stefanucci A, Maccallini C, Mirzaie S, Novellino E, Mollica A. Hemopressin peptides as modulators of the endocannabinoid system and their potential applications as therapeutic tools. Protein Pept Lett 2016;23(12):1045–51.

[143] Wei F, Zhao L, Jing Y. Signaling molecules targeting cannabinoid receptors: hemopressin and related peptides. Neuropeptides 2020;79:101998.

[144] Wei F, Zhao L, Jing Y. Hemoglobin-derived peptides and mood regulation. Peptides 2020;127:170268.

[145] Schoenheimer R, Ratner S, Rittenberg D. Studies in protein metabolism: VII. The metabolism of tyrosine. J Biol Chem 1939;127(1):333−44.

[146] Schoenheimer R. The dynamic state of body constituents. Cambridge, MA: Harvard Universitty Press; 1946. x + 78 pp.

[147] de Duve C. The lysosome turns fifty. Nat Cell Biol 2005;7(9):847−9.

[148] Ciechanover A. Intracellular protein degradation: from a vague idea thru the lysosome and the ubiquitin-proteasome system and onto human diseases and drug targeting. Best Pract Res Clin Haematol 2017;30(4):341−55.

[149] Choi SG, Olivet J, Cassonnet P, Vidalain PO, Luck K, Lambourne L, et al. Maximizing binary interactome mapping with a minimal number of assays. Nat Commun 2019;10(1):3907.

[150] Heimann AS, Giardini AC, Sant'Anna MB, Dos Santos NB, Gewehr MCF, Munhoz CD, et al. NFKF is a synthetic fragment derived from rat hemopressin that protects mice from neurodegeneration. Neurosci Lett 2020;721:134765.

[151] Xapelli S, Agasse F, Grade S, Bernardino L, Ribeiro FF, Schitine CS, et al. Modulation of subventricular zone oligodendrogenesis: a role for hemopressin? Front Cell Neurosci 2014;8(FEB).

[152] Toniolo EF, Maique ET, Ferreira Jr WA, Heimann AS, Ferro ES, Ramos-Ortolaza DL, et al. Hemopressin, an inverse agonist of cannabinoid receptors, inhibits neuropathic pain in rats. Peptides 2014;56.

[153] Gomes I, Dale CS, Casten K, Geigner MA, Gozzo FC, Ferro ES, et al. Hemoglobin-derived peptides as novel type of bioactive signaling molecules. AAPS J 2010;12(4):658−69.

[154] Gomes I, Grushko JS, Golebiewska U, Hoogendoorn S, Gupta A, Heimann AS, et al. Novel endogenous peptide agonists of cannabinoid receptors. FASEB J 2009;23 (9):3020−9.

[155] Heimann AS, Gomes I, Dale CS, Pagano RL, Gupta A, de Souza LL, et al. Hemopressin is an inverse agonist of CB1 cannabinoid receptors. Proc Natl Acad Sci USA 2007;104 (51):20588−93.

[156] Petrucci V, Chicca A, Glasmacher S, Paloczi J, Cao Z, Pacher P, et al. Pepcan-12 (RVD-hemopressin) is a CB2 receptor positive allosteric modulator constitutively secreted by adrenals and in liver upon tissue damage. Sci Rep 2017;7(1):9560.

[157] Hofer SC, Ralvenius WT, Gachet MS, Fritschy JM, Zeilhofer HU, Gertsch J. Localization and production of peptide endocannabinoids in the rodent CNS and adrenal medulla. Neuropharmacology 2015;98:78−89.

[158] Dodd GT, Worth AA, Hodkinson DJ, Srivastava RK, Lutz B, Williams SR, et al. Central functional response to the novel peptide cannabinoid, hemopressin. Neuropharmacology 2013;71:27−36.

[159] Haynes CM, Yang Y, Blais SP, Neubert TA, Ron D. The matrix peptide exporter HAF-1 signals a mitochondrial UPR by activating the transcription factor ZC376.7 in C. elegans. Mol Cell 2010;37(4):529−40.

[160] Kondo T, Plaza S, Zanet J, Benrabah E, Valenti P, Hashimoto Y, et al. Small peptides switch the transcriptional activity of Shavenbaby during *Drosophila embryogenesis*. Science 2010;329(5989):336−9.

[161] Parada A Carolina, et al. Cells 2022;24(11):385. Available from: https://doi.org/10.3390/cells11030385.

Chapter 12

Applications of computational three-dimensional structure prediction for antimicrobial peptides

Marlon Henrique Cardoso[1,2,3], Octavio Luiz Franco[1,2] and William Farias Porto[1,4]

[1]*S-Inova Biotech, Pós Graduação em Biotecnologia, Universidade Católica Dom Bosco, Campo Grande, Mato Grosso do Sul, Brazil,* [2]*Centro de Analises Proteomicas e Bioquimicas, Programa de Pós-Graduação em Ciências Genômicas e Biotecnologia, Universidade Católica de Brasília, Brasília, Distrito Federal, Brazil,* [3]*Instituto de Biociências (INBIO), Universidade Federal de Mato Grosso do Sul, Cidade Universitária, Campo Grande, Mato Grosso do Sul, Brazil,* [4]*Porto Reports, Brasília, Distrito Federal, Brazil*

12.1 Introduction

Excessive use of antibiotics has become a public concern, leading to the development of multidrug-resistance in microorganisms [1]. Particularly in hospitals and communities where the arsenal of antibiotics is no longer effective to counter bacterial infections, this situation is heading toward a crisis of antibiotic resistance [2]. Thus, efforts have been undertaken by the research community to develop favorable alternatives to restrain such infections. In this context, antimicrobial peptides (AMPs) represent promising candidates to overcome antibiotic resistance, since they belong to a class of evolutionarily ancient and persistent multifunctional molecules that have been identified from diverse sources, including microorganisms, plants and animals. AMPs play an important role in the innate immune system and are an early line of defense to protect internal and external surfaces of the host [3,4], and, overall, AMPs have an amphipathic and cationic structure, with a positive net charge between +3 and +9, and their size can range from 12 to 100 amino acid residues.

In protein engineering and drug design, the description of molecular systems at the atomic level has become increasingly important [5,6]. Structural

Peptide and Peptidomimetic Therapeutics. DOI: https://doi.org/10.1016/B978-0-12-820141-1.00012-1

information is crucial for understanding several aspects of AMPs, including evolutionary origins, mechanisms of action and rational design. However, currently, there are almost 17,000 known AMP nonredundant sequences [7], but until January 2020, only approximately 300 sequences are structurally characterized, according to the antimicrobial peptides database (APD) [8]. Understanding how AMPs are arranged at the structural level is important to further elucidate their mechanisms of action toward specific targets. This correlation has represented a focal point in AMP studies and has been reported as the "AMP structure–function relationship." Over the years, reports have shown that AMP functions are not straightforward, but are rather a complex map involving structure–function relationships, allowing AMPs to adapt and develop new functions [9]. Thus, different functions may be performed by orthologues with similar structures, and the same structure may perform multiple activities [10,11]. In addition, studies have shown that despite variation in the sequences, in many cases, the structure is conserved. Furthermore, several reports showed that some AMP classes resemble others [12–15].

Studies focusing on the development of more efficient AMPs with improved antimicrobial activity have proposed a series of rational design strategies, which mainly involve amino acid substitutions and modifications of physicochemical properties. In this context, diverse methodologies have helped researchers to understand the AMP structure based on experimental techniques and computational simulations.

Computational techniques allow a faster three-dimensional structure prediction than experimental techniques, despite being less accurate in some cases. Computational methods have been used for diverse purposes, including rational design of drugs, protein function analysis, molecular interactions, antigenic behavior, and redesign of proteins and/or peptides with greater stability and specificity [9,16–18]. Therefore, computational strategies for predicting protein three-dimensional structure, including comparative and ab initio molecular modeling, represent essential tools for the development of effective AMPs. This chapter presents applications of protein three-dimensional structure prediction to the development of AMPs.

12.2 What is a protein three-dimensional model?

Borrowing from engineering's definition, a model is a simplification of reality intended to promote understanding and reasoning [19]. In other words, a model is a representation of reality with a high level of abstraction. Thus, even the primary sequence of a protein or peptide could be considered a model and maybe the model with maximum abstraction.

In light of abstraction, even the protein three-dimensional structures solved by experimental methods (Table 12.1) could be considered models, as they are representations of reality with reduced complexity that could be interpreted as a picture of a protein in a particular condition. However, here,

TABLE 12.1 Experimental methods for protein structure determination and their applications to antimicrobial peptides.

Experimental method	Principles	Application to AMPs
X-ray crystallography	This is the most favored experimental technique in protein structure determination due to there being no limit to protein molecular weight. The goal of X-ray crystallography is to obtain a three-dimensional molecular structure from a crystal based on the distribution of the electron density. Thus, the first requirement for protein structure determination by X-ray crystallography is to obtain protein crystals diffracting at high resolution. The atom positions of a polypeptide chain in the crystal can be determined by means of an electron density map, generated by X-ray diffraction on the crystal structure [20,21].	A large number of AMPs have had their structures determined by X-ray crystallography. X-ray studies with human beta-defensins, for example, have shown evidence of high-order oligomerization in this class of AMPs [22]. Similarly, plant defensins have been structurally characterized, revealing a conserved dimeric cationic grip conformation to mediate microbial cell lysis [23]. Interestingly, disulfide-rich AMPs can be notoriously difficult to crystallize. In this context, researchers have explored the use of racemic crystallization procedures to determine the structure of the cyclotide kalata-B1 [24] and snakin-1 [15].
Nuclear magnetic resonance	This is a spectroscopic technique to observe local magnetic fields around atomic nuclei, such as ^{1}H, ^{13}C and ^{15}N. The chemical environment of atomic nuclei is affected by other nuclei, and they can be induced to flip between nuclear spin-states of different energy by the application of radio waves. Therefore, state changes can be assigned to resonance signals, which in turn are used to derive constraints for the distances separating them. Thus, the distance restraints can be used to deduce the three-dimensional structure [25]. Specifically for proteins, NMR covers the smaller	Among the NMR strategies employed for AMP structural studies, we can cite solution NMR and solid-state NMR. The first approach has been largely used to determine the three-dimensional structure of AMPs in diverse solutions, including water [26], mixtures of water with fluorinated organic solvents [27], micelles (e.g., sodium dodecyl sulfate (SDS) and dodecylphosphocholine (DPC) micelles) [28]. On the other hand, solid-state NMR represents the most suitable technique by which to study AMP structure, dynamics, and topology in liquid-disordered lipid bilayers, and depending on which nucleus is

(Continued)

TABLE 12.1 (Continued)

Experimental method	Principles	Application to AMPs
	sequences, but, despite being considered a limitation, NMR is suitable for AMP structure determination due to their typical lengths.	investigated, useful information may be acquired for both lipids and peptides [29].
Cryogenic electron microscopy	Cryo-EM is a form of cryogenic electron microscopy. The cryo-EM sample preparation process is relatively more instantaneous and easier than crystallography. The protein sample for cryo-EM analysis is cooled to cryogenic temperatures using liquid ethane [30]. After fast freezing, samples are ready for visualization under EM [30]. While the development of the technique began in the 1970s, recent advances in detector technology and software algorithms have allowed the determination of biomolecular structures at a near-atomic resolution [31].	When it comes to AMP structural analysis, cryo-EM has been used for supporting studies of complex stability for AMPs loaded into liquid crystalline gels and nanoparticles [32]. The study of complex systems such as those including human cells, bacterial biofilms and AMPs has also been explored, where Cryo-EM was used to visualize the activity of the peptide P60.4Ac (derived from LL-37) on *Staphylococcus aureus* methicillin-resistant strain LUH14616 biofilms on human epidermal models [33]. Analysis of peptide-membrane interaction can also be performed, where magainin 2 was used for studies of AMP optimization, targeting Gram-negative bacteria using synthetic vesicles composed of different proportions of POPC/POPG and DMPC/POPG in contact with the peptide [34].

a model represents a protein three-dimensional structure generated using in silico methods. Among these methods, we can cite comparative modeling, which requires an experimentally derived structure to generate models, or ab initio modeling, which is based only on theoretical calculations. Despite being less accurate than experimentally derived structures, even coarse models with low resolution can reveal some aspects of biological function. These in silico methods and their applications for designing AMPs will be discussed in the next sections.

12.3 Comparative modeling

Comparative modeling relies on the alignment between the target sequence and at least one protein of known structure solved by experimental techniques (Table 12.1), which serves as the template for modeling (Fig. 12.1). Thus, identifying the most appropriate templates is critical to this technique, which in turn depends on the purpose of the comparative modeling. Overall,

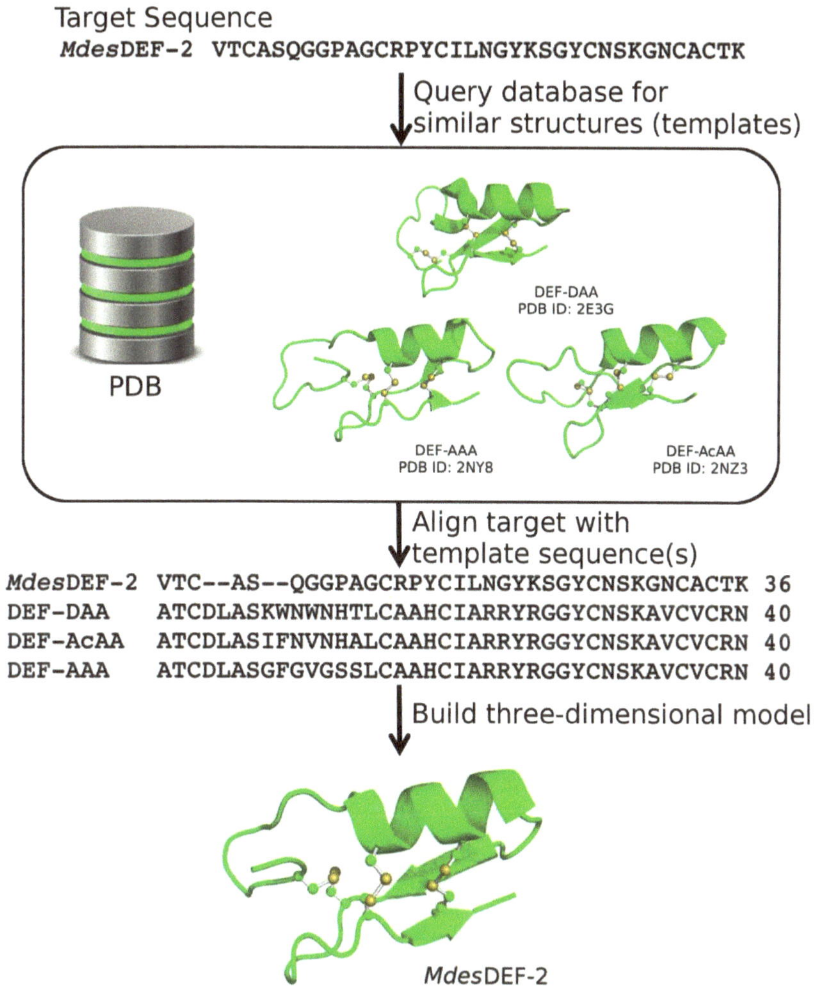

FIGURE 12.1 General workflow for comparative modeling. The modeling of *Mdes*DEF-2 [36] is depicted. The first step is to identify the template in the Protein Data Bank, using a local alignment search tool (direct modeling) or using a fold recognition tool (threading). After the selection, the query sequence is aligned to the template, and the resulting alignment is used to extract special restraints and finally build the model.

high sequence identity is the golden rule for selecting a template. However, this rule can be overridden in specific cases such as protein–ligand complexes or active site analysis, where a template that contains the ligand or a high-resolution template is desirable, respectively. In fact, the use of several templates approximately equidistant from the target sequence generally increases the final model's accuracy [35].

Direct modeling strictly follows the golden rule, where local alignment searches are the most convenient way to identify a template. For this purpose, the main tools are BLAST [37] and FASTA [38], with BLAST being the most popular tool. Local alignments are suitable for this kind of modeling because alignments with more than 30% of identity generate reliable models, as this kind of alignment is normally correct [39]. However, as the sequence similarity decreases, alignments contain an increasingly large number of gaps and alignment errors. When the sequence identity is below 30% (also known as the "twilight zone"), the alignment becomes less accurate, and therefore the detection of structural similarity has low statistical significance [40]. Thus, if simple comparative modeling is applied to these proteins the result would be models with poor quality.

As an alternative way to overcome this obstacle, peptide sequences in the twilight zone can be predicted by using threading techniques. These techniques predict the three-dimensional structure of a query sequence using experimentally determined structures as templates, without relying on sequence similarity since the structural similarity of distant homologous proteins is often preserved even with low sequence identity [41]. However, due to the absence of a direct correlation between structure and function, the identification of a sequence with a similar AMP structure does not mean this sequence is actually an AMP.

Comparative modeling has also been used to gain insights into novel AMPs, including those identified using classical purification processes, molecular biology, transcriptome data, or even databases. For instance, Konno et al. [42] isolated an AMP named anoplin from *Anoplius samariensis*, which was subjected to comparative modeling using mastoparan-X atomic coordinates [42]. Both peptides presented a similar helical conformation that was further related to similar biological activities (e.g., mast cell degranulation and broad spectrum antimicrobial activity against Gram-positive and -negative bacteria) between mastoparan-X and anoplin [42]. Cunha et al. [43] applied comparative modeling to compare two cyclotides from South American plants identified by molecular biology techniques, showing the implications of amino acid replacements between them [43]. Similarly, Cândido et al. [44] used comparative modeling to characterize the putative AMPs in the spathe transcriptome of *Zantedeschia aethiopica* [44]. Porto et al. [45] modeled the interactions of hevein-like peptides identified from the NCBI's nonredundant protein database and chitin fragments, indicating that those sequences are lectins with putative antifungal activity [45].

Besides characterization itself, comparative modeling has also been used for rational design of novel variants with improved activity, stability and/or reduced length. For instance, Irazazabal et al. [46] performed comparative modeling of a mastoparan-L analog (I^5R^8), showing that I^5R^8 was capable of forming an amphipathic α-helix that favored membrane perturbation [46]. Similarly, the peptide medusin-PT was subjected to threading modeling, and the three-dimensional model was used to perform modifications to the peptide, generating a variant with improved helicity by a single target modification Pro6Lys, which also improved the analog's activity [47]. Moreover, Bao et al. [48] used threading to study the effects of substitution of Rana box (CGLXGLC) from the nigrocin-HL peptide with a phenylalanine residue, showing that this replacement increased the helix formation, reduced the peptide length and improved its activity [48].

12.4 Ab initio modeling

Although comparative modeling uses threading techniques capable of predicting three-dimensional structures for sequences in the twilight zone, there are some cases where there is no template available and thus comparative modeling is unlikely to succeed. In these cases, ab initio modeling, also known as *de novo* or free modeling, is capable of predicting the protein structure from scratch using a designed energy function to guide the conformational search [49]. The overall workflow of ab initio modeling is depicted in Fig. 12.2. While comparative modeling depends on a sequence alignment with a template, ab initio modeling depends on the length of the target sequence. The prediction of a larger protein through ab initio modeling is extremely difficult because increasing the sequence length also increases the conformational phase space of sampling, resulting in a loss of accuracy [50,51]. Thus, for sequences up to 100 amino acid residues, ab initio modeling has success rates between 20% and 25% for constructing good quality models [36,52]. Fortunately, in the last few years, this length limitation has been overcome, with successful predictions of sequences of approximately 150 amino acid residues [53]. Because the length of an AMP sequence is often <100 amino acid residues, ab initio modeling may be applied. Regardless, a three-dimensional model does not have to be absolutely perfect to be helpful in biology.

The case of snakin-1 is, perhaps, the most emblematic example of one of the limits of ab initio modeling. The first structural insights were made by ab initio molecular modeling along with disulfide bridge prediction, where the structure was predicted to contain an N-terminal helix-turn helix motif and a C-terminal extended loop [13]. The main drawback of snakin-1, however, is due to the presence of 12 cysteine residues that could establish 66 possible disulfide connections. Therefore, the possibility of an erroneous prediction of all disulfides is high (98.5%). Thus, the establishment of a correct disulfide

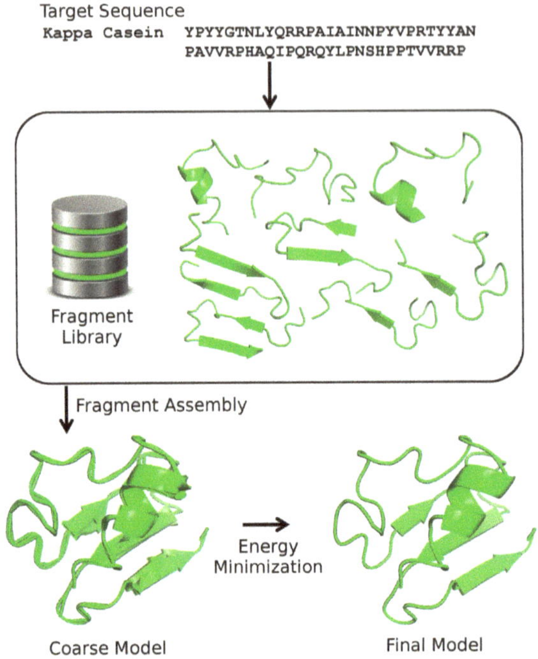

FIGURE 12.2 General workflow for ab initio protein structure prediction. The modeling of kappa casein [54] is depicted. The query sequence is fragmented by a sliding window algorithm. Three-dimensional structures for each fragment are generated, guided by atomic force fields. Then the fragments are assembled in full structures and undergo energy minimization steps until they generate the final models.

bond pattern was considered very difficult and could compromise the accuracy of the final theoretical model. In a later work, three disulfide bridges were determined by mass spectrometry, with only one matching the previous ab initio prediction [55]. A few years later, the full structure of snakin-1 was elucidated, showing that only two ab initio predicted disulfide bonds were correct [15]. However, it is important to highlight that the crystal and the predicted structures share the same helix-turn-helix motif, indicating that the prediction was partially correct. Taken together, these findings emphasize the importance of combining computational and experimental structural techniques to achieve high-quality three-dimensional structures for constrained peptides.

Ab initio modeling has also played a crucial role in AMP structural investigation. In this context, Mandal et al. [56] have reported the ab initio modeling of AMPs isolated from green coconut water, revealing that the arrangement of amino acids along the peptides' sequence favors the formation of helices and, thus, interferes with their action [56]. In addition, Porto et al. [57] reported the structural comparison of a glycine-rich natural peptide

(Pg-AMP1) isolated from *Psidium guajava* with its recombinant analog. In this work, the authors showed that both molecules present α-helical conformations with glycine-rich regions, which confer structural flexibility. In contrast, different conformations were reported due to the histidine tag, which might be a key point for their bactericidal activities [58].

Indeed, ab initio modeling has shown great advantages for predicting AMP structure, as sequence length is no longer an imperative limitation and experimental atomic coordinates are not needed to build the models, which is the case with comparative modeling. Since there are families and single sequences without known structures, ab initio modeling could also support the identification of novel putative folds. Using large scale ab initio prediction of 184 AMP sequences ranging from 20 to 120 amino acid residues, 12 putative novel AMP folds were identified [59]. These include ten all-α folds, one all-β fold and one mixed αβ fold. Interestingly, 11 out of 12 putative novel AMP folds have no matches against any solved structure deposited in the Protein Data Bank (PDB). Common AMP folds for cysteine-stabilized peptides were observed for sequences lacking the characteristic intramolecular disulfide bonds [59]. Thus, despite requiring experimental validation (Table 12.1), these findings could represent an expansion of current knowledge in the AMP structural universe and may contribute to the development of more potent AMPs using protein design [60].

12.5 Contact-assisted modeling

In the last decade, a new kind of prediction method has emerged, known as contact-assisted modeling. Given the initial classification of modeling methods and the fact that it does not required template structures, contact-assisted modeling has been classified as an ab initio modeling strategy [61]. The main principle of this technique is that correlated mutations can carry information about distance relationships in protein structure (Fig. 12.3). In other words, contact-assisted modeling uses a similar principle to threading, with related sequences sharing the same or similar contacts. The technique uses multiple sequence alignments with enough sequences (hundreds to thousands) to deduce the correlated mutations, which are used to construct a pairwise residue contact map; then a very coarse three-dimensional model is created from the contact map, and finally, the model is refined by standard molecular modeling techniques [61].

Contact-assisted modeling accuracy depends on the precision of contact maps, and therefore the technique was originally thought appropriate for application to genomic data [61,62] and later to metagenomic data [63]. Interestingly, despite the fact that contact-assisted modeling emerged as a new ab initio method, this technique ended up incorporating the threading and "classical" ab initio modeling methods. Since then, a significant improvement has been observed in folding performance, mainly for ab initio

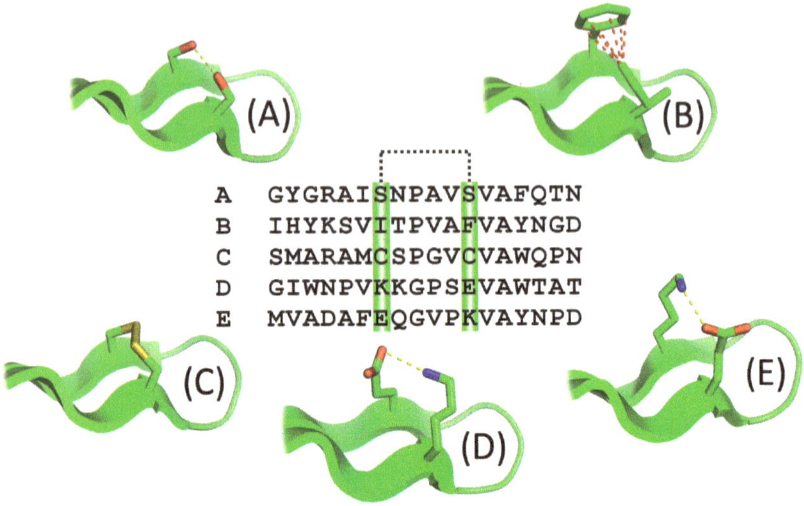

FIGURE 12.3 Correlated mutations carry information about distance relationships in protein structure. In this example, two positions are correlated, with contacts between them, which could vary from hydrogen bonds (A), C—H—π interactions (B), disulfide bridges (C) or saline bridges (D and E). Electrostatic interactions are shown as yellow dotted lines in (A), (D), and € and hydrophobic interactions as red dotted lines in (B). This approach has not been applied to AMPs yet. Currently, multiple deep sequence alignments, with hundreds to thousands of sequences, should be constructed to obtain the correlation data, which is converted into an amino acid contact map with distances <8 Å and used as additional restraints in both comparative and ab initio modeling.

targets [64]. Nevertheless, the confidence of contact predictions should be taken into account because false-positive contacts could reduce the modeling accuracy. Fortunately, the precision of contact prediction algorithms has recently improved drastically with the introduction of deep learning algorithms along with metagenomic data [63,64].

Metagenomic data, accurate contact maps and ab initio modeling were the three pillars for the discovery of 137 novel folds not represented in the PDB [63], highlighting the importance of ab initio contact-assisted modeling. Nonetheless, contact-assisted modeling has not been applied to AMPs yet. This could be attributed to many factors, including the availability of template structures for the main peptide families [65], which facilitates the use of comparative modeling techniques; moreover, when a new class is discovered, it might not have enough homologous sequences to deduce an accurate contact map, which favors the use of pure ab initio modeling [12,58].

12.6 Conclusions

Knowledge about three-dimensional structure has increased in relevance for the study of AMPs, given the need to understand their structure to gain insights into mechanisms of action [66]. Unfortunately, experimental techniques (Table 12.1) are less accessible and harder to apply on a large scale. Therefore, computational methods have gained attention. The success of computational methods for structure prediction is a product of two factors; structures of small proteins are easier to predict than larger ones, and homologous structures can be explored to predict other protein structures [67]. Hence, these computational approaches are widely applied to study AMPs, which have short sequences and a lot of homologous sequences. Overall, the models are coarse, and additional experiments are desirable to evaluate and add more reliability, including the subsequent use of molecular dynamics simulations and/or circular dichroism analysis or even combining NMR secondary chemical shifts and temperature coefficient analyses with molecular modeling and dynamics simulations [17]. At this point, integrating automated structural results from different sources is interesting, as demonstrated by the combination of metagenomic data, contact maps and ab initio modeling to discover new peptide scaffolds [63]. The algorithms applied in bioinformatics will be of great value for integrating the structural biochemical information generated, making it possible to bring together the knowledge engendered by different techniques and to apply this to the rational design of new AMPs. In the next few years, the application of contact maps to refine AMP models are expected, allowing more accurate analysis of data deposited in public databases.

Acknowledgments

This work was supported by CNPq, CAPES, FUNDECT and FAPDF.

References

[1] Cao H, et al. 266 Prevalence and risk factors for antimicrobial resistance on U.S. dairy operations. J Anim Sci 2017;95(suppl. 4):131–2. Available from: https://doi.org/10.2527/asasann.2017.266.

[2] Xie J, et al. Clinical features and antimicrobial resistance profiles of important Enterobacteriaceae pathogens in Guangzhou representative of Southern China, 2001–2015. Microb Pathogen 2017;107:206–11. Available from: https://doi.org/10.1016/j.micpath.2017.03.038.

[3] Nissen-Meyer J, Nes IF. Ribosomally synthesized antimicrobial peptides: their function, structure, biogenesis, and mechanism of action. Arch Microbiol 1997;167(2–3):67–77. Available from: https://doi.org/10.1007/s002030050418.

[4] Silva ON, et al. Exploring the pharmacological potential of promiscuous host-defense peptides: from natural screenings to biotechnological applications. Front Microbiol 2011;2:232. Available from: https://doi.org/10.3389/fmicb.2011.00232.

[5] Fjell CD, Jenssen H, et al. Optimization of antibacterial peptides by genetic algorithms and cheminformatics. Chem Biol Drug Des 2011;77(1):48–56. Available from: https://doi.org/10.1111/j.1747-0285.2010.01044.x.

[6] Vriend G. WHAT IF: a molecular modeling and drug design program. J Mol Graph 1990;8(1):52–6. Available from: https://doi.org/10.1016/0263-7855(90)80070-V.

[7] Aguilera-Mendoza L, et al. Overlap and diversity in antimicrobial peptide databases: compiling a non-redundant set of sequences. Bioinforma (Oxford, Engl) 2015;31(15):2553–9. Available from: https://doi.org/10.1093/bioinformatics/btv180.

[8] Wang G, Li X, Wang Z. APD3: the antimicrobial peptide database as a tool for research and education. Nucl Acids Res 2016;44(D1):D1087–93. Available from: https://doi.org/10.1093/nar/gkv1278.

[9] Fjell CD, Hiss JA, et al. Designing antimicrobial peptides: form follows function. Nat Rev Drug Discov 2011;11(1):37–51. Available from: https://doi.org/10.1038/nrd3591.

[10] Franco OL. Peptide promiscuity: an evolutionary concept for plant defense. FEBS Lett 2011;585(7):995–1000. Available from: https://doi.org/10.1016/j.febslet.2011.03.008 Federation of European Biochemical Societies.

[11] Phoenix DA, Dennison SR, Harris F. Antimicrobial peptides: their history, evolution, and functional promiscuity. Antimicrobial peptides. Weinheim, Germany: Wiley-VCH Verlag GmbH & Co. KGaA; 2013. p. 1–37. Available from: http://doi.org/10.1002/9783527652853.ch1.

[12] Mandal SM, et al. The attack of the phytopathogens and the trumpet solo: identification of a novel plant antifungal peptide with distinct fold and disulfide bond pattern. Biochimie 2013;95(10):1939–48. Available from: https://doi.org/10.1016/j.biochi.2013.06.027 Elsevier Masson SAS.

[13] Porto WF, Franco OL. Theoretical structural insights into the snakin/GASA family. Peptides 2013;44:163–7. Available from: https://doi.org/10.1016/j.peptides.2013.03.014 Elsevier Inc.

[14] Singh PK, et al. The intramolecular disulfide-stapled structure of laterosporulin, a class IId bacteriocin, conceals a human defensin-like structural module. FEBS J 2014;. Available from: https://doi.org/10.1111/febs.13129.

[15] Yeung H, et al. Radiation damage and racemic protein crystallography reveal the unique structure of the GASA/snakin protein superfamily. Angew Chem (Int ed Engl) 2016;. Available from: https://doi.org/10.1002/anie.201602719.

[16] Cardoso MH, et al. Computer-aided design of antimicrobial peptides: are we generating effective drug candidates? Front Microbiol 2020. Available from: https://doi.org/10.3389/fmicb.2019.03097.

[17] Porto WF, Pires AS, Franco OL. Computational tools for exploring sequence databases as a resource for antimicrobial peptides. Biotechnol Adv 2017;35(3). Available from: https://doi.org/10.1016/j.biotechadv.2017.02.001.

[18] Porto WF, Silva ON, Franco OL. Prediction and rational design of antimicrobial peptides. In: Faraggi E, editor. Protein Structure. InTech; 2012. p. 377–96. Available from: http://doi.org/10.5772/2335.

[19] Kramer J. Is abstraction the key to computing? Commun ACM 2007;50(4):36–42. Available from: https://doi.org/10.1145/1232743.1232745.

[20] Ilari A, Savino C. Protein structure determination by X-ray crystallography. 2008, pp. 63–87. Available from: https://doi.org/10.1007/978-1-60327-159-2_3.

[21] Smyth MS, Martin JH. X ray crystallography. Mol Pathol MP 2000;53(1):8–14. Available from: https://doi.org/10.1136/mp.53.1.8.

[22] Hoover DM, et al. The structure of human β-defensin-2 shows evidence of higher order oligomerization. J Biol Chem 2000;275(42):32911–18. Available from: https://doi.org/10.1074/jbc.M006098200.

[23] Baxter AA, et al. The tomato defensin TPP3 binds phosphatidylinositol (4,5)-bisphosphate via a conserved dimeric cationic grip conformation to mediate cell lysis. Mol Cell Biol 2015;35(11):1964–78. Available from: https://doi.org/10.1128/MCB.00282-15.

[24] Wang CK, et al. Racemic and quasi-racemic X-ray structures of cyclic disulfide-rich peptide drug scaffolds. Angew Chem Int Ed 2014;53(42):11236–41. Available from: https://doi.org/10.1002/anie.201406563.

[25] Pellecchia M, Sem DS, Wüthrich K. NMR in drug discovery. Nat Rev Drug Discov 2002;1(3):211–19. Available from: https://doi.org/10.1038/nrd748.

[26] Rosengren KJ, et al. Twists, knots, and rings in proteins. Structural definition of the cyclotide framework. J Biol Chem 2003;278(10):8606–16. Available from: https://doi.org/10.1074/jbc.M211147200.

[27] Resende JM, et al. Solution NMR structures of the antimicrobial peptides phylloseptin-1, -2, and -3 and biological activity: the role of charges and hydrogen bonding interactions in stabilizing helix conformations. Peptides 2008;29(10):1633–44. Available from: https://doi.org/10.1016/j.peptides.2008.06.022.

[28] Campagna S, et al. Structure and mechanism of action of the antimicrobial peptide piscidin. Biochemistry 2007;46(7):1771–8. Available from: https://doi.org/10.1021/bi0620297.

[29] Fillion M, Auger M. Oriented samples: a tool for determining the membrane topology and the mechanism of action of cationic antimicrobial peptides by solid-state NMR. Biophys Rev 2015;7(3):311–20. Available from: https://doi.org/10.1007/s12551-015-0167-5.

[30] Saibil HR. Macromolecular structure determination by cryo-electron microscopy. Acta Crystallogr Sect D Biol Crystallogr 2000;56(10):1215–22. Available from: https://doi.org/10.1107/S0907444900010787.

[31] Cheng Y, et al. A primer to single-particle cryo-electron microscopy. Cell 2015;161 (3):438–49. Available from: https://doi.org/10.1016/j.cell.2015.03.050.

[32] Boge L, et al. Lipid-based liquid crystals as carriers for antimicrobial peptides: phase behavior and antimicrobial effect. Langmuir 2016;32(17):4217–28. Available from: https://doi.org/10.1021/acs.langmuir.6b00338.

[33] Haisma EM, et al. Antimicrobial peptide P60.4Ac-containing creams and gel for eradication of methicillin-resistant *Staphylococcus aureus* from cultured skin and airway epithelial surfaces. Antimicr Agents Chemother 2016;60(7):4063–72. Available from: https://doi.org/10.1128/AAC.03001-15.

[34] Han M, et al. Characterization of antibiotic peptide pores using cryo-EM and comparison to neutron scattering. Biophys J 2009;97(1):164–72. Available from: https://doi.org/10.1016/j.bpj.2009.04.039.

[35] Martí-Renom MA, et al. Comparative protein structure modeling of genes and genomes. Annu Rev Biophys Biomol Struct 2000;29(1):291–325. Available from: https://doi.org/10.1146/annurev.biophys.29.1.291.

[36] Porto WF, Fensterseifer GM, Franco OL. In silico identification, structural characterization, and phylogenetic analysis of MdesDEF-2: a novel defensin from the Hessian fly, Mayetiola destructor. J Mol Model 2014;20(7):2339. Available from: https://doi.org/10.1007/s00894-014-2339-9.

[37] Altschul SF, et al. Gapped BLAST and PSI-BLAST: a new generation of protein database search programs. Nucleic Acids Res 1997;25(17):3389–402. Available from: https://doi.org/10.1093/nar/25.17.3389.

[38] Pearson WR. Rapid and sensitive sequence comparison with FASTP and FASTA. Meth Enzymol 1990;183:63–98. Available at: http://www.ncbi.nlm.nih.gov/pubmed/2156132 (Accessed: 21 December 2014).

[39] Saqi MA, Russell RB, Sternberg MJ. Misleading local sequence alignments: implications for comparative protein modelling. Protein Eng Des Selection 1998;11(8):627–30. Available from: https://doi.org/10.1093/protein/11.8.627.

[40] Koehl P, Levitt M. Sequence variations within protein families are linearly related to structural variations. J Mol Biol 2002;323(3):551–62. Available from: https://doi.org/10.1016/S0022-2836(02)00971-3.

[41] Gille C, et al. Conservation of substructures in proteins: interfaces of secondary structural elements in proteasomal subunits. J Mol Biol 2000;299(4):1147–54. Available from: https://doi.org/10.1006/jmbi.2000.3763.

[42] Konno K, et al. Anoplin, a novel antimicrobial peptide from the venom of the solitary wasp Anoplius samariensis. Biochimica et Biophysica Acta (BBA) - Protein Struct Mol Enzymol 2001;1550(1):70–80. Available from: https://doi.org/10.1016/S0167-4838(01)00271-0.

[43] Cunha NBD, et al. Cloning and characterization of novel cyclotides genes from South American plants. Biopolymers 2016;106(6). Available from: https://doi.org/10.1002/bip.22938.

[44] Cândido E, de S, Fernandes G, da R, et al. Shedding some light over the floral metabolism by Arum Lily (*Zantedeschia aethiopica*) spathe de novo transcriptome assembly. PLoS One 2014;9(3):e90487. Available from: https://doi.org/10.1371/journal.pone.0090487 Edited by F. Zhou. Public Library of Science.

[45] Porto WF, et al. In silico identification of novel hevein-like peptide precursors. Peptides 2012;38(1):127–36. Available from: https://doi.org/10.1016/j.peptides.2012.07.025 Elsevier Inc.

[46] Irazazabal LN, et al. Selective amino acid substitution reduces cytotoxicity of the antimicrobial peptide mastoparan. Biochimica et biophysica acta 2016;1858(11):2699–708. Available from: https://doi.org/10.1016/j.bbamem.2016.07.001.

[47] Gao Y, et al. Targeted modification of a novel amphibian antimicrobial peptide from phyllomedusa tarsius to enhance its activity against MRSA and microbial biofilm. Front Microbiol 2017;8. Available from: https://doi.org/10.3389/fmicb.2017.00628.

[48] Bao K, et al. Modification targeting the "rana box" motif of a novel nigrocin peptide from hylarana latouchii enhances and broadens its potency against multiple bacteria. Front Microbiol 2018;9. Available from: https://doi.org/10.3389/fmicb.2018.02846.

[49] Lee J, Wu S, Zhang Y. Ab initio protein structure prediction. In: Rigden DJ, editor. From protein structure to function with bioinformatics. Springer; 2009. p. 3–25.

[50] Floudas CA, et al. Advances in protein structure prediction and de novo protein design: a review. Chem Eng Sci 2006;61(3):966–88. Available from: https://doi.org/10.1016/j.ces.2005.04.009.

[51] Wu S, Skolnick J, Zhang Y. Ab initio modeling of small proteins by iterative TASSER simulations. BMC Biol 2007;5(1):17. Available from: https://doi.org/10.1186/1741-7007-5-17.

[52] Rigden D. Ab initio modeling led annotation suggests nucleic acid binding function for many DUFs. OMICS J Integr Biol 2011;15(7–8):431–8. Available from: https://doi.org/10.1089/omi.2010.0122.

[53] Kinch LN, et al. Evaluation of free modeling targets in CASP11 and ROLL. Proteins Struct Funct Bioinforma 2016;84:51–66. Available from: https://doi.org/10.1002/prot.24973.

[54] Mandal SM, et al. Identification of multifunctional peptides from human milk. Peptides 2014;56:84–93. Available from: https://doi.org/10.1016/j.peptides.2014.03.017 Elsevier Inc.

[55] Harris PWR, et al. Plant antimicrobial peptides snakin-1 and snakin-2: chemical synthesis and insights into the disulfide connectivity. Chem (Weinh an der Bergstrasse, Ger) 2014;20(17):5102–10. Available from: https://doi.org/10.1002/chem.201303207.

[56] Mandal SM, et al. Identification and structural insights of three novel antimicrobial peptides isolated from green coconut water. Peptides 2009;30(4):633–7. Available from: https://doi.org/10.1016/j.peptides.2008.12.001.

[57] Porto WF, et al. Screening and functional prediction of conserved hypothetical proteins from *Escherichia coli*. J Proteom Bioinforma 2014;07(07):203–13. Available from: https://doi.org/10.4172/jpb.1000321.

[58] Porto WF, Nolasco DO, Franco OL. Native and recombinant Pg-AMP1 show different antibacterial activity spectrum but similar folding behavior. Peptides 2014;55:92–7. Available from: https://doi.org/10.1016/j.peptides.2014.02.010 Elsevier Inc.

[59] Kozic M, et al. Large scale ab initio modeling of structurally uncharacterized antimicrobial peptides reveals known and novel folds. Proteins Struct Funct Bioinforma 2018;86 (5):548–65. Available from: https://doi.org/10.1002/prot.25473.

[60] Bhardwaj G, et al. Accurate de novo design of hyperstable constrained peptides. Nature 2016;538(7625):329–35. Available from: https://doi.org/10.1038/nature19791.

[61] Marks DS, et al. Protein 3D structure computed from evolutionary sequence variation. PLoS ONE 2011;6(12):e28766. Available from: https://doi.org/10.1371/journal.pone.0028766 Edited by A. Sali.

[62] Sulkowska JI, et al. Genomics-aided structure prediction. Proc Natl Acad Sci 2012;109 (26):10340–5. Available from: https://doi.org/10.1073/pnas.1207864109.

[63] Ovchinnikov S, et al. Protein structure determination using metagenome sequence data. Science 2017;355(6322):294–8. Available from: https://doi.org/10.1126/science.aah4043.

[64] Zheng W, et al. Deep-learning contact-map guided protein structure prediction in CASP13. Proteins Struct Funct Bioinforma 2019;87(12):1149–64. Available from: https://doi.org/10.1002/prot.25792.

[65] de S. Cândido E, de S. Cardoso MH, et al. The use of versatile plant antimicrobial peptides in agribusiness and human health. Peptides 2014;55:65–78. Available from: https://doi.org/10.1016/j.peptides.2014.02.003.

[66] Cardoso MH, et al. The structure/function relationship in antimicrobial peptides: what can we obtain from structural data? 2018, pp. 359–84. doi:10.1016/bs.apcsb.2018.01.008.

[67] Zhang Y, Skolnick J. Tertiary structure predictions on a comprehensive benchmark of medium to large size proteins. Biophys J 2004;87(4):2647–55. Available from: https://doi.org/10.1529/biophysj.104.045385.

Section 4

Therapeutic applications

Chapter 13

Knottin peptidomimetics as therapeutics

Tatiana Galochkina[1,2] and Jean-Christophe Gelly[1,2]

[1]*Université Paris Cité, INSERM, BIGR, Paris, France,* [2]*Laboratoire d'Excellence GR-Ex, Paris, France*

13.1 Historical context

In 1982 Rees and Lipscomb determined a three-dimensional structure of the complex between carboxypeptidase A and a natural inhibitor from the potato PCI (potato carboxypeptidase A inhibitor) containing 38 amino acids, including six cysteines forming three disulfide bonds. When describing the molecule, the authors noted the presence of an original and remarkable knot-like topology formed by the "threading of the disulfide bond between CysP18 and CysP34 through a loop closed by the other two-disulfide bridges" [1]. Seven years after, the first structural studies of tryptic inhibitors from *Cucurbitaceae* revealed an architecture similar to that observed earlier in PCI, namely, an antiparallel triple beta-sheet stabilized by a cystine knot [2–4]. This cystine knot consists of a macrocycle formed by two-disulfide bridges and a fragment of the protein main chain, which is further crossed by a third cysteine bridge. A few years later, this particular architecture was also detected in animal toxins, conotoxins from marine shellfish [5] and agatoxins from spider venom [6]. Interestingly, this structural family incorporated a human hormone [7–9], as well as a range of insect toxins [10], scorpion toxins [11], antifungal [12,13], and antibacterial peptides [14].

As early as 1994, a large number of homologous macrocyclic proteins with the same topological arrangement were isolated from plants of the *Rubiaceae* and *Violaceae* families, which have been named cyclotides for "cyclic peptides" [15]. Later, more and more new families of macrocyclic proteins sharing the same topology were identified in *Solanaceae*, *Cucurbitaceae*, and *Fabaceae* plant families [16]. Nearly 500 cyclotides are

Peptide and Peptidomimetic Therapeutics. DOI: https://doi.org/10.1016/B978-0-12-820141-1.00019-4

currently described. As of today, more than 350 structures of knottins have been resolved and nearly 3500 sequences are known [17–20].

13.2 Why "knottin"?

Due to the presence of a pseudo-knot formed by disulfide bridges, in 1990 the term "knottin" was proposed to designate the proteins belonging to this structural family [21]. Nevertheless, several alternative names were proposed subsequently. Since initially all the discovered proteins were inhibitors of proteases (trypsin, carboxypeptidase) or the ion channels (animal toxins), Pallaghy et al. proposed to name them "inhibitor cystine knot" or ICK [22]. Nevertheless, this term seems to us a little restrictive, since for a certain number of the cysteine macrocyclic proteins the mechanism of action is still unknown. Therefore, we prefer to keep the initial term "knottin."

The term "knottin" defines a very particular cysteine macrocycle peptide topology. Indeed, more general terms such as "T-knot" [23–25] or "cystine knot" [26] are not specific for knottins and include other structural families, in particular the growth factors of the transforming growth factor family. Similar to knottins, these growth factors (growth factor cystine knots) have six cysteines forming three disulfide bridges with I–IV, II–V, III–VI connectivity as well as a macrocycle formed by two-disulfide bridges and crossed by a third disulfide bridge. However, several fundamental properties distinguish them from knottins: (1) the disulfide bridge penetrating the macrocycle consists of cysteines I and IV (instead of III and VI for knottins), (2) differences in the connectivity and spatial arrangement of the disulfide bridges eliminates any possibility of structural superposition with knottins, and (3) these proteins have much larger sizes on average and exist mostly as dimers. Therefore these growth factors form a distinct structural family and are not considered in the current work.

To represent in s comprehensive way the general structure of knottins, and to facilitate their selection and classification according to their architecture, a nomenclature based on the length of the loops between the knot-forming cysteines has been proposed [17]. The loops and the number of residues between the cysteines are labeled successively by the letters a, b, c, d, and e. To clarify the topology description, the two loops involved in the formation of a macrocycle together with two-disulfide bridges are indicated in parentheses. In the case of cyclic knottins (cyclotides, cyclic squash inhibitors), the additional loop connecting the last cysteine of the knot with the first one indicated by the letter [f] in square brackets. Schematic representation of the comparative topologies of knottins and cystine-knotted growth factors, as well as the arrangement of the loops in the three-dimensional structure, is shown in Figs. 13.1 and 13.2.

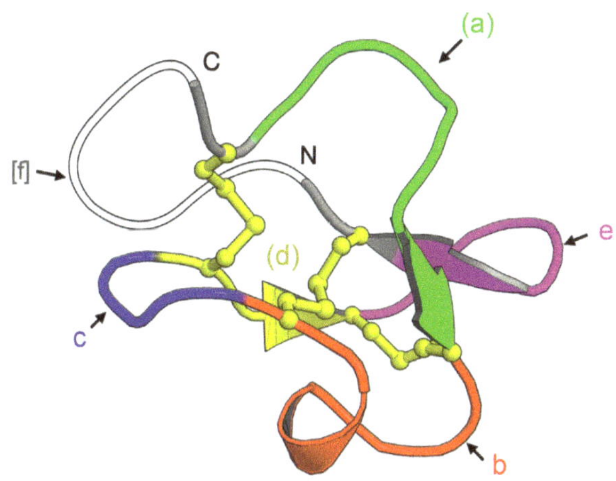

FIGURE 13.1 Schematic representation of a knottin with different colors indicating the a–f loops. The disulfide bonds appear as sticks.

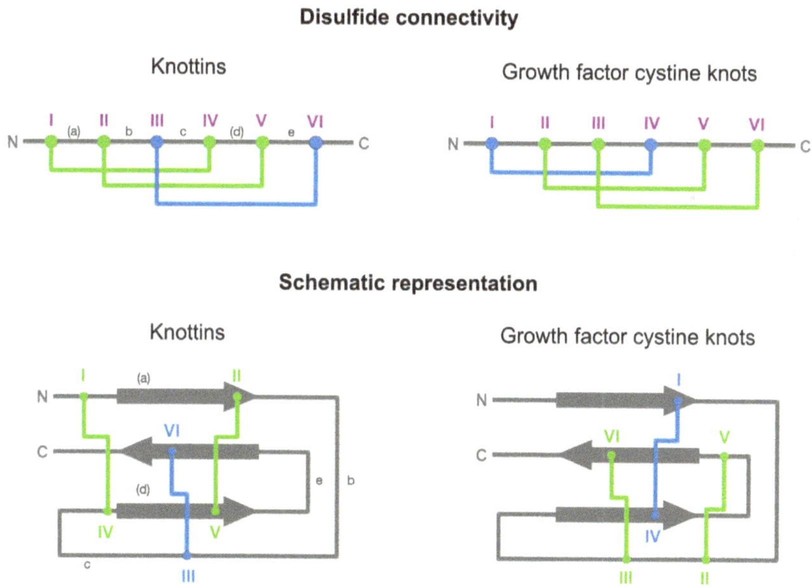

FIGURE 13.2 Schematic representations of the topologies of knottins and cystine knot growth factors. Loops between cysteines are noted (a), b, c, (d), e.

13.3 The highly conserved basic structural cystine-stabilized beta-sheet motif

Despite the strong topological similarity between "squash" inhibitors and the potato carboxypeptidase inhibitors (PCI) detected as early as 1989, their further structural comparison revealed an important difference. Indeed, the disulfide bridge between cysteines I and IV was not strictly conserved between two knottin families thus suggesting that the best conserved structural motif contained only disulfide bridges II–V and III–VI [27]. This observation was quickly supported by the isolation and structural study of a folding intermediate of the squash inhibitor EETI-II [28]. This folding intermediate, stable and having a quasinative structure contained only disulfide bridges II–V and III–VI, while cysteines I and IV remained free. A few years later, synthesis and subsequent structural and biophysical characterization of a truncated peptide with removed I–IV disulfide bridge (Min-23) finally confirmed that the structural motif with two-disulfide bridges is indeed an autonomous folding unit [29]. This elementary motif was named CSB for cystine-stabilized beta-sheet, by analogy with the already described disulfide-bridged motifs CSH [30] found in endothelin, as well as apamin and sarafotoxins [31] found in insect defensins and in some scorpion toxins. In Fig. 13.3, we show the relationships between the CSB motif, knottins, and cyclic peptides, as well as the introduced nomenclature. A slightly modified version of the basic CSB structural motif was later proposed by other authors under the name DDH, for disulfide-directed hairpin [32].

It is important to note that the CSB motif is not only present in knottins but also in many other proteins making it the most common structural motif in small disulfide-bridged proteins. In this regard, it should be emphasized that the term "knottin" is currently used differently by different authors. For instance, the SCOP structural database contains a fold type called knottin, which actually corresponds to CSB motif proteins and thus includes many proteins in which no knots or pseudo-knots are present (e.g., epidermal growth factor [EGF] motifs, some scorpion toxins, hevein, and elafin).

It is particularly interesting that the CSB motif with two-disulfide bridges demonstrates significant thermo stability (Tm ~100°C for Min-23 [29]), although it is still lower than that of the corresponding knottins with three disulfide bridges. Furthermore, a two-disulfide toxin from the venom of scorpion *Liocheles australasiae* LaIT1 has been shown to display a pseudo-knottin fold [33]. Although two-disulfide knottin-like domains were predicted to be stable many years ago [28,29], this is, however, the first evidence of a wild two-disulfide knottin-like miniprotein without additional domains or disulfide bridges. As there is no cystine knot in this toxin, it has not been classified as a knottin. The CSB motif could also correspond to ancestral proteins that have largely disappeared, with the exception of LaIT1, and that would have been the origin of different structural families such as knottins, EGF motifs, and scorpion toxins by divergent evolution. However, it remains possible that the CSB motif appeared in these diverse families as a result of convergent evolution. In this case, rather than an ancestral motif, the CSB motif would appear as a stable arrangement common to small

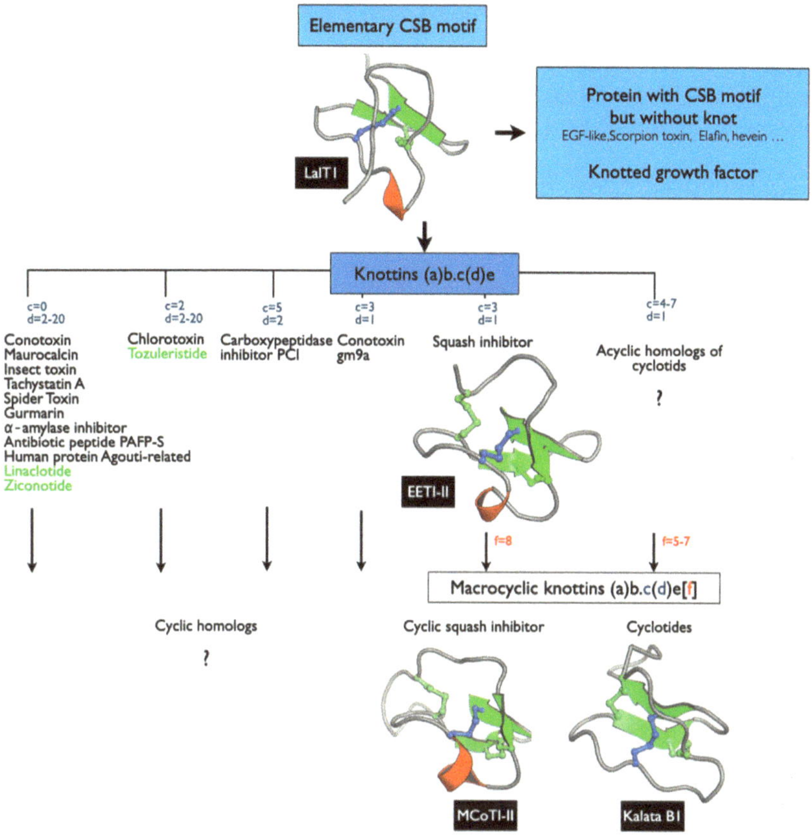

FIGURE 13.3 General diagram showing the relationship between knottins, CSB elemental structural motif and natural cyclic proteins. The nomenclature a–f, explained above, indicates the number of amino acids in each loop. *CSB*, Cystine-stabilized beta-sheet.

disulfide-bridged proteins, in the same way, that the Greek key or helix clusters are frequently observed in globular proteins.

Recently, a high-throughput production and screening of cystine-dense peptides (CDPs) served as a basis for classification based on the sequence and structure protein properties [34]. Interestingly, the statistical analysis of the cystine connectivity type and sequence/structure relationship revealed a very low correlation between CDP sequence and structure type. This CDP property can therefore significantly hinder the mapping of the new sequences to existing CDPs structures necessary for the new protein design. Remarkably, this study also showed that the knotted topology is the most frequently observed. Moreover, the 1–4, 2–5, 3–6 connectivity scheme was by far the most commonly observed for both knotted and nonknotted topologies.

13.4 Knottins for drug design and engineering

The use of peptides in pharmacology has often been discussed. The main difficulties generally considered to hinder successful peptide-based drug development are the following [35]:

- low bioavailability, especially through oral absorption
- poor metabolic stability
- potential immunogenicity

Nevertheless, the following unique knottin properties make them promising scaffolds for the development of peptide drugs.

13.4.1 Very diverse functions and sequences

First, knottins identified so far have very diverse functions among which protease inhibitors, toxins, neurotransmitters, antimicrobials, insecticides, fungicides, hemolytic properties, etc. Second, knottins are found in various eukaryotic organisms: plants, fungi, molluscs, arthropods, and vertebrates including humans. In addition to this plurality of functions and location, knottins demonstrate great sequence variability at the level of the loops connecting the cysteines. While such variability hinders comparison between the proteins and increases the difficulties in interpretation of the sequence–structure relationships, it also shows that nature has used the knottin architecture extensively for various purposes. Therefore, such architecture, highly tolerant to sequence variations, must be an important candidate to be used as a structural support for the implementation of various functions in drug design or protein engineering.

Knottin synthesis was already successfully applied in several drug-design studies [36–38]. For example, several chimeric proteins have been obtained from the squash inhibitor EETI-II by insertion of active sites [18,21,27,39–41]. The same knottin has been used for ligand selection from combinatorial libraries [42,43].

13.4.2 Stable structures accessible to chemical synthesis

Another important feature of knottins further supporting their potential for drug design is their small size that makes these proteins easily accessible to chemical synthesis, which opens the way to chemical modifications (pseudopeptides) to limit immunogenic reactions. Moreover, knottins are very stable and demonstrate high resistance to proteolysis thanks to their disulfide bridges and their pseudo-knot topology.

The chemical synthesis of cyclic knottins has been the subject of several works [44–46]. The value of the cyclization observed in cyclotides and cyclic squash inhibitors is not obvious *a priori*, since the noncyclic analogs already show quite remarkable stability and structuring. The additional property that cyclization may provide is probably an insensitivity of knottin interactions with exoproteases due to the removal of the N- and C-termini.

Finally, the last argument concerns the discovery of a human knottin, the Agouti-related protein [9]. This discovery suggests that humanization approaches, similar to those developed for antibodies, could be considered when grafting active sites onto a knottin architecture.

13.4.3 Knottin-based therapies

The potential of knotting used for therapeutic development is also strongly supported by already existing successful applications. The most striking example, which has been put forward several times in consideration of knottins for medical applications, concerns the use of knottins in traditional African medicine. An infusion prepared from the leaves of the tropical plant *Oldenlandia affinis* is used in the Congo to accelerate vaginal delivery [47]. The uterotonic activity of the decoction could be attributed to Kalata B1, a cyclotide-like cyclic knottin. The activity of this peptide, after one hour of boiling in water and ingestion, shows the very high thermal stability and resistance to proteases of these particular proteins. The oral absorption of the decoction shows that due to its high resistance to proteases, this protein can pass the intestinal barrier.

Several later studies also highlight the potential of knottins to treat many different diseases [38]:

- potential anticancer drugs [48,49]
- cardiovascular drugs, in particular for myocardial infarction [50]
- drug delivery and imaging vectors [51]
- immunological modulators [52,53]

It is also worth mentioning that an ecologically friendly product (SeroX) containing a cyclic knottin is used as an insecticide [54].

Finally, several synthetic knottins are already successfully commercialized for the use in clinical applications. Since 2004, the derivative of the ωω-conotoxin from the cone snail *Conus magus* ziconotide (SNX-111; Prialt) is used as a medication for chronic pain [55]. Acting as a selective inhibitor of the voltage-gated calcium (CaV) channels in the spinal cord, ziconotide is efficient even for patients that do not respond to first-line analgesic drugs and is 1000 times more potent than morphine. Another potentially efficient application of knottins to chronic pain treatment concerns the capacity of spider venom knottins to modulate the function of NaV sodium channels. A range of spider knottins are developed as potential analgesic leads and are the subject of intensive studies to enhance their potency and selectivity [56].

Linaclotide (Linzess) is another example of the synthetic peptide drug using a knottin architecture as a scaffold. Derived from bacterial heat-stable enterotoxin, linaclotide acts as guanylate cyclase-C agonist and together with its modifications is used for the treatment of irritable bowel syndrome [57]. In addition, tozuleristide (tumor paint) is a tumor-homing, fluorophore conjugate of the scorpion knottin chlorotoxin (CTX). Tozuleristide is in phase III clinical trials [58].

13.5 Summary

Knottins represent a unique structural class of peptide and peptidomimetic molecules. These scaffolds are being increasingly leveraged for therapeutic programs and have already demonstrated their potential in successful development of several approved drugs and compounds. We expect to see increasing use of knottins and their derivatives in the nearest future.

References

[1] Rees DC, Lipscomb WN. Refined crystal structure of the potato inhibitor complex of carboxypeptidase A at 2.5-A resolution. J Mol Biol 1982;160:475–98.

[2] Heitz A, Chiche L, Le-Nguyen D, Castro B. 1H 2D NMR and distance geometry study of the folding of *Ecballium elaterium* trypsin inhibitor, a member of the squash inhibitors family. Biochemistry 1989;28:2392–8.

[3] Bode W, Greyling HJ, Huber R, Otlewski J, Wilusz T. The refined 2.0-A X-ray crystal structure of the complex formed between bovine beta-trypsin and CMTI-I, a trypsin inhibitor from squash seeds (*Cucurbita maxima*). Topological similarity of the squash seed inhibitors with the carboxypeptidase A inhibitor from potatoes. FEBS Lett 1989;242:285–92.

[4] Chiche L, Gaboriaud C, Heitz A, Mornon JP, Castro B, Kollman PA. Use of restrained molecular dynamics in water to determine three-dimensional protein structure: prediction of the three-dimensional structure of *Ecballium elaterium* trypsin inhibitor II. Proteins 1989;6:405–17.

[5] Davis JH, Bradley EK, Miljanich GP, Nadasdi L, Ramachandran J, Basus VJ. Solution structure of omega-conotoxin GVIA using 2-D NMR spectroscopy and relaxation matrix analysis. Biochemistry 1993;32:7396–405.

[6] Yu H, Rosen MK, Saccomano NA, Phillips D, Volkmann RA, Schreiber SL. Sequential assignment and structure determination of spider toxin omega-Aga-IVB. Biochemistry 1993;32:13123–9.

[7] Jackson PJ, McNulty JC, Yang YK, Thompson DA, Chai B, Gantz I, et al. Design, pharmacology, and NMR structure of a minimized cystine knot with agouti-related protein activity. Biochemistry 2002;41:7565–72.

[8] McNulty JC, Thompson DA, Bolin KA, Wilken J, Barsh GS, Millhauser GL. High-resolution NMR structure of the chemically-synthesized melanocortin receptor binding domain AGRP(87–132) of the agouti-related protein. Biochemistry 2001;40:15520–7.

[9] Bolin KA, Anderson DJ, Trulson JA, Thompson DA, Wilken J, Kent SB, et al. NMR structure of a minimized human agouti related protein prepared by total chemical synthesis. FEBS Lett 1999;451:125–31.

[10] Bernard C, Corzo G, Mosbah A, Nakajima T, Darbon H. Solution structure of Ptu1, a toxin from the assassin bug *Peirates turpis* that blocks the voltage-sensitive calcium channel N-type. Biochemistry 2001;40:12795–800.

[11] Mosbah A, Kharrat R, Fajloun Z, Renisio JG, Blanc E, Sabatier JM, et al. A new fold in the scorpion toxin family, associated with an activity on a ryanodine-sensitive calcium channel. Proteins 2000;40:436–42.
[12] Gao GH, Liu W, Dai JX, Wang JF, Hu Z, Zhang Y, et al. Molecular scaffold of a new pokeweed antifungal peptide deduced by 1H nuclear magnetic resonance. Int J Biol Macromol 2001;29:251–8.
[13] Gao GH, Liu W, Dai JX, Wang JF, Hu Z, Zhang Y, et al. Solution structure of PAFP-S: a new knottin-type antifungal peptide from the seeds of *Phytolacca americana*. Biochemistry 2001;40:10973–8.
[14] Fujitani N, Kawabata S, Osaki T, Kumaki Y, Demura M, Nitta K, et al. Structure of the antimicrobial peptide tachystatin A. J Biol Chem 2002;277:23651–7.
[15] Craik DJ, Daly NL, Bond T, Waine C. Plant cyclotides: a unique family of cyclic and knotted proteins that defines the cyclic cystine knot structural motif. J Mol Biol 1999;294:1327–36.
[16] Wang CK, Kaas Q, Chiche L, Craik DJ. CyBase: a database of cyclic protein sequences and structures, with applications in protein discovery and engineering. Nucleic Acids Res 2008;D206–10.
[17] Gelly JC, Gracy J, Kaas Q, Le-Nguyen D, Heitz A, Chiche L. The KNOTTIN website and database: a new information system dedicated to the knottin scaffold. Nucleic Acids Res 2004;32:D156–9.
[18] Chiche L, Heitz A, Gelly JC, Gracy J, Chau PTT, Ha PT, et al. Squash inhibitors: from structural motifs to macrocyclic knottins. Curr Protein Pep Sci 2004;5(5):341–9 9.
[19] Postic G, Gracy J, Périn C, Chiche L, Gelly JC. KNOTTIN: the database of inhibitor cystine knot scaffold after 10 years, toward a systematic structure modeling. Nucleic Acids Res 2018;46(D1):D454–8.
[20] Gracy J, Le-Nguyen D, Gelly J-C, Kaas Q, Heitz A, Chiche L. KNOTTIN: the knottin or inhibitor cystine knot scaffold in 2007. Nucleic Acids Res 2008;36:D314–19. Available from: https://doi.org/10.1093/nar/gkm939.
[21] Le Nguyen D, Heitz A, Chiche L, Castro B, Boigegrain RA, Favel A, et al. Molecular recognition between serine proteases and new bioactive microproteins with a knotted structure. Biochimie 1990;72:431–5.
[22] Pallaghy PK, Nielsen KJ, Craik DJ, Norton RS. A common structural motif incorporating a cystine knot and a triple-stranded beta-sheet in toxic and inhibitory polypeptides. Protein Sci 1994;3:1833–9.
[23] Lin SL, Nussinov R, Polticelli F, Pascarella S, Bordo D, Bolognesi M, et al. A disulphide-reinforced structural scaffold shared by small proteins with diverse functions the T-knot motif revisited. A structural superfamily of growth factors containing a cystine knot motif. Nat Struct Biol 1995;2:835–7.
[24] McDonald NQ, Hendrickson WA. A structural superfamily of growth factors containing a cystine knot motif. Cell 1993;73:421–4.
[25] Polticelli F, Pascarella S, Bordo D, Bolognesi M, Ascenzi P, McDonald NQ, et al. The T-knot motif revisited a structural superfamily of growth factors containing a cystine knot motif. Biol Chem 1999;380:1247–50.
[26] Narasimhan L, Singh J, Humblet C, Guruprasad K, Blundell T. Snail and spider toxins share a similar tertiary structure and "cystine motif". Nat Struct Biol 1994;1:850–2.
[27] Chiche L, Heitz A, Padilla A, Le-Nguyen D, Castro B. Solution conformation of a synthetic bis-headed inhibitor of trypsin and carboxypeptidase A: new structural alignment between the squash inhibitors and the potato carboxypeptidase inhibitor. Protein Eng 1993;6:675–82.

[28] Le-Nguyen D, Heitz A, Chiche L, el Hajji M, Castro B. Characterization and 2D NMR study of the stable [9–21, 15–27] 2 disulfide intermediate in the folding of the 3 disulfide trypsin inhibitor EETI II. Protein Sci 1993;2:165–74.

[29] Heitz A, Le-Nguyen D, Chiche L. Min-21 and min-23, the smallest peptides that fold like a cystine-stabilized beta-sheet motif: design, solution structure, and thermal stability. Biochemistry 1999;38:10615–25.

[30] Tamaoki H, Miura R, Kusunoki M, Kyogoku Y, Kobayashi Y, Moroder L. Folding motifs induced and stabilized by distinct cystine frameworks. Protein Eng 1998;11:649–59.

[31] Cornet B, Bonmatin JM, Hetru C, Hoffmann JA, Ptak M, Vovelle F. Refined three-dimensional solution structure of insect defensin A. Structure 1995;3:435–48.

[32] Wang X, Connor M, Smith R, Maciejewski MW, Howden ME, Nicholson GM, et al. Discovery and characterization of a family of insecticidal neurotoxins with a rare vicinal disulfide bridge. Nat Struct Biol 2000;7:505–13.

[33] Horita S, Matsushita N, Kawachi T, et al. Solution structure of a short-chain insecticidal toxin LaIT1 from the venom of scorpion *Liocheles australasiae*. Biochem Biophys Res Comm 2011;411(3):738–44.

[34] Correnti CE, Gewe MM, Mehlin C, et al. Screening, large-scale production and structure-based classification of cystine-dense peptides. Nat Struct Mol Biol 2018;25:270–8.

[35] Latham PW. Therapeutic peptides revisited. Nat Biotechnol 1999;17(8):755–7. Available from: https://doi.org/10.1038/11686.

[36] Craik DJ, Daly NL, Waine C. The cystine knot motif in toxins and implications for drug design. Toxicon 2001;39:43–60.

[37] Craik DJ, Simonsen S, Daly NL. The cyclotides: novel macrocyclic peptides as scaffolds in drug design. Curr Opin Drug Discov Devel 2002;5:251–60.

[38] Kintzing JR, Cochran JR. Engineered knottin peptides as diagnostics, therapeutics, and drug delivery vehicles. Curr Opin Chem Biol 2016;34:143–50.

[39] Christmann A, Walter K, Wentzel A, Kratzner R, Kolmar H. The cystine knot of a squash-type protease inhibitor as a structural scaffold for *Escherichia coli* cell surface display of conformationally constrained peptides. Protein Eng 1999;12:797–806.

[40] Ay J, Hilpert K, Krauss N, Schneider-Mergener J, Hohne W. Structure of a hybrid squash inhibitor in complex with porcine pancreatic elastase at 1.8-A resolution. Acta Crystallogr D Biol Crystallogr 2003;59:247–54.

[41] Heitz A, Le-Nguyen D, Dumas C, Chiche L. Engineering potential inhibitors of the interaction between the HIV-1 NEF protein and kinase SH3 domains. In: Martinez J, Fehrentz JA, editors. Peptides 2000. Paris, France: Editions EDK; 2000. p. 415–6.

[42] Baggio R, Burgstaller P, Hale SP, Putney AR, Lane M, Lipovsek D, et al. Identification of epitope-like consensus motifs using mRNA display. J Mol Recognit 2002;15:126–34.

[43] Wentzel A, Christmann A, Kratzner R, Kolmar H. Sequence requirements of the GPNG beta-turn of the *Ecballium elaterium* trypsin inhibitor II explored by combinatorial library screening. J Biol Chem 1999;274:21037–43.

[44] Daly NL, Love S, Alewood PF, Craik DJ. Chemical synthesis and folding pathways of large cyclic polypeptides: studies of the cystine knot polypeptide kalata B1. Biochemistry 1999;38:10606–14.

[45] Le Nguyen D, Barry LG, Tam JP, Heitz A, Chiche L, Hernandez JF, et al. Synthesis of MCoTI-I, a cyclic trypsin inhibitor from *Momordica cochinchinensis*. In: Benedetti E, Pedone C, editors. Peptides 2002. Napoli, Italy: Edizioni Ziino; 2002. p. 182–3.

[46] Tam J, Lu Y-A. Synthesis of large cyclic cystine-knot peptide by orthogonal coupling strategy using unprotected precursor. Tetrahedron Lett 1997;38:5599–602.

[47] Gran L, Sandberg F, Sletten K. Oldenlandia affinis (R&S) DC. A plant containing uteroactive peptides used in African traditional medicine. J Ethnopharmacol 2000;70:197–203.

[48] Cox N, Kintzing JR, Smith M, Grant GA, Cochran JR. Integrin-targeting knottin peptide-drug conjugates are potent inhibitors of tumor cell proliferation. Angew Chem Int Ed Engl 2016;55(34):9894–7.

[49] Huang YH, Henriques S, Wang C, et al. Design of substrate-based BCR-ABL kinase inhibitors using the cyclotide scaffold. Sci Rep 2015;5:12974.

[50] Aboye T, Meeks CJ, Majumder S, Shekhtman A, Rodgers K, Camarero JA. Design of a MCoTI-based cyclotide with angiotensin (1–7)-like activity. Molecules 2016;21:12974.

[51] Dardevet L, Rani D, Aziz TA, Bazin I, Sabatier JM, Fadl M, et al. Chlorotoxin: a helpful natural scorpion peptide to diagnose glioma and fight tumor invasion. Toxins (Basel) 2015;7:1079–101.

[52] Gründemann C, Stenberg KG, Gruber CW. T20K: an immunomodulatory cyclotide on its way to the clinic. Int J Pept Res Ther 2019;25:9–13.

[53] Wong CT, Rowlands DK, Wong CH, Lo TW, Nguyen GK, Li HY, et al. Orally active peptidic bradykinin B1 receptor antagonists engineered from a cyclotide scaffold for inflammatory pain treatment. Angew Chem Int Ed Engl 2021;51:5620–4.

[54] Oguis GK, Gilding EK, Huang YH, Poth AG, Jackson MA, Craik DJ. Insecticidal diversity of butterfly pea (*Clitoria ternatea*) accessions. Ind Crop Products 2020;147:112214.

[55] Schmidtko A, Lötsch J, Freynhagen R, Geisslinger G. Ziconotide for treatment of severe chronic pain. Lancet 2010;375(9725):1569–77.

[56] Dongol Y, Cardoso FC, Lewis RJ. Spider knottin pharmacology at voltage-gated sodium channels and their potential to modulate pain pathways. Toxins 2019;11(11):626.

[57] Bassotti G, Usai-Satta P, Bellini M. Linaclotide for the treatment of chronic constipation. Expert Opin Pharmacother 2018;19(11):1261–6.

[58] Leary S, Blatt JE, Cohen AR, et al. A phase II/III randomized, blinded study of tozuleristide for fluorescence imaging detection during neurosurgical resection of pediatric primary central nervous system (CNS) tumors: PNOC012 (Pacific Pediatric Neuro-oncology Consortium). J Clin Onc 2020;38(15_suppl):TPS2575.

Chapter 14

Venom peptides and peptidomimetics as therapeutics

Suhail Akhtar and Zulfiqar Ahmad
Department of Biochemistry, Kirksville College of Osteopathic Medicine, A.T. Still University, Kirksville, MO, United States

14.1 Introduction

Antimicrobial resistance (AMR) has become an existential threat to humanity, and our ability to treat microbial infections with antibiotics is challenged by new emerging microbial strains. As a result, even modest microbial infections are becoming untreatable with existing antibiotics. According to the World Health Organization (WHO) and the United States Centers for Disease Control and Prevention (CDC) data, the future forecast regarding AMR is dire with predictions of large numbers of deaths and huge economic impacts by 2050 [1–6]. To provide affordable and effective healthcare to the citizens of the world, it is of paramount importance to identify and use additional and alternative ways of combating microbial infections. In this chapter, we will focus on the importance of peptidomimetics and peptides for combating AMR and other disease conditions.

Peptidomimetics are compounds that mimic a natural peptide or protein. They not only interact with the biological target of a natural peptide or protein but also produce the same biological effect [7,8]. Peptidomimetics can be generated from alterations of an existing peptide, or they can be designed to mimic peptides. A good example of a designed peptidomimetic is peptoid, or poly-*N*-substituted glycine [9]. The side chains of a peptoid are attached to the nitrogen atom of the peptide backbone instead of the α-carbons (which is usually the case with amino acids). The R groups in amino acids of native peptides are represented by 20 different substitutions. In the case of peptoids, however, the possibilities of different R groups are potentially limitless [9,10]. Whatever the approach, the ultimate goal of generating a peptidomimetic is to modify the molecular properties of a peptide to our advantage to get a transformed chemical structure that is highly stable and has better biological activity. These advantageous modifications in peptide structure may

Peptide and Peptidomimetic Therapeutics. DOI: https://doi.org/10.1016/B978-0-12-820141-1.00018-2

have a very important role in the development of drugs and other therapeutics.

Traditionally, peptidomimetics have been categorized into three classes based on their structural and functional characteristics: type I mimetics, or structural mimetics; type II mimetics, or functional mimetics; and type III mimetics, or functional-structural mimetics [11,12]. However, in recent years, peptidomimetics have been grouped into four different classes: Class A, B, C, and D [12]. This classification of peptidomimetics is based on the similarity of peptidomimetics with their precursor peptide. Class A mimetics are peptides with minimal modifications to the peptide side chains and backbone [12]. Class B mimetics are modified forms of class A mimetics. While they are still peptide in nature, class B mimetics have nonnatural amino acids with backbone and side chain modifications [12]. Peptidomimetics in class C have highly modified structures that completely replace the entire peptide backbone with a small molecule character [12]. Class D mimetics have the least similarity with their precursor peptide. They are small molecule drugs that mimic the mode of action of a bioactive peptide without a direct link to its side chains [12]. Some class D mimetics are in clinical trials, and some are available in the market.

14.2 Peptide categories

14.2.1 Natural

To date, there are an estimated 7000 peptides/peptidomimetics identified from natural resources. These naturally occurring peptides include hormones, neurotransmitters, growth factors, and antiinflammatory agents [13]. Many of these peptides have been successfully extracted from the skin of toads and frogs [14]. Venom peptides from spiders, snakes, snails, and other predatory animals are being used to treat neurological or cardiovascular diseases or to alleviate pain [15]. Exenatide is one such example of a natural peptide that was originally discovered in the saliva of a venomous lizard native to the southwestern United States and northwestern Mexican state of Sonora. This peptide, structurally similar to a glucagon-like peptide, is used in the treatment of type 2 diabetes mellitus. A synthetic version of the peptide is commercially available and used as a drug for the treatment of type 2 diabetes [16].

Although biological diversity in the oceans has decreased dramatically since industrialization began in the 19th century, marine sources still represent the majority of global biodiversity. Many marine invertebrates, for example, cnidarian, tunicates, mollusks, or sponges, serve as hosts to different symbiotic microorganisms. These associated microorganisms metabolize various biologically active compounds that can be used for the discovery of potentially novel peptides [17]. Peptides isolated from marine sources show high potential and therapeutic efficacy against several infectious agents. Further, they are being utilized by the pharmaceutical industry in the development of new drugs and for the treatment of several infections [18].

A large number of antimicrobial peptides (AMPs) are produced by different marine organisms [19]. Also known as host defense peptides, AMPs are short less than 50 amino acids which are ubiquitously present in nature [20]. More than 2700 naturally occurring or synthetic AMPs have been identified so far [21,22]. These AMPs have activity not only against Gram-positive and Gram-negative bacteria but also against fungi and enveloped viruses [23,24]. Most of these peptides are positively charged (cationic) α-helical molecules that have an important role in host defense. These cationic AMPs bind and electrostatically interact with negatively charged (anionic) bacterial cytoplasmic membranes, subsequently damaging the membrane and leading to bacterial cell death [20] (Fig. 14.1). The binding of cationic AMPs to a bacterial cytoplasmic membrane is preferred over the zwitterionic membrane of mammalian cells [25]. AMPs have also been shown to pass the bacterial membrane and target intracellular DNA, RNA, and protein synthesis [26]. When the AMP reaches the lipid membrane of the bacterial cytoplasmic membrane, it acquires amphipathic conformation, which helps the AMP integrate into or move across the membrane [27].

The rapid and continued emergence of resistant bacteria worldwide threatens the efficacy of antibiotics. The overuse and misuse of antibiotics have led to this current crisis. In 2013, the CDC declared that the human race is now in the postantibiotic era [28], and The WHO warned that AMR is becoming dire [28]. AMPs hold great potential for combating bacterial

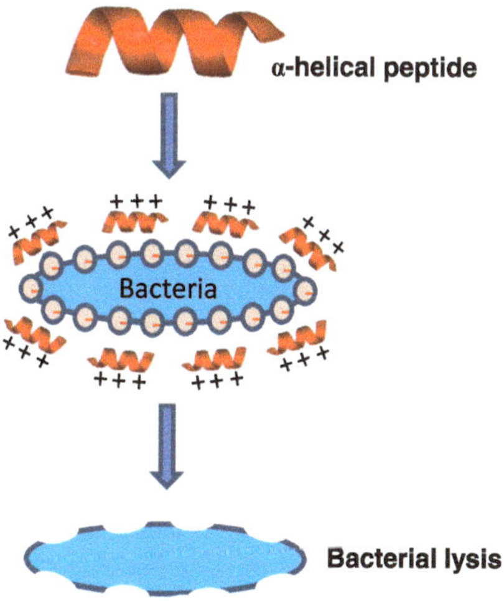

FIGURE 14.1 Positively charged α-helical antimicrobial peptide binding to the bacterial membrane through electrostatic interactions, leading to disruption of the bacterial membrane.

resistance and are promising next-generation antibiotics. However, AMPs have some intrinsic drawbacks, such as susceptibility to enzymatic degradation and a short half-life [20]. Thus, for clinical applications, it is necessary and important to develop long-lasting AMP analogs that can overcome the disadvantages of their natural peptides.

14.2.2 Synthetic

Synthetic peptides are exceptional protein mimics. Synthetic peptides can be generated with diverse chemical modifications, including the introduction of nonnatural amino acids or modification of the peptide backbone. These diverse structural and chemical modifications of synthetic peptides increase their stability against proteolysis and enhance their use in biological research and applications. Synthetic peptides can also be used as drugs to treat different diseases, such as bacterial and viral infections, autoimmune disorders, and cancer [29–31].

The use of small molecules as inhibitors of protein–protein interactions is usually less effective because of the involvement of larger interfaces in these interactions. Further, protein-based drugs are not very successful because of the immunogenic nature of proteins, which results in their immunological clearance before they even reach their target site. Many diseases involve specific protein–protein interactions. The use of synthetic peptides, which mimic the inhibitor of such protein–protein interactions, is a good alternative to both small molecules and protein-based drugs. These peptides not only have high specific binding but also have low immunogenicity [13].

Synthetic peptides can also be used as mimics of various cellular receptors. Viruses interact with cellular receptors to enter cells. Further, these receptors have an important role in different signal transduction pathways. Synthetic peptides mimicking cellular receptors can be used to develop new strategies to block the entry of the virus into cells, such as for HIV [32,33]. Novel synthetic peptides are also being developed for the treatment of Coronavirus disease 2019 (COVID-19). Researchers worldwide are searching for peptides that can either block severe acute respiratory syndrome-Coronavirus 2 (SARS-CoV-2) entry and replication into the cell or treat conditions once the cell is invaded by the virus. The spike protein of the SARS virus uses the angiotensin-converting enzyme 2 (ACE2) receptor of the host cell to enter cells [34]. Peptides that block the interactions of spike and ACE2 proteins offer a potential strategy to restrict viral entry into cells. There are 21 peptide drugs currently under development for the treatment of COVID-19. Among these, five peptides are being used in different phases of clinical trials, specifically, four in phase II and one in phase I (*Source*: global data healthcare).

Because cancer is a leading cause of death worldwide, the main focus of synthetic peptide development has been the treatment of cancer. These peptides have been developed to work directly as anticancer drugs or to redirect toxic anticancer drugs away from normal cells and toward cancerous cells [29,35,36]. As such, cancer immunotherapy is another major area where peptides may have a key role in future. Peptides can be used as vaccines presenting tumor-associated antigens, thereby eliciting an immune response against cancer cells in the patient [37]. One of the challenges of the delivery of anticancer drugs is the discrimination between self and nonself. This results in the clearance of the cancer drug by the body before the drug reaches its target cell. The membrane protein CD47 is an established marker of self that prevents clearance by phagocytes. In mice, it increases the circulation time of nanoparticles, which may provide a new opportunity for enhanced delivery of drugs [38].

A similar idea was used for the delivery of the anticancer drug paclitaxel, which was loaded into nanoparticles. These nanoparticles expressed the self-peptide originating from CD47 on their surface, preventing phagocytosis of nanoparticles in mice and thus prolonging their circulation. Because of the delayed clearance, treatment of mice with lung adenocarcinoma epithelial tumors with self-peptide expressing nanobeads resulted in a more efficient reduction of tumor size when compared with mice treated with beads without the peptide. Even though it is not a bioactive compound, this peptide provides an excellent strategy for drug delivery to cancerous cells [38].

14.3 ATP synthase as a potent molecular drug target

ATP synthase is the main source of energy in almost all organisms from bacteria to humans. As the energy currency of cells, ATP (adenosine triphosphate) is generated by combining nucleotide ADP and Pi at catalytic sites of ATP synthase. The structural and functional integrity of ATP synthase is essential in health and disease conditions. Malfunction of this enzyme complex has been implicated in a wide variety of diseases, including cancer, tuberculosis, neuropathy, Alzheimer's, Parkinson's, and a group of devastating diseases known as mitochondrial myopathies. ATP synthase has also been used and suggested as a potent molecular drug target for the treatment of such diseases as tuberculosis, AMR infections, cancer, heart disease, mitochondrial diseases, immune deficiency, cystic fibrosis, diabetes, ulcers, Parkinson's, and Alzheimer's [39–42].

14.4 Peptides as selective inhibitors of ATP synthase

A variety of plant- and animal-based products, including many natural and modified peptides, bind and inhibit ATP synthase [3,39,43–45]. As shown in Fig. 14.2, ATP synthase has a peptide binding site known as the βDELSEED-motif. This peptide binding site is formed by the β-subunit residues 380–386 [3,41,46,47]. Cationic peptides are known to interact with the

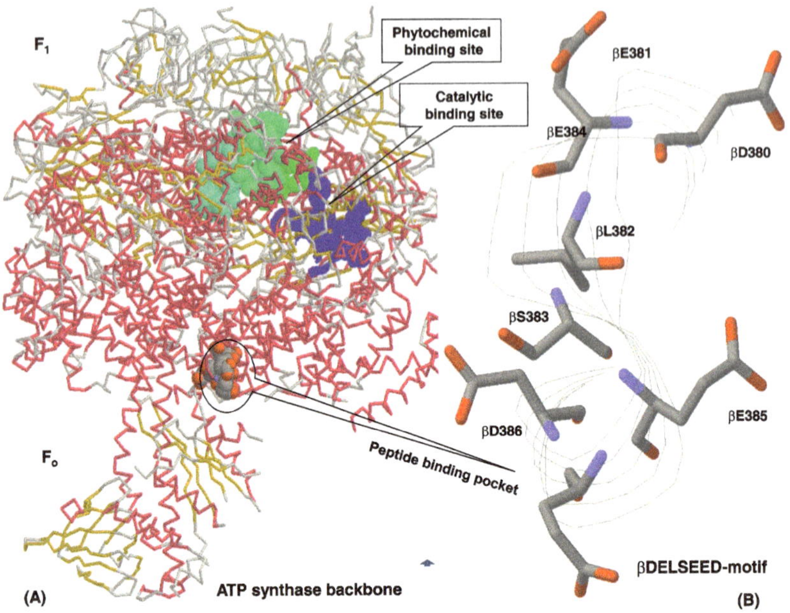

FIGURE 14.2 Structure of *Escherichia coli* ATP synthase and peptide binding site βDELSEED: (A) *E. coli* ATP synthase F1 sector in backbone form showing phytochemical, catalytic, and peptide binding sites in space fill form. (B) Enlarged peptide binding site in wireframe form identifying the involved residues. *Figure adapted from Ahmad Z, Okafor F, Azim S, Laughlin TF. Curr Med Chem 2013;20:1956–73, with permission.*

negatively charged residues of the βDELSEED-motif, leading to potent inhibition of ATP synthase. Of the identified AMPs to date, about 84% possess antibacterial activity, about 36% possess antifungal activity, about 3% possess antiparasitic activity, about 7% possess anticancer activity, and about 6% possess antiviral activity [48].

The inhibitory potency of peptides varies by the number and charge of amino acids and by the complexity of the α-helix (Fig. 14.3). The helical wheel plots of a selection of venom peptides, generated with HeliQuest (http://heliquest.ipmc.cnrs.fr/), are shown in Fig. 14.4. The downward arrow (↓) indicates the level of hydrophobicity, which makes these peptides useful as ATP synthase inhibitors. The size of the one-letter code circles is proportional to the size of the amino acid side chain. The hydrophobic and hydrophilic faces of the helical wheel plots are in yellow and blue colors, respectively.

Several peptides—melittin, melittin-related peptide, ascaphin, aurein, carein, citropin, dermaseptin, maculatin, maganin, XT-7, cathelicidin, lycotoxin, anoplin, cupiennin, latarcin, and pandinin—cause variable inhibition of ATP synthase [3,41,45,50]. Potent and selective inhibition of target molecules can be achieved by structural modification of peptides. As shown in Fig. 14.5,

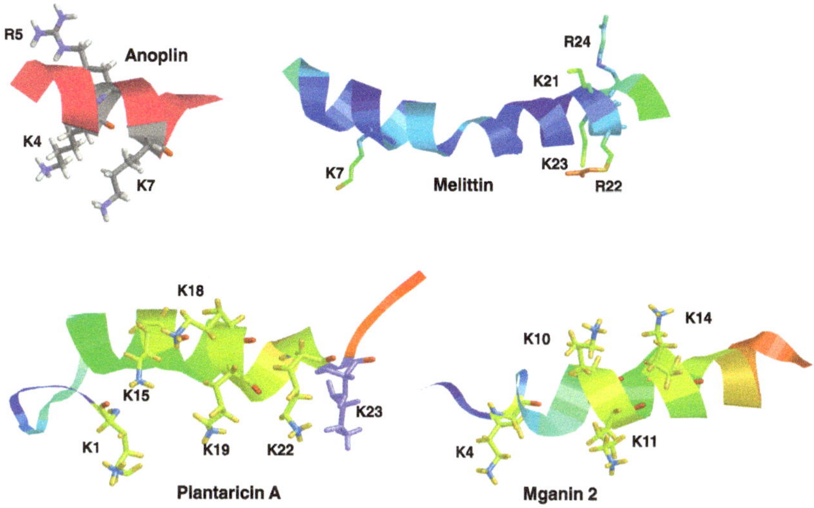

FIGURE 14.3 X-ray crystallographic structures of select peptides showing the positively charged residues expected to interact with the negatively charged βDELSEED-motif residues of ATP synthase. *Figure adapted from Ahmad Z, Okafor F, Azim S, Laughlin TF. Curr Med Chem 2013;20: 1956–1973, with permission.*

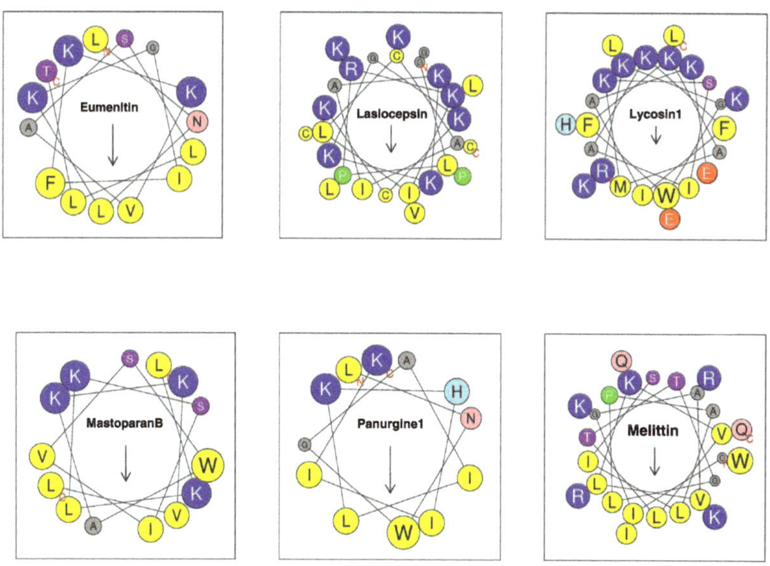

FIGURE 14.4 Helical wheel plots of selected venom. The length of downward arrow ↓ indicates the level of hydrophobicity. The size of the one-letter code circles is proportional to the size of the amino acid side chain. The hydrophobic and hydrophilic faces of the helical wheel plots are in yellow and blue colors, respectively. For color version see the reference [49]. *Figure adapted from Syed H, Tauseef M, Ahmad Z. Int J Biol Macromol 2018; 119:23–31; Amini A, Raheem S, Steiner A, Deeba F, Ahmad Z. Int J Biol Macromol 2020;150:23–30; Steiner A, Raheem S, Ahmad Z. Int J Biol Macromol 2020;165:2588–97, with permission.*

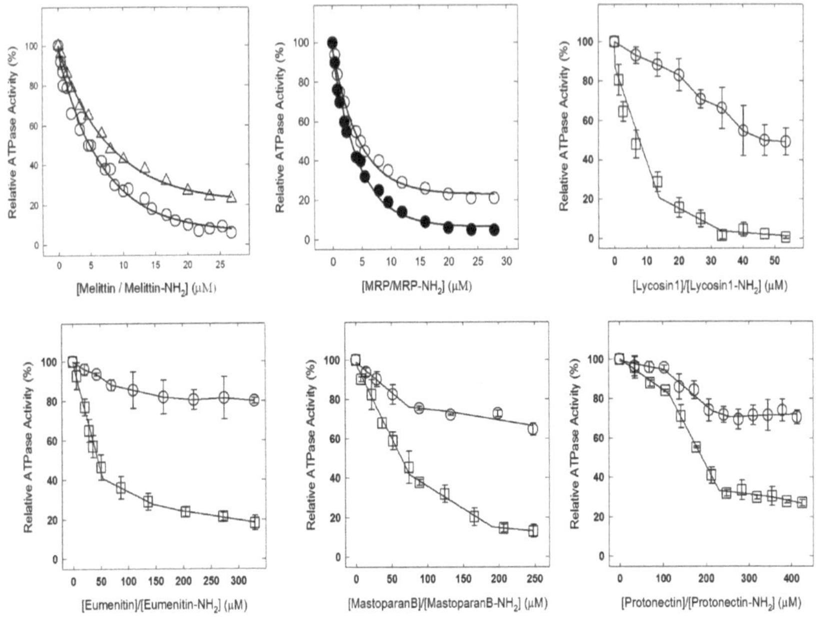

FIGURE 14.5 Inhibition of membrane-bound F_1F_o ATP synthase by melittin, melittin-related peptide, lycosin, eumenitin, mastoparanB, protonectin, and their C-terminal -NH_2 variants. *Figure adapted from Syed H, Tauseef M, Ahmad Z. Int J Biol Macromol 2018;119: 23–31, with permission.*

addition of the C-terminal -NH_2 group to peptides causes an augmented inhibition of ATP synthase [3,41,42].

Peptides also inhibit the ATP synthase in *Streptococcus mutans*. *S. mutans* is a primary microbial agent in the pathogenesis of dental caries through biofilm formation and acid production. Therefore, inhibition of ATP synthase of *S. mutans* inhibits biofilm formation and acid production [51,52]. In *Mycobacterium* ATP synthase, two mutations in the c-subunit (D32V and A63P) have been shown to confer resistance in the new tuberculosis drug diarylquinoline [53,54]. Thus, a better understanding of ATP synthase inhibition may aid in the treatment of these and other diseases.

14.5 Conclusion

Peptidomimetics represent a promising field that has opened unlimited opportunities in the area of peptide-based therapeutics. Peptides and peptidomimetics work on a variety of molecular targets. ATP synthase is proving to be a valuable molecular drug target in search for new molecular targets. Initially isolated from animal venom, many peptides have tremendous therapeutic potential in the pharmaceutical sciences. New peptide drugs are being used to treat different diseases, and many more are being used in clinical trials. Continuous advancement in peptide research has led to the development of peptides that are more stable with increased half-life and enhanced pharmaceutical properties. Modern experimental approaches such as genomics, metabolomics, proteomics, and transcriptomics have greatly helped in discovering peptides with distinct structural properties. These unique and advanced peptides have the potential to play an important role in combating future disease outbreaks, such as the COVID-19 pandemic.

References

[1] O'Neill J. Antimicrobial Resistance: tackling a crisis for the health and wealth of nations. Review on antimicrobial resistance, 2016, pp. 1–20.

[2] World Health Organization, 2017, pp. 1–232.

[3] Syed H, Tauseef M, Ahmad Z. A connection between antimicrobial properties of venom peptides and microbial ATP synthase. Int J Biol Macromol 2018;119:23–31.

[4] Shen W, He P, Xiao C, Chen X. From antimicrobial peptides to antimicrobial poly (α-amino acid)s. Adv Healthc Mater 2018;7:e1800354.

[5] Center for Disease Control and Prevention (USA), 2021.

[6] Prestinaci F, Pezzotti P, Pantosti A. Antimicrobial resistance: a global multifaceted phenomenon. Pathog Glob Health 2015;109:309–18.

[7] Kumar MS. Peptides and peptidomimetics as potential antiobesity agents: overview of current status. Front Nutr 2019;6:11.

[8] Vagner J, Qu H, Hruby VJ. Peptidomimetics, a synthetic tool of drug discovery. Curr Opin Chem Biol 2008;12:292–6.

[9] Murphy JE, Uno T, Hamer JD, Cohen FE, Dwarki V, Zuckermann RN. A combinatorial approach to the discovery of efficient cationic peptoid reagents for gene delivery. Proc Natl Acad Sci USA 1998;95:1517–22.

[10] Schneider JA, Craven TW, Kasper AC, Yun C, Haugbro M, Briggs EM, et al. Design of peptoid-peptide macrocycles to Inhibit the β-catenin TCF interaction in prostate cancer. Nat Commun 2018;9:4396.

[11] Marshall GR, Ballante F. Limiting assumptions in the design of peptidomimetics. Drug Dev Res 2017;78:245–67.

[12] Pelay-Gimeno M, Glas A, Koch O, Grossmann TN. Structure-based design of inhibitors of protein-protein interactions: mimicking peptide binding epitopes. Angew Chem Int Ed Engl 2015;54:8896–927.

[13] Fosgerau K, Hoffmann T. Peptide therapeutics: current status and future directions. Drug Discov Today 2015;20:122–8.

[14] Li J, Xu X, Xu C, Zhou W, Zhang K, Yu H, et al. Anti-infection peptidomics of amphibian skin. Mol Cell Proteom 2007;6:882–94.

[15] Uhlig T, Kyprianou T, Martinelli FG, Oppici CA, Heiligers D, Hills D, et al. The emergence of peptides in the pharmaceutical business: from exploration to exploitation. EuPA OPEN Proteom 2014;4:58–69.

[16] Bond A. Exenatide (Byetta) as a novel treatment option for type 2 diabetes mellitus. Proc (Bayl Univ Med Cent) 2006;19:281–4.

[17] Liu J, Jung JH, Liu Y. Antimicrobial compounds from marine invertebrates-derived microorganisms. Curr Med Chem 2016;23:2892–905.

[18] Anjum K, Abbas SQ, Akhter N, Shagufta BI, Shah SAA, Hassan SSU. Emerging biopharmaceuticals from bioactive peptides derived from marine organisms. Chem Biol Drug Des 2017;90:12–30.

[19] Sable R, Parajuli P, Jois S. Peptides, peptidomimetics, and polypeptides from marine sources: a wealth of natural sources for pharmaceutical applications. Mar Drugs 2017;15.

[20] Lei J, Sun L, Huang S, Zhu C, Li P, He J, et al. The antimicrobial peptides and their potential clinical applications. Am J Transl Res 2019;11:3919–31.

[21] Li J, Koh JJ, Liu S, Lakshminarayanan R, Verma CS, Beuerman RW. Membrane active antimicrobial peptides: translating mechanistic insights to design. Front Neurosci 2017;11:73.

[22] Bormann N, Koliszak A, Kasper S, Schoen L, Hilpert K, Volkmer R, et al. A short artificial antimicrobial peptide shows potential to prevent or treat bone infections. Sci Rep 2017;7:1506.

[23] Wimley WC, Hristova K. Antimicrobial peptides: successes, challenges and unanswered questions. J Membr Biol 2011;239:27–34.

[24] Reddy KV, Yedery RD, Aranha C. Antimicrobial peptides: premises and promises. Int J Antimicrob Agents 2004;24:536–47.

[25] Nguyen LT, Haney EF, Vogel HJ. The expanding scope of antimicrobial peptide structures and their modes of action. Trends Biotechnol 2011;29:464–72.

[26] Xiong YQ, Bayer AS, Yeaman MR. Inhibition of intracellular macromolecular synthesis in *Staphylococcus aureus* by thrombin-induced platelet microbicidal proteins. J Infect Dis 2002;185:348–56.

[27] Teixeira V, Feio MJ, Bastos M. Role of lipids in the interaction of antimicrobial peptides with membranes. Prog Lipid Res 2012;51:149–77.

[28] Michael CA, Dominey-Howes D, Labbate M. The antimicrobial resistance crisis: causes, consequences, and management. Front Public Health 2014;2:145.

[29] Kaspar AA, Reichert JM. Future directions for peptide therapeutics development. Drug Discov Today 2013;18:807–17.

[30] Bunschoten A, Ippel JH, Kruijtzer JA, Feitsma L, de Haas CJ, Liskamp RM, et al. A peptide mimic of the chemotaxis inhibitory protein of *Staphylococcus aureus*: towards the development of novel anti-inflammatory compounds. Amino Acids 2011;40:731–40.

[31] Shankaramma SC, Athanassiou Z, Zerbe O, Moehle K, Mouton C, Bernardini F, et al. Macrocyclic hairpin mimetics of the cationic antimicrobial peptide protegrin I: a new family of broad-spectrum antibiotics. Chembiochem 2002;3:1126–33.

[32] Wu B, Chien EY, Mol CD, Fenalti G, Liu W, Katritch V, et al. Structures of the CXCR4 chemokine GPCR with small-molecule and cyclic peptide antagonists. Science 2010;330:1066–71.

[33] Drakopoulou E, Vizzavona J, Vita C. Engineering a CD4 mimetic inhibiting the binding of the human immunodeficiency virus-1 (HIV-1) envelope glycoprotein gp120 to human lymphocyte CD4 by the transfer of a CD4 functional site to a small natural scaffold. Claudio Vita Lett Peptide Sci 1998;5:241–5.

[34] Yang J, Petitjean SJL, Koehler M, Zhang Q, Dumitru AC, Chen W, et al. Molecular interaction and inhibition of SARS-CoV-2 binding to the ACE2 receptor. Nat Commun 2020;11:4541.

[35] Stiuso P, Caraglia M, De Rosa G, Giordano A. Bioactive peptides in cancer: therapeutic use and delivery strategies. J Amino Acids 2013;2013:568953.

[36] Qvit N, Rubin SJS. Peptide therapeutics: scientific approaches, current development trends, and future directions. Curr Top Med Chem 2020;20:2903.

[37] Miller MJ, Foy KC, Kaumaya PT. Cancer immunotherapy: present status, future perspective, and a new paradigm of peptide immunotherapeutics. Discov Med 2013;15:166–76.

[38] Rodriguez PL, Harada T, Christian DA, Pantano DA, Tsai RK, Discher DE. Minimal "Self" peptides that inhibit phagocytic clearance and enhance delivery of nanoparticles. Science 2013;339:971–5.

[39] Hong S, Pedersen PL. ATP synthase and the actions of inhibitors utilized to study its roles in human health, disease, and other scientific areas. Microbiol Mol Biol Rev 2008;72:590–641.

[40] Ahmad Z, Laughlin TF. Medicinal chemistry of ATP synthase: a potential drug target of dietary polyphenols and amphibian antimicrobial peptides. Curr Med Chem 2010;17:2822–36.

[41] Laughlin TF, Ahmad Z. Inhibition of *Escherichia coli* ATP synthase by amphibian antimicrobial peptides. Int J Biol Macromol 2010;46:367–74.

[42] Ahmad Z, Okafor F, Azim S, Laughlin TF. ATP synthase: a molecular therapeutic drug target for antimicrobial and antitumor peptides. Curr Med Chem 2013;20:1956–73.

[43] Weber J, Senior AE. Effects of the inhibitors azide, dicyclohexylcarbodiimide, and aurovertin on nucleotide binding to the three F1-ATPase catalytic sites measured using specific tryptophan probes. J Biol Chem 1998;273:33210–15.

[44] Zheng J, Ramirez VD. Inhibition of mitochondrial proton F0F1-ATPase/ATP synthase by polyphenolic phytochemicals. Br J Pharmacol 2000;130:1115–23.

[45] Amini A, Raheem S, Steiner A, Deeba F, Ahmad Z. Insect venom peptides as potent inhibitors of *Escherichia coli* ATP synthase. Int J Biol Macromol 2020;150:23–30.

[46] Ahmad Z, Tayou J, Laughlin TF. Asp residues of βDELSEED-motif are required for peptide binding in the *Escherichia coli* ATP synthase. Int J Biol Macromol 2015;75:37–43.

[47] Azim S, Ahmad Z. Glu residues of βDELSEED-motif are essential for peptide binding in *Escherichia coli* ATP synthase. Int J Biol Macromol 2018;116:977–82.

[48] Wang G, Li X, Wang Z. APD3: the antimicrobial peptide database as a tool for research and education. Nucleic Acids Res 2016;44:D1087–93.

[49] Steiner A, Raheem S, Ahmad Z. Significance of Leu and Ser in the βDELSEED-loop of *Escherichia coli* ATP synthase. Int J Biol Macromol 2020;165:2588–97.

[50] Azim S, McDowell D, Cartagena A, Rodriguez R, Laughlin TF, Ahmad Z. Venom peptides cathelicidin and lycotoxin cause strong inhibition of *Escherichia coli* ATP synthase. Int J Biol Macromol 2016;87:246–51.

[51] Percival RS, Devine DA, Duggal MS, Chartron S, Marsh PD. The effect of cocoa polyphenols on the growth, metabolism, and biofilm formation by *Streptococcus mutans* and *Streptococcus sanguinis*. Eur J Oral Sci 2006;114:343–8.

[52] Duarte S, Gregoire S, Singh AP, Vorsa N, Schaich K, Bowen WH, et al. Inhibitory effects of cranberry polyphenols on formation and acidogenicity of Streptococcus mutans biofilms. FEMS Microbiol Lett 2006;257:50–6.

[53] Andries K, Verhasselt P, Guillemont J, Gohlmann HW, Neefs JM, Winkler H, et al. A diarylquinoline drug active on the ATP synthase of Mycobacterium tuberculosis. Science 2005;307:223–7.

[54] Cole ST, Alzari PM. Microbiology. TB—a new target, a new drug. Science 2005;307:214–15.

Further reading

Gewirtz DA, Bristol ML, Yalowich JC. Curr Opin Investig Drugs 2010;11:612–14.

Majeed H, Gupta V. Adverse Effects of Radiation Therapy. Treasure Island (FL): StatPearls; 2021.

Alban TJ, Chan TA. Nat Rev Clin Oncol 2021;.

Garcia-Fernandez C, Saz A, Fornaguera C, Borros S. Cancer Gene Ther 2021;.

Flower DR. Nat Chem Biol 2013;9:749–53.

Kaumaya PT, Foy KC. Future Oncol 2012;8:961–87.

Timur SS, Gursoy RN. J Drug Target 2021;1–15.

Serafino A, Moroni N, Psaila R, Zonfrillo M, Andreola F, Wannenes F, et al. Biochim Biophys Acta 2012;1822:1004–18.

Takahashi H, Okamoto M, Shimodaira S, Tsujitani S, Nagaya M, Ishidao T, et al. D.C.-v.s.g.a.t. J.S.o.I.C. Therapy Eur J Cancer 2013;49:852–9.

Rothenberg ML, Moore MJ, Cripps MC, Andersen JS, Portenoy RK, Burris 3rd HA, et al. Ann Oncol 1996;7:347–53.

Nishida S, Koido S, Takeda Y, Homma S, Komita H, Takahara A, et al. J Immunother 2014;37:105–14.

Masuzawa T, Fujiwara Y, Okada K, Nakamura A, Takiguchi S, Nakajima K, et al. Int J Oncol 2012;41:1297–304.

Ishikawa H, Imano M, Shiraishi O, Yasuda A, Peng YF, Shinkai M, et al. Gastric Cancer 2014;17:173–80.

Dhillon S. Drugs 2014;74:675–86.

Amiri-Kordestani L, Blumenthal GM, Xu QC, Zhang L, Tang SW, Ha L, et al. Clin Cancer Res 2014;20:4436–41.

Gaffo A, Saag KG, Curtis JR. Am J Health Syst Pharm 2006;63:2451–65.

Bossaller L, Rothe A. Expert Opin Biol Ther 2013;13:1257–72.

Gokhale A, Weldeghiorghis TK, Taneja V, Satyanarayanajois SD. J Med Chem 2011;54:5307–19.

Yoshihara Y, Nakamura H, Obata K, Yamada H, Hayakawa T, Fujikawa K, et al. Ann Rheum Dis 2000;59:455–61.

Nam EJ, Kang JH, Sung S, Sa KH, Kim KH, Seo JS, et al. Arthritis Rheum 2013;65:1753–63.

Waters ST, McDuffie M, Bagavant H, Deshmukh US, Gaskin F, Jiang C, et al. J Exp Med 2004;199:255–64.

Bloom O, Cheng KF, He M, Papatheodorou A, Volpe BT, Diamond B, et al. Proc Natl Acad Sci U S A 2011;108:10255–9.

Koutsokeras T, Healy T. Nat Rev Drug Discov 2014;13:173–4.

Monneaux F, Muller S. Adv Exp Med Biol 2007;601:105–12.

Page N, Gros F, Schall N, Decossas M, Bagnard D, Briand JP, et al. Ann Rheum Dis 2011;70:837–43.

Muller S, Monneaux F, Schall N, Rashkov RK, Oparanov BA, Wiesel P, et al. Arthritis Rheum 2008;58:3873–83.

McFarland HF, Martin R. Nat Immunol 2007;8:913–19.

Badawi AH, Siahaan TJ. Clin Immunol 2012;144:127–38.

Srinivasan M, Janardhanam S. J Biol Chem 2011;286:44799–810.

Calenoff E. ISRN Neurol 2012;2012:851541.

Chapter 15

Therapeutic peptides targeting protein kinase: progress, challenges, and future directions, featuring cancer and cardiovascular disease

Nir Qvit
The Azrieli Faculty of Medicine in the Galilee, Bar-Ilan University, Safed, Israel

15.1 Introduction

Kinases derived their name from the Greek word kinein meaning "to move" and are a major class of regulatory enzymes that catalyze phosphorylation. This entails the transfer of a gamma phosphate group from the molecule adenosine triphosphate (ATP) onto hydroxyl groups of specific amino acids in a recipient molecule. In eukaryotes, these are usually serine, threonine and tyrosine residues that act as a substrate (Fig. 15.1). Histidine phosphorylation is common in prokaryotes and occurs to some extent (6% of total phosphorylation) in eukaryotes. However, due to its rapid hydrolysis histidine phosphorylation is less observed [1]. Protein kinases exist in balance with phosphatases, which remove a phosphate group from amino acids (catalyze the transfer of the phosphate from a phosphoprotein to a water molecule) and modulate numerous activities of proteins in a cell by regulating protein function, mainly turning them "on" and "off." The functions of protein

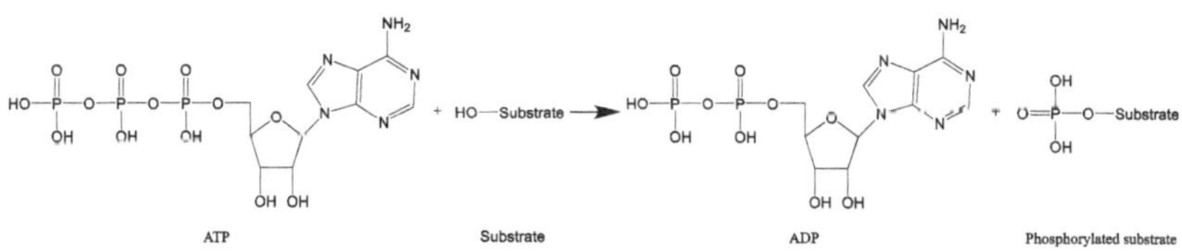

FIGURE 15.1 General phosphorylation reaction catalyzed by kinases.

Peptide and Peptidomimetic Therapeutics. DOI: https://doi.org/10.1016/B978-0-12-820141-1.00005-4

kinases and phosphatases are mediated by their target substrates, which are diverse and include lipids, carbohydrates, nucleotides and proteins. However, herein we will focus on the mechanism and function of protein kinase phosphorylation of substrate proteins on serine or threonine and tyrosine residues. Protein phosphorylation is the most widespread type of posttranslational modification used in signal transduction, and it is also one of the most extensively and diligently studied. While phosphorylation is essential for normal cellular processes, abnormal phosphorylation has been inseparably linked with various human diseases.

The field of phosphorylation began with the discovery of phosphorylated vitellin in 1906 [2]. Subsequently, in the 1940s, the active and inactive forms of the phosphorylase enzymes that transfer inorganic phosphate to acceptor molecules and are involved in the process of glycogen metabolism were identified by Gerty and Carl Cori [3]. Cori's research was awarded a Nobel Prize in Physiology or Medicine in 1947 and provided the foundation for understanding the process of reversible phosphorylation. In 1954, the process of enzymatic phosphorylation was demonstrated for the first time by transfer of a phosphate group from ATP to protein in rat liver mitochondria catalyzed by protein phosphokinase [4], and in 1955, the Nobel Prize was awarded to Fischer and Krebs for revealing the importance of this reversible process as a major biological regulatory mechanism [5].

Protein kinases represent one of the largest families of genes in eukaryotes, which accounts for nearly 2% of the human genome and includes approximately 538 kinases [6,7]. The human kinome has been classified into two major classes, the serine/threonine kinases (about 388 protein kinases) and the tyrosine protein kinases (about 90 protein kinases), which share extensive sequence and structure homologies. However, several protein kinases share the consensus kinases motifs of both serine/threonine and tyrosine kinases and can phosphorylate serine/threonine as well as tyrosine residues [8]. In addition, there are atypical protein kinases that have biochemical kinase activity but lack sequence similarity to the conventional eukaryotic kinases (about 40 protein kinases) [9]. The eukaryotic protein kinases are further subdivided into eight major kinase groups, namely, the CaMK (calcium/calmodulin-dependent kinases) group, the AGC [protein kinase A (PKA), protein kinase G (PKG), and protein kinase C (PKC)] group, the collectively CMGC [cyclin-dependent kinases (CDK)], mitogen-activated protein kinases, glycogen synthase kinase (GSK) and CDC-like kinases (CLK) group, the TK (tyrosine kinases) group, the STE (homologs of yeast sterile 7) group, the CK1 (casein kinases) group, the TKL (tyrosine kinase-like) group and the RCG (receptor guanylatecyclases) group [10]. In addition, there is a group of pseudokinases, kinases that are considered catalytically inactive (about 10% of all human kinases). Despite their lack of catalytic activity pseudokinases have essential signaling functions by acting as scaffolding proteins and allosteric regulators of catalytically active kinases [11].

Protein kinases are highly regulated molecular switches that can adopt at least two extreme conformations: an "on" state that is maximally active, and an "off" state that has minimal activity. They are turned "on" and "off" by various biological signals, such as hormone, neurotransmitter, nutrient deprivation, or other triggers either initiated by an intra- or extra-cellular signal and affects almost all basic cellular processes, including metabolism, growth, division, differentiation, motility, organelle trafficking, membrane transport, muscle contraction, immunity, learning and memory [10,12]. Many proteins may be phosphorylated by a single protein kinase and multiple protein kinases may phosphorylate a single protein, thus creating a highly complex though synchronized cascade of cell programming in response to a stimulus. It has been estimated that 30% of all cellular proteins are regulated through some kinase activity [10,13]. Phosphorylation of protein substrates may occur at a single site or multiple sites and can have profound effects, such as enzyme activation, enzyme inhibition, cellular location, the creation of recognition sites for recruitment of other proteins, and transitions in protein state from order to disorder or disorder to order. Phosphorylation regulates protein functions by inducing conformational changes or by disruption and creation of protein-protein interaction surfaces [14,15].

Studies on protein kinases structure and conformational changes have provided insights into their complex mechanisms that regulate their activation and into the exquisite substrate phosphorylation site specificity shown by them, which ensures fidelity in their function. The first protein kinase structure was solved in 1991 of the PKA catalytic subunit [16], and since then the number of kinase structures in the protein data bank has risen rapidly. In general, protein kinases share a conserved kinase domain, but there is considerable diversity among kinases within the additional flanking domains and sequences that may regulate processes such as catalytic function, substrate phosphorylation, or serve as regulatory domains. The protein kinase domain is composed of two structurally dissimilar lobes, N-lobe and a C-lobe, joined by a peptide coil often referred to as the "linker," and the adenine ring of ATP is buried at the base of the cleft between the two lobes. Protein kinases can adopt different structures between the active ("on") and inactive ("off") states, in response to specific signaling events, which are transduced via kinase-associated regulatory domains in *cis* or *trans*, and/or by phosphorylation. Phosphorylation imparts conformational changes in the protein thereby modifying its affinity for another protein, activates or deactivates an enzyme protein, or facilitates its degradation by ubiquitin-proteasome complex. Despite the well-conserved organization of the ATP-binding and catalytic site, kinases have quite diverse functions depending on the substrate that is phosphorylated. Some kinases may phosphorylate and regulate only a couple of substrates, whereas other kinases may phosphorylate hundreds of different substrates.

Phosphorylation, on one hand, is vital for normal cellular function under physiological conditions, while, on the other hand, dysregulation of protein kinases is one of the prime causes of structural and functional alteration leading to abnormal phosphorylation resulting in numerous pathological conditions. Therefore it is not surprising that protein kinases have become major targets for basic research and therapy, as well as protein kinase structures have had a significant impact on the development of selective and specific targeted therapies.

15.2 Protein kinase architecture

Eukaryotic protein kinases are enzymes that belong to a very extensive family of proteins that share a conserved catalytic domain, while there is considerable diversity among kinases within the additional flanking domains and sequences that may regulate processes such as catalytic function, substrate phosphorylation, or serve as regulatory domains in this family. The emergence of various protein kinases crystal structures from many research groups gives a deep insight to understand the catalysis and regulation of different protein kinases, indicating how kinases transfer phosphate to serine, threonine, or tyrosine residues of their protein substrates. The conserved kinase domain (aka catalytic domain) is comprised of 250–300 amino acids, and it has a bilobed three-dimensional structure consisting of conserved amino-(N) and carboxy-(C) terminal lobes that are coordinated in their movement in relation to each other depending on kinase activity. The typical protein kinase contains a conserved ATP-binding site, substrate interaction sites in the C-terminal lobe, and activation sites [17].

15.2.1 Protein kinase catalytic domains

The tertiary structure and overall folding of the kinase catalytic domain are highly conserved throughout the kinase family. In general, the kinase catalytic domain fold consists of two domains, also called lobes (N-terminal lobe (N-lobe) and C-terminal lobe (C-lobe)), which are connected to each other via a flexible hinge region. The N-lobe, the smaller lobe, contains five β-strands (β1–β5) and a single α-helix (the "C helix"). The larger C-lobe is consisting mostly of α-helices and loops. Both lobes form a hydrophobic cleft which serves as a binding site for ATP. The N-terminal lobe consists of the ATP-binding and catalytic site that serves the function of transferring the terminal phosphate (PO_4^{3-}) from magnesium-ATP and catalyzing its transfer to the hydroxyl group (OH^-) of serine, threonine, or tyrosine residue of the substrate protein. The Mg^{2+} helps stabilize and position the negatively charged phosphate on ATP for transfer onto the substrate. A conserved C-terminal glycine-rich motif and lysine residue in the N-terminal lobe also coordinate the ATP phosphates. Although protein kinases structural information is an important source that provides helpful guidance in designing substrate-competitive inhibitors, other strategies including bioinformatics, computational modeling, and high-throughput screening are often employed for developing specific protein kinase inhibitors, vide infra.

15.3 Kinase inhibitors

Over the last several decades, a vast amount of research has discovered protein kinases and their functions in regulating specific physiological functions demonstrating that overexpression and/or dysregulation of protein kinase activity is directly involved in the pathogenesis of many human diseases including diabetes mellitus, autoimmune disease, inflammatory disease and neurological diseases as well as cardiovascular disease (CVD) and cancer. Thus the development of specific drugs to inhibit protein kinases has been one of the most important targets of academic and pharmaceutical industry research in the 21st century, and currently, the human protein kinase superfamily accounts for nearly 25% of all current drug discovery research and development efforts, and 35% of all new targets screened by the global pharmaceutical industry [18–20]. While many drug discovery programs are aimed to develop kinase-specific drug therapies, in which the main therapeutic objective is to inhibit constitutively active kinase, there are also programs to develop inhibitors that target multiple kinases simultaneously or polypharmacologic properties that are uniquely effective against several dysregulated kinases associated with specific diseases, such as cancer [21].

Kinase inhibitors are broadly classified into six categories that are grouped largely based on the structural interactions between the inhibitor and target kinase. Currently, the most common approach to target kinases is by occupying the ATP site either by preventing interactions with ATP when the enzyme is in an active (Type I) or inactive (Type II) state. Concurrently by solely occupying an allosteric site in the kinase domain but do not affect ATP-binding (Type III), or kinase inhibitors that target allosteric sites outside the kinase domain and are generally designed to interfere with the interactions (Type IV). In addition, there are bivalent compounds that target both the ATP-binding site and unique allosteric sites outside the kinase domain (Type V), or compounds that form covalent interactions with amino acids (e.g., cysteine) in the ATP-binding site or other regions of the kinase (Type VI) [22].

Contemporary strategies for inhibiting protein kinase activity fall into three main categories: The first one using small molecules to compete for the nucleotide-binding site of the protein kinase was considered for many years the most effective approach to inhibit protein kinase activity, and currently many kinase inhibitors are available are small molecules, however, several challenges still exist with this strategy, mainly target selectivity and resistance. The second approach, antibody-based approaches, is by clearing the kinase through antibody-mediated deactivation, however, significant technical and physiological drawbacks limit the antibody-based approaches in drug discovery and development. Finally, peptides can be used to regulate protein kinase activity which we will discuss vide infra.

15.3.1 Small molecules targeting protein kinases

Small-molecule kinase inhibitors (SMKIs) are predominantly designed based on the competition for the ATP-binding pocket, blocking the catalytic function of protein kinase by preventing the phosphorylation of downstream substrates and therefore mitigating protein kinase-induced disease. The ROCK serine/threonine kinase inhibitor fasudil that was approved in Japan in 1995 for the treatment of cerebral vasospasm was the first approved SMKI for clinical use [23]. Next, mTOR serine/threonine kinase inhibitor sirolimus (brand name Rapamune) was approved in 1999 for use during renal transplants in combination with cyclosporine to prevent organ rejection. The first tyrosine kinase inhibitor developed, imatinib (brand name Gleevec or Glivec), was approved by the FDA in 2001 and was considered a groundbreaking discovery due to the way it was developed and the results it achieved. Imatinib is an orally administered SMKI, which inhibits tyrosine kinases, specifically BCR-ABL, c-KIT, and PDGFRA [24].

Up to January 1, 2021, the FDA approved a total of 62 SMKIs, nearly all of which are orally effective [25]. While in some cases SMKIs are highly effective, SMKIs have also several limitations. First, given that protein kinases share similar ATP-binding sites, it was considered highly challenging to develop compounds that inhibit just one kinase. Second, it was thought that these compounds would have to compete with high levels of intracellular ATP (millimolar ATP concentrations). Third, they have to distinguish among the ATP-binding sites of approximately 2,000 proteins utilizing ATP [26]. Finally, the development of resistance has always been and will probably always be the greatest problem limiting the use of kinase inhibitors. From a clinical point of view in addition to the resistance to kinase inhibitors, SMKI-induced toxicity (e.g., cardiotoxicity [27]) remains a major limitation to SMKI-based therapies [28].

15.3.2 Antibodies targeting protein kinases

One alternative strategy to SMKIs has been to develop antibodies against protein kinases. Indeed, monoclonal antibodies emerged as an alternative tool since they are designed to complement a relatively large surface area of the protein, thereby enhancing target selectivity. Antibodies differ from SMKIs in their modes of action at the target level, since antibodies are usually too large to penetrate cell membranes, so they are designed to bind to epitopes expressed on the cell surface [29,30]. Their mechanism of action

(MoA) is either interfering with signal transduction from the cell surface, acting as a stimulatory agent to induce cellular events (e.g., apoptosis), or provoking an immune response. Antibodies may target protein kinases with a high level of selectivity, specifically and efficiency [31,32]. In general, antibodies have a longer half-life than SMKIs and are usually administered intravenously and not orally.

Thc first antibodies were generated by hybridoma technology and were nonhuman [33]; the nature of these nonhuman antibodies and their repeated administration stimulated inflammatory reactions and the development of human antimurine antibodies in recipients. Accordingly, recombinant technologies were used to engineer antibodies with partially human sequences in viruses or yeast, and later fully humanized antibodies, such as adalimumab and panitumumab, that do not contain any nonhuman components were generated, resulting in antibodies with reduced immunogenicity problems and with increased specificities for target epitopes [34,35].

The first monoclonal antibody protein kinase inhibitor (PKI) to be approved was trastuzumab (brand name: Herceptin) in 1998 by Genentech. Trastuzumab targets the extracellular domain of human epidermal growth factor receptor 2 (HER2, aka ErbB2) for the treatment of ErbB2-positive breast cancer, which represents a significant number of breast cancers [36,37]. Trastuzumab binds to the HER2 receptor sites on the surface of breast cancer cells, blocking the receptor sites, and preventing tumor growth by interrupting the growth signal to intracellular domains [38]. Several other successful examples include cetuximab (brand name: Erbitux), a recombinant human/mouse chimeric monoclonal antibody developed by ImClone that blocks the epidermal growth factor receptor (EGFR; ErbB-1; HER1 in humans). Another solution for infusion that contains the active substance of cetuximab was approved in 2004 for the treatment of metastatic colorectal cancer [39–41]. Bevacizumab (brand name: Avastin) is a humanized monoclonal antibody developed by Genentech, which was approved by the FDA in 2006. Bevacizumab works by slowing the growth of new blood vessels by inhibiting vascular endothelial growth factor A (VEGF-A) as an antiangiogenic anticancer drug and as first-line therapy for metastatic colorectal cancer. Bevacizumab has a long terminal half-life of 17 to 21 days in humans [42,43]. Pertuzumab (brand name: Omnitarg/Perjeta) is a humanized monoclonal antibody also developed by Genentech, which was approved by the FDA in 2012 for the treatment of metastatic HER2-positive breast cancer. In total there are more than 30 FDA-approved monoclonal antibodies developed to block the activity of mostly receptor and nonreceptor tyrosine kinases involved in disease.

However, there are several significant technical and physiological drawbacks that hinder antibody-based approaches to drug development. Full-length antibodies are relatively large biomolecules; their production is often costly, and recombinant expression of antibodies can result in end-product heterogeneity. In addition, a major drawback of antibody therapies is their inability to access the intracellular space, which effectively limits their application to targeting serum proteins or proteins that contain extracellular domains. However, numerous protein kinases are intracellular proteins that lack extracellular domains and are unable to be targeted using antibody-based therapeutics. Finally, resistance and/or immunogenicity are largely inevitable for antibody-based therapeutics, at least when dosed chronically.

15.4 Peptides targeting protein kinases

Protein kinases have a canonical catalytic domain that mediates the enzymatic function with a remarkable structural degree of homology. However, despite their high degree of structural conservation at the catalytic site, protein kinases exhibit significant specificity in their ability to recognize individual proteins that are targeted and phosphorylated by them, as well as other binding partners that might serve as inhibitors or scaffold proteins. Multiple mechanisms have evolved that contribute to this exquisite specificity based on unique substrate recognition motifs of protein kinases that are composed of amino acids which form topological surfaces that are complementary to the kinase-binding domains of cognate substrates. Therefore it has been reasoned that peptides that target these sites may act as a specific regulator of kinase interactions. Several recognition sites are used to develop peptides as specific regulators, including, the substrate-binding region located along with the N- and C-terminal lobes that exhibit remarkable diversity; the pseudosubstrate domain that holds the kinase in an inactive state; the distal docking site that is used as a specific binding domain on the kinase and increases the kinase/substrate affinity; and the interaction regions with anchor proteins that regulate the trafficking of kinases throughout the cell. The biomimetic nature and sequence specificity of peptides advanced their development as protein kinase regulators. Over the past 50 years, researchers have been developing synthetic peptides as a means to inhibit protein kinase activity. Herein we will focus on the development of peptide-based kinase regulators that have shown success in inhibiting kinase activity. We will also cover applications of such peptide-based kinase inhibitors as potential therapeutics [20,44,45].

15.4.1 Peptide-based kinase inhibitors targeting kinase-substrate sites

It is estimated that eukaryotic cells contain about 10,000 different proteins, that each protein is an average of about 400 amino acids, and that the frequency of the commonly phosphorylated amino acid in proteins is about 17% [Ser ($\sim$8%), Thr ($\sim$6%), or Tyr ($\sim$3%)] [46], thus there are approximately 700,000 potential phosphorylation sites for any given kinase.

Nevertheless, under physiological conditions, a kinase may only phosphorylate several specific phosphorylation sites. The specificity of a protein kinase is primarily controlled by the substrate-binding region located along with the N- and C-terminal lobes. The substrate-kinase binding regions are highly diverse, forming a shallow surface area that serves as a complementary binding site for the target substrate and has proven to be difficult to target using small molecules. One promising approach to inhibit kinase function has been to inhibit such interactions using peptide-based mimetics of kinase-substrate interaction domains.

Protein kinases selectivity for their substrates is primarily determined by the "substrate recognition motif," a sequence that contains particular amino acids surrounding the phosphorylation site with physicochemical features that influence substrate selection that is critical in substrate recognition by the protein kinase. Substrate-based kinase inhibitors are typically designed based on the amino acid sequences around the sites of phosphorylation in the substrate, mimetic of the direct substrate consensus sequence, and usually include in addition to the phosphorylation site, several additional residues that are critical for specific kinase recognition [47,48].

15.4.1.1 Peptide-based kinase inhibitors targeting pseudosubstrate sites

The pseudosubstrate domain is an inhibitory sequence in the regulatory region initially identified in PKC by Kemp and coworkers [49,50]. The pseudosubstrate domain is an autoinhibitory region that acts by mimicking the features of the substrate. Structurally, the pseudosubstrate domain mimics the protein substrate phosphorylation site of the enzyme of the protein kinase and extends into the binding cavity of the catalytic domain. This short sequence of amino acids resembled that of a substrate except an alanine occupied the phosphor-acceptor position. The pseudosubstrate is a structure that folds back into the active site turning the protein kinase to be inactive by mimicking the substrate-like recognition motif. An activator bound to the protein kinase removes the pseudosubstrate region from the active site, resulting in the expression of protein kinase activity [51]. Therefore pseudosubstrate-based kinase inhibitors have been designed to disrupt interactions between protein kinases and their own A-loop. The pseudosubstrate site can either be located within the full-length kinase (e.g., the regulatory domain of PKC) or on a separate protein (e.g., R-subunits of PKA).

Various findings established the hypothesis that the pseudosubstrate occupied the substrate-binding cavity, therefore maintaining the enzyme in an inactive state. For example, a study identified peptides based on this sequence that were effective competitive PKC inhibitors, as well as peptides modified to have a serine at the putative phosphor-acceptor position that were good substrates [50,52]. In another study an antibody that was generated against this sequence activated PKC even in the absence of cofactors, likely due to withdrawing the pseudosubstrate from the active site, unmasking the enzyme [53]. Finally, it was demonstrated that protein kinase activation is accompanied by the release of the pseudosubstrate sequence from the kinase core (e.g., PKC [54]), as well as that the pseudosubstrate is resistant to proteolysis in the inactive conformation but susceptible to proteolysis at active conformation (e.g., PKC [55]). Interestingly, mutations or deletions within the pseudosubstrate site may result in protein kinases that are constitutively active [56,57]. Therefore the concept of developing pseudosubstrate-based peptides as kinase regulators has been explored. Initially, peptides designed to mimic pseudosubstrate domains were used mainly to study the pseudosubstrate roles in regulating protein kinase activity. Recent studies developed such peptides to directly regulate the protein kinases activity, and it has been reasoned that these peptides may also serve as potential therapeutics in the treatment of kinase-related diseases.

Myosin light chain kinase (MLCK, aka MYLK, or MLCK) is a calcium/calmodulin-dependent serine/threonine protein kinase that belongs to the immunoglobulin superfamily. It phosphorylates the regulatory myosin light chains of myosin II leading to the activation of myosin so that it can interact with actin. It plays a central role in regulating the actin-myosin interaction of smooth muscles. MLCK autoinhibitory sequence is located within the C-terminal [58], and the autoinhibited state of the C-terminal extends over the surface of the cleft between the two lobes of the kinase domain, masking the substrate-binding site and the ATP-binding pocket. Calmodulin binding to the pseudosubstrate region results in a conformational change that removes the autoinhibitory sequence from the binding site [59]. A peptide that was designed based on the pseudosubstrate region was shown to inhibit MLCK bioactivity (Ki 46 nM) [60]. However, since the MLCK pseudosubstrate peptide sequence overlaps with the calmodulin-binding region, it also inhibited calmodulin-dependent protein kinases (CaMKII). Further screening of peptide libraries with related sequences resulted in the identification of "Peptide 18" which is specific to MLCK (IC_{50} 50 nM) [61].

15.4.1.1.1 Peptides as substrate-based inhibitors of protein kinases in cardiovascular disease

CVD generally refers to diseases that are related to heart or blood vessels, and are the leading cause of mortality and morbidity worldwide, with very few new medications approved for intervention/treatment. Therefore new approaches to treating CVD are urgently required. PKC is a family consisting of 10 structurally-related serine/threonine protein kinases that are critical regulators of cell proliferation and survival and their expression or activity is altered in many human diseases. Relevant to CVD, many PKC isozymes play a vital role in the initiation and progression of CVD, including regulating cell proliferation [62–64], heart failure [65], heart attack [66], angiogenesis [67], and the immune response [68].

PKC contains an N-terminal autoinhibitory pseudosubstrate sequence that was initially described in 1987. In the mature but inactive conformation of PKC, the pseudosubstrate site is lodged in the kinase active site mediates the binding of the regulatory domain to the catalytic domain, thus maintaining PKC in an inactive conformation, preventing it from phosphorylating substrates. A peptide-based on this sequence was found to inhibit PKC bioactivity (Ki 150 nM). Agonist stimulation induces PKC to translocate to intracellular membranes, where membrane engagement of its regulatory domains allosterically removes the pseudosubstrate from the active site, enabling it to phosphorylate substrates. However, due to the similarity of the pseudosubstrate sequence of the α, β, and γ isoforms, this peptide inhibits these three isoforms with comparable potency [49]. Braun et al. demonstrated that inhibition of endogenous ζPKC in neonatal rat cardiac fibroblasts with ζ-pseudosubstrate reduced basal growth and transforming growth factor-β1 (TGFβ1)-induced growth, indicating that ζPKC stimulates proliferation of neonatal rat cardiac fibroblasts [64].

15.4.1.1.2 Peptides as substrate-based inhibitors of protein kinases in cancer

Dysregulation of protein kinases is implicated in various processes of carcinogenesis. Numerous studies revealed the role of deranged phosphorylation of proteins in the manifestation of cancer. Deregulated phosphorylation elicits one of the most crucial changes associated with cancer. The first substrate-competitive inhibitor was discovered in the early 1970s by Ashby et al., who identified a 76-amino acid protein that coprecipitates and inhibits PKA activity (aka PKI). This protein has two functional regions, a nuclear export sequence that transports PKA out of the nucleus and a pseudosubstrate region that inhibits PKA catalytic activity [69]. Based on this non-ATP-competitive regulator Cheng et al. developed a series of short peptides derived from the canonical RRNSL PKI consensus sequence that contained an Alanine residue in place of the phosphoracceptor Serine, RRN*A*L, (italic) that were effective PKA inhibitors (Ki ~ 0.2 nM) [70]. Glass et al. further performed structure-activity relationship studies and identified the residues within the consensus sequence that are important for kinase recognition and specificity [71]. Currently, the PKI peptide termed IP20 is commercially available as a peptide-based PKA selective and potent inhibitor (Ki ~ 2 nM) that serves as a pseudosubstrate inhibitor for PKA. While this peptidc is an effective PKA binder, it was demonstrated that only one-third of PKA substrates contain this entire sequence [52].

PKI-amide can revert the proliferative effect of oxytocin-treated MDA-MB231 breast carcinoma cells by inhibiting the cAMP-PKA pathway [72].

A recent study showed that inhibiting PKA by PKI can divert G-protein-coupled receptor signals toward cell growth in cancer cells [73]. In addition, PKA inhibition could influence the response of cancer cells to taxol and taxane therapy as was demonstrated by PKA inhibition using PKI-amide that significantly reduced their toxicity to prostate cancer cells [74].

15.4.2 Peptide-based kinase inhibitors targeting docking sites

Protein kinases form complexes with their substrates that are bridged by a third protein, vide infra, or involve direct, high-affinity interaction between a part of the kinase and a short sequence of the substrate, known as a docking site. These binding motifs increase the efficiency of the protein kinase with a specific substrate both in vitro and in cells by many times. The sites are modular often located separately from the kinase active site and the substrate phosphorylation site; in addition, several docking sites can be present in a single substrate, increasing affinity for a kinase [13,75,76]. Docking sites have been identified in numerous kinases, such as TGFβ receptor [77], cyclin-dependent kinase 2 (CDK2) [78], glycogen synthase kinase-3 (GSK3) [79], and c-Jun N-terminal kinases (JNKs) [80] to name a few, indicating the generality of these interactions.

15.4.2.1 Peptide-based kinase inhibitors targeting docking sites in cardiovascular disease

δPKC is a pleiotropic kinase that phosphorylates many substrates, including heat shock protein 27 (HSP27), myristoylated alanine-rich C-kinase substrate (MARCKS), signal transducer and activator of transcription (STAT), glyceraldehyde 3-phosphate dehydrogenase (GAPDH), troponin I, and pyruvate dehydrogenase kinase (PDK) [81], and its signaling was demonstrated to be vital in many models of human diseases, including cancer [82], diabetes [83], neurodegenerative diseases [84] and CVDs [85]. A series of peptides derived from the docking sites of δPKC and various substrates were developed. These peptides demonstrate high specificity and high bioactivity in vitro, in cells and in an animal model of human diseases.

Using rational design approaches [86–88], Qvit and coworkers developed a peptide corresponding to the PDK substrate docking site on δPKC [89]. PDK is a protein kinase that participates in the regulation of the pyruvate dehydrogenase complex by inactivating the enzyme pyruvate dehydrogenase via phosphorylating it. They hypothesized that an inhibitor of the PDK docking site should selectively inhibit the phosphorylation of only PDK without affecting the phosphorylation of the other δPKC substrates. Based on this approach they developed a highly bioactive peptide that selectively inhibits only δPKC-mediated PDK phosphorylation. The peptide binds to δPKC in vitro in a time-dependent manner (Kd 53 nM), and treatment with the peptide in a model of heart attack *ex vivo* decreased infarct size by over 60% when delivered only during reperfusion, and treatment with the peptide in vivo, immediately after 30 min ischemia reduced infarct size by approximately 50%, indicating the efficacy of the peptide. Finally, the peptide is highly specific (IC_{50} 5 nM), without affecting the phosphorylation of other δPKC substrates even at 1 μM [90].

Next, Qvit et al. studied the mechanism by which oxidative stress inhibits the protective GAPDH-mediated elimination of damaged mitochondria. GAPDH is a δPKC substrate that plays a role in inducing mitochondrial elimination under oxidative stress. However, the direct role of GAPDH in this process is still an enigma. Using rational design, they developed a highly selective δPKC inhibitor based on GAPDH docking site on δPKC. In vitro the peptide inhibited δPKC/GAPDH PPIs and subsequent GAPDH phosphorylation by over 70% at concentrations as low as 1 μM. Interestingly, the peptide inhibits GAPDH oligomerization and GAPDH-mediated glycolytic activity. The peptide was highly specific and did not affect the phosphorylation of other δPKC substrates in whole-cell lysates. Finally, the conjugated peptide acted as an inhibitor of the elimination of damaged mitochondria in cardiac H9C2 cells following oxidative stress [91]. This approach is general and maybe a promising approach suggesting the use of antidocking site compounds as a therapeutic strategy to counteract various CVDs as well as other human diseases.

15.4.2.2 Peptide-based kinase inhibitors targeting docking sites in cancer

The hepatocyte growth factor (HGF)/scatter factor is a multifunctional protein that has been implicated in embryo development, tissue repair, and cancer growth. The factor is produced and secreted by stromal and mesenchymal cells, and it stimulates epithelial cell proliferation, motility, morphogenesis, and angiogenesis in various organs via tyrosine phosphorylation of its cognate receptor, Met. The HGF-Met signaling pathway contributes in a paracrine manner to the development of epithelial organs, exerts regenerative effects, and promotes the regression of fibrosis. Importantly, it is correlated with the biology of cancer types, neurons and immunity. Cantelmo et al. developed peptides that mimic the docking site of the Met carboxyl-terminal tail. These peptides were highly bioactive in vitro inhibiting ligand-dependent endothelial cell proliferation, motility, invasion, and morphogenesis. Importantly the peptides were less toxic than a small-molecule Met inhibitor, PHA-665752, and demonstrated higher bioactivity. The peptides were also highly bioactive in vivo, inhibiting HGF-induced angiogenesis and reducing xenograft tumor growth and vascularization [92].

15.4.2.3 Peptide-based kinase inhibitors targeting anchoring proteins

Localization of protein kinases to specific locations at the cell level provides spatial control over receptor signaling, and sequestering them within various organelles allows their association with localized substrates to be tightly controlled [93]. Subcellular localization of protein kinases can be mediated through the interaction with anchor proteins that regulate kinase trafficking. Therefore a common approach to inhibiting kinase activity has been to develop peptides that target these interactions by mimicking anchor protein interaction domains. The most common examples for the development of peptides that disrupt interactions between protein kinases and their associated anchor proteins are in PKC and PKA, and in many cases, these peptides have shown efficacy in inhibiting kinase function.

PKC isozyme subcellular localizations are determined by lipid interactions and by their interactions with scaffold proteins, such as Receptor for Activated C Kinases (RACKs). RACKs belong to the tryptophan-aspartate 40 (WD40) motif repeat family regulating protein-protein interactions and they are homologous to the β-subunit of heterotrimeric G-proteins. RACK structurally unrelated adaptor or scaffold proteins that are functionally defined by their ability to bind and activate specific PKC isozymes were first identified by Mochly-Rosen et al. [94]. The binding of a specific PKC isozyme to its RACK provides access to and phosphorylation of PKC substrates; hence, the PKC and RACK interaction and subsequently PKC anchoring to RACK is a required step for PKC function [95]. RACK1 is a member of the WD-repeat family of proteins and it adopts a seven-bladed β-propeller structure which facilitates protein binding. RACK1 is neither a substrate nor a PKC inhibitor. It binds specifically to βPKC, it is not phosphorylated by PKC (not a substrate), and it does not cause inhibition of substrate phosphorylation by PKC (e.g., histone III, not an inhibitor). RACK1 increases histone III phosphorylation, indicating its ability to stabilize the active conformation of PKC [96]. Ron et al. developed a βPKC-derived peptide that interferes with the inhibitory intramolecular interaction within the enzymes which is a βPKC activator. This peptide acts as an agonist of PKC function and binds to PKC, exposing the RACK-binding site, enabling the association of the enzyme with its anchoring RACK to the plasma membrane near its localized substrates even in the absence of PKC activators, both in vitro and in vivo [97].

Similar approaches have been used to inhibit the activity of PKA which is also regulated through interactions with the anchor protein family of A-kinase anchor proteins (AKAPs), which contain a helix binding motif that anchors the PKA-regulatory subunit (R) [98]. For example, Rosenmund et al. demonstrated that peptides derived from the conserved amphipathic helix kinase-binding region of AKAPs inhibit PKA-mediated phosphorylation of

AMPA/kainite channels in cultured neurons. Furthermore, treatment of cells inhibited PKA-mediated regulation of AMPA/kainate currents, indicating for the first time the role of PKA-anchoring proteins in the regulation of synaptic function [99]. Carlson et al. developed a high-affinity regulatory subunit anchoring disruptor (RIAD) peptide-based on bioinformatics analyses and peptide-screening arrays, which was highly specific [100].

15.4.2.4 Peptide-based kinase inhibitors targeting anchoring proteins in cardiovascular disease

PKC isozymes are involved in a variety of chronic cardiac diseases as well as in acute cardiac injuries and preconditioning. Ron et al. developed peptides derived from the RACK1 and βPKC binding sites as specific inhibitors of PKC translocation and function. The C2 region of βPKC contains part of the RACK-binding site on the enzyme [101]. A series of short peptides derived from the C2 region was developed and several peptides were highly bioactive inhibitors of hormone-induced translocation and functions of the C2-containing βPKC isozymes, these peptides specifically inhibited phorbol ester-induced translocation of the C2-containing isozymes in cardiac myocytes [102].

15.4.2.5 Peptide-based kinase inhibitors targeting anchoring proteins in cancer

Many of the defining characteristics of cancer, including uncontrolled growth, survival, neovascularization, metastasis and invasion, result from perturbation of regulatory signaling pathways, which are normally under tight control. JNKs, a family of serine/threonine protein kinases, are stress-activated proteins that upon activation phosphorylate several transcription factors, and cellular substrates implicated in cell survival and proliferation, resulting in the modulation of several pathways implicated in a variety of disease states. JNKs bind to scaffold proteins, such as JNK-interacting protein-1 (JIP1), containing a D-domain [103]. JIP1 is a scaffolding protein that enhances JNK signaling by creating a proximity effect between JNK and upstream kinases. Bonny et al. identified peptides that inhibit JNK phosphorylation of c-Jun in vitro and reduce c-Jun expression in cells. Barr et al. identified the minimal region necessary for JNK inhibition, and a peptidomimetic of the native peptides, which were generated by synthesizing the peptide in reverse order out of D-amino acids, inhibited IL-1β-induced apoptosis in pancreatic β-cells [104]. This peptide could bind JNK directly and inhibit JNK activity [105]. Gao et al. assessed the effect of the D-amino acid version of the JIP-1-derived peptide (D-JNKI-1) in the melanoma murine model. In this model, cells were injected into the hind paw of the mouse and the peptidomimetic bioactivity was evaluated on tumor growth and tumor-induced pain. D-JNKI-1 prevented tumor growth entirely during treatment and reduced tumor growth [106].

15.4.3 Additional approaches to develop peptide-based kinase inhibitors

15.4.3.1 Development of protein kinase inhibitors using bioinformatics and structural data

KinAce is a technology that combines bioinformatics and crystallographic data to identify variable regions of structurally conserved catalytic protein kinase domains to develop peptides that regulate the protein kinase-dependent signal transduction in a specific manner. The KinAce strategy analyzed protein kinase sequence, three-dimensional structure, and involvement in protein-protein interaction, using these tools it identified regions of interest, such as subdomains of various protein kinases (e.g., subdomain V (the αD region), the loop between subdomains IX and X (the HJ loop), and subdomain X (the αG helix)). Next, this information is used to develop leads as specific peptide-based specific kinase inhibitors [107]. For example, using this approach the authors developed peptides derived from the αD region of the tyrosine kinases c-Kit and Lyn that selectively inhibited SCF-induced transphosphorylation (IC_{50} 7 μM). In another example, they developed peptides derived from the HJ-αG region of the serine/threonine kinases 3-phosphoinositide-dependent kinase-1 (PDK1) that reduced Akt/protein kinase B (PKB) phosphorylation in prostate cancer cells and inhibited cell proliferation at (IC_{50} low- to mid-μM). Finally, a peptide derived from the Lyn subdomain III of HJ-αG region was bioactive reducing the phosphorylation of Syk and Lyn in a dose-dependent manner, yielding 80%–90% inhibition at concentrations of 5 μM [107].

15.4.3.2 Bisubstrate and bivalent approaches

Another promising strategy is the bisubstrate and bivalent approaches that consist of two conjugated fragments, each targeted to a different binding site of the enzyme. This is an established and effective approach to enhance the potency and specificity of the inhibitors [108]. For example, a PKC-directed bisubstrate inhibitor consisting of a pseudosubstrate peptide containing a single serine followed by six arginine residues, which targets the protein substrate site of PKC, linked to an ATP-competitive 5-isoquinolinylsulfonyl inhibitor, which targets the ATP-binding site. Both fragments, the peptide and ATP-competitive ligands, were covalently bound using a linker whose length was calculated using bioinformatic studies. In an in vitro activity assay, it was demonstrated that the bisubstrate inhibitor (IC_{50} 0.3 μM) was more potent than the 5-isoquinolinylsulfonyl ATP-competitive inhibitor alone (IC_{50} 20 μM). However, the potency of the bisubstrate inhibitor was

100-fold greater for PKA than for PKC [109]. A bivalent inhibitor of JNK1 was generated by linking an ATP-competitive inhibitor of the JNK family to a peptide that targets JNK1's docking groove. The small-molecule ATP mimetic, SP600125, is a highly potent but not very selective JNK inhibitor. An analog of SP600125 was conjugated to a minimal sequence derived from the D-domain of the JNK1 scaffolding protein, JIP1, which does not inhibit JNK1 activity (IC_{50} > 50 mM), through a glycine-glycine-diaminopropyl linkage. This inhibitor was highly bioactive against JNK1 in vitro (IC_{50} 0.7 nM), a 20,000-fold increase in JNK1 inhibition compared to the ATP-competitive inhibitor alone (IC_{50} 14 μM) [110].

15.5 Conclusions and perspectives

The phosphorylation of proteins in response to both intracellular and extracellular signals fulfills a pivotal role in a wide variety of cellular signal transduction pathways, and protein kinases have therefore attracted much attention as potential drug targets. The majority of kinase inhibitors are ATP-competitive small-molecule inhibitors that target the ATP-binding pocket; however, selectivity remains a major challenge due to the conservation of the binding pocket across the kinase superfamily. Therefore such inhibitors are restricted to indications such as cancer, where a low degree of selectivity may be an advantage (e.g., sunitinib, a growth factor receptor kinase inhibitor, which blocks signaling through multiple oncogenic pathways). In many other cases, small molecules or monoclonal antibodies have mostly failed to produce effective or sustained clinical responses despite the significant evidence supporting kinases as key drivers of disease, emphasizing the need for alternative therapeutics.

Peptides demonstrate many advantages for targeting protein complexes compared to small molecules, such as conformational flexibility and increased selectivity; in addition, they are easier and less expensive to manufacture compared to antibodies. However, peptides have limitations as therapeutics, such as high protease sensitivity and rapid plasma clearance. While in several indications, such as an acute treatment of ST-elevation myocardial infarction by intravenous infusion, these pharmaceutical properties (e.g., rapid clearance) have the advantage of minimizing potential toxicity and improving the dose proportionality of exposure, in other cases, these limitations hamper their development into successful drugs. Nevertheless, these limitations can be overcome using peptidomimetics, molecules generated to mimic and maintain the biological function of the natural peptide, while simultaneously removing their liabilities [111–113]. Several strategies are used to convert peptides into peptidomimetics, including local modification (e.g., using *N*-methyl and *D*-amino acids) and global modification (e.g., backbone cyclization) [114,115], leading to a sustained increase in the number of peptides and peptidomimetics entering clinical studies, such as CVD

[116] and cancer [117]. Another advantage of peptides and peptidomimetics is that they can be further enhanced by conjugating peptides to cell-penetrating peptides to facilitate their penetration into cells and tissues.

There are currently several approaches for inhibiting protein kinase activity with peptides, including peptides that target the kinase-substrate interaction domains (e.g., pseudosubstrate site and kinase-substrate specific interaction sites); peptides that target allosteric protein kinase binding sites (e.g., docking interacting sites, sites that are distinct from the kinase active site that facilitates interaction with protein partners); and peptides that target the interactions between protein kinases and cell-targeting and adaptor proteins (e.g., anchoring proteins, proteins that regulate the trafficking of kinases throughout the cell). These domains are more diverse than the ATP-binding site. Therefore competitive inhibitor peptide-mimetics based on these sites are expected to be highly selective. Currently, one of the most promising classes of PKIs to be discovered are peptide and peptidomimetics of naturally occurring protein interactions, as the characterization of these targets and their signaling pathways has led to the identification of key interaction domains.

Funding

This research was supported by a grant from the United States-Israel Binational Science Foundation (BSF), Jerusalem, Israel; a grant from the Ministry of Health (MoH), Jerusalem, Israel; and a grant from the Israel Science Foundation (ISF), Israel, all to N.Q.

References

[1] Adam K, Hunter T. Histidine kinases and the missing phosphoproteome from prokaryotes to eukaryotes. Lab Investig 2018;98(2):233–47.

[2] Levene PA, Alsberg CL. The cleavage products of vitellin. J Biol Chem 1906;2(1):127–33.

[3] Cori CF, Cori GT. Glycogen formation in the liver from d-and l-lactic acid. J Biol Chem 1929;81(2):389–403.

[4] Burnett G, Kennedy EP. The enzymatic phosphorylation of proteins. J Biol Chem 1954;211(2):969–80.

[5] Fischer EH, Krebs EG. Conversion of phosphorylase b to phosphorylase a in muscle extracts. J Biol Chem 1955;216(1):121–32.

[6] Pinna LA, Ruzzene M. How do protein kinases recognize their substrates? Biochim Biophys Acta 1996;1314(3):191–225.

[7] Cohen P. The regulation of protein function by multisite phosphorylation—a 25 year update. Trends Biochem Sci 2000;25(12):596–601.

[8] Dhanasekaran N, Reddy EP. Signaling by dual specificity kinases. Oncogene 1998;17(11):1447–55.

[9] Endicott JA, Noble MEM, Johnson LN. The structural basis for control of eukaryotic protein kinases. Annu Rev Biochem 2012;81(1):587–613.

[10] Manning G, Whyte DB, Martinez R, Hunter T, Sudarsanam S. The protein kinase complement of the human genome. Science 2002;298(5600):1912–34.

[11] Taylor SS, Shaw A, Hu J, Meharena HS, Kornev A. Pseudokinases from a structural perspective. Biochem Soc Trans 2013;41(4):981–6.

[12] Manning G, Plowman GD, Hunter T, Sudarsanam S. Evolution of protein kinase signaling from yeast to man. Trends Biochem Sci 2002;27(10):514–20.

[13] Ubersax JA, Ferrell Jr. JE. Mechanisms of specificity in protein phosphorylation. Nat Rev Mol Cell Biol 2007;8(7):530–41.

[14] Holt LJ, Tuch BB, Villén J, Johnson AD, Gygi SP, Morgan DO. Global analysis of Cdk1 substrate phosphorylation sites provides insights into evolution. Science 2009;325 (5948):1682–6.

[15] Serber Z, Ferrell Jr. JE. Tuning bulk electrostatics to regulate protein function. Cell 2007;128(3):441–4.

[16] Knighton DR, Zheng JH, Ten Eyck LF, Ashford VA, Xuong NH, Taylor SS, et al. Crystal structure of the catalytic subunit of cyclic adenosine monophosphate-dependent protein kinase. Science 1991;253(5018):407–14.

[17] Hanks SK, Quinn AM, Hunter T. The protein kinase family: conserved features and deduced phylogeny of the catalytic domains. Science 1988;241(4861):42–52.

[18] Cohen P. Protein kinases—the major drug targets of the twenty-first century? Nat Rev Drug Discovery 2002;1(4):309–15.

[19] Roskoski R. A historical overview of protein kinases and their targeted small molecule inhibitors. Pharmacol Res 2015;100:1–23.

[20] Holub JM. Novel peptide-based inhibitors of protein kinases. Next generation kinase inhibitors. Springer; 2020. p. 169–206.

[21] Rao S, Du G, Hafner M, Subramanian K, Sorger PK, Gray NS. A multitargeted probe-based strategy to identify signaling vulnerabilities in cancers. J Biol Chem 2019;294 (21):8664–73.

[22] Roskoski R. Classification of small molecule protein kinase inhibitors based upon the structures of their drug-enzyme complexes. Pharmacol Res 2016;103:26–48.

[23] Shibuya M, Suzuki Y, Sugita K, et al. Effect of AT877 on cerebral vasospasm after aneurysmal subarachnoid hemorrhage. Results of a prospective placebo-controlled double-blind trial. J Neurosurg 1992;76(4):571–7.

[24] Iqbal N, Iqbal N. Imatinib: a breakthrough of targeted therapy in cancer. Chemother Res Pract 2014;2014:357027.

[25] Roskoski R. Properties of FDA-approved small molecule protein kinase inhibitors: a 2021 update. Pharmacol Res 2021;165:105463.

[26] Fischer P. The design of drug candidate molecules as selective inhibitors of therapeutically relevant protein kinases. Curr Med Chem 2004;11(12):1563–83.

[27] Jin Y, Xu Z, Yan H, He Q, Yang X, Luo P. A comprehensive review of clinical cardiotoxicity incidence of FDA-approved small-molecule kinase inhibitors. Front Pharmacol 2020;11:891.

[28] Bhullar KS, Lagarón NO, McGowan EM, Parmar I, Jha A, Hubbard BP, et al. Kinase-targeted cancer therapies: progress, challenges and future directions. Mol Cancer 2018;17(1):48 48.

[29] Imai K, Takaoka A. Comparing antibody and small-molecule therapies for cancer. Nat Rev Cancer 2006;6(9):714–27.

[30] Dassonville O, Bozec A, Fischel JL, Milano G. EGFR targeting therapies: monoclonal antibodies vs tyrosine kinase inhibitors: similarities and differences. Crit Rev Oncol/ Hematol 2007;62(1):53–61.

[31] Fauvel B, Yasri A. Antibodies directed against receptor tyrosine kinases: current and future strategies to fight cancer. mAbs 2014;6(4):838–51.

[32] Swain SM, Kim S-B, Cortés J, et al. Pertuzumab, trastuzumab, and docetaxel for HER2-positive metastatic breast cancer (CLEOPATRA study): overall survival results from a randomised, double-blind, placebo-controlled, phase 3 study. Lancet Oncol 2013;14 (6):461–71.

[33] Köhler G, Milstein C. Continuous cultures of fused cells secreting antibody of predefined specificity. 1975. J Immunol 2005;174(5):2453–5.

[34] Nelson AL, Dhimolea E, Reichert JM. Development trends for human monoclonal antibody therapeutics. Nat Rev Drug Discov 2010;9(10):767–74.

[35] Scott AM, Wolchok JD, Old LJ. Antibody therapy of cancer. Nat Rev cancer 2012;12 (4):278–87.

[36] Valabrega G, Montemurro F, Aglietta M. Trastuzumab: mechanism of action, resistance and future perspectives in HER2-overexpressing breast cancer. Ann Oncol 2007;18 (6):977–84.

[37] Hudis CA. Trastuzumab—mechanism of action and use in clinical practice. N Engl J Med 2007;357(1):39–51.

[38] Slamon DJ, Leyland-Jones B, Shak S, et al. Use of chemotherapy plus a monoclonal antibody against HER2 for metastatic breast cancer that overexpresses HER2. N Engl J Med 2001;344(11):783–92.

[39] Mitchell P. Erbitux diagnostic latest adjunct to cancer therapy. Nature Publishing Group; 2004.

[40] Saltz LB, Meropol NJ, Loehrer Sr. PJ, Needle MN, Kopit J, Mayer RJ. Phase II trial of cetuximab in patients with refractory colorectal cancer that expresses the epidermal growth factor receptor. J Clin Oncol 2004;22(7):1201–8.

[41] Adams GP, Weiner LM. Monoclonal antibody therapy of cancer. Nat Biotechnol 2005;23 (9):1147–57.

[42] Ferrara N, Hillan KJ, Gerber H-P, Novotny W. Discovery and development of bevacizumab, an anti-VEGF antibody for treating cancer. Nat Rev Drug discovery 2004;3 (5):391–400.

[43] Roskoski Jr R. The ErbB/HER receptor protein-tyrosine kinases and cancer. Biochem Biophys Res Commun 2004;319(1):1–11.

[44] Kaidanovich-Beilin O, Eldar-Finkelman H. Peptides targeting protein kinases: strategies and implications. Physiol (Bethesda) 2006;21:411–18.

[45] Helton LG, Limaye AJ, Bendzunas GN, Kennedy EJ. Novel stabilized peptide inhibitors of protein kinases. Next generation kinase inhibitors. Springer; 2020. p. 141–67.

[46] Echols N, Harrison P, Balasubramanian S, Luscombe NM, Bertone P, Zhang Z, et al. Comprehensive analysis of amino acid and nucleotide composition in eukaryotic genomes, comparing genes and pseudogenes. Nucleic Acids Res 2002;30(11):2515–23.

[47] Eldar-Finkelman H, Eisenstein M. Peptide inhibitors targeting protein kinases. Curr Pharm Des 2009;15(21):2463–70.

[48] Han KC, Kim SY, Yang EG. Recent advances in designing substrate-competitive protein kinase inhibitors. Curr Pharm Des 2012;18(20):2875–82.

[49] House C, Kemp BE. Protein kinase C contains a pseudosubstrate prototope in its regulatory domain. Science 1987;238(4834):1726–8.

[50] House C, Kemp BE. Protein kinase C pseudosubstrate prototope: structure-function relationships. Cell Signal 1990;2(2):187–90.

[51] Kemp BE, Pearson RB, House CM. Pseudosubstrate-based peptide inhibitors in Methods enzymology 1991;287–304.

[52] Kemp BE, Parker MW, Hu S, Tiganis T, House C. Substrate and pseudosubstrate interactions with protein kinases: determinants of specificity. Trends Biochem Sci 1994;19 (11):440–4.

[53] Makowske M, Rosen OM. Complete activation of protein kinase C by an antipeptide antibody directed against the pseudosubstrate prototope. J Biol Chem 1989;264(27):16155–9.

[54] Orr JW, Keranen LM, Newton AC. Reversible exposure of the pseudosubstrate domain of protein kinase C by phosphatidylserine and diacylglycerol. J Biol Chem 1992;267 (22):15263–6.

[55] Orr JW, Newton AC. Requirement for negative charge on "activation loop" of protein kinase C. J Biol Chem 1994;269(44):27715–18.

[56] Pears CJ, Kour G, House C, Kemp BE, Parker PJ. Mutagenesis of the pseudosubstrate site of protein-kinase-C leads to activation. Eur J Biochem 1990;194(1):89–94.

[57] Zhao J, Renner O, Wightman L, Sugden PH, Stewart L, Miller AD, et al. The expression of constitutively active isotypes of protein kinase C to investigate preconditioning. J Biol Chem 1998;273(36):23072–9.

[58] Lei J, Tang X, Chambers TC, Pohl J, Benian GM. Protein kinase domain of twitchin has protein kinase activity and an autoinhibitory region. J Biol Chem 1994;269 (33):21078–85.

[59] Allen BG, Walsh MP. The biochemical basis of the regulation of smooth-muscle contraction. Trends Biochem Sci 1994;19(9):362–8.

[60] Kemp BE, Pearson RB, Guerriero Jr. V, Bagchi IC, Means AR. The calmodulin binding domain of chicken smooth muscle myosin light chain kinase contains a pseudosubstrate sequence. J Biol Chem 1987;262(6):2542–8.

[61] Lukas TJ, Mirzoeva S, Slomczynska U, Watterson DM. Identification of novel classes of protein kinase inhibitors using combinatorial peptide chemistry based on functional genomics knowledge. J Med Chem 1999;42(5):910–19.

[62] Li W, Jiang YX, Zhang J, Soon L, Flechner L, Kapoor V, et al. Protein kinase C-delta is an important signaling molecule in insulin-like growth factor I receptor-mediated cell transformation. Mol Cell Biol 1998;18(10):5888–98.

[63] Kiley SC, Clark KJ, Duddy SK, Welch DR, Jaken S. Increased protein kinase C delta in mammary tumor cells: relationship to transformation and metastatic progression. Oncogene 1999;18(48):6748–57.

[64] Braun MU, Mochly-Rosen D. Opposing effects of delta- and zeta-protein kinase C isozymes on cardiac fibroblast proliferation: use of isozyme-selective inhibitors. J Mol Cell Cardiol 2003;35(8):895–903.

[65] Bowling N, Walsh RA, Song G, et al. Increased protein kinase C activity and expression of Ca2 + -sensitive isoforms in the failing human heart. Circulation 1999;99(3):384–91.

[66] Inagaki K, Hahn HS, Dorn GW, Mochly-Rosen D. Additive protection of the ischemic heart ex vivo by combined treatment with delta-protein kinase C inhibitor and epsilon-protein kinase C activator. Circulation 2003;108(7):869–75.

[67] Teicher BA, Menon K, Alvarez E, Shih C, Faul MM. Antiangiogenic and antitumor effects of a protein kinase C beta inhibitor in human breast cancer and ovarian cancer xenografts. Investig N Drugs 2002;20(3):241–51.

[68] Isakov N, Altman A. Protein kinase C(theta) in T cell activation. Annu Rev Immunology 2002;20:761–94.

[69] Ashby CD, Walsh DA. Characterization of the interaction of a protein inhibitor with adenosine 3′, 5′-monophosphate-dependent protein kinases: I. Interaction with the catalytic subunit of the protein kinase. J Biol Chem 1972;247(20):6637–42.

[70] Cheng H-C, Kemp B, Pearson R, Smith AJ, Misconi L, Van Patten SM, et al. A potent synthetic peptide inhibitor of the cAMP-dependent protein kinase. J Biol Chem 1986;261 (3):989–92.

[71] Glass DB, Cheng HC, Mende-Mueller L, Reed J, Walsh DA. Primary structural determinants essential for potent inhibition of cAMP-dependent protein kinase by inhibitory peptides corresponding to the active portion of the heat-stable inhibitor protein. J Biol Chem 1989;264(15):8802–10.

[72] Cassoni P, Sapino A, Fortunati N, Munaron L, Chini B, Bussolati G. Oxytocin inhibits the proliferation of MDA-MB231 human breast-cancer cells via cyclic adenosine monophosphate and protein kinase A. Int J Cancer 1997;72(2):340–4.

[73] Hoy JJ, Salinas Parra N, Park J, Kuhn S, Iglesias-Bartolome R. Protein kinase A inhibitor proteins (PKIs) divert GPCR-Gαs-cAMP signaling toward EPAC and ERK activation and are involved in tumor growth. Faseb j 2020;34(10):13900–17.

[74] Zynda ER, Matveev V, Makhanov M, Chenchik A, Kandel ES. Protein kinase A type II-α regulatory subunit regulates the response of prostate cancer cells to taxane treatment. Cell Cycle 2014;13(20):3292–301.

[75] Holland PM, Cooper JA. Protein modification: docking sites for kinases. Curr Biol 1999;9 (9):R329–31.

[76] Cunningham AD, Qvit N. Specific kinase inhibition using peptides to target kinase-substrate docking. Chim Oggi/Chem Today 2016;34(5):22–5.

[77] Chen YG, Hata A, Lo RS, Wotton D, Shi Y, Pavletich N, et al. Determinants of specificity in TGF-beta signal transduction. Genes Dev 1998;12(14):2144–52.

[78] Schulman BA, Lindstrom DL, Harlow E. Substrate recruitment to cyclin-dependent kinase 2 by a multipurpose docking site on cyclin A. Proc Natl Acad Sci USA 1998;95 (18):10453–8.

[79] Williams DD, Marin O, Pinna LA, Proud CG. Phosphorylated seryl and threonyl, but not tyrosyl, residues are efficient specificity determinants for GSK-3β and Shaggy. FEBS Lett 1999;448(1):86–90.

[80] Kallunki T, Su B, Tsigelny I, Sluss HK, Dérijard B, Moore G, et al. JNK2 contains a specificity-determining region responsible for efficient c-Jun binding and phosphorylation. Genes & Dev 1994;8(24):2996–3007.

[81] Qvit N, Kornfeld OS, Mochly-Rosen D. Engineered substrate-specific delta PKC antagonists to enhance cardiac therapeutics. Angew Chem Int Ed Engl 2016;55(50):15672–9.

[82] Kim J, Koyanagi T, Mochly-Rosen D. PKCdelta activation mediates angiogenesis via NADPH oxidase activity in PC-3 prostate cancer cells. Prostate 2011;71(9):946–54.

[83] Pereira S, Park E, Mori Y, et al. FFA-induced hepatic insulin resistance in vivo is mediated by PKCdelta, NADPH oxidase, and oxidative stress. Am J Physiol Endocrinol Metab 2014;307(1):E34–46.

[84] Qi X, Inagaki K, Sobel RA, Mochly-Rosen D. Sustained pharmacological inhibition of delta PKC protects against hypertensive encephalopathy through prevention of blood-brain barrier breakdown in rats. J Clin Investig 2008;118(1):173–82.

[85] Chen L, Hahn H, Wu G, et al. Opposing cardioprotective actions and parallel hypertrophic effects of delta PKC and epsilon PKC. Proc Natl Acad Sci USA 2001;98(20):11114–19.

[86] Churchill EN, Qvit N, Mochly-Rosen D. Rationally designed peptide regulators of protein kinase C. Trends Endocrinol Metab 2009;20(1):25–33.
[87] Mochly-Rosen D, Qvit N. Peptide inhibitors of protein-protein interactions. Chim Oggi/CHEM today 2010;28(1):14–16.
[88] Qvit N. Development and therapeutic applications of oligonucleotides and peptides. Chim Oggi/CHEM today 2011;29(2):4–7.
[89] Qvit N, Mochly-Rosen D. The many hats of protein kinase C delta: one enzyme with many functions. Biochem Soc Trans 2014;42(6):1529–33.
[90] Qvit N, Disatnik MH, Sho J, Mochly-Rosen D. Selective phosphorylation inhibitor of delta protein kinase C-pyruvate dehydrogenase kinase protein-protein interactions: application for myocardial injury in vivo. J Am Chem Soc 2016;138(24):7626–35.
[91] Qvit N, Joshi AU, Cunningham AD, Ferreira JC, Mochly-Rosen D. GAPDH (glyceraldehyde-3-phosphate dehydrogenase) protein-protein interaction inhibitor reveals a non-catalytic role for GAPDH oligomerization in cell death. J Biol Chem 2016;291(26):13608–21.
[92] Cantelmo AR, Cammarota R, Noonan DM, Focaccetti C, Comoglio PM, Prat M, et al. Cell delivery of Met docking site peptides inhibit angiogenesis and vascular tumor growth. Oncogene 2010;29(38):5286–98.
[93] Alto N, Carlisle Michel JJ, Dodge KL, Langeberg LK, Scott JD. Intracellular targeting of protein kinases and phosphatases. Diabetes 2002;51(Suppl 3):S385–8.
[94] Mochly-Rosen D, Khaner H, Lopez J. Identification of intracellular receptor proteins for activated protein kinase C. Proc Natl Acad Sci USA 1991;88(9):3997–4000.
[95] Qvit N, Mochly-Rosen D. Highly specific modulators of protein kinase C localization: Applications to heart failure. Drug Discov Today: Dis Mech 2010;7(2):e87–93.
[96] Ron D, Chen CH, Caldwell J, Jamieson L, Orr E, Mochly-Rosen D. Cloning of an intracellular receptor for protein kinase C: a homolog of the beta subunit of G proteins. Proc Natl Acad Sci USA 1994;91(3):839–43.
[97] Ron D, Mochly-Rosen D. An autoregulatory region in protein-kinase-C—the pseudoanchoring site. Proc Natl Acad Sci USA 1995;92(2):492–6.
[98] Beene DL, Scott JD. A-kinase anchoring proteins take shape. Curr Opin Cell Biol 2007;19(2):192–8.
[99] Rosenmund C, Carr DW, Bergeson SE, Nilaver G, Scott JD, Westbrook GL. Anchoring of protein kinase A is required for modulation of AMPA/kainate receptors on hippocampal neurons. Nature 1994;368(6474):853–6.
[100] Carlson CR, Lygren B, Berge T, Hoshi N, Wong W, Taskén K, et al. Delineation of type I protein kinase A-selective signaling events using an RI anchoring disruptor. J Biol Chem 2006;281(30):21535–45.
[101] Mochly-Rosen D, Miller KG, Scheller RH, Khaner H, Lopez J, Smith BL. p65 fragments, homologous to the C2 region of protein kinase C, bind to the intracellular receptors for protein kinase C. Biochemistry 1992;31(35):8120–4.
[102] Ron D, Luo J, Mochly-Rosen D. C2 region-derived peptides inhibit translocation and function of beta protein kinase C in vivo. J Biol Chem 1995;270(41):24180–7.
[103] Yang SH, Whitmarsh AJ, Davis RJ, Sharrocks AD. Differential targeting of MAP kinases to the ETS-domain transcription factor Elk-1. EMBO J 1998;17(6):1740–9.
[104] Bonny C, Oberson A, Negri S, Sauser C, Schorderet DF. Cell-permeable peptide inhibitors of JNK: novel blockers of beta-cell death. Diabetes 2001;50(1):77–82.
[105] Barr RK, Kendrick TS, Bogoyevitch MA. Identification of the critical features of a small peptide inhibitor of JNK activity. J Biol Chem 2002;277(13):10987–97.

[106] Gao YJ, Cheng JK, Zeng Q, Xu ZZ, Decosterd I, Xu X, et al. Selective inhibition of JNK with a peptide inhibitor attenuates pain hypersensitivity and tumor growth in a mouse skin cancer pain model. Exp Neurol 2009;219(1):146–55.

[107] Niv MY, Rubin H, Cohen J, et al. Sequence-based design of kinase inhibitors applicable for therapeutics and target identification. J Biol Chem 2004;279(2):1242–55.

[108] Lavogina D, Enkvist E, Uri A. Bisubstrate inhibitors of protein kinases: from principle to practical applications. ChemMedChem 2010;5(1):23–34.

[109] Ricouart A, Gesquiere J, Tartar A, Sergheraert C. Design of potent protein kinases inhibitors using the bisubstrate approach. J Med Chem 1991;34(1):73–8.

[110] Stebbins JL, De SK, Pavlickova P, et al. Design and characterization of a potent and selective dual ATP- and substrate-competitive subnanomolar bidentate c-Jun N-terminal kinase (JNK) inhibitor. J Med Chem 2011;54(18):6206–14.

[111] Rubin S, Qvit N. Cyclic peptides for protein-protein interaction targets: applications to human disease. Crit Rev Eukaryot Gene Expr 2016;26(3):199–221.

[112] Cunningham AD, Qvit N, Mochly-Rosen D. Peptides and peptidomimetics as regulators of protein-protein interactions. Curr OpStruct Biol 2017;44:59–66.

[113] Qvit N, Rubin SJS, Urban TJ, Mochly-Rosen D, Gross ER. Peptidomimetic therapeutics: scientific approaches and opportunities. Drug Discov Today 2017;22(2):454–62.

[114] Rubin SJ, Tal-Gan Y, Gilon C, Qvit N. Conversion of protein active regions into peptidomimetic therapeutic leads using backbone cyclization and cycloscan–how to do it yourself!. Curr Top Med Chem 2018;18(7):556–65.

[115] Rubin SJS, Qvit N. Backbone-cyclized peptides: a critical review. Curr Top Med Chem 2018;18(7):526–55.

[116] Lerner Y, Hanout W, Ben-Uliel SF, Gani S, Leshem MP, Qvit N. Natriuretic peptides as the basis of peptide drug discovery for cardiovascular diseases. Curr Top Med Chem 2020;.

[117] Rubin SJ, Qvit N. Engineering "antimicrobial peptides" and other peptides to modulate protein-protein interactions in cancer. Curr Top Med Chem 2020;.

Chapter 16

Therapeutic peptidomimetics for infectious diseases

Shikun Ge[1,2], Long Xu[1,2] and Xiaoying Zhang[1,2]

[1]*Department of Biology, Centre of Molecular and Environmental Biology, University of Minho, Braga, Portugal,* [2]*Chinese-German Joint Laboratory for Natural Product Research, State Key Laboratory of Biological Resources and Ecological Environment of Qinba Areas, College of Biological Science and Engineering, Shaanxi University of Technology, Hanzhong, P.R. China*

16.1 Introduction

Healthcare is facing diversified challenges of infectious diseases along with the increase of antibiotic resistance as well as evolution and spread of new zoonoses [i.e., Coronavirus disease 2019 (COVID-19)]. Pharmaceutical companies are facing major challenges in cost-effective drug research and development, which has caused a substantial delay in launching new antiinfection agents to the markets in the past decades. This phenomenon is not only related to human healthcare but also observed in veterinary medicine and agriculture applications [1]. As a strategy to overcome these issues, peptide and peptidomimetics (modified peptides) has gained increasing interest and have been increasingly appeared on the market in the past two decades, mostly owing to their advantages in biocompatibility and target selectivity compared to conventional drugs [2].

Peptidomimetics are organic molecules composed of 2–10 amino acids to mimic some properties of natural peptide ligands. They are designed and applied as novel antiinfection agents to overcome the barriers posed by natural peptides, such as premature proteolysis, low availability and less receptor selectivity [2].

16.2 Emerging peptidomimetic technologies

Novel strategies in peptidomimetic design and generation are developing along with the maturation and utilization of novel technologies, such as antigen epitope analysis, immunoassays, overlapping peptide synthesis and peptide structure analysis, next-generation gene sequencing, peptide simulation technology, and bioinformatics [3]. These technologies allow a better

Peptide and Peptidomimetic Therapeutics. DOI: https://doi.org/10.1016/B978-0-12-820141-1.00006-6

modeling approach for simplified peptide design, as well as three-dimensional pharmacophore model exploration and reassembling, to better match with the receptor of desire [2].

16.2.1 Antibody mimetics

Antibody mimetics are peptidomimetics derived from the functional antibody fragments [e.g., single-chain variable fragment (scFv), antigen-binding fragment (Fab) or variable fragment (Fv)] by protein-directed evolution or complementarity determining regions (CDRs) fusion through framework region (FR) of Fab in different sequences.

Directed evolution is currently used to harness the power of natural selection to evolve proteins with desired properties, which involves four major steps: (1) Identification: the suitable sequence is chosen based on its perceived proximity to the desired function and its evolvability; (2) Diversification: the parent sequence is subjected to diversification by error-prone PCR and DNA shuffling; (3) Selection: the generated library is tested for the presence of mutants by screening or selection; (4) Amplification: the variants are screened, selected and replicated multiple-fold to achieve a variant with the desired properties [4]. Vast combinatorial libraries have been constructed by randomizing amino acid positions in structurally variable loops of proteins or by exon-shuffling and phage display [5]. The mutant libraries are subjected to phage display or ribosome display selection against a target of interest.

CDR-FR peptides are constructed by fusing two CDRs through cognate FR. Protein antigens are usually recognized by all six CDRs from the both VL and VH domains of the intact antibody combining site(s). The CDRH3 loop is usually considered the most indispensable part of the mimetic, as it is often the most accessible of the CDR loops and is routinely involved in antigen binding to the greatest extent due to its greater sequence diversity. The C-terminus of the chosen CDR1 or CDR2 loop and the N-terminus of the selected CDRH3 loop are joined with a FR selected from VH or VL [6].

16.2.2 Tripeptide analogs

The proteasome is an essential nonlysosomal intracellular protease found in all eukaryotic cells. Dysregulation of the proteasome can lead to altered protein activity and buildup of misfolded protein complexes in mammalian cells, resulting in cancer and a range of other diseases [7]. A series of drug-like molecules have been synthesized with different specificities [8]. One of the well-known tripeptide aldehyde, Leupeptin, inhibits the trypsin-like (TL) activity of the proteasome and is also reported to bind to calpain [9]. The classic tripeptide aldehyde inhibitor is active against all caspase-like (CPL), TL and chymotrypsin-like (CTL) activities of the proteasome but shows diminished activity against calpain. Comparison of inhibitor substrate preference indicates that the N-terminal group of tripeptide aldehydes confers a degree of specificity against protease targets [9].

The Openeye suite of bioinformatic tools was utilized to undertake docking simulations to investigate the binding modes of inhibitors and tripeptide aldehyde inhibitors to better understand the observed selectivity profiles [8]. Inhibitors were docked into the active site of the crystal structures of calpain 1 (PDB 2G8J) and calpain 2 (PDB 3BOW) [9]. The selected "Compound 1" (IC_{50} 51 nM for calpain 1) docked into calpain 1 with all three critical hydrogen bonds present a β-Strand backbone conformation of the inhibitor binding to Gly 208 and Gly 271 in the active site. The docked poses for "Compound 2" (IC_{50} 8 nM for calpain 1), also revealed a β-Strand conformation of the tripeptide aldehyde inhibitor, but with only two of the three hydrogen bonds apparently interacted. Inhibitors 1 (IC_{50} 6 nM) and 2 (IC_{50} 11 nM) were also docked with the calpain 2 enzyme, with both showing a β-Strand backbone conformation with "Compound 1" interacting via two of the three hydrogen bonds and inhibitor 2 with all three hydrogen bonds [9].

16.2.3 Cyclic heptapseudopeptides

To minimize some limits of conventional peptides such as low bioavailability and low conformational flexibility, it is necessary to the radical redesign of peptide structures. Carbohydrates, particularly the cyclic oligosaccharides, can be used to produce novel peptides and peptidomimetics as these hybrid molecules, chimeras, have properties arising largely from the combination of structural characteristics of carbohydrates with the functional group diversity of peptides. The successful applications of carbohydrate scaffolds for non-peptide peptidomimetics of cyclic peptides demonstrate the general utility of carbohydrates in peptide and protein architecture [10]. A new family of peptidomimetics-cyclic heptapseudopeptides was generated, inspired by a natural bacterial peptide (Fig. 16.1). Recently, two peptidomimetics (Fig. 16.1) are found to be effective against methicillin-resistant *Staphylococcus aureus* (MRSA) in mild and severe sepsis mouse models without exhibiting toxicity on human erythrocytes and kidney cells, zebra fish embryos, and mice. These two peptidomimetics also exhibited significant inhibitory effects on *Psuedomonas aeruginosa* [1].

The activity of heptapseudopeptides was explained by the ability of unnatural amino acids to strengthen dynamic association with bacterial lipid bilayers and to induce membrane permeability, leading to bacterial death. Based on structure determination, domains surrounded by an extended hydrophobic core could improve bactericidal activity [1]. The ultrastructure changes were examined by scanning electron microscopy (SEM) and transmission electron microscopy (TEM) in MRSA by each peptidomimetic at the minimum inhibitory concentration (MIC). The treated MRSA showed

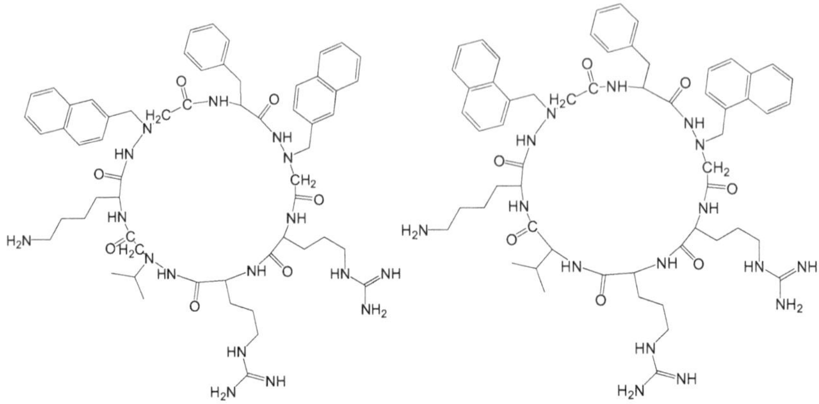

FIGURE 16.1 Two cyclic heptapseudopeptides structures. *Adapted from Nicolas, I, Bordeau, V, Bondon, A, Baudy-Floc'h, M, & Felden, B. Novel antibiotics effective against gram-positive and -negative multi-resistant bacteria with limited resistance. PLoS Biol 2019;17(7): e3000337.*

inflated and distorted shapes, and their cell walls seemed eroded by these two peptidomimetics. TEM micrographs suggested membrane deformation and detachment, with internal lateral expansion and stacks. SEM and TEM were also examined in Gram-negative cells exposed to each pseudopeptide at their respective MIC, evidencing membrane and cytoplasmic deformations, as well as cell division defects. To explore their ability to permeate Gram-negative outer and inner membranes, leakage assays were also conducted on *Escherichia coli* cells. All pseudopeptides permeated the outer and inner *E. coli* membranes. Thus, these new drugs may act by membrane permeation of both Gram-negative bacteria (GNB) and Gram-positive bacteria (GPB), with perturbations inside the cells [1].

16.3 Peptidomimetics in the therapy of infectious diseases

As a rapid developing drug design technology, a series of peptidomimetics molecules have been approved and applied for antimicrobial therapeutics, and more peptidomimetics are in pipeline of drug development, to better overcome the drawbacks of conventional drugs [11,12] A remarkable fact is that in the recent years, most Food and Drug Administration-approved antiviral agents are peptidomimetics and macrocyclic compounds that interact with the active site of a targeted protease [12].

16.3.1 Plasmepsin inhibitors with potent antimalarial activity

Malaria is a life-threatening disease caused by *Plasmodium* parasites and transmitted by mosquitoes. The currently used drugs, including artemisinin, are largely impeded by the widespread resistance of the parasites [13]. In a large-scale cell-based (phenotypic) high through screening (HTS) assay conducted by GlaxoSmithKline (GSK), a number of antimalarial compounds were identified, including a drugable hydroxyethyl amine derivative compound named 1a [14], which is an inhibitor of the *P. falciparum* aspartic proteases-plasmepsin subtypes Plm I, Plm II, and Plm IV, with particularly high potency against Plm IV. Compounds 1b and 1c (Fig. 16.2), as well as potent inhibitors of Plm IV and structurally simplified analogs of 1a, which remain highly effective in inhibiting the growth of *Plasmodium falciparum* [13]. Based on sequence alignment of Plm IV and Cat D, putative selectivity inducing structural motifs were sought in S3 and S4 subpocket-occupying substituents of the inhibitors. The potent antimalarial activity did not correlate with the relatively weak inhibition of Plm I and II; in contrast, there was a much better correlation with Plm IV inhibition [14]. A more detailed investigation of the action mechanism of the selected compounds showed that they interfered with parasite egress and maturation of the parasite serine protease SUB1, indicating a strong link between antimalarial activity, inhibition of the no hemoglobinase plasmepsin and SUB1 maturase Plm X. Further studies should clarify whether cooperative Plm IV and Plm X inhibition or only Plm X inhibition is necessary to achieve optimal antimalarial activity [13].

16.3.2 Application of peptidomimetics for targeting multidrug-resistant organisms

With the increased cases of antibiotic resistance, the control of infectious diseases with antibiotics is still perilously fragile. Peptidomimetics is of particular interest due to their potent antimicrobial activity, versatile design, and convenient optimization via assembly by standard solid-phase procedures [15,16].

P. aeruginosa is one of the most common nosocomial pathogens and no drug is yet available for the treatment of infections caused by *P. aeruginosa* [17]. A new family of potent peptidomimetic antibiotics (POL7001 and

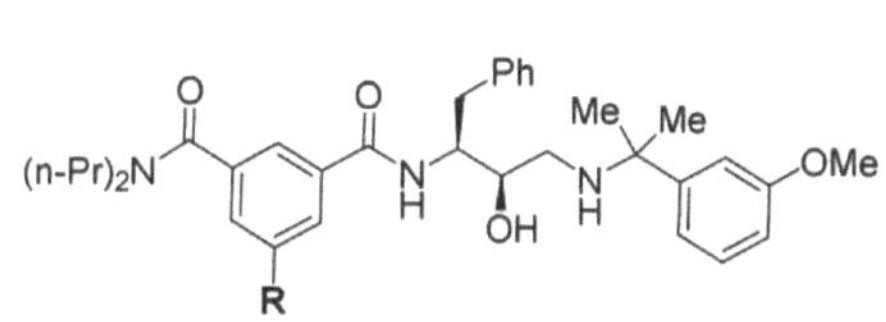

1b

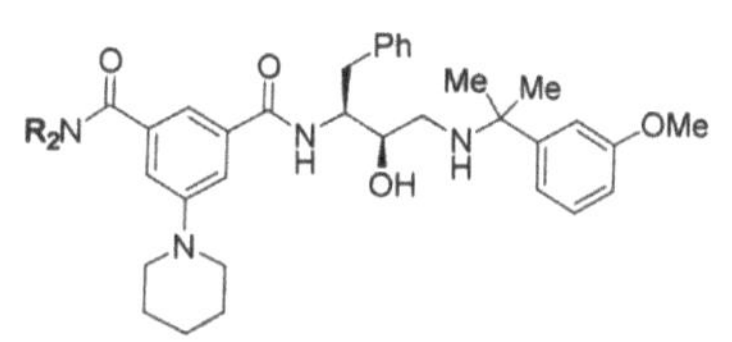

1c

FIGURE 16.2 Representative Plm inhibitors 1b and 1c. *Adapted from Zogota R, Kinena L, Withers-Martinez C, Blackman M, Bobrovs R, Panteļejevs T, et al. Peptidomimetic plasmepsin inhibitors with potent anti-malarial activity and selectivity against cathepsin D. Eur J Medicinal Chem 2018;163(1): 344–352.*

POL7080), patterned on the naturally occurring cationic β-hairpin antibiotic protegrin I (PG-I), has recently been described [16]. The novel compounds POL7001 and POL7080, which are effective against a wide range of strains, display potent and selective action against more than one hundred *P. aeruginosa* clinical strains, including many isolates from cystic fibrosis (CF) patients. POL7001 and POL7080 are active in the nanomolar range against Gram-negative *Pseudomonas sp.* but are largely inactive against other GNB and GPB [18]. Biochemical and genetic studies showed that the peptidomimetics has a nonmembrane-lytic mechanism of action. Instead, these novel compounds interact with and functionally inhibit the Pseudomonas outer membrane (OM) protein LptD, which serves an essential lipopolysaccharide (LPS) transport function [16].

16.3.3 Antiinflammatory peptidomimetics

The activation of leukocytes and the subsequent immune cascade play essential roles in infectious inflammation. Dysregulation of these immune responses or excess leukocyte activation can induce tissue damage, organ dysfunction and mortality [19]. Formyl peptide receptors (FPRs) are functionally diverse pattern recognition receptors responsible for recognizing different endogenous damage-associated molecular patterns or exogenous pathogen-associated molecular patterns, FPRs mediate leukocyte activation during inflammation [20]. FPR1 antagonists and FPR2 agonists have demonstrated significant antiinflammatory effects based on in vitro and in vivo studies [21].

A different strategy was invented by Cidara Therapeutics, Inc. to link an antifungal agent with an FPR ligand-inducing chemotaxis. The use of the terms "modulator" and "modulation" to classify agent interactions with FPRs is not limited to receptor agonists or antagonists. Consequently, FPR modulators are used primarily for the treatment of acute immune cell-dominant and chronic inflammatory lung diseases, such as chronic obstructive pulmonary disease, asthma, emphysema, CF, and acute lung injury [21]. Advanced techniques of genetic engineering have been increasingly used to discover the diverse physiological and pathological roles of the FPR family via gene knockout studies. The corresponding ligands have been isolated from compound library screening. The FPR family also affects malignancy progression, nociceptive processing, autoimmune disease, and intestinal epithelial homeostasis [22]. These alternative therapeutic indications of FPR modulators are still uncertain but may emerge in the future. Among the FPR family members, FPR2 is an appealing therapeutic target as FPR2 is involved in many physiological functions and pathogeneses and FPR2 interacts with structurally diverse ligands [19]. Library screening studies largely focused on the identification of novel compounds capable of FPR2 modulation. An increasing number of synthesized compounds targeting FPRs, especially potential FPR1 antagonists and FPR2 agonists, have been disclosed in patents [23].

16.3.4 Peptidomimetics targeting oral bacteria

A variety of approaches have been pursued to develop novel therapeutics that can effectively combat or target oral bacteria. The structure and function of naturally occurring antimicrobial peptides from various sources have been exploited to develop numerous peptide mimetics that show greater activity, stability and other desirable characteristics compared to their parent peptides [24]. Many of them have demonstrated potent activity in preclinical investigations and animal model study of oral disease. However, for most peptidomimetics, their activities against the more complex oral microbial communities that occur in vivo have not been thoroughly investigated [25]. Future studies to further develop these active agents will likely begin to focus on developing suitable formulations for oral delivery, additional toxicity testing and evaluating stability and pharmacokinetics in the oral environment.

The specifically targeted antimicrobial peptide "C16G2 STAMP," is capable of selectively killing cariogenic *Streptococcus mutans*, and has already been formulated in an oral gel, in a dental varnish and in oral strips. The initial clinical evaluation of these formulations indicated that a single varnish application outperformed multiple gel applications and resulted in a significant reduction in *S. mutans* populations. Further Phase II clinical trial is ongoing. Additional peptide and/or small molecule mimetics will likely undergo clinical evaluation in the future [26]. C16G2 is relatively stable in phosphate-buffered saline (PBS) for over 20 h at 4°C and for at least 4 h at 25°C and exhibited a half-life of approximately 19 min in human saliva [25], suggesting that C16G2 could be formulated in a basic PBS mouth rinse to evaluate its activity in vivo in humans. In a pilot clinical study, patients using the placebo rinse showed a daily increase in *S. mutans* levels relative to baseline during the 4-day period whereas, in the 0.04% C16G2 rinse group, there was no increase in *S. mutans* despite the daily sucrose challenges [24].

16.3.5 Application of peptidomimetics in microbial vaginosis

Antimicrobial resistance and infection recurrence are critical issues in treating vaginosis, which is the most common vaginal disorder found in women of reproductive age. The abnormal vaginal microbiome, or vaginal dysbiosis, which includes bacterial (e.g., *Gardnerella vaginalis*) or fungal (e.g., *Candida albicans*) vaginosis, has been linked to adverse pregnancy outcomes, pelvic inflammatory disease, and increased risk of sexually transmitted infections and other reproductive health problems [27,28].

Novel alternatives, such as peptidomimetics, to traditional antibiotics or antifungals have the potential to address this challenge.

Although the ability to cause vaginitis is likely a complex process involving multiple interactions between *C. albicans* and the host, secreted aspartic proteinases (SAPs) appear to be a major virulence factor in these mucosal infections. *C. albicans* strains express at least eight distinct genes (SAP1−8) at different stages of infection, indicating that different SAPs have distinct functions. In particular, SAP2 is one of the most expressed enzymes implicated in host invasion, and it has been recognized as a crucial virulence factor. SAP2 includes a class of bicyclic compounds, which have become a target for screening peptide inhibitors. The screening of a library of small molecule peptidomimetics toward SAP2 allowed for the identification of two potential hit compounds, namely APG12 and APG19 (Fig. 16.3) [30]. In the in vitro enzyme inhibitory assay, APG12 and APG19 showed comparable inhibitory potency to pepstatin A, a low molecular weight inhibitor of acid proteases. The in vivo trials showed that the two candidate compounds maintain the inhibitory potency, whereas fluconazole failed, demonstrating the inhibition of SAP2 as a valid approach for the development of new drugs against drug-resistant *C. albicans* [29]. The effect of APG12 and APG19 creams on fungal clearance in rat vitreous was similar to that of natural SAP2 inhibitors pepstatin A and fluconazole. Importantly, APG cream treatment was able to control infection caused by fluconazole-resistant and caspofungin-resistant *C. albicans* strains, a finding of special interest in light of increasing fungal resistance [31].

Cationic gemini-like amphiphiles (CAms) have been shown to effectively target *G. vaginalis* biofilms with notable minimal inhibitory and bactericidal effects on planktonic lactobacilli and biofilms [32]. CAm compounds synthesized include Ether-(series 1) and Ester-(series 2) linked CAms with varying cationic linker lengths (nC, where n is the total carbon linker length) (Fig. 16.4). Two new series of CAms were developed with varying levels of flexibility between the hydrophilic head groups and the hydrophobic domains

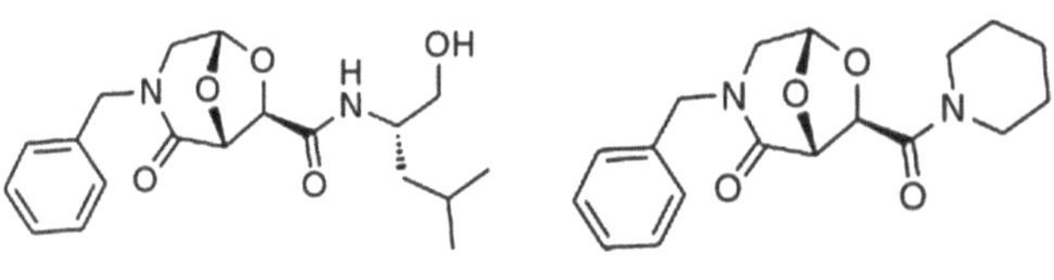

APG12 APG19

FIGURE 16.3 Chemical structure of lead compounds APG12 and APG19. *Structure adapted from DE Bernardis F., Arancia S., Tringali G., Greco M., Ragazzoni E., Calugi C., et al. Evaluation of efficacy, pharmacokinetics and tolerability of peptidomimetic aspartic proteinase inhibitors as cream formulation in experimental vaginal candidiasis. J Pharm Pharmacology 2014;66(8): 1094−1101.*

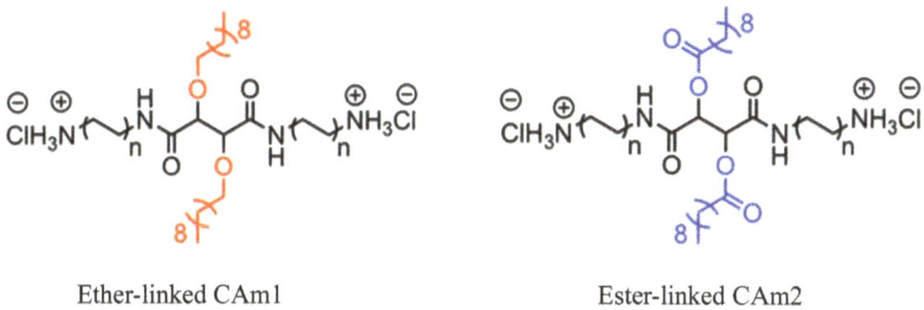

FIGURE 16.4 Chemical compositions of Cationic gemini-like amphiphiles [33]. *Structure adapted from Weeks RM, Moretti A, Song S, Uhrich KE, Karlyshev AV, & Chikindas ML. Cationic amphiphiles against* Gardnerella vaginalis *resistant strains and bacterial vaginosis-associated pathogens. Pathog Dis 2019;77(8).*

[34]. The cationic nature of the CAms may lead to coating of the negatively charged attachment surface, thereby preventing initial attachment of bacterial cells and subsequent biofilm formation. CAms have also been demonstrated to act via insertion and disruption of the bacterial membrane, leading to cell rupture and eventual cell death [32].

16.3.6 Application of peptidomimetics to diphtheria toxin

Diphtheria is a severe infection caused by the exotoxin-producing *Corynebacterium diphtheriae*. This disease is characterized by psuedomembrane formation in the pharynx and swollen lymph nodes and can lead to life-threatening complications like myocarditis or polyneuritis [35]. Although effective vaccines are available, diphtheria is still a dangerous disease in certain developing countries. The released toxin is cleaved by furin into two subunits (A and B) [36]. Furin cleaves and thereby activates several bacterial toxin precursors, including anthrax, diphtheria, or Shiga toxins [37]. Therefore the effective inhibition of furin represents a potential strategy for the short-time treatment of numerous severe infectious diseases.

Novel elongated and shortened derivatives of the peptidomimetic furin inhibitor phenylacteyl-Arg-Val-Arg-4-amidinobenzylamide were synthesized, contain additional basic residues at the N-terminus and inhibit furin in the low picomolar range (Fig. 16.5). Furthermore, the p5-moiety was removed to decrease the molecular weight of this inhibitor type [37]. To inhibit the furin-catalyzed cleavage of diphtheria toxin in early endosomes, only nanomolar inhibitor concentrations were needed to obtain significant effects on toxin activation. Multibasic inhibitors have a tendency to accumulate at the negatively charged cell surface followed by endocytotic uptake, comparable to polybasic cell-penetrating peptides [38]. This mechanism could explain the improved potency of these inhibitors when addressing endosomal targets.

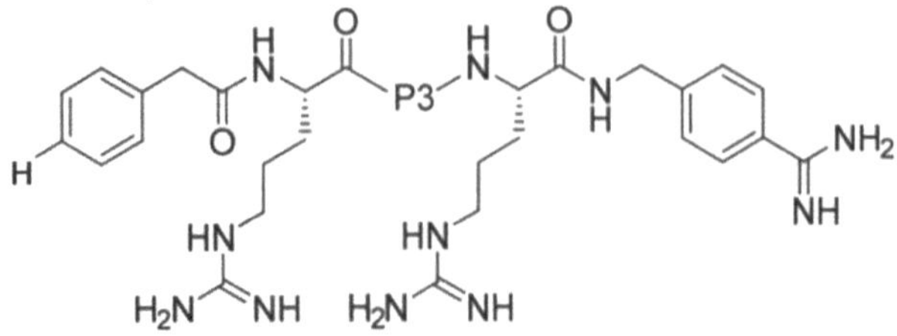

FIGURE 16.5 Furin inhibitors. *Structure adapted from Hardes K, Ivanova T, Thaa B, Mcinerney GM, Klokk TI, Sandvig K, et al. Elongated and shortened peptidomimetic inhibitors of the proprotein convertase furin. ChemMedChem 2017;12(8): 613–620.*

All presently known and more potent synthetic furin inhibitors contain at least three or even more basic residues, which limits their cellular bioavailability and makes them unsuitable for oral administration. Nevertheless, the parenteral application should be possible with this inhibitor type, for example, for the prophylaxis and treatment of eye infections caused by *P. aeruginosa* [39].

16.4 Peptidomimetics in vaccine design

Peptidomimetics vaccine utilizes the advances in microbial genomics and proteomics to enable high-throughput production and testing of recombinant microbial proteins, to identify potential protein-based vaccine candidates. According to different preparation methods, the peptidomimetics vaccine is mainly divided into polyantigen peptide vaccine, synthetic peptide vaccine and lipopeptide vaccine; according to different functions, it can be divided into preventive vaccines and therapeutic vaccines. Compared with traditional attenuated or inactivated vaccines, peptidomimetics vaccine has lower production cost, higher specificity, more safety and easy standardization. In particular, the synthetic peptide vaccine prepared by chemical synthesis of antigen epitope amino acid sequence is considered to be the most ideal vaccine at present [40]. Peptidomimetics vaccine has been used in the fields of bacteria and parasites, but mainly in viruses, especially hepatitis C virus (HCV), human papillomavirus (HPV), and human immunodeficiency virus (HIV), which pose great public health threats.

HCV is susceptible to mutate and no vaccine yet can effectively prevent HCV infection. HPV is the culprit causing genital wart lesions, cervical epithelial neoplasia and cervical cancer. However, the HPV and HCV peptidomimetics vaccines described above need to match a specific human leukocyte antigen (HLA) type and require an adjuvant to enhance the immune response. If a longer peptidomimetics fragment with an affinity for the HLA tissue type is used, the application scope of this vaccine may be expanded.

At present, highly active antiretroviral therapy (HAART) is effective in treating HIV infection, but it can cause myelosuppression in 42.6% of patients during the treatment [41]. As an alternative, the vac-4x therapeutic peptidomimetics vaccine, consisting of four synthetic peptides based on the conserved region of HIV core protein p24, is one of the emerging approaches. The vaccine was well tolerated in phase II B clinical trials. A significant reduction in viral load was observed despite there was no statistically significant difference in CD4 count between the trial group and the placebo group [42].

16.5 Peptidomimetics in diagnosis of infectious disease

Peptidomimetic-based materials hold great potential in the diagnosis of infectious diseases. It is tempting to suggest that methods for the de novo prediction of interacting sequences can be used to mimic both antigen–antibody interactions and the complex (even multivalent) interactions in general [43]. At present, peptidomimetics antigens are mainly applied in the diagnosis of infectious diseases in virus-related fields, especially when the serological detection of a virus is prone to have cross-reactivity, or when a virus has multiple serotypes that are difficult to distinguish. Peptidomimetics can be generated against a range of biomarkers associated with specific infectious diseases for the development of electronic and other formats of multiplex biosensors, reagents for detection in routine immunological analysis.

Peptidomimetics antigens are also used in the diagnosis of parasites and bacteria. In a diagnosis of malaria [3], seven specific peptides were screened from an overlapping polypeptide library, and further developed into peptidomimetics that constructed and combined them into microarray to detect *Plasmodium falciparum* and *Plasmodium vivax*, with total sensitivity and specificity of 95.6% and 99.1%, respectively. In the diagnosis of tuberculosis and tuberculous meningitis, four peptides were screened and further developed into peptidomimetics for *Mycobacterium tuberculosis* antigen 85 complex detection and the establishment of antibody capture ELISA, showed that the sensitivity and specificity of three peptidomimetics for the diagnosis of tuberculosis and tubercular meningitis exceeded 85% and 84%, respectively [44].

16.6 Future prospects of peptidomimetics in infectious diseases

Owing to high target retention, rapid tissue penetration and blood clearance, peptidomimetics are gaining importance as therapeutics and diagnostics for infectious diseases. Compared to macro biological molecules such as antibodies, peptidomimetics have lower molecular weights and higher stability, but shorter half-life as they cannot bind host transport receptors like FcRn.

Conjugation strategies using proline–alanine–serine (PASylation) or albumin can extend the half-life of peptidomimetics [5]. There are clear unmet needs where use of peptidomimetics is critical for combating infectious diseases; further study and improvement of peptidomimetic compounds is a priority.

References

[1] Nicolas I, Bordeau V, Bondon A, Baudy-Floc'h M, Felden B. Novel antibiotics effective against gram-positive and -negative multi-resistant bacteria with limited resistance. PLoS Biol 2019;17(7):e3000337.

[2] Shah A, Barathi B, Nair A, Rajam C. Peptidomimetics as a cutting edge tool for advanced healthcare. Int J Pharm Sci Res 2017;8(12):4992–5000.

[3] Lu Y, Li Z, Teng H, Xu H, Qi S, He JA, et al. Chimeric peptide constructs comprising linear B-cell epitopes: application to the serodiagnosis of infectious diseases. Sci Rep 2015;21(5):13364–75.

[4] Romero P, Arnold F. Exploring protein fitness landscapes by directed evolution. Nat Rev Mol Cell Biol 2009;10(12):866–76.

[5] Baloch A. Antibody mimetics: promising complementary agents to animal-sourced antibodies. Crit Rev Biotechnol 2014;36(2):1–8.

[6] Qiu XQ, Wang H, Cai B, Wang LL, Yue ST. Small antibody mimetics comprising two complementarity-determining regions and a framework region for tumor targeting. Nat Biotechnol 2007;25(8):921–9.

[7] Mani A, Gelmann EP. The ubiquitin-proteasome pathway and its role in cancer. J Clin Oncol 2005;23(21):4776–89.

[8] Kelley B, Brown S, Warren G, Muchmore S. POSIT: flexible shape-guided docking for pose prediction. J Chem Inf Modeling 2015;55(8):1771–80.

[9] Pehere A, Nguyen S, Garlick S, Wilson D, Hudson I, Sykes M, et al. Tripeptide analogues of MG132 as protease inhibitors. Bioorganic & Medicinal Chem 2019;27(2):436–41.

[10] Jensen KJ, Brask J. Carbohydrates in peptide and protein design. Biopolymers 2005;80 (6):747–61.

[11] Gomes B, Augusto M, Romão, Felício M, Hollmann A, Franco O, et al. Designing improved active peptides for therapeutic approaches against infectious diseases. Biotechnol Adv 2018;36(2):415–29.

[12] Chang K, Kim Y, Lovell S, Rathnayake AD, Groutas WC. Antiviral drug discovery: norovirus proteases and development of inhibitors. Viruses 2019;11(2):197–211.

[13] Zogota R, Kinena L, Withers-Martinez C, Blackman M, Bobrovs R, Panteļejevs T, et al. Peptidomimetic plasmepsin inhibitors with potent anti-malarial activity and selectivity against cathepsin D. Eur J Medicinal Chem 2018;163(1):344–52.

[14] Jaudzems K, Tars K, Maurops G, Ivdra N, Otikovs M, Leitans J, et al. Plasmepsin inhibitory activity and structure-guided optimization of a potent hydroxyethylamine-based antimalarial hit. ACS Medicinal Chem Lett 2014;5(4):373–7.

[15] Molchanova N, Hansen PR, Franzky H. Advances in development of antimicrobial peptidomimetics as potential drugs. Molecules 2017;22(9):1430–90 Basel.

[16] Srinivas N, Jetter P, Ueberbacher B, Werneburg M, Zerbe K, Steinmann J, et al. Peptidomimetic antibiotics target outer-membrane biogenesis in *Pseudomonas aeruginosa*. Science 2010;327(5968):1010–13.

[17] Page M, Heim J. Prospects for the next anti-Pseudomonas drug. Curr OpPharmacology 2009;9(5):558–65.

[18] Bragonzi A. Fighting back: peptidomimetics as a new weapon in the battle against antibiotic resistance. Sci Transl Med 2010;2(21) 21-21.

[19] Tsai YF, Shun-Chin Y, Hwang TL. Formyl peptide receptor modulators: a patent review and potential applications for inflammatory diseases (2012–2015). Expert Opin Therapeutic Pat 2016;26(10):1139–56.

[20] Sachdeva S. Peptides as 'drugs': the journey so far. Int J Peptide Res Therapeutics 2016;23(1):49–60.

[21] Chew MF, Tham HW, Mohamed, Ibrhaim MR, Hassan S. Anti-dengue virus serotype 2 activity and mode of action of a novel peptide. J Appl Microbiology 2015;119 (4):1170–80.

[22] Zhang H, Jin R, Yao C, Zhang T, Wang M, Xia W, et al. Combination of long-acting HIV fusion inhibitor albuvirtide and LPV/r showed potent efficacy in HIV-1 patients. AIDS Res Ther 2016;13(1):8–12.

[23] Edwards IA, Elliott A, Kavanagh A, Zuegg J, Blaskovich M, Cooper M. Contribution of amphipathicity and hydrophobicity to the antimicrobial activity and cytotoxicity of β-hairpin peptides. ACS Infect Dis 2016;2(6):442–50.

[24] Tan J, Patil P, Luzzio F, Demuth D. In vitro and in vivo activity of peptidomimetic compounds that target the periodontal pathogen *Porphyromonas gingivalis*. Antimicrobial Agents Chemotherapy 2018;62(7) 10.1128/AAC.00400-18.

[25] Patil P, Tan J, Demuth D, Luzzio F. 1,2,3-Triazole-based inhibitors of *Porphyromonas gingivalis* adherence to oral streptococci and biofilm formation. Bioorganic & Medicinal Chem 2016;24(21):5410–17.

[26] Sztukowska M, Roky M, Demuth D. Peptide and non-peptide mimetics as potential therapeutics targeting oral bacteria and oral biofilms. Mol Oral Microbiology 2019;34(5):169–82.

[27] Bagnall P, Rizzolo D. Bacterial vaginosis: a practical review. J Am Acad Physician Assistants 2017;30(12):15–21.

[28] Paavonen J, Brunham RC. Bacterial vaginosis and desquamative inflammatory vaginitis. N Engl J Med 2018;379(23):2246–54.

[29] DE Bernardis F, Arancia S, Tringali G, Greco M, Ragazzoni E, Calugi C, et al. Evaluation of efficacy, pharmacokinetics and tolerability of peptidomimetic aspartic proteinase inhibitors as cream formulation in experimental vaginal candidiasis. J Pharm Pharmacology 2014;66(8):1094–101.

[30] Trabocchi A, Gloria M, Guarna F, Machetti F, Scarpi D, Guarna A. Design, synthesis, and applications of 3-aza-6,8-dioxabicyclo[3.2.1]octane-based scaffolds for peptidomimetic chemistry. ChemInform 2006;37(20):0331–53.

[31] Sanglard D, Odds F. Resistance of Candida species to antifungal agents: molecular mechanisms and clinical consequences. Lancet Infect Dis 2002;2(2):73–85.

[32] Algburi A, Zehm S, Netrebov V, Bren AB, Chistyakov V, Chikindas ML. Subtilosin prevents biofilm formation by inhibiting bacterial quorum sensing. Probiotics Antimicrobial Proteins 2017;9(1):81–90.

[33] Weeks RM, Moretti A, Song S, Uhrich KE, Karlyshev AV, Chikindas ML. Cationic amphiphiles against *Gardnerella vaginalis* resistant strains and bacterial vaginosis-associated pathogens. Pathog Dis 2019;77(8) pii: ftz059.

[34] Moretti A, Weeks RM, Chikindas M, Uhrich KE. Cationic amphiphiles with specificity against gram-positive and gram-negative bacteria: chemical composition and architecture combat bacterial membranes. Langmuir Acs J Surf & Colloids 2019;35(16):5557–67.

[35] Hadfield TL, Mcevoy P, Polotsky Y, Tzinserling VA, Yakovlev AA. The pathology of diphtheria. J Infect Dis 2000;181(Suppl 1):S116–20.

[36] Tsuneoka M, Nakayama K, Hatsuzawa K, Komada M, Kitamura N, Mekada E. Evidence for involvement of furin in cleavage and activation of diphtheria toxin. J Biol Chem 1993;268(35):26461–5.

[37] Hardes K, Ivanova T, Thaa B, Mcinerney GM, Klokk TI, Sandvig K, et al. Elongated and shortened peptidomimetic inhibitors of the proprotein convertase furin. ChemMedChem 2017;12(8):613–20.

[38] Fuchs SM, Raines RT. Internalization of cationic peptides: the road less (or more?) traveled. Cell Mol Life Sci 2006;63(16):1819–22.

[39] Karicherla P, Hobden JA. Nona-D-arginine therapy for *Pseudomonas aeruginosa* keratitis. Investig Ophthalmol & Vis Sci 2009;50(1):256–62.

[40] Li W, Joshi MD, Singhania S, Ramsey KH, Murthy AK. Peptide *Vaccine: Progress and Challenges*. Vaccines 2014;2(3):515–36 Basel.

[41] Lu DY, Wu HY, Yarla NS, Xu B, Ding J, Lu TR. HAART in HIV/AIDS treatments: future trends. Infect Disord Drug Targets 2018;18(1):15–22.

[42] Pollard R, Rockstroh J, Pantaleo G, Asmuth D, Peters B, Lazzarin A, et al. Safety and efficacy of the peptide-based therapeutic vaccine for HIV-1, Vacc-4x: A phase 2 randomised, double-blind, placebo-controlled trial. Lancet Infect Dis 2014;14(4):291–300.

[43] Cretich M, Gori A, D'annessa I, Chiari M, Colombo G. Peptides for infectious diseases: from probe design to diagnostic microarrays. Antibodies 2019;8(1):23–38.

[44] Kashyap R, Shekhawat S, Nayak A, Purohit H, Taori G, Daginawala H. Diagnosis of tuberculosis infection based on synthetic peptides from *Mycobacterium tuberculosis* antigen 85 complex. Clin Neurol Neurosurg 2013;115(6):678–83.

Chapter 17

Antiparasitic therapeutic peptidomimetics

Wessal Hanout and Nir Qvit
The Azrieli Faculty of Medicine in the Galilee, Bar-Ilan University, Safed, Israel

17.1 Introduction to parasites

Parasite infections are a serious and devastating public health problem worldwide and are accountable for high mortality and morbidity rates, affecting 3.5 billion people leading to 450 million illnesses [1]. The name parasite originates from the Greek word parasitos, from pará, "beside," and from sîtos, "food," meaning "one who eats at the table of another and repays him with flattery" [2]. The biological definition of parasitism is a type of symbiosis where a parasite is an organism that lives at the expense of its human or animal host organism by consuming host nutrients.

Some parasites do not harm their host organism (e.g., *Chilomastix mesnili*) [3], while other parasites can invade and harm the host organism, causing disease that can manifest with a wide spectrum of signs and symptoms (e.g., *Toxoplasma gondii*) [4]. In addition, the relationship between the parasite and the host organism can be mutual, in which both sides benefit from their interaction (e.g., *Paramecium tetraurelia* and *Caedibacter*) [5]. In nature, there are several forms and categories of parasitism, and a parasite can be categorized into more than one group depending on the relationship, severity of parasitism, as well as other characteristics. One type of parasites are the facultative parasites, which are capable of living and reproducing individually, inside or outside of host cells, and they do not rely on the host organism to complete their life cycle (e.g., *Entamoeba histolytica*) [6]. Obligate parasites, on the other hand, are necessarily dependent on the host organisms to complete their life cycle and have no other alternative. An example of obligate parasites is *Plasmodium*, which are intracellular parasites including the species responsible for malaria in humans [7].

Parasites are further classified into permanent parasites and temporary parasites. Permanent parasites perform parasitic activity throughout their life cycle and development, whereas temporary parasites are parasitic in certain

Peptide and Peptidomimetic Therapeutics. DOI: https://doi.org/10.1016/B978-0-12-820141-1.00015-7

phases of their life cycle. Another two distinct classifications are ectoparasites and endoparasites. Ectoparasites are parasites that live on the surfaces of host organisms or in the skin and do not invade the inner body, such as *Rhipicephalus sanguineus*, or the brown dog tick [8]. The brown dog tick is a dog infecting parasite, although it can also be transmitted to other hosts including humans and can lead to many serious diseases [8]. Endoparasites are parasites that live inside the host; well-known examples are the tapeworms inhabiting the gut lumen of vertebrates, such as *Taenia saginata* (beef tapeworm) and *Taenia solium* (pork tapeworm).

Parasites that can cause diseases in humans comprise three main classes: protozoa, helminths, and ectoparasites. Protozoan parasites are microscopic single-celled organisms, which can be free-living or parasitic and have the ability to multiply in host organisms. They usually have a flagellum, a long, whip-like structure that helps them move, such as in the case of kinetoplastida protozoan parasites. Protozoan parasites cause high morbidity and mortality and are responsible for many life-threatening diseases, including malaria, toxoplasmosis, leishmaniasis, and others. While there are drugs available that may be effective against these parasites, the emergence of drug resistance and the lack of vaccines to prevent these diseases represent major public health concerns.

Helminth parasites are giant multicellular organisms and can be seen by the unaided eye. Unlike protozoan parasites, helminths cannot multiply in the host organism. In addition to living as a parasite, helminths can also live host free. Examples of helminths include filarial worms, schistosomes, nematodes, and cestodes. Helminth infections are distributed worldwide and cause several serious diseases, such as soil-transmitted helminthiases, lymphatic filariasis, and schistosomiasis. Helminth infections occur mainly due to the consumption of contaminated food or water. In addition, some helminth infections are transmitted by mosquitoes and flies. Approximately one billion people are infested with helminth worms. The symptoms of helminth infection vary depending on the type. The most common symptoms include nausea, diarrhea and abdominal pain; some infections can lead to life-threatening complications, such as neurocysticercosis, which is an infection of the brain by *Taenia solium.* Treatment with benzimidazoles is most frequently used against helminth parasites, although some strains may be resistant [9].

The final class of parasites that cause major diseases in humans are ectoparasites. Unlike protozoa and helminth parasites, ectoparasites can live on the skin of the host organism, and they include lice and fleas. Ectoparasites are also responsible for causing many devastating diseases, such as pediculosis, which is an infestation of lice. The *Ixodes scapularis* ectoparasitic tick can also transmit *Borrelia bugdorferi* bacteria, which causes Lyme disease in the human host.

In this chapter, we focus on kinetoplastida parasites, covering their life cycle, the diseases they cause, disease burdens, and limitations associated with disease treatment. Finally, we discuss the potential therapeutic abilities of peptides and peptidomimetics (modified peptides) in addressing therapeutic challenges for kinetoplastida parasites.

17.2 Kinetoplastida parasites

Kinetoplastida is a large group of flagellated protozoan parasites that can infect animals and plants and may also be free-living. Kinetoplastida is distinguished by having unique kinetoplasts, comprising mainly deoxyribonucleic acid (kDNA) present in their only mitochondrion and considered to be the most complex mitochondrial DNA in nature. The first observation of a kinetoplast was a century ago when a basophilic granule associated with a flagellum was observed through a light microscope and recognized as a structure located in the mitochondrion that stains with basic dyes [10]. However, only after the development of transmission electron microscope (TEM), this structure was unraveled and characterized in greater detail [11,12].

The DNA kinetoplast is a structure found adjacent to the basal body of the flagellum and comprises about 30% of the total cell DNA. This kDNA is circular and forms a unique network of interlocked rings. In addition, kDNA contains two different types of DNA rings that differ in size: (1) maxicircles (20–40 kb) that encode rRNAs (ribonucleic acid) and proteins of the respiratory chain, such as nicotinamide adenine dinucleotide dehydrogenase subunits, ATPase subunit 6, cytochrome *b*, and cytochrome oxidase subunits; and (2) minicircles (0.5–10 kb), which encode guide RNAs that enable the RNA editing process and assist the posttranscriptional modifications of the mitochondrial proteins [12]. Another feature of kinetoplastida parasites is that they are flagellated at all of their life cycle stages. While the primary functions of the flagella are locomotion and facilitation of attachment to insect epithelia, their flagella also play important roles in sensing and signal transduction [13].

The most common human diseases caused by kinetoplastida include African trypanosomiasis (African sleeping sickness) caused by infection of *Trypanosoma brucei* (*T. brucei*), Chagas disease caused by infection with *Trypanosoma cruzi* (*T. cruzi*), and various forms of leishmaniasis caused by infection with different species of *Leishmania*. These kinetoplastid pathogens are typically transmitted from insects to humans. Although they are distinct diseases and cause diverse clinical symptoms, they are caused by related kinetoplastid protozoan pathogens and are mainly and disproportionately distributed in developing countries, afflicting populations with limited access to health services. These diseases are classified as neglected tropical diseases (NTDs) due to limited research focus and funds dedicated to these conditions. These diseases also continue to be a tremendous burden and threat to humankind and cause a high rate of mortality and morbidity, claiming the lives of over 50,000 people annually [14]. To date, there is no effective vaccine for the prevention of these diseases, and the drugs that are being used for treatment currently are accompanied by severe toxicity and side effects. In addition, parasite drug resistance is beginning to raise concerns. Therefore a search for novel therapies to combat these devastating diseases is required [15]. Despite the fact that these parasites cause clinically different diseases and are transmitted by different invertebrates, they share many biological features (Fig. 17.1). In this chapter, we focus on the kinetoplastid diseases and review their characteristics as well as therapeutic agents and their roles in combatting these diseases.

Overview of the transmission, infection, and life cycles of kinetoplastid protozoa. The life cycle of kinetoplastida consists of several different morphological stages that develop in both invertebrate and vertebrate hosts. In leishmaniasis, the promastigotes (Pro) form is injected into the skin of the vertebrate host during a blood meal of the phlebotomine sandfly vector. In the vertebrate hosts, the parasites are phagocytized by macrophages and transform into amastigotes (Amas), a proliferative form that can replicate and infect other tissues. During a blood meal with a new vector, the amastigotes are digested and transformed into promastigotes in the midgut of the new vector, which continue the cycle during a new blood meal. In Chagas disease, metacyclic (Meta) trypomastigote (Trypo) forms of *T. cruzi* are injected into vertebrate hosts by the bite of infected triatomine bugs. In the

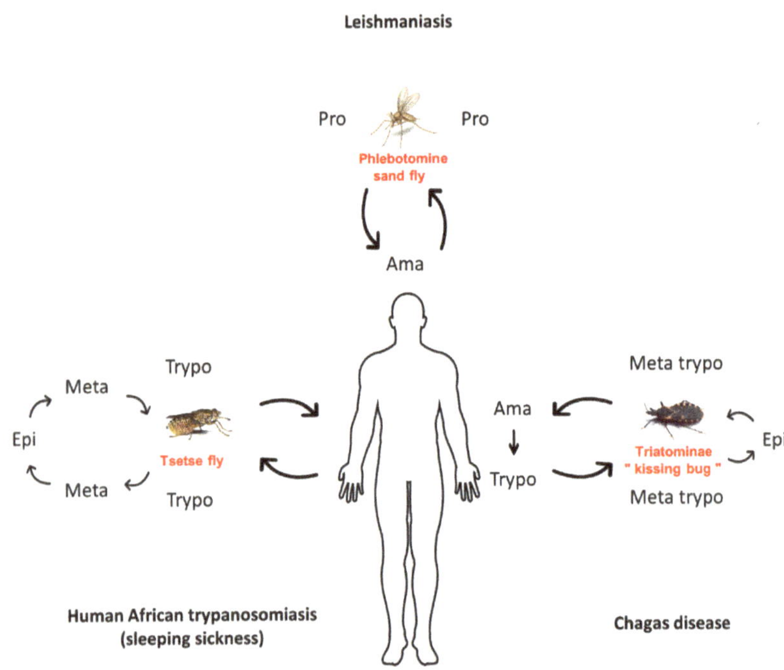

FIGURE 17.1 Kinetoplastid parasites infection and life cycles.

vertebrate host, trypomastigotes transform into the intracellular Amas proliferative stage and replicate before transforming back into trypomastigotes and diffusing into the blood circulation. Upon the blood meal of a new triatomine bug of the infected vertebrate, the trypomastigotes from the bloodstream are ingested and transform into the epimastigote (Epi) proliferative stage in the vector's midgut before differentiating back into metacyclic trypomastigote forms to readily continue the infectious cycle in human. In African trypanosomiasis, during a blood meal upon bite of the tsetse fly (Glossinidae), *T. brucei* parasites are injected in the Trypos from the saliva of the vector and multiply as trypomastigotes in the bloodstream of the vertebrate host. During a new blood-feeding on infected host blood, a new vector ingests trypomastigotes that differentiate into procyclic trypomastigotes in the midgut of the vector. In the midgut, they transform into Epis forms and then reach the fly's salivary glands and multiply before finally transforming into Meta trypomastigotes.

17.3 Human African trypanosomiasis

Human African trypanosomiasis (HAT, or African sleeping sickness), is a parasitic NTD caused by two subspecies of *T. brucei* protozoa, namely *T. brucei gambiense* and *T. brucei rhodesiense*. Infection by these parasites is chronic, and in the absence of treatment, results in death. The disease is transmitted by tsetse flies of the genus *Glossina*; thus HAT is distributed in the 36 sub-Saharan African countries where the tsetse fly is found. The disease is a major health concern with an estimated 70 million people at risk of infection, 10,000 new cases per year, and when left untreated the mortality rate of African sleeping sickness is close to 100% [16].

Metacyclic trypomastigote *T. brucei* is injected into the skin tissue of the mammalian host upon the bite of an infected tsetse fly during a blood meal. The parasites invade the lymphatic system and the bloodstream. The parasites enter the bloodstream in trypomastigote form and replicate by binary fission and diffuse to many sites and tissues including the lymph and spinal fluid. During the blood meal of a new tsetse fly, the fly ingests the parasites and they are transformed into procyclic trypomastigotes in the midgut of the fly. After leaving the midgut, they are transformed into epimastigotes, then reach the salivary glands, and transform to the metacyclic trypomastigotes, thus continuing their life cycle (Fig. 17.2).

The early stage of infection, in which the trypanosomes have diffused throughout the bloodstream and the lymphatic system, is accompanied by nonspecific signs and symptoms, such as fever, headaches, lymphadenopathy, joint pain, stiffness, weakness, heart failure, pulmonary edema, and lymphadenopathy. In the late stage when trypanosomes have crossed the blood–brain barrier (BBB) and invaded the central nervous system, there are various neurological manifestations, including loss of sensation, sleep abnormalities, coma, and finally death [17]. The two subspecies of *T. brucei*

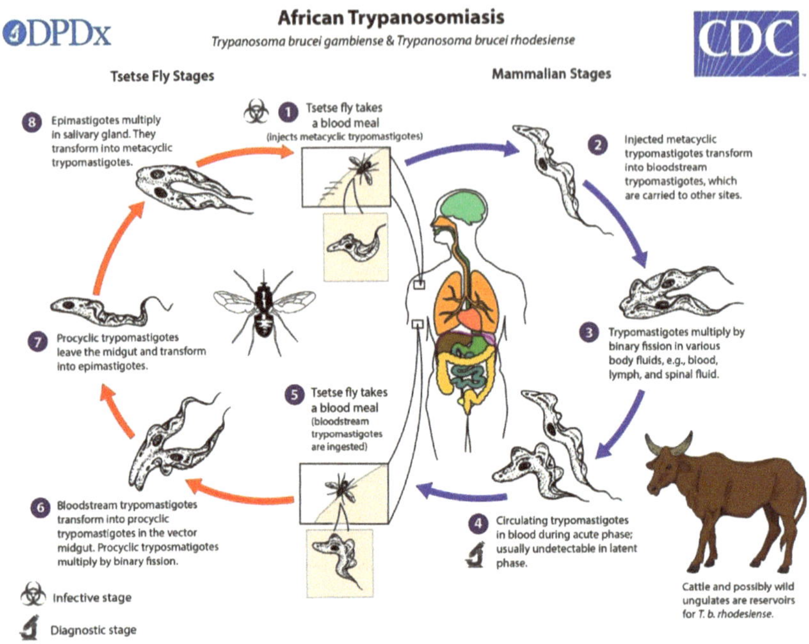

FIGURE 17.2 Diagrammatic representation of the *Trypanosoma brucei* life cycle in the human and the tsetse fly. *Image credit: Centers for Disease Control Public Health Image Library.*

differ in disease progression; however, their disease stages and symptoms are similar.

T. brucei rhodesiense infections are distributed in eastern and central Asia and cause an acute disease that progresses rapidly, though it accounts for only 2% of the infections worldwide. Conversely, *T. brucei gambiense* infections are distributed in western and central Asia and cause a chronic disease that progresses slowly and are responsible for 98% of total infections [16]. The disease is curable, and treatment is available depending on the stage of the disease and the type of the infection. For the early stage of the disease where the parasite has diffused throughout the lymphatic system and the bloodstream, suramin and pentamidine are used for both *T. brucei gambiense* and *T. brucei rhodesiense* infections. Suramin is a symmetrical polysulfonated polyaromatic urea that interacts with trypanosomal glycolytic enzymes to disrupt glycolysis, whereas pentamidine is an antiprotozoal diamidine that disrupts the structure of trypanosomal mitochondrial DNA. In the second late-stage after parasite invasion of the central nervous system, the arsenical drug melarsoprol is used to treat *T. brucei rhodesiense* infections by disrupting parasite energy production, and the combination of the nitroheterocyclic drugs nifurtimox and eflornithine (difluoromethylornithine) cause DNA breakdown and target ornithine decarboxylase [18].

African trypanosomiasis is lethal if left untreated. Therefore diagnosis of HAT is crucial. For the diagnosis of *T. brucei gambiense*, three steps are practiced, which include screening, diagnostic confirmation, and identification of the disease stage (first stage or second stage). There are many serological methods for screening. The most adapted method in the endemic areas is the Card Agglutination Test for Trypanosomiasis (CATT/*T*), which is rapid and highly sensitive. Other screening methods exist but are less sensitive and include the latex agglutination test (LATEX/*T. brucei gambiense*) and immunofluorescence assays. The subsequent diagnostic confirmation is made through the observation of the parasites in the lymph nodes, blood, or cerebrospinal fluid (CSF); however, standard parasitological methods for diagnostic confirmation lack sensitivity and 20%–30% of infected people are not diagnosed. The staging of the disease is performed via investigation of the cerebrospinal fluid [19].

The diagnosis of *T. brucei rhodesiense* infection is more challenging than *T. brucei gambiense*. While the CATT/*T*-test is used for the detection of *T. brucei gambiense* infection, there is no similar test for the detection of *T. brucei rhodesiense*. Therefore screening mostly depends on clinical signs and symptoms [19]. Although the disease is curable and several drugs are available, these therapies are often accompanied by severe reactions and side effects (Table 17.1). For instance, high doses of suramin can result in nausea, fatigue, anemia, neuropathy, anaphylactic shock, and renal problems [20]. Nevertheless, suramin has a high success rate, especially for *T. brucei rhodesiense* first stage infection, though it is less active against *T. brucei gambiense*. Pentamidine is another therapeutic agent, which is the preferred treatment for the first stage of *T. brucei gambiense* infection, but is unable to cross the BBB and is thus ineffective for the second stage of the disease [20]. Compared to suramin, pentamidine is better tolerated and accompanied by fewer side effects, although it can cause nausea, vomiting, hypotension, hypoglycemia, and nephrotoxicity [20]. In addition, resistance to the drug has been reported [21], limiting its effectiveness.

Drugs that are used to treat the second stage of the disease must be able to cross the BBB. Few drugs are available, such as melarsoprol as well as eflornithine and nifurtimox-eflornithine combination therapy (NECT). However, these drugs are highly toxic and their administration is more complex. Melarsoprol is used for *T. brucei gambiense* and *T. brucei rhodesiense* infections and is accompanied by severe side effects. About 5%–10% of patients encounter posttreatment reactive encephalopathy (PTRE) after melarsoprol and experience high fever, headaches, nausea, dizziness, and vomiting. Moreover, PTRE leads to death in 50% of patients who develop the condition [20]. In addition, parasite resistance has been reported through genetic alternations related to drug uptake [27]. Eflornithine administration is considered to be a better alternative to melarsoprol; however, the drawbacks of eflornithine include lack of activity against *T. brucei rhodesiense*, difficult administration [28], and

TABLE 17.1 Names and mechanisms of action of some drugs currently used in the therapy of human African trypanosomiasis.

Subspecies	Stage	Drug	Mechanism of action	Side effects	Drug resistance reported	Sources
Gambiense	Stage 1	Pentamidine	Disruption of mitochondrial processes	Nausea, vomiting, nephrotoxicity, hypotension, and hypoglycemia	Yes	[20,22,23]
		Suramin	Disruption of glycolysis	Nausea, fatigue, anemia, neuropathy, anaphylactic shock, and nephrotoxicity	Yes	[20,24,25]
	Stage 2	Melarsoprol	Disruption of glycolysis and redox metabolism	Post-treatment reactive encephalopathy (PTRE), high fever, headaches, nausea, dizziness, and vomiting	Yes	[20,22]
		Eflomithine	Inhibition of proliferation	Hypertension, fever, gastrointestinal problems, and convulsions	Yes	[20,26]
		Nifurtimox-eflornithine combination therapy (NECT)	Inhibition of proliferation and induction of oxidative stress	Nausea, vomiting, and neurological disturbances	No	[20]
Rhodesiense	Stage 1	Suramin	Disruption of glycolysis	Nausea, fatigue, anemia, neuropathy, anaphylactic shock, and nephrotoxicity	Yes	[20,24,25]
	Stage 2	Melarsoprol	Disruption of glycolysis and redox metabolism	Post-treatment reactive encephalopathy (PTRE), high fever, headaches, nausea, dizziness, and vomiting	Yes	[20,22]

severe side effects including hypertension, fever, gastrointestinal problems, and convulsions [29]. Recently, the nifurtimox-eflornithine combination has emerged for the treatment of *T. brucei rhodesiense* infections and is better tolerated, safer, and easier to administer than melarsoprol or eflornithine monotherapy. Nonetheless, the combination is still ineffective for *T. brucei rhodesiense* infections [30], is accompanied by diverse side effects, such as nausea and vomiting in over 60% of the patients, and may lead to neurological disturbances [31].

The aforementioned drug toxicity, drug resistance, and ability of the parasites to bypass drug mechanisms of action highlight the unmet need for the development of new compounds to combat this devastating disease [32]. Recently, the development of peptides to treat these diseases is receiving growing attention and interest. Several peptides have shown antitrypanosome activity and are under investigation and development [33]. We elaborate on these peptides in the following section.

17.4 Chagas disease

Chagas disease (American trypanosomiasis) is a lethal and life-long parasitic neglected disease caused by infection of the kinetoplastid flagellate *T. cruzi*. The disease was discovered in 1909 by the Brazilian hygienist and bacteriologist Carlos Chagas [34]. However, tissue specimens of mummies revealed that Chagas disease originated from as early as 9000 years ago [35].

Similar to the *T. brucei* species, *T. cruzi* also has a complex life cycle that is completed between the vector and the mammalian host organism. *T. cruzi* are injected as trypomastigotes into the mammalian host, transformed into intracellular amastigotes, then transformed back into trypomastigotes, and diffuse into the blood circulation. The trypomastigotes in the bloodstream infect other cells and tissues and then transform back into intracellular amastigotes. The intracellular amastigotes can replicate, while the trypomastigotes cannot, and are responsible for the infection process. When a new vector ingests trypomastigotes, it becomes infected and the trypomastigotes transform into epimastigotes in the midgut before finally being transformed into metacyclic trypomastigotes in the hindgut (Fig. 17.3).

Presently, Chagas disease affects approximately 18–20 million people throughout Central and South America as well as the United States and accounts for 12,500 deaths annually. There are several modes transmission, most commonly through the feces of infected bloodsucking insects known as kissing bugs, of the subfamily triatomine bugs. During the night, the insects bite the sleeping human host on the face and deposit parasite-ridden feces nearby. The parasite cannot penetrate the healthy skin; however, it can penetrate through the mucous membranes or through opened wounds. A person may unconsciously scratch the itchy bite region, which paves the way for the parasite to enter the body. Other transmission mechanisms may be through

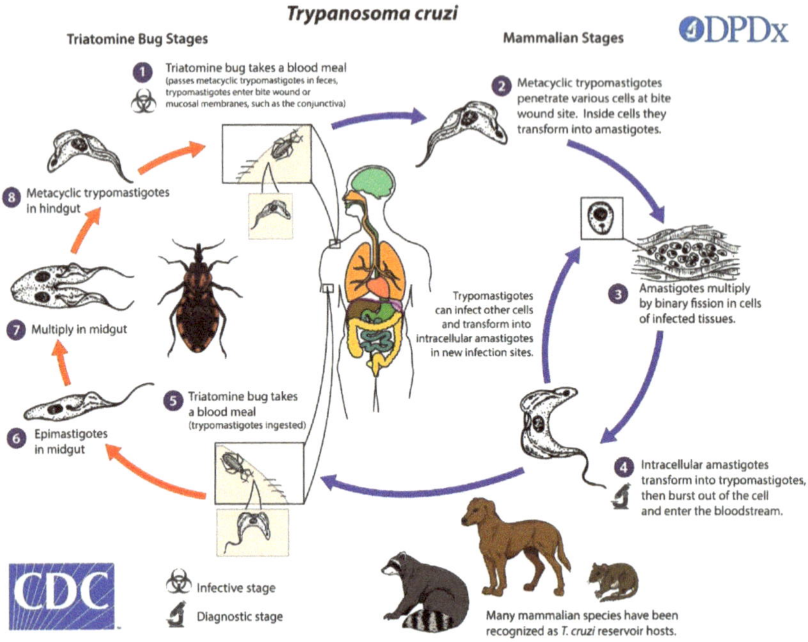

FIGURE 17.3 Diagrammatic representation of the *Trypanosoma cruzi* parasite lifecycle. *Image credit: Centers for Disease Control Public Health Image Library.*

vertical transmission from mother at birth, blood transfusion, and organ transplant [36].

Chagas disease includes two stages. The acute phase lasts for 40–60 days. Typically, this phase is asymptomatic or causes mild symptoms, such as fever, fatigue, headache, loss of appetite, vomiting, and diarrhea. These symptoms are general and similar to many other illnesses; thus recognition of the disease in the acute phase is often challenging. In some cases, a chagoma, or swelling that may develop at the site of parasite inoculation, may appear at the site of the bite. In addition, enlargement of the spleen and liver may be noticed. Importantly, *T. cruzi* parasites can be detected by laboratory methods in the acute phase. The gold standard for the detection of trypomastigotes in peripheral blood involves visualization using conventional microscopy [37].

Following the acute phase, the patient with Chagas disease enters the chronic phase, which may extend for several years. During this phase, the patient shows no clinical symptoms and parasite replication is suppressed by a strong and effective immune response. Hence, the vast majority of the infected people are unaware of their chronic infection, and less than 1% are diagnosed. Special and expensive laboratory methods are required to detect *T. cruzi* parasites in the chronic phase. Blood parasite levels in the chronic stage are very low, and direct parasitological assays are limited by insufficient sensitivity. Therefore immunological and serological tests are generally used to detect the existence of anti-*T. cruzi* antibodies. The most widely used methods are indirect immunofluorescence, indirect hemagglutination, and enzyme-linked immunosorbent assay. However, a single one of these tests alone is usually inadequate to confirm the presence of the parasite [38].

The ability of the immune system to control parasite replication is limited. After up to 20 asymptomatic years of the infection, 30%−40% of the patients may encounter classical pathological complications of Chagas disease, including cardiomyopathy, digestive syndromes, and damage to the nervous system, depending on the tissue invaded [35]. There are two available drugs that are used for the treatment of the disease, namely benznidazole (BZ) and nifurtimox (NFX), which are used during both acute and chronic phases (Table 17.2). These drugs are more effective in the acute phase than the chronic phase; however, they are extremely toxic and are accompanied by a wide range of side effects. Treatment with nifurtimox can lead to sleepiness, psychic alterations, excitability, weight loss, anorexia, intestinal colic, diarrhea, nausea, and vomiting. In addition, benznidazole can cause skin manifestations, such as dermatitis with cutaneous eruptions, hypersensitivity, fever, generalized edema, lymphadenopathy, and articular and muscular pain [39]. In summary, diagnosis of the disease in the acute phase is difficult, and these drugs are less effective after disease has progressed. In addition, long-term use of these drugs increases the chance of developing drug resistance. Therefore novel compounds are needed to combat and overcome the devastating disease.

17.5 Leishmaniasis

Leishmaniasis includes a group of diseases with a broad clinical spectrum of symptoms, caused by different species of obligate intracellular flagellated protozoa in the *Leishmania* genus of the trypanosomatida order of the kinetoplastea class. *Leishmania* is often transmitted by the bite of an infected female phlebotomine sandfly. Leishmaniasis is a neglected parasitic disease and is distributed in many tropical and subtropical areas around Europe, Africa, Asia, and the Americas, including both developed and underdeveloped countries. The *Leishmania donovani* species has a long history, with DNA fragments found in ancient Egyptian and Christian Nubian mummies from 3500 to 2800 BCE [44]. There are three primary forms of the disease depending on the infectious species, classified as cutaneous, mucosal, and visceral leishmaniasis (VL). In addition to these main disease types, less serious forms are known to exists, such as post kala-azar dermal leishmaniasis (PKDL, a complication of VL) and diffuse cutaneous leishmaniasis (CL) [45].

The life stages of the *Leishmania* parasite are less complex than the *Trypanosoma* parasites. Upon injection of the promastigote stage parasite

TABLE 17.2 Names and mechanisms of action of some drugs currently used in the therapy of Chagas disease.

Pathogen	Stage	Drug	Mechanism of action	Side effects	Resistance reported	Sources
Trypanosoma cruzi	Acute and chronic stages	Benznidazole (BZ)	Inhibition of DNA, RNA and proteins synthesis	Dermatitis with cutaneous eruptions, hypersensitivity, fever, generalized edema, lymphadenopathy, and articular and muscular pain	Yes	[39–41]
		Nifurtimox (NFX)	Activation of nitroreductase enzymes, leading to production of reactive metabolites	Sleepiness, psychic alterations, excitability, loss weight, anorexia, intestinal colic, diarrhea, nausea, and vomiting	Yes	[39,42,43]

from a vector, the parasite transforms into an amastigote after being phagocytized by macrophages. *Leishmania* then replicates and infect other cells and tissues. A new vector ingests the amastigotes through blood-feeding, after which they transform back to promastigotes in the midgut (Fig. 17.4) [46].

The most common form of leishmaniasis is CL, a disease caused by 15 different species of the *Leishmania* parasite. An estimated 2–2.5 million new cases occur annually and are distributed worldwide. The different species are classified into *old world* species and *new world* species. The old world CL is mainly caused by *Leishmania* (*L.*) *major*, *L. infantum*, and *L. tropica* species, while the new world CL is caused by parasites belonging to *L. peruviana*, *L. guyanensis*, *L. braziliensis*, and *L. mexicana* species. Although CL itself is not a lethal disease and causes painless chronic skin lesions at the infection site, it can lead to social stigma. In addition, in some cases, CL can progress to for more serious disease. Mucosal leishmaniasis (ML) results from hematogenous or lymphatic dissemination from the skin to the naso-oropharyngeal mucosa. Because of its heterogeneous presentation, ML is often confused or underestimated by clinicians. VL is the most serious form of leishmaniasis. Every year, 0.2–0.4 million new VL cases appear. As the disease progresses, it is accompanied by fever, weight loss, malaise, splenomegaly, and is life-threatening if untreated [45].

The diagnosis of leishmaniasis is often challenging due to the broad spectrum of *Leishmania* species that can cause disease and different clinical

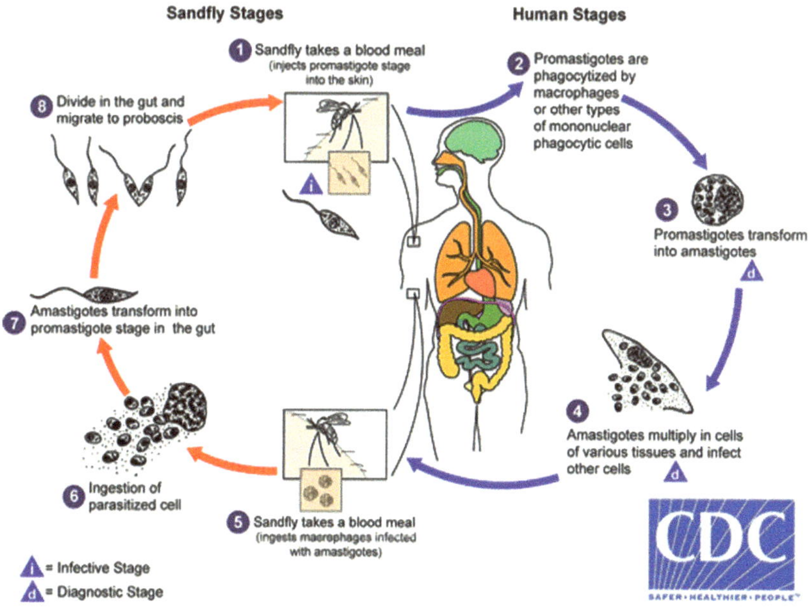

FIGURE 17.4 Overview of the *Leishmania* parasite life cycle. *Image credit: Centers for Disease Control Public Health Image Library.*

manifestations. In addition, other diseases may show similar clinical features, such as malaria and schistosomiasis, making it difficult to differentiate and diagnose the infection [47]. Despite these challenges, several laboratory methods have been developed to detect *Leishmania* infection. The gold standard is a microscopic examination of Giemsa-stained biopsy slides. The tissue from which the biopsy is taken depends on the manifestation suspected. For example, for CL the biopsy is taken from the lesions, and for VL the biopsy can be taken from spleen aspirates or from the bone marrow [47]. Many other immunological techniques and molecular approaches are available to support microscopic examination [47].

No effective vaccinations exist for leishmaniasis, but several drug treatments are available depending on the disease form (Table 17.3). Pentavalent antimonial compounds have long been used as first-line therapy for VL and CL; however, their effectiveness is limited due to their toxicity, including severe cardiotoxicity and hepatotoxicity, as well as drug resistance. Miltefosine is an oral alkyl phospholipid drug used for the treatment of VL and CL. Despite this convenient mode of administration, miltefosine is costly and associated with increasing treatment failure [48].

Amphotericin B is another drug used for the treatment for leishmaniasis also accompanied by severe adverse reactions. Amphotericin B lipid formulations have been developed to minimize the side effects, although still not ideal [62]. Paromomycin is a relatively new drug used for the treatment of VL that was introduced in 2006. Despite the high cure rate of paromomycin, the route of administration (intramuscular) injection is painful and the treatment period is long and associated with toxicity to the kidney and ear. In addition, paromomycin shows poor cell permeability and therefore is only used for local infections [58]. Furthermore, drug resistance has been reported for paromomycin [58]. Pentamidine is wildly used for the treatment of CL, although this drug is associated with the development of diabetes. Consequently, drug toxicity, high costs of treatment, long treatment periods, and the development of drug resistance are the main barriers to effective treatment for leishmaniasis [63].

17.6 Peptides as drug candidates

Peptides are chains of amino acids linked together by peptides bonds. Unlike proteins, peptides are short and range from about two to 50 amino acids. Peptides are sometimes classified according to their sizes, such as dipeptides (peptides comprised of only two amino acids), tripeptides (peptides comprised of three amino acids), oligopeptides (peptides comprised of up to about ten amino acids), polypeptides (peptides comprised of around 11–50 amino acids), and polypeptides over about 50 amino acids in length that are considered to be proteins. Importantly, peptides can exist in linear or cyclic forms.

In recent years, peptides have emerged as attractive therapeutic candidates for various diseases due to their unique properties, such as high

TABLE 17.3 Names and mechanisms of action of some drugs currently used in the therapy of leishmaniasis.

Pathogen	Drug	Mechanism of action	Side effects	Resistance reported	Sources
Leishmania	Pentavalent antimonials	Inhibition of type I DNA topoisomerase	Nausea, vomiting, weakness, myalgia, abdominal colic, diarrhea, skin rashes, hepatotoxicity, and cardiotoxicity	Yes	[49–51]
	Amphotericin B	Membrane distribution	Nausea, vomiting, rigors, fever, hypertension, hypotension, hypoxia, and nephrotoxicity	Yes	[52,53]
	Miltefosine	Induction of apoptosis and disruption of lipid-dependent cell signaling pathways	Vomiting, diarrhea, nausea, nephrotoxicity, and hepatotoxicity	Yes	[54–56]
	Paromomycin	Inhibition of protein synthesis	Nausea, abdominal cramps, diarrhea, nephrotoxicity, hepatotoxicity, and ototoxicity	Yes	[57–60]
	Pentamidine	Disruption of mitochondrial processes	Nausea, vomiting, hypotension, hypoglycemia, and nephrotoxicity	Yes	[20,22,23,61]

selectivity (e.g., for targeting protein-protein interactions that were considered "undruggable") as well as their ability to mimic and modulate natural signaling pathways [64,65]. Peptide advantages over antibodies (also used to modulate protein–protein interactions) include cell permeability and relative affordability. Historically, the first peptide developed as a drug was insulin, which is a peptide hormone composed of 51 amino acids. Insulin was isolated and used in diabetic patients who did not produce sufficient insulin naturally. The success of insulin therapy led to the beginning of a peptide-based drug era.

Cyclic peptides are chains of amino acids which contain a circular sequence of peptide bonds, either naturally occurring or synthetically produced. Cyclic peptides have been used widely and clinically as immune suppression, cancer therapy, and antibiotics, amongst other applications. The advantages that cyclic peptides offer over linear peptides is relative rigidity and resistance to hydrolysis by *endo*- and *exo*-peptidases, which leads to increased biostability. In addition, in many cases, cyclic peptides show increased binding affinity and selectivity toward their target compared to linear peptides. Furthermore, cyclic peptides have demonstrated improved bioavailability due to increased membrane penetration. An example of a naturally occurring cyclic peptide used as a pharmaceutical agent is cyclosporine A. Cyclosporine is a cyclic endecapeptide first discovered in 1972 that originated from the two fungi *Trichoderma polysporum* and *Cylindrocarpon lucidum*. The drug is used as an immunosuppressive agent to enable successful organ transplantation [66].

Due to the numerous advantages of cyclic peptides, efforts are being made to develop peptide therapeutic agents with antiparasitic activity. Numerous natural cyclic peptides have been identified with antitrypanosomal activity. Additionally, synthetic cyclic peptides have been developed for the treatment of kinetoplastid diseases. Over the last five decades, the field of peptide-based drug discovery has grown tremendously worldwide, which has led to massive progress in the discovery of new therapeutic peptides, and currently, more than 60 peptides have been approved to treat various diseases and disorders ranging from cancer to various infections, to metabolic diseases and endocrine conditions [67].

The past years have witnessed significant medicinal chemistry and pharmaceutical advances with discoveries of therapeutic agents ranging from small molecules to antibodies as well as peptides and peptidomimetics. The following sections provide a discussion of some recent studies on peptides and peptidomimetics for treating parasitic diseases. There is an increasing interest in peptide drug therapy, and although there are no currently approved peptides to treat protozoan diseases, several peptides are under development and evaluation in clinical trials. We discuss applications of different classes of peptides and peptidomimetics as well as their implications for treating parasitic diseases.

17.7 Antimicrobial peptides

Antimicrobial peptides (AMPs, also host defense peptides, or HDPs) are small AMPs of up to 100 amino acids that are components of the innate immune system in various eukaryote organisms. These peptides exhibit diversity in structure and mechanisms of action and also possess broad antibacterial, antiviral and antifungal activity.

The observation that organisms such as insects can be infected without showing disease symptoms has stimulated research regarding their natural defense mechanisms against parasites. Several studies showed that insects express AMPs to combat infectious diseases and confer resistance to parasites. For example, one study investigated the expression of genes before and after trypanosome infection of the tsetse fly vector responsible for transmitting *Trypanosoma*. Genes encoding the AMPs defensin, attacin, and cecropin were markedly upregulated [68]. In addition, the knockdown of these genes increased the susceptibility of the tsetse fly to disease symptoms [69]. Therefore the use of AMPs suggests a promising therapeutic avenue against parasitic infections.

AMPs are characterized by their amphipathicity and high cationicity at physiological pH, which promotes their interaction with the negatively charged cell membrane, thus causing a change in the electrochemical potential on the bacterial cell membrane [70]. AMPs can function via several mechanisms of action, including membrane rupture and permeabilization, inhibition of cell wall synthesis, regulation of intracellular targets, and modulation of the immune system [71]. On the other hand, natural AMPs can also exhibit host cell toxicity and thus may have limited therapeutic potential due to their characteristic cytolytic activity. Construction of peptide synthetic analogs or peptide hybrids based on AMPs can be used to overcome these obstacles by improving pathogen selectivity. Inspired by the properties and the capacity of AMPs, several researchers have synthesized AMP-derived peptides and examined their potential against kinetoplastida parasites. The following sections discuss some of the most potent natural AMPs and AMP analogs being investigated for the treatment of parasite infections.

17.7.1 Defensins

One of the most extensively investigated classes of AMPs is the defensins. Defensins are a class of small cationic AMPs. They are cysteine-enriched and are produced by a broad spectrum of organisms including invertebrates, vertebrates, plants, and fungi. The production of defensins is a crucial part of the innate immune system. Defensins are categorized into three classes, namely the α-, β- and θ-defensins, and they exhibit wide antimicrobial activity against pathogens through divergent modes of action, targeting both the cell membrane as well as intracellular targets. The positive charge of the peptide enables it to bind to the pathogen's negatively charged membrane and exert its antimicrobial activity [72].

Besides their antimicrobial activity, several studies have demonstrated the potential therapeutic abilities of defensins against kinetoplastid parasites. In one in vitro study, a synthetic D-amino acid peptide derived from the putative active site of an insect defensin peptide (RLYLRIGRR-NH_2) showed antiparasitic activity against *T. brucei* and suppressed its growth in mammalian cells, with minimal toxicity [73]. The trypanocidal effect was traced to the disruption of parasite membrane potential. In addition, a parallel study demonstrated similar results for synthetic nonameric peptides derived from insect defensin, which were shown to suppress African trypanosomes in axenic culture by disrupting parasite membrane potential [74].

Belonging to the family of defensins are the human neutrophil peptides (HNP) 1–4 produced by neutrophils during infections. These molecules exhibit potent antimicrobial activities, trigger host cytokine production and activate the immune system. HNP-1 (ACYCRIPACIAGERRYGTCIYQGRLWAFCC) has been reported to significantly reduce *L. major* promastigote and amastigotes load. This effect was found due to boosting the production of the proinflammatory cytokine TNF-α, which in turn enhances neutrophil microbicidal capacity [75]. In a different study, HNP was linked to antitrypanosomal activity against *T. cruzi* trypomastigotes and amastigotes. In this study, HNP1 was shown to be toxic to parasites due to the induction of membrane pore formation and DNA fragmentation, thus leading to their destruction [76].

Another member of the defensin family is brevinin 2R, a naturally occurring nonhemolytic antimicrobial polypeptide extracted from the skin of the *Rana ridibunda* marsh frog consisting of 25 amino acids (KLKNFAKGVAQSLLNKASCKLSGQC). Brevinin 2R is well known for its antibacterial and antifungal activity and was shown to have potent activity against multidrug resistance bacteria [77]. In addition, brevinin 2R exerts antitumor activity on various malignant cells via disturbance of mitochondrial membrane potential and induction of apoptosis, triggering the secretion of interleukin-6 (IL-6) and IL-1β cytokines [78]. Brevinin 2R was thus proposed to have antiparasitic activity [78], and in 2019 a group tested this hypothesis by studying the impact of both brevinin 2R and lauric acid-conjugated brevinin 2R treatment for *L. major* infection. In vitro results showed efficient brevinin 2R inhibition of *L. major* promastigote proliferation. Conjugation of brevinin 2R to lauric acid enhanced the inhibitory effect. This inhibition was found mainly due to induction of *L. major* promastigote necrosis by the destruction of cell membrane potential and cell morphology. Despite toxic effects in vitro, the peptide showed no side effects in animal studies and potently maintained low parasite levels in animal models [79].

Defensins are also found in plants, protecting them against various pathogens, such as bacteria and fungi. PvD_1 is a plant defensin isolated from

Phaseolus vulgaris French bean seeds known to exert its effect by disrupting the plasma membrane, triggering the formation of reactive oxygen species (ROS) and nitric oxide (NO) [80]. PvD_1 was therefore proposed to have antiparasitic effects. In a 2015 study, PvD_1 was isolated and investigated for antileishmanial activity against *L. amazonensis*. At a concentration of 300 μg/mL, PvD_1 suppressed parasite proliferation by 70% after 24 h of incubation. Furthermore, PvD_1 reduced the number of protozoa by disrupting their plasma membrane, leading to the formation of cytoplasmic vacuoles in the parasites and causing cytoplasmic fragmentation [81]. These results demonstrated that PvD_1 can be harnessed to control *Leishmania* parasites.

17.7.2 Temporins

Another family of AMPs are the temporins, which are hydrophobic short peptides (8–17 amino acids) that contain one positively charged amino acid and an α-amidated C-terminus. These peptides are released by granular glands and secreted from the skin of some North American and Eurasian frogs. Temporins represent crucial protective elements of the frog's innate immune system defense against a broad spectrum of invasive infections, exhibiting antifungal activity and antibacterial activity against Gram-positive bacteria, as well as some activity against Gram-negative bacteria. These peptides were first discovered in 1996 in frog skin [82] and later were identified also in wasp venom [83].

Temporins adapt a helical conformation in a hydrophobic environment, and their catalytic function arises from their ability to perturb the cell membrane. Temporins gained much attention due to their powerful antimicrobial and chemotactic characteristics [84]. Consequently, temporins were expected to have promising trypanolytic activities and many researchers are investigating these molecules as potential therapeutic agents against parasitic infections [85,86]. As expected, several studies have shown antiparasitic activity of natural and synthetic temporins. One example is the temporin-SHd originating from the North African ranid frog, *Pelophylax saharicus*. Temporin-SHd is a 17-residue peptide (FLPAALAGIGGILGKLF) effective against both the promastigote and amastigote forms of *Leishmania* that cause visceral, cutaneous and mucocutaneous leishmaniasis [87]. In addition, temporin-SHd has shown activity against the trypanosome parasites *T. brucei* and *T. cruzi*. Temporin-SHd activity is attributed to membrane disruption by pore formation [85].

Another study investigated the potency of two temporin-derived hybrid peptides named temporizin and temporizin 1 against *T. cruzi*. Temporizin was designed with an N-terminal region consisting of a pore-forming region from temporin A and a C-terminal region containing alternating leucine and lysine residues. Temporizin 1 is almost the same, but with a modification that reduces the extent of insertion into membranes. Both peptides displayed activity against *T. cruzi* parasites with no or only slight toxicity in mammalian cells, suggesting that these peptides could be potential drugs for Chagas disease [86].

17.7.3 Cathelicidins

Cathelicidins are a family of AMPs which were first isolated from bone marrow myeloid cells of mammals and are known as "myeloid AMPs" (MAPs), though they are also found in nonmyeloid cells, such as epithelial cells. They vary in sequence and structure, are stored in neutrophil and macrophage granules, and possess a variety of antimicrobial activities against many pathogens. These peptides and their activity present a crucial part of the innate immune system.

One cathelicidin called bovine myeloid AMP 28 (BMAP-28) is found in bovine neutrophils, protects against bacterial infection by induction of mitochondrial membrane pores and was thus suggested to have antiparasitic activity. A study in 2011 tested this hypothesis and investigated the effectiveness of BMAP-28 as well as its D-amino acid form (D-BMAP-28) and retro-inverso form (RI-BMAP-28) against promastigote and amastigote forms of *L. major* parasites. These peptides showed potency against *Leishmania* parasites, acting via cell membrane disruption and the induction of caspase-independent cell death [88].

eCATH1 is another Cathelicidin found in horses. eCATH1 is composed of 26 amino acid residues and adapts α-helical structure. This peptide was found to exhibit potent antimicrobial activity against bacteria. Further study in 2016 demonstrated the trypanocidal activity of eCATH1 against *Trypanosoma* parasites by disturbing their mitochondrial membrane potential and causing permeabilization with an IC_{50} of 9.5 μM [89].

17.7.4 Melittin

Melittin is a cationic, hemolytic linear AMP found in bee venom (*Apis mellifera*) and consists of 26 amino acids. Melittin has antileishmanial activity against promastigotes and intracellular amastigotes of *L. infantum* with an IC_{50} of 28.3 μg/mL against promastigotes and 1.4 μg/mL against intracellular amastigotes. However, the peptide exhibited cell cytotoxicity in mice peritoneal macrophages, with an IC_{50} of 5.7 μg/mL. Melittin acts via the induction of macrophage immunomodulatory effects [90]. In a different study, melittin was also found to inhibit the growth of *T. cruzi* parasites, inducing cell death [91]. These studies show the important therapeutic value of melittin and its possible use as the basis for the development of peptides to treat parasitic diseases, as further optimization is needed to reduce observed native peptide host cell cytotoxicity.

To overcome the toxicity of natural AMPs, short and hybrid AMPs have been synthesized and evaluated for therapeutic use. In a 2018 study, a hybrid of melittin C and cecropin A AMPs was investigated for activity against amastigote forms of *L. major*. This combination merged the hydrophobic C-terminal sequence from melittin (residues 6–9) with the basic N-terminal sequence of cecropin A (residues 2–8) to form a short cationic AMP (WKLFKKILKVL-NH_2) termed CM11. The results demonstrated antileishmanial activity by CM11. At a concentration of 8 μM the peptide CM11 inhibited the growth of *L. major* in fibroblast cells for 24 h and did not show any cytotoxicity; after 48 h the IC_{50} was 9.58 μM [92].

17.7.5 Dermaseptins

Dermaseptins (DSs) are cationic, AMPs that consist of 24–34 residues, isolated from skin secretions of the frog genus *Phyllomedusa*. DSs have demonstrated antimicrobial activity against many microbes, such as Gram-negative and Gram-positive bacteria, viruses, and fungi. In addition, studies have shown that DSs exhibit robust activity against kinetoplastid parasites [93].

DS 01 (GLWSTIKQKGKEAAIAAA-KAAGQAALGAL-NH_2) is a cationic peptide belonging to the DS family and is extracted from the skin secretions of amphibians. Many studies have shown potent activity of this peptide. For example, one study showed the strong interaction between DS 01 and the Langmuir negatively charged monolayers of a natural lipid extract from *L. amazonensis* phospholipid membranes [94]. A more recent study has shown potential antileishmanial activity of DS 01 using atomic force microscopy and scanning electron microscopy to observe cell morphological changes, such as membrane damage and alterations to flagella [95].

Historically, many natural substances have been used even since ancient times, demonstrating therapeutic effects in wide range of disorders and diseases. These substances gained much interest due to the possibility of developing successful drugs inspired by nature. One of these compounds is *Royal Jelly*. Royal Jelly is a highly nutritious milky secretion produced by the hypopharyngeal glands of the young worker honey bee (*Apis mellifera*) and is rich in carbohydrates, lipids, proteins, and vitamins. Royal Jelly is used to feed the bee larvae and those selected to become queens. Recently researches demonstrated several therapeutic effects of Royal Jelly, including lifespan extension [96], reduction of inflammation, reducing free radicals, antioxidant functions, and antitumor activities [97,98]. Further, Royal Jelly possesses antibacterial activity via destruction of the bacterial cell wall [99]. This inhibitory effect was traced to Jelleins, short positive AMPs derived from Royal Jelly [99]. Due to the potent inhibitory effects of Jelleins, they were proposed to also have antiparasitic properties. A recent study has evaluated the therapeutic uses of a synthetic peptide Jellein (PFKISIHL) against *L. major* parasites. The peptide suppressed the growth of the parasite at

6.25 μg/mL, mediating pore formation and disrupting membrane potential [100]. These results demonstrate the promising applications of Jellein in kinetoplastid diseases.

17.7.6 Bacteriocins

Bacteriocins are a class of ribosomal AMPs produced by Gram-positive and Gram-negative bacteria and are classified into many groups. These peptides exhibit diverse antimicrobial activities against a wide spectrum of microorganisms. Additionally, the bacteria colonizing the human gastrointestinal tract produce bacteriocin to protect against many pathogens. Although they were described many years ago, interest in the clinical use of bacteriocins is recent. Bacteriocins are mainly used in the food service industry as food preservatives. However, they are now being studied and developed as potential drugs for infectious diseases and may address the problem of antibiotic resistance [101]. More recently, bacteriocins have gained considerable attention for their possible role as antiparasitic agents.

The first bacteriocin reported to inhibit the growth of parasites was enterocin AS-48, a circular peptide produced by *Enterococcus faecalis* bacteria that is both hydrophobic and cationic. A 2017 study reported that AS-48 suppressing the growth of *Leishmania* promastigotes and intracellular amastigotes with minimal toxicity. The decrease in leishmanial parasites was due to several mechanisms, including permeabilization of the plasma membrane, depolarization of mitochondria, and ROS production [102]. In another study, AS-48 activity was also assessed for *Trypanosoma* parasites. The peptide was lethal for *T. brucei gambiense*, *T. brucei rhodesiense*, and *T. brucei brucei* with an EC_{50} of 1.7 to 3.12 nM. Interestingly, in contrast to the previous study, AS-48 was not shown to cause membrane permeabilization, but a novel autophagy-like cell death mechanism was reported for this peptide [103]. AS-48 appears to interact with the variant surface glycoprotein (VSG) on the parasite surface, after which the bound VSG is endocytosed, finally triggering autophagic cell death [103]. In summary, bacteriocins may be highly beneficial for the treatment of *Leishmania* and *Trypanosoma* infections.

17.7.7 Histatins

Up until this point, most AMPs mentioned in this chapter originated from plants, bacteria, insects, and other nonhuman organisms. AMPs are also fundamental components of the innate immune system in humans (e.g., HNPs), and are produced abundantly by various cells and tissues to protect against infectious pathogens. Among the most important AMPs in humans are defensins, cathelicidins and histatins. Here we focus on the latter.

Histatins are a family of 12 structurally related, low molecular weight, cationic antimicrobial polypeptides rich in histidine and secreted by the parotid and submandibular salivary glands in humans. Histatins play pivotal roles in protecting the oral cavity against invading pathogens and have been shown to exert antiviral, antimicrobial, and anticandidal activities [104]. Histatins also possess wound healing properties [105]. Moreover, histatins are considered to maintain low toxicity, as they originate from humans and therefore are being evaluated widely for therapeutic applications [106].

Histatins are being developed to combat *Candida albicans* (pathogenic yeast) infections [107] and multidrug resistance ESKAPE pathogens, which are six bacterial pathogens that carry antibiotic resistance genes (*Enterococcus faecium, Staphylococcus aureus, Klebsiella pneumoniae, Acinetobacter baumannii, Pseudomonas aeruginosa*, and *Enterobacter species*) [108]. Histatins have also been proposed as potential antiprotozoal agents.

One of the major and most potent members of the histatin family is histatin 5. This peptide is known to target fungal mitochondria [109]. As the inhibition of the respiratory chain in *Leishmania* promastigotes results in a metabolic arrest, this peptide was further investigated for its ability to inhibit leishmanial growth [110]. A study in 2008 demonstrated the potential antileishmanial activity of histatin 5 against the growth of *Leishmania* promastigotes and amastigotes. The peptide caused the destruction of the parasite plasma membrane, translocated to the cytoplasm, accumulated in the mitochondria, and inhibited the F(1)F(0)-ATPase, leading to a decrease in critical ATP synthesis. This human antimicrobial and cell-penetrating peptide (CPP) vide infra, represents a powerful, nontoxic tool to inhibit the growth of parasites [111].

17.7.8 Combination therapy of antimicrobial peptides

So far, each study presented has investigated the impact of a single AMP on suppressing parasite counts. However, native AMPs in source organisms such as insects work synergistically. Thus a mixture of AMPs was proposed to confer a more potent effect. A 2016 study investigated the impact on *Trypanosoma* of a peptide mixture containing three synthetic AMPs originating from bumblebees, namely abaecin, a defensin peptide, and hymenoptaecin. Notably, these peptides are the only AMPs found to be expressed upon bumblebee infection. The combination demonstrated greater suppressive activity toward eight different strains of *Crithidia bombi* trypanosomatidae compared to one AMP alone [112]. An additional study in 2010 showed the synergistic cytotoxicity of apidaecin, magainin II, melittin, and cecropin A AMPs. The combination improved activity against *T. cruzi* [113]. These data suggest that a combination of AMPs may be an improved strategy to combat infectious disease.

17.8 Marine peptides

A wide variety of organisms living in the oceans survive their environments using intrinsic antifungal, antibacterial, anti-inflammatory, anticancer, antimicrobial, and neuroprotective mechanisms. Thus marine organisms are a source of many potential natural therapeutic compounds to fight a wide range of diseases.

Several drugs originating from marine organisms have been developed and approved for the treatment of human diseases. One example is ziconotide, a peptidic analgesic drug originally isolated from the venom of the *C. magus* marine snail, a fish-hunting species [114]. Another example of a marine-derived peptide drug is vidarabine, which is used to treat viral infections such as varicella-zoster viruses and herpes simplex. This marine-derived peptide originates from the Caribbean sponge *Tethya crypta* [115]. In addition, trabectedin was found in the sea squirt *Ecteinascidia turbinate* and is used as an anticancer agent [116].

A 2010 study reported the isolation of a marine peptide consisting of three units of D-α-hydroxyisovaleryl-D-valyl-L-lactoyl-L-valyl joined together to form a cyclic depsipeptide. The depsipeptide, now known as valinomycin, was isolated from *Streptomyces sp.* strains from Mediterranean sponges. The peptide functions as an ionophore that facilitates the transport of ions, especially potassium, across membranes. In this study, valinomycin exhibited significant inhibitory activity against both *L. major* ($IC_{50} < 0.11\ \mu M$) and *T. brucei* (IC_{50} 0.0032 μM). However, this compound also exhibited general cytotoxicity against mammalian cells, and therefore structural optimization may be necessary to decrease the host cell cytotoxicity [117].

Another cyclic peptide isolated from a marine source is the lipopeptide viridamide A, which was isolated from the cyanobacterium Oscillatoria nigro-viridis. The peptide displayed antitrypanosomal activity (IC_{50} 1.1 μM) against *T. cruzi* and antileishmanial activity (IC_{50} 1.5 μM) against *L. Mexicana*. Although viridamide A also showed slight toxicity against mammalian cells, it remains an example of a potent compound against parasitic pathogens [118].

A 2013 study reported the identification of a marine Kunitz-type serine protease inhibitor, a 55-amino acid peptide isolated from the Caribbean sea anemone *Stichodactyla helianthus*. This peptide selectively inhibited serine proteases, reduced viability of *T. cruzi*, induced morphological changes and efficiently killed metacyclic trypomastigote and epimastigote forms at a 10 μM concentration [119]. Continuous and extensive research is being carried out to screen and identify additional therapeutic candidates from the rich marine environment.

17.9 Peptides targeting vital proteins and their interactions

17.9.1 Glycoprotein gp63

To survive and proliferate, parasites have developed many sophisticated ways to evade the host immune system. One vital virulence factor is the zinc-metalloprotease glycoprotein gp63, which is expressed on the surface of both amastigote and promastigote *Leishmania* [120]. Gp63 plays an essential role in the attachment of the parasite to macrophages of the host [121], suppressing the host immune system, thus promoting parasite survival [122].

Two proteins are known to be substrates of the glycoprotein gp63 protein, namely myristoylated alanine-rich C kinase substrate (MARCKS) and MARCKS-related protein (MRP). These proteins are extensively expressed in a wide range of cells, including macrophages. In MRP and MARCKS proteins, a basic 24–25 amino acid sequence known as the effector domain (ED) binds to calmodulin, actin, and acidic phospholipids. This ED also contains a serine cleavage site, and phosphorylation of this residue protects the protein from degradation by gp63 [123]. Based on these findings, peptides derived from the ED were proposed to bind glycoprotein gp63 and inhibit its activity. Such peptides were synthesized in a 2002 study for evaluation of their inhibitory activity against gp63. The results showed that ED-derived peptidomimetics were able to modulate the ED interaction and inhibit its activity [124].

In a more recent study screening peptides that bind and target leishmania proteins, molecular docking analysis identified one peptide termed P2 to bind and interact with the gp63 metalloprotease domain. The P2 peptide inhibited *L. major* growth in vitro and in vivo and reduced cutaneous lesions in infected mice [125]. Thus these studies suggest an additional potential target for inhibiting the activity of *Leishmania* parasites.

17.9.2 *Leishmania* receptor for activated C-kinase and *Trypanosoma* receptor for activated C-kinase

The *Leishmania* parasite has developed several sophisticated strategies to inhabit its host organism, which require several pivotal proteins. Among these proteins, two scaffold proteins in *Leishmania* and *Trypanosoma* were been found to play crucial roles in parasite invasion, namely *Leishmania* receptor for activated C-kinase (LACK) and its homolog *Trypanosoma* receptor for activated C-kinase (TRACK). These proteins are plasminogen binding proteins that assist in the adhesion of the parasites to host cells. Parasites lacking these proteins have shown lower viability and limited invasion capacities. Taking advantage of this mechanism, peptides were designed for antiparasitic activity based on the conserved sequence of LACK and TRACK proteins but not the human homolog, receptor for activated C-kinase (RACK). One peptide, p4a (RNGQCQRKGGYGRKKRRQRRR), showed antiparasitic activity, reducing the viability of *L. donovani* promastigotes and *L. amazonensis* promastigotes by 60% at 48 h and approximately 100% after 24 h, respectively. In addition, treatment of *T. cruzi* trypomastigotes with the p4a peptide reduced parasite viability in culture by approximately 75%. Using the same rational design, cyclic peptides were designed and synthesized. The cyclic peptide p4d had improved activity compared to the linear peptide, p4a, with a lower IC_{50} value in culture. Next, the peptides were evaluated for their host cell toxicity, and were found to have no effect on the viability of mammalian cells. Furthermore, the p4d peptide was found to reduce the number of parasites in infected mice. In conclusion, specific LACK and TRACK inhibitor peptides hold great promise toward developing effective nontoxic drugs to treat kinetoplastid diseases [126,127].

17.9.3 Gp85-cytokeratin interaction

Parasites find their way into the host cell by adhering to host cell components. Following a 2001 study in which an interaction between *T. cruzi* and cytokeratin of the host cell was described [128], efforts have been made to better understand and inhibit this interaction.

Sequence analysis of *T. cruzi* proteins and cytokeratin revealed a shared motif between *T. cruzi* gp85 and host cell cytokeratin filament protein, a sequence named TS9 (FTLVASVTI). The TS9 peptide was synthesized and evaluated for binding to host cell cytokeratin and to modulate the *T. cruzi* gp85 interaction with host cell cytokeratin. The epithelial cell line LLC-MK2 was treated with synthetic peptide TS9. Treatment with 200 μM of TS9 for 15 min prior to trypomastigote infection decreased infection by 78% after 48 h. Furthermore, the peptide was nontoxic for host cells and had no impact on cell viability. The antiparasitic activity and minimal host cell cytotoxicity indicate the promise of the TS9 peptide in further development and optimization as a drug to fight *T. cruzi* parasites responsible for Chagas disease [129].

17.9.4 *N*-mirystoyltransferase enzyme

Protein *N*-myristoylation is a lipid cotranslational modification process, where a 14-carbon saturated fatty acid myristate is attached to the N-terminal glycine of substrate proteins. This process is facilitated by the enzyme *N*-myristoyltransferase (NMT), which ensures proper protein trafficking, promotes attachment to cell membranes, and enables many crucial cellular functions. NMT is expressed in *Leishmania* and *Trypanosoma*, an essential protein for their survival, and thus a potential drug target [130]. A recent study showed that the inhibition of *L. donovani* parasite NMT led to a decline in parasite number by 52% [131]. In addition, another study in 2016 suggested NMT be a drug target for *T. cruzi* [132]. Many compounds are under development to specifically target and inhibit the parasitic NMT without targeting the human NMT. In another study, antifungal NMT inhibitors were used as the basis for developing peptidomimetic inhibitors of *Leishmania* NMT. Among this series of compounds, the most potent inhibitor showed submicromolar potency against *L. donovani*, but only moderate selectivity over the human NMT. Therefore further optimization is needed to improve target selectivity [133]. With further investigation and selectivity optimization, NMT represents a potential target for treatment of parasitic infections.

17.9.5 Breast cancer 2

Homologous recombination (HR) is crucial process that assists DNA replication, mostly repairing DNA, via DNA double-strand break repair mechanisms. HR is characterized by a series of pathways that promote the exchange of nucleotide sequences among sister chromatids or homologous chromosomes. This mechanism is crucial for gene stability and diversity. HR is a critical pathway in parasites as well, allowing the parasites to acclimate with environmental conditions and bypass humoral immunity of host organisms. In *T. brucei* parasites, HR pathways facilitate the alternation of cell surface antigens from a homogeneous VSG coating, allowing successful evasion of the host humoral immune responses. The importance of the HR process for the infection, proliferation and persistence of parasites makes it an attractive mechanism to target and disrupt [134].

HR processes are catalyzed by many proteins, the most important of which is RAD51, which is responsible for the recognition of homology and pairing between the homologous DNA strands. Another critical protein is the multifunctional scaffolding protein, breast cancer 2 (BRCA2). BRCA2 has been shown to interact with RAD51, recruiting it to double-stranded DNA (dsDNA). Taking advantage of this pivotal interaction, a 2011 study characterized the interaction of *T. brucei* BRCA2 (TbBRCA2) with the RAD51 paralogues from *T. brucei*. The investigators revealed that a repeating BRC motif in TbBRCA2 is responsible for the interaction with RAD51. Based on these findings, a 44 amino acid peptide aptamer representing the BRC motif was expressed in *T. brucei* to evaluate its effects. Cells with the peptide aptamer were more sensitive to methylmethane sulfonate, a DNA damaging agent used in cancer treatment, and their growth and proliferation were restricted. These results imply that the peptide aptamer inhibited the interaction between TbBRCA2 and *T. brucei* RAD51 paralogues, thus impairing the DNA damage repair mechanism [135]. This research shed light on an attractive therapeutic target, not only for *T. brucei* but also for *T. cruzi* and *Leishmania*, as they possess similar mechanisms [134].

17.9.6 Pseudomonas exotoxin

All pathogens develop unique virulence and self-defense mechanisms in order to facilitate their survival and adaption to their environments. Pseudomonas exotoxin (PE) is a major virulence factor and enzyme toxin produced by the Gram-negative, rod-shaped bacterium *Pseudomonas aeruginosa*, which can cause severe acute and chronic infections in humans. This enzyme, once in the cell, inhibits protein synthesis by ADP ribosylation modification to elongation factor 2, an important element in protein production, leading to cell death [136]. PE A is a 638 amino acid protein, which upon endocytosis undergoes processing to finally be released as a 613 amino acid mature toxin. The protein consists of three major domains, namely a receptor-binding domain at the N-terminus, a translocation domain, and a catalytic domain at the C-terminus (PE40) responsible for the inactivation of elongation factor 2.

A study involving point mutations has shown that the last amino acids at the C-terminal of (PE40), namely REDLK, are important for the cytotoxic activity of PE [137]. A peptide derived from this sequence was hypothesized to exert antimicrobial activity. In a 2019 study, investigators synthesized the KDEL peptide, and its activity against *L. tarentolae* promastigotes and amastigotes was evaluated. The results showed remarkable therapeutic potential for KDEL peptides. The KDEL peptide disrupted the surface membrane integrity and induced parasite apoptosis, reducing parasite numbers [138]. Although the molecules were also slightly more toxic to host cells, the increase in toxicity was much greater against the target parasite. Further studies are recommended to evaluate the efficiency of KDEL and REDLK peptides in the elimination of other pathogenic parasites. In addition, a recent study synthesized a REDLK peptide and evaluated its activity against *L. tarentolae* promastigote growth. The results showed that REDLK facilitates antileishmanial activity in a dose-dependent manner, through the induction of apoptosis and disruption of cell membrane integrity [139].

17.10 Peptides that target critical pathways

17.10.1 Endoplasmic reticulum-associated degradation pathway

A quality control system in eukaryotic cells is the endoplasmic reticulum-associated degradation (ERAD) pathway, an ubiquitin-proteasome system that is responsible for the degradation of misfolded or misassembled proteins or peptides. Parasites such as those that cause malaria, toxoplasmosis, trypanosomiasis, and leishmaniasis also possess this pathway. However, compared to higher eukaryotic cells, the ERAD network in parasites is simpler. Bioinformatic analysis has shown that in parasites there are fewer proteins involved in the ERAD pathway. Therefore due to the simplicity of the ERAD system in parasites, the effect of any change in the ERAD pathway is expected to be much more significant.

A 2012 study investigated the impact of inhibiting a signal peptide peptidase (SPP), which is involved in the ERAD pathway. The study mainly investigated the impact of inhibiting SPP on the growth of *P. falciparum*, a parasite that causes malaria. The researchers also investigated the inhibition of *T. cruzi* and *T. gondii*, as these parasites have SPP orthologues.

The investigators developed NITD731, a small molecule that directly targets the SPP. The molecule was shown to be potent against *P. falciparum* parasites and also showed antiparasitic activity against *T. cruzi* and *T. gondii*. The study introduced an attractive and selective target for the inhibition of protozoan parasites [140].

17.10.2 Glycolysis pathway

Trypanosoma are unable to store carbohydrates and they lack the Krebs cycle and oxidative phosphorylation pathways. Glycolysis of glucose obtained from the host organism is the only source of adenosine triphosphate (ATP) to sustain the metabolic needs of *Trypanosoma*. Therefore the glycolysis pathway serves as a "sweet spot" to selectively target parasitic metabolism [141].

T. brucei hexokinase 1 (TbHK1) and *T. brucei* hexokinase 2 (TbHK2) are two trypanosomal proteins localized in peroxisome-like organelles, and they initiate the first step of glycolysis. Specific RNAi targeting those proteins was shown to be toxic to the parasite [142], making them an attractive potential target. Previously, benzamidobenzoic acid (BABA) derivatives were validated to be potent inhibitors of TbHK1 but showed weak trypanocidal activity. To optimize the trypanocidal activity of TbHK1 inhibitors, the tripeptide peroxisomal targeting sequence SKL was used as a delivery agent for the BABA compound to the peroxisome-like organelles of *Trypanosoma* to improve the inhibitor efficacy. SKL-conjugated BABA remarkably improved delivery of BABA to the target and increased the inhibitory effect against the growth of *T. brucei*. These data demonstrate the use of peptide sequences to facilitate the transport of poorly penetrating compounds to their target and thus improve their therapeutic efficacy [143].

17.11 Peptides that target proteases

Proteases are proteolytic enzymes that cleave and degrade proteins by peptide bond hydrolysis. Proteases play key roles in the pathogenesis of the protozoan parasite infections, are expressed in all parasite life stages, are involved in many critical parasite life-sustaining pathways, and contribute to parasite invasion and immune evasion [144]. Parasites express proteases homologous to mammalians that show structural similarities, but still exhibit some differences, making them attractive molecular drug targets. In this section, we discuss the implications of protease inhibitors in kinetoplastid diseases and their suitability as antimicrobial drugs.

17.11.1 Cysteine peptidase B

Cysteine peptidase B (CPB) belongs to the papain family expressed in *L. Mexicana* and is involved in parasite immune evasion mechanisms [145].

The mechanism may involve regulation of gp63, a glycosyl-phosphatidyl-inositol (GPI)-anchored metalloprotease which plays a key role in host exploitation by parasites [146]. To discover peptide substrates of CPB, a 2001 study investigated a combinatorial fluorescence quenched peptide library. This method allows the identification of specific substrates for proteolytic enzymes and led to the identification of a lead peptide, FAARYLNGA. This peptide was resistant to hydrolysis by the enzyme and had an inhibitory activity with a Ki value of 4 μM. The peptide was further optimized by extension and addition of an ortho-aminobenzoic acid group (Abz-FAARYLNGAQ) to yield a more potent peptide with Ki value of 30 nM [147].

Cruzipain is a cathepsin L-like cysteine protease and a member of the papain family, expressed in all *T. cruzi* developmental stages. This protease plays pivotal role in cell invasion and parasite replication in the setting of Chagas disease [148]. Therefore cruzipain represents a target for antitrypanosomal drug development. A previous study identified a novel peptide that bound and inhibited the mammalian lysosomal cysteine proteinase cathepsin L with a Ki value of 1.7 pM. This peptide is a 65 amino acid fragment derived from p41 of the invariant chain (Ii), which associates and binds to the peptide-binding groove of the major histocompatibility complex (MHC) class II molecule [149]. The peptide was evaluated for its inhibitory activity and binding to the parasitic cysteine protease cruzipain. Results showed potent binding to the enzyme ($Ki = 5.8 \times 10^{-11}$ M) [150]. These findings demonstrate the potential use of the P41 invariant chain for inhibition of the pivotal parasitic enzyme.

A high-throughput screening method was used in a 2001 study to identify peptides inhibitors of a recombinant cysteine protease from *L. Mexicana* using a "one-bead-two-compounds" library assay approach. In this assay, every bead was bound to a different putative protease inhibitor and an accompanying fluorescence-quenched protease substrate. The library was screened with a recombinant cysteine protease from *L. Mexicana* called CPB2.8 DeltaCTE. The screened peptides were based on Diels-Alder reactions, with two classes of inhibitors identified, namely bicyclic Diels-Alder product isosters and N-terminal alpha, beta-unsaturated ketone Michael acceptors. These inhibitors were resynthesized and evaluated for their inhibitory effects. The results demonstrated an inhibitory range between high nanomolar to low micromolar concentrations [151].

1H-[1,2,3]-triazole molecules are five-membered rings that consist of two carbon atoms and three nitrogen atoms. Triazole molecules have gained much attention due to their pharmacological properties such as anti-inflammatory, anticancer, antiviral, antifungal, and antimicrobial activities [152–154]. The unique structure of the 1H-[1,2,3]-triazole molecules results in a rigid and stable molecule under in vivo conditions, which makes them favorable for drug discovery and development. A combinatorial library of peptidomimetics was synthesized with the addition of regiospecific copper(I) to catalyze the synthesis of 1H-[1,2,3]-triazoles, forming peptidotriazoles. The peptidotriazole library was screened for inhibitory activity against CPB2.8ΔCTE, a recombinant cysteine protease from *L. Mexicana*. Twenty-three compounds were identified as inhibitors of the protease and were resynthesized separately and examined for their inhibitory activities against *L. Mexicana* using in-solution assays. The results exhibited potent restriction activity with low micromolar inhibition. The most potent compound displayed a Ki value of 76 nM [155]. These results demonstrate the potential antiparasitic effects of peptidotriazoles.

A similar one-bead-two compound combinational chemistry screening method was used in an additional study to screen for octapeptide inhibitors of the recombinant cysteine protease B from *L. Mexicana*. The identified peptides were evaluated for inhibitory activity against the enzyme and displayed micromolar to nanomolar Ki values. The peptide DKHF(CH_2NH) LLVK displayed a Ki value of 1 μM and exhibited antileishmanial activity with an IC_{50} of 50 μM [156]. These studies suggest cysteine proteases as potential targets in drug development for the treatment of kinetoplastid diseases due to their importance in the pathogenesis of parasites.

17.12 Dipeptides

Dipeptides are short peptides containing only two amino acids. Their attractive properties include rapid absorption in the body and flexible penetration through membranes and the BBB. Therefore dipeptides and dipeptide hybrids are being developed for many disorders including parasite infections.

One study in 2018 described the synthesis of triazole-based hybrid dipeptides and their effects on the growth of *L. major*. Triazoles are bioactive heterocycles widely used in infectious disease drugs like pentavalent antimonials, amphotericin B and miltefosine for leishmaniasis. To increase the lipophilicity and cytotoxicity, triazole-based hybrid dipeptides containing 1-(2-Amino-phenyl)-1H- [1,2,3] triazole-4-carboxylic acid at the N-terminus and Leu/Phe/Val/Gly at the C-terminus were synthesized and assessed for antileishmanial activity. Among these peptides, a dipeptide containing a Leu residue exhibited the most potent inhibitory activity, with an IC_{50} value of 11 μg/mL for *L. major* promastigotes. This was due to higher lipophilicity and the free intramolecular hydrogen bond sites of the dipeptide structure, which mediated intracellular interactions with the target. Furthermore, the dipeptide did not show any toxic effects on mammalian macrophages, demonstrating a safe and effective antileishmanial drug lead [157].

Another study showed the potency of dipeptide nitriles in the suppression of *T. brucei* growth. One of these dipeptides specifically targeted Rhodesian, the cathepsin L-like protease of *T. brucei*. Rhodesain is required for the degradation of host cell proteins and is crucial for the survival of parasites. In the study, many dipeptide nitrile compounds were evaluated against

T. brucei. A 3-bromo-Phe residue and biphenyl fragment compound, as well as a compound comprising a Leu residue and a triarylic moiety with thiophene, a 1,2,4-oxadiazole and a phenyl ring, exhibited promising antitrypanosomal activity with an EC_{50} value of 8 μM. These compounds were found to fit into the S2-S3 pockets of Rhodesian and exhibited high selectivity and affinity for Rhodesian. Further optimization will be helpful for improved antiparasitic activity [158].

17.13 Peptoids

Peptoids are mimics of α-peptides, in which their side chains are appended to the nitrogen atom of the peptide backbone, rather than to the α-carbons. This change in the structure enhances their permeability and protects the peptoids against proteolytic degradation. Thus peptoids offer a promising alternative to highly degradable traditional peptides and represent an emerging peptidomimetic format in the drug discovery field [159].

The first study showing antiparasitic activity of linear peptoids was a 2015 study on *L. Mexicana* parasites, which cause the cutaneous form of leishmaniasis. The peptoids displayed an EC_{50} value of 7 and 17 μM against promastigotes and amastigotes, respectively. This study led to further investigation of peptoids as antiparasitic agents [160]. An extended study in 2016 expanded the peptoid library, and some of the identified peptoids showed more potent activity than in the previous study. In addition, structure–activity relationship studies revealed a possible membrane perturbation mode of action for the peptoids [161]. A 2019 study described an easy and efficient method to synthesize peptoids and their evaluation against *L. Amazonensis* promastigotes in vitro. The results showed even greater parasite inhibitory potential with improved IC_{50} values for some of the compounds (IC_{50} 2.6–7.9 μM) [162]. In conclusion, the results from these studies demonstrate that the structural rigidity and stability of peptoids can be harnessed for the development of a promising new class of antiparasitic agents.

17.14 Peptides as drug delivery agents

Peptide sequences can also be utilized as delivery agents to target other compounds to parasites. The antiparasitic activity of peptides or drugs is dependent on their binding capability to the parasite. To affect the parasite, a peptide may either bind to parasite cell surfaces or directly penetrate their cells.

One recent study involved the use of a random phage library screening to identify peptides that bind *T. cruzi*; one peptide was identified from a bacteriophage and termed EPI18 (NMYKSPDYCHKAKSSDP) [163]. Biochemical assays were carried out to determine and confirm the binding of EPI18 to epimastigote surfaces. Although this study did not reveal antiparasitic activity, the binding capability and the specificity of EPI18 peptide toward the epimastigotes of *T. cruzi* suggest it may represent a potential parasite targeting or drug delivery peptide [163].

CPPs are peptides that can penetrate directly through the plasma membrane. Thus CPPs can be utilized to deliver various therapeutic substances that are poorly passed through the biological cell membrane [164]. We mentioned above the emerging drug resistance of kinetoplastid infections. One method to bypass the resistance may be to conjugate a known drug to a CPP, thus providing an alternative mode of entry into kinetoplastid parasites.

Miltefosine is an orally administered drug used to treat VL, for which parasite resistance has been reported [55]. Studies have shown that this resistance mostly arises from reduced uptake of the drug due to the inactivation of *L. donovani* miltefosine transporter (LdMT), and more specifically the B subunit of LdMT called LRos3. In addition, the drug is depleted from the cell due to overexpressed ATP-binding cassette (ABC) transporters [165]. By elucidating these pathways, a group in 2012 aimed to overcome resistance mechanisms by conjugating miltefosine to the transactivator of transcription (TAT) CPP (48–60) to facilitate its uptake. This miltefosine analog was evaluated against promastigote and intracellular amastigote forms of miltefosine-resistant *L. donovani* parasites. The results showed that the cell-penetrating analog retained its accumulation ability in parasites and reduced parasite resistance [166].

Another example of potential use for CPPs is the delivery of the antileishmanial drug paromomycin. Paromomycin is an aminoglycoside antibiotic used for the treatment of many infections, including leishmaniasis [167]. However, paromomycin is poorly absorbed, and several studies attempted to improve its bioavailability. In a 2017 study, cell-penetrating TAT (48–60) peptide conjugated to a fluorescent dye was used to deliver the drug. Confocal fluorescence microscopy visualization showed that the drug was successfully delivered into *Leishmania* cells [168]. In summary, these studies demonstrate that peptides may be used as powerful tools in the delivery of molecules to their targets, thus overcoming challenges associated with poorly delivered drugs.

17.15 From antibody to peptides

Killer peptide (KP) is an engineered peptide composed of ten amino acids (AKVTMTCSAS) and derived from the single-chain recombinant antibody of fungicidal yeast killer toxin that targets the β-glucan cell wall receptors. The KP was shown to exert antifungal, antibacterial, and antiviral activity against many species [169].

A 2006 study sought to test the inhibitory effect of KP against the growth of *L. major* and *L. infantum* promastigotes. The results showed that the peptide has potential antileishmanial and antiproliferative activities at concentrations in the micromolar range. Ultrastructural analysis by TEM revealed that the KP does not cause DNA fragmentation nor plasma-membrane destruction, but rather intracellular damage [170].

17.16 Peptides as diagnostic tools for parasitic diseases

Part of treatment efficiency for parasitic diseases relies on their early detection. However, diagnosing the diseases, for example, the early stages of Chagas disease, is often challenging because patients show nearly no symptoms in the acute and even sometimes in the chronic phases. After development into later stages, the disease cannot be successfully treated. Therefore early detection and diagnosis of the disease improve the chances of successful treatment. Most laboratory methods to detect parasites are invasive, demand expertise, and/or require complex devices and equipment that may not be available in every laboratory. Therefore simpler and more convenient diagnostic methods are in need. Discovering new parasite antigens represents a rapid noninvasive alternative diagnostic method. Therefore numerous efforts are being invested in detecting new candidates.

While peptides have emerged as a potential therapeutic approach, peptides are also being investigated for their detection of parasites. One example was described in a 2017 study, where attempts were made to identify peptide antigens for *Leishmania* parasites. Peptide phage display library screening was performed using antibodies from *L. braziliensis* patients. Three peptide candidates were selected (P1, P2, and P3), synthesized, and further evaluated for their efficiency in CL serological diagnosis. The peptides individually and in combination demonstrated *Leishmania* detection capability. P1 showed 72% sensitivity, and the mix of the three peptides showed 79% sensitivity, with 78%–100% specificity, which is clearly in the range of clinical utility. This study provided a novel, simple and nondemanding method for detecting *Leishmania* parasite infection [171].

In 2019, researchers described a rapid method for detecting VL antibodies. The concept of this method was to immobilize a peptide-based probe of the promastigote surface antigen PSA-38S on electrospun polyamide-6 chitosan nanofibers. The relative change in impedance was measured using electrochemical impedance spectroscopy upon incubating serum samples with the PSA-38S peptide. The change in impedance was proportional to the PSA-38S peptide concentrations. Moreover, this method appeared to differentiate between negative and positive VL [172].

A recent method for the detection of various biomarkers of diseases is based on serum peptide profiling. The concept behind this method is relatively simple. A serum peptide profile of the patient is analyzed using matrix-assisted laser desorption/ionization (MALDI)-time-of-flight (TOF) mass spectrometry (Ms) and compared to a known disease positive serum peptide profile. This method enables the early detection of disease biomarkers that could be difficult to detect using traditional methods [173,174].

A group of researchers in 2012 applied this method to detect *L. donovani* infection in a mouse model. Serum peptides from a *L. donovani* mouse model were profiled by MALDI-TOF Ms. This method showed high accuracy, sensitivity and specificity in detecting early *L. donovani* infection [175]. These results demonstrated a novel and powerful proof of concept method to detect the early asymptotic infections caused by parasites.

17.17 Concluding remarks

We provided a broad overview of parasitic diseases and focused on the applications of peptides to fight life-threatening conditions, such as HAT, Chagas disease, and various forms of leishmaniasis, which are caused by kinetoplastid parasites. Until now, no effective vaccines have been developed for the elimination of these diseases. Currently used drugs are accompanied by severe toxicity and side effects, and emerging drug resistance of certain parasites is raising concerns regarding the effectiveness of these treatments in the future.

Peptides have gained considerable attention as therapeutics over recent years and can be used to fight or assist in fighting parasite infections. Their various advantages over currently used small molecule drugs, such as reduced toxicity, increased specificity, high affinity for their target, and the ability to penetrate cell membranes make them desirable as alternative therapeutic agents. A wide range of species, such as plants, bacteria, fungi, mammals, and insects express peptides that assist their defense mechanisms against pathogens. Inspired by these mechanisms, these peptides have been extracted or synthesized for use against kinetoplastid parasites. Further modification either in peptide sequence or in chemical structure to produce peptidomimetics, such as cyclic peptides or peptoids, can improve the efficacy and potential use for fighting these challenging diseases.

Peptides have played several roles in fighting parasitic diseases, serving as therapeutic agents, drug delivery agents, and as diagnostic agents. Their effectiveness has often been enhanced by combining different peptides to provide a multitarget effect. Peptide development holds great promise for identifying lead peptidic candidates to diagnose and treat parasitic diseases, and they represent valuable future tools to augment commonly used drugs.

References

[1] Tigabu A, Taye S, Aynalem M, Adane K. Prevalence and associated factors of intestinal parasitic infections among patients attending Shahura Health Center, Northwest Ethiopia. BMC Res Notes 2019;12(1):333.

[2] Burton J, Tam D. Towards a parasitic ethics. Theory Cult Soc 2016;33(4):103–25.

[3] Liao CW, Fu CJ, Kao CY, Lee YL, Chen PC, Chuang TW, et al. Prevalence of intestinal parasitic infections among school children in capital areas of the Democratic Republic of São Tomé and Príncipe, West Africa. Afr Health Sci 2016;16(3):690–7.

[4] Mendez OA, Koshy AA. Toxoplasma gondii: entry, association, and physiological influence on the central nervous system. PLoS Pathog 2017;13(7):e1006351.
[5] Vannier-Santos MA, Lenzi HL. Parasites or cohabitants: cruel omnipresent usurpers or creative "éminences grises? J Parasitol Res 2011;2011:214174.
[6] Tharmaratnam T, Kumanan T, Iskandar MA, D'Urzo K, Gopee-Ramanan P, Loganathan M, et al. Entamoeba histolytica and amoebic liver abscess in northern Sri Lanka: a public health problem. Trop Med Health 2020;48:2.
[7] Antinori S, Galimberti L, Milazzo L, Corbellino M. Biology of human malaria plasmodia including *Plasmodium knowlesi*. Mediterr J Hematol Infect Dis 2012;4(1): e2012013.
[8] Dantas-Torres F. Biology and ecology of the brown dog tick, *Rhipicephalus sanguineus*. Parasit Vectors 2010;3:26.
[9] Furtado LFV, de Paiva Bello ACP, Rabelo ÉML. Benzimidazole resistance in helminths: from problem to diagnosis. Acta Trop 2016;162:95–102.
[10] Meyer H, Porter KR. A study of *Trypanosoma cruzi* with the electron microscope. Parasitology 1954;44(1–2):16–23.
[11] Meyer H, De Oliveira Musacchio M, De Andrade Mendonca I. Electron microscopic study of *Trypanosoma cruzi* in thin sections of infected tissue cultures and of blood-agar forms. Parasitology 1958;48(1–2):1–8.
[12] Cavalcanti DP, de Souza W. The kinetoplast of Trypanosomatids: from early studies of electron microscopy to recent advances in atomic force microscopy. Scanning 2018; 2018:9603051.
[13] Landfear SM, Tran KD, Sanchez MA. Flagellar membrane proteins in kinetoplastid parasites. IUBMB Life 2015;67(9):668–76.
[14] Khare S, Nagle AS, Biggart A, Lai YH, Liang F, Davis LC, et al. Proteasome inhibition for treatment of leishmaniasis, Chagas disease and sleeping sickness. Nature 2016;537 (7619):229–33.
[15] Collett CF, Kitson C, Baker N, Steele-Stallard HB, Santrot MV, Hutchinson S, et al. Chemogenomic profiling of antileishmanial efficacy and resistance in the related kinetoplastid parasite *Trypanosoma brucei*. Antimicrob Agents Chemother 2019; 63(8).
[16] Franco JR, Simarro PP, Diarra A, Jannin JG. Epidemiology of human African trypanosomiasis. Clin Epidemiol 2014;6:257–75.
[17] Stich A, Abel PM, Krishna S. Human African trypanosomiasis. BMJ 2002;325(7357): 203–6.
[18] Keating J, Yukich JO, Sutherland CS, Woods G, Tediosi F. Human African trypanosomiasis prevention, treatment and control costs: a systematic review. Acta Trop 2015; 150:4–13.
[19] Chappuis F, Loutan L, Simarro P, Lejon V, Büscher P. Options for field diagnosis of human African trypanosomiasis. Clin Microbiol Rev 2005;18(1):133–46.
[20] Babokhov P, Sanyaolu AO, Oyibo WA, Fagbenro-Beyioku AF, Iriemenam NC. A current analysis of chemotherapy strategies for the treatment of human African trypanosomiasis. Pathog Glob Health 2013;107(5):242–52.
[21] Baker N, de Koning HP, Mäser P, Horn D. Drug resistance in African trypanosomiasis: the melarsoprol and pentamidine story. Trends Parasitol 2013;29(3):110–18.
[22] Bernhard SC, Nerima B, Mäser P, Brun R. Melarsoprol- and pentamidine-resistant *Trypanosoma brucei* rhodesiense populations and their cross-resistance. Int J Parasitol 2007;37(13):1443–8.

[23] Yang G, Choi G, No JH. Antileishmanial mechanism of diamidines involves targeting kinetoplasts. Antimicrob Agents Chemother 2016;60(11):6828–36.

[24] Zoltner M, Campagnaro GD, Taleva G, Burrell A, Cerone M, Leung KF, et al. Suramin exposure alters cellular metabolism and mitochondrial energy production in African trypanosomes. J Biol Chem 2020;295(24):8331–47.

[25] Fantini J, Rognoni JB, Roccabianca M, Pommier G, Marvaldi J. Suramin inhibits cell growth and glycolytic activity and triggers differentiation of human colic adenocarcinoma cell clone HT29-D4. J Biol Chem 1989;264(17):10282–6.

[26] Vincent IM, Creek D, Watson DG, Kamleh MA, Woods DJ, Wong PE, et al. A molecular mechanism for eflornithine resistance in African trypanosomes. PLoS Pathog 2010;6(11): e1001204.

[27] Fairlamb AH, Horn D. Melarsoprol resistance in African trypanosomiasis. Trends Parasitol 2018;34(6):481–92.

[28] Burri C, Brun R. Eflornithine for the treatment of human African trypanosomiasis. Parasitol Res 2003;90(Supp 1):S49–52.

[29] Priotto G, Pinoges L, Fursa IB, Burke B, Nicolay N, Grillet G, et al. Safety and effectiveness of first line eflornithine for *Trypanosoma brucei* gambiense sleeping sickness in Sudan: cohort study. BMJ 2008;336(7646):705–8.

[30] Priotto G, Kasparian S, Mutombo W, Ngouama D, Ghorashian S, Arnold U, et al. Nifurtimox-eflornithine combination therapy for second-stage African *Trypanosoma brucei* gambiense trypanosomiasis: a multicentre, randomised, phase III, non-inferiority trial. Lancet 2009;374(9683):56–64.

[31] Burri C, Blum J. In: Cohen J, Powderly WG, Opal SM, editors. Human African trypanosomiasis. Infectious diseases. 4th ed. Elsevier; 2017, p. 966–70. e961.

[32] Barrett MP, Vincent IM, Burchmore RJ, Kazibwe AJ, Matovu E. Drug resistance in human African trypanosomiasis. Future Microbiol 2011;6(9):1037–47.

[33] Baker CH, Welburn SC. The long wait for a new drug for human African trypanosomiasis. Trends Parasitol 2018;34(10):818–27.

[34] Chagas C. Nova espécie mórbida do homem produzida por um trypanosoma (*Trypanosoma cruzi*). Nota prévia. Bras Médico 1909;230:161.

[35] Aufderheide AC, Salo W, Madden M, Streitz J, Buikstra J, Guhl F, et al. A 9,000-year record of Chagas' disease. Proc Natl Acad Sci USA 2004;101(7):2034–9.

[36] Moncayo A, Silveira AC. Current epidemiological trends for Chagas disease in Latin America and future challenges in epidemiology, surveillance and health policy. Mem Inst Oswaldo Cruz 2009;104(Suppl 1):17–30.

[37] Balouz V, Agüero F, Buscaglia CA. Chagas disease diagnostic applications: present knowledge and future steps. Adv Parasitol 2017;97:1–45.

[38] Santos FL, de Souza WV, Barros Mda S, Nakazawa M, Krieger MA, Gomes YDM. Chronic chagas disease diagnosis: a comparative performance of commercial enzyme immunoassay tests. Am J Trop Med Hyg 2016;94(5):1034–9.

[39] Castro JA, de Mecca MM, Bartel LC. Toxic side effects of drugs used to treat Chagas' disease (*American trypanosomiasis*). Hum Exp Toxicol 2006;25(8):471–9.

[40] Trochine A, Creek DJ, Faral-Tello P, Barrett MP, Robello C. Benznidazole biotransformation and multiple targets in *Trypanosoma cruzi* revealed by metabolomics. PLoS Negl Trop Dis 2014;8(5):e2844.

[41] Campos MC, Leon LL, Taylor MC, Kelly JM. Benznidazole-resistance in *Trypanosoma cruzi*: evidence that distinct mechanisms can act in concert. Mol Biochem Parasitol 2014;193(1):17–19.

[42] Wilkinson SR, Taylor MC, Horn D, Kelly JM, Cheeseman I. A mechanism for cross-resistance to nifurtimox and benznidazole in trypanosomes. Proc Natl Acad Sci USA 2008;105(13):5022–7.

[43] Boiani M, Piacenza L, Hernández P, Boiani L, Cerecetto H, González M, et al. Mode of action of nifurtimox and N-oxide-containing heterocycles against *Trypanosoma cruzi*: is oxidative stress involved?". Biochem Pharmacol 2010;79(12):1736–45.

[44] Steverding D. The history of leishmaniasis. Parasit Vectors 2017;10(1):82.

[45] Torres-Guerrero E, Quintanilla-Cedillo MR, Ruiz-Esmenjaud J, Arenas R. Leishmaniasis: a review. F1000Res 2017;6:750.

[46] Wheeler RJ, Gluenz E, Gull K. The cell cycle of Leishmania: morphogenetic events and their implications for parasite biology. Mol Microbiol 2011;79(3):647–62.

[47] Reithinger R, Dujardin JC. Molecular diagnosis of leishmaniasis: current status and future applications. J Clin Microbiol 2007;45(1):21–5.

[48] Monge-Maillo B, López-Vélez R. Miltefosine for visceral and cutaneous leishmaniasis: drug characteristics and evidence-based treatment recommendations. Clin Infect Dis 2015;60(9):1398–404.

[49] Rojas R, Valderrama L, Valderrama M, Varona MX, Ouellette M, Saravia NG. Resistance to antimony and treatment failure in human Leishmania (Viannia) infection. J Infect Dis 2006;193(10):1375–83.

[50] Chakraborty AK, Majumder HK. Mode of action of pentavalent antimonials: specific inhibition of type I DNA topoisomerase of Leishmania donovani. Biochem Biophys Res Commun 1988;152(2):605–11.

[51] Frezard F, Demicheli C, Ribeiro RR. Pentavalent antimonials: new perspectives for old drugs. Molecules 2009;14(7):2317–36.

[52] Al Balushi A, Khamis F, Klaassen CHW, Gangneux JP, Van Hellemond JJ, Petersen E. Double infection with *Leishmania tropica* and L. major in an HIV patient controlled with high doses of amphotericin B. Open Forum Infect Dis 2018;5(12):ofy323.

[53] Mbongo N, Loiseau PM, Billion MA, Robert-Gero M. Mechanism of amphotericin B resistance in *Leishmania donovani* promastigotes. Antimicrob Agents Chemother 1998;42(2):352–7.

[54] More B, Bhatt H, Kukreja V, Ainapure SS. Miltefosine: great expectations against visceral leishmaniasis. J Postgrad Med 2003;49(1):101–3.

[55] Srivastava S, Mishra J, Gupta AK, Singh A, Shankar P, Singh S. Laboratory confirmed miltefosine resistant cases of visceral leishmaniasis from India. Parasit Vectors 2017;10(1):49.

[56] Sundar S, Olliaro PL. Miltefosine in the treatment of leishmaniasis: clinical evidence for informed clinical risk management. Ther Clin Risk Manag 2007;3(5):733–40.

[57] Rahman R, Goyal V, Haque R, Jamil K, Faiz A, Samad R, et al. Safety and efficacy of short course combination regimens with AmBisome, miltefosine and paromomycin for the treatment of visceral leishmaniasis (VL) in Bangladesh. PLoS Negl Trop Dis 2017;11(5):e0005635.

[58] Wiwanitkit V. Interest in paromomycin for the treatment of visceral leishmaniasis (kala-azar). Ther Clin Risk Manag 2012;8:323–8.

[59] Chawla B, Jhingran A, Panigrahi A, Stuart KD, Madhubala R. Paromomycin affects translation and vesicle-mediated trafficking as revealed by proteomics of paromomycin -susceptible -resistant *Leishmania donovani*. PLoS One 2011;6(10):e26660.

[60] Jhingran A, Chawla B, Saxena S, Barrett MP, Madhubala R. Paromomycin: uptake and resistance in *Leishmania donovani*. Mol Biochem Parasitol 2009;164(2):111–17.

[61] Basselin M, Denise H, Coombs GH, Barrett MP. Resistance to pentamidine in *Leishmania mexicana* involves exclusion of the drug from the mitochondrion. Antimicrob Agents Chemother 2002;46(12):3731–8.
[62] Dupont B. Overview of the lipid formulations of amphotericin B. J Antimicrob Chemother 2002;49(Suppl 1):31–6.
[63] Zulfiqar B, Shelper TB, Avery VM. Leishmaniasis drug discovery: recent progress and challenges in assay development. Drug Discov Today 2017;22(10):1516–31.
[64] Qvit N, Rubin SJS, Urban TJ, Mochly-Rosen D, Gross ER. Peptidomimetic therapeutics: scientific approaches and opportunities. Drug Discov Today 2017;22(2):454–62.
[65] Rubin S, Qvit N. Cyclic peptides for protein-protein interaction targets: applications to human disease. Crit Rev Eukaryot Gene Expr 2016;26(3):199–221.
[66] Laupacis A, Keown PA, Ulan RA, McKenzie N, Stiller CR. Cyclosporin A: a powerful immunosuppressant. Can Med Assoc J 1982;126(9):1041–6.
[67] Fosgerau K, Hoffmann T. Peptide therapeutics: current status and future directions. Drug Discovery Today 2015;20(1):122–8.
[68] Boulanger N, Brun R, Ehret-Sabatier L, Kunz C, Bulet P. Immunopeptides in the defense reactions of Glossina morsitans to bacterial and *Trypanosoma brucei* infections. Insect Biochem Mol Biol 2002;32(4):369–75.
[69] Hu C, Aksoy S. Innate immune responses regulate trypanosome parasite infection of the tsetse fly *Glossina morsitans*. Mol Microbiol 2006;60(5):1194–204.
[70] Yeaman MR, Yount NY. Mechanisms of antimicrobial peptide action and resistance. Pharmacol Rev 2003;55(1):27–55.
[71] Kumar P, Kizhakkedathu JN, Straus SK. Antimicrobial peptides: diversity, mechanism of action and strategies to improve the activity and biocompatibility in vivo. Biomolecules 2018;8(1).
[72] Cytryńska M, Zdybicka-Barabas A. Defense peptides: recent developments. Biomol Concepts 2015;6(4):237–51.
[73] Yamage M, Yoshiyama M, Grab DJ, Kubo M, Iwasaki T, Kitani H, et al. Characteristics of novel insect defensin-based membrane-disrupting trypanocidal peptides. Biosci Biotechnol Biochem 2009;73(7):1520–6.
[74] Kitani H, Naessens J, Kubo M, Nakamura Y, Iraqi F, Gibson J, et al. Synthetic nonamer peptides derived from insect defensin mediate the killing of African trypanosomes in axenic culture. Parasitol Res 2009;105(1):217–25.
[75] Dabirian S, Taslimi Y, Zahedifard F, Gholami E, Doustdari F, Motamedirad M, et al. Human neutrophil peptide-1 (HNP-1): a new anti-leishmanial drug candidate. PLoS Negl Trop Dis 2013;7(10):e2491.
[76] Madison MN, Kleshchenko YY, Nde PN, Simmons KJ, Lima MF, Villalta F. Human defensin alpha-1 causes *Trypanosoma cruzi* membrane pore formation and induces DNA fragmentation, which leads to trypanosome destruction. Infect Immun 2007;75(10): 4780–91.
[77] Conlon JM, Ahmed E, Condamine E. Antimicrobial properties of brevinin-2-related peptide and its analogs: efficacy against multidrug-resistant *Acinetobacter baumannii*. Chem Biol Drug Des 2009;74(5):488–93.
[78] Ghavami S, Asoodeh A, Klonisch T, Halayko AJ, Kadkhoda K, Kroczak TJ, et al. Brevinin-2R(1) semi-selectively kills cancer cells by a distinct mechanism, which involves the lysosomal-mitochondrial death pathway. J Cell Mol Med 2008;12(3):1005–22.
[79] Zahedifard F, Lee H, No JH, Salimi M, Seyed N, Asoodeh A, et al. Anti-leishmanial activity of Brevinin 2R and its Lauric acid conjugate type against L. major: in vitro

mechanism of actions and in vivo treatment potentials. PLoS Negl Trop Dis 2019;13(2):e0007217.

[80] Mello EO, Ribeiro SF, Carvalho AO, Santos IS, Da Cunha M, Santa-Catarina C, et al. Antifungal activity of PvD1 defensin involves plasma membrane permeabilization, inhibition of medium acidification, and induction of ROS in fungi cells. Curr Microbiol 2011;62(4):1209–17.

[81] do Nascimento VV, Mello Ede O, Carvalho LP, de Melo EJ, Carvalho Ade O, Fernandes KV, et al. PvD1 defensin, a plant antimicrobial peptide with inhibitory activity against *Leishmania amazonensis*. Biosci Rep 2015;35(5).

[82] Simmaco M, Mignogna G, Canofeni S, Miele R, Mangoni ML, Barra D. Temporins, antimicrobial peptides from the European red frog Rana temporaria. Eur J Biochem 1996;242(3):788–92.

[83] Rollins-Smith LA, Carey C, Conlon JM, Reinert LK, Doersam JK, Bergman T, et al. Activities of temporin family peptides against the chytrid fungus (*Batrachochytrium dendrobatidis*) associated with global amphibian declines. Antimicrob Agents Chemother 2003;47(3):1157–60.

[84] Saviello MR, Malfi S, Campiglia P, Cavalli A, Grieco P, Novellino E, et al. New insight into the mechanism of action of the temporin antimicrobial peptides. Biochemistry 2010;49(7):1477–85.

[85] Raja Z, André S, Abbassi F, Humblot V, Lequin O, Bouceba T, et al. Insight into the mechanism of action of temporin-SHa, a new broad-spectrum antiparasitic and antibacterial agent. PLoS One 2017;12(3):e0174024.

[86] Souza AL, Faria RX, Calabrese KS, Hardoim DJ, Taniwaki N, Alves LA, et al. Temporizin and temporizin-1 peptides as novel candidates for eliminating *Trypanosoma cruzi*. PLoS One 2016;11(7):e0157673.

[87] Abbassi F, Raja Z, Oury B, Gazanion E, Piesse C, Sereno D, et al. Antibacterial and leishmanicidal activities of temporin-SHd, a 17-residue long membrane-damaging peptide. Biochimie 2013;95(2):388–99.

[88] Lynn MA, Kindrachuk J, Marr AK, Jenssen H, Panté N, Elliott MR, et al. Effect of BMAP-28 antimicrobial peptides on *Leishmania major* promastigote and amastigote growth: role of leishmanolysin in parasite survival. PLoS Negl Trop Dis 2011;5(5):e1141.

[89] Cauchard S, Van Reet N, Buscher P, Goux D, Grotzinger J, Leippe M, et al. Killing of trypanozoon parasites by the equine cathelicidin eCATH1. Antimicrob Agents Chemother 2016;60(5):2610–19.

[90] Pereira AV, de Barros G, Pinto EG, Tempone AG, Orsi Rde O, Dos Santos LD, et al. Melittin induces in vitro death of Leishmania (Leishmania) infantum by triggering the cellular innate immune response. J Venom Anim Toxins Incl Trop Dis 2016;22:1.

[91] Adade CM, Oliveira IR, Pais JA, Souto-Padrón T. Melittin peptide kills Trypanosoma cruzi parasites by inducing different cell death pathways. Toxicon 2013;69:227–39.

[92] Khalili S, Mohebali M, Ebrahimzadeh E, Shayan E, Mohammadi-Yeganeh S, Moosazadeh Moghaddam M, et al. Antimicrobial activity of an antimicrobial peptide against amastigote forms of Leishmania major. Vet Res Forum 2018;9(4):323–8.

[93] Bartels EJH, Dekker D, Amiche M. Dermaseptins, multifunctional antimicrobial peptides: a review of their pharmacology, effectivity, mechanism of action, and possible future directions. Front Pharmacol 2019;10:1421.

[94] Salay LC, Nobre TM, Colhone MC, Zaniquelli ME, Ciancaglini P, Stabeli RG, et al. Dermaseptin 01 as antimicrobial peptide with rich biotechnological potential: study of

peptide interaction with membranes containing *Leishmania amazonensis* lipid-rich extract and membrane models. J Pept Sci 2011;17(10):700–7.

[95] Eaton P, Bittencourt CR, Costa Silva V, Véras LM, Costa CH, Feio MJ, et al. Antileishmanial activity of the antimicrobial peptide DRS 01 observed in *Leishmania infantum* (syn. *Leishmania chagasi*) cells. Nanomedicine 2014;10(2):483–90.

[96] Kunugi H, Mohammed Ali A. Royal jelly and its components promote healthy aging and longevity: from animal models to humans. Int J Mol Sci 2019;20(19).

[97] Miyata Y, Sakai H. Anti-cancer and protective effects of royal jelly for therapy-induced toxicities in malignancies. Int J Mol Sci 2018;19(10).

[98] Aslan Z, Aksoy L. Anti-inflammatory effects of royal jelly on ethylene glycol induced renal inflammation in rats. Int Braz J Urol 2015;41(5):1008–13.

[99] Brudzynski K, Sjaarda C. Honey glycoproteins containing antimicrobial peptides, Jelleins of the major royal jelly protein 1, are responsible for the cell wall lytic and bactericidal activities of honey. PLoS One 2015;10(4):e0120238.

[100] Zahedifard F, Lee H, No JH, Salimi M, Seyed N, Asoodeh A, et al. Comparative study of different forms of Jellein antimicrobial peptide on Leishmania parasite. Exp Parasitol 2020;209:107823.

[101] Meade E, Slattery MA, Garvey M. Bacteriocins, potent antimicrobial peptides and the fight against multi drug resistant species: resistance is futile? Antibiotics (Basel) 2020;9(1).

[102] Abengozar MA, Cebrian R, Saugar JM, Garate T, Valdivia E, Martinez-Bueno M, et al. Enterocin AS-48 as evidence for the use of bacteriocins as new leishmanicidal agents. Antimicrob Agents Chemother 2017;61(4).

[103] Cebrián R, Rodríguez-Cabezas ME, Martín-Escolano R, Rubiño S, Garrido-Barros M, Montalbán-López M, et al. Preclinical studies of toxicity and safety of the AS-48 bacteriocin. J Adv Res 2019;20:129–39.

[104] Khurshid Z, Naseem M, Sheikh Z, Najeeb S, Shahab S, Zafar MS. Oral antimicrobial peptides: types and role in the oral cavity. Saudi Pharm J 2016;24(5):515–24.

[105] Oudhoff MJ, Bolscher JG, Nazmi K, Kalay H, van't W, Hof AV, et al. Histatins are the major wound-closure stimulating factors in human saliva as identified in a cell culture assay. Faseb J 2008;22(11):3805–12.

[106] Khurshid Z, Najeeb S, Mali M, Moin SF, Raza SQ, Zohaib S, et al. Histatin peptides: pharmacological functions and their applications in dentistry. Saudi Pharm J 2017;25 (1):25–31.

[107] Zambom CR, da Fonseca FH, Crusca Jr. E, da Silva PB, Pavan FR, Chorilli M, et al. A novel antifungal system with potential for prolonged delivery of histatin 5 to limit growth of *Candida albicans*. Front Microbiol 2019;10:1667.

[108] Du H, Puri S, McCall A, Norris HL, Russo T, Edgerton M. Human salivary protein histatin 5 has potent bactericidal activity against ESKAPE pathogens. Front Cell Infect Microbiol 2017;7:41.

[109] Petruzzelli R, Clementi ME, Marini S, Coletta M, Di Stasio E, Giardina B, et al. Respiratory inhibition of isolated mammalian mitochondria by salivary antifungal peptide histatin-5. Biochem Biophys Res Commun 2003;311(4):1034–40.

[110] Van Hellemond JJ, Tielens AG. Inhibition of the respiratory chain results in a reversible metabolic arrest in *Leishmania promastigotes*. Mol Biochem Parasitol 1997;85(1): 135–8.

[111] Luque-Ortega JR, van't Hof W, Veerman EC, Saugar JM, Rivas L. Human antimicrobial peptide histatin 5 is a cell-penetrating peptide targeting mitochondrial ATP synthesis in Leishmania. FASEB J 2008;22(6):1817–28.

[112] Marxer M, Vollenweider V, Schmid-Hempel P. Insect antimicrobial peptides act synergistically to inhibit a trypanosome parasite. Philos Trans R Soc Lond B Biol Sci 2016;371(1695).

[113] Fieck A, Hurwitz I, Kang AS, Durvasula R. *Trypanosoma cruzi*: synergistic cytotoxicity of multiple amphipathic anti-microbial peptides to *T. cruzi* and potential bacterial hosts. Exp Parasitol 2010;125(4):342–7.

[114] McGivern JG. Ziconotide: a review of its pharmacology and use in the treatment of pain. Neuropsychiatr Dis Treat 2007;3(1):69–85.

[115] Sagar S, Kaur M, Minneman KP. Antiviral lead compounds from marine sponges. Mar Drugs 2010;8(10):2619–38.

[116] Ganjoo KN, Patel SR. Trabectedin: an anticancer drug from the sea. Expert Opin Pharmacother 2009;10(16):2735–43.

[117] Pimentel-Elardo SM, Kozytska S, Bugni TS, Ireland CM, Moll H, Hentschel U. Antiparasitic compounds from *Streptomyces* sp. strains isolated from Mediterranean sponges. Mar Drugs 2010;8(2):373–80.

[118] Simmons TL, Engene N, Urena LD, Romero LI, Ortega-Barria E, Gerwick L, et al. Viridamides A and B, lipodepsipeptides with antiprotozoal activity from the marine cyanobacterium *Oscillatoria nigro*-viridis. J Nat Prod 2008;71(9):1544–50.

[119] de Almeida Nogueira NP, Morgado-Díaz JA, Menna-Barreto RFS, Paes MC, da Silva-López RE. Effects of a marine serine protease inhibitor on viability and morphology of *Trypanosoma cruzi*, the agent of Chagas disease. Acta Trop 2013;128(1):27–35.

[120] Frommel TO, Button LL, Fujikura Y, McMaster WR. The major surface glycoprotein (GP63) is present in both life stages of Leishmania. Mol Biochem Parasitol 1990;38(1):25–32.

[121] Russell DG, Wilhelm H. The involvement of the major surface glycoprotein (gp63) of *Leishmania promastigotes* in attachment to macrophages. J Immunol 1986;136(7):2613–20.

[122] Isnard A, Shio MT, Olivier M. Impact of Leishmania metalloprotease GP63 on macrophage signaling. Front Cell Infect Microbiol 2012;2:72.

[123] Corradin S, Ransijn A, Corradin G, Roggero MA, Schmitz AA, Schneider P, et al. MARCKS-related protein (MRP) is a substrate for the Leishmania major surface protease leishmanolysin (gp63). J Biol Chem 1999;274(36):25411–18.

[124] Corradin S, Ransijn A, Corradin G, Bouvier J, Delgado MB, Fernandez-Carneado J, et al. Novel peptide inhibitors of *Leishmania* gp63 based on the cleavage site of MARCKS (myristoylated alanine-rich C kinase substrate)-related protein. Biochem J 2002;367(Pt 3):761–9.

[125] Rhaiem RB, Houimel M. Targeting Leishmania major parasite with peptides derived from a combinatorial phage display library. Acta Trop 2016;159:11–19.

[126] Qvit N, Kornfeld OS. Development of a backbone cyclic peptide library as potential antiparasitic therapeutics using microwave irradiation. J Visual Exp 2016;107:e53589.

[127] Qvit N, Schechtman D, Pena DA, Berti DA, Soares CO, Miao Q, et al. Scaffold proteins LACK and TRACK as potential drug targets in kinetoplastid parasites: development of inhibitors. Int J Parasitol: Drugs Drug Resist 2016;6:74–84.

[128] Magdesian MH, Giordano R, Ulrich H, Juliano MA, Juliano L, Schumacher RI, et al. Infection by *Trypanosoma cruzi*. Identification of a parasite ligand and its host cell receptor. J Biol Chem 2001;276(22):19382–9.

[129] Teixeira AA, de Vasconcelos Vde C, Colli W, Alves MJ, Giordano RJ. *Trypanosoma cruzi* binds to cytokeratin through conserved peptide motifs found in the laminin-G-Like domain of the gp85/trans-sialidase proteins. PLoS Negl Trop Dis 2015;9(9):e0004099.

[130] Price HP, Menon MR, Panethymitaki C, Goulding D, McKean PG, Smith DF. Myristoyl-CoA:protein N-myristoyltransferase, an essential enzyme and potential drug target in kinetoplastid parasites. J Biol Chem 2003;278(9):7206–14.

[131] Corpas-Lopez V, Moniz S, Thomas M, Wall RJ, Torrie LS, Zander-Dinse D, et al. Pharmacological validation of N-myristoyltransferase as a drug target in *Leishmania donovani*. ACS Infect Dis 2019;5(1):111–22.

[132] Herrera LJ, Brand S, Santos A, Nohara LL, Harrison J, Norcross NR, et al. Validation of N-myristoyltransferase as potential chemotherapeutic target in mammal-dwelling stages of *Trypanosoma cruzi*. PLoS Negl Trop Dis 2016;10(4):e0004540.

[133] Olaleye TO, Brannigan JA, Roberts SM, Leatherbarrow RJ, Wilkinson AJ, Tate EW. Peptidomimetic inhibitors of N-myristoyltransferase from human malaria and leishmaniasis parasites. Org Biomol Chem 2014;12(41):8132–7.

[134] Kelso AA, Waldvogel SM, Luthman AJ, Sehorn MG. Homologous recombination in protozoan parasites and recombinase inhibitors. Front Microbiol 2017;8:1716.

[135] Hall 3rd M, Misra S, Chaudhuri M, Chaudhuri G. Peptide aptamer mimicking RAD51-binding domain of BRCA2 inhibits DNA damage repair and survival in *Trypanosoma brucei*. Microb Pathog 2011;50(5):252–62.

[136] Pollack M. The role of exotoxin A in pseudomonas disease and immunity. Rev Infect Dis 1983;5(Suppl 5):S979–84.

[137] Seetharam S, Chaudhary VK, FitzGerald D, Pastan I. Increased cytotoxic activity of *Pseudomonas exotoxin* and two chimeric toxins ending in KDEL. J Biol Chem 1991;266 (26):17376–81.

[138] Cao L, Jiang W, Cao S, Zhao P, Liu J, Dong H, et al. In vitro leishmanicidal activity of antimicrobial peptide KDEL against *Leishmania tarentolae*. Acta Biochim Biophys Sin (Shanghai) 2019;51(12):1286–92.

[139] Yu Y, Zhao P, Cao L, Gong P, Yuan S, Yao X, et al. A novel anti-microbial peptide from pseudomonas, REDLK induced growth inhibition of *Leishmania tarentolae* promastigote in vitro. Korean J Parasitol 2020;58(2):173–9.

[140] Harbut MB, Patel BA, Yeung BK, McNamara CW, Bright AT, Ballard J, et al. Targeting the ERAD pathway via inhibition of signal peptide peptidase for antiparasitic therapeutic design. Proc Natl Acad Sci USA 2012;109(52):21486–91.

[141] Verlinde CL, Hannaert V, Blonski C, Willson M, Périé JJ, Fothergill-Gilmore LA, et al. Glycolysis as a target for the design of new anti-trypanosome drugs. Drug Resist Updat 2001;4(1):50–65.

[142] Coley AF, Dodson HC, Morris MT, Morris JC. Glycolysis in the african trypanosome: targeting enzymes and their subcellular compartments for therapeutic development. Mol Biol Int 2011;2011:123702.

[143] Gordhan HM, Milanes JE, Qiu Y, Golden JE, Christensen KA, Morris JC, et al. A targeted delivery strategy for the development of potent trypanocides. Chem Commun (Camb) 2017;53(62):8735–8.

[144] Ferreira LG, Andricopulo AD. Targeting cysteine proteases in trypanosomatid disease drug discovery. Pharmacol Ther 2017;180:49–61.

[145] Denise H, McNeil K, Brooks DR, Alexander J, Coombs GH, Mottram JC. Expression of multiple CPB genes encoding cysteine proteases is required for *Leishmania mexicana* virulence in vivo. Infect Immun 2003;71(6):3190–5.

[146] Casgrain PA, Martel C, McMaster WR, Mottram JC, Olivier M, Descoteaux A. Cysteine peptidase B regulates *Leishmania mexicana* virulence through the modulation of GP63 expression. PLoS Pathog 2016;12(5):e1005658.

[147] Alves LC, St Hilaire PM, Meldal M, Sanderson SJ, Mottram JC, Coombs GH, et al. Identification of peptides inhibitory to recombinant cysteine proteinase, CPB, of *Leishmania mexicana*. Mol Biochem Parasitol 2001;114(1):81–8.

[148] Meirelles MN, Juliano L, Carmona E, Silva SG, Costa EM, Murta AC, et al. Inhibitors of the major cysteinyl proteinase (GP57/51) impair host cell invasion and arrest the intracellular development of *Trypanosoma cruzi* in vitro. Mol Biochem Parasitol 1992;52(2):175–84.

[149] Bevec T, Stoka V, Pungercic G, Dolenc I, Turk V. Major histocompatibility complex class II-associated p41 invariant chain fragment is a strong inhibitor of lysosomal cathepsin L. J Exp Med 1996;183(4):1331–8.

[150] Bevec T, Stoka V, Pungercic G, Cazzulo JJ, Turk V. A fragment of the major histocompatibility complex class II-associated p41 invariant chain inhibits cruzipain, the major cysteine proteinase from *Trypanosoma cruzi*. FEBS Lett 1997;401(2–3):259–61.

[151] Graven A, St Hilaire PM, Sanderson SJ, Mottram JC, Coombs GH, Meldal M. Combinatorial library of peptide isosters based on Diels-Alder reactions: identification of novel inhibitors against a recombinant cysteine protease from *Leishmania mexicana*. J Comb Chem 2001;3(5):441–52.

[152] Naaz F, Preeti Pallavi MC, Shafi S, Mulakayala N, Shahar Yar M, Sampath Kumar HM. 1,2,3-triazole tethered Indole-3-glyoxamide derivatives as multiple inhibitors of 5-LOX, COX-2 & tubulin: their anti-proliferative & anti-inflammatory activity. Bioorg Chem 2018;81:1–20.

[153] Petrova KT, Potewar TM, Correia-da-Silva P, Barros MT, Calhelha RC, Ćiric A, et al. Antimicrobial and cytotoxic activities of 1,2,3-triazole-sucrose derivatives. Carbohydr Res 2015;417:66–71.

[154] Peyton LR, Gallagher S, Hashemzadeh M. Triazole antifungals: a review. Drugs Today (Barc) 2015;51(12):705–18.

[155] Tornøe CW, Sanderson SJ, Mottram JC, Coombs GH, Meldal M. Combinatorial library of peptidotriazoles: identification of [1,2,3]-triazole inhibitors against a recombinant *Leishmania mexicana* cysteine protease. J Comb Chem 2004;6(3):312–24.

[156] St Hilaire PM, Alves LC, Herrera F, Renil M, Sanderson SJ, Mottram JC, et al. Solid-phase library synthesis, screening, and selection of tight-binding reduced peptide bond inhibitors of a recombinant *Leishmania mexicana* cysteine protease B. J Med Chem 2002;45(10):1971–82.

[157] Maji K, Abbasi M, Podder D, Datta R, Haldar D. Potential antileishmanial activity of a triazole-based hybrid peptide against Leishmania major. ChemistrySelect 2018;3(36):10220–5.

[158] Schirmeister T, Schmitz J, Jung S, Schmenger T, Krauth-Siegel RL, Gutschow M. Evaluation of dipeptide nitriles as inhibitors of rhodesain, a major cysteine protease of *Trypanosoma brucei*. Bioorg Med Chem Lett 2017;27(1):45–50.

[159] Simon RJ, Kania RS, Zuckermann RN, Huebner VD, Jewell DA, Banville S, et al. Peptoids: a modular approach to drug discovery. Proc Natl Acad Sci USA 1992;89(20):9367–71.

[160] Eggimann GA, Bolt HL, Denny PW, Cobb SL. Investigating the anti-leishmanial effects of linear peptoids. ChemMedChem 2015;10(2):233–7.

[161] Bolt HL, Eggimann GA, Denny PW, Cobb SL. Enlarging the chemical space of anti-leishmanials: a structure–activity relationship study of peptoids against *Leishmania mexicana*, a causative agent of cutaneous Leishmaniasis. MedChemComm 2016;7(5):799–805.

[162] Previdi D, Rodrigues S, Coelho MG, Candido ACB, Magalhães LG, Donate PM. Synthesis and antileishmanial activity of some functionalized peptoids. J Braz Chem Soc 2019;30(6):1334–40.

[163] Saenz-Garcia JL, Yamanaka IB, Pacheco-Lugo LA, et al. Targeting epimastigotes of *Trypanosoma cruzi* with a peptide isolated from a phage display random library. Exp Parasitol 2020;210:107830.

[164] Ramsey JD, Flynn NH. Cell-penetrating peptides transport therapeutics into cells. Pharmacol Ther 2015;154:78–86.

[165] Ponte-Sucre A, Gamarro F, Dujardin JC, Barrett MP, López-Vélez R, García-Hernández R, et al. Drug resistance and treatment failure in leishmaniasis: A 21st century challenge. PLoS Negl Trop Dis 2017;11(12):e0006052.

[166] Luque-Ortega JR, de la Torre BG, Hornillos V, Bart JM, Rueda C, Navarro M, et al. Defeating Leishmania resistance to miltefosine (hexadecylphosphocholine) by peptide-mediated drug smuggling: a proof of mechanism for trypanosomatid chemotherapy. J Control Rel 2012;161(3):835–42.

[167] Davidson RN, den Boer M, Ritmeijer K. Paromomycin. Trans R Soc Trop Med Hyg 2009;103(7):653–60.

[168] Defaus S, Gallo M, Abengózar MA, Rivas L, Andreu D. A synthetic strategy for conjugation of paromomycin to cell-penetrating Tat(48–60) for delivery and visualization into Leishmania parasites. Int J Pept 2017;2017:4213037.

[169] Paulone S, Ardizzoni A, Tavanti A, Piccinelli S, Rizzato C, Lupetti A, et al. The synthetic killer peptide KP impairs *Candida albicans* biofilm in vitro. PLoS One 2017;12 (7):e0181278.

[170] Savoia D, Scutera S, Raimondo S, Conti S, Magliani W, Polonelli L. Activity of an engineered synthetic killer peptide on Leishmania major and Leishmania infantum promastigotes. Exp Parasitol 2006;113(3):186–92.

[171] Link JS, Alban SM, Soccol CR, Pereira GV, Thomaz Soccol V. Synthetic peptides as potential antigens for cutaneous Leishmaniosis diagnosis. J Immunol Res 2017;2017: 5871043.

[172] Liberato MS, Mancini RSN, Factori IM, Ferreira FF, de Oliveira VL, Carnielli JBT, et al. Peptide-based assemblies on electrospun polyamide-6/chitosan nanofibers for detecting visceral Leishmaniasis antibodies. ACS Appl Electron Mater 2019;1(10): 2086–95.

[173] Rarick MU, Montgomery T, Groshen S, Sullivan-Halley J, Jamin D, Mazumder A, et al. Intravenous immunoglobulin in the treatment of human immunodeficiency virus-related thrombocytopenia. Am J Hematol 1991;38(4):261–6.

[174] Karimi A, Amanati A. Matrix-assisted laser desorption/ionization time of flight mass spectrometry: a new guide to infectious disease. Arch Pediatr Infect Dis 2016;4:e31816.

[175] Li L, Li J, Jin H, Shang L, Li B, Wei F, et al. Detection of Leishmania donovani infection using magnetic beads-based serum peptide profiling by MALDI-TOF MS in mice model. Parasitol Res 2012;110(3):1287–90.

Chapter 18

Peptides and antibiotic resistance

Ekaterina Umnyakova, Dmitriy Orlov, and Olga Shamova
Institute of Experimental Medicine, St. Petersburg, Russia

18.1 Introduction

Antimicrobial peptides (AMPs) or host defense peptides (HDPs) are mostly cationic amphipathic polypeptides with low molecular weight (<10 kDa) that inactivate bacteria, fungi, viruses, protozoa, and parasites entering the host organism. In the 1950s–60s, cationic peptides from neutrophils were shown to be capable of killing bacteria through an oxygen-independent mechanism [1]. Moreover, this particular mechanism was clearly not associated with an adaptive immune system. In 1962, the peptide of animal origin—bombinin—from the frog *Bombina variegate* was described. This was the first described case in the history of AMPs formed during classical translation on the ribosomes of eukaryotic cells [2].

AMPs are expressed by all living organisms from bacteria and archaea to mammals [3]. According to the Collection of Antimicrobial Peptides (CAMP), more than 8000 sequences of AMPs of microbial, plant, and animal origin are already known [4]. At present, three-dimensional structures of more than 750 peptides are known, and the database continues to grow. Some antibiotic peptides of microbial origin have been used as therapeutics for a long time, for instance, gramicidin S—a cyclodecapeptide produced by the Gram-positive bacterium *Bacillus brevis*, discovered by G. Gause and M. Brazhnikova in 1944. Within only 1 year after this discovery, gramicidin S became broadly applied in hospitals to treat infections [5]. Another microbial peptide—daptomycin, from *Streptomyces roseosporus*—was discovered by researchers at Eli Lilly and Company in the late 1980s which has been used since 2003 in the clinic [6]. Nowadays, AMPs of the innate immune system are also considered to be attractive candidates for novel tools to combat multidrug-resistant (MDR) bacteria [7,8].

18.2 Classification of antimicrobial peptides

There are several classifications of AMPs, although it is rather difficult to classify this group due to the wide structural diversity of these substances. In

Peptide and Peptidomimetic Therapeutics. DOI: https://doi.org/10.1016/B978-0-12-820141-1.00025-X

1995, Hans Boman proposed a classification for AMPs based on their structural features [9]. The first group of peptides that do not contain disulfide bonds consists of linear peptides, which mainly have α-helical structures: human LL-37/hCAP-18 [10], bovine BMAP-27 and BMAP-28 [11], magainins from the skin of the spur frog [12], buforin from the skin glands of the toad [13], and cecropins of insects [14]. The peptides of the second group also do not contain intramolecular disulfide bonds and are characterized by a high content of certain amino acids including, for example, arginine- and proline-rich bactenecins ChBac 3.4 and ChBac5 from goat neutrophils [15,16], porcine PR-39 and profenin [17,18], and tryptophan-rich bovine indolicidin [19]. The third group of AMPs is peptides with one disulfide bond. Dodecapepide from bovine leukocytes is an example of a member of this group [20]. The fourth group of AMPs is cationic peptides with two or more disulfide bonds that have mainly β-strands in their structure. These peptides include protegrins from porcine neutrophils [21] and arenicins from coelomocytes of marine polychaeta *Arenicola marina* [22]. Further, the family of defensins is characterized by the presence of three intramolecular disulfide bonds [23–25]. Depending on how the cysteine residues are located, groups of α- and β-defensins are distinguished (at the time this classification was created, the macrocyclic peptides from the θ-defensin family were not yet described). The fifth and last group of AMPs includes precursors, which are larger polypeptide molecules with other known functions. An example is lactoferricin, the precursor of which is lactoferrin, the main iron-transporting protein of leukocyte origin.

Another AMP classification is based on three-dimensional structures and was adopted for the APD3 database [26,27]:

1. **Alpha family:** Peptides with helical structures, for example, human cathelicidin LL-37.
2. **Beta family:** Peptides with β-strands, for example, human α-defensins.
3. **Alpha beta family:** Peptides with both α-helical and β-strand structures in the same three-dimensional fold, for example, human β-defensins.
4. **Nonalpha beta family:** Peptides with neither α-helices nor β-strands, for example, indolicidin, which is isolated from cattle.

Today, three-dimensional structures for more than 750 AMPs are known, and the previous approach to AMP classification is not sufficient. Other classifications based on properties such as covalent bonding patterns, hydrophobicity, net charge, or molecular targets are required [27].

The structures listed above are typical for processed mature peptides, as most AMPs from both multi- and unicellular organisms are derived from precursor sequences. Consequently, they display a number of posttranslational modifications that include: amidation, incorporation of D-amino acids, glycosylation, halogenation, phosphorylation, and even cyclization [28].

Some AMPs can also be synthesized nonribosomely, for example, like several bacteriocins, AMPs of bacteria.

AMPs can be synthetized constitutively (e.g., human β-defensin-1 in keratinocytes), and also the synthesis can be induced (e.g., human LL-37 or β-defensin-2 in barrier epithelial cells), in response to infection or stimulation with pro-inflammatory cytokines [29]. The structure of AMP genes is different in various groups of animals. There is an assumption that peptides with antimicrobial activity appeared in different groups independently and that it happened several times during evolution [3].

Peptides are expressed on body surfaces that are in contact with the environment: epithelium of the airways, genitourinary tract, as well as the epithelium of the gastrointestinal tract and integument epithelium. Peptides also accumulate in the granules of phagocytes and the endosomes of epithelial cells; when the contents of the granules are released, the peptides can participate in the destruction of microorganisms.

Despite the fact that the family of AMPs is heterogeneous, members of the group are characterized by the following: peptides are predominantly cationic, contain L-amino acids in their chains, and these peptides are amphipathic, that is, in their three-dimensional structure hydrophobic and hydrophilic sites are spatially separated.

The main determining structural parameters of AMPs are molecular size, conformation, charge, hydrophobicity, amphipathicity, and polar angle. It is important to note that these parameters are closely interrelated and often a change in one of the parameters leads to a compensatory modification of the others [30].

18.2.1 Size

AMPs are relatively small polypeptide molecules: they contain from 8 to 60 amino acid residues. However, larger AMPs are also found (e.g., porcine profenin).

18.2.2 Conformation

The secondary structure of most AMPs is represented by α-helices and β-strands. It is also known that other AMP molecules are rich in proline, arginine, and tryptophan. For example, PR-39 peptide, profenin, and some bactenecins have the structure of a type II polyproline helix.

18.2.3 Charge

Most known AMPs are characterized by a positive net charge (from +2 to +9) due to the presence in the primary structure of positively charged amino acids: arginine and lysine.

18.2.4 Amphipathicity

AMP amphipathicity is realized due to the fact that amino acids with hydrophobic and hydrophilic side groups are located in space on opposite sides of the molecule. α-Helical peptides are characterized by a hydrophobic moment, the vector sum of the hydrophobicity of individual amino acids normalized in an ideal helix. An increase in the hydrophobic moment leads to an increase in the ability to bind to lipid membranes and disrupt their integrity.

18.2.5 Polar angle

This is a parameter of the relative proportions of the polar and nonpolar surfaces of the peptide with an amphipathic helical structure. For a helix, in which one surface consists only of hydrophilic residues and the other consists of hydrophobic residues, this angle will be 180 degrees.

18.2.6 Hydrophobicity

Hydrophobicity depends on the percentage of hydrophobic amino acid residues. Hydrophobicity is a property that allows water-soluble peptides to integrate into the bilipid layer of the plasma membrane of cells.

18.3 Antimicrobial peptide mechanisms of action

AMPs are able to act either directly on microorganisms, interacting with heterogenous cells, or indirectly, modulating the immune mechanisms of the host. There are two main types of action of peptides on bacterial cells: one that occurs due to disruption of the plasma membrane through the formation of pores and the second that happens due to the ability of AMPs to disturb various intracellular processes in microbial cells.

Bacterial cell membranes contain acidic phospholipids such as phosphatidylglycerol (PG) and cardiolipin. The negative charge of the membrane of Gram-positive bacteria is partly determined by the presence of lipoteichoic acids. A component of the outer membrane of Gram-negative bacteria is a lipopolysaccharide (LPS). All of the above compounds are negatively charged and contribute to the formation of a common negative charge on the bacterial surface. This total negative charge is more than the negative charge of the host cell membrane. Therefore, AMPs have a higher affinity for bacterial membranes compared to the surfaces of eukaryotic cells [30]. The membrane of a bacterial cell differs from the membrane of eukaryotic cells by the value of the membrane potential (in mammals from −80 to −100 mV, in bacteria from −130 to −150 mV). It is quite possible that this difference also contributes to the fact that the cytotoxicity of AMPs to the host cells is not as significant [31].

There is a lot of convincing data showing that AMPs can destroy lipid bilayers in vitro. However, there is still no agreement among scientists regarding the hypothetical mechanisms described by each model, and there is no clear boundary between all these mechanisms. The first step is considered to involve electrostatic interactions between the positively charged amino acids in AMPs and the negatively charged bacterial membrane, resulting in AMP adsorption on the surface [32].

After reaching some threshold concentration, AMPs can aggregate on the bacterial membrane and incorporate into it. That is the mechanism of pore formation that can occur in different ways. Several models for such pore formation have been suggested:

1. The barrel-stave model proposes that peptides associate together and insert perpendicularly in the phospholipid bilayer, forming a pore. That mechanism is typical for the peptides with an α-helical structure. The hydrophobic region of AMPs interacts with the hydrophobic region of the membrane bilayer and the hydrophilic surface of the peptide is oriented inwards, forming a hydrophilic channel. In this way, the formed peptide complex forms a pore inserted into the membrane. This process alters the membrane structure and membrane potential, and provokes the leakage of water, ions, and different solutes and eventually, it results in cell lysis [33].
2. The toroidal pore mechanism suggests peptides inserted perpendicularly in the bilayer and inducing a local membrane curvature with the pore that is partly composed of peptides and partly of phospholipid head groups. Initially, the peptide orients itself parallel to the plasma membrane which is typical for AMPs at the first stages of this process. When the peptide–lipid molar ratio is exceeded, the peptides are reoriented perpendicular to the bilayer and they invert themselves in conjunction with several surrounding lipids toward the hydrophobic region. The formation of the dynamic supramolecular peptide–lipid complex occurs in this way, which causes irreversible rupture of the membrane [34].
3. The carpet mechanism, according to which AMPs adsorb to the bilayer and after reaching sufficient density of coverage, act like detergent that disintegrates the membrane without entering it. At the beginning of the interaction, the AMPs orient themselves parallel to the membrane. Then, when the AMP reaches a critical concentration, the monomers reorient toward the hydrophobic core of membrane phospholipids. That causes the formation of micelles and disintegration of bilayer which actually happens when the detergent is added to the phospholipid membrane. The pore formation is mainly governed by hydrophobic interactions between the AMPs and the hydrophobic part of the lipid bilayer [35]. This model

explains the action of AMPs with a size of less than 24 amino acid residues that are considered not to cross the membrane [33]. It can also be noted that at high concentrations all AMPs act similar to detergent, regardless of how the specific peptide acts at lower concentrations [36].

As has been shown, these are not the only mechanisms of AMP action on bacterial cells; many peptides affect intracellular processes [37]. Human α-defensins, HNP-1 and -2, can interfere with the synthesis of DNA, RNA, and protein in *E. coli* cells [38]. Proline-rich AMPs can bind to the chaperone of bacterial DnaK, as has been shown, for example, for insect AMPs—drosocin, apidaecin, etc., disrupting protein folding in microbial cells [39] as well as interacting with the bacterial ribosome, inhibiting translation (bovine Bac 7 fragments, oncocin, and others) [40,41]. Buforin II and PR-39 are able to bind to nucleic acids: DNA and RNA [42]. Indolicidin also has the ability to bind to DNA with sequence preference and can inhibit topoisomerase 1, which cuts one strand of double-stranded DNA, relaxing the strands. This also may contribute to indolicidin's antimicrobial action [43,44].

AMPs have a wide spectrum of activity against different microorganisms, but their biological activity is not limited only to (cyto)toxic function. It has been shown mainly for human AMPs that they also play an important role in the regulation of the immune system, acting as chemoattractants for immune cells [45–52], as an LPS binding agent [53,54], as well as participating in angiogenesis [55,56] and collagen synthesis [57]. In particular, for some of them and not only for human AMPs the ability to interact with complement proteins and to regulate the complement system has been described [58–62]. Human AMPs were also shown to participate in autoimmune processes [63]. Moreover, it was discovered that some AMPs can demonstrate corticostatic activity and used to be known as corticostatins [64,65].

18.4 Advantages of antimicrobial peptides compared to conventional antibiotics

The main function of conventional antibiotics is to inactivate bacteria. The long-term and frequent use of these pharmaceuticals can lead to bacterial mutations that provoke bacterial drug resistance. This has already been reported to result in serious problems all over the world because there is a threat of the absence of effective antibiotics due to bacterial adaptation to them. That is why efforts should be maintained in developing new antibacterial pharmaceutical agents and this task is becoming increasingly difficult.

Despite some limitations that exist now for using AMPs in clinical practice, there is a tendency for developing new antibiotics utilizing natural AMPs as a template for improved and designed peptides that will be more effective against microbes. In fact, it was shown by different scientific groups that many AMPs are more effective against MDR bacteria strains [66–70].

Intriguingly, there are some FDA–approved AMPs for treating human infections caused by MDR strains of Gram-positive bacteria [71]. These are mainly bacterial AMPs or bacteriocins: vancomycin, that is synthesized by the soil bacterium *Amycolatopsis orientalis*, and its derivatives – teicoplanin, dalbavicin, and telavancin; daptomycin – alipopeptide antibiotic, that is a naturally occurring compound found in the soil saprotroph *Streptomyces roseosporus*; bacitracin, that is a mixture of related cyclic peptides produced by *Bacillus subtilis* var Tracy. These peptides are synthetized by nonribosomal peptide synthetases (NRPSs) and they are rather toxic for humans [71].

Moreover, the FDA has also approved an antibacterial peptide named nisin, a bacteriocin from Gram-positive bacteria that belongs to *Lactococcus* and *Streptococcus* species, for use in the food industry as a preservative mainly to extend the shelf lives of meat and dairy products [72]. Nisin has a broad spectrum of activity against Gram-positive bacteria, antibiofilm properties, and is generally noncytotoxic to mammals [73].

Still, there are a lot of limitations to using AMPs in clinical practice that are mainly connected to the low stability of these antimicrobial molecules in the human body. Nevertheless, there are some ways to overcome these difficulties including rational design and modification of natural AMPs, and there are some reasons why these peptides are still so pervasive. First, in contrast to the majority of traditional antibiotics, AMPs interact with the bacterial cell membranes. Due to their positive net charge, peptides selectively interact with negatively charged bacterial membrane and further permeabilize membranes, which causes cell death very rapidly at relatively low bactericidal concentrations, reducing the possibility of bacterial drug resistance [74]. These peptides act more efficiently than conventional antibiotics. They display many advantages over antibiotics that are now used in medical and veterinary practice with broad-spectrum antibacterial, antifungal, and antiviral activities [74,75]. AMPs also have synergistic effects with traditional antibiotics to neutralize endotoxin [74,76]. Second, besides their broad spectrum of antimicrobial activities, AMPs have high thermal stability, good water solubility [77], low immunogenicity, a wide range of structure–activity relationship study approaches, and weak or low sensitization [78,79].

Since some of the AMPs (usually human AMPs) act as chemoattractants for immune cells, they can also be used as adjuvants for vaccination. Adjuvants themselves often do not have immunogenic properties, but they serve to enhance the immune response in relation to the components of the vaccine, thus reducing the dose of the vaccine and prolonging its effect [80]. Moreover, many studies have demonstrated that cancer cells are more sensitive to AMPs than normal cells because cationic AMPs selectively interact with the acidic phospholipids on the outer surfaces of cancer cells [74].

Thus, peptides easily insert into the membranes and form pores to destroy the cancer cells or result in leakage of cell contents [81,82].

Currently, many pharmaceutical companies have begun to develop chemically synthesized analogs of AMPs. The most successful examples are the following peptides [83]:

1. Omiganan—Omiganan-MX-226 ("Migenix/BioWest therapeutics") is a synthetic analog of indolicidin with antimicrobial and immunostimulating properties. It is undergoing phase III of clinical trials as a drug for external use in the treatment of atopic dermatitis, acne, and to prevent infections when using catheters.
2. Magainin/Peksiganan acetate ("Dipexium Pharma") is a linear AMP consisting of 22 amino acid residues, an analog of AMP from the *Xenopus laevis* frog. The compound is currently in phase III clinical trials as a topical antimicrobial tool as a therapy for diabetic foot ulcer infections.
3. OP-145 ("OctoPlus") is an LL-37 derivative capable of binding LPS very efficiently. It is intended to be used to treat chronic otitis media. It has passed phase II of clinical trials.
4. hLF1−11 ("AM-Pharma") is a derivative of the cationic human AMP-lactoferricin. This peptide has passed phase I/II of clinical trials.
5. Arenicins (Adenium Biotech) are several forms of modified Ar-3 from the sea polychaeta *Arenicola marina.* Preclinical trials are underway as drugs for the treatment of infections caused by MDR strains of Gram-positive bacteria.

Thus, membrane-permeabilizing AMPs can be widely used in the food industry and in medicine for treating infections including diseases that are caused by MDR bacteria. They present the potential for a new therapy against microbes and may be clinically applied as a strategy to overcome the resistance of microorganisms to conventional antibiotics. However, despite success in creating antibiotics using AMPs as templates, there are still a lot of questions, and the road to novel drugs for the treatment of severe bacterial infections is rather long.

18.5 Mechanisms of bacterial resistance to antimicrobial peptides

Despite the interest in AMPs as novel antibiotic drugs, only a limited number of studies on the development of bacterial resistance have been reported. Scientists previously believed that resistance to AMPs is difficult to acquire because the main target of AMPs is a bacterial membrane. However, there is evidence that bacteria can develop resistance to AMPs [84].

Bacterial resistance to AMPs can occur via constitutive (passive) or adaptive (inducible) mechanisms. Passive mechanisms refer to inherent properties of a microorganism and are usually expressed even in the absence of AMPs or antimicrobial proteins. On the contrary, inducible resistance mechanisms include those triggered in response to antimicrobials or the target cell stresses it causes. It is known that some AMPs in subinhibitory concentrations can modulate the expression of genes involved in bacterial virulence and resistance [85].

Passive resistance occurs in some bacterial species as a result of a decrease in the negative charge of lipid A, which is typical for such species as *Burkholderia, Proteus, Providencia, Serratia, Morganella* [86,87]. This reduces the possibility of AMPs interacting with the membrane of microorganisms because they lose the affinity to bacterial membrane. Inducible (or adaptive) resistance caused by molecular modifications in both Gram-negative and Gram-positive bacteria. The most typical mechanism of such resistance is the incorporation of positively charged molecules in the cell surface that reduces the interaction with cationic AMPs [30,88–90]. For example, an addition of aminoarabinose into *K. pneumonia* LPS in the presence of low concentrations of polymyxin B is occurring followed by activation of the expression of genes involved in LPS modification and leading to the increase of resistance to this peptide [91]. The resistance to AMPs can be raised upon addition of phosphoethanolamines causing a conformational rearrangement of LPS and affecting the binding of polymyxin B to the bacterial outer membrane [92].

18.5.1 Proteolysis

One of the mechanisms to overcome the antimicrobial activity of peptides is to digest them using proteolytic agents. For example, the PgtE protein was identified in *Salmonella* as an outer membrane endopeptidase [93]. This protease cleaves peptides between basic residue pairs and the carboxy-terminal part of basic amino acid residues preceding nonpolar residues. There are also some more proteases from different microorganisms that protect them from lactoferricin B, LL-37, etc. [94], for instance, the aspartate proteases of the Omptin family cleave α-helical AMPs reached the outer membrane of enterobacteria [93,95], while other proteases such as ZmpB of *Bulcholderia cenocepacia* can digest beta-sheet peptides including beta-defensins [96].

18.5.2 Capsular polysaccharides

This protective layer, that is formed from the anionic monomer subunits, is typically negatively charged and can be widely heterogeneous. It is not only a physical barrier for peptides but it also binds them with great avidity. The absorbed AMPs lose their activity because of the limitation of their access to internal targets. These mechanisms of AMP resistance are typical for a respiratory tract or other mucosal pathogens that form biofilms and have capsules, including *Streptococcus*, *Haemophilus*, *Staphylococcus*, *Klebsiella*,

Legionella species [97]. Other bacterial components such as membrane vesicles demonstrate a similar ability to trap AMPs. *E. coli* membrane vesicles are released after incubation of bacteria with subinhibitory concentrations of polymyxine B resulting in a higher resistance to this antibiotic peptide [98].

18.5.3 Cell wall

The incorporation of D-amino acids (e.g., D-Ala) into teichoic acids of the cell wall of *Staphylococcus aureus* leads to a decrease in the negative charge due to the formation of ester bonds using free COO^- groups, and the remaining free NH_2^+ groups increase the positive charge of the molecule. Therefore, there is a decrease in AMP affinity to the membrane [94].

18.5.4 Plasma membrane

Some bacteria develop resistance by changing the composition of the plasma membrane. Such modifications alter the charge and/or the membrane potential. For example, *Pseudomonas fluorescens* adaptively decreases levels of anionic phospholipids [phosphatidylethanolamine (PE), PG, and cardiolipin] and increases cationic ornithine-amide constituents to resist cationic AMPs [99]. The incorporation of L-lysine into PG by membrane protein multipeptide resistance factor (MprF) is also a strategy that leads to diminishing of the negative surface charge in Gram-positive bacteria and causes a higher AMP resistance [100].

18.5.5 Efflux pumps

AMPs are considered not only to destroy bacterial envelopes but also to permeabilize the cytoplasmic membrane and to interact with intracellular targets. Hence, bacteria use mechanisms that help to evacuate the AMPs from the cells, in particular, they exploit the efflux pumps. For instance, in *Neisseria gonorrhoeae* it has been shown that an energy-dependent efflux system named *mtr* mediates resistance to different AMPs [101].

18.5.6 Biofilm formation

Biofilm formation is one of the most successful mechanisms of defense of pathogenic bacteria from natural or synthetic antimicrobials. Densely packed communities of microorganisms, embedded within extracellular polymeric substances and growing on living or inert surfaces, become up to 1000-fold more resistant to conventional antibiotics than planktonic forms and therefore represent a global problem for the treatment of infectious diseases. The antimicrobial activity of AMPs is usually decreased toward bacterial biofilms in comparison with their effects on the planktonic microbes due to binding of the peptides to the exopolysaccarides of bacterial biofilms. However, it is known that some peptides can inhibit biofilm formation by decreasing the number of bacteria at the early stage of biofilm formation as well as by inhibiting bacterial quorum sensing. It was shown for a small cationic peptide 1037, that possesses a weak antimicrobial activity, but effectively prevents biofilm formation ($>50\%$ reduction in cell biomass) by the Gram-negative bacteria *Pseudomonas aeruginosa*, *Burkholderia cenocepacia* and the Gram-positive pathogen *Listeria monocytogenes*, at the concentration of 1/30th of its minimal inhibitory concentration [102]. The same was observed for the peptide GL13K [103], subtilosin [104], and other AMPs [70].

18.6 Approaches for overcoming bacterial resistance to antimicrobial peptides

Despite bacteria developing an array of mechanisms of AMP resistance, the prospects of practical use of the peptides are based on several approaches allowing this problem to be avoided. One approach is to design AMPs de novo with decreased sensitivity to bacterial proteases using chemical modification of their structure via cyclization, N-terminal acetylation, substituting some, mainly cationic, or all amino acid residues by D-enantiomers, or creating peptidomimetic molecules [83,105]. Incorporation of D-enantiomers in some cases helps to reduce the resistance caused by D-alanylation of Gram-positive bacteria cell wall teichoic and lipoteichoic acids: D-enantiomer of a peptide GL13K effectively killed an *Enterococcus faecalis* strain that carried such the modification, suggesting that this peptide is able to bypass the D-alanylation resistance mechanism, as well as that a peptide's chiral structure can be a crucial determinant of the interaction with the cell wall [106].

Inhibition of D-alanylation of teichoic acids results in altering of the pathogenicity of Gram-positive bacteria, therefore the inhibitors of DltA, a protein engaged in the initial step of D-alanylation, may serve as potential therapeutics, particularly when used in combination with other antimicrobials [107]. A protein MprF, responsible for the incorporation of L-lysine into PG, also may be a good target, since it exists in both Gram-positive as well as Gram-negative pathogens, and the inhibitors of MprF can be considered as tools to prevent AMP resistance [108].

To avoid the action of bacterial efflux pumps some authors suggest creating immobilized nano/microparticles with AMPs [109]. The incorporation of AMPs or antimicrobial proteins into varied nanomaterials can also prevent their proteolytic degradation and improve the antibacterial activity [110–114]. Conjugation and functionalization of nanoparticles with AMPs expands the spectrum of their applications as promising tools for drug delivery as well as for diagnostics [110].

To combat bacterial growth in a biofilm form, particular strategies have focused on the peptides that are specifically designed to prevent the biofilm formation or eliminate existing biofilms [115]. Some of these peptides specifically attack cellular mechanisms involved in biofilm formation (e.g., cell attachment) and thus may cause less bacterial resistance than bactericidal AMPs [116]. The antibiofilm action of AMPs can occur via different mechanisms. For example, lactoferrin's antibiofilm activity is proposed due to the iron-chelating features of this protein, which increase the surface motility and facilitate the formation of thin, flat biofilms by surface-wandering bacterial cells, causing higher susceptibility of these cells to detachment [117]. The human cathelicidin LL-37 decreases an initial attachment and creation of biofilm in *Staphylococcus epidermidis*, inhibits the attachment of *P. aeruginosa* cells onto the surface of medical devices, promotes the motility mediated by type IV pili, and downregulates the Las and Rhl quorum sensing system [118,119].

One of the most promising ways to overcome the resistance of bacteria to AMPs is a combined application of AMPs with conventional antibiotics or other antimicrobials that results in a synergistic action. It was shown that synergistic combinations of AMPs with varied conventional antibiotics as well as with silver nanoparticles are effective tools against MDR bacteria (*Staphylococcus aureus*, *E. coli* ESBL, *Pseudomonas aeruginosa*, *Klebsiella pneumoniae* ESBL, *Acinetobacter baumannii*, etc.) including carbapenem-resistant strains causing severe hospital infections [120–122]. On the other hand, some clinically used antibiotics may enhance the antimicrobial activity of endogenous AMPs. A synergistic effect has been revealed for the combinations of hBD-3 with gentamicin against *E. coli*, *A. baumannii*, hBD-2 with rifampicin and polymyxin B against *A. baumannii*, and aHNP-1 or hBD-3 with rifampicin against MRSA ATCC 33591 [120]. Understanding the interaction between conventional antibiotics and endogenous AMPs is important for the elaboration of cure strategies in the therapy of infections caused by MDR pathogens. Administration of beta–lactam antibiotics to meticillin-resistant *S. aureus* or vancomycin-resistant *Enterococcus* spp. potentiate the antimicrobial action of human cathelicidin LL-37 or cationic peptide antibiotic daptomycin in synergy studies in vitro and pilot clinical studies in patients with previously recalcitrant infections [123,124].

A synergistic effect on bacterial biofilm formation has been observed for the combined action of AMPs and varied antimicrobial substances [125,126]. However, despite the undeniable advantages of the combined use of peptides and antibiotics, it is necessary to consider the possibility of increasing the toxic effects of these combinations on host cells, as well and to comprehensively explore these effects in vitro and in vivo to avoid side effects upon a practical application of AMPs [127].

Except for direct antimicrobial action, AMPs exert a potent immunomodulatory activity, they can modulate the immune response through the activation of different cell types. Taking into account this feature of AMPs, a unique class of peptide immunomodulators has been developed on a base of bovine neutrophil AMP bactenecin 1 (dodecapeptide) and named innate defense regulator (IDR) peptides, which can enhance host innate immune responses against microbial pathogens including drug-resistant strains [128,129]. IDRs selectively increase chemokine production and polarize cellular differentiation while suppressing/balancing the pro-inflammatory response [130]. They exert a variety of biological effects, for instance, a small synthetic peptide IDR-1018 demonstrates antiinfective, antiinflammatory, wound–healing, and antibiofilm activities [131]. It selectively targets bacterial biofilms, formed by MDR bacteria, and shows synergy with conventional antibiotics to both prevent biofilm formation and treat preexisting biofilms. Using AMPs as immunomodulators is a promising approach to combat antibiotic-resistant infections by applying an adjunctive therapy [131].

18.7 Conclusion

AMPs of the innate immune system are natural potent antiinfective molecules and attractive candidates for the development of novel therapeutics to combat MDR bacteria. AMPs possess a rapid antimicrobial action and kill microorganisms within several minutes by disrupting their membranes, as well as affecting various intracellular targets. Due to the fast and multitargeted action of AMPs, pathogenic bacteria hardly develop resistance to the peptides in comparison with that to common antibiotics. Nevertheless, some pathogens have evolved varied inducible or intrinsic AMP–defense mechanisms to enable their survival and colonization. These mechanisms include proteolytic degradation of AMPs, altering the cell surface net charge, trapping the peptides by specific molecules or bacterial membrane vesicles, using efflux pumps, formation of biofilms, etc. However, any such attempts of bacteria to evade the AMP action can be diminished by using varied strategies: rational design of AMPs with improved resistance to bacterial proteases via replacement of L-amino acids with their D-enantiomers, immobilization of AMPs on special carriers, using nanocontainers, creating peptidomimetics, and employing AMPs with particular inhibitors of proteins involved in the realization of bacterial resistance.

One of the most promising strategies for overcoming AMP resistance is a combined use of AMP-based therapeutics with conventional antibiotics which target different sites in bacterial cells. This approach seems to be very promising considering the ability of some peptides to exert a synergistic action with common antibiotics in the prevention of biofilm formation as well as eradication of preexisting biofilms. Biofilm-related infections cause a serious health problem nowadays as biofilms are more resistant to antibiotics that are routinely used in clinics compared to planktonic bacteria, and there is an urgent need to find potent antibiofilm tools pointing to the prospects of AMP–based therapy.

While there are many positive aspects of a potential application of AMPs as antiinfective therapeutics, there is concern about the development of cross-resistance against host AMPs as a result of the administration of exogenous peptides [132]. On the other hand, there was no significant resistance found in a course of long-time usage of nisin for food conservation as well as for polymyxins in medicine and veterinary uses. Despite the presence of pathogenic microorganisms, having successfully evolved AMP-resistance mechanisms, resistance to a broad range of AMPs has not yet occurred [133]. In higher vertebrates, endogenous AMPs participate in host defense in cooperation with many other defensive molecules as well as in the interplay with adaptive immunity factors; this may prevent the development of essential resistance to the peptides. AMP-based antiinfective therapeutics with combined antimicrobial and immunomodulatory properties may represent the most promising approach for the treatment of antibiotic-resistant infections [134]. A further comprehensive study of AMP-resistance mechanisms and host–pathogen interactions is also required to determine the optimal way of using AMPs to combat the increasing emergence of MDR bacteria.

References

[1] Hirsch JG. Phagocytin: a bactericidal substance from polymorphonuclear leucocytes. J Exp Med 1956;103(5):589–611.

[2] Michl H, Bachmayer H. Über das Giftsekret der Gelbbauchunke Bombina variegata L. (2. Mitt.) Monatshefte für Chemie; 1963.

[3] Tennessen JA. Molecular evolution of animal antimicrobial peptides: widespread moderate positive selection. J Evol Biol 2005;18(6):1387–94.

[4] Biomedical Informatics Centre, NIRRH, Mumbai. Cationic Collection of Antimicrobial Peptides (CAMP). http://www.camp.bicnirrh.res.in/index.php; 2020 [accessed 18.06.20].

[5] Gause GF, Brazhnikova MG. Gramicidin S and its use in the Treatment of Infected Wounds. Nature 1944;154(3918):703.

[6] Raybak MI. The efficacy and safety of daptomycin: first in a new class of antibiotics for Gram-positive bacteria. Clin Microbiol Infect 2006;12(S1):24–32.

[7] Nuti R, Goud NS, Saraswati AP, Alvala R, Alvala M. Antimicrobial peptides: a promising therapeutic strategy in tackling antimicrobial resistance. Curr Med Chem 2017;24:4303.

[8] Pachón-Ibáñez ME, Smani Y, Pachón J, Sánchez-Céspedes J. Perspectives for clinical use of engineered human host defense antimicrobial peptides. FEMS Microbiol Rev 2017;41 (3):323–42.

[9] Boman HG. Peptide antibiotics and their role in innate immunity. Annu Rev Immunol 1995;13(1):61–92.

[10] Agerberth B, Gunne H, Odeberg J, Kogner P, Boman HG, Gudmundsson GH. FALL-39, a putative human peptide antibiotic, is cysteine-free and expressed in bone marrow and testis. Proc Natl Acad Sci USA 1995;92(1):195–9.

[11] Skerlavaj B, Gennaro R, Bagella L, Merluzzi L, Risso A, Zanetti M. Biological characterization of two novel cathelicidin-derived peptides and identification of structural requirements for their antimicrobial and cell lytic activities. The J Biol Chem 1996;271:28375–81.

[12] Zasloff M. Magainins, a class of antimicrobial peptides from Xenopus skin: isolation, characterization of two active forms, and partial cDNA sequence of a precursor. Proc Natl Acad Sci USA 1987;84(15):5449–53.

[13] Park CB, Kim HS, Kim SC. Mechanism of action of the antimicrobial peptide buforin II: buforin II kills microorganisms by penetrating the cell membrane and inhibiting cellular functions. Biochem Biophys Res Commun 1998;244(1):253–7.

[14] Steiner H, Hultmark D, Engström Å, Bennich H, Boman HG. Sequence and specificity of two antibacterial proteins involved in insect immunity. Nature 1981;292:246–8.

[15] Shamova O, Orlov D, Stegemann C, Czihal P, Hoffmann R, Brogden K, et al. ChBac3.4: A novel proline-rich antimicrobial peptide from goat leukocytes. Int J Pept Res Ther 2009;15(1):31–42.

[16] Shamova O, Brogden KA, Zhao C, Nguyen T, Kokryakov VN, Lehrer RI. Purification and properties of proline-rich antimicrobial peptides from sheep and goat leukocytes. Infect Immun 1999;67(8):4106–11.

[17] Agerberth B, Lee J-Y, Bergman T, Carlquist M, Boman HG, Mutt V, et al. Amino acid sequence of PR-39: isolation from pig intestine of a new member of the family of proline-arginine-rich antibacterial peptides. Eur J Biochem 1991;202:849–54.

[18] Harwig SS, Kokryakov VN, Swiderek KM, Aleshina GM, Zhao C, Lehrer RI. Prophenin-1, an exceptionally proline-rich antimicrobial peptide from porcine leukocytes. FEBS Lett 1995;362(1):65–9.

[19] Selsted ME, Novotny MJ, Morris WL, Tang YQ, Smith W, Cullor JS. Indolicidin, a novel bactericidal tridecapeptide amide from neutrophils. J Biol Chem 1992;267(7):4292–5.

[20] Romeo D, Skerlavaj B, Bolognesi M, Gennaro R. Structure and bactericidal activity of an antibiotic dodecapeptide purified from bovine neutrophils. J Biol Chem 1988;263:9573–5.

[21] Kokryakov VN, Harwig SS, Panyutich EA, Shevchenko AA, Aleshina GM, Shamova OV, et al. Protegrins: leukocyte antimicrobial peptides that combine features of corticostatic defensins and tachyplesins. FEBS Lett 1993;327(2):231–6.

[22] Ovchinnikova TV, Aleshina GM, Balandin SV, Krasnosdembskaya AD, Markelov ML, Frolova EI, et al. Purification and primary structure of two isoforms of arenicin, a novel antimicrobial peptide from marine polychaeta Arenicola marina. FEBS Lett 2004;577 (1–2):209–14.

[23] Selsted ME, Harwig SS. Purification, primary structure, and antimicrobial activities of a guinea pig neutrophil defensin. Infect Immun Am Soc Microbiol 1988;55(9):2281–6.

[24] Daher KA, Lehrer RI, Ganz T, Kronenberg M. Isolation and characterization of human defensin cDNA clones. Proc Natl Acad Sci USA 1988;85(19):7327–31.

[25] Eisenhauer PB, Harwig SS, Szklarek D, Ganz T, Selsted ME, Lehrer RI. Purification and antimicrobial properties of three defensins from rat neutrophils. Infect Immun 1989;57 (7):2021–7.

[26] Dept of Pathology & Microbiology, UNMC. The Antimicrobial Peptide Database (ADP). http://aps.unmc.edu/AP/; 2020 [accessed 18.06.20].

[27] Wang G, Li X, Wang Z. APD3: the antimicrobial peptide database as a tool for research and education. Nucleic Acids Res 2016;44(D1):D1087–93.

[28] Wang G. Post-translational modifications of natural antimicrobial peptides and strategies for peptide engineering. Curr Biotechnol 2012;1(1):72–9.

[29] Zhang L, Gallo RL. Antimicrobial peptides. Curr Biol 2016;26:R1–21.

[30] Yeaman MR, Yount NY. Mechanisms of antimicrobial peptide action and resistance. Pharmacol Rev 2003;55(1):27–55.

[31] Hancock RE. Peptide antibiotics. Lancet 1997;349(9049):418–22.

[32] Epand RM, Vogel HJ. Diversity of antimicrobial peptides and their mechanisms of action. Biochim Biophys Acta 1999;1462:11–28.

[33] Zhao H, Rinaldi A, Rufo A, Bozzi A, Kinnunen P, Di Giulio A. Structural and charge requirements for antimicrobial peptide insertion into biological and model membranes. In: Menestrina G;, Dalla-Serra M, Lazarovici P, editors. Cellular and molecular mechanism of toxin action. Pore-forming peptides and protein toxins. Taylor & Francis Group; 2003. p. 151–77.

[34] Huang H. Action of antimicrobial peptides: two-state model. Biochemistry 2000;39 (29):8347–52.

[35] Papo N, Shai Y. Can we predict biological activity of antimicrobial peptides from their interactions with model phospholipid membranes? Peptides 2003;24(11):1693–703.

[36] Chang WK, Wimley WC, Searson PC, Hristova K, Merzlyakov M. Characterization of antimicrobial peptide activity by electrochemical impedance spectroscopy. Biochim Biophys Acta 2008;1778(10):2430–6.

[37] Brogden KA. Antimicrobial peptides: pore formers or metabolic inhibitors in bacteria? Nat Rev Microbiol 2005;3(3):238–50.

[38] Lehrer RI, Barton A, Daher KA, Harwig SS, Ganz T, Selsted ME. Interaction of human defensins with Escherichia coli. Mechanism of bactericidal activity. J Clin Invest 1989;84 (2):553–61.

[39] Hansen A, Schäfer I, Knappe D, Seibel P, Hoffmann R. Intracellular toxicity of proline-rich antimicrobial peptides shuttled into mammalian cells by the cell-penetrating peptide penetratin. Antimicrob Agents Chemother 2012;56(10):5194–201.

[40] Seefeldt AC, Graf M, Pérébaskine N, Nguyen F, Arenz S, Mardirossian M, et al. Structure of the mammalian antimicrobial peptide Bac7(1–16) bound within the exit tunnel of a bacterial ribosome. Nucleic Acids Res 2016;44(5):2429–38.

[41] Roy RN, Lomakin IB, Gagnon MG, Steitz TA. The mechanism of inhibition of protein synthesis by the proline-rich peptide oncocin. Nat Struct Mol Biol 2015;22(6):466–9.

[42] Lee HS, Park CB, Kim JM, Jang SA, Park IY, Kim MS, et al. Mechanism of anticancer activity of buforin IIb, a histone H2A-derived peptide. Cancer Lett 2008;271(1):47–55.

[43] Hsu CH, Chen C, Jou ML, Lee AY, Lin YC, Yu YP, et al. Structural and DNA-binding studies on the bovine antimicrobial peptide, indolicidin: evidence for multiple conformations involved in binding to membranes and DNA. Nucleic Acids Res 2005;33:4053–64.

[44] Marchand C, Krajewski K, Lee HF, Antony S, Johnson AA, Amin R, et al. Covalent binding of the natural antimicrobial peptide indolicidin to DNA abasic sites. Nucleic Acids Res 2006;34:5157–65.

[45] Yang D, Chen Q, Schmidt AP, Anderson GM, Wang JM, Wooters J, et al. LL-37, the neutrophil granule- and epithelial cell-derived cathelicidin, utilizes formyl peptide receptor-like 1 (FPRL1) as a receptor to chemoattract human peripheral blood neutrophils, monocytes, and T cells. J Exp Med 2000;192:1069–74.

[46] Yang D, Chertov O, Bykovskaia SN, Chen Q, Buffo MJ, Shogan J, et al. β-Defensins: linking innate and adaptive immunity through dendritic and T cell CCR6. Science 1999;286:525–8.

[47] Niyonsaba F, Hirata M, Ogawa H, Nagaoka I. Epithelial cell-derived antibacterial peptides human beta-defensins and cathelicidin: multifunctional activities on mast cells. Curr Drug Targets Inflamm Allergy 2003;2(3):224–31.

[48] Niyonsaba F, Iwabuchi K, Someya A, Hirata M, Matsuda H, Ogawa H, et al. A cathelicidin family of human antibacterial peptide LL-37 induces mast cell chemotaxis. Immunology 2002;106(1):20–6.

[49] Territo MC, Ganz T, Selsted ME, Lehrer R. Monocyte-chemotactic activity of defensins from human neutrophils. J Clin Invest 1989;84:2017–20.

[50] Grigat J, Soruri A, Forssmann U, Riggert J, Zwirner J. Chemoattraction of macrophages, T lymphocytes, and mast cells is evolutionarily conserved within the human α-defensin family. J Immunol 2007;179:3958–65.

[51] Niyonsaba F, Ogawa H, Nagaoka I. Human beta-defensin-2 functions as a chemotactic agent for tumour necrosis factor-alpha-treated human neutrophils. Immunology 2004;111 (3):273–81.

[52] Tjabringa GS, Ninaber DK, Drijfhout JW, Rabe KF, Hiemstra PS. Human cathelicidin LL-37 is a chemoattractant for eosinophils and neutrophils that acts via formyl-peptide receptors. Int Arch Allergy Immunol 2006;140:103–12.

[53] Suzuki K, Murakami T, Kuwahara-Arai K, Tamura H, Hiramatsu K, Nagaoka I. Human anti-microbial cathelicidin peptide LL-37 suppresses the LPS-induced apoptosis of endothelial cells. Int Immunol 2011;23(3):185–93.

[54] Scott A, Weldon S, Buchanan PJ, Schock B, Ernst RK, McAuley DF, et al. Evaluation of the ability of LL-37 to neutralise LPS in vitro and ex vivo. PLoS One 2011;6(10):e26525.

[55] Suarez-Carmona M, Hubert P, Gonzalez A, Duray A, Roncarati P, Erpicum C. ΔNp63 isoform-mediated β-defensin family up-regulation is associated with (lymph) angiogenesis and poor prognosis in patients with squamous cell carcinoma. Oncotarget 2014;5:1856–68.

[56] Chavakis T, Cines DB, Rhee JS, Liang OD, Schubert U, Hammes HP, et al. Regulation of neovascularization by human neutrophil peptides (alpha-defensins): a link between inflammation and angiogenesis. FASEB J 2004;18:1306–8.

[57] Oono T, Shirafuji Y, Huh W-K, Akiyama H, Iwatsuki K. Effects of human neutrophil peptide-1 on the expression of interstitial collagenase and type I collagen in human dermal fibroblasts. Arch Dermatol Res 2002;294:185–9.

[58] Umnyakova ES, Gorbunov NP, Zhakhov AV, Krenev IA, Ovchinnikova TV, Kokryakov VN, et al. Modulation of human complement system by antimicrobial peptide arenicin-1 from *Arenicola marina*. Marine Drugs 2018;16:e480.

[59] Van den Berg RH, Faber-Krol MC, van Wetering S, Hiemstra PS, Daha MR. Inhibition of activation of the classical pathway of complement by human neutrophil defensins. Blood 1998;92(10):3898–903.

[60] Berlov MN, Umnyakova ES, Leonova TS, Milman BL, Krasnodembskaya AD, Ovchinnikova TV, et al. Interaction of arenicin-1 with C1q protein. Russ J Bioorgan Chem 2015;41:597–601.

[61] Panyutich AV, Szolda OH, Peon PH, Tseng Y, Ganz T. Identification of defensin binding to C1 complement. FEBS Lett 1994;356(2–3):169–73.

[62] Prohászka Z, Német K, Csermely P, Hudecz F, Mezõ G, Füst G. Defensins purified from human granulocytes bind C1q and activate the classical complement pathway like the transmenbrane glycoprotein gq41 of HIV-1. Mol Immunol 1997;34(11):809–16.

[63] Umnyakova ES, Zharkova MS, Berlov MN, Shamova OV, Kokryakov VN. Human antimicrobial peptides in autoimmunity. Autoimmunity 2020;. Available from: https://doi.org/10.1080/08916934.2020.1711517.

[64] Shamova OV, Lesnikova MR, Kokryakov VN, Shkhinek EK, Korneva EA. Effect of defensins on the blood level of corticosterone and the immune response during stress. Bull Exp Biol Med 1993;115:728–73.

[65] Zhu Q, Singh A, Bateman A, Esch F, Solomon S. The corticostatin anti-ACTH and cytotoxic activity of peptides isolated from fetal, adult and tumor-beating lung. J Steroid Biochem 1987;27:1017–423.

[66] Bayramov D, Li Z, Patel E, Izadjoo M, Kim H, Neff J. A novel peptide-based antimicrobial wound treatment is effective against biofilms of multi-drug resistant wound pathogens. Military Med 2018;183(1):481–6.

[67] Straus SK, Hancock REW. Mode of action of the new antibiotic for Gram-positive pathogens daptomycin: comparison with cationic antimicrobial peptides and lipopeptides. Biochim Biophys Acta Biomembr 2006;1758:1215–23.

[68] Uhlig T, Kyprianou T, Martinelli FG, Oppici CA, Heiligers D, Hills D, et al. The emergence of peptides in the pharmaceutical business: from exploration to exploitation. EuPA Open Proteomics 2014;4:58–69.

[69] Zharkova MS, Kopeykin PM, Afinogenov GE, Orlov DS, Artamonov AY, Safiullina KE, et al. Effects of proline-rich peptides of the innate immune system on drug-resistant bacterial strains. Med Immunol (Russia) 2018;20(1):107–14.

[70] De la Fuente-Núñez C, Hancock RE. Using anti-biofilm peptides to treat antibiotic-resistant bacterial infections. Postdoc J 2015;3(2):1–8.

[71] Lei J, Sun L, Huang S, Zhu C, Li P, He J, et al. The antimicrobial peptides and their potential clinical applications. Am J Transl Res 2019;11(7):3919–31.

[72] Juturu V, Wu JC. Microbial production of bacteriocins: Latest research development and applications. Biotechnol Adv 2018;36(8):2187–200.

[73] Maher S, McClean S. Investigation of the cytotoxicity of eukaryotic and prokaryotic antimicrobial peptides in intestinal epithelial cells in vitro. Biochem Pharmacol 2006;71(9):1289–98.

[74] Mahlapuu M, Hakansson J, Ringstad L, Bjorn C. Antimicrobial peptides: an emerging category of therapeutic agents. Front Cell Infect Microbiol 2016;6:194.

[75] Amso Z, Hayouka Z. Antimicrobial random peptide cocktails: a new approach to fight pathogenic bacteria. Chem Commun (Camb) 2019;55:2007–14.

[76] Bahar AA, Ren D. Antimicrobial peptides. Pharmaceuticals (Basel) 2013;6:1543–75.

[77] Rathinakumar R, Walkenhorst WF, Wimley WC. Broad-spectrum antimicrobial peptides by rational combinatorial design and high-throughput screening: the importance of interfacial activity. J Am Chem Soc 2009;131:7609–17.

[78] Lee DL, Hodges RS. Structure-activity relationships of de novo designed cyclic antimicrobial peptides based on gramicidin S. Biopolymers 2003;71:28–48.

[79] Raguse TL, Porter EA, Weisblum B, Gellman SH. Structure-activity studies of 14-helical antimicrobial beta-peptides: probing the relationship between conformational stability and antimicrobial potency. J Am Chem Soc 2002;124:12774–85.

[80] Nicholls EF, Madera L, Hancock REW. Immunomodulators as adjuvants for vaccines and antimicrobial therapy. Ann N Y Acad Sci 2010;1213(1):46–61.

[81] Wang KR, Zhang BZ, Zhang W, Yan JX, Li J, Wang R. Antitumor effects, cell selectivity and structure-activity relationship of a novel antimicrobial peptide polybia-MPI. Peptides 2008;29:963–8.

[82] Jakel CE, Meschenmoser K, Kim Y, Weiher H, Schmidt-Wolf IG. Efficacy of a proapoptotic peptide towards cancer cells. In Vivo 2012;26:419–26.

[83] Molchanova N, Hansen PR. Advances in development of antimicrobial peptidomimetics as potential drugs. Molecules 2017;22(9):1430.

[84] Andersson D, I., Hughes D, Kubicek-Sutherland JZ. Mechanisms and consequences of bacterial resistance to antimicrobial peptides. Drug Resist Updates 2016;26:43–57.

[85] Strempel N, Neidig A, Nusser M, Geffers R, Vieillard J, Lesouhaitier O, et al. Human host defense peptide LL-37 stimulates virulence factor production and adaptive resistance in Pseudomonas aeruginosa. PLoS ONE 2003;8:e82240.

[86] Viljanen P, Vaara M. Susceptibility of Gram-negative bacteria to polymyxin B nonapeptide. Antimicrob Agents Chemother 1984;25:701–5.
[87] Manniello JM, Heymann H, Adair FW. Resistance of spheroplasts and whole cells of *Pseudomonas cepacia* to polymyxin B. Antimicrob Agents Chemother 1978;14:500–4.
[88] Anaya-Lopez JL, Lopez-Meza JE, Ochoa-Zarzosa A. Bacterial resistance to cationic antimicrobial peptides. Crit Rev Microbiol 2013;39:180–95.
[89] Matamouros S, Miller SI. S. Typhimurium strategies to resist killing by cationic antimicrobial peptides. Biochim Biophys Acta 2015;1848:3021–5.
[90] Nawrocki KL, Crispell EK, McBride SM. Antimicrobial peptide resistance mechanisms of gram-positive bacteria. Antibiotics (Basel) 2014;3:461–92.
[91] Llobet E, Campos MA, Giménez P, Moranta D, Bengoechea JA. Analysis of the networks controlling the antimicrobial-peptide-dependent induction of *Klebsiella pneumoniae* virulence factors. Infect Immun 2011;79:3718–32.
[92] Salazar J, Alarcón M, Huerta J, Navarro B, Aguayo D. Phosphoethanolamine addition to the Heptose I of the lipopolysaccharide modifies the inner core structure and has an impact on the binding of polymyxin B to the *Escherichia coli* outer membrane. Arch Biochem Biophys 2017;620:28–34.
[93] Guina T, Yi EC, Wang H, Hackett M, Miller SI. A PhoP-regulated outer membrane protease of Salmonella enterica serovar typhimurium promotes resistance to alpha-helical antimicrobial peptides. J Bacteriol 2000;182:4077–86.
[94] Yount NY, Bayer AS, Xiong YQ, Yeaman MR. Advances in antimicrobial peptide immunobiology. Curr Trends Peptide Sci 2006;84:435–458.
[95] Thomassin JL, Brannon JR, Gibbs BF, Gruenheid S, Le Moual H. OmpT outer membrane proteases of enterohemorrhagic and enteropathogenic Escherichia coli contribute differently to the degradation of human LL-37. Infect Immun 2012;80:483–92.
[96] Kooi C, Sokol PA. Burkholderia cenocepacia zinc metalloproteases influence resistance to antimicrobial peptides. Microbiology 2009;155:2818–25.
[97] Campos MA, Vargas MA, Regueiro V, Llompart CM, Alberti S, Bengoechea JA. Capsule polysaccharide mediates bacterial resistance to antimicrobial peptides. Infect Immunol 2004;72:7107–14.
[98] Manning AJ, Kuehn MJ. Contribution of bacterial outer membrane vesicles to innate bacterial defense. BMC Microbiol 2011;11:258.
[99] Dorrer E, Teuber M. Induction of polymyxin resistance in *Pseudomonas fluorescens* by phosphate limitation. Arch Microbiol 1977;114:87–9.
[100] Andra J, Goldmann T, Ernst CM, Peschel A, Gutsmann T. Multiple peptide resistance factor (MprF)-mediated resistance of *Staphylococcus aureus* against antimicrobial peptides coincides with a modulated peptide interaction with artificial membranes comprising lysylphosphatidylglycerol. J Biol Chem 2011;286:18692–700.
[101] Shafer WM, Qu X, Waring AJ, Lehrer RI. Modulation of *Neisseria gonorrhoeae* susceptibility to vertebrate antibacterial peptides due to a member of the resistance/nodulation/division efflux pump family. Proc Natl Acad Sci USA 1998;95:1829–33.
[102] De la Fuente-Núñez C, Korolik V, Bains M, Nguyen U, Breidenstein EB, Horsman S, et al. Inhibition of bacterial biofilm formation and swarming motility by a small synthetic cationic peptide. Antimicrob Agents Chemother 2012;56:2696–704.
[103] Hirt H, Gorr SU. Antimicrobial peptide GL13K is effective in reducing biofilms of *Pseudomonas aeruginosa*. Antimicrob Agents Chemother 2013;57(10):4903–10.

[104] Algburi A, Zehm S, Netrebov V, Bren AB, Chistyakov V, Chikindas ML. Subtilosin prevents biofilm formation by inhibiting bacterial quorum sensing. Probiot Antimicrob Proteins 2017;9:81–90.

[105] De la Fuente-Núñez C, Reffuveille F, Mansour SC, Reckseidler-Zenteno SL, Hernández D, Brackman G, et al. D-enantiomeric peptides that eradicate wild-type and multidrug-resistant biofilms and protect against lethal *Pseudomonas aeruginosa* infections. Chem Biol 2015;22(2):196–205.

[106] Hirt H, Hall JW, Larson E, Gorr SU. A D-enantiomer of the antimicrobial peptide GL13K evades antimicrobial resistance in the Gram-positive bacteria *Enterococcus faecalis* and *Streptococcus gordonii*. PLoS One 2018;13(3):e0194900.

[107] May JJ, Finking R, Wiegeshoff F, Weber TT, Bandur N, Koert U, et al. Inhibition of the D-alanine:D-alanyl carrier protein ligase from Bacillus subtilis increases the bacterium's susceptibility to antibiotics that target the cell wall. FEBS 2005;272:2993–3003.

[108] Koprivnjak T, Peschel A. Bacterial resistance mechanisms against host defense peptides. Cell Mol Life Sci 2011;68(13):2243–54.

[109] Reinhardt A, Neundorf I. Design and application of antimicrobial peptide conjugates. Int J Mol Sci 2016;17(5):E701.

[110] Mohid SA, Bhunia A. Combining antimicrobial peptides with nanotechnology: an emerging field in theranostics. Curr Protein Pept Sci 2020;21(4):413–28.

[111] Pham TN, Loupias P, Dassonville-Klimpt A, Sonnet P. Drug delivery systems designed to overcome antimicrobial resistance. Med Res Rev 2019;39(6):2343–96.

[112] Golubeva OY, Shamova OV, Yakovlev AV, Zharkova MS. Synthesis and study of the biologically active lysozyme–silver nanoparticles–montmorillonite K10 complexes. Glass Phys Chem 2016;42(1):84–91.

[113] Golubeva OY, Golubkov VV, Yukhnev VA, Shamova OV. Synthesis of silver nanoparticles with lysozyme and sodium dioctyl sulfosuccinate and comparison of their biological activities. Glass Phys Chem 2017;43(1):63–9.

[114] Golubeva OY, Brazovskaya EY, Shamova OV. Biological activity and sorption ability of synthetic montmorillonitemodified by silver/lysozyme nanoparticles. Appl Clay Sci 2018;163:56–62.

[115] Pletzer D, Coleman SR, Hancock RE. Anti-biofilm peptides as a new weapon in antimicrobial warfare. Curr Opin Microbiol 2016;33:35–40.

[116] Chung PY, Khanum R. Antimicrobial peptides as potential anti-biofilm agents against multidrug-resistant bacteria. J Microbiol, Immunol Infection 2017;50(4):405–10.

[117] Jorge P, Lourenco A, Pereira MO. New trends in peptide-based anti-biofilm strategies: a review of recent achievements and bioinformatic approaches. Biofouling 2012;28 (10):1033–61.

[118] Hell E, Giske CG, Nelson A, Römling U, Marchini G. Human cathelicidin peptide LL-37 inhibits both attachment capability and biofilm formation of *Staphylococcus epidermidis*. Lett Appl Microbiol 2010;50:211–15.

[119] Overhage J, Campisano A, Bains M, Torfs EC, Rehm BH, Hancock RE. Human host defense peptide LL-37 prevents bacterial biofilm formation. Infect Immun 2008;76:4176–82.

[120] Zharkova MS, Orlov DS, Golubeva OY, Chakchir OB, Eliseev IE, Grinchuk TM, et al. Application of antimicrobial peptides of the innate immune system in combination with conventional antibiotics—a novel way to combat antibiotic resistance? Front Cell Infect Microbiol 2019;. Available from: https://doi.org/10.3389/fcimb.2019.00128.

[121] Amani J, Barjini KA, Moghaddam MM, Asadi A. In vitro synergistic effect of the CM11 antimicrobial peptide in combination with common antibiotics against clinical isolates of six species of multidrug-resistant pathogenic bacteria. Protein Pept Lett 2015;22:940–51.

[122] Panteleev PV, Bolosov IA, Kalashnikov AA, Kokryakov VN, Shamova OV, Emelianova AA, et al. Combined antibacterial effects of goat cathelicidins with different mechanisms of action. Front Microbiol 2018;9(2983):1–19.

[123] Sakoulas G, Bayer AS, Pogliano J, Tsuji BT, Yang S-J, Mishra NN, et al. Ampicillin enhances daptomycin-and cationic host defense peptide-mediated killing of ampicillin and vancomycin-resistant *Enterococcus faecium*. Antimicrob Agents Chemother 2012;56:838–44.

[124] Dhand A, Bayer AS, Pogliano J, Yang SJ, Bolaris M, Nizet V, et al. Use of antistaphylococcal beta-lactams to increase daptomycin activity in eradicating persistent bacteremia due to methicillin-resistant *Staphylococcus aureus*: role of enhanced daptomycin binding. Clin Infect Dis 2011;53:158–63.

[125] Berditsch M, Jäger T, Strempel N, Schwartz T, Overhage J, Ulrich AS. Synergistic effect of membrane-active peptides polymyxin B and gramicidin S on multidrug-resistant strains and biofilms of *Pseudomonas aeruginosa*. Antimicrob Agents Chemother 2015;59:5288–96.

[126] Pletzer D, Mansour SC, Hancock REW. Synergy between conventional antibiotics and anti-biofilm peptides in a murine, sub-cutaneous abscess model caused by recalcitrant ESKAPE pathogens. PLoS Pathog 2018;14(6):e1007084.

[127] Pletzer D, Hancock RE. Is synergy the key to treating high-density infections? Future Microbiol 2018;13:1629–32.

[128] Niyonsaba F, Madera L, Afacan N, Okumura K, Ogawa H, Hancock RE. The innate defense regulator peptides IDR-HH2, IDR-1002, and IDR-1018 modulate human neutrophil functions. J Leukoc Biol 2013;94(1):159–70.

[129] Scott MG, Dullaghan E, Mookherjee N, Glavas N, Waldbrook M, Thompson A, et al. An anti-infective peptide that selectively modulates the innate immune response. Nat Biotechnol 2007;25(4):465–72.

[130] Madera L, Hancock RE. Anti-infective peptide IDR-1002 augments monocyte chemotaxis towards CCR5 chemokines. Biochem Biophys Res Commun 2015;464(3):800–6.

[131] Mansour SC, de la Fuente-Núñez C, Hancock RE. Peptide IDR-1018: modulating the immune system and targeting bacterial biofilms to treat antibiotic-resistant bacterial infections. J Pept Sci 2015;21(5):323–9.

[132] Duperthuy M. Antimicrobial peptides: virulence and resistance modulation in gram-negative bacteria. Microorganisms 2020;8(2):280. Available from: https://doi.org/10.3390/microorganisms8020280.

[133] Cole JN, Nizet V. Bacterial evasion of host antimicrobial peptide defenses. Microbiol Spectr 2016;4(1) 10.1128/microbiolspec.VMBF-0006–2015.

[134] Silva ON, de la Fuente-Núñez C, Haney EF, Fensterseifer IC, Ribeiro SM, Porto WF, et al. An anti-infective synthetic peptide with dual antimicrobial and immunomodulatory activities. Sci Rep 2016;6:35465.

Chapter 19

Antimicrobial peptides and the skin and gut microbiomes

Adrienne Joseph[1,*], Meital Nuriel-Ohayon[2,*], Shai Bel[2] and Tamia A. Harris[1,3]

[1]Department of Dermatology, University of Texas Southwestern Medical Center, Dallas, TX, United States, [2]Azrieli Faculty of Medicine, Bar Ilan University, Safed, Israel, [3]Department of Immunology, University of Texas Southwestern Medical Center, Dallas, TX, United States

19.1 Introduction

Animals live in a microbial-dominant world. This means they are in constant contact with a large diversity of microorganisms and spend a considerable amount of energy in an effort to keep these microbes out of internal tissues. Nowhere is this more evident than on the skin and in the gastrointestinal tract, where concentrations of bacteria can reach a staggering $10^{11}/m^2$ of skin and $10^{11}/1$ mL of colon content [1]. To protect themselves from bacterial invasion, animals have developed sophisticated barriers. These are not just static physical barriers, but also dynamic chemical barriers that actively kill and sequester potential invaders. This is mainly achieved by secreting a large range of antimicrobial proteins and peptides (AMPs) [2].

In addition to the presence of bacterial pathogens, the skin and intestine are colonized by beneficial bacteria (often termed commensals) that serve various functions for the host. Commensal species produce nutrients and prevent the colonization of potentially invasive pathogens [3]. Thus, it is undesirable to indiscriminately kill bacteria that reside in and on the host. AMPs are an evolutionary adaptation to facilitate this selective symbiosis.

AMPs are short amino acids chains (up to 50 amino acids) that kill or inactivate microorganisms. These AMPs are produced by all animals and constitute an evolutionarily conserved mechanism of immune defense. The means by which AMPs disable bacteria are numerous and likely reflect the large diversity of bacteria that colonize the host. Importantly, these peptides

* These authors contributed equally.

Peptide and Peptidomimetic Therapeutics. DOI: https://doi.org/10.1016/B978-0-12-820141-1.00016-9

must remain active in the harsh environment of the intestine and skin (pH and proteases) while also not harming the host's own cells [2].

In this chapter, we review the molecular mechanism of AMP antimicrobial activity in the skin and intestine and describe several notable examples (Table 19.1). We then detail how these AMPs are regulated at the transcriptional and posttranslational levels, and how they affect the microbiome. Lastly, we discuss how AMP dysfunction is involved in human disease. Importantly, AMPs also have a role in directly shaping host immune response [4], antiviral immunity [5], and some AMPs are also produced by the bacteria themselves [6], although these latter proteins will not be the focus of this chapter.

19.2 Mechanism of action

Antimicrobial proteins kill bacteria primarily through binding and dismantling the bacterial cell wall. Gram-positive bacteria have a two-layered cell wall consisting of an inner membrane and an outer peptidoglycan layer. Gram-negative bacteria have an additional outer membrane surrounding the peptidoglycan layer. When AMPs bind membranes they create pores in the bacterial cell wall leading to loss of membrane potential and ultimately cell lysis [6]. The specific mechanism underlying this disruption is often driven by the structure of the AMP. AMPs tend to fall into structural families composed predominantly of alpha helices or beta sheets (Fig. 19.1 and Table 19.1).

19.2.1 Alpha helices

In the skin, two notable AMPs have an alpha-helical structure, cathelicidin and dermcidin, while in the intestine only members of the cathelicidin alpha-helical family are expressed. The cathelicidins are AMPs comprised of cationic proteins that assume an α helical structure in the presence of cell membranes. Cathelicidin proteins bind and insert into negatively charged cell surfaces, leading to membrane disruption and cell death [16,17]. The cathelicidins can also cross cell membranes, inciting protein degradation and disrupting intracellular processes, such as nucleic acid synthesis and enzymatic activity [17].

Dermcidin is a negatively charged protein secreted by eccrine sweat glands. As most AMPs are cationic and make use of charge interactions with negatively charged bacterial membranes for their mechanism of action, the anionic charge of dermcidin makes this peptide unique in comparison to other AMPs. The structure of dermcidin is made up of elongated α helices which form a hexameric channel [18]. Although the mechanism through which dermcidin kills microbes is unclear, it is thought that dermcidin also leads to pore formation and disruption of bacterial cell membranes [19].

TABLE 19.1 Notable antimicrobial peptides and their characteristics.

Family (representative protein)	Structure	Size (aa)	Target organism	Mechanism of action	Distribution	Specific targets
Cathelicidins (LL-37)	α-helix	37aa	Gram-positive bacteria, Gram-negative bacteria, fungi, viruses, parasites	Bind and insert into negatively charged cell surfaces, leading to membrane disruption and cell death	Lungs, large intestine, skin	*E. coli* [7] *S. aureus* [8] *Aspergillus fumigatus* [9] vaccinia virus [10] *Entamoeba histolytica* [11]
Dermcidin (DCD-1L)	α-helix	47aa	Gram-positive bacteria, Gram-negative bacteria, fungi, viruses	Membrane perforating (Chemoatractant)	Human sweat	*S. aureus* *Mycobacterium tuberculosis* [12]
α-defensins (Human Defensin 6 (HD6))	β-sheet	32aa	Bacteria and fungi	Performing "nanonets" that entrap microbes and prevent pathogens translocation from lumen into the host cells	Small intestine (Paneth cells, macrophages, neutrophils)	*C. albicans* [13]
β-defensin (hBD1–3)	β-sheet	> 50aa	Gram-positive bacteria, Gram-negative bacteria, fungi, viruses, protozoa	Interacting with negatively charged phospholipids in cell membrane	Large intestine, skin	*S. aureus* [14], *Bacillus cereus, Bacillus thuringiensis, Bacillus licheniformis* [15], *E. coli* [7]

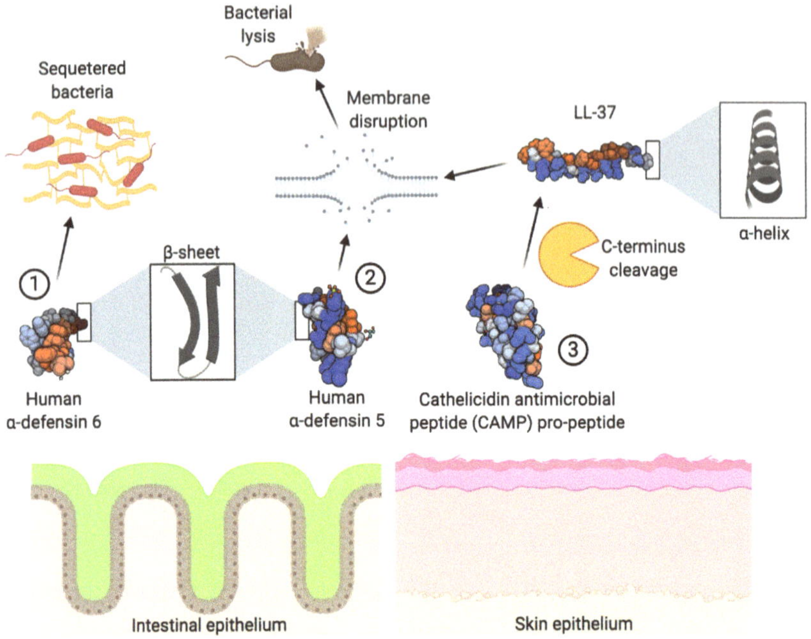

FIGURE 19.1 AMP mechanisms of action. While human α-defensin 6 and human α-defensin 5 both have a α-sheet structure, their antimicrobial mechanisms of action differ. (1) Human α-defensin 6 assembles into a multimer nanofiber structure that binds and sequsters bacteria, thus denying them access to host tissues. (2) Human α-defensin 5 binds to bacterial membranes, creating pores that lead to bacterial lysis. (3) Cathelicidin undergoes cleavage in its C-terminus, creating multiple AMPs. The most studied member of the Cathelicidin family is LL-37, which disrupts bacterial membranes, leading to bacterial lysis.

19.2.2 Beta sheets

AMPs that have a β-sheet structure all fall within the defensin family. This diverse and evolutionarily conserved family of peptides has a broad spectrum of antimicrobial activities and can target bacteria, fungi, viruses, and even protozoa. The defensin subfamilies α and β are classified based on cysteine residue-pairing in their structure. The third subfamily of defensins are the θ-defensins [20], which have a circular structure.

α-Defensins kill bacteria by dimerizing via cysteine bonds and directly interacting with bacterial membranes, while they also sequester lipid II, which inhibits bacterial cell wall synthesis [6]. An exception to this is the α-defensin human defensin 6 (HD6), which does not kill, but rather sequesters bacteria into clumps by oligomerizing them into long fibrils. These fibrils of HD6 and bacteria create "nanonets," preventing bacterial invasive potential [21].

Like the cathelicidins, β-defensins have a net positive charge. The β-defensins typically have a free N-terminus and three antiparallel β-sheets.

They are also cysteine-rich and characterized by a 6-cysteine motif stabilized by three disulfide bonds between cysteines 1–5, 2–4 and 3–6 [20,22]. The β-defensins carry out their antimicrobial properties by binding to negatively charged phospholipids in the cell membrane and forming defensin pores that result in bacterial cell lysis [16,22].

19.2.3 Antimicrobial proteins

Though not classified as AMPs (as they are composed of >50 amino acids), antimicrobial proteins also provide the host with crucial protection against invading microbes. While antimicrobial proteins are not the focus of this chapter, we will briefly describe a few notable examples. Like cathelicidins and defensins, antimicrobial proteins often kill bacteria via pore formation. These include the regenerating (Reg) protein family of lectins. Of note is RegIIIα (also known as HIP/PAP) and its mouse ortholog RegIIIγ, which maintain the intestinal mucus barrier and preserve homeostasis [23]. Other antimicrobial proteins kill bacteria by enzymatic activity. Lysozyme is one such protein that kills bacteria by hydrolyzing peptidoglycan in the bacterial cell wall [24]. Another protein, secretory phospholipase A2 (sPLA2), hydrolyzes phospholipids in the bacterial membrane, leading to loss of membrane potential [25]. Antimicrobial proteins also provide protection by mechanisms other than cell wall disruption. For example, angiogenin 4 is an RNase that targets both Gram-positive and -negative bacteria, though it is not clear if this enzymatic activity is needed for its bacteriocidic function [26]. Another example is lipocalin-2, which is secreted into the colon and sequesters iron, thus effectively starving bacteria of a key component of the respiratory chain [27].

19.3 Regulation of antimicrobial peptides

Disruption of bacterial membranes is a frequently shared mechanism of action of AMPs. Though some specificity is often linked to the structural and charge properties of these peptides, there is considerable risk imposed to the host by producing short peptides capable of producing damage to host membranes. Thus, the production of active AMPs is highly regulated.

19.3.1 Regulation of cathelicidins

Human cathelicidin is encoded by a single gene, *CAMP*, consisting of four exons on chromosome 3p21 [17,20]. The *CAMP* gene encodes a single precursor protein, cathelicidin AMP (CAMP). The CAMP propeptide is then posttranslationally cleaved at the C-terminus to generate multiple mature AMPs, including the most studied human cathelicidin, LL-37 [20,28]. RK-31 and KS-30 are other mature AMPs produced by the cleavage of the human cathelicidin precursor protein and have the ability to induce host cytokines.

Thus, postenzymatic processing is important in dictating the activity and function of cathelicidin proteins [20].

Cathelicidin expression is also regulated at the transcriptional level. In healthy skin and intestine, LL-37 expression is low; however, in the setting of injury, infection, or inflammation, LL-37 expression is upregulated [16,29]. This upregulation in response to infection is secondary to response elements to AP-1 and cAMP in the cathelicidin promoter [30]. Cathelicidin is also under transcriptional control of a vitamin D response element. Thus, increased levels of 1,25-dihydroxyvitamin D3 leads to increased expression of cathelicidin mRNA [20].

19.3.2 Regulation of dermcidin

Dermcidin is constitutively secreted in human sweat. Similar to cathelicidin, dermcidin is regulated posttranslationally after secretion. Proteolytic processing of dermcidin produces a series of truncated peptides varying in length from 25 to 48 amino acids [31]. The mature peptides are distributed to the skin surface via human sweat and have activity against a variety of pathogenic microbes [32].

19.3.3 Regulation of β-defensins

As described above, there are three defensin subfamilies, namely α-, β-, and circular θ-defensins [20]. Beta-defensins are critical AMPs in the human skin and intestine. The β-defensins are encoded by 90 different genes that each have unique activity and expression patterns [16]. β-Defensin genes mainly consist of two exons. The first exon encodes a leucine-rich signal and prosequence, while the second exon encodes the mature peptide. Among the β-defensins, exon 1 is highly conserved, while exon 2 is less well conserved [22]. hBD1−4 are the most well-known β-defensins. These genes cluster on chromosome 8p23.1, a site of frequent chromosomal rearrangement. The copy number of defensins varies between individuals and correlates with the level of mRNA transcript. Defensins are generated as proproteins, and then the mature peptide is cleaved out via posttranslational processing [20]. The β-defensins have a broad range of antimicrobial activity and tissue distribution [22]. In the human skin and intestine, β-defensins are upregulated during periods of inflammation and infection [33].

19.4 Antimicrobial peptides in the skin

19.4.1 Overall function of antimicrobial peptides in the skin

The major function of AMPs in the skin is to protect the skin from infectious agents. Thus, bacteria can trigger the expression of AMP expression in keratinocytes through the engagement of toll-like receptors (TLRs) and other pattern recognition receptors [34,35]. The development of mice lacking cathelicidin was the first direct evidence of AMPs defending against skin pathogens [16]. Many pathogens, including Group A *Streptococcus, Staphylococcus* species, Vaccinia virus and *Leishmania* species have shown susceptibility to AMPs [16,20].

AMPs have activity against a wide number of clinically relevant targets. The human β-defensins (hBD1−3) and human cathelicidin (LL-37) have exhibited antimicrobial activity against *Escherichia coli* and methicillin-resistant *S. aureus* (MRSA) [34]. LL-37 can also kill Group A *Streptococcus* and *Candida albicans* (*C. albicans*) and can inhibit biofilm formation of *S. aureus* and *Pseudomonas aeruginosa* [34]. In vitro, dermcidin has exhibited antimicrobial activity against *E. coli, Enterococcus faecalis, S. aureus,* and *C. albicans* [29].

AMPs also serve a barrier function in the skin. Cathelicidin related AMP, CRAMP, in mice is similar to the human cathelicidin, LL-37. This AMP is stored in lamellar bodies in the granular layer of skin keratinocytes and is then secreted between the stratum corneum and stratum granulosum [36]. Barrier dysfunction has been demonstrated during cellular assays in CRAMP-deficient mice [36]. In comparison to wild-type mice, mice deficient in murine CRAMP experience more severe necrotic lesions after subcutaneous injection with *S. pyogenes* [37].

19.4.2 Antimicrobial proteins and skin microbiota

Endogenous AMPs help our skin maintain healthy microbiota. In addition to endogenous AMPs, the commensal skin microbiota secretes bactericidal proteins that promote skin health. Bacteriocins are secreted by Gram-positive bacteria, such as *Lactococcus*, *Streptococcus* and *Streptomyces* species. These AMPs inhibit the growth of pathogenic bacterial strains and species that compete for similar resources. *S. epidermidis*, a major commensal bacterial species in the skin microbiome produces AMPs known as lantibiotics. These lantibiotics contain characteristic thioether amino acids, lanthionine and/or methyllanthionine. *S. epidermidis* also produces phenol-soluble modulin-γ and phenol-soluble modulin-δ. Similar to endogenous AMPs, these proteins result in cell membrane destruction [16,38]. A thiolactone peptide produced by *S. epidermidis* has been found to block the *S. aureus* system that controls the production of virulence factors [16,39]. The biofilms of *S. aureus* are also inhibited by specific strains of *S. epidermidis* that produce serine proteases [16,40]. Coagulase-negative staphylococci, *S. epidermidis* and *S. hominis*, have been found to produce AMPs that inhibit *S. aureus* overgrowth. When these AMPs were applied to lesional skin of atopic dermatitis (AD) patients, *S. aureus* colonization decreased [41,42]. This highlights the importance of maintaining a healthy skin microbiota for fighting off potential pathogens.

Although endogenous AMPs and AMPs produced by commensal microbes provide sophisticated and robust protection against infectious microorganisms, pathogenic bacteria have developed mechanisms to resist antimicrobial proteins. Gram-positive and Gram-negative bacteria have the ability to increase their net positive surface charge to electrostatically repel cationic AMPs, resulting in evasion of cytoplasmic membrane disruption by AMPs [37]. For example, *S. aureus* can increase the positive charge on its cell surface via D-alanylation of cell wall teichoic acid. Gram-negative bacteria decrease the net negative surface charge via the ability to modify lipopolysaccharide (LPS) [37]. Other AMP evasion mechanisms employed by pathogenic bacteria include AMP trapping via surface-associated or secreted proteins and polysaccharides, efflux mechanisms, proteolytic degradation, transcriptional regulation of bacterial AMP resistance factors, and suppression of host AMP expression [37].

19.4.3 Environmental effects on skin antimicrobial peptides

The atmosphere that our skin experiences also can alter AMPs in the skin. UV radiation is one way in which the environment alters antimicrobial expression. UV rays induce AMP production, which triggers activation of the innate immune system. UV radiation has been shown to increase the expression of AMPs, cathelicidin, β-defensin, ribonuclease 7 (RNase7), and S100A7 (psoriasin) by keratinocytes in vitro and in vivo [43,44]. The mechanism by which UV radiation induces AMPs is thought to involve the production of vitamin D, resulting in immunomodulatory effects in the skin [43].

19.4.4 Antimicrobial peptides in skin disorders

AMPs play an important function in a number of skin conditions. AD is one skin condition where alterations in AMPs may be a factor in determining disease burden. AD is a chronic skin condition that results in itchy and dry skin. Filaggrin, a protein responsible for the maintenance of the skin barrier, is deficient in many AD patients. AD skin expresses less β-defensins, dermcidin and cathelicidin than healthy skin [29,34,45]. Lack of AMP promoting cytokines such as IL-1, IL-17 and IL-22 as well as excess production of Th-2 cytokines IL-4, IL-10 and IL-13 likely contribute to the decreased AMP expression in AD [29,34]. Reduced amounts of dermcidin in AD sweat has also been reported [20]. This alteration in AMP expression likely contributes to a skin infection in patients with AD. For example, *S. aureus* invades the skin epidermis via proteolysis, as well as by the failure of the AMPs or physical skin barrier to fight infection [46]. LL-37 has also shown activity against *S. pyogenes* and has the ability to work synergistically with β-defensins against *S. aureus* [33]. The presence of *S. aureus* in AD skin is correlated with decreased LL-37, hBD-2, hBD-3, and dermcidin, supporting the notion that lack of AMPs contribute to *S. aureus* infection [34]. The use of vitamin D in AD increases cathelicidin expression and improve outcomes in AD [46].

Like AD, low levels of AMP expression and filaggrin expression is seen in cutaneous T cell lymphoma (CTCL). Similar to patients with AD, patients with CTCL frequently suffer from infection with *S. aureus*. Decreased levels of AMPs and filaggrin serves as a plausible mechanism for increased infection and dry skin secondary to barrier dysfunction in CTCL [47]. Indeed, altered expression of AMPs and barrier function are closely linked. In mouse studies, disruption of the skin barrier via tape stripping, acetone treatment, and essential fatty acid-deficient diet resulted in an increase of AMPs expression. In contrast, the use of occlusive dressings significantly reduced AMP expression [34]. Further, hBD-3 was also found to contribute to skin barrier maintenance via increased expression of tight junction proteins [34].

AMPs are also implicated in rosacea, a chronic skin condition characterized by facial erythema, papules, pustules, and telangiectasias. In contrast to AD, patients with rosacea secrete elevated levels of cathelicidin [45]. Abnormal TLR2 or polymorphisms in the vitamin D receptor may lead to increased transcription of cathelicidin [16]. Cathelicidin has also been shown to promote angiogenesis and thus could explain the redness and increased telangiectasias seen in rosacea.

Psoriasis patients upregulate a number of AMPs. In fact, many AMPs like hBD-2, hBD-3 and psoriasin were discovered in psoriasis skin [33,45]. Psoriasis is a skin condition that results in skin hyperproliferation. Interestingly, psoriasis is associated with reduced incidence of superinfections of the skin compared to other chronic inflammatory skin conditions, possibly related to upregulation of AMPs. It has also been shown that the cathelicidin LL-37 may contribute to psoriasis pathogenesis. In psoriasis, LL-37 binds to self-DNA in the dermis and triggers cutaneous inflammation via activation of dermal plasmacytoid dendritic cells. In contrast to its proinflammatory role in the dermis, in the epidermis, LL-37 binds cytosolic DNA in keratinocytes, blocking the proinflammatory activity of cytosolic DNA. This possibly explains why vitamin D reduces inflammation in psoriatic lesions [29].

AMPs also contribute to the pathogenesis of acne. In the most common type of acne, acne vulgaris, *Cutibacterium acnes* (formerly *Propionibacterium acnes*) induces the expression of AMPs. hBD-2 and IL-8 are increased by *C. acnes* via TLR-2 and TLR-4 activation. Increased hBD-2 expression in the pilosebaceous unit acts as a chemoattractant for inflammatory cells promoting the inflammation and cyst formation seen in acne. Psoriasin and LL-37 are also upregulated in the pilosebaceous units of acne vulgaris patients and may contribute to killing *C. acnes* [29]. Thus, AMPs may contribute to the pathogenesis of acne vulgaris by increasing inflammation and through direct effects on the microbiome of the pilosebaceous unit.

AMPs expression is also altered in hidradenitis suppurativa (HS). The expression of LL-37, hBD-2, psoriasin, and calprotectin are increased in HS.

LL-37 is thought to lead to increased inflammatory activity in HS [29,48]. Unlike other AMPs, dermcidin is downregulated in HS. This is possibly due to decreased sweat secretion in HS or the result of an overall decreased number of eccrine sweat glands in the condition [49]. Although AMPs have shown altered expression in HS, the role that they play in the pathogenesis of the disorder is not well understood.

19.4.5 Antimicrobial peptides in regulation of epithelial damage

In addition to their antimicrobial functions, AMPs participate in wound healing after epithelial damage. Many AMPs are increased at sites of skin injury. For example, hBD-2 and LL-37 are induced at sites of skin injury via EGFR activation [34,50]. hBD-3 is also upregulated during times of wound healing via growth factors that are important for wound repair. These AMPs act in various ways to aid in the healing process. First, AMPs directly kill microbes, and increased microbial colonization has been shown to delay wound healing. Second, AMPs have the ability to bind and inactivate proinflammatory molecules, such as LPS. Finally, AMPs promote cellular migration and proliferation, which is crucial for skin restoration [50].

LL-37 has also exhibited the ability to induce angiogenesis, which is important in tissue repair. LL-37 inhibitory antibodies have been found to impede postwound reepithelization in a concentration-dependent manner [33]. Additionally, chronically wounded skin shows a reduced expression of AMPs, highlighting their critical role in wound healing. Cathelicidin expression is low in chronic ulcers, while expression of hBD-2 is drastically decreased in burn wounds [33].

Interestingly, the cathelicidin gene, *CAMP*, is under transcriptional control of a vitamin D response element. After experiencing an insult or infection, 25-OH vitamin D3 is hydroxylated to 1,25 OH vitamin D3, the active form of vitamin D, via cytochrome p450 enzyme cyp27B1 [16]. This reaction is mediated via activation of TLR2 and local cytokines such as TNF or type I IFNs. Increased vitamin D is associated with a reduction in diabetic foot ulcers, as well as decreased postoperative surgical site infections, signifying an important role for vitamin D in skin restoration, protection and maintenance [51,52].

19.4.6 Emerging skin antimicrobial proteins

In recent years many other skin proteins have shown antimicrobial properties. Proteins in the resistin family can directly kill bacteria via membrane disruption [53,54]. Mouse skin expresses two resistin family proteins, including resistin and resistin-like molecule α (RELMα). Human skin expresses the protein resistin at the skin surface [54]. Bacterial colonization, vitamin A and vitamin A analogs all trigger the expression of RELMα in mouse skin.

Additionally, sebocytes treated with vitamin A showed increased human resistin expression in vitro. Mouse skin devoid of RELMα is also more susceptible to *S. pyogenes* skin infection [54].

Filaggrin has also exhibited antimicrobial properties, preferentially against *P. aeruginosa* [55]. The C-terminus of filaggrin-2 likely exhibits antimicrobial properties via membrane blebbing, not pore formation. Filaggrin peptide fragments may also interfere with bacterial replication by impairing DNA polymerase [55]. This suggests that in addition to filaggrin's importance in the preservation of the skin barrier, it may also play a role in preventing skin infections.

Recent data also indicate that cationic and intrinsically disordered proteins expressed throughout the epidermis have direct antimicrobial activity. Cationic intrinsically disordered antimicrobial peptides (CIDAMPs) [56] have a high percentage of disorder-promoting amino acids and a low percentage of order-promoting amino acids and thus generate cationic linear intrinsically disordered peptides of variable lengths. Hornerin (HRNR) is a protein that generates CIDAMPs in the skin. HRNR derived CIDAMPs have shown activity against *P. aeruginosa*. S-palmitoylation of these fragments increase their activity toward *S. aureus*. Synthetically derived CIDAMPs have also shown microbicidal activity, suggesting a therapeutic role for these peptides in the future [56].

19.5 Antimicrobial peptides in the intestine

The mammalian intestines are colonized by a staggering number of bacteria, fungi, protozoa, and viruses, which we collectively term the microbiome. Unlike the skin, the intestinal epithelium is composed of a single layer of live epithelial cells that covers the intestinal connective tissue, the lamia propria. As these epithelial cells express a wide variety of microbial sensors that can go on to activate an immune response, there is a need for proper separation between the microbiome and the epithelium [57]. AMPs that are secreted by intestinal epithelial cells diffuse into the intestinal mucus layer to create a chemical "minefield" that maintains this barrier. The intestine is also a prime invasion site for foodborne pathogens. AMPs prevent these pathogens from invading by killing and sequestering them as they travel from the intestinal lumen to the tissue itself [16].

19.5.1 α-Defensins in the intestine

Only two members of the alpha-defensin family are known to be expressed in the human gut, HD5 and HD6. HD5 and HD6 expression are specifc to the small intestinal Paneth cells. They are secreted as inactive propeptides and then activated via trypsin proteolysis in the intestinal lumen [16]. As these AMPs are not produced by mice, their importance for host defense was not clear until transgenic mouse models were developed. Mice that are genetically engineered to express HD5 are resistant to intestinal infection by the enteric pathogen *Salmonella enterica* serovar Typhimurium (*S. typhimurium*). However, these mice remained susceptible to systemic infection, thus highlighting the tissue-specific role of HD5 [58]. While HD5 directly kills bacteria, HD6 was found to sequester bacteria via the formation of long fibrils. In a transgenic mouse model, expression of HD6 also limited intestinal infection by *S. typhimurium* [21]. Notably, HD6 also limits the growth of other pathogens such as *Yersinia enterocolitica* and *L. monocytogenes* in vitro [59].

19.5.2 β-Defensins in the intestine

Unlike α-defensins, β-defensins are expressed by enterocytes in the intestinal epithelium and the colon. Of the 28 β-defensins encoded by the human genome, eight are expressed in the intestine [6]. These exhibit a wide range of antimicrobial activity in vitro against a variety of enteric pathogens. Their specificity to the intestine is exemplified in their activity under the unique physiological conditions of the intestine. The intestinal environment is largely anaerobic and acidic. The human β-defensin hBD-1 has negligible bacteriocidic activity under aerobic conditions. However, under anaerobic conditions, it undergoes a conformational shift and becomes a potent inhibitor of Gram-positive bacteria [60]. Mice express four β-defensins that exhibit a wide range of antimicrobial activity in vitro.

19.5.3 Cathelicidins in the intestine

Cathelicidins are expressed by the epithelium in both the small intestine and the colon. The expression of cathelicidin is upregulated during enteric infection [61]. When the human cathelicidin hCAP18 is cleaved, it produces the active LL-37 peptide, which is crucial for the prevention of enterohemorrhagic *E. coli* (EHEC) infection [62]. In the intestine, cathelicidins also limit infection by *Shigella* and other Gram-negative and -positive bacteria, as well as fungi [62].

19.5.4 Antimicrobial peptides and the intestinal microbiome

An intricate relationship exists between the host and its microbiome. The microbiome often regulates the production and secretion of AMPs by the host, and in turn, the AMPs shape the intestinal microbiome. As these are complicated feedback loops, investigators are just starting to unravel this ecosystem [63]. An example of the interplay between host-derived AMPs and the microbiome is seen in the effect of HD5 on the microbiome. Transgenic mice expressing HD5 harbor an altered microbiome, characterized by reduced numbers of segmented filamentous bacteria (SFB), a commensal from the *Clostridiaceae* family [64]. These SFBs are immunomodulatory bacteria, as they affect host levels of T helper 17 cells (Th17) [65]. As the expression of defensins is affected by IL-17, which is produced by Th17 cells, this regulation of SFB levels by HD5 then affects its own expression. Thus, HD5 expression creates a negative feedback loop via intestinal bacteria (Fig. 19.2).

The microbiome produces a variety of short-chain fatty acids (SCFAs), which are absorbed by the host in the colon and regulate immune function. SCFA have been shown to upregulate cathelicidin expression [63]. The enteric pathogen *Shigella*, on the other hand, downregulates the expression of cathelicidin in the colon. Administering butyrate, a microbiome-produced SCFA, to infected mice reduces *Shigella* pathogenicity by increasing cathelicidin production in the colon [66]. Thus, the microbiome can control AMP

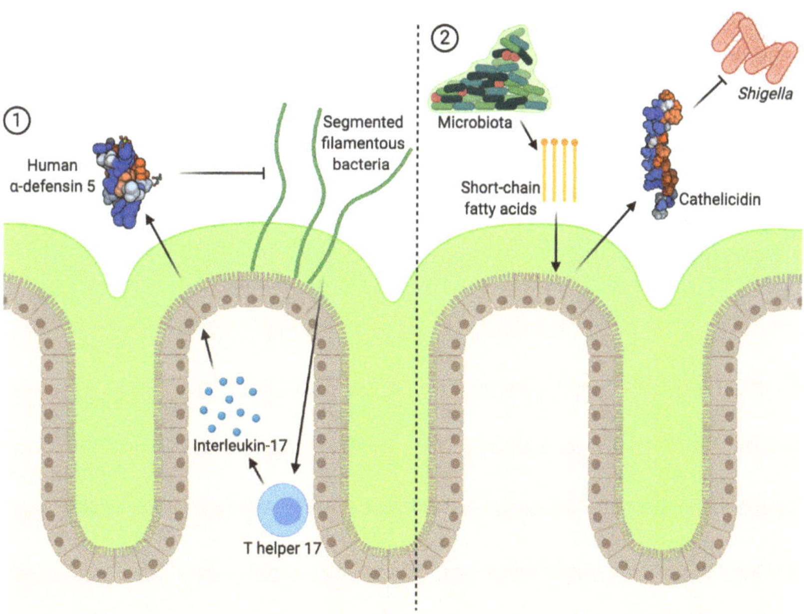

FIGURE 19.2 Crosstalk between AMPs and the microbiome. (1) When expressed in mice, human α-defensin 5 targets segmented filamentous bacteria (SFB). These SFB induce T helper 17 cell differentiation and release of interleukin-17 (IL-17). IL-17 then signals intestinal epithelial cells to produce more AMPs. Thus, human α-defensin 5 expression creates a negative feedback loop via intestinal bacteria. (2) Short-chain fatty acids (SCFAs) are produced by the microbiota in the colon. These SCFAs have immunomodulatory properties that induce the expression of cathelicidin that protects the host from Shigella infection. Host tissue is shown in shades of brown and mucus in green.

function via the production of immunomodulatory molecules that aid the host by limiting bacterial invasion and pathogenesis.

19.5.5 Antimicrobial peptides in inflammatory bowel diseases

Inflammatory bowel diseases (IBD), including Crohn's disease and ulcerative colitis, are persistent illnesses characterized by chronic inflammation in the intestine(s). Though research points to a link between genetics and environment, the pathogenesis of these conditions are poorly characterized and there is currently no lasting cure. At the base of IBD is an inappropriate immune response to the microbiome [3].

Clinical evidence indicates that AMP production is reduced in patients with IBD. In IBD patients, bacteria can penetrate the intestinal mucus layer and cause inflammation [67]. Additionally, IBD patients have large numbers of bacteria associated with their intestinal epithelial surfaces, suggesting a breakdown of the intestinal barrier related in part to AMPs [23]. There are several hypothesized reasons for this reduction in AMP secretion. First, mutations in the bacterial sensor NOD2 and the transcription factor TCF4, which have been linked to IBD, both lead to reduced production of AMP by the intestinal epithelium [23]. Additionally, dysfunction of Paneth cells, the major producers of AMPs in the intestine, are a hallmark of Crohn's disease. Mutations in the autophagy gene ATG16L1 are a risk allele for Crohn's disease, and when cloned into transgenic mice, these mutations hamper AMP secretion by Paneth cells [68,69]. Thus, maintaining a functional barrier by secretion of AMPs seems to be crucial for intestinal homeostasis, and the breakdown of this barrier can lead to IBD development.

19.6 Therapeutic opportunities

There are a number of therapeutic opportunities through which AMPs and the microbiome can be utilized to promote skin and intestinal health. The ability to synthetically derive AMPs have practical clinical implications. Their broad spectrum of activity under a variety of conditions make these peptides promising therapeutics. The use of topical AMPs for microbial infections has been reported with some success [34].

In the era of antibiotic resistance, AMPs serve as promising anti-infective agents [70]. In synergy studies and a small clinical series, beta-lactam antibiotics have shown the ability to sensitize methicillin-resistant *S. aureus* (MRSA) and vancomycin-resistant *Enterococcus* spp. (VRE) to killing by LL-37 and the cationic antibiotic daptomycin [37]. Further investigation into the efficacy of AMPs in eradicating infection, as well as the extent of bacterial resistance to AMPs, should be pursued. If found to be successful in eradicating infections, AMPs may serve as a novel tool in treating infectious diseases.

The maintenance of a healthy microbiome also has important implications for preserving a robust repertoire of AMPs for host defense. Topically applied probiotics have the ability to produce AMPs that may benefit the skin's immune response and eradicate pathogens. Prebiotics can be applied to the skin and selectively promote the growth and activity of commensal microorganisms. The growth of favorable skin microbes supports healthy skin and can have favorable effects on various skin disorders [71].

Vitamin D and Vitamin A increase the expression of AMPs, making these fat-soluble molecules promising treatments for certain skin conditions [54]. The use of vitamin D has proved to be promising in AD and psoriasis [29]. This has important implications for inflammatory skin conditions and wound healing. This also emphasizes the significance that environmental and dietary effects may have on AMP composition.

References

[1] Sender R, Fuchs S, Milo R. Revised estimates for the number of human and bacteria cells in the body. PLoS Biol 2016;14.

[2] Lu W. Antimicrobial peptides. Semin Cell Dev Biol 2019;88:105–6.

[3] Harris-Tryon TA, Bel S, Harris TA, Bel S. Gene/environment interaction and autoimmune disease. In: Teperino R, editor. Beyond our genes: pathophysiology of gene and environment interaction and epigenetic inheritance. Springer International Publishing; 2020, p. 139–56. Available from: http://doi.org/10.1007/978-3-030-35213-4_8.

[4] Lee EY, Lee MW, Wong GCL. Modulation of toll-like receptor signaling by antimicrobial peptides. Semin Cell Dev Biol 2019;88:173–84.

[5] Shartouny JR, Jacob J. Mining the tree of life: Host defense peptides as antiviral therapeutics. Semin Cell Dev Biol 2019;88:147–55.

[6] Chung LK, Raffatellu MGI. Pros: antimicrobial defense in the gastrointestinal tract. Semin Cell Dev Biol 88;129–137.

[7] Chen X, et al. Synergistic effect of antibacterial agents human beta-defensins, cathelicidin LL-37 and lysozyme against *Staphylococcus aureus* and *Escherichia coli*. J Dermatol Sci 2005;40.

[8] Herman A, Herman AP. Antimicrobial peptides activity in the skin. Skin Res Technol 2019;25.

[9] Luo X-L, et al. LL37 Inhibits *Aspergillus fumigatus* infection via directly binding to the fungus and preventing excessive inflammation. Front Immunol 2019;10:283.

[10] Howell MD, et al. Selective killing of vaccinia virus by LL-37: implications for eczema vaccinatum. J Immunol 2004;172:1763–7.

[11] Rico-Mata R, De Leon-Rodriguez LM, Avila EE. Effect of antimicrobial peptides derived from human cathelicidin LL-37 on *Entamoeba histolytica* trophozoites. Exp Parasitol 2013;133:300–6.

[12] Paulmann M, et al. Structure-activity analysis of the dermcidin-derived peptide DCD-1L, an anionic antimicrobial peptide present in human sweat. J. Biol Chem 2012;287.

[13] Chairatana P, Chiang I-L, Nolan EM. Human α-defensin 6 self-assembly prevents adhesion and suppresses virulence traits of *Candida albicans*. Biochemistry 2017;56.

[14] Sass V, et al. Human beta-defensin 3 inhibits cell wall biosynthesis in Staphylococci. Infect Immun 2010;78:2793–800.

[15] Yadava P, Zhang C, Sun J, Hughes JA. Antimicrobial activities of human beta-defensins against bacillus species. Int J Antimicrob Agents 2006;28.
[16] Gallo RL, Hooper LV. Epithelial antimicrobial defence of the skin and intestine. Nat Rev Immunol 2012;12:503–16.
[17] Harten RM Van, Woudenbergh E Van, Dijk A Van, Haagsman HP. Cathelicidins: immunomodulatory antimicrobials. Vaccines 2018;6:63.
[18] Song C, et al. Crystal structure and functional mechanism of a human antimicrobial membrane channel. Biophys Comput Biol 2013;110:4586–91.
[19] Schittek B. The multiple facets of dermcidin in cell survival and host defense. J Innate Immun 2012;4:349–60.
[20] Kenshi Y, Richard G. Antimicrobial peptides. Hum Skin Dis 2009;18:11–21.
[21] Chu H, et al. Human α-defensin 6 promotes mucosal innate immunity through self-assembled peptide nanonets. Sci (80-) 2012. Available from: https://doi.org/10.1126/science.1218831.
[22] Semple F, Dorin JR. β-Defensins: multifunctional modulators of infection, inflammation and more? J Innate Immun 2012;4:337–48.
[23] Mukherjee S, Hooper LV. Antimicrobial defense of the intestine. Immunity 2015. Available from: https://doi.org/10.1016/j.immuni.2014.12.028.
[24] Ragland SA, Criss AK. From bacterial killing to immune modulation: recent insights into the functions of lysozyme. PLoS Pathog 2017. Available from: https://doi.org/10.1371/journal.ppat.1006512.
[25] Dore E, Boilard E. Roles of secreted phospholipase A 2 group IIA in inflammation and host defense. Biochimica et Biophysica Acta - Mol Cell Biol Lipids 2019. Available from: https://doi.org/10.1016/j.bbalip.2018.08.017.
[26] Hooper LV, Stappenbeck TS, Hong CV, Gordon JI. Angiogenins: a new class of microbicidal proteins involved in innate immunity. Nat Immunol 2003. Available from: https://doi.org/10.1038/ni888.
[27] Deriu E, et al. Probiotic bacteria reduce salmonella typhimurium intestinal colonization by competing for iron. Cell Host Microbe 2013. Available from: https://doi.org/10.1016/j.chom.2013.06.007.
[28] Kosciuzukm EM, et al. Cathelicidins: family of antimicrobial peptides. A review. Mol Biol Rep 2012;10957–70. Available from: https://doi.org/10.1007/s11033-012-1997-x.
[29] Marcinkiewicz M, Majewski S. The role of antimicrobial peptides in chronic inflammatory skin diseases. Adv Dermatol Allergol 2016;33:6–12.
[30] Chakraborty K, Maity PC, Sil AK, Takeda Y, Das S. cAMP stringently regulates human cathelicidin antimicrobial peptide expression in the mucosal epithelial cells by activating cAMP-response element-binding protein, AP-1, and inducible cAMP early repressor. J Biol Chem 2009;284:21810–27.
[31] Senyurek I, et al. Dermcidin-derived peptides show a different mode of action than the cathelicidin LL-37 against *Staphylococcus aureus*. Antimicrob Agents Chemother 2009;53:2499–509.
[32] Schittek B, et al. Dermcidin: a novel human antibiotic peptide secreted by sweat glands. Nat Immunol 2001;2:1133–7.
[33] Braff MH, Bardan A, Nizet V, Gallo RL. Cutaneous defense mechanisms by antimicrobial peptides. Soc Investig Dermatol 2004.
[34] Herman A, Herman AP. Antimicrobial peptides activity in the skin. Ski Res Technol 2019;25:111–17.
[35] Aranow C. Vitamin D and the immune system. J Investig Med 2014;59:881–6.

[36] Natsuga K. Epidermal barriers. Cold Spring Harb Perspect Med 2014;4:a018218 –a018218.
[37] Cole JN, et al. Bacterial evasion of host antimicrobial peptide defenses. Microbiol Spectr 2016;4.
[38] Cogen AL, et al. Selective antimicrobial action is provided by phenol-soluble modulins derived from staphylococcus epidermidis, a normal resident of the skin. J Invest Dermatol 2010;130:192–200.
[39] Otto M, Sussuth R, Vuong C, Gunther J, Gotz F. Inhibition of virulence factor expression in Staphylococcus aureus by the *Staphylococcus epidermidis* agr pheromone and derivatives. FEBS Lett 1999;450:257–62.
[40] Iwase T, et al. *Staphylococcus epidermidis* Esp inhibitors *Staphylococcus aureus* biofilm formation nasal colonization. Nature 2010;465.
[41] Steiman C, Gern J. Antimicrobials from human skin commensal bacteria protect against *Staphylococcus aureus* and are deficient in atopic dermatitis. Pediatrics 2017;140:2015–18.
[42] Nakatsuji T, et al. Antimicrobials from human skin commensal bacteria protect against *Staphylococcus aureus* and are deficient in atopic dermatitis. Sci Transl Med 2017;9:eaah4680.
[43] Patra V, Byrne SN, Wolf P. The skin microbiome: is it affected by UV-induced immune suppression? Front Microbiol 2016;7:1235.
[44] Patra V, Wagner K, Arulampalam V, Wolf P. Skin microbiome modulates the effect of ultraviolet radiation on cellular response and immune function. iScience 2019;15:211–22.
[45] Brandwein M, Bentwich Z, Steinberg D. Endogenous antimicrobial peptide expression in response to bacterial epidermal colonization. Front Immunol 2017;8:1637.
[46] Nakatsuji T, et al. *Staphylococcus aureus* exploits epidermal barrier defects in atopic dermatitis to trigger cytokine expression. J Invest Dermatol 2016;136:2192–200.
[47] Suga H, Sugaya M, Miyagaki T, Ohmatsu H, Kawaguchi M. Skin barrier dysfunction and low antimicrobial peptide expression in cutaneous T-cell lymphoma. Clin Cancer Res 2014. Available from: https://doi.org/10.1158/1078-0432.CCR-14-0077.
[48] Almansouri D, Zouboulis CC. Review article calprotectin functions in the skin: an overview. Hong Kong J Dermatol Venereol 2017;25:115–21.
[49] Coates M, et al. The skin transcriptome in hidradenitis suppurativa uncovers an antimicrobial and sweat gland gene signature which has distinct overlap with wounded skin. PLoS One 2019;14:e0216249.
[50] Mangoni ML, McDermott AM, Zasloff M. Antimicrobial peptides and wound healing: biological and therapeutic considerations. Exp Dermatol 2016;25:167–73.
[51] Dai J, Jiang C, Chen H, Chai Y. Vitamin D and diabetic foot ulcer: a systematic review and meta-analysis. Nutr Diabetes 2019. Available from: https://doi.org/10.1038/s41387-019-0078-9.
[52] Abdehgah AG, Monshizadeh A, Tehrani MM, Afhami S. ScienceDirect relationship between preoperative 25-hydroxy vitamin D and surgical site infection. J Surg Res 2019;245:338–43.
[53] Propheter DC, Chara AL, Harris TA, Ruhn KA, Hooper LV. Resistin-like molecule beta is a bactericidal protein that promotes spatial segregation of the microbiota and the colonic epithelium. Proc Natl Acad Sci USA 2017. Available from: https://doi.org/10.1073/pnas.1711395114.
[54] Harris TA, et al. Resistin-like molecule a provides vitamin-A-dependent antimicrobial protection in the skin article resistin-like molecule a provides vitamin-A-dependent antimicrobial. Cell Host Microbe 2019;25:1–12.
[55] Hansmann B, Schröder J, Gerstel U. Skin-derived C-terminal filaggrin-2 fragm are *Pseudomonas aeruginosa*. Dir Antimicr Target Bacter Repl 2015;1–21. Available from: https://doi.org/10.1371/journal.ppat.1005159.

[56] Latendorf T, et al. Cationic intrinsically disordered antimicrobial peptides (CIDAMPs) represent a new paradigam of innate defense with a potential for novel anti-infectives. Sci Rep 2019;1–25. Available from: https://doi.org/10.1038/s41598-019-39219-w.

[57] Hooper LV, Littman DR, Macpherson AJ. Interactions between the microbiota and the immune system. Sci (80-) 2012;336:1268–73.

[58] Salzman NH, Ghosh D, Huttner KM, Paterson Y, Bevins CL. Protection against enteric salmonellosis in transgenic mice expressing a human intestinal defensin. Nature 2003;422:522–6.

[59] Chairatana P, Nolan EM. Molecular basis for self-assembly of a human host-defense peptide that entraps bacterial pathogens. J Am Chem Soc 2014. Available from: https://doi.org/10.1021/ja5057906.

[60] Schroeder BO, et al. Reduction of disulphide bonds unmasks potent antimicrobial activity of human β 2-defensin 1. Nature 2011. Available from: https://doi.org/10.1038/nature09674.

[61] Hase K, Eckmann L, Leopard J, N. V-I. undefined. Cell differentiation is a key determinant of cathelicidin LL-37/human cationic antimicrobial protein 18 expression by human colon epithelium. Am Soc Microbiol 2002.

[62] Chromek M, Arvidsson I, Karpman D. The antimicrobial peptide cathelicidin protects mice from *Escherichia coli* O157: H7-mediated disease; 2012. ncbi.nlm.nih.gov.

[63] Thaiss CA, Zmora N, Levy M, Elinav E. The microbiome and innate immunity. Nature 2016;535:65–74.

[64] Salzman NH, et al. Enteric defensins are essential regulators of intestinal microbial ecology. Nat Immunol 2010. Available from: https://doi.org/10.1038/ni.1825.

[65] Bel S, Hooper LV. Immunology: a bacterial nudge to T-cell function. Nature 2015. Available from: https://doi.org/10.1038/526328a.

[66] Raqib R, et al. Improved outcome in shigellosis associated with butyrate induction of an endogenous peptide antibiotic. Proc Natl Acad Sci USA 2006. Available from: https://doi.org/10.1073/pnas.0602888103.

[67] van der Post S, et al. Structural weakening of the colonic mucus barrier is an early event in ulcerative colitis pathogenesis. Gut gutjnl- 2019;2018–317571. Available from: https://doi.org/10.1136/gutjnl-2018-317571.

[68] Bel S, et al. Paneth cells secrete lysozyme via secretory autophagy during bacterial infection of the intestine. Science 2017;357(6355):1047–52. Available from: https://doi.org/10.1126/science.aal4677.

[69] Bel S, Hooper LV. Secretory autophagy of lysozyme in paneth cells. Autophagy 2018;14:719–21.

[70] Mahlapuu M, Håkansson J, Ringstad L, Björn C. Antimicrobial peptides: an emerg category therapeutic agents. 6, 1–12 (2016).

[71] Al-Ghazzewi FH, Tester RF. Impact of prebiotics and probiotics on skin health. Benef Microbes 2014. Available from: https://doi.org/10.3920/BM2013.0040.

Chapter 20

Peptide and peptidomimetic-based vaccines

Abhishek Singh[1,2] and Mukesh Thakur[1]
[1]*Zoological Survey of India, New Alipore, Kolkata, West Bengal, India,* [2]*Department of Forensic Biotechnology, National Forensic Sciences University, Gandhinagar, Gujarat, India*

20.1 Vaccines

The human immune system has an extraordinary ability to remember the encounter of any pathogen or antigen, making it a highly complex and effective biological system. This complexity and extent of effectiveness have long been exploited since the discovery of the first vaccine in 1796 [1]. Vaccines are biological products that can be used to boost or induce the immune response against subsequent exposure to any pathogen or disease [2]. Each pathogen consists of a unique genetic makeup and thus the immune response against the pathogen also differs in its extent and severity. For vaccines to effectively induce defense against the pathogen, they must contain components either from the pathogen itself or synthetically produced antigens representing the pathogen [2]. Therefore, vaccines are designed and produced considering the antigenic efficacy of the antigen targeted to induce the immune response. Earlier, vaccines were classified based on live (attenuated) and nonlive (inactivated) pathogens [3]. Vaccines based on live pathogens have the potential to produce both cellular and humoral immune responses with high immunogenicity but are unstable due to their dependence on consistent cold temperatures [4]. They also pose a risk of severe complications in an immune-compromised individual due to their ability to undergo secondary mutation. Vaccines based on killed or inactivated pathogens are stable but generally produce a weaker immune response [5]. Currently, most vaccines contain protein antigens; however, vaccines based on polysaccharides have also proven effective for the cases of pneumonia and meningitis [2]. Over the past few decades, several platforms based on nucleic acids and virus-like particles have been developed for vaccine production. Vaccine formulation is one of the major factors that determine the use, storage, and

Peptide and Peptidomimetic Therapeutics. DOI: https://doi.org/10.1016/B978-0-12-820141-1.00008-X

administration of vaccines. Furthermore, the safety of vaccines should always be paramount in light of the preventative nature of vaccine use.

20.2 Vaccine formats

Based on the type of antigen used for the preparation of vaccines, four types of vaccines—that is, live attenuated vaccine (LAV), inactivated (killed antigen), subunit (purified antigen), and toxoid (inactivated toxins)—have been recognized globally considering their efficiency and safety [6] (Fig. 20.1). In a broader sense, vaccines are also classified as mono/polyvalent and combination based on the number of strains and the antigens. Vaccines with a single strain of single-antigen are called monovalent vaccines, whereas those having two or more strains of the same antigen are called polyvalent vaccines. Combination vaccines contain multiple strains of pathogen causing similar diseases [7].

20.2.1 Live attenuated vaccines

LAVs contain disease-causing pathogens which have been weakened to reduce their pathogenicity. On vaccination, they grow inside the individual but due to their weakened nature, they are unable to cause disease or any

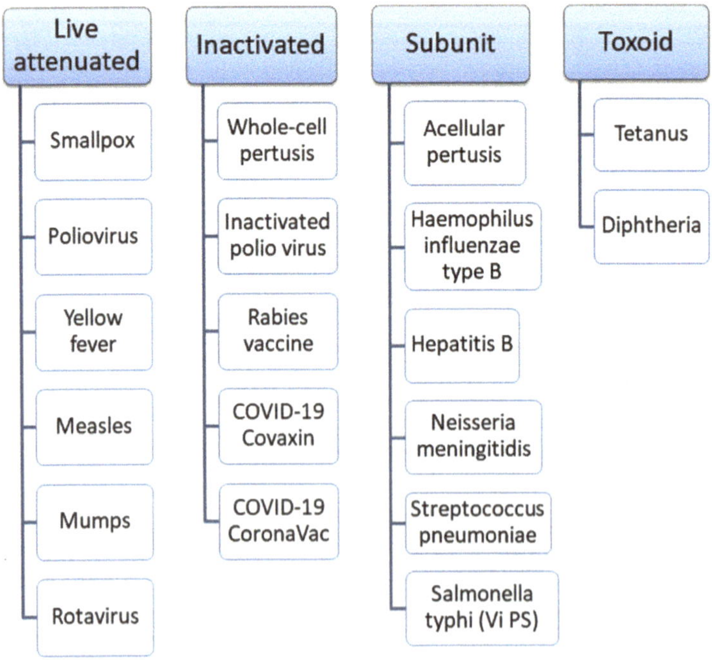

FIGURE 20.1 Format of vaccines developed against their respective pathogens.

serious discomfort. Since LAVs are derived from live pathogens, they stimulate the immune response as well as wild-type pathogens, which provides sufficient time for the immune system to produce the memory cells [8]. Usually, LAVs do not undergo any reversion to their pathogenic form except in cases of vaccine-associated paralytic poliomyelitis and vaccine-derived poliovirus [9]; however, some exceptions pose safety risks. Another major concern with LAVs is the requirement of cold chains to sustain their effectiveness. One of the most widely used LAVs derived from bacteria for tuberculosis disease is Bacillus Calmette–Guérin vaccine [10]. Several other LAVs which are derived from viruses are oral polio vaccine, rotavirus, measles, and yellow fever.

20.2.2 Inactivated vaccines

Inactivated vaccines are derived from whole organisms which have been killed through several physical and chemical processes. These vaccines may not induce a strong immune response; hence, several doses may be required for sufficient immunogenic response. They are considered more stable than LAVs due to the absence of live organisms or their derivative [11]. Some of the most widely used inactivated vaccines are whole-cell pertussis and inactivated poliovirus.

20.2.3 Subunit vaccines

Subunit vaccines are derived from the antigenic parts of the pathogen sufficient to elicit an effective immune response [12]. The extent of the immune response depends on several factors such as the antigenic property of the components, the combination of the component used, and to what extent these components are inducing an immunological memory. Therefore, extensive research and trials are required to develop an effective subunit vaccine. Based on the type of component/fragment of pathogen used to derive the vaccines, subunit vaccines are categorized into three types, that is, protein, polysaccharide, and conjugate.

20.2.3.1 Protein

Protein-based subunit vaccines are derived from the specific protein of the pathogen with proven capability to generate an effective immune response [3]. The stability of proteins is highly dependent on factors like temperature and pH. In the case of temperature fluctuation, isolated proteins may denature and/or bind to other molecules. Some of the common protein-based subunit vaccines are acellular pertussis (aP) vaccine and hepatitis B vaccine [13].

20.2.3.2 Polysaccharide

Polysaccharide-based vaccines are specially designed for those bacteria that when invading the body of an individual are protected by a polysaccharide capsule. The polysaccharide molecules derived from bacteria are so small that they are unable to induce a strong immunogenic response and hence produce short-term immunity. Vaccines available for pneumococcal disease and meningococcal disease are polysaccharide-based vaccines [13].

20.2.3.3 Conjugate

Conjugate subunit vaccines are similar to polysaccharide-based subunit, vaccines as both are derived from the polysaccharide of the pathogen. However, conjugate vaccines require an additional carrier protein that binds to the polysaccharide and produces a long-term immunogenic response. Therefore, they can provide long term protection against bacterial disease for which normal polysaccharide vaccines fail to perform. Some of the common conjugate-based subunit vaccines are haemophilus influenzae type b (Hib), pneumococcal conjugate, and meningococcal A vaccines [13].

20.2.4 Toxoid vaccines

In certain diseases, pathogens like bacteria produce toxins that when entering the bloodstream cause specific symptoms. Toxoid vaccines use those toxins as an antigen to elicit an effective immune response. These toxins are weakly immunogenic and hence an adjuvant is required to boost the immune response. These vaccines are highly stable as they are not dependent on external factors like temperature, humidity, etc. Some of the commonly used toxoid vaccines are tetanus toxoid and diphtheria toxoid [14].

20.3 Peptides or peptidomimetics as potential immunogens

Traditional vaccines have been known for their excellent efficiency since 1796 after the discovery of the smallpox vaccine by Edward Jenner [1]. These vaccines have the potential to generate a robust immune response because they contain both B-cell and T-cell binding epitopes in the required conformation. However, several difficulties remain with the manufacturing and administration of these vaccines. To deal with these difficulties, immunotherapeutic techniques have been developed in the last few decades for alternative approaches to immunization.

The most acceptable approach is to find a peptide epitope with the capability to induce the required immune response and produce the epitope synthetically without the use of the pathogen. The epitopes are selected considering several parameters to avoid side effects and since the peptides are synthetically made, they do not undergo mutation leading to pathogenic effects. Synthetic peptides are also highly stable, as they are produced considering external factors that may hamper their function. Antigens derived from natural sources have a limitation in the production capacity but synthetically made peptides may be produced according to the demand [15]. For a peptide to effectively induce an immune response, it must be presented by major histocompatibility complex (MHC) molecules on antigen-presenting cells (APCs), which are recognized by the T-cells. The source of antigen/peptide predominantly determines the type of MHC-antigen complex, that is, MHC class-I and MHC class-II. MHC class-I induces a cytotoxic T-cell response on the encounter of proteins or peptides which are derived from proteins inside the host cell (e.g., viral proteins), whereas MHC class-II deals with exogenous proteins/peptides (e.g., bacterial excretions). This further induces CD4 T-helper cell responses which in turn elicit a humoral or cell-mediated immune response [4]. In this context, vaccines based on peptides deliver specific epitopes to induce the desired immune response, thus decreasing the dependency on natural antigens.

Despite all these advantages and successful production, peptide-based vaccines have failed to produce the desired potential. One of the major factors attributing to this failure is the requirement of adjuvant to boost the immune response. The use of adjuvants in humans poses serious health risks which not only hampers the efficiency of the vaccine but may induce various health issues in immune-compromised individuals [16]. Furthermore, peptides are more susceptible to enzymatic degradations and reduced *in vivo* bioavailability. In an attempt to overcome these issues, the field of peptidomimetics was introduced in the early 1990s [17]. The process of chemical modification of a peptide to change the inherent properties while imitating the original compound is called peptidomimetics [16]. A peptidomimetic candidate vaccine is based on the modification of native peptides to decrease the degradation by peptidase, enhance the bioavailability, reduce the rate of clearance, and/or improve the transport through the blood–brain barrier (BBB) [18]. These modifications are achieved through adjustment in various factors, that is, increase in the half-life of peptides, increase in affinity with MHC molecules, and achieving the structural requirement for efficient T-cell recognition. Therefore, chemical alteration of the peptides may improve the immunogenicity and stability of the vaccines, thus providing more efficacious candidates.

One of the most widely reported and preferable approaches for the generation of peptidomimetics is the altered solid-phase peptide synthesis (SPPS) method [19]. Peptide synthesis using SPPS is carried out through automation to insert nonnatural amino acids and achieve the posttranslational modifications of the backbone. Some common examples of peptidomimetics are peptoids, peptide sulfonamides, urea peptidomimetics, D-amino acid substitutions, azapeptides and β-peptides. In the present scenario, peptidomimetics have wide applications in cancer therapy [20–22] and in the modification of drugs for HIV [23–25]. There have been many attempts to design novel peptidomimetic-based vaccines for COVID-19 caused by SARS-COV-2 [26] but to date, no such vaccine is approved for human use.

20.4 Peptide-based vaccines

Peptide-based vaccines are composed of specific epitopes of 20–30 amino acids, which are produced considering the targeted antigen of any disease. These peptides are designed in such a way that each of them is recognized as an invading pathogen concerning the targeted disease by the pattern recognition receptors (PRRs). The immune system is equipped with APCs like dendritic cells and macrophages, which recognize the invading pathogen through the PRRs and other molecules present on their surface [27]. The captured pathogens then undergo phagocytosis or receptor-mediated endocytosis, where the pathogens are expressed as small fragments or peptides to their MHC complex. The peptide MHC complex expresses the epitopes as an antigen to the T-cell for generating an effective humoral or cell-mediated immune response [28,29]. Therefore, the pathogens are represented to our immune system in small fragments or as peptides. This clearly defines the reason behind the use of specific peptides instead of the whole organism in the development of peptide vaccines. The use of specific immunoactive peptides also reduces the generation of antibodies against ineffective epitopes, which increases the efficiency of the immunogenic response upon vaccination, and can even be used to reduce side effects in certain cases. The immune active peptides representing the antigen of the target organism are designed using in silico methods following chemical synthesis of peptides and conjugation with the carrier protein. Constructs are designed and tested based on their immunogenicity, antigen specificity, and extent of efficacy in animals and humans (Table 20.1).

20.4.1 Epitope selection

Epitope selection is a crucial step in vaccine design and development [30]. Epitopes are scrutinized on target proteins using several computational approaches. The primary approach is to retrieve the structural and sequence information of the target protein and assess the antigenicity of the protein using in silico tools. Commonly, two major epitopes, for B-cell and CD8 T-cell interaction, are identified, as vaccines having B-cell epitopes will act against the extracellular pathogen, and CD8 T-cell epitopes will act against intracellular pathogens [4]. Since B-cells constitute humoral immunity and T-cells are responsible for cellular immunity, identification of both B-cell and T-cell epitopes is often necessary for an effective vaccine.

The selection of the type of immune cell epitopes also depends on the type of pathogen. B-cell epitopes are required to maintain their native

TABLE 20.1 Peptide-based vaccines in advanced stages of development.

Company	Phase	Vaccine	Comments
Vaxxinity	II	UB-311	Alzheimer's Disease
Araclon Biotech	II	ABvac 40	Alzheimer's Disease
GemVax & KAEL Bio	II	GV1001	Alzheimer's Disease
M.D. Anderson Cancer Center	III	Proteinase 1 (PR1)	Leukemia
Bionor Immuno	II/III	Vacc-4x	HIV p24 peptide
Immune Response BioPharma	II/III	NeuroVax	Multiple Sclerosis
Vaxxinity	II/III	UB-612	COVID-19
Federal Budgetary Research Institution	Complete	EpiVacCorona	COVID-19

conformation for effectively inducing humoral immunity and are hence called conformational epitopes [15]. For example, the B-cell epitope derived from the APR-1 protein of *Necator americanus* failed to induce APR-1 selective antibodies due to the absence of native β-sheet conformation [31]. The length of the peptide epitopes could vary significantly; longer peptides are incorporated in peptide-based vaccines to maintain their native conformation. Generally, peptides of shorter length are found more antigenic and hence these short peptide epitopes could be flanked with sequences inducing the desired secondary structure. Good and colleagues introduced the sequences of yeast GCN4 protein to flank the C- and N-termini of B-cell epitopes to achieve the desired conformation [32]. While selecting epitopes, proteolytic degradation of the peptides is one of the major concerns in the development of the peptide-based vaccine. To resolve this issue, artificial chemical bonds were introduced between the side chains of the amino acids lying on the same face of the peptide helix to form a stapled peptide. These modifications are obtained by several chemical processes, such as cysteine cross-linking, hydrocarbon stapling, cycloaddition reaction, double-click approach, etc. [33]. This chemical modification not only helps the peptides to maintain their secondary structure but also protects them from proteolytic degradation. Peptides are examined carefully after the modification for any change in their immunological properties [34].

In contrast, to induce cellular immunity, CD8 T-cell epitopes are not required to maintain their native conformation; the processing of longer peptides into shorter peptides is crucial for effective functioning [15]. Therefore, no modification is necessary to maintain their conformational structure; hence computational prediction of final CD8 T-cell epitopes is relatively easy.

The only difficulty with CD8 T-cell epitopes is that they might lose their immunogenicity upon conjugation to a delivery system [15]. Hence, careful examination is required while assessing CD8 T-cell immunogenicity.

In addition to B-cells and CD8 T-cell, CD4 T-helper cells also play crucial roles in immunity. Vaccines with only B-cell and CD8 T-cell can induce an immune response, but inducing T-helper cells enhances the memory response across heterogeneous populations [35]. Usually, T-helper epitopes are derived from the targeted protein of the pathogen (disease-specific), but there is another class of synthetic T-helper epitopes, that is, the Pan-Dr helper T-lymphocyte epitope (PADRE) [36]. PADRE (AKFVAAWTLKAAA) is a synthetic peptide that is specially designed to bind to most of the human HLA-Dr receptors and provide a universal immune response in heterogeneous populations. Therefore, an effective peptide-based vaccine should ideally consist of B-cell, CD8 T-cell, and CD4 T-helper cell epitopes.

20.4.2 Characterization of epitopes

Assessment of interaction between selected epitopes and MHC proteins is a crucial step to understand the efficiency of vaccines in diverse populations. It has been long known that MHCs have a specific signature for a particular population and since MHCs are encoded by HLA genes, HLA genes differ from population to population [37]. This clearly explains the reason why an epitope recognized by specific MHCs of a particular population might not be recognized by the MHC of another population. Therefore, it is essential to assess the population coverage of selected epitopes based on the interaction analysis with specific and universal HLA types. There are several computational tools such as the Immune Epitope Database (IEDB) population coverage analysis tool, which represents the population coverage of an epitope [38]. Another important aspect of characterization is the assessment of autoimmunity induced by selected epitopes. If any of the epitopes selected show similarity with any human proteins, then the vaccine might induce an autoimmune response [39]. Therefore, using computational methods, it is essential to analyze and compare all epitopes to the human proteome.

20.4.3 Immunostimulants and vaccine delivery

20.4.3.1 Immunostimulants

Immunostimulants or adjuvants are agents used to enhance the immune response of the peptide epitopes against the targeted antigen [40]. The term adjuvant is originated from the Latin word "*adjuvare*" meaning to help [4]. Peptide epitopes possess a weaker signature for the immune system to induce the necessary immune response and thus require a stimulant [15]. A whole pathogen vaccine usually has a relatively strong ability to stimulate the immune response, but peptide-based vaccines generally require an adjuvant to enhance immunogenicity.

There are several adjuvants discovered and designed so far, but the first clue of any substance to act as an immune stimulant was given by Alexander T. Glenny and colleagues in 1926 [41]. They observed precipitation of antigens over insoluble particles of potash alum ("aluminum potassium sulfate"), which induced better antibody responses than the antigen alone. Alum was also the first adjuvant approved by the US Food and Drug Administration (FDA) for use in humans [42]. Alum is well known for its efficacy in prophylactic vaccines designed for diseases like hepatitis A, hepatitis B, meningococcal and pneumococcal infections, diphtheria-tetanus-pertussis (DTaP, Tdap), anthrax, *Haemophilus influenzae* type B, etc. [4]. However, in subsequent years due to problems in manufacturing reproducibility, alum was replaced by several other aluminum salts, such as Adju-Phos (aluminum phosphate) and Alhydrogel (aluminum hydroxide) [41].

In peptide-based vaccines, alum fails to induce cellular immunity and is thus considered a weak stimulant [43]. This poor adjuvanting potency for peptide epitopes was resolved by the conjugation of peptides to a protein carrier. The vaccine developed against group-A *streptococcus* was designed in such a way that the conserved B-cell epitopes were conjugated to the diphtheria toxin (DT) protein [44]. This conjugation of conserved epitopes with the carrier protein allowed the vaccine to induce an immune response with alum as an adjuvant. Several other adjuvants, such as the squalene-based emulsions AS03, AF03, MF59, and monophosphoryl lipid A containing AS04 have been approved by specific countries for specific diseases [45]. Some of these adjuvants were reported to induce undesired inflammatory responses, although some of the evidence is inconclusive. Most of these adjuvants were developed and tested for whole pathogen and protein-based vaccines. Hence, knowledge of adjuvanting potency for peptide-based vaccines is still very limited (Table 20.2). Therefore, many potent adjuvants such as complete Freund's adjuvant, incomplete Freund's adjuvant, and lipid A have been tested for peptide-based vaccines in animal models [46]. Despite being able to induce humoral immunity, these adjuvants proved too toxic for human application. This stimulated ongoing development in the field of adjuvant design and construction.

Among the variety of adjuvants from bacterial components to chemical compounds, modification of already known natural adjuvants to reduce their toxicity has been a reasonable approach. One of the promising examples of this approach is dephosphorylation of lipid A to monophosphoryl lipid A, which significantly reduced the toxicity of lipid A. This monophosphoryl lipid A in combination with aluminum hydroxide was approved for use in the GSK *human papillomavirus* vaccine [15]. In a recent study from 2020, aluminum hydroxide was used as an adjuvant in the EpiVacCorona vaccine designed against SARS-COV-2 [47]. This is the first peptide-based vaccine

TABLE 20.2 Adjuvants used in protein- and peptide-based vaccines.

Adjuvant	Type	Adjuvant composition	Types of vaccine	References
Aluminum hydroxide	Mineral salt	Aluminum hydroxide	Inactivated, toxoid, peptide, conjugate	[47]
Virosome	Liposome	Inactivated strain of hepatitis A virus adsorbed to the surface of liposomes; Liposomes carrying viral envelope proteins	Inactivated and protein-based	[48,49]
ISA-51	Montanide	Mineral oil with a mannide monooleate emulsifier	Protein-based	[50]

COVID-19, which was developed and evaluated in Russia and approved for human use in several countries.

20.4.3.2 *Delivery system*

Delivery systems and adjuvants are critical components of vaccine development. Both have the responsibility to protect the antigen from degradation and to stimulate an effective immune response. In addition, a delivery system has the responsibility to transport other components of the vaccine-like adjuvant to the desired tissue. Sometimes, APCs recognize the antigens but fail to identify adjuvants that induce tolerance to the antigen. Therefore, the selection and use of delivery systems based on the route of administration and the targeted pathogen is an important factor for efficient vaccine development. In some cases, the delivery systems themselves act as an adjuvant by presenting the target vaccine to APCs to boost immune response [51]. These types of delivery systems are called self-adjuvanting delivery systems [15]. Recently, Skwarczynski et al. [52] proposed a self-adjuvanting "poly (amino acids)" delivery system for peptide-based vaccines, which proved more effective and safer in comparison to the conventional delivery systems.

20.4.3.2.1 Route of administration

Usually, vaccines are administered through the three most common parenteral routes, that is, subcutaneous, intramuscular, and intravenous [53]. For peptide-based vaccines, mucosal routes such as intranasal and oral routes offer many advantages as well. These routes do not require highly skilled personnel and also offer a needle-free, economical and convenient approach [15]. However, the gastrointestinal represents a particular challenge for vaccine delivery via oral administration. The low pH, proteolytic enzymes, and presence of bile salts in the stomach may disrupt vaccine function [54]. The delivery system needs to protect the peptide-based vaccine against the adverse conditions of the gastrointestinal tract and should also accelerate the uptake of antigens before they are discharged or digested. Therefore, to improve the stability of peptide-based vaccines, they are encapsulated within the delivery system in such as way that the antigens should have sufficient time to cross the epithelial membrane. The large surface area of the gastrointestinal tract usually dilutes the vaccine, and thus higher doses of vaccines are required to effectively reach the target immune cells [55]. However, the administration of large doses of vaccine through the oral route might produce oral tolerance, which may lead to the altered delayed-type hypersensitivity and a decrease in proliferation of T-cells. To overcome the obstacle of oral tolerance and proteolytic degradation, nasal routes are recommended, as nasal mucosae are enriched with immune cells and bypass the first-pass metabolism in comparison to the oral route. Upon increasing doses of vaccine, the absorption of aqueous solution vaccines may still be low, and hence an effective delivery system is required to enhance the bioavailability of the vaccine formulation.

20.5 Advantages of peptide and peptidomimetic vaccines

The most crucial advantage of peptide-based vaccines is that they can be designed by computationally targeting the pathogen and assessed based on several designs before synthesizing chemically. Lack of infectious materials from the pathogen decreases the probability of the emergence of disease. The selection of specific peptides as the epitopes excludes the inclusion of deleterious sequences from whole protein which may be oncogenic, such as Epstein–Barr virus nuclear antigens [56]. The use of small peptides for immune stimulation decreases the unnecessary antigenic load and allergenicity of the epitopes. In a peptide-based vaccine, other macromolecules like carbohydrate, lipid, and phosphate can be used in a controlled way to improve the immunogenicity and stability of the vaccines [57]. This provides an advantage over the other conventional methods of vaccine design where strategic designing is very limited and majorly depends on the immunogenicity of the pathogen. Many pathogens are difficult to culture in a controlled environment, but synthetic peptides can be produced on large scales without the involvement of an actual pathogen. These vaccines can be stored in dried conditions and thus the storage, transportation, and distribution of vaccines do not require maintaining the cold-chain procedure. Peptide-based vaccines can be designed to produce multiple epitopes from the same or different pathogens, targeting multiple antigens simultaneously. Another prominent advantage of the use of peptides as antigens in the peptide-based vaccine is that these vaccines show more drug-like properties compared to conventional vaccines [57]. Additionally, peptidomimetic techniques have improved the stability and efficiency of these vaccines. This widens the scope of development for peptide- and peptidomimetic-based vaccines and their delivery systems.

20.6 Current and future perspectives

Currently, no peptide-based vaccine is approved for human use on a widespread global scale. EpiVacCorona was designed to counter SARS-COV-2 infection and was given limited approval as of 2021. In addition to EpiVacCorona, many other peptide-based vaccines entered clinical trials and are still under evaluation. With technological advances, the scope of more sophisticated methods to identify patient-specific or population-specific epitopes based on HLA haplotypes has increased dramatically. The use of synthetic peptides has bypassed the complexity involved with the processing and delivery of conventional vaccines. Most importantly, the drug-like properties of peptides and the flexibility of chemical modification significantly widens the potential scope of development and application for peptide-based vaccines.

References

[1] Riedel S. Edward Jenner and the history of smallpox and vaccination. Baylor Univ Med Center Proc 2005;18(1):21–5. Available from: https://doi.org/10.1080/08998280.2005.11928028.

[2] Pollard AJ, Bijker EM. A guide to vaccinology: from basic principles to new developments. Nat Rev Immun 2021;21(2):83–100. Available from: https://doi.org/10.1038/s41577-020-00479-7.

[3] Kallerup RS, Foged C. Classification of Vaccines; 2015. p. 15–29. Available from: https://doi.org/10.1007/978-1-4939-1417-3_2.

[4] Nevagi RJ, Toth I, Skwarczynski M. Peptide-based vaccines. Peptide Applications in Biomedicine, Biotechnology and Bioengineering. Elsevier Inc; 2018. p. 327–58. Available from: https://doi.org/10.1016/B978-0-08-100736-5.00012-0.

[5] Clem AS. Fundamentals of vaccine immunology. J Glob Infect Dis 2011;3(1):73–8. Available from: https://doi.org/10.4103/0974-777X.77299.

[6] MODULE 2. Types of vaccine. WHO Vaccine Safety Basics. (n.d.). <https://vaccine-safety-training.org/types-of-vaccine.html> [accessed 23.05.21].

[7] Schlingmann B, Castiglia KR, Stobart CC, Moore ML. Polyvalent vaccines: High-maintenance heroes. PLoS Pathog 2018;14(4). Available from: https://doi.org/10.1371/journal.ppat.1006904.

[8] MODULE 2. Live attenuated vaccines (LAV). WHO Vaccine Safety Basics. (n.d.). <https://vaccine-safety-training.org/live-attenuated-vaccines.html> [accessed 23.05.21].

[9] Polio Vaccine: Vaccine-derived poliovirus. CDC. (n.d.). <https://www.cdc.gov/vaccines/vpd/polio/hcp/vaccine-derived-poliovirus-faq.html> [accessed 23.05.21].

[10] Fuerst TR, Stover CK, De la Cruz VF. Development of BCG as a live recombinant vector system: potential use as an HIV vaccine. Biotechnol Ther 1991;2(1−2):159−78. Available from: https://pubmed.ncbi.nlm.nih.gov/1845119/.

[11] Dumpa N, Goel K, Guo Y, McFall H, Pillai AR, Shukla A, et al. Stability of Vaccines. AAPS PharmSciTech 2019;20(2):1−11. Available from: https://doi.org/10.1208/s12249-018-1254-2.

[12] Moyle P, Toth I. Self-Adjuvanting Lipopeptide Vaccines. Curr Med Chem 2008;15 (5):506−16. Available from: https://doi.org/10.2174/092986708783503249.

[13] MODULE 2. Subunit vaccines. WHO Vaccine Safety Basics. (n.d.). <https://vaccine-safety-training.org/subunit-vaccines.html> [accessed 24.05.21].

[14] Yadav DK, Yadav N, Khurana SMP. Vaccines: present status and applications. Animal Biotechnology: Models in Discovery and Translation. Elsevier Inc; 2013. p. 491−508. Available from: https://doi.org/10.1016/B978-0-12-416002-6.00026-2.

[15] Skwarczynski M, Toth I. Peptide-based synthetic vaccines. Chem Sci 2016;7(2):842−54. Available from: https://doi.org/10.1039/c5sc03892h.

[16] Croft NP, Purcell AW. Peptidomimetics: modifying peptides in the pursuit of better vaccines. Exp Rev Vacc 2011;10(2):211−26. Available from: https://doi.org/10.1586/erv.10.161.

[17] Vanpatten S, He M, Altiti A, Cheng KF, Ghanem MH, Al-Abed Y. Evidence supporting the use of peptides and peptidomimetics as potential SARS-CoV-2 (COVID-19) therapeutics. Future Med Chem 2020;12(18):1647−56. Available from: https://doi.org/10.4155/fmc-2020-0180.

[18] Vagner J, Qu H, Hruby VJ. Peptidomimetics, a synthetic tool of drug discovery. Curr Opin Chem Biol 2008;12(3):292−6. Available from: https://doi.org/10.1016/j.cbpa.2008.03.009.

[19] Avan I, Dennis Hall C, Katritzky AR. Peptidomimetics via modifications of amino acids and peptide bonds. Chem Soc Rev 2014;43(10):3575−94. Available from: https://doi.org/10.1039/c3cs60384a.

[20] Boltjes A, Huang Y, Van De Velde R, Rijkee L, Wolf S, Gaugler J, et al. Fragment-based library generation for the discovery of a peptidomimetic p53-Mdm4 inhibitor. ACS Combin Sci 2014;16(8):393−6. Available from: https://doi.org/10.1021/co500026b.

[21] Brown CJ, Cheok CF, Verma CS, Lane DP. Reactivation of p53: from peptides to small molecules Trends in Pharmacol Sci, 32. Elsevier Ltd.; 2011. p. 53−62Issue 1. Available from: https://doi.org/10.1016/j.tips.2010.11.004.

[22] Obrecht D, Chevalier E, Moehle K, Robinson JA. β-Hairpin protein epitope mimetic technology in drug discovery. Drug Discov Today Technol 2012;9(1):e63−9. Available from: https://doi.org/10.1016/j.ddtec.2011.07.006.

[23] Collier AC, Coombs RW, Schoenfeld DA, Bassett RL, Timpone J, Baruch A, et al. Treatment of human immunodeficiency virus infection with saquinavir, zidovudine, and zalcitabine. N Engl J Med 1996;334(16):1011−18. Available from: https://doi.org/10.1056/nejm199604183341602.

[24] Kempf DJ, Marsh KC, Kumar G, Rodrigues AD, Denissen JF, McDonald E, et al. Pharmacokinetic enhancement of inhibitors of the human immunodeficiency virus protease by coadministration with ritonavir. Antimicrob Agents Chemother 1997;41(3):654−60. Available from: https://doi.org/10.1128/aac.41.3.654.

[25] Kosel BW, Beckerman KP, Hayashi S, Homma M, Aweeka FT. Pharmacokinetics of nelfinavir and indinavir in HIV-1-infected pregnant women. AIDS 2003;17(8):1195−9. Available from: https://doi.org/10.1097/00002030-200305230-00011.

[26] Alagumuthu M, Rajpoot S, Baig MS. Structure-based design of novel peptidomimetics targeting the SARS-CoV-2 spike protein. Cell Mol Bioeng 2021;14(2):177–85. Available from: https://doi.org/10.1007/s12195-020-00658-5.

[27] De Temmerman ML, Rejman J, Demeester J, Irvine DJ, Gander B, De Smedt SC. Particulate vaccines: on the quest for optimal delivery and immune response. Drug Discov Today 2011;16(13–14):569–82. Available from: https://doi.org/10.1016/j.drudis.2011.04.006.

[28] Neefjes J, Jongsma MLM, Paul P, Bakke O. Towards a systems understanding of MHC class i and MHC class II antigen presentation. Nat Rev Immunol 2011;11(12):823–36. Available from: https://doi.org/10.1038/nri3084.

[29] Spellberg B, Edwards JE. Type 1/type 2 immunity in infectious diseases. Clin Infect Dis 2001;32(1):76–102. Available from: https://doi.org/10.1086/317537.

[30] Sette A, Fikes J. Epitope-based vaccines: an update on epitope identification, vaccine design and delivery. Curr Opin Immunol 2003;15(4):461–70. Available from: https://doi.org/10.1016/S0952-7915(03)00083-9.

[31] Fuaad AAHA, Pearson MS, Pickering DA, Becker L, Zhao G, Loukas AC, et al. Lipopeptide nanoparticles: development of vaccines against hookworm parasite. ChemMedChem 2015;10(10):1647–54. Available from: https://doi.org/10.1002/cmdc.201500227.

[32] Cooper JA, Hayman W, Reed C, Kagawa H, Good MF, Saul A. Mapping of conformational B cell epitopes within alpha-helical coiled coil proteins. Mol Immunol 1997;34(6):433–40. Available from: https://doi.org/10.1016/S0161-5890(97)00056-4.

[33] Lau YH, De Andrade P, Wu Y, Spring DR. Peptide stapling techniques based on different macrocyclisation chemistries. Chem Soc Rev 2015;44(1):91–102. Available from: https://doi.org/10.1039/c4cs00246f.

[34] Pentier JM, Sewell AK, Miles JJ. Advances in T-cell epitope engineering. Front Immunol 2013;4(JUN):133. Available from: https://doi.org/10.3389/fimmu.2013.00133.

[35] Alexander J, Sidney J, Southwood S, Ruppert J, Oseroff C, Maewal A, et al. Development of high potency universal DR-restricted helper epitopes by modification of high affinity DR-blocking peptides. Immunity 1994;1(9):751–61. Available from: https://doi.org/10.1016/S1074-7613(94)80017-0.

[36] Alexander J, del Guercio M-F, Maewal A, Qiao L, Fikes J, Chesnut RW, et al. Linear PADRE T helper epitope and carbohydrate B cell epitope conjugates induce specific high titer IgG antibody responses. The J Immunol 2000;164(3):1625–33. Available from: https://doi.org/10.4049/jimmunol.164.3.1625.

[37] Borghans JAM, Keşmir C, De Boer RJ. MHC diversity in individuals and populations. In Silico Immunology. Springer US; 2007. p. 177–95. Available from: https://doi.org/10.1007/978-0-387-39241-7_10.

[38] Bui HH, Sidney J, Dinh K, Southwood S, Newman MJ, Sette A. Predicting population coverage of T-cell epitope-based diagnostics and vaccines. BMC Bioinformatics 2006;7. Available from: https://doi.org/10.1186/1471-2105-7-153.

[39] Singh A, Thakur M, Sharma LK, Chandra K. Designing a multi-epitope peptide based vaccine against SARS-CoV-2. Sci Rep 2020;10(1):16219. Available from: https://doi.org/10.1038/s41598-020-73371-y.

[40] Di Pasquale A, Preiss S, Da Silva FT, Garçon N. Vaccine adjuvants: From 1920 to 2015 and beyond. Vaccines 2015;3(2):320–43. Available from: https://doi.org/10.3390/vaccines3020320.

[41] Marrack P, McKee AS, Munks MW. Towards an understanding of the adjuvant action of aluminium. Nat Rev Immunol 2009;9(4):287–93. Available from: https://doi.org/10.1038/nri2510.

[42] Gupta RK. Aluminum compounds as vaccine adjuvants. Adv Drug Deliv Rev 1998;32(3):155–72. Available from: https://doi.org/10.1016/S0169-409X(98)00008-8.

[43] Petrovsky N, Aguilar JC. Vaccine adjuvants: current state and future trends. Immunol Cell Biol 2004;82(5):488–96. Available from: https://doi.org/10.1111/j.0818-9641.2004.01272.x.

[44] Batzloff MR, Hayman WA, Davies MR, Zeng M, Pruksakorn S, Brandt ER, et al. Protection against group a streptococcus by immunization with J8-diphtheria toxoid: Contribution of J8- and diphtheria toxoid-specific antibodies to protection. J Infect Dis 2003;187(10):1598–608. Available from: https://doi.org/10.1086/374800.

[45] Shah RR, Brito LA, O'Hagan DT, Amiji MM. Emulsions as Vaccine Adjuvants; 2015. p. 59–76. <https://doi.org/10.1007/978-1-4939-1417-3_4>.

[46] Reed SG, Bertholet S, Coler RN, Friede M. New horizons in adjuvants for vaccine development. Trends Immunol 2009;30(1):23–32. Available from: https://doi.org/10.1016/j.it.2008.09.006.

[47] Ryzhikov AB, Ryzhikov EA, Bogryantseva MP, Danilenko ED, Imatdinov IR, Nechaeva EA, et al. Immunogenicity and protectivity of the peptide candidate vaccine against SARS-CoV-2. Ann Russ Acad Med Sci 2021;76(1):5–19. Available from: https://doi.org/10.15690/vramn1528.

[48] Bovier PA, Epaxal. a virosomal vaccine to prevent hepatitis A infection. Expert Rev Vaccines 2008;7:1141–50.

[49] Mischler R, Metcalfe IC. Inflexal V a trivalent virosome subunit influenza vaccine: production. Vaccine 2002;20:B17–23.

[50] van Doorn E, Liu H, Huckriede A, Hak E. Safety and tolerability evaluation of the use of Montanide ISA™51 as vaccine adjuvant: A systematic review. Human vaccines & immunotherapeutics 2016; 12:159–69.

[51] Sioud M, Skorstad G, Mobergslien A, Sæbøe-Larssen S. A novel peptide carrier for efficient targeting of antigens and nucleic acids to dendritic cells. FASEB J 2013;27(8):3272–83. Available from: https://doi.org/10.1096/fj.12-224758.

[52] Skwarczynski M, Zhao G, Boer JC, Ozberk V, Azuar A, Cruz JG, et al. Poly(amino acids) as a potent self-adjuvanting delivery system for peptide-based nanovaccines. Sci Adv 2020;6(5):eaax2285. Available from: https://doi.org/10.1126/sciadv.aax2285.

[53] Jin JF, Zhu LL, Chen M, Xu HM, Wang HF, Feng XQ, et al. The optimal choice of medication administration route regarding intravenous, intramuscular, and subcutaneous injection. Patient Preference Adherence 2015;9:923–42. Available from: https://doi.org/10.2147/PPA.S87271.

[54] Marasini N, Skwarczynski M, Toth I. Oral delivery of nanoparticle-based vaccines. Expert Rev Vacc 2014;13(11):1361–76. Available from: https://doi.org/10.1586/14760584.2014.936852.

[55] Vela Ramirez JE, Sharpe LA, Peppas NA. Current state and challenges in developing oral vaccines. Adv Drug Deliv Rev 2017;114:116–31. Available from: https://doi.org/10.1016/j.addr.2017.04.008.

[56] Moss DJ, Schmidt C, Elliott S, Suhrbier A, Burrows S, Khanna R. Strategies involved in developing an effective vaccine for EBV-associated diseases. Adv Cancer Res 1996; 69:213–45. Available from: https://doi.org/10.1016/s0065-230x(08)60864-7.

[57] Purcell AW, McCluskey J, Rossjohn J. More than one reason to rethink the use of peptides in vaccine design. Nat Rev Drug Discov 2007;6(5):404–14. Available from: https://doi.org/10.1038/nrd2224.

Chapter 21

Therapeutic peptidomimetics for cancer treatment

Carmen Avendaño and J. Carlos Menéndez
Departamento de Química Orgánica y Farmacéutica, Universidad Complutense, Madrid, Spain

21.1 Introduction

Cancer displays unique cellular events in its initiation, progression, and metastases, with the participation of a large number of transcription factors and signaling proteins. Many stages of signaling pathways, gene transcription and translation and cell growth and repair depend on protein–protein interactions (PPIs). Therefore the design of specific inhibitors of aberrant PPIs to realign a system toward homeostasis is a highly relevant goal in anticancer drug discovery.

Peptidomimetics (or peptide mimics) are very interesting as modulators of PPIs because, although they do not possess a natural peptide backbone structure, they may have essential functional groups in a required three-dimensional pattern complementary to the interface of PPIs and can be designed to mimick the amino acid residues that are critical points ("hot-spots") of this interface for mutual recognition. PPIs may occur between similarly sized proteins or protein domains, such as those involved in homo- or heterodimers, or may imply the tight binding of a small ligand to special pockets within its receptor. If the ligand is rigid and the receptor is flexible, this interaction may change the receptor conformation to allow an induced fit of the ligand, while if the ligand is flexible and the receptor is rigid, the flexible nature of the ligand would require to determine which is the most energetically favorable conformation for binding with the receptor [1,2]. Besides mimicking the amino acid residues that are "hotspots" in critical PPIs involved in aberrant cell growth, the structure of rationally designed peptide mimics studied as anticancer agents have been based on secondary structural motifs of proteins, mostly α-helices, because α-helices are more common than β-pleated sheets at interfaces of PPIs.

Nonpeptidic-based peptidomimetics, which have drastic modifications at the peptide backbone and amino acid side chains but still mimic the targeted

Peptide and Peptidomimetic Therapeutics. DOI: https://doi.org/10.1016/B978-0-12-820141-1.00010-8

biological activity of a peptide, are generally found by a random screening of synthetic or natural peptide libraries using conventional methods or phage display, a method of screening a library of candidate peptides by displaying them on the surface of bacteriophage. In addition, nonpeptidic-based peptidomimetics may be obtained by synthesis of proteolytically stable nonpeptidic scaffolds consisting of a repeating structural unit amenable to an iterative construction that presents functionalized peptide-like side chains to mimic the conformation of native peptide substrates. These structural scaffolds are further modified to achieve the desired specificity, potency, efficacy, and stability. For instance, the polypyrrolinone scaffold may be used to generate β-strand/β-sheet, β-turn, and helical peptidomimetics [3].

21.2 Peptidomimetics targeting proteasomal protein regulation

21.2.1 The role of the ubiquitin–proteasome system in cancer

In eukaryotic cells, the main mechanism for the control of protein homeostasis is extralysosomal protein degradation by the ubiquitin–proteasome system (UPS). Each human cell contains 20,000–30,000 proteasomes, which are located in the cytoplasm and in the nucleus and their role is to remove proteins that are not functional or are no longer needed.

To prevent undesired hydrolysis, proteins are targeted for UPS degradation through the incorporation of polyubiquitin chains at the ε-NH_2 group of Lys residues [4]. The protein is subsequently deubiquitinated, unfolded and degraded at the 20S core of the proteasome, a complex of 28 subunits that are organized in four stacked heptameric rings, creating a cylindrical structure. The proteins at the top and bottom rings are called α subunits, while the ones in the two inner rings are called β. The β_1, β_2, and β_5 subunits contain proteolytic sites, formed by the N-terminal threonine residues that face the central cavity of the 20S complex [5]. To be functional in vivo, the 20S proteasome must be capped at both ends by the 19S regulatory complex (Fig. 21.1). Proteasomes are amino-terminal nucleophile (Ntn) hydrolases whose catalytic cycle, involving two tetrahedral transition states, is summarized in Fig. 21.2. The orchestrated processes of ubiquitinylation and hydrolysis by the proteasome regulates many cellular functions including cell cycle progression, development, apoptosis, signal transduction, and antigen presentation.

Because the proteasome controls the levels of several regulatory proteins that are essential for the progression of the cell cycle and apoptosis, [6] its inhibition has received much attention as a strategy for cancer treatment [7–9]. Constitutive proteasomes are upregulated in several types of cancer, reflecting increased demand for protein degradation and therefore a higher

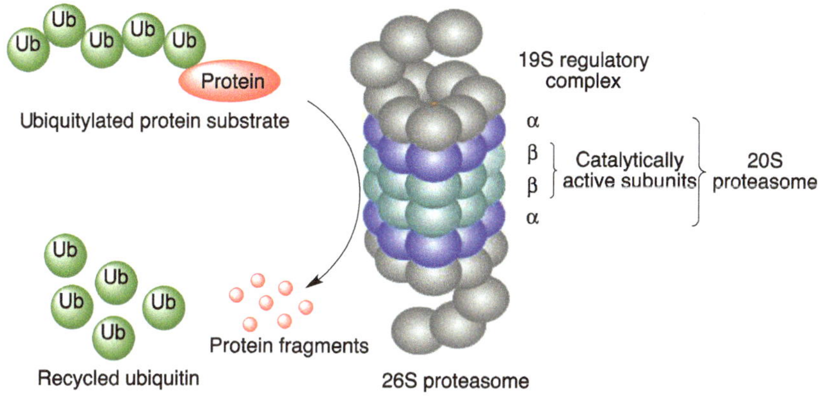

FIGURE 21.1 Structure and function of the proteasome. *Ub*: ubiquitin.

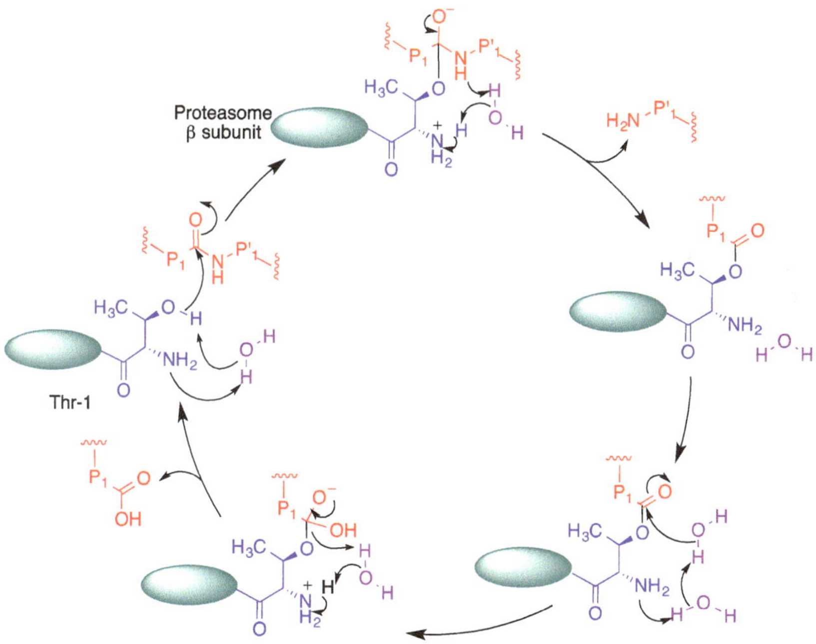

FIGURE 21.2 Hydrolytic catalytic cycle of the proteasome.

metabolic activity. This makes it an attractive target for anticancer therapy, especially in multiple myeloma (MM) and mantle cell lymphoma [10–12].

Two important examples of proteins subject to proteasomal inhibition are the MDM2 protein, whose overexpression contributes to tumor development through inhibition of p53 activity, [13] and the IκB inhibitor proteins that regulate the activity of nuclear factor-κB, a family of transcription factors that play critical roles in inflammation, immunity, cell proliferation, differentiation, and survival. Proteasome inhibition also prevents completion of the cell cycle and the mitotic proliferation of cancer cells because the cyclins involved in this processess cannot be degraded.

21.2.2 Anticancer peptidomimetics targeting the ubiquitin–proteasome system

Most known proteasome inhibitors are peptidomimetics containing an electrophilic functional group that is normally placed at one end of the molecule and reacts with the hydroxyl of the N-terminal threonine of the proteasome β catalytic subunits.

The natural product lactacystin, isolated from *Streptomyces lactacystinaeus*, is an example of serendipity in drug discovery [14]. Its skeleton contains *N*-acetylcysteine and a novel pyroglutamic acid derivative. It functions as a prodrug in vivo that spontaneously generates the cell-permeable peptidomimetic omuralide (clasto-lactacystin-β-lactone), which covalently modifies the N-terminal threonine residues at the proteasome β-subunits. The structurally related salinosporamine A (marizomib), a naturally occuring analog of omuralide isolated from the marine actinomycete *Salinospora tropica*, is currently being studied in phase III for glioblastoma (Fig. 21.3) [15].

Besides fused β-lactone moieties, which are probably opened by the threonine hydroxyl to give esters (Fig. 21.4A), other electrophiles that have been incorporated to peptidomimetic proteasome inhibitors including α,β-unsaturated sulfones (Fig. 21.4B), boronic acids, which interact reversibly with the threonine hydroxyl due to the ability of the latter group to form stable, reversible tetrahedral intermediates by electron donation to an empty p-orbital on boron (Fig. 21.4C) and α-epoxy ketones that generate hemiacetals by addition of the threonine hydroxyl to the carbonyl group, followed by attack of the threonine free amino group onto the more hindered epoxide carbon atom (Fig. 21.4D) [16].

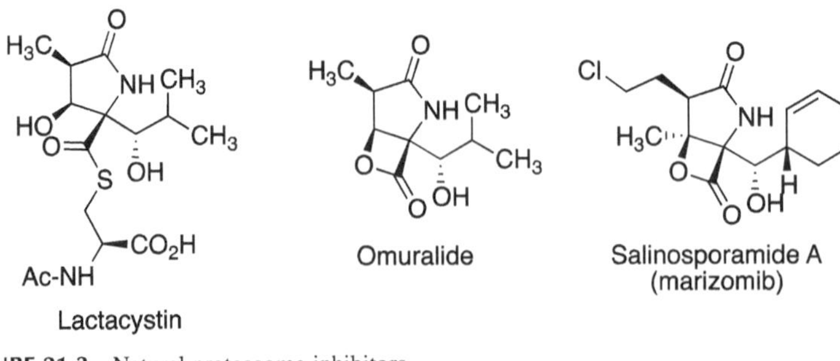

FIGURE 21.3 Natural proteasome inhibitors.

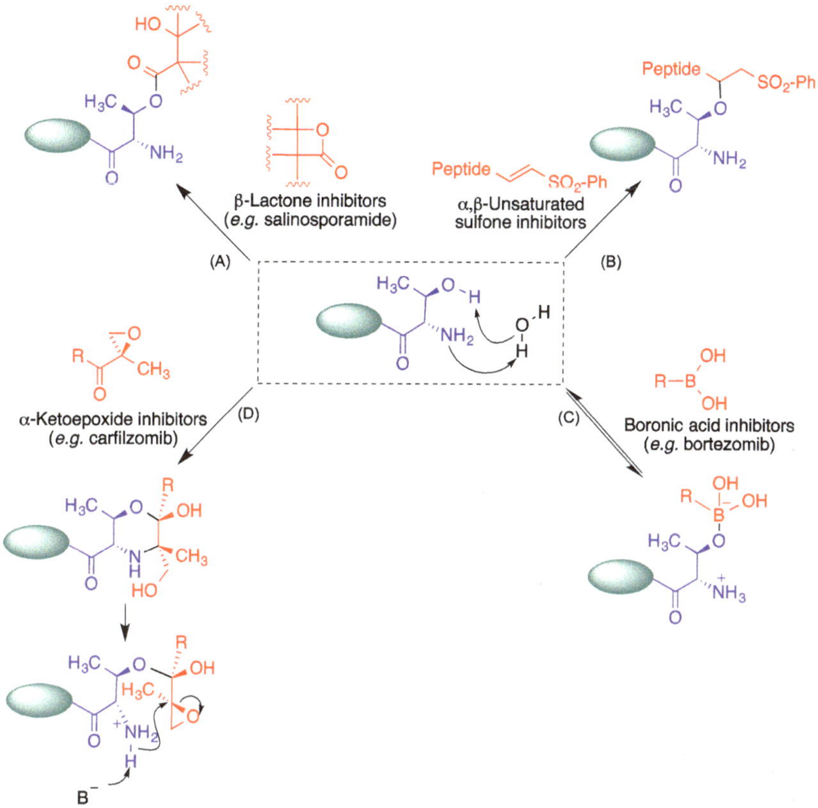

FIGURE 21.4 (A–D) Several structural types of peptidomimetic proteasome inhibitors and their modes of action.

The dipeptide boronic acid bortezomib (PS-341, MLN-341, Velcade) inhibits the chymotrypsin-like (CT-L) activity of the 26S proteasome by preferentially binding to the active site of the β_5 subunit. It affects multiple signaling cascades in cancer cells (e.g., NF-κB) and induces G/M phase arrest followed by apoptosis. Several studies have also shown that bortezomib has a broad off-target inhibitory effect on other proteases, which are likely to contribute to its multiple clinical side effects. This drug was approved in 2003 for the treatment of MM, the second most common hematological cancer. Ixazomib (Ninlaro, MLN-9708) and delanzomib (CEP-18770) are additional orally bioavailable boronic acid-based reversible proteasome inhibitors. Ixazomib is a prodrug whose cyclic boronate ester structure is immediately hydrolyzed when exposed to aqueous solutions to yield its active form, the boronic acid MLN2238. It preferentially inhibits the β_5 and β_{5i} subunits and has a significantly reduced cytotoxicity compared to delanzomib, which inhibits both β_5 and β_1. Ixazomib was granted FDA

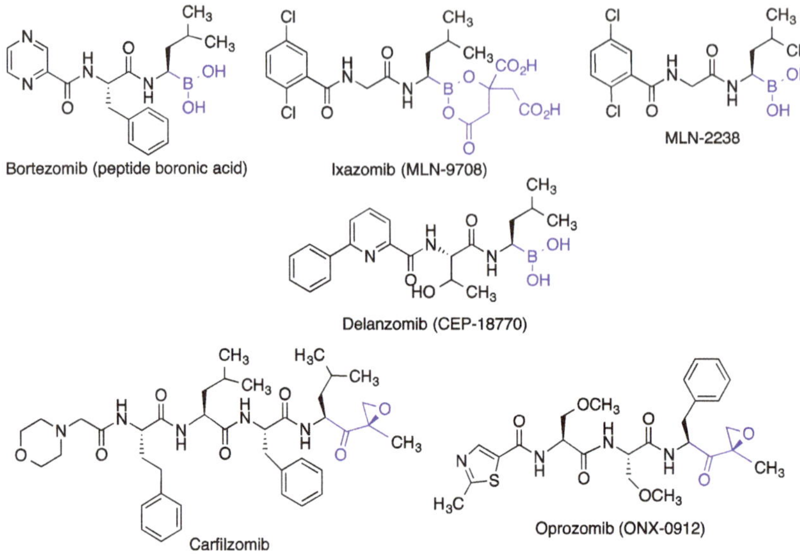

FIGURE 21.5 Boronic acid-derived and epoxyketone-derived proteasome inhibitors.

approval in 2015 for the treatment of MM, but the development of delanzomib was halted due to its toxicity. Among epoxy ketones, the tetrapeptide carfilzomib (Krypolis) was FDA-approved for MM in 2012. It is highly selective for the β_5 and β_{5i} subunits, with minimal cross reactivity to other proteases, allowing more sustained and specific inhibition of the 20 S proteasome. However, although carfilzomib has reduced side effects compared to bortezomib, its C-terminal epoxyketone function is highly unstable in vivo, resulting in a short plasma half life and a low tissue drug distribution. Oprozomib (ONX-0912, PR-047) [17] is an orally bioavailable tripetide epoxyketone related to carfilzomib that has entered several early phase trials, alone or in combination with pomalidomide and dexamethasone [18] (Fig. 21.5).

21.3 Peptidomimetic matrix metalloproteinase inhibitors

Many members of the matrix metalloproteinase family (MMPs) are involved in primary tumor growth and in the proteolytic degradation of extracellular matrix and basement membrane components required for metastasis. Furthermore, some antiangiogenic drugs act by inhibiting MMPs because they are necessary to initiate the breakdown of the extracellular matrix (ECM), allowing the migration of proliferating endothelial cells to form new vascular lumens. Importantly, tumor growth beyond about 2 mm^3 requires the generation of a network of blood vessels [19].

The hydrolytic activity of MMPs is based on the enhanced acidity of a water molecule as a consequence of coordination of its oxygen atom with a zinc cation that is also coordinated with three imidazole side chains of the protein histidine residues. All inhibitors replace this water molecule and coordinate with zinc in a monodentate or bidentate fashion (Fig. 21.6) [20,21].

Peptidomimetics that contain a hydroxamic acid moiety are among the most potent known MMP inhibitors owing to the fact that hydroxamic acid anion is a very efficient 1,4-bidentate Zn^{2+} ligand that binds with two contacts and creates a distorted trigonal bipiramidal geometry around the metal.

Batimastat (BB-94), a potent broad-spectrum MMP inhibitor that showed antineoplastic and antiangiogenic activity in various tumor models, reached Phase III clinical trials but it was not orally active and is no longer being considered. Marimastat, the first orally bioavailable MMP inhibitor that entered clinical trials in the field of oncology, underwent several Phase III studies but showed poor performance. The second generation of MMP inhibitors includes more specific peptidomimetics, designed on the basis of structural studies of the MMP active site by NMR and X-ray crystallography. The first subgroup of these inhibitors, that includes prinomastat (AG-3340) and MMI-270, still bears an hydroxamic function. Prinomastat was studied in Phase III trials for advanced prostate and lung cancers, but its development was halted due to lack of efficacy. Clinical studies with MMI-270 (CGS-27023A) were advanced to Phase II, but they were interrupted due to poor tolerance (Fig. 21.7).

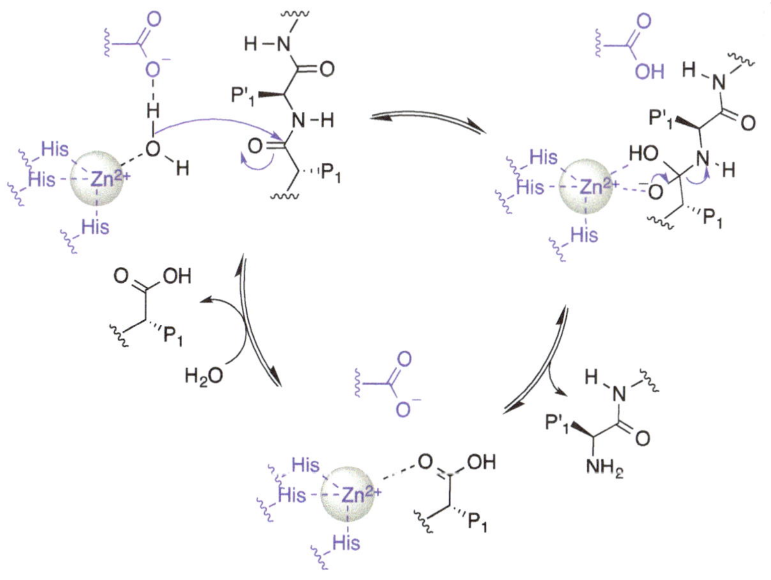

FIGURE 21.6 Catalytic cycle of zinc metalloproteinases.

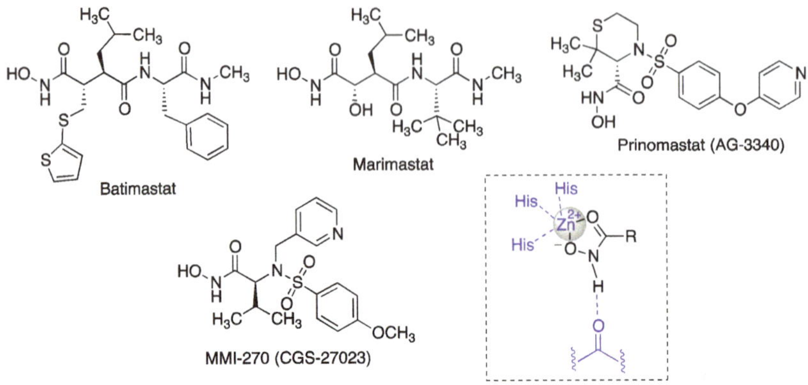

FIGURE 21.7 Hydroxamic acid-derived peptidomimetic MMP inhibitors. The MMP catalytic zinc therapeutic target is shown in the box on the right.

Other substructures that can interact with the Zn^{2+} cation in MMPs are the carboxy and mercapto functional groups. Tanomastat (BAY 12–9566) is a biphenyl carboxylic acid-based compound that inhibits MMP-2, MMP-3, and MMP-9 and was generally well tolerated, but failed during Phase III studies for treatment of advanced small-cell lung cancers. Another carboxylic acid-based MMP inhibitor is S-3304, which entered Phase III trials for the treatment of solid tumors. Rebimastat is a thiol-based compound that inhibits MMP 1, 2, 8, 9, and 14 that underwent Phase III clinical trials for nonsmall-cell lung cancer. Finally, some tetracyclines, which are well-known chelating agents, are MMP inhibitors. One example is incyclinide (COL-3, Metastat), which entered Phase II trials for Kaposi's sarcoma and advanced brain tumors (Fig. 21.8). MMP inhibitors that do not bind Zn^{2+} in the active site of MMPs are also known, including a considerable number of natural products [22].

At present, it has become apparent that some MMPs, especially MMP-3, -8, and -9, when genetically eliminated in knockout animal models, induce enhanced tumorigenesis and metastasis. Consequently, the broad-spectrum activity of most of studied MMP inhibitors (MMPIs) might block these MMPs and result in tumor progression. Furthemore, although MMPs are involved in the early stages of tumor progression, MMPIs have normally been tested in patients with advanced disease, beyond the stage when these compounds could be effective. Therefore future success will depend on the design of new trials in early stage tumors [23].

21.4 Aminopeptidase inhibitors

21.4.1 Aminopeptidase N structure and functions

Aminopeptidases are proteolytic enzymes that cleave an amino acid residue or a peptide chain from the N-terminus of a substrate. The human

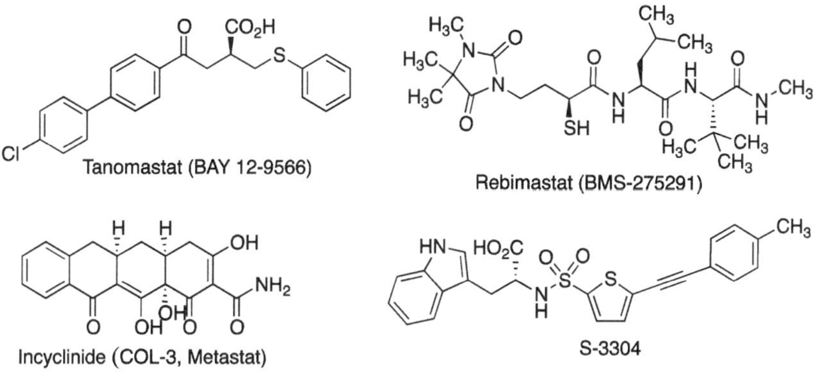

FIGURE 21.8 Representative nonhydroxamic MMP inhibitors.

aminopeptidase N (APN) is a ubiquitous transmembrane glycoprotein, with a large extracellular region and a small intracellular NH_2-terminal membrane anchor that may lead to an arch-like arrangement on the cell surface, forming and open or a closed conformation [24]. It is a member of the M1 zinc metallopeptidase family that shows structural homology to related enzymes, including MMPs, and its active site is similar to that of matrix metalloprotease 2 (MMP-2). The dimeric structure of both enzymes contains two hydrophobic cavities (the S1 and S1′ pockets) around the catalytic Zn^{2+} ion.

APN may act as an enzyme in peptide cleavage, a receptor in endocytosis, and a signaling molecule in signal transduction. Acting as an enzyme, APN is involved in the degradation of the extracellular matrix in cancer cells, affecting cell survival, cell migration and angiogenesis. APN also plays a major role in modulating several signaling pathways of a variety of cancer cells through the binding of a specific substrate followed by activation of mitogen-activated protein kinase (MAPK), phosphoinositide 3-kinases (PI3Ks)/protein kinase B (PKB, also known as Akt), and nuclear factor kappa B (NF-κB) kinases, triggering of MMP-2 and MMP-9 to promote migration, and invasion [25]. Its misregulation in almost all types of cancer and the correlation found between the levels of APN expression and the invasive capacity of cells make APN an attractive therapeutical target for cancer treatment [26].

Besides its direct effects, the inhibition of aminopeptidases avoids the hydrolysis of peptide fragments derived from the action of proteasome on proteins, a step necessary for the efficient recycling of amino acids. The subsequent amino acid deprivation prevents cancer cell survival and proliferation [27].

21.4.2 Aminopeptidase N inhibitors

Basic research on aminopeptidase inhibitors started with the isolation of the peptidomometic bestatin (ubenimex), an inhibitor of B and N

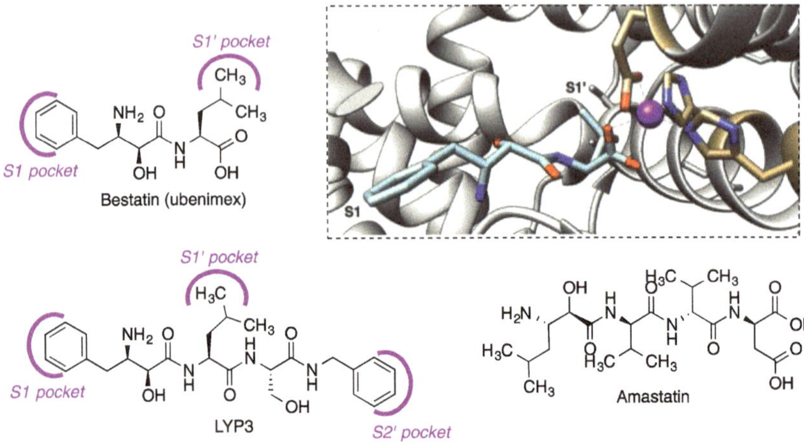

FIGURE 21.9 Bestatin and its analogs. The structure of human APN bound to bestatin was generated from PDB 4FYR.

aminopeptidases, from *Streptomyces olivoreticuli* [28]. It entered several clinical trials for a variety of malignancies, including cancers of the stomach, esophagus, and lung, melanoma, lymphomas, leukemias, nonsmall-cell lung carcinoma (NSCLC), and squamous cell carcinoma [29].

To improve the water solubility of bestatin, its structure was elongated at the C-terminal position by incorporating a L-serine benzylamide unit to obtain LYP3 [30]. This compound, in which the additional fragment may be involved in an interaction with the S2′ pocket of APN, showed a higher activity than the parent compound [31]. Amastatin is a representative example of the many natural aminopeptidase competitive inhibitors that can be viewed as bestatin analogs. The binding of these compounds to the aminopeptidase active site is exemplified by the bestatin-aminopeptidase complex (Fig. 21.9). The phenylalanine residue of the drug is deeply inserted into the S1 subsite, the leucine residue is directed toward the S1′ pocket and the leucine carboxylic acid is coordinated with the catalytic Zn^{2+} [24].

Tosedostat (CHR-2797), which bears an hydroxamic acid moiety as a Zn-chelating unit, was 300-fold more potent than bestatin as an APN inhibitor. This compound is an ester prodrug that liberates by hydrolysis the poorly membrane permeable active metabolite CHR-79888 (Fig. 21.10). It showed encouraging clinical responses in patients with acute leukemia and several other blood-related cancers, and it is also under clinical study for solid tumors [32].

Many additional compounds have been found to have potent APN inhibitory activity. Their scaffolds include the leucine ureido function, leucinethiol, 3-amino-2-tetralone, pyrrolidine, cyclic imide, AHPA, and indoline-2,3-dione moieties.

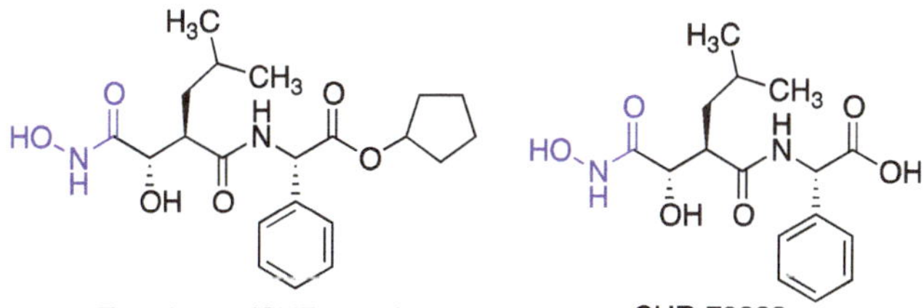

FIGURE 21.10 Todesostat and its metabolite, CHR-79888.

21.5 Peptidomimetics acting on the Ras-Raf-MAPK pathway

21.5.1 The Ras-Raf-MAPK pathway

Rat sarcoma (Ras) proteins are GTPases that bind to guanosine triphosphate (GTP) and hydrolyze it to guanosine diphosphate (GDP). The main three members of the Ras family are the H-Ras, K-Ras and N-Ras proteins. The Ras-Raf-MAPK pathway (Fig. 21.11) regulates diverse physiological processes, coordinating a large variety of cellular responses. About 25% of human cancers, including at least 30% of colon, thyroid and lung tumors and almost all pancreatic tumors, show a mutation that leads to Ras proteins remaining locked in an active state [33].

Ras activation requires its binding to GTP, allowing the subsequent recruitment of effectors. Membrane-bound Ras alternates between the inactive ("off") GDP-bound and the active ("on") GTP-bound forms, which are interconverted via a conformational change that alters two loops of the protein that are placed close to the nucleotide binding site, known as switch 1 and switch 2.

Due to the high affinity of Ras to GDP, the exchange of this nucleotide by GTP requires binding of Ras to the effector Son of Sevenless (Sos), which inserts its catalytic subunit between the switch 1 and 2 regions, opening the nucleotide binding pocket and releasing GDP, thereby allowing Ras to bind to GTP. The membrane-bound Ras-GTP complex activates the Raf protein through its phosphorylation, although other PPIs also play key roles in rapidly accelerated fibrosarcoma (Raf) activation [34]. Active Raf induces MAPK/Erk kinase (MEK) protein kinase kinases to phosphorylate MAPK that, finally, stimulate several transcription factors and gene expression (Fig. 21.12A-C).

21.5.2 Direct Ras inhibitors

Because of the key role of the Ras-Raf-MAPK pathway, its blockade has considered relevant for anticancer drug development for more than twenty years. However, the Ras-Sos interface has a very large interacting surface (3600 Å^2) that lacks deep pockets suitable for binding small-molecule

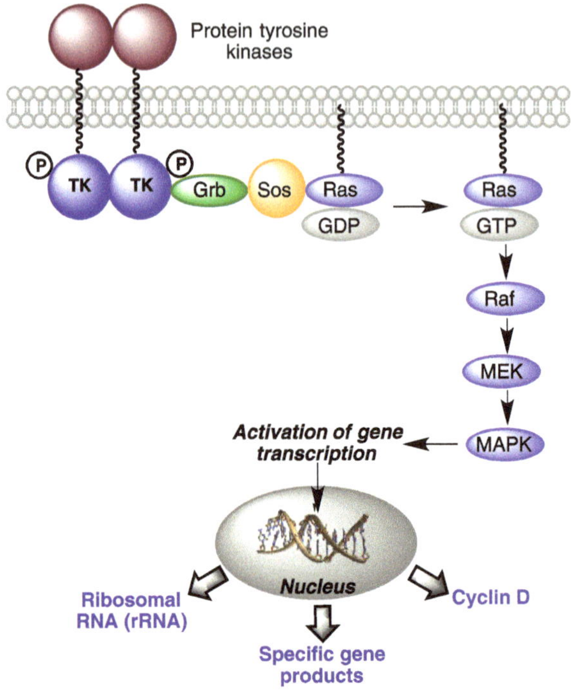

FIGURE 21.11 The Ras-Raf-MAPK pathway. (TK—tyrosine kinases; Grb—growth factor receptor-bound protein 2; SoS—Son of Sevenless; Ras—rat sarcoma protein; GDP—guanosine diphosphate; GTP—guanosine triphosphate; Raf—rapidly accelerated fibrosarcoma protein; MEK—mitogen-activated protein kinase kinase; MAPK—mitogen-activated protein kinase).

inhibitors, and therefore Ras itself has long been considered an undruggable target. Fortunately, in recent years covalent binding has been shown to provide a viable approach to overcome the lack of protein-inhibitor contacts. Following the identification of the K-Ras (G12C) mutation in about 10%–20% of all Ras-driven cancers, efforts were made to exploit the presence of this additional Cys residue to design covalent inhibitors endowed with selectivity for the inactive state of the mutated form of K-Ras [35]. Electrophilic moieties were incorporated to $\alpha_3\beta$ chimeric mimic of the Sos helix to provide proteolytically stable peptidomimetic structures that covalently target oncogenic Ras (G12C) [36]. Small molecules can also serve the same purpose while having better drug-like profiles, and the vinylsulfonamide derivative **1** was discovered using a fragment screening assay. This molecule binds covalently to an allosteric pocket, that has become known as the switch 1/2 pocket, inhibiting nucleotide exchange (Fig. 21.12D) [37]. Its optimization has led to additional promising compounds, such as ARS-853 [38].

Several allosteric covalent K-Ras (G12C) inhibitors have reached clinical trials in patients with this mutation [39]. MRTX-849, from Mirati

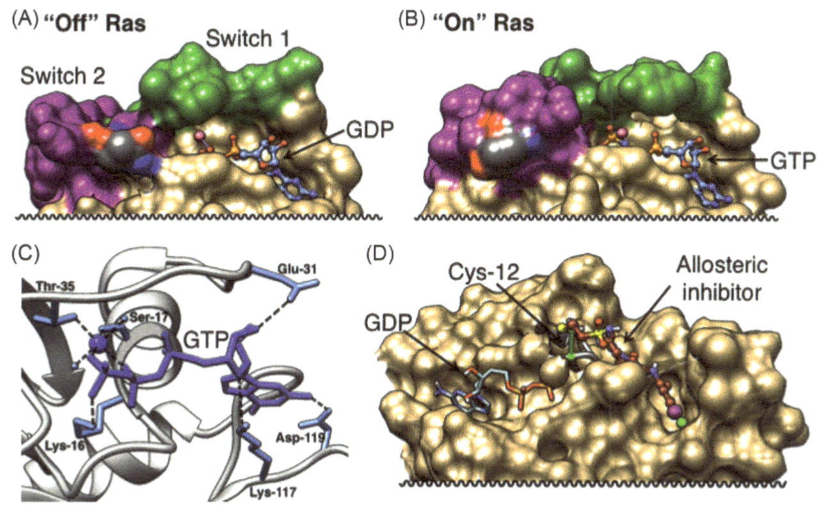

FIGURE 21.12 (A) Structure of inactive ("off") Ras, from PDB 5P21. (B) Structure of active ("on") Ras, from PDB 4Q21. (C) Details of the interaction of GTP with the active site of "on"-Ras, from PDB 4Q21. (D) Binding of compound **1**, a covalent allosteric inhibitor of K-Ras (G12C).

Therapeutics, is in a Phase I/II study, and sotorasib (AMG-510), developed by Amgen, has been approved to begin Phase II clinical trials for the treatment of metastatic NSCLC and additional solid tumors with this mutation. Due to restricted rotation around a N-Ar bond, this compound exists as two atropisomers, one of which is more potent. Other K-Ras (G12C) covalent inhibitors that have reached the clinic include Johnson & Johnson's JNJ-74699157 (previously known as ARS-3248) and Eli Lilly's LY3499446. We will finally mention BI-2852, which is a noncovalent inhibitor designed by fragment-based techniques and displayed a nanomolar inhibitory activity at the switch 1/2 pocket (Fig. 21.13) [40].

All the progress described in this section notwithstanding, it is currently understood that in K-Ras-mutant cancers the disease normally is not exclusively dependent on the dysregulated K-Ras signaling, and for this reason, future clinical research in this area will probably proceed via combination therapy.

21.5.3 Indirect Ras inhibitors: peptidomimetic inhibitors of Ras processing by farnesyltranferase

In this section, we will focus on peptidomimetics that indirectly inhibit the Ras-Raf-MAPK pathway by preventing the attachment of cytosolic Ras to the cell membrane [41,42]. This process requires the post-translational structural modification of the Ras C-terminus by incorporation of three highly

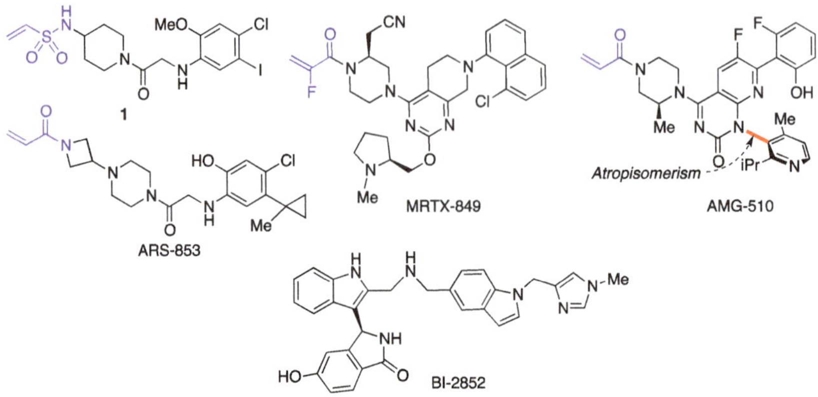

FIGURE 21.13 Major direct Ras inhibitors.

lipophilic moieties. Farnesyltransferase (FTase) is a zinc metalloprotein that recognizes the terminal CAAX sequence of Ras (the CAAX box), where C represents cysteine, A is an aliphatic amino acid (Leu, Ileu, Val) and X is Met, Ser, Leu or Gln. The cysteine sulfur is coordinated to the zinc cation in the active site, which lowers the pKa of the thiol and facilitates its *S*-farnesylation. A subsequent proteolytic step, esterification of the new carboxyl terminus and introduction of two palmitoyl groups by a palmitoyl CoA transferase, complete the preparation of Ras for attachment to the cell membrane (Fig. 21.14).

In the design of peptidomimetics targeted at the terminal CAAX sequence of Ras, the tetrapeptide Cys-Val-Phe-Met was used as a lead for systematic structural modification to achieve analogs, retaining the zinc-coordinating thiol group. Some of the labile peptide bonds were replaced by stable methylenamino or methylenoxy links (e.g., L-739750), and replacement of the central amino acid by *p*-aminobenzoic acid was also assayed (FTI-276). To enhance oral absorption, ester prodrugs (L-744832, FTI-277) and a double prodrug (AZD-3409) were designed and some of these compounds reached clinical trials, [43] but were later discontinued (Fig. 21.15).

Some heterocyclic compounds with activity as FTIs have been discovered through screening approaches [42]. Because they act by a peptide-competitive mechanism, they can be considered peptidomimetics. Some representative examples are BMS-214662, tipifarnib (R-115777, Zarnestra), L-778123, lonafarnib (SCH-66336) and SCH-226374. BMS-214662 is a benzodiazepine derivative active against tumor cells with and without RAS mutations that may reverse the malignant phenotype of H-Ras-transformed cells. Tipifarnib and L-778123 contain imidazole rings that are able to coordinate the catalytic zinc cation, competing with the cysteine unit at the CAAX motif in Ras. SCH-226374 was designed as an imidazole-bearing analog of lonafarnib. ABT-100 was particularly potent as a FTase inhibitor and was characterized

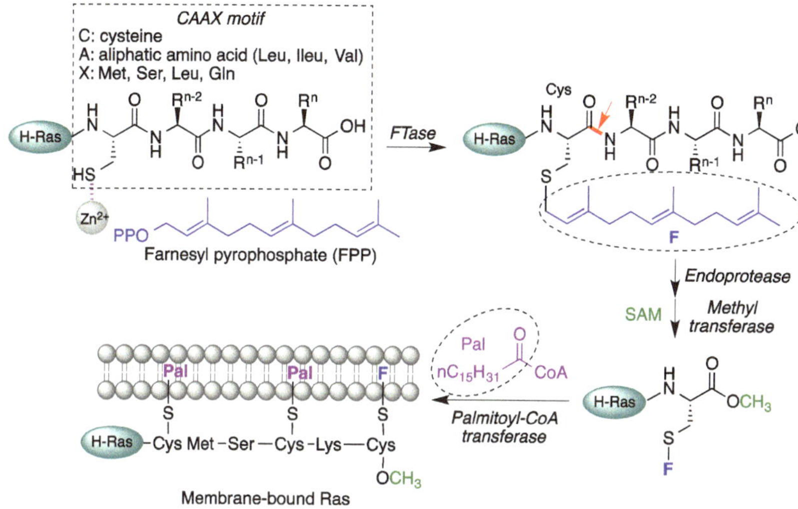

FIGURE 21.14 Steps required for the anchoring of Ras to the cell membrane.

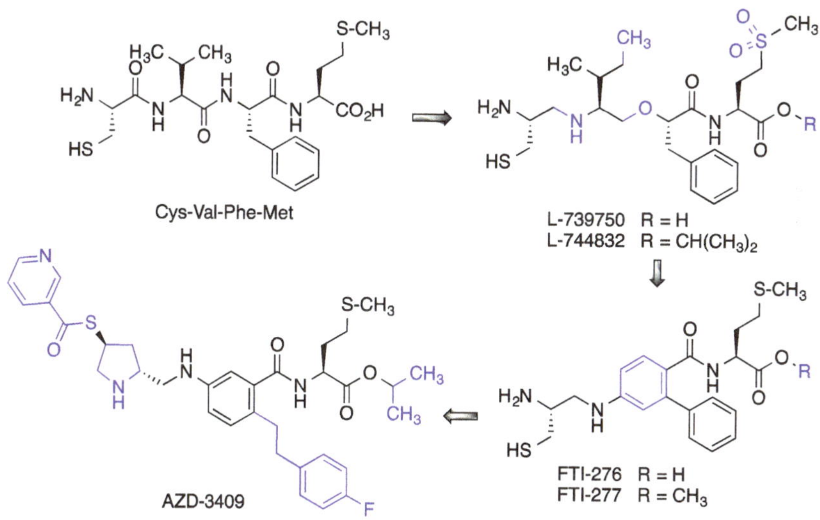

FIGURE 21.15 Design of peptidomimetics targeted at the terminal CAAX.

as a liver cancer chemopreventive agent. Both tifiparnib and lonafarnib are orally bioavailable, while BMS-24662 and L-7781123 have been studied as intravenous formulations. BMS-24662, tifiparnib, L-778123 and lonafarnib have been under clinical studies against a variety of cancers.

X-ray diffraction studies of BMS-214662 and tipifarnib complexed with farnesyl transferase showed that they bind to a hydrophobic cleft at the

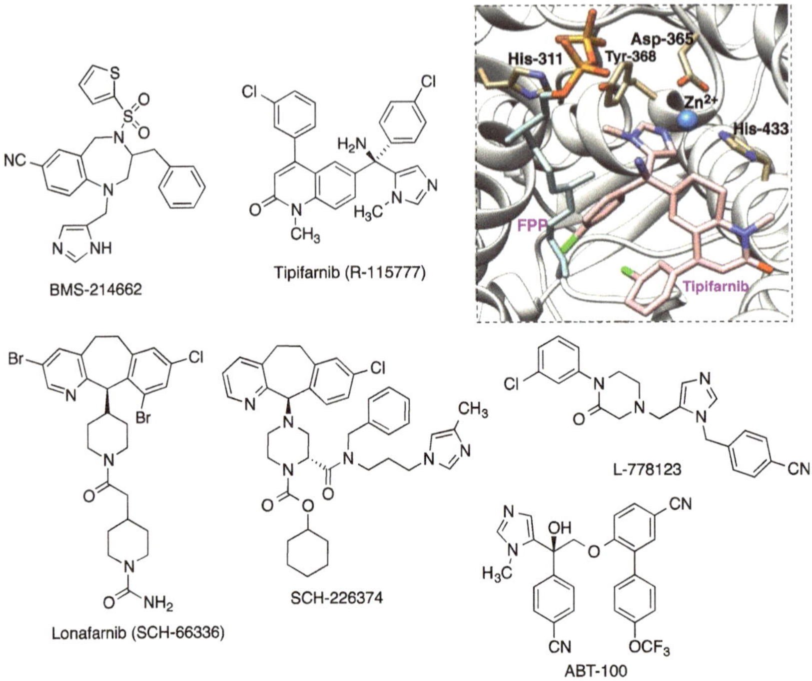

FIGURE 21.16 Several heterocyclic FTase inhibitors. The complex of tipifarnib with *Aspergillus fumigatus* farnesyltransferase (top right) was generated from PDB 4LNG.

interface of the α and β subunits of the enzyme, binding the catalytic zinc cation and forming a ternary complex with the FPP substrate and the enzyme. Aromatic stacking interactions are important, including those between the 4-chlorophenyl unit and the farnesyl moiety, the quinoline unit and Tyr-361β, and the 3-chlorophenyl ring and Trp-102β and Trp-106β (Fig. 21.16). Similarly, the imidazole ring in BMS-214662 binds to the zinc cation in the active site and the union is stabilized by several π-stacking interactions [44].

21.6 Anticancer peptidomimetics targeting HER2, HER2-HER3, and HER2-VEGF

21.6.1 HER2 as an anticancer target

The tyrosine kinase receptor HER2 (ErbB2) is an important therapeutic target that is overexpressed in approximately 20%–30% of patients with aggressive breast, colon, bladder, and biliary cancers. The introduction of HER2-targeted therapy has led to major advances in the treatment of HER2-positive breast cancer.

HER2 amplification or activating mutations results in activated oncogenic downstream signaling pathways, such as PI3K/Akt/mammalian target of rapamycin (mTOR) or MAPK, and ultimately promotion of cellular proliferation, survival and angiogenesis. HER2 (ErbB2) is a member of a tyrosine kinase receptor family that also includes HER1 (ErbB1, EGFR), HER3 (ErbB3), and HER4 (ErbB4). These are also often overexpressed, amplified, or mutated in many forms of cancer, making them the focus of cancer immunotherapeutic strategies [45]. Besides being present in the nucleus acting as transcriptional regulator kinases, all these receptors are located at the membrane, and are activated following ligand binding and receptor dimerization, being switched off through dephosphorylation, ubiquitination or removal through endosomal sorting and lysosomal degradation [46].

In addition to HER2 amplification and overexpression, over the past decade several activating HER2 mutations of potential therapeutic relevance have been recognized, and the number of new therapeutics targeting HER2, including small-molecule inhibitors, monoclonal antibodies (Mabs), antibody-drug conjugates, and bispecific antibodies, has grown rapidly. In spite of the efficiency of the HER2-targeted Mabs, their high cost, short half life, toxicity, and resistance development, has motivated new targeted aproaches, among which the use of stable, less expensive, and nontoxic peptidomimetics that block this signaling is an interesting alternative.

21.6.2 Peptidomimetic inhibitors of HER2-EGFR and HER2-HER3 heterodimerization

Generally speaking, epidermal growth factor receptors (EGFRs) undergo the conformational changes leading to dimerization upon binding to an EGF ligand, but in the absence of ligands HER2 overexpression drives its heterodimerization with other receptors of the family. Blocking of these PPIs with antibodies such as trastuzumab or pertuzumab is therapeutically useful because it prevents kinase signaling [47]. These antibodies bind to HER2 on the C-terminal portion of extracelular domain IV near the transmembrane domain, therefore the blocking of any protein interaction in this domain can disrupt the signaling pathways that promote cancer.

Some peptidomimetics targeted to the receptor to modulate HER2-mediated signaling disrupt EGFR-HER2 and HER2-HER3 PPIs. Starting from the X-ray crystal structures of human HER2 complexed with the antigen-binding fragments (Fabs) of trastuzumab [48] or pertuzumab [49], the peptidemometic HERP5 was designed [50]. It interacts specifically with HER2 extracellular domain (ECD) and disrupts the dimerization of EGFRs, showing antiproliferative activity against HER2-overexpressing breast cancer cell lines [51]. By using enzyme fragment complementation and proximity ligation assays, peptidomimetic **2** (Fig. 21.17) was also designed, and was shown to target the extracellular domain of HER2 and

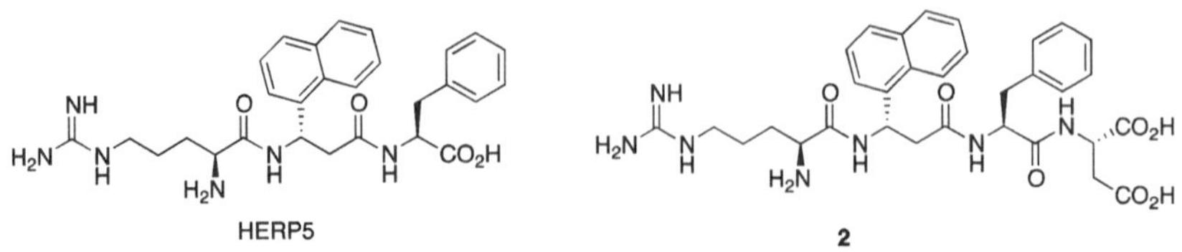

FIGURE 21.17 Several peptidomimetics that disrupt EGFR-HER2 and HER2-HER3 PPIs.

inhibit the HER2-HER3 interaction. The results of in vivo studies support further work in this direction [47].

21.7 Inhibitors of insulin-like growth factor-1 receptors

21.7.1 Insulin-like growth factor-1 receptors

Insulin-like growth factors (IGFs) are peptides that are part of a complex system often referred to as the IGF "axis." They have a high sequence homology with insulin and an important role in promotion of cell proliferation and inhibition of apoptosis. The insulin-like growth factor-1 receptor (IGF1R) is a membrane tyrosine kinase receptor with 70% homology to the insulin receptor. When it is activated by its ligands IGF1, IGF2, or by insulin, it transmits a signal to two adaptor proteins: the *insulin receptor substrate-1* (IRS1) and the *Src homologous and collagen* (Shc) [52]. Both adaptor proteins lack intrinsic enzymatic activity, but they possess various domains that allow them to recruit different signaling molecules. Shc is recruited by activated cell surface receptors (like IGF1R) or by integrins, then being phosphorylated by receptor tyrosine kinases (e.g., EGFR) or cytosolic tyrosine kinases. Phosphorylated Shc can then interact with Grb2, leading to the activation of different pathways such as MAPK, PI3K/Akt and Ras homolog family member A (RhoA) that depend on the activation of Ras (Fig. 21.18).

IGF1R is particularly important in the establishment and maintenance of the transformed phenotype, in mediating proliferation, and in the survival of tumor cells with anchorage-independent growth. IGF1R also exerts antiapoptotic activity and has a substantial influence on the control of the cell and body size, enabling transformed cells to form macroscopic tumors and to survive the process of detachment required for metastasis [53].

Various strategies, such as anti-IGF-1R antibodies, IGF-1 mimetic peptides, antisense oligonucleotides, and IGF-1R-specific peptide aptamers, as well as the induction of targeted degradation of IGF-1R and expression of dominant-negative IGF-1R mutants, have been explored to inhibit IGF-1R signaling. Targeting the intracellular kinase domain of IGF-1R with small molecules (most commonly ATP antagonists) has gained considerable attention, but because of the high sequence homology of the kinase domains of

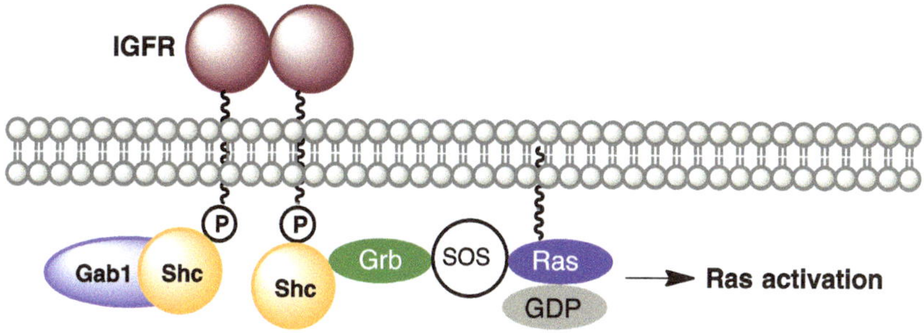

FIGURE 21.18 Activation of Ras by IGFR via Shc recruitment (IGF1R—insulin-like growth factor-1 receptor; Shc—Src homologous and collagen; GAB1—GRB2-associated-binding protein 1; Grb—growth factor receptor-bound protein 2; SoS—Son of Sevenless; Ras—rat sarcoma protein; GDP—guanosine diphosphate).

the IGF-1R and insulin receptor (IR) this strategy may be associated with metabolic adverse effects.

Since 2000, several agents targeting IGF1-IGF1R signaling have shown antitumor activity in preclinical studies which led to the evaluation of different drug candidates in clinical trials [54]. Although antibody-based anti-IGF1 signaling therapeutic development was at the forefront, many efforts were directed to design and develop small-molecule IGF1R kinase inhibitors [55]. Since these compounds cannot be considered peptidomimetics, they are outside the scope of this chapter.

21.7.2 Anti-IGF1R peptidomimetics

A peptide or peptidomimetic is considered to inhibit IGF1R activity if it inhibits any biological functions by binding to IGF1R or its targets, thereby interfering with IGFR1-ligand binding. Alternatively, or in addition, it may inhibit the dimerization of IGF1R. These activities were found in several partial sequences (for instance the sequence RXGNGV), containig at least six contigous amino acids of the juxtamembrane regions of IGF1Rs from several different species [56].

Following a different approach, two novel peptide vaccines and two B cell epitope mimics that specifically target the IGF1R and inhibit IGF1R-dependent signaling pathways were designed to interact with the contact sites between IGF1 and IGF1R on the basis of crystal structure and mutagenesis studies. The 56−81 and 233−251 B cell epitopes proved to be the best candidates, either as vaccines or peptide mimics, eliciting the most significant inhibitory effects both in vitro and in vivo [57]. Additionally, additive antitumor effects were found for the combination treatment of the IGF1R 56−81 epitope with HER1−418 and HER2−597 epitopes.

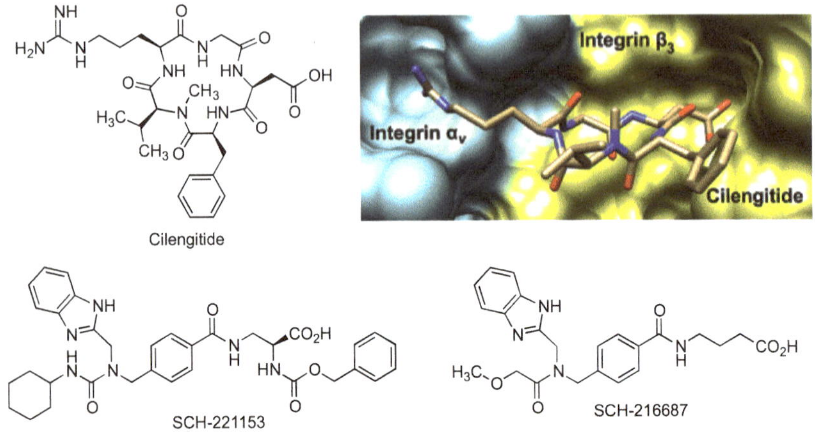

FIGURE 21.19 Cilengitide and its interaction with the integrin α_ν-β_3 interface (generated from PDB 1L5G) and additional peptidomimetic integrin inhibitors.

21.8 Peptidomimetic integrin inhibitors

Integrins have a key role in cell-cell and cell-extracellular matrix protein adhesion processes and therefore their inhibition constitutes an interesting approach to antiangiogenic therapy since their overexpression in endothelial cells of the tumor vasculature involves a high rate of neovascularization and invasive potential.

Integrins comprise an α- and a β-subunits and they often recognize their binding partners through an Arg-Gly-Asp (RGD) sequence. For this reason, several RGD-derived peptides and peptidomimetics have been studied as integrin antagonists. The first successful compound in this area was the cyclic peptide cilengitide, cyclo-(Arg-Gly-Asp-D-Phe-D-NMeVal), which binds at the interface between the α_ν and β_3 integrin subunits (Fig. 21.19). Cilengitide received orphan drug status from EMA in 2008, although a subsequent phase III trial with glioblastoma patients, in addition to the standard temozolomide and radiotherapy therapy, did not increase survival in patients with newly diagnosed glioblastoma and methylated MGMT promoter status. Other compounds that mimic the RGD motif and are able to inhibit angiogenesis and tumor growth are SCH-221153 and SCH-216687, which are dual inhibitors of the α_ν β_3 and α_ν β_5 integrins [58].

21.9 Peptidomimetics acting on transcriptional regulation

Defects in the modulation of several signaling pathways involved in transcriptional regulation such as the STAT, Notch, Wnt and Hedgehog pathways, are important in the genesis and growth of several types of cancer. Some stages in these pathways involve PPIs that can be disrupted by peptides [59] or peptidomimetic compounds, which will be our focus here.

21.9.1 Peptidomimetics acting on the signal transducer and activator of transcription pathway

Cytokine receptors do not have intrinsic kinase activity and rely on Janus kinases (JAKs), which are able to intracelularly activate downstream signaling proteins by phpsphorylation. JAKs phosphorylate signal transducer and activator of transcription (STAT) proteins, including STAT3, for instance, which are transcription factors that regulate many aspects of cell growth, survival, and differentiation and have a role in the activity of the immune system. Two molecules of the activated STAT3 dimerize by a reciprocal phosphotyrosine-SH2 domain interaction and the dimer then translocates to the cell nucleus, where it regulates gene transcription and thereby induce cell proliferation. Alternatively, STAT3 can be phosphorylated by the receptors of epidermal growth factors (Fig. 21.20).

The JAK/STAT pathway was identified as an anticancer target following the discovery that many patients with breakpoint cluster region (Bcr)-Abl1-negative chronic myeloproliferative cancers showed a mutation in a JAK. Aberrant regulation of the *STAT3* oncogene was also found to contribute to

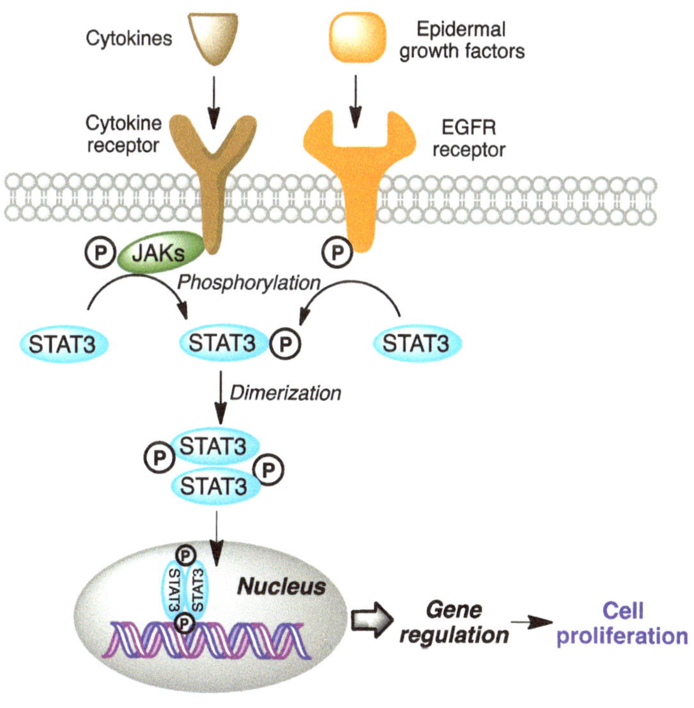

FIGURE 21.20 The JAK-STAT pathway using STAT3 as an example and its crosstalk with EGFR signaling (STAT—Signal transducer and activator of transcription, JAK—Janus kinase, EGFR—Epidermal growth factor).

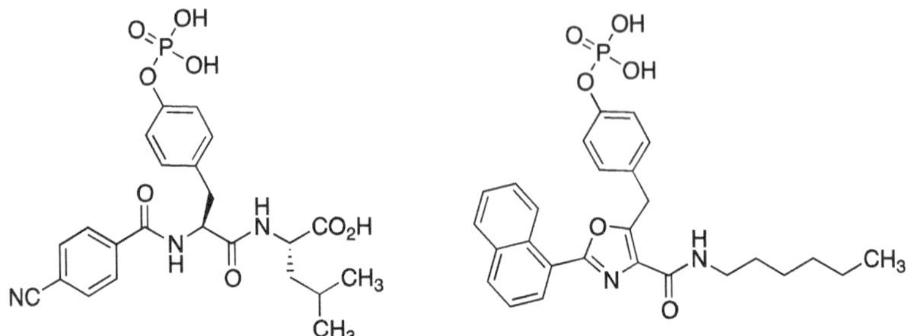

FIGURE 21.21 Peptidomimetic STAT3-targeting anticancer agents.

the formation and progression of many cancers, including head and neck squamous cell carcinoma (HNSCC).

STAT3-targeting anticancer agents may act by a variety of mechanisms [60], among which we will focus on peptidomimetic strategies aimed at disrupting the phosphorylated STAT3 (pSTAT3) dimer [61]. The starting point was the amino acid sequence close to phosphotyrosine 705 that takes part in STAT3 dimerization (Pro-**P**Tyr^{705}-Leu-Lys-Thr-Lys). Some interesting compounds derived from these studies are ISS-610, whose interaction with the Src Homology 2 (SH2) domain of STAT3, whose inhibition may block the STAT3 activity, was studied in depth, guiding the development of further analogs such as the oxazole-based S3I-M2001, which additionally exhibited inhibition of STAT3 phosphorylation and tumor growth in human breast tumor xenografts (Fig. 21.21).

21.9.2 Peptidomimetics acting at the Notch, TGFβ, and β-catenin pathways

The Notch1 receptor is a transcriptional activator of multiple genes, and its upregulation activates many signals connected to apoptosis inhibition and proliferation activation. Upregulation of Notch receptors and ligands is frequently observed in cancer and has been correlated with increased tumor cell growth and survival in hematological malignancies and solid tumors such as breast, lung and adenoid cystic carcinoma [62]. Furthermore, Notch inhibition downregulates other oncogenic pathways such as transforming growth factor beta (TGFβ) via the interaction between the Notch intracelular domain (NICD) and Smads, which are TGFβ downstream signaling mediators. It is also connected to the Wnt-β-catenin pathway, a key feature of epithelial cancers that is also critical for metastasis. Activation of the Wnt receptor leads to sequestration of the β-catenin phosphorylation complex, and the consequent suppression of β-catenin phosphorylation allows its translocation to the nucleus, where it promotes the transcription of a number of proteins, including TGFβ (Fig. 21.22).

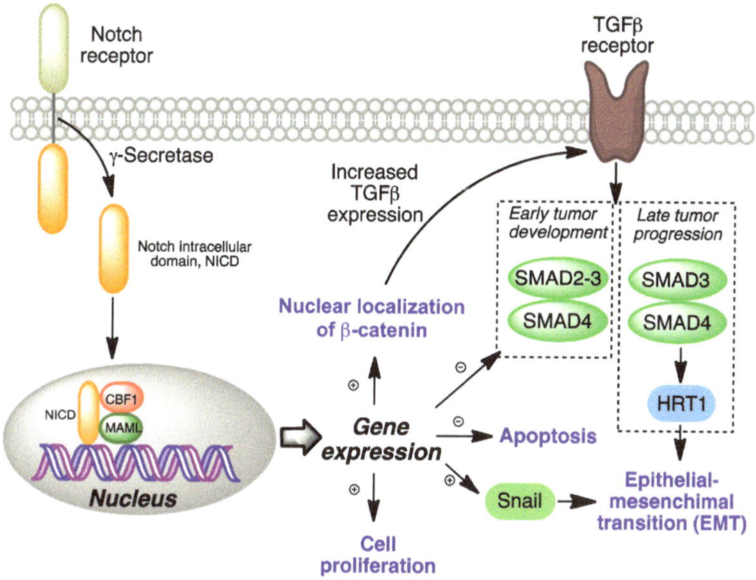

FIGURE 21.22 The Notch pathway and its crosstalk with TGF-β signaling.

The Notch intracellular domain activates transcription by forming a trimeric complex with the coactivator proteins of the MAML family and DNA-bound CBF1. The α-helical binding domain of MAML was used as the basis for the design of the (i − i + 4) stapled peptide SAHM1, which potently inhibited the trimer formation and reduced the transcription of Notch-related genes in leukemia cells [63].

The direct targeting of β-catenin is also a relevant goal in cancer treatment. The formation of the β-catenin/T-cell factor (Tcf) complex is the major driving force for the initiation, progression, and metastasis of triple-negative breast cancer. Compound **3** disrupts the β-catenin/Tcf PPI, which has a critical role in the β-catenin signaling pathway and was shown to inhibit the viability of Wnt/β-catenin hyperactive cancer cells [64]. Similarly, cell-penetrating peptidomimetic helical foldamers derived from sulfono-γ-AApeptides of general structure **4** were found to disrupt β-catenin/BCL9 PPIs with high potency and specificity due to their ability to mimic the BCL9 helix (Fig. 21.23) [65].

21.10 Anticancer peptidomimetics that modulate hormone action

In castration-resistant prostate cancer cells, the estrogen 17β-estradiol (E2) induces the formation of ternary complexes with androgen receptor (AR) and the transcriptional coactivator proline-, glutamic acid- and leucine-rich

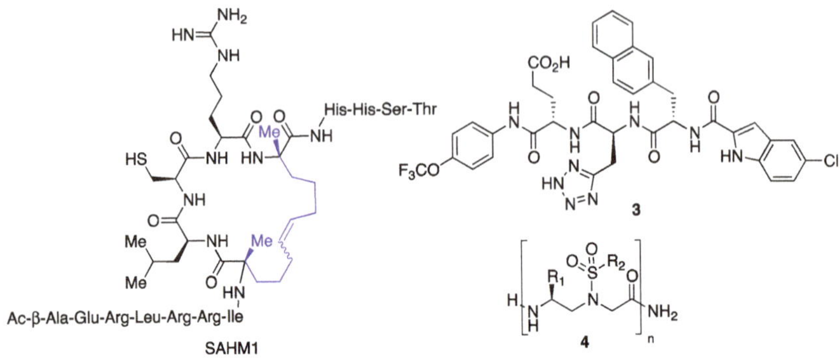

FIGURE 21.23 Peptidomimetic compounds targeting β-catenin protein–protein interactions.

protein 1 (PELP-1), enabling the transcription of AR-responsive genes [66]. PELP-1 has several LXXLL motifs (L, leucine; X, any amino acid) that adopt a helical structure and participate in many PPIs associated with transcription activation or repression. In the interaction of PELP-1 with the androgen receptor, the leucine side chains of the LXXLL motifs fit into the hydrophobic pocket.

As an interesting example of the use of small-molecule scaffolds to generate α-helix mimetics, the design of the peptidomimetic known as D2 was based on the structural modification of oligobenzamides **5** containing two or three 3-alkoxy-4-aminobenzoic acids. The two isobutyl groups of D2 occupy the leucine pockets at the i and $i+4$ positions of the LXXLL motif (Fig. 21.24A) [67]. D2 is stable, nontoxic and efficiently taken up by prostate cancer cells, where it disrupts androgen receptor-coregulator interactions. Structure-activity relationships (SARs) in a series of D2 derivatives, through modification of X, Y, R_1, and N_2 substituents (Fig. 21.24B), showed that a nitro group is critical to inhibit androgen-induced proliferation of prostate cancer cells in vivo, whereas compounds possessing either no substituent or an amine were inactive [68]. Using similar strategies, a number of peptidomimetics have been developed that target the estrogen receptor-α-coactivator interaction for breast cancer therapy [69].

21.11 Peptidomimetics targeting regulation of apoptosis

The balance between cell proliferation and apoptosis, or programmed cell death, is controlled by several pro- and antiapoptotic proteins, and the loss of this balance in favor of proliferation is a feature of many types of cancer. Apoptotic cell death requires the concerted action of sensor and effector proteins and comprises the extrinsic and intrinsic pathways. The former starts with the activation of cell surface death receptors (e.g., FAS and TNF-R1), while the latter is initiated when DNA damage or signaling imbalance are

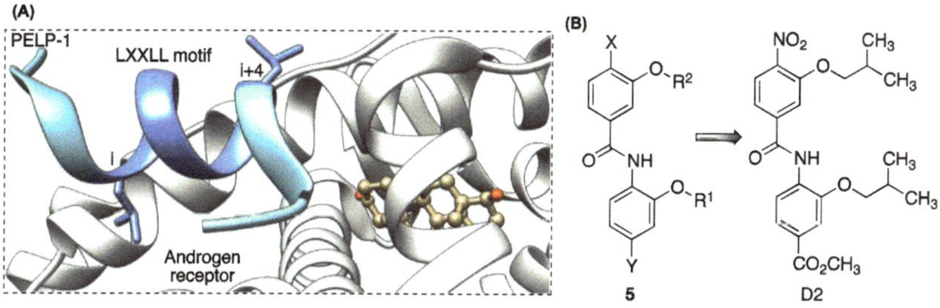

FIGURE 21.24 (A) PELP-1 (left) bound to the androgen receptor (right), generated from PDB 1t63. The LXXLL motif and the side chains of the *i* and *i* + 4 Leu residues are shown in a darker shade. Dihydrotestosterone is shown as a ball-and-stick molecule. (B) Structure of D2 and the structure-activity relationship study at the N/C-terminus (X and Y) and side chains (R1 and R2) that led to compound D2.

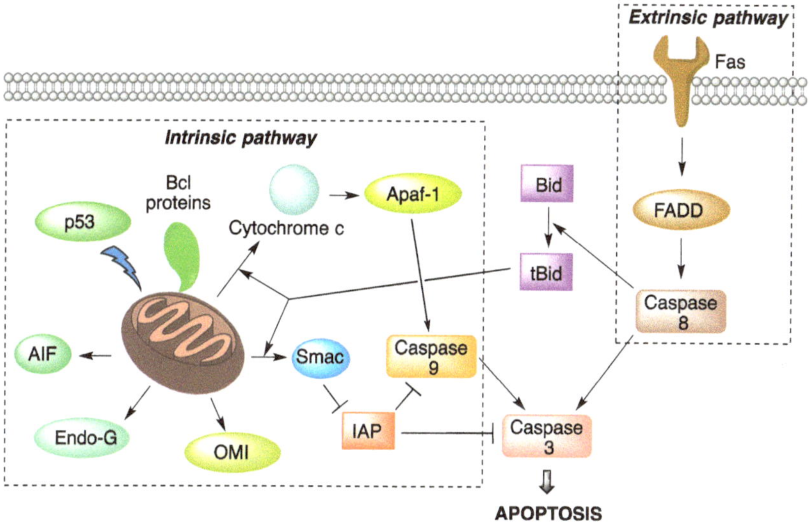

FIGURE 21.25 An outline of apoptosis. Pointed arrows indicate activation; flat arrows indicate inhibition. *FADD*—Fas-associated protein with death domain; *AIF*—apoptosis inducing factor; *IAP*—inhibitor of apoptosis protein; *APAF1*—apoptotic protease activating factor 1; *Bid*—BH3-interacting domain death agonist; *tBid*—truncated Bid.

detected by intracellular sensors. Both pathways require the proteolytic activation of procaspases to caspases, which are the effectors of apoptosis. On the other hand, several families of proteins, such as Bcls and IAPs, are inhibitors of apoptosis (Fig. 21.25).

Many cancer cells are able to evade apoptotic signals by a variety of mechanisms, including (1) upregulation of a nonsignaling FAS receptor that competes with the functional FAS; (2) loss of the proapoptotic p53 protein via mutation of the p53 tumor suppressor gene; (3) overexpression of Bcl-2;

(4) high levels of XIAP, one of the members of the IAP family of proteins and therefore an apoptosis inhibitor; and (5) low levels of Smac, an endogenous inhibitor of IAP. Targeting these mechanisms is an interesting strategy for anticancer therapy, and in this section, we focus on the development of Smac mimics for this purpose [70]. Smac binds competitively to the BIR3 domain of XIAP, which is mainly responsible for its interaction with caspase 9. The interaction of Smac with the XIAP surface involves its N-terminal tetrapeptide (Ala-Val-Pro-Ile, AVPI), and is shown in Fig. 21.26A and B.

The first generation of Smac mimetics was generated by manipulation of the AVPI tetrapeptide by N-methylation of the N-terminal Ala residue, replacement of the Val or Pro residues with unnatural amino acids and capping of the C-terminal residue with aromatic or heteroaromatic substituents, which occupy the pocket corresponding to the Ile side chain. Some compounds of interest arising from these studies include GDC-0152 **6**, the first IAP inhibitor to reach clinical trials, and GDC-0917 **7**, which had the advantage of being orally bioavailable due to its higher lipophilicity. In efforts to improve metabolic stability, additional modifications were designed using a structure-based approach, culminating in SM-406 (DeBio-1143, **8**), which is undergoing clinical trials for cancer treatment (Fig. 21.26C) [71].

The observation that dimeric Smac can interact with two different domains of XIAP, namely BIR2 and BIR3, led to the idea of preparing bivalent mimetics, containing two AVPI mimetic residues connected by a linker. The main representative of this family of compounds is birinapant (Fig. 21.27), which in 2019 entered a Phase I study in combination with radiation therapy for the treatment of recurrent head and neck squamous cell carcinoma.

The Bcl2 protein family is also key to the apoptosis process and takes part in a large number of interactions with both pro-apoptotic and antiapoptotic proteins. Therefore modulation of these proteins is of great interest in anticancer drug discovery [72].

The p53 protein, often described as the "guardian of the genome," mediates the expression of genes involved in apoptosis and also in cytoprotective processes such as DNA repair and cell cycle arrest. This transcription factor is downregulated by the MDM2 and MDMX proteins, which act either by mediating the ubiquitylation of p53 and its subsequent degradation by the proteasome, as mentioned in Section 2.1 or by directly antagonizing p53. The latter mechanism can be blocked by PPI inhibitors, opening up the possibility of using peptidomimetics to upregulate the activity of p53. Many peptide-like compounds have been designed with that purpose, and one representative example is the stapled peptide SAH-p53−8, a modified p53 designed to have a similar α-helix structure [73]. Small molecules able to interfere with the MDM-p53 interaction have also been developed, exemplified by nutlin-2.

In the p53-MDM interaction, the Phe-19, Trp-23, and Leu-26 residues in p53 occupy hydrophobic pockets of the MDM protein, as shown by X-ray

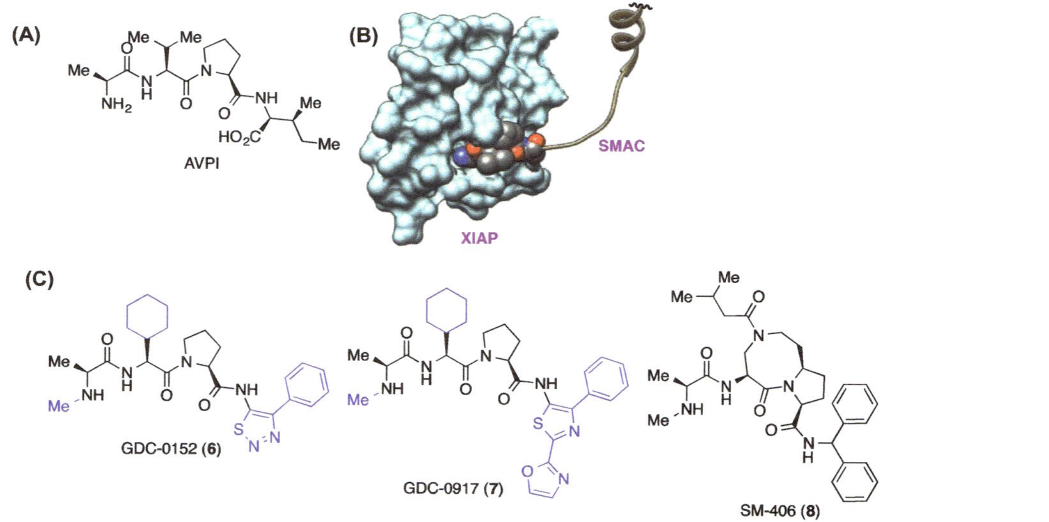

FIGURE 21.26 (A) The N-terminal Ala-Val-Pro-Ile tetrapeptide of Smac. (B) Interaction between XIAP and Smac (generated from PDB 1g73). (C) Representative first-generation Smac mimetics.

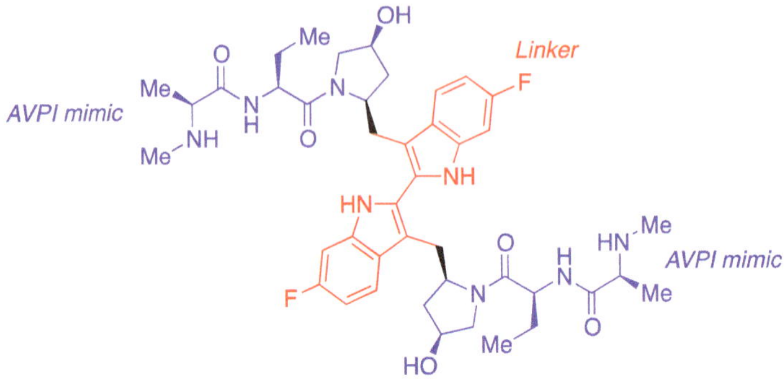

Birinapant

FIGURE 21.27 Structure of birinapant.

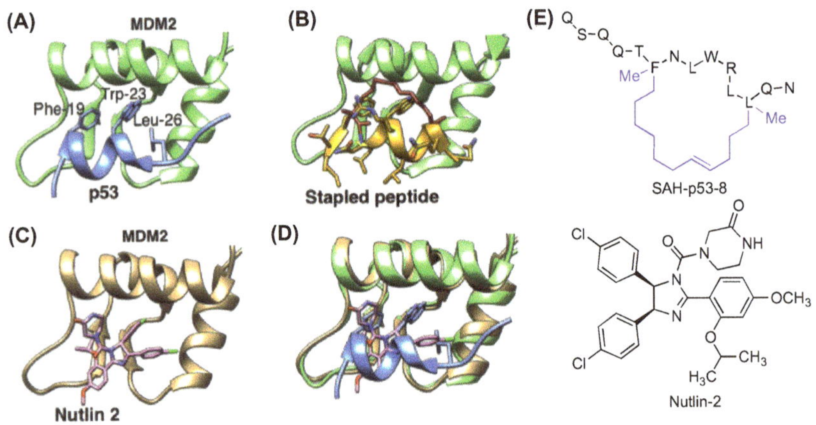

FIGURE 21.28 (A–C) Interaction of MDM2 with p53 (PDB 1YCR), the stapled peptide SAH-p53–8 (PDB 3V3B), and nutlin-2 (PDB 4HG7), respectively. (D) Superimposition of p53 and nutlin-2 at the MDM2 binding site. (E) Structures of SAH-p53–8 and nutlin.

studies (Fig. 21.28A). In a similar fashion, SAH-p53–8 fills the binding site in a similar manner via the same Phe-19, Trp-23 and Leu-26 residues (Fig. 21.28B). Regarding small-molecule inhibitors, the ethoxy and bromophenyl groups of nutlin-2 bind to the pockets corresponding to Phe-19, Trp-23, and Leu-26 of p53 and prevents its interaction with the N-terminal domain of MDM2 (Fig. 21.28C and D).

21.12 Peptidomimetics targeting tubulin

Microtubules have a key role in mitosis and for this reason, they represent one of the main targets for anticancer drugs. Among microtubule targeting agents, paclitaxel is the one most commonly used in the clinic (Fig. 21.29).

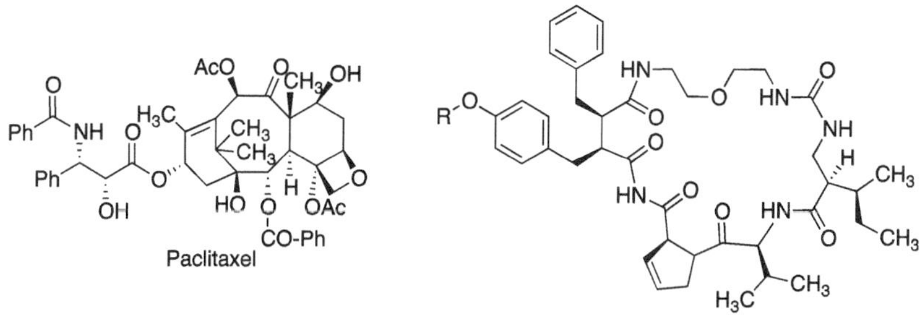

9 R = H
10 R = Bn

FIGURE 21.29 Structures of paclitaxel and cyclic peptidomimetics that mimick the interaction of Nur77 with tubulin and Bcl-2.

Taxanes, besides targeting β-tubulin, also behave as Bcl-2 molecular switchers, inducing apoptosis. Paclitaxel behaves as a peptidomimetic of the endogenous protein Nur77, which interacts with both paclitaxel targets on the same binding sites. In the course of this work, some linear peptides that mimicked the interaction of Nur77 with tubulin and Bcl-2 were developed [74]. From this starting point, several cyclic peptides and peptidomimetics, represented by compounds **9** and **10**, were synthesized, which interacted with tubulin and induced apoptosis (Fig. 21.29) [75].

21.13 Conclusion

PPI interference and modulation is an emerging field in drug discovery that has brought to light targets previously viewed as undruggable. The use of peptidomimetics for targeting PPIs in cancer therapy has yielded drug candidates in clinical trials. The approach is sufficiently mature and is poised to become an indispensable tool in drug discovery.

References

[1] Akram ON, DeGraff DJ, Sheehan JH, Tilley WD, Matusik RJ, Ahn JM, et al. Tailoring peptidomimetics for targeting protein-protein interactions. Mol Cancer Res 2014; 12:967–78.

[2] Mabonga L, Kappo AP. Peptidomimetics: a synthetic tool for inhibiting protein–protein interactions in cancer. Int J Peptide Res Therap 2020;26:225–41.

[3] Smith AB, Charnley AK, Hirschmann R. Pyrrolinone-based peptidomimetics. "Let the enzyme or receptor be the judge.". Acc Chem Res 2011;44:180–93.

[4] Jacobson AD, Zhang NY, Xu P, Han K-J, Noone S, Cheng J, et al. The lysine 48 and lysine 63 ubiquitin conjugates are processed differently by the 26 S proteasome. J Biol Chem 2009;284:35485–94.

[5] Dick TP, Nussbaum AK, Deeg M, Heinemeyer W, Groll M, Schirle M, et al. Contribution of proteasomal beta-subunits to the cleavage of peptide substrates analyzed with yeast mutants. J Biol Chem 1998;273:25637–46.

[6] Myung J, Kim KB, Crews CM. Proteasome inhibition: mechanism and inhibitors. Med Res Rev 2001;21:245–73.

[7] Kisselev AF, van der Linden WA, Overkleeft HS. Proteasome inhibitors: an expanding army attacking a unique target. Chem Biol 2012;19:99–115.

[8] Johnson DE. The ubiquitin-proteasome system: opportunities for therapeutic intervention in solid tumors. Endocr Relat Cancers 2015;22:T1–17.

[9] Moreau P, Richardson PG, Cavo M, Orlowski RZ, San Miguel JF, Palumbo A, et al. Proteasome inhibitors in multiple myeloma: 10 years later. Blood 2012;120:947–59.

[10] Manasanch EE, Orlowski RZ. Proteasome inhibitors in cancer therapy. Nat Rev Clin Oncol 2017;14:417–33.

[11] Morozov AV, Karpov VL. Proteasomes and several aspects of their heterogeneity relevant to cancer. Front Oncol 2019;9:761.

[12] Grigoreva TA, Tribulovich VG, Garabadzhiu AV, Melino G, Barlev NA. The 26S proteasome is a multifaceted target for anti-cancer therapies. Oncotarget 2015;6:24733–49.

[13] Karni-Schmidt O, Lokshin M, Prives C. The roles of MDM2 and MDMX in cancer. Annu Rev Pathol: Mech Dis 2016;11:617–44.

[14] Ōmura S, Crump A. Lactacystin: first-in-class proteasome inhibitor still excelling and an exemplar for future antibiotic research. J Antibiot 2019;72:189–201.

[15] https://clinicaltrials.gov/ct2/show/NCT03345095.

[16] Avendaño C, Menéndez JC. Chapter 11 Medicinal chemistry of anticancer drugs. 2nd Edition Elsevier; 2015.

[17] Zhou HJ, Aujay MA, Bennett MK, Dajee M, Demo SD, Fang Y, et al. Design and synthesis of an orally bioavailable and selective peptide epoxyketone proteasome inhibitor (PR-047). J Med Chem 2009;52:3028–38.

[18] Shah J, Usmani S, Stadtmauer EA, Rifkin RM, Berenson JR, Berdeja JG, et al. Oprozomib, pomalidomide, and dexamethasone in patients with relapsed and/or refractory multiple myeloma. Clin Lymphoma Myeloma Leukemia 2019;19:570–8.

[19] Jain RK. Antiangiogenesis strategies revisited: From starving tumors to alleviating hypoxia. Cancer Cell 2014;26:605–22.

[20] Rao BG. Recent developments in the design of specific matrix metalloproteinase inhibitors aided by structural and computational studies. Curr Pharm Des 2005;11:295–322.

[21] Comunanza V, Busolino F. Therapy for cancer: strategy of combining anti-angiogenic and target therapies. Front Cell Develop Biol 2017;5 article 101.

[22] Kumar GB, Nair BG, Perry JP, Martin DBC. Recent insights into natural product inhibitors of matrix metalloproteinases. Med Chem Commun 2019;10:2024–37.

[23] Winer A, Sylvia Adams S, Mignatti P. Mol Cancer Therap 2018;17:1147–55.

[24] Wong AH, Zhou D, Rini JM. The X-ray crystal structure of human aminopeptidase N reveals a novel dimer and the basis for peptide processing. J Biol Chem 2012;287:36804–13.

[25] Liang W, Gao B, Xu G, Weng D, Xie M, Qian Y. Possible contribution of aminopeptidase N (APN/CD13) to migration and invasion of human osteosarcoma cell lines. Int J Oncol 2014;45:2475–85.

[26] Amin SA, Adhikari N, Jha T. Design of aminopeptidase N inhibitors as anti-cancer agents. J Med Chem 2018;61:6468–90.

[27] Krige D, Needham LA, Bawden LJ, Flores N, Farmer H, Miles LEC, et al. CHR-2797: an antiproliferative aminopeptidase inhibitor that leads to amino acid deprivation in human leukemic cells. Cancer Res 2008;68:6669–79.

[28] Scornik OA, Botbol V. Bestatin as an experimental tool in mammals. Curr Drug Metab 2001;2:67–85.
[29] Ota K, Uzuka Y. Clinical trials of bestatin for leukemia and solid tumors. Biotherapy 1992;4:205–14.
[30] Luan Y, Ma C, Sui Z, Wang X, Feng J, Liu ND, et al. LYP3, a new bestatin derivative for aminopeptidase N inhibition. Med Chem 2011;7:32–6.
[31] Gao JJ, Gao ZH, Zhao CR, Yuan Y, Cui SX, Zang XF, et al. LYP, a novel bestatin derivative, inhibits cell growth and suppresses APN/CD13 activity in human ovarian carcinoma cells more potently than bestatin. Investig N Drugs 2011;29:574–82.
[32] Lee S, Desai P, Edirisinghe B, Pianello S, Singer JW, Samuel M, et al. Phase II clinical study of the clinical efficacy and safety of tosedostat in patients with myelodysplastic syndromes (MDS) after failure of hypomethylating agent-based therapy. Blood 2017;130 (Supplement 1):2966.
[33] Simanshu DK, Nissley DV, McCormick F. RAS proteins and their regulators in human disease. Cell 2017;170:17–33.
[34] Leicht DT, Balan V, Kaplun A, Singh-Gupta V, Kaplun L, Dobson M, et al. Raf kinases: function, regulation and role in human cancer. Biochimica et Biophysica Acta 2007;1773:1196–212.
[35] Ni D, Li X, He X, Zhang H, Zhang J, Lu S. Drugging K-Ras^{G12C} through covalent inhibitors: mission possible? Pharmacol Therap 2019;202:1–17.
[36] Yoo DY, Hauser AD, Joy ST, Bar-Sagi D, Arora PS. Covalent targeting of Ras G12C by rationally designed peptidomimetics. ACS Chem Biol 2020;15:1604–12.
[37] Ostrem JM, Peters U, Sos ML, Wells JA, Shokat KM. K-Ras(G12C) inhibitors allosterically control GTP affinity and effector interactions. Nature 2013;503:548–51.
[38] Patricelli MP, Janes MR, Li L-S, Hansen R, Peters U, Kessler LV, et al. Cancer Discov 2016;6:316–29.
[39] Goebel L, Müller MP, Goody RS, Rauh D. KRasG12C inhibitors in clinical trials: a short historical perspective. RSC Med Chem, 11. 2020. p. 760–70.
[40] Kessler D, Gmachl M, Mantoulidis A, Martin LJ, Zoephel A, Mayer M, et al. Drugging an undruggable pocket on KRAS. Proc Natl Acad Sci 2019;116:15823–9.
[41] Moorthy NSHN, Sousa SF, Ramos MJ, Fernandes PA. Farnesyltransferase inhibitors: a comprehensive review based on quantitative structural analysis. Curr Med Chem 2013;20:4888–923.
[42] Wang J, Yao X, Huang J, Wang J, Yao X, Huang J. New tricks for human farnesyltransferase inhibitor: cancer and beyond. Medchemcomm. 2017;8:841–54.
[43] Le DT, Shannon KM. Ras processing as a therapeutic target in hematologic malignancies. Curr OpHematol 2002;9:308–15.
[44] Reid TS, Beese LS. Crystal structures of the anticancer clinical candidates R115777 (Zarbestra®/tipifarnib) and BMS-214662 complexed with protein farnesyltransferase suggest mechanism of FTI selectivity. Biochemistry 2004;43:6877–84.
[45] Meric-Bernstam F, Johnson AM, Dumbrava EEI, Raghav K, Balaji K, Bhatt M, et al. Advances in HER2-targeted therapy: novel agents and opportunities beyond breast and gastric cancer. Clin Cancer Res 2019;25:2033–241.
[46] Arteaga CL, Engelman JA. ERBB receptors: from oncogene discovery to basic science to mechanism-based cancer therapeutics. Cancer Cell 2014;25:282–303.
[47] Kanthala S, Banappagari S, Gokhale A, Liu Y-Y, Xin G, Zhao Y, et al. Novel peptidomimetics for inhibition of HER2:HER3 heterodimerization in HER2-positive breast cancer. Chem Biol Drug Des 2015;85:702–14.

[48] Cho HS, Mason K, Ramyar KX, Stanley AM, Gabelli SB, Denney DW, et al. Structure of the extracellular region of HER2 alone and in complex with the herceptin Fab. Nature 2003;421:756–60.

[49] Franklin MC, Carey KD, Vajdos FF, Leahy DJ, de Vos AM, Sliwkowski MX. Insights into ErbB signaling from the structure of the ErbB2-pertuzumab complex. Cancer Cell 2004;5:317–28.

[50] Satyanarayanajois S, Villalba S, Jianchao L, Lin GM. Design, synthesis, and docking studies of peptidomimetics based on HER2-herceptin binding site with potential antiproliferative activity against breast cancer cell lines. Chem Biol Drug Des 2009;74:246–57.

[51] Banappagari S, Ronald S, Satyanarayanajois SD. A conformationally constrained peptidomimetic binds to the extracellular region of HER2 protein. J Biomol Struct Dyn 2010;28:289–308.

[52] Samrein BM, Ahmed SBM, Prigent SA. Insights into the Shc family of adaptor proteins. J Mol Signal 2017;12 article 2.

[53] Haisa M. The type 1 insulin-like growth factor receptor signalling system and targeted tyrosine kinase inhibition in cancer. J Int Med Res 2013;41:253–64.

[54] Osborne R. Commercial interest waxes for IGF-1 blockers. Nat Biotechnol 2008; 26:719–20.

[55] Li R, Pourpak A, Morris SW. Inhibition of the insulin-like growth factor-1 receptor (IGF1R) tyrosine kinase as a novel cancer therapy approach. J Med Chem 2009; 52:4981–5004.

[56] Tarasova NI, Tarasov SG. Synthetic analogs of the juxtamembrane domain of igf1r and uses thereof. United States Patent Application Publication No: US 201/019591 A1, **2011**.

[57] Foy JC, Miller MJ, Overholser J, Donnelly SM, Nahta R, Kaumaya PTP. IGF-1R peptide vaccines/mimics inhibit the growth of BxPC3 and JIMT-1 cancer cells and exhibit synergistic antitumor effects with HER-1 and HER-2 peptides. Oncoimmunology 2014;3: e956005.

[58] Kumar CC, Malkowski M, Yin Z, Tanghetti E, Yaremko B, Nechuta T, et al. Inhibition of angiogenesis and tumor growth by SCH221153, a dual the α_ν β_3 and α_ν β_5 integrin receptor antagonist. Cancer Res 2001;61:2232–8.

[59] Inamoto I, Shin JA. Peptide therapeutics that directly target transcription factors. Peptide Sci 2019;111:e24048.

[60] Geiger JL, Grandis JR, Bauman JE. The STAT3 pathway as a therapeutic target in head and neck cancer: barriers and innovations. Oral Oncol 2016;56:84–92.

[61] Morlacchi P, Robertson FM, Klostergaard J, McMurray JS. Targeting SH_2 domains in breast cancer. Future MedChem 2014;6:1909–26.

[62] Moore G, Annett S, McClements L, Robson T. Top Notch targeting strategies in cancer: a detailed overview of recent insights and current perspectives. Cells 2020;9:1503.

[63] Moellering RE, Cornejo M, Davis TN, Del Bianco C, Aster JC, Blacklow SC, et al. Direct inhibition of the NOTCH transcription factor complex. Nature 2009;462:182–8.

[64] Wang Z, Zhang M, Wang J, Ji H. Optimization of peptidomimetics as selective inhibitors for the β-catenin/T-cell factor protein – protein interaction. J Med Chem 2019;62:3617–35.

[65] Sanga P, Zhang M, Shi Y, Li C, Abdulkadir S, Li Q, et al. Inhibition of β-catenin/B cell lymphoma 9 protein – protein interaction using α-helix–mimicking sulfono-γ-AApeptide inhibitors. Proc Natl Acad Sci 2019;116:10757–62.

[66] Yang L, Ravindranathan P, Ramanan M, Kapur P, Hammes SR, Hsieh J-T, et al. Central role for PELP1 in nonandrogenic activation of the androgen receptor in prostate cancer. Mol Endocrinol 2012;26:550–61.

[67] Ravindranathan P, Lee T-K, Yang L, Centenera MM, Butler L, Tiley WD, et al. Peptidomimetic targeting of critical androgen receptor−coregulator interactions in prostate cancer. Nat Commun 2013;4 article 1923.
[68] Lee T-K, Ravindranathan P, Sonavane R, Raj GV, Ahn J-M. A structure-activity relationship study of bis-benzamides as inhibitors of androgen receptor-coactivator interaction. Molecules 2019;24:2783.
[69] Qina W, Xiea M, Qin X, Fang Q, Yin F, Li Z. Recent advances in peptidomimctics antagonists targeting estrogen receptor a-coactivator interaction in cancer therapy. Bioorg Med Chem Lett 2018;28:2827−36.
[70] For a review, see Elsayed NMY, Abou El Ella DA, Serya RAT, Abouzid KAM. Targeting apoptotic machinery as approach for anticancer therapy: Smac mimetics as anticancer agents. Future J Pharm Sci 2015;1:16−21.
[71] Hurwitz HI, Smith DC, Pitot HC, Brill JM, Chugh R, Rouits E, et al. Safety, pharmacokinetics, and pharmacodynamic properties of oral DEBIO1143 (AT-406) in patients with advanced cancer: results of a first-in-man study. Cancer Chemother Pharmacol 2015;75:851−9.
[72] Moldoveanu T, Follis AV, Kriwacki RW, Green DR. Many players in BCL-2 family affairs. Trends Biochem Sci 2014;39:101−11.
[73] Baek S, Kutchukian PS, Verdine GL, Huber R, Holak TA, Lee KW, et al. Structure of the stapled p53 peptide bound to Mdm2. J Am Chem Soc 2012;134:103−6.
[74] Ferlini C, Cicchillitti L, Raspaglio G, Bartollino S, Cimitan S, Bertucci C, et al. Paclitaxel directly binds to Bcl-2 and functionally mimics activity of Nur77. Cancer Res 2009;69:6906−14.
[75] Brindisi M, Maramai S, Brogi S, Fanigliulo E, Butini S, Guarino E, et al. Development of novel cyclic peptides as pro-apoptotic agents. Eur J Med Chem 2016;117:301−20.

Chapter 22

Immunomodulatory peptidomimetics for multiple sclerosis therapy—the story of glatiramer acetate (Copaxone)

Rina Aharoni
Department of Immunology, Weizmann Institute of Science, Rehovot, Israel

22.1 Introduction

Multiple sclerosis (MS) is a chronic disease of the central nervous system (CNS), and one of the most prevalent neurological disorders leading to chronic disability among young adults [1]. The disease is heterogeneous in its clinical manifestations as well as in its pathological mechanisms. The most typical clinical progression pattern is an initial phase of relapsing and remitting symptoms (relapsing-remitting MS, RRMS) that can develop to a progressive disease course (secondary progressive MS, SPMS) in some patients. A fraction of patients show disease progression from the disease onset (primary progressive MS, PPMS), which represents a somewhat different pathology [2]. The definite etiology of MS is still unknown, but it is established that complex interactions between environmental factors and multiple genes are involved [1].

Traditionally, MS is considered an autoimmune disease in which the immune system reacts against the body's own constituents, in this case against the myelin that surrounds the axons, initiating a vicious inflammatory cascade in the CNS [1,3,4]. The disease is mediated by autoreactive effector T-cells, T-helper (Th)-1 and Th-17, that recognize specific myelin antigens. Immune cells of both the adaptive and innate immune systems, including cytotoxic T-cells, B-cells and monocytes/macrophages, are further involved in the inflammatory network and inflict myelin damage. In addition to inflammation and demyelination, axonal and neuronal pathologies are crucial components of this disease, in particular during the progressive stage in which degeneration prevails [2]. The disease is modulated by antiinflammatory

Peptide and Peptidomimetic Therapeutics. DOI: https://doi.org/10.1016/B978-0-12-820141-1.00020-0

constituents of the immune system, such as T-cells of the Th-2 pathway, regulatory T-cells (Tregs) and regulatory B-cells (Bregs) [4–6]. Repair processes such as remyelination and neuroprotection are further stimulated in an attempt to overcome myelin damage and degeneration [4,7]. Therapeutic approaches aim to suppress the detrimental inflammation and enhance counteracting antiinflammatory and neuroprotective repair processes in the CNS. The main pathological and modulatory pathways in MS are depicted in Fig. 22.1.

Essential data on MS has been obtained using the animal model of experimental autoimmune encephalomyelitis (EAE), which is induced by immunization with myelin antigens such as myelin basic protein (MBP), myelin oligodendrocyte glycoprotein (MOG), or myelin proteolipid protein (PLP) [4,8]. The encephalitogenic regions (epitopes) of these myelin antigens were identified, and the disease can be induced by immunization with the corresponding peptides. EAE induction by different encephalitogenic peptides in susceptible animal strains leads to the development of different disease forms

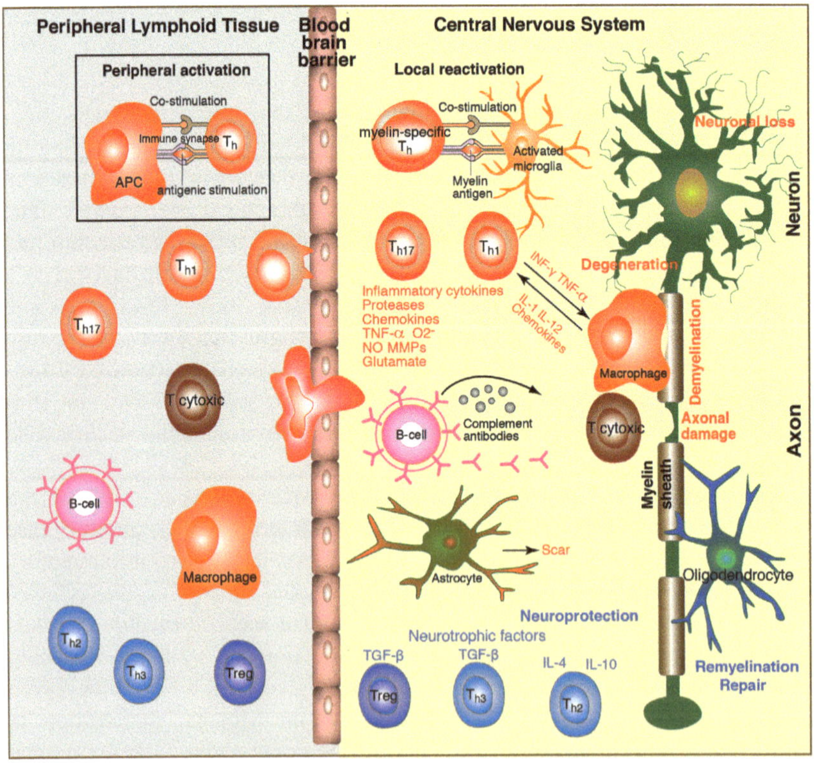

FIGURE 22.1 The main pathological and modulatory pathways in multiple sclerosis and experimental autoimmune encephalomyelitis.

that resemble various MS patterns. For instance, characteristic clinical symptoms manifested as ascending paralysis occurs in an acute monophasic manner in SJL/J mice immunized by the peptide corresponding to amino acids 89–101 of MBP, a chronic manner in C57BL/6 mice immunized by the peptide corresponding to amino acids 35–55 of MOG, and in a relapsing-remitting manner in SJL/J mice immunized by the peptide corresponding to amino acids 139–151 of PLP [9]. An immune response to these peptides is manifested in the blood and the cerebrospinal fluid (CSF) of MS patients [10], suggesting that these epitopes are also involved in human disease.

Interactions between effector T-cells and antigen-presenting cells (APCs) govern the immune response in EAE as well as in MS [11]. Specific interactions occur between encephalitogenic peptides presented by major histocompatibility complex (MHC) class II molecules on APCs to the T-cell receptor (TCR), whereas nonspecific signals occur between pairs of costimulatory molecules expressed on these cells. Several approaches to develop peptidomimetic analogs for the various components participating in the disease have been explored in attempts to modulate the disease. The following are examples of such strategies.

22.2 Peptidomimetics for modulation of experimental autoimmune encephalomyelitis / multiple sclerosis

22.2.1 Antigen-specific approaches

Antigen based MS therapies aim at targeting the specific encephalitogenic response to myelin and inducing "restricted tolerance," or protective antiinflammatory immune responses, without disabling the entire immune system. In the EAE model, several myelin proteins are effective in preventing the disease [12]. However, in MS patients, a trial testing oral administration of the whole MBP and PLP proteins resulted in negative findings [13]. Synthetic peptides with a high specificity could offer more effective and selective therapy, but the use of natural endogenously occurring myelin peptides as such is hazardous, due to their encephalitogenic potential. In this context, peptidomimetic molecules designed to mimic existing peptides could constitute a promising class of therapeutics.

Therapeutic interventions aiming to modulate specific pathogenic T-cell responses have been attempted using altered peptide ligands (APLs) [14]. These peptides are created by applying substitutions of one or several amino acids to the native encephalitogenic sequences recognized by autoreactive T-cells. Depending on the substitution at the TCR contact residues, APLs prevented and even reversed EAE. Subsequently, several clinical trials tested various APLs corresponding to the principal encephalitogenic epitope of MBP (amino acids 83–99), with D-amino acid substitutions in the TCR contact site [15]. Reduced inflammation and elevated antiinflammatory responses against the native antigen were detected. However, trials were suspended due to severe adverse events, such as hypersensitivity reactions and disease exacerbations, probably due to cross-reactivity between the APLs and the native agonists.

The current consensus is that there is not one, but several autoantigens involved in MS, and there is heterogeneity in patient responses as well as epitope spreading [16]. Whereas in the initial disease stages autoreactivity may be specific to a particular antigen, eventually tissue damage induces priming and shifting to multiple encephalitogenic determinants. The possibility to overcome such a multifaceted disease by targeting a single specific epitope is therefore questionable.

22.3 Approaches for blocking activation signals

Several peptidomimetic strategies for blocking signals leading to nonspecific effector T-cell activation have been designed, and their effects were demonstrated in vitro and animal models. Major efforts have focused on the disruption of the interactions between T-cells and APCs.

One approach is to target MHC class II on APCs. Peptidomimetic compounds that bind to MHC class II molecules are expected to compete with antigen presentation on this haplotype (MHC blocking), including the autoantigens presentation to pathogenic T-cells, and thereby interrupt the disease process. Indeed, several such compounds were shown to inhibit antigen presentation and suppress EAE in transgenic mice carrying the relevant human class II alleles (HLA-DR) [17].

Another approach is to target molecules on T-cells. The T-cell transmembrane glycoprotein CD4 plays a central role in T-cell function, interacting with MHC class II molecules on the APC and facilitating cell-to-cell interactions. Therefore, there is interest in CD4 modifiers for potential therapeutic use. Several aromatically modified exocyclic peptides derived from the CDR3 region of the CD4 molecule were shown to affect the T-cell response [18]. Although these peptides had similar structures, they presented different biological functions. Whereas CD4-Ser and CD4-Cys mimetics exhibited classical antagonistic activity and reduced EAE severity, the CD4-Met mimetic revealed properties of an agonist and exacerbated the disease, emphasizing the hazardous potential of this approach. Peptidomimetics for blocking the costimulatory interactions of CD28/CTLA-4:B7 and CD40: CD40L, which are required for full T-cell activation, have also been explored and have demonstrated efficacy in treating EAE [19].

The above peptidomimetics provided insights into the complex interactions mediating the immune response. However, their use for human therapy requires thorough preclinical assessment due to their hazardous potential to act in an agonistic manner and exacerbate the disease. Furthermore, whereas antigen-specific peptidomimetics might be "too specific" to combat the multiple autoantigens involved in MS pathology, peptidomimetics that block overall activation signals might nonspecifically disable the immune system broadly. So far, small peptides fulfilling the "classical" definition of peptidomimetics targeting a single aspect of MS pathophysiology have not been fruitful in counteracting this complex disease. A wider definition of peptidomimetics is exemplified by the case of glatiramer acetate (GA or Copaxone), which was originally designed as a mimetic to the whole autoantigen MBP in an attempt to create an experimental model but was found to be one of the first disease-modifying therapies (DMTs) for MS.

22.4 The story of glatiramer acetate (Copaxone)

In the late 1960s, Sela, Arnon, and Teitelbaum at the Weizmann Institute in Israel, conducting basic research on the immunological properties of synthetic polymers and copolymers, made a serendipitous drug discovery [20]. They hypothesized that synthetic polypeptides with amino acid analogous to those of the autoantigen MBP would induce EAE and designed several such copolymeric mixtures. None of these copolymers proved to be encephalitogenic. Counterintuitively, some of the molecules protected animals against EAE. Copolymer 1 (Cop 1), randomly composed of L-alanine, L-lysine, L-glutamic acid, and L-tyrosine, in a molar ratio of 4.2:3.4:1.4:1.0 proved to have the greatest activity in ameliorating EAE. This activity was not restricted to a particular species or encephalitogen, as Cop 1 ameliorated EAE induced by several myelin antigens in guinea pigs, rabbits, various mouse strains, and in two kinds of monkeys [21]. Extensive clinical trials and long-term studies established the efficacy and particular safety of this copolymer as a drug for MS, which was named GAr (Copaxone) and approved for the treatment of RRMS in 1997 [22–25]. GA at a subcutaneous dose of 20 mg daily or 40 mg three times weekly alters the course of RRMS by reducing relapse rate and disability. These consequences are consistent with magnetic resonance imaging (MRI) findings from various clinical centers [26]. Currently, GA is still one of the most widely prescribed first-line treatments for RRMS, although several novel treatments have recently been approved [27].

GA is the first, and so far the only, therapeutic agent to have a copolymer as its active ingredient. Due to the complexity and variability of its polypeptide mixture, a clear definition of its active component has not been established. The notion is that the overall "pool" of various amino acid sequences is required for its therapeutic activity, engendering multiple peptidomimetic species and facilitating several modes of action. Over the last four decades, findings from our and other laboratories have delineated GA mechanisms of action [28]. An overview of GA effects in EAE/MS is discussed below and depicted in Fig. 22.2.

Peripheral immunomodulation	Competition for MHC	Promiscuous binding to various MHC class II molecules, displacement of myelin antigens from the MHC binding groove.	
	Alteration of the innate immune response	Inhibitory effect on monocytes reactivity, deviation of dendritic cells and monocytes to produce less TNF-α and IL-12, more IL-10 and TGF-β, and to stimulate Th2 anti-inflammatory responses.	
	T-cell receptor antagonism	Inhibition of the activation of T-cells specific to the 82-100 epitope of MBP.	
	T-cell deviation	Induction of specific Th2/3-cells that secrete high amounts of IL-4, IL-5, IL-10, and TGF-β. Elevation of the prevalence and function of T-regulatory cells, activation of the transcription factor Foxp3. Reduction of Th-17 cells and their transcription factors RORγt. Improvement of the regulatory function of $CD8^+$ T-cells.	
	Modification of B-cells	Induction of antibodies with beneficial rather than neutralizing activity. Bias towards production of anti-inflammatory cytokines such as IL-10. Down-regulating of chemokine receptors.	
Immunomodulation in the CNS	Secretion of anti-inflammatory cytokine	GA-specific Th2/3 cells cross the BBB and secrete *in situ* anti-inflammatory cytokines. Bystander expression of IL-10 and TGF-β by resident astrocyte and microglia. Reduction in the overall expression of IFN-γ.	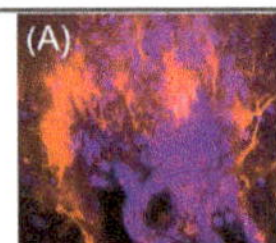 (A)
	Th-17 and T-regulatory cells	Decrease in the amount of Th-17 cells. Increase in T-regulatory cells.	 (B)
Neuroprotection	Elevation of neurotrophic factors	GA-specific T-cells express BDNF in the brain. Restoration of the impaired expressions of BDNF, NT-3, NT4, IGF-1, and IGF-2.	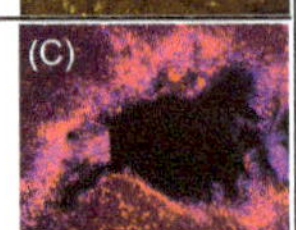 (C)
	Reduced CNS injury	Prevention of demyelination. Preservation of retinal ganglion cells. Inhibition of motor neuron loss. Preservation of brain tissue integrity by the MRI parameters MTR and DTI. Reduced formation of "black holes." Increase in NAA:Cr ratio.	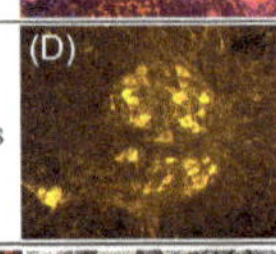 (D)
	Remyelination	Augmented remyelination. Increased proliferation, maturation and survival of oligodendrocyte progenitor cells and their accumulation in the lesions.	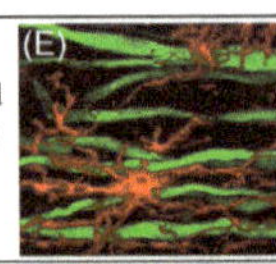 (E) 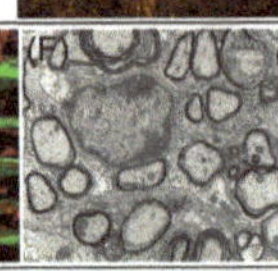 (F)
	Neurogenesis	Elevated proliferation, migration and differentiation of neuronal progenitor cells and their recruitment into injury sites.	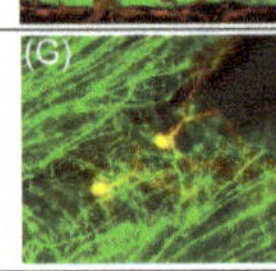 (G) 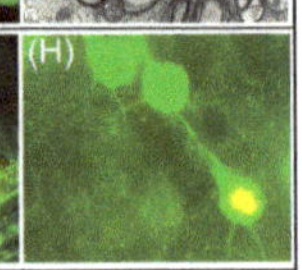 (H)

FIGURE 22.2 Immunomodulatory and neuroprotective effects of glatiramer acetate in multiple sclerosis and experimental autoimmune encephalomyelitis. Inserts demonstrate the *in situ* consequences of GA in the CNS. (A) GA-specific T-cells (*blue*) expressing IL-10 (*red*); (B) Infiltration of Foxp3 expressing T-cells (*yellow*); (C) GA-specific T-cells (*blue*) expressing BDNF (*red*); (D) Preservation of motor neurons; (E) Oligodendrocyte progenitor cells (*red*) extending processes between transected neuronal fibers (*green*); (F) Remyelination zone with newly myelinated axons surrounding an oligodendrocyte; (G) Neuronal progenitor cells (*yellow*) in a lesion site following GA treatment; (H) A BrdU expressing neuron (*yellow*) born during BrdU/GA injection, also expressing a mature neuronal marker in the cortex (*green*).

22.5 Immunomodulatory mechanisms

Various studies indicate that GA induces an immunomodulatory effect on several aspects of the immune system. The initial prerequisite step involves the binding of GA to MHC class II molecules responsible for antigen presentation. An in vitro study showed that GA rapidly and efficiently binds to various murine and human MHC class II molecules and even displaces other peptides from the MHC binding groove [29]. This competitive binding can prevent the presentation of various antigens and hinder T-cell activation (MHC blocking). In addition, several groups have demonstrated that GA induces generalized alterations of APCs, such as dendritic cells and monocytes, so that they preferentially induce protective antiinflammatory responses [30,31]. This modulation on the innate immune system level is the least specific step in the immunological processes affected by GA and can be beneficial for inhibiting responses to multiple myelin antigens. In addition, for the immunodominant encephalitogenic epitope of MBP (comprising amino acids 82–100), GA acts in an antigen-specific manner as an APL by TCR antagonism [32].

Most studies attribute the primary mechanism of GA activity to its ability to shift the T-cell response from proinflammatory to antiinflammatory pathways. It has been long known that GA-treated animals develop specific T-cells in their immune system [28,33]. These cells act as modulatory suppressor cells, as they inhibit the response of MBP-specific effector cells in vitro and adoptively transfer protection against EAE *in vivo*. T-cell lines induced by GA progressively polarize toward the Th-2 subset, secreting high amounts of antiinflammatory cytokines such as interleukin (IL)-4, IL-5, IL-10, and transforming growth factor-β (TGF-β), until they completely lose the ability to secrete proinflammatory Th-1 cytokines such as interferon-γ (IFN-γ) [33]. Antiinflammatory cytokine secretion by GA-specific T-cell lines was observed in response to either GA or MBP. Other myelin antigens such as PLP and MOG do not activate the GA-specific T-cells, yet EAE induced by PLP and MOG can be suppressed by GA as well as by GA-induced T-cells [34]. These findings are indicative of the "bystander suppression mechanism," which is especially important in the context of the epitope spreading occurring in MS and EAE [16]. A shift from the proinflammatory Th-1-biased cytokine profile toward an antiinflammatory Th-2-biased profile was also observed in GA-treated MS patients [35,36], indicating that such cells are involved in the therapeutic effect of GA in MS.

The effect of GA on T-cells is not restricted to the Th-2 versus Th-1 pathway. Several studies have demonstrated the effect of GA on Th-17 and on Treg cells, which are pivotal mediators of disease exacerbation and suppression, respectively [5,37]. In vitro exposure of peripheral CD4 T-cells from humans or mice to GA resulted in elevated levels of Tregs, through activation of the transcription factor forkhead box P3 (Foxp3) [38].

Furthermore, GA treatment leads to increased Foxp3 expression in CD4 T-cells of MS patients, whose baseline Foxp3 level is low. GA treatment of EAE mice increased levels of Tregs and reduced Th-17 cells, as demonstrated by the detection of their specific transcription factors, Foxp3 and RORγt, respectively, on both the mRNA and the protein levels [39]. Thus, GA modulates the immune response through a diverse array of mechanisms.

Therapeutic efficacy can be evaluated by the effect on the diseased organ, in MS by the ability to modulate pathological processes in the CNS. The initial immunological activity of GA apparently occurs in the periphery, at the injection sites and in the draining lymph nodes. An indication for dendritic cell uptake of GA and its delivery to the CNS has been demonstrated [40]. However, since GA is rapidly degraded in the periphery, it is unlikely that sufficient amounts reach the CNS and effectively compete with myelin antigens or initiate specific immune responses locally. The majority of researchers agree that the therapeutic effect of GA is mediated by GA-induced T-cells that penetrate the CNS from the periphery. The presence of GA-specific T-cells in the CNS was demonstrated by their isolation from the brains of sensitized mice [41]. Furthermore, the ability of GA-induced cells to cross the blood–brain barrier (BBB) and accumulate in the CNS was confirmed by injecting labeled GA-specific T-cells into the periphery of EAE-mice and their subsequent detection in the brain [40]. There is currently a consensus that the brain is not an immune privileged site and that activated T cells, regardless of their specificity, penetrate the CNS, especially in the course of MS/EAE when the BBB is disrupted [3,4].

While the cross-reactivity of GA-specific T-cells with MBP is not essential for entering the CNS, it may enable their *in situ* reactivation. In the CNS of EAE-mice, GA-specific T-cells express the potent regulatory antiinflammatory cytokines IL-10 and TGF-β [42]. Of special interest is the finding that IL-10 and TGF-β are expressed not only by the GA-specific T-cells but also by CNS resident cells in their vicinity, such as astrocytes and microglia. In contrast, the overall expression of the inflammatory cytokine IFN-γ is reduced. In the CNS of GA-treated EAE mice, there is also a drastic reduction of proinflammatory Th-17 cells, and a parallel elevation of Tregs [39]. Importantly, analysis of GA-reactive T-cells from the CSF of GA-treated MS patients revealed a pronounced antiinflammatory profile [43]. These cumulative results indicate that GA induces an overall immunomodulatory effect in the CNS and generates an *in situ* antiinflammatory cytokine shift, thus restraining the proinflammatory disease pathology.

22.6 Neuroprotective repair mechanisms

An essential challenge for MS therapy is to target not only the inflammatory characteristics of the disease but also its neuroaxonal pathology by inducing neuroprotection. Cumulative findings indicate that GA treatment generates neuroprotective consequences in the CNS [7,28,44].

The initial indication for GA neuroprotective activity was the ability of murine and human GA-specific T-cells to secrete brain-derived neurotrophic factor (BDNF) [42,45,46]. Furthermore, GA treatment augments BDNF expression in the brains of EAE mice [47]. A similar elevation was found for additional neurotrophic factors such as neurotrophin (NT)-3 and NT-4 [47]. Members of the neurotrophin family such as BDNF are important regulators of neuronal function and survival [48]. Besides their role in neuronal development, neurotrophins promote axonal outgrowth, remyelination and regeneration. Of special significance is the elevation of neurotrophic factor levels even when treatment starts late, in the chronic disease phase, when levels are diminished [47]. Furthermore, in Mecp2 knockout mice in which BDNF levels are low, GA treatment elevated BDNF to levels found in healthy mice [49]. The relevance of this effect to human therapy is evident by the GA-induced restoration of serum and CSF BDNF levels in MS patients [50].

The neuroprotective effect of GA is manifested by CNS preservation and reduction of typical MS tissue damage. Several studies utilizing immunohistochemistry and electron microscopy in different EAE models demonstrated the protective outcome of GA treatment on myelin [50,51]. Furthermore, in mice with MOG-induced EAE in which a chronic disease with extensive neurodegeneration is typically manifested, GA treatment reduced neuroaxonal damage [52,53]. In a study that employed MRI measurement of magnetization transfer ratio and diffusion tensor imaging to assess the whole brain as well as to detect specific affected regions, GA restored all MRI parameters in both chronic and relapsing-remitting EAE models [54]. These effects were obtained by various regimens of GA administration. When treatment was applied before the appearance of clinical manifestations, blocking the development of the pathological processes (prevention regimen), mice displayed nearly no damage. When treatment was applied in a therapeutic schedule after disease exacerbation or even late in the chronic phase once injury was already manifested (suppression regimen), a significant reduction in myelin and neuroaxonal damages was observed. This suggests that GA induces repair mechanisms.

The ability of GA to augment myelin repair via remyelination was established by applying transmission electron microscopy that facilitates the visualization of newly myelinated axons [9]. Ultrastructural quantitative analysis in the spinal cord of mice induced with relapsing-remitting EAE provided evidence for a significant increase in remyelination after GA treatment. This was manifested by seven- and threefold increases in relative remyelination compared to demyelination, over untreated mice, when GA was applied during the first or the second disease exacerbation, respectively. The mode of action in this system is attributed to increased proliferation and survival of oligodendrocyte progenitor cells and their recruitment to sites of injury [51].

Current opinions claim that stem cells with the potential to give rise to new neurons reside in different regions of the mammalian brain and contribute to repair after injury [55]. EAE induction triggers neuroprogenitor proliferation in the neuroproliferative zones, but this effect is of short duration and subsequently declines to levels below that of naïve mice. GA treatment augments neuronal proliferation to a higher level than that observed in EAE mice, and this effect persists for a prolonged duration [53]. Furthermore, neuronal progenitors were seen diverging from classic migratory streams and spreading to damaged sites that do not normally undergo neurogenesis. Neurons "born" during GA treatment were found in the cortex of treated mice, expressing mature neuronal markers and displaying mature morphology. These newly formed neurons could constitute a pool for replacing dead or dysfunctional cells and/or inducing a growth-promoting environment that supports neuroprotection and axonal growth. The latter activity is evidenced by the extensive BDNF expression manifested by these new neurons.

Findings from human studies support the notion that GA confers neuroprotection in MS patients. GA treatment reduces the formation of permanent T1 hypointense lesions that evolve into "black holes," which are associated with irreversible neurological disability [56]. Since T1 hypointensity is also associated with reduced myelin content, the finding of fewer black holes in GA-treated patients can be related to less axonal damage and/or improved myelin repair. Quantitative MRI analysis revealed that GA treatment for one year leads to a significant increase in the *N*-acetylaspartate (NAA): creatine (Cr) ratio compared to pretreatment values, implying an axonal metabolic recovery [57]. Altogether, these findings may explain the long-term beneficial effect of GA in patients followed for over 15 years [24,25].

22.7 Concluding remarks

MS is a multifaceted disease mediated by an immune response to different myelin antigens, eliciting inflammation, demyelination and degeneration in the CNS. Several approaches to develop peptidomimetic analogs for various components participating in the pathological process have been used in attempts to suppress the disease. These include antigen-specific peptidomimetics (APLs) and peptidomimetics that block activation signals. However, the complexity and heterogeneity of MS pathology make the prospect of a single peptidomimetic to control the disease unlikely. Whereas antigen-specific peptidomimetics might be "too specific" to combat the multiple autoantigens involved in MS pathology, peptidomimetics that block nonspecific activation signals might excessively disable the immune system. Application of peptidomimetics for human therapy requires thorough preclinical assessment due to their hazardous potential to act in an agonistic exacerbating manner. Although multiple peptidomimetic strategies to treat MS have been designed, so far small peptides fulfilling the "classical" definition of peptidomimetics have been effective only in vitro or in animal models.

A widened perspective on peptidomimetics is the case of GA (Copaxone), designed as a mimetic to the autoantigen MBP and found to be one of the first MS-modifying therapies. The unique mechanism of action facilitates immunomodulatory effects on different components of the immune system. These include competition for binding to APCs and to the TCR, steering dendritic cells and monocytes toward antiinflammatory responses, induction of Th-2/3 and regulatory T cells, and downregulation of both Th-1 and Th-17 cells. Antiinflammatory immune cells induced by GA can reach the inflamed CNS and secrete *in situ* antiinflammatory cytokines that alleviate pathological processes. Furthermore, these cells enhance neurotrophic factor secretion, remyelination and neurogenesis, supporting the notion that neuroprotective repair processes can be upregulated by immunomodulatory peptidomimetics.

References

[1] Compston A, Coles A. Multiple sclerosis; 2008.

[2] Lassmann H, Van Horssen J, Mahad D. Progressive multiple sclerosis: pathology and pathogenesis. Nat Rev Neurol 2012;8(11):647.

[3] Hohlfeld R, Wekerle H. Autoimmune concepts of multiple sclerosis as a basis for selective immunotherapy: from pipe dreams to (therapeutic) pipelines. Proc Natl Acad Sci 2004;101(suppl 2):14599–606.

[4] Aharoni R. New findings and old controversies in the research of multiple sclerosis and its model experimental autoimmune encephalomyelitis. Exp Rev Clin Immunol 2013; 9(5):423–40.

[5] Bettini M, Vignali DA. Regulatory T cells and inhibitory cytokines in autoimmunity. Curr Opin Immunol 2009;21(6):612–18.

[6] Staun-Ram E, Miller A. Effector and regulatory B cells in multiple sclerosis. Clin Immunol 2017;184:11–25.

[7] Aharoni R, Arnon R. Linkage between immunomodulation, neuroprotection and neurogenesis. Drug News Perspect 2009;22(6):301–12.

[8] Batoulis H, et al. Experimental autoimmune encephalomyelitis–achievements and prospective advances. APMIS 2011;119(12):819–30.

[9] Aharoni R, et al. Distinct pathological patterns in relapsing–remitting and chronic models of experimental autoimmune enchephalomyelitis and the neuroprotective effect of glatiramer acetate. J Autoimmun 2011;37(3):228–41.

[10] Grau-López L, et al. Myelin peptides in multiple sclerosis. Autoimmun Rev 2009;8 (8):650–3.

[11] Santana M, Rosenstein Y. What it takes to become an effector T cell: the process, the cells involved, and the mechanisms. J Cell Physiol 2003;195(3):392–401.

[12] Sospedra M, Martin R. Antigen-specific therapies in multiple sclerosis. Int Rev Immunol 2005;24(5–6):393–413.

[13] Panitch H, Francis G. Clinical results of a phase III trial of oral myelin in relapsing-remitting multiple sclerosis. Ann Neurol. Philadelphia, PA: Lippincott-Raven Publ 227 East Washington Sq; 1997.

[14] Correale J, Farez M, Gilmore W. Vaccines for multiple sclerosis. CNS Drugs 2008;22(3): 175–98.

[15] Kappos L, et al. Induction of a non-encephalitogenic type 2 T helper-cell autoimmune response in multiple sclerosis after administration of an altered peptide ligand in a placebo-controlled, randomized phase II trial. Nat Med 2000;6(10):1176.

[16] Tuohy VK, Kinkel RP. Epitope spreading: a mechanism for progression of autoimmune disease. Autoimmunity. Springer; 2001. p. 39–48.

[17] Rosloniec EF, et al. Second-generation peptidomimetic inhibitors of antigen presentation effectively treat autoimmune diseases in HLA-DR-transgenic mouse models. J Autoimmun 2006;27(3):182–95.

[18] Horie T, et al. Study of disabling T-cell activation and inhibiting T-cell-mediated immunopathology reveals a possible inverse agonist activity of CD4 peptidomimetics. Exp Mol Pathol 2002;73(2):93–103.

[19] Allen S, et al. Therapeutic peptidomimetic strategies for autoimmune diseases: costimulation blockade. The J Peptide Res 2005;65(6):591–604.

[20] Teitelbaum D, et al. Suppression of experimental allergic encephalomyelitis by a synthetic polypeptide. Eur J Immunol 1971;1(4):242–8.

[21] Sela M, Teitelbaum D. Glatiramer acetate in the treatment of multiple sclerosis. Expert Opin Pharmacother 2001;2(7):1149–65.

[22] Bornstein MB, et al. A pilot trial of Cop 1 in exacerbating–remitting multiple sclerosis. N Engl J Med 1987;317(7):408–14.

[23] Johnson K, et al. Copolymer 1 reduces relapse rate and improves disability in relapsing-remitting multiple sclerosis: results of a phase III multicenter, double-blind, placebo-controlled trial. Neurology 1995;45(7):1268–76.

[24] Ford C, et al. Continuous long-term immunomodulatory therapy in relapsing multiple sclerosis: results from the 15-year analysis of the US prospective open-label study of glatiramer acetate. Multiple Sclerosis J 2010;16(3):342–50.

[25] Miller A, et al. Long-term (up to 22 years), open-label, compassionate-use study of glatiramer acetate in relapsing—remitting multiple sclerosis. Multiple Sclerosis J 2008;14(4):494–9.

[26] Comi G, et al. European/Canadian multicenter, double-blind, randomized, placebo-controlled study of the effects of glatiramer acetate on magnetic resonance imaging–measured disease activity and burden in patients with relapsing multiple sclerosis. Ann Neurol 2001;49(3):290–7.

[27] Rommer PS, et al. Immunological aspects of approved MS therapeutics. Front Immunol 2019;10:1564.

[28] Aharoni R. The mechanism of action of glatiramer acetate in multiple sclerosis and beyond. Autoimmun Rev 2013;12(5):543–53.

[29] Fridkis-Hareli M, et al. Direct binding of myelin basic protein and synthetic copolymer 1 to class II major histocompatibility complex molecules on living antigen-presenting cells–specificity and promiscuity. Proc Natl Acad Sci 1994;91(11):4872–6.

[30] Vieira PL, et al. Glatiramer acetate (copolymer-1, copaxone) promotes Th2 cell development and increased IL-10 production through modulation of dendritic cells. The J Immunol 2003;170(9):4483–8.

[31] Weber MS, et al. Type II monocytes modulate T cell–mediated central nervous system autoimmune disease. Nat Med 2007;13(8):935.

[32] Aharoni R, et al. Copolymer 1 acts against the immunodominant epitope 82–100 of myelin basic protein by T cell receptor antagonism in addition to major histocompatibility complex blocking. Proc Natl Acad Sci 1999;96(2):634–9.

[33] Aharoni R, et al. Copolymer 1 induces T cells of the T helper type 2 that crossreact with myelin basic protein and suppress experimental autoimmune encephalomyelitis. Proc Natl Acad Sci 1997;94(20):10821–6.

[34] Aharoni R, et al. Bystander suppression of experimental autoimmune encephalomyelitis by T cell lines and clones of the Th2 type induced by copolymer 1. J Neuroimmunol 1998;91(1–2):135–46.

[35] Neuhaus O, et al. Multiple sclerosis: comparison of copolymer-1-reactive T cell lines from treated and untreated subjects reveals cytokine shift from T helper 1 to T helper 2 cells. Proc Natl Acad Sci 2000;97(13):7452–7.

[36] Duda PW, et al. Glatiramer acetate (Copaxone®) induces degenerate, Th2-polarized immune responses in patients with multiple sclerosis. J Clin Invest 2000;105(7):967–76.

[37] Becher B, Segal BM. TH17 cytokines in autoimmune neuro-inflammation. Curr Opin Immunol 2011;23(6):707–12.

[38] Hong J, et al. Induction of CD4+ CD25+ regulatory T cells by copolymer-I through activation of transcription factor Foxp3. Proc Natl Acad Sci 2005;102(18):6449–54.

[39] Aharoni R, et al. Glatiramer acetate reduces Th-17 inflammation and induces regulatory T-cells in the CNS of mice with relapsing–remitting or chronic EAE. J Neuroimmunol 2010;225(1–2):100–11.

[40] Liu J, et al. T cell independent mechanism for copolymer-1-induced neuroprotection. Eur J Immunol 2007;37(11):3143–54.

[41] Aharoni R, et al. Specific Th2 cells accumulate in the central nervous system of mice protected against experimental autoimmune encephalomyelitis by copolymer 1. Proc Natl Acad Sci 2000;97(21):11472–7.

[42] Aharoni R, et al. Glatiramer acetate-specific T cells in the brain express T helper 2/3 cytokines and brain-derived neurotrophic factor in situ. Proc Natl Acad Sci 2003;100 (24):14157–62.

[43] Hestvik A, et al. Multiple sclerosis: glatiramer acetate induces anti-inflammatory T cells in the cerebrospinal fluid. Multiple Sclerosis J 2008;14(6):749–58.

[44] Aharoni R. Immunomodulation neuroprotection and remyelination–The fundamental therapeutic effects of glatiramer acetate: a critical review. J Autoimmun 2014;54:81–92.

[45] Kipnis J, et al. T cell immunity to copolymer 1 confers neuroprotection on the damaged optic nerve: possible therapy for optic neuropathies. Proc Natl Acad Sci 2000;97(13):7446–51.

[46] Ziemssen T, et al. Glatiramer acetate-specific T-helper 1-and 2-type cell lines produce BDNF: implications for multiple sclerosis therapy. Brain 2002;125(11):2381–91.

[47] Aharoni R, et al. The immunomodulator glatiramer acetate augments the expression of neurotrophic factors in brains of experimental autoimmune encephalomyelitis mice. Proc Natl Acad Sci 2005;102(52):19045–50.

[48] Lessmann V, Gottmann K, Malcangio M. Neurotrophin secretion: current facts and future prospects. Prog Neurobiol 2003;69(5):341–74.

[49] Ben-Zeev B, et al. Glatiramer acetate (GA, Copolymer-1) an hypothetical treatment option for Rett syndrome. Med Hypotheses 2011;76(2):190–3.

[50] Azoulay D, et al. Lower brain-derived neurotrophic factor in serum of relapsing remitting MS: reversal by glatiramer acetate. J Neuroimmunol 2005;167(1–2):215–18.

[51] Aharoni R, et al. Demyelination arrest and remyelination induced by glatiramer acetate treatment of experimental autoimmune encephalomyelitis. Proc Natl Acad Sci 2008;105 (32):11358–63.

[52] Maier K, et al. Effects of glatiramer acetate and interferon-β on neurodegeneration in a model of multiple sclerosis: a comparative study. Am J Pathol 2006;169(4):1353–64.

[53] Aharoni R, Arnon R, Eilam R. Neurogenesis and neuroprotection induced by peripheral immunomodulatory treatment of experimental autoimmune encephalomyelitis. J Neurosci 2005;25(36):8217–28.
[54] Aharoni R, et al. Magnetic resonance imaging characterization of different experimental autoimmune encephalomyelitis models and the therapeutic effect of glatiramer acetate. Exp Neurol 2013;240:130–44.
[55] Magavi SS, Leavitt BR, Macklis JD. Induction of neurogenesis in the neocortex of adult mice. Nature 2000;405(6789):951.
[56] Filippi M, et al. Glatiramer acetate reduces the proportion of new MS lesions evolving into "black holes.". Neurology 2001;57(4):731–3.
[57] Khan O, et al. Axonal metabolic recovery and potential neuroprotective effect of glatiramer acetate in relapsing-remitting multiple sclerosis. Multiple Sclerosis J 2005;11(6):646–51.

Chapter 23

Therapeutic peptidomimetics in metabolic diseases

Worood Sirhan and Ron Piran

The Laboratory for Diabetes and Regenerative Medicine, Azrieli Faculty of Medicine, Bar-Ilan University, Safed, Israel

23.1 Introduction

Cells are dependent on glucose as their main energy source. Higher-ordered organisms need to adequately control glucose uptake by individual cells, and the insulin–glucagon endocrine system is the glucose regulation mechanism. When blood glucose levels are high, insulin facilitates effective glucose entry into the cells, while glucagon acts in opposition, balancing low blood glucose levels. As a result, insulin deficiency prevents glucose entry into the cells, leading to glucose build up in the bloodstream and to diabetes. All types of diabetes are characterized by inefficient insulin signaling mechanisms, which can be the result of insufficient insulin supply as in the case of type I diabetes and the progressive stages of type II diabetes, or the result of an insufficient response to insulin (also known as insulin resistance). In essence, diabetes is a disease of starved tissues, unable to absorb glucose (as well as other nutrients), with one symptom being high blood glucose levels. Prior to the discovery of insulin, diabetic patients suffered from glucose malabsorption (Fig. 23.1 and Table 23.1).

23.2 Key regulators of glucose homeostasis

23.2.1 Insulin

Insulin is a peptide hormone encoded by the *INS* gene [1], and it is produced and secreted by endocrine pancreatic β-cells in response to high blood glucose levels. Insulin is critical for glucose entry into different body tissues and plays a major role in energetic homeostasis [2]. Insulin is synthesized as a 51-amino-acid protein. The protein has a molecular weight of 5808 Daltons (Da). The primary translational product of the *INS* gene is

Peptide and Peptidomimetic Therapeutics. DOI: https://doi.org/10.1016/B978-0-12-820141-1.00023-6

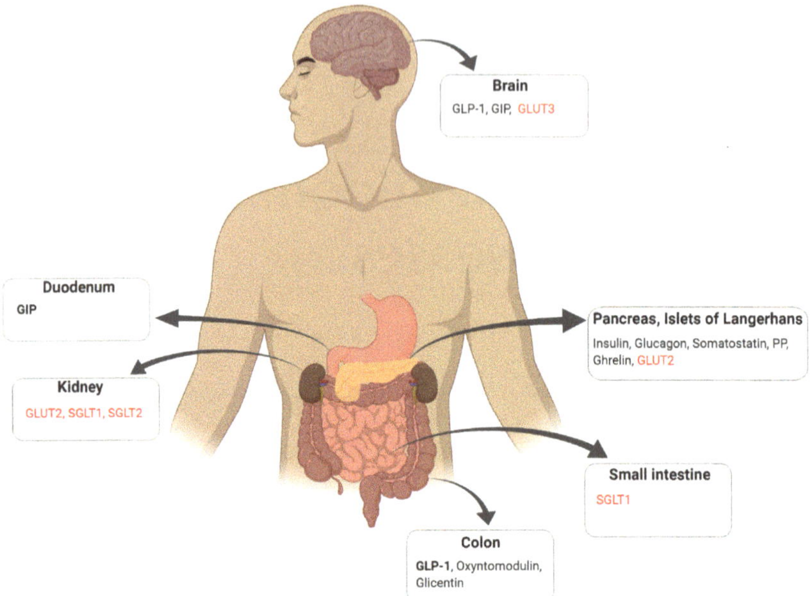

FIGURE 23.1 Peptide hormones and glucose transporters are discussed in this chapter. The different hormones (*black*) and the glucose transporters (*red*) covered in the chapter are organized by the tissues in which they are expressed.

preproinsulin. This is a 109-amino-acid single-chain protein, with a molecular weight of 11,100 Da [3], The preproinsulin sequence consists of proinsulin and a signal peptide chain, which is required for preproinsulin entry through the rough endoplasmic reticulum (rER) [4]. When the signal peptide in the N-terminus of the preproinsulin chain is proteolytically cleaved in the rER [5], proinsulin, an 86-amino-acid peptide, is created. Proinsulin is composed of three chains, including the A-chain, B-chain, and C-peptide, which are joined by disulfide bridges. In the Golgi apparatus, a protease cleaves the proinsulin chain to form the mature insulin alongside the C-peptide (Fig. 23.2). Until they are metabolically needed, insulin and C-peptide are stored within glucose-regulated secretory vesicles as zinc insulin hexamers [6,7].

Insulin binds and activates the transmembrane insulin receptor (IR), which is encoded by the *INSR* gene [2]. Two peptide chains can constitute the IR in different combinations: IR-A and IR-B (Fig. 23.3) [8]. The IR-A chain is predominantly expressed in the kidneys and brain, while the IR-B chain is the major isoform in the liver [8–11]. Insulin can bind the different hybrids of IR-A and IR-B, and in addition, can bind insulin-like growth factor receptors (IGFIR) in low affinity (Fig. 23.3). Additional molecules can interact with the different chain hybrids of IR, like insulin-like growth factor

TABLE 23.1 Peptide hormones and their analogs presented in this chapter. Sequences are organized according to the figures in which they are presented. The colors correspond to the colors in the text. X = α-aminoisobutyric acid.

Amino acid sequence	Peptide name	Relevant figure
MALWMRLLPLLALLALWGPDPAAAFVNQHLCGSHLVEALYLVCGERGFFYTPKTRREAEDLQVGQVE LGGGPGAGSLQPLALEGSLQKRGIVEQCCTSICSLYQLENYCN	Preproinsulin	23.2A
FVNQHLCGSHLVEALYLVCGERGFFYTPKTRREAEDLQVGQVELGGGPGAGSLQPLALEGSLQKRGIVEQ CCTSICSLYQLENYCN	Proinsulin	23.2B
GIVEQCCTSICSLYQLENYCN	Insulin A-chain	23.2C
FVNQHLCGSHLVEALYLVCGERGFFYTPKT	InsulinB-chain	23.2C
RREAEDLQVGQVELGGGPGAGSLQPLALEGSLQKR	C-peptide	23.2C
LKVKMHTMSSSHLFYLALCLLTFTSSATAGPETLCGAELVDALQFVCGDRGFYFNKPTGYGSSSRRAPQTGI VDECCFRSCDLRRLEMYCAPLKPAKSARSVRAQRHTDMPKTQKYQPPSTNKNTKSQRRKGSTFEERK	IGF-I	23.3
AYRPSETLCGGELVDTLQFVCGDRGFYFSRPASRVSRRSRGIVEECCFRSCDLALLETYCATPAKSE	IGF-II	23.3
HSQGTFTSDYSKYLDSRRAQDFVQWLMNT	Glucagon	23.4
HSQGTFTSDYSKYLDSRRAQDFVQWLMNTKRNKNNIA	Oxyntomodulin	23.4
RSLQDTEEKSRSFSASQADPLSDPDQMNEDKRHSQGTFTSDYSKYLDSRRAQDFVQWLMNTKRNKNNIA	Glicentin	23.4
HAEGTFTSDVSSYLEGQAAKEFIAWLVKGRG	GLP-1	23.4, 23.5
HADGSFSDEMNTILDNLAARDFINWLIQTKITD	GLP-2	23.4
FVNQHLCGSHLVEALYLVCGERGFFYTKPT(A-chain is identical to the insulin A-chain)	Lispro	23.6B
FVKQHLCGSHLVEALYLVCGERGFFYTPET(A-chain is identical to the insulin A-chain)	Glulisine	23.6C

(*Continued*)

TABLE 23.1 (Continued)

Amino acid sequence	Peptide name	Relevant figure
FVNQHLCGSHLVEALYLVCGERGFFYTDKT(A-chain is identical to the insulin A-chain)	Aspart	23.6D
FVNQHLCGSHLVEALYLVCGERGFFYTPKTRRGIVEQCCTSICSLYQLENYCG	Glargine	23.6E
FVNQHLCGSHLVEALYLVCGERGFFYTPK(A-chain is identical to the insulin A-chain)	Detemir	23.6F
HGEGTFTSDLSKQMEEEAVRLFIEWLKNGGPSSGAPPPS	Extendin-4	23.7A
HGEGTFTSDLSKQMEEEAVRLFIEWLKNGGPSSGAPPPSKKKKKK	Lixisenatide	23.7B
HAEGTFTSDVSSYLEGQAAKEEFIAWLVRGRG	Liraglutide	23.7C
HXEGTFTSDVSSYLEGQAAKEEFIAWLVRGRG	Semaglutide	23.7D
HGEGTFTSDVSSYLEGQAAKEFIAWLVKGR HGEGTFTSDVSSYLEGQAAKEFIAWLVKGR	Albiglutide	23.7E
HGEGTFTSDVSSYLEEQAAKEFIAWLVKGGGGGGKVLWAIFEKAAQEELYSSVDSTFTGEGH	Dulaglutide	23.7F
GSSFLSPEHQRVQQRKESKKPPAKLQPR	Ghrelin	Not in figure
YAEGTFISDYSIAMDKIHQQDFVNWLLAQKGKKNDWKHNITQ	GIP	Not in figure
APLEPVYPGDNATPEQMAQYAADLRRYINMLTRPRY	Pancreatic polypeptide	Not in figure
AGCKNFFWKTFTSC	Somatostatin (14 amino acids)	Not in figure
SANSNPAMAPRERKAGCKNFFWKTFTSC	Somatostatin (28 amino acids)	Not in figure

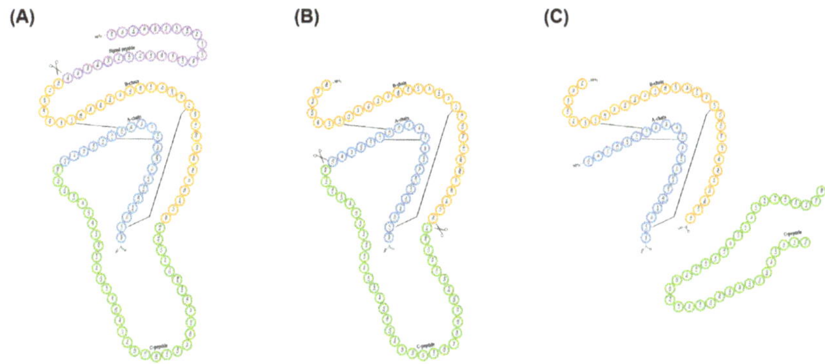

FIGURE 23.2 The insulin synthesis pathway. (A) Preproinsulin in its translated form. Preproinsulin is docked to the rER by the signal peptide (*purple*). (B) Preproinsulin is cleaved from its signal peptide to form proinsulin. (C) The mature insulin, formed from two peptide chains, the A-chain in blue and B-chain in yellow together with the C-peptide (*green*) that was cleaved from proinsulin. These two molecules are encapsulated in vesicles. When blood glucose levels rise, both insulin and C-peptide are secreted together at a 1:1 ratio.

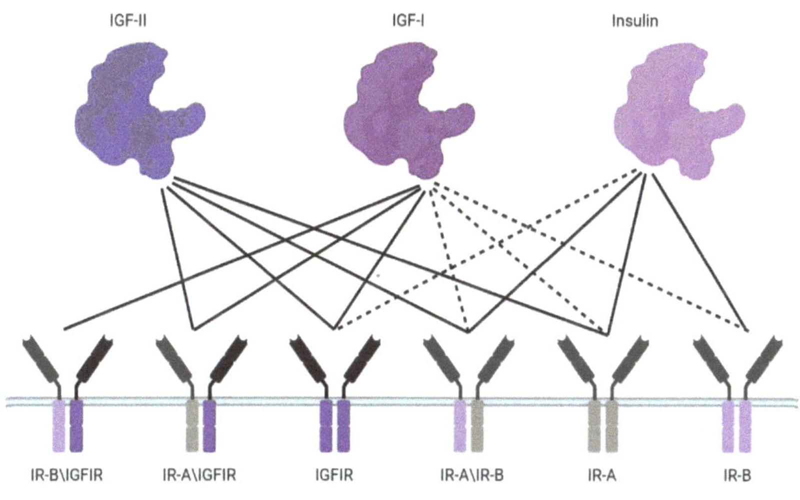

FIGURE 23.3 Various receptors bind insulin, IGF-I, and IGF-II. Insulin, IGF-I, and IGF-II bind receptors that are composed of two chains. These receptor chains can be IR-A, IR-B, and IGFIR. The receptor chains can form homodimers and heterodimers. Insulin binds the IR-A and IR-B homodimers, as well as to their heterodimer and to IGFIR homodimer (although in lesser affinity). IGFI binds all dimers, but it prefers binding the IGFIR chain in homodimer or heterodimer. IGFII binds all possible dimers except for the IR-B homodimer and the IR-B/IGFIR heterodimer.

I (IGFI) and IGFII. IGFI can bind all different hybrids. It binds the IR-A, IR-B, and IR-A/IR-B in low affinity, and binds the other three hybrid receptors in high affinity. IGFII however, binds only the IR-A, IR-A/IR-B, IGFIR, and IR-A/IGFIR in high affinity [8,12–15]. Research of sequence analysis revealed a significant homology between insulin, IGFI, and IGFII. The IGFI peptide is encoded by the *IGFI* gene [16,17] and has a molecular weight of 7649 Da [18]. The IGFII peptide is encoded by the *IGFII* gene and has a molecular weight of 7505 Da [19].

23.2.2 The glucagon gene and its products

23.2.2.1 Glucagon

Glucagon is the insulin counter hormone. Its role is to release glucose from the liver's glycogen reservoirs and to promote glycogenesis [20]. In addition, it promotes lipolysis in the adipose tissues [21,22]. It is secreted from the pancreatic α-cells, and is a single-chain endocrine hormonal peptide composed of 29 amino acids. As with insulin, glucagon is first expressed as the preproglucagon chain from the glucagon gene, termed *GCG* [23]. This single peptide is then cleaved to form proglucagon followed by a subsequent cleavage by prohormone convertase to produce the separated mature peptides: glucagon, glucagon-like peptide 1 (GLP-1), glucagon-like peptide 2 (GLP-2), oxyntomodulin, glicentin, and glicentin-related pancreatic polypeptide [24], (Fig. 23.4). Glucagon has a molecular weight of 3483 Da [25]. It has one receptor, encoded by the *GCGR* gene. This 54,009 Da receptor is a member of the G-protein coupled receptor (GPCR) family and is expressed in many tissues including the adipose tissues, the liver, and additional tissues such as the kidney, pancreas, heart, spleen, thymus, adrenal glands, cerebrum (specifically in the cerebral cortex), and gastrointestinal tract [26].

23.2.2.2 GLP-1 and GLP-2

The GLP-1 and GLP-2 hormones are secreted from the gut L-intestinal endocrine cells in response to nutrient ingestion [27–29]. These hormones can decrease blood glucose levels. GLP-1 is a 31-amino-acid long peptide; it is glucose-dependent and insulinotropic. GLP-1 is mostly known for its function in slowing stomach emptying and inducing satiety effects [30]. Furthermore, GLP-1 decreases postmeal intestinal chylomicron production and triglyceride levels [29]. It has a stimulatory role in β-cell-induced insulin

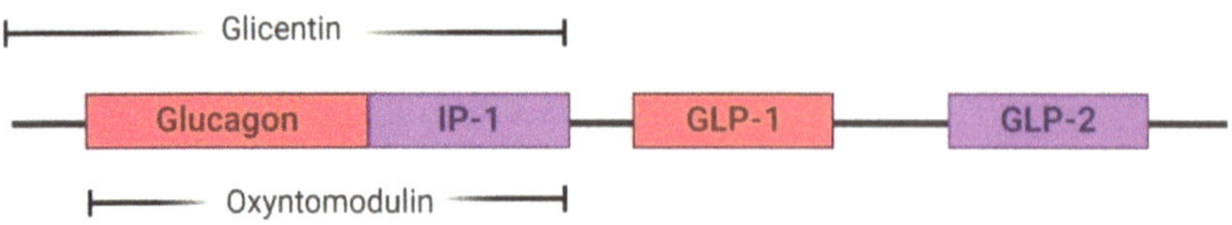

FIGURE 23.4 The *GCG* gene and its translated hormones. The *GCG* gene can be translated to generate different hormone peptides. Glucagon, GLP-1, and GLP-2 have their own genetic sequence on the gene, while glicentin and oxyntomodulin are splice variants of the glucagon segment, fused to additional sequences.

secretion and it prolongs β-cell life and function [31]. GLP-2 is a 33-amino-acid long peptide. It has a number of functions, including the enhancement of intestinal functions, intestinal growth, reduction of bone breakdown, neuro-protection, activation of the sodium-dependent glucose transporter-1 (SGLT-1) [32], and activation of glucose transporter-2 (GLUT-2) [33]. GLP-1 and GLP-2 have some contradictory roles, particularly on intestinal lipid delivery [28,29].

23.2.2.3 Glicentin and oxyntomodulin

These two hormones are produced and secreted to the colon by the enteroendocrine gut L-cells [34]. Both are expressed in the central nervous system [35]. Glicentin is 69 amino acids long [36], and oxyntomodulin is 37 amino acids long [34]. These two peptides are known as inhibiting hormones, but their specific targets remain unexplored [37], possibly because no diseases, as yet, has been associated with their signaling. Oxyntomodulin is known to suppress appetite. It binds both the GLP-1 receptor and the glucagon receptor. However, it has not yet been determined if its biological effects are due to the activation of one of these receptors or by an unidentified receptor [38]. The biological functions of glicentin include insulin secretion stimulation, gut growth and motility regulation, and gastric acid secretion inhibition [39].

23.3 The cephalic phase

Before food is even tasted and digested, the body reacts to its near coming. This is called the cephalic phase. The meaning of cephalic is found near, in, or on the head [40]. This is achieved by the sensory identification of the food by smelling it, seeing it, or even thinking about it, and later by the taste of the food [40–46]. In Pavlov's classical experiment, dogs were given food together with an auditory stimulus of a bell. This habituation was followed for a while [47,48]. The dogs started to salivate when the bell was sound. Later, Pavlov was able to demonstrate that only the sound of the bell was enough to stimulate the dogs' salivation, a phenomenon that was later coined as Pavlovian conditioning [47,48].

Here, we use the cephalic phase term to describe the reaction of the body to food before it enters the mouth. This phase induces some hormonal and neural signaling [49]. Cephalic stimulation eventually triggers the vagus nerve, induces different autonomic responses including the secretion of hormones, digestive enzymes, saliva, and gastric acid [45,50–57], as well as nonsecretory responses, such as increased gastric motor activity, increased blood pressure, changes in respiratory measures, changes in cardiac measures, and postprandial thermogenesis [58–63].

23.3.1 Cephalic phase insulin release

Of the many affected responses to the cephalic phase, and the most investigated, is insulin release. It comes as a response to oral cavity stimulation. This can be achieved by mastication (chewing), food tasting, or ingestion [42,64]. Different studies have indicated that when the insulin response is inhibited during cephalic phase insulin release (CPIR), inadequate food absorption is observed, which is accompanied by poor glycemic control [65]. Insulin release prior to the actual arrival of glucose to the bloodstream is thought to facilitate optimal postprandial glucose homeostasis by inhibiting blood glucose sharp increases and also preventing sudden insulin secretion spikes [64]. Moreover, CPIR inhibits liver gluconeogenesis and adipose lipolysis [66,67]. Insulin-resistant obese and type II diabetics (T2D) display reduced CPIR and present an approximately 40% increase in postprandial blood glucose levels [65,68]. Small amounts of insulin infusion prior to or during the first minutes of a carbohydrate-rich meal have no effect in lean patients, but present a significant advantage in both type I diabetic (T1D), T2D, and obese patients [69–71], therefore highlighting the importance of CPIR to maintain efficient normoglycemia.

The pancreatic polypeptide (PP) hormone is secreted mostly by PP cells in the islets of Langerhans in response to the ingestion of food (see below). There is robust PP secretion following mastication [72–74] and, therefore, it is commonly used as a marker for cephalic–vagal stimulation [75,76].

23.4 The incretin effect, the duodenum, and the small intestine

In a series of articles, McIntyre et al. demonstrated that oral glucose administration produced much higher blood insulin levels in comparison to the same glucose amounts that were injected intravenously [77,78], thus confirming an already hypothesized theory of the incretin hormones [79].

23.4.1 Glucose-dependent insulinotropic polypeptide, or gastric inhibitory polypeptide (GIP)

Gastric inhibitory polypeptide (GIP) was the first incretin hormone to be discovered. The name GIP was recently modified to a glucose-dependent insulinotropic polypeptide, to highlight its role in stimulating insulin secretion [80]. It was isolated from porcine small intestine crude extracts in 1970 [81–83]. GIP peptide is encoded by the *GIP* gene, it is 42 amino acids long [84], and is formed from a precursor of a 153-amino-acid long peptide that is cleaved to generate the mature hormone. GIP is expressed in many tissues like the central nervous system, intestinal enteroendocrine K cells, which are primarily found in the proximal duodenum, and the pancreatic α-cells, which may indicate a possible intra-islet influence of α- on β-cell insulin secretion [85–87]. Furthermore, GIP was demonstrated to promote pancreatic β-cell survival, since GIP administration led to an increase of β-cell mass in several studies [88]. In addition, GIP increased fat accumulation in an insulin-independent manner in adipose tissues, and adipocytes were found to express the GIP receptor (GIPR). To highlight GIP central function, GIPR knock-out mice did not respond to a high-fat diet and their weight did not increase [1,89,90].

23.4.2 GLP-1

The second incretin hormone to be discovered was GLP-1 (Fig. 23.5A). Interestingly, this discovery was a byproduct of the sequencing of the glucagon gene (*GCG*, see above) [83,91,92]. GLP-1 is a substrate of the target enzyme dipeptidyl peptidase-4 (DPP-4) [93]. DPP-4 is so efficient in GLP-1 catalysis that only 25% of the newly secreted hormone succeeds in leaving the gastrointestinal tract [94,95]. As a result, GLP-1 has a short half-life of 1–2 min [93]. As GLP-1 has such valuable functions, many efforts have been aimed at inhibiting its DPP-4 rapid degradation. This trait is relevant to T2D patients, in which insulin production is still possible, especially in patients who suffer from overweight and obesity. In these patients, DPP-4 inhibitors (DPP-4i) are known to lower fasting and satiety glycemic levels [96].

23.4.3 DPP-4

DPP-4 is a protein expressed in different tissues in the body, including the intestinal and renal brush border membranes, vascular endothelium, liver, pancreas, and cells of the immune system [97]. It is first expressed as a membrane protein, which has a large extracellular catalytic domain (aa 49–766), but it can be processed to discharge the soluble extracellular

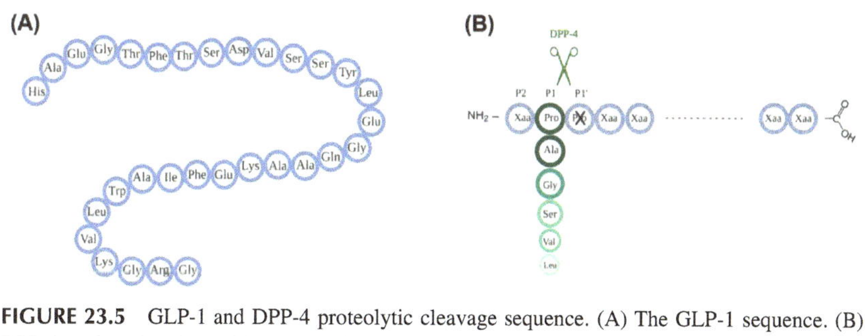

FIGURE 23.5 GLP-1 and DPP-4 proteolytic cleavage sequence. (A) The GLP-1 sequence. (B) DPP-4 preferred amino acid sequence for cleavage. DPP-4 prefers to cut two sequential prolines, but the first proline can be replaced with other amino acids as presented in (B), where alanine maintains the best catalytic activity and leucine is the weakest replacement.

domain. It is not yet determined if soluble DPP-4 or the membrane-anchored protein is the active enzyme.

The activity of DPP-4 is manifested by cutting the two most N-terminal amino acids from a target peptide, mainly in the proline or alanine positions. DPP-4 can cleave additional amino acid residues such as glycine, serine, or valine, though this catalysis is significantly slower (Fig. 23.5B).

GIP and GLP-1 are found among the DPP-4 substrates [98,99]. As GLP-1 is so important in decreasing food intake, many efforts have been aimed at preventing its rapid degradation by using DPP-4i as a therapeutic approach, especially with T2D patients and patients who suffer from overweight and obesity. The administration of DPP-4i can stabilize endogenic GLP-1 and, as a result, will lead to lower fasting and satiety blood glucose levels [96].

23.4.4 Sodium-glucose cotransporter 1

Sodium-glucose cotransporters (SGLT) translocate glucose with sodium ions. They are found mainly in the small intestine (SGLT1) and in the renal tube (proximal tube, SGLT1 and SGLT2, see below). SGLT3, another member of the family, was recently characterized in humans, and therefore, its role is only superficially understood [100]. SGLT1 is a high-affinity, low-capacity glucose transporter. In the intestine, it is responsible for the active uptake of glucose across the apical membrane into the bloodstream in a sodium-dependent manner. SGLT1 is found in many tissues such as intestinal cells, the heart, and the central nervous system [101]. Since SGLT1 has wide tissue distribution, its inhibition suffers from substantial side effects and as a result was not chosen as a specific drug candidate. However, recently drugs were developed that act as SGLT1/2 coinhibitors, and were approved or are in clinical trials [102].

23.5 Transport from the bloodstream into tissues

Once glucose enters the bloodstream it can be consumed by different tissues. Glucose is the main energy source for all cells and especially for the brain, which under normal conditions does not exploit other nutrients. From the large family of GLUTs, GLUT 1–4 are the most significant [103] (Table 23.2). GLUT1 is the most abundant, and in some tissues is coexpressed alongside GLUT4. GLUT1 is usually constitutively expressed and its kinetic parameters [Kcat (Kcat represents the limiting rate of an enzyme-catalyzed reaction at saturation)/km (km represents the substrate concentration at which the reaction rate is half of Vmax)] are relatively low, having a km of about 2 mM [104]. GLUT2, is expressed in pancreatic β-cells, hepatocytes, and renal tubular cells. GLUT2 has a higher km of about 17 mM [91].

TABLE 23.2 The main human GLUT isoforms and the major tissues in which they are found.

GLUT isoform	km for glucose (approximated, mM)	Main tissue expression
GLUT1[a]	2	Throughout the body
GLUT2[b]	17	Liver, kidney, pancreatic β-cells
GLUT3[c]	1–2	Brain
GLUT4[d]	5	Cardiac and skeletal muscles, adipose tissue

[a]*Data from Nishimura H, Pallardo FV, Seidner GA, Vannucci S, Simpson IA, Birnbaum MJ. Kinetics of GLUT1 and GLUT4 glucose transporters expressed in Xenopus oocytes. J Biol Chem 1993;268(12):8514–8520.*
[b]*Data from Uldry M, Ibberson M, Hosokawa M, Thorens B. GLUT2 is a high affinity glucosamine transporter. FEBS Lett 2002;524(1–3):199–203.*
[c]*Data from Leino RL, Gerhart DZ, van Bueren AM, McCall AL, Drewes LR. Ultrastructural localization of GLUT 1 and GLUT 3 glucose transporters in rat brain. J Neurosci Res 1997;49 (5):617–626.*
[d]*Data from Nishimura H, Pallardo FV, Seidner GA, Vannucci S, Simpson IA, Birnbaum MJ. Kinetics of GLUT1 and GLUT4 glucose transporters expressed in Xenopus oocytes. J Biol Chem 1993;268(12):8514–8520; Huang S, Czech MP. The GLUT4 glucose transporter. Cell Metab 2007;5(4):237–252.*

In fact, GLUT2 acts as a glucose indicator for β-cells as it starts transporting glucose only when blood glucose levels rise above a certain threshold. In the kidneys, the second step of glucose uptake following SGLT1 and SGLT2 preliminary reabsorption is mediated by GLUT2 [105]. The brain specifically expresses GLUT3, which is thus also termed neuronal GLUT. GLUT3 km is significantly lower than for the other GLUTs (1–2 mM), which results in higher glucose affinity compared with other organs, allowing glucose consumption priority over other tissues. This trait is important, especially when blood glucose levels are low [106]. Three principal tissues express GLUT4: the cardiac muscles, the skeletal muscles, and the adipose tissues. GLUT4 has a km of 5 mM. GLUT4 is the only GLUT whose expression is dependent on insulin signaling and that is a high-capacity glucose transporter [107]. GLUT4 is brought up to the cell membrane following insulin and allows for significant glucose consumption. Thus, GLUT4 expression levels can increase during exercise [108].

23.6 Reaching the endocrine pancreas

The pancreas is a slender and long organ, which is composed of exocrine and endocrine cells [6]. The exocrine cell population is composed of acinar and ductal cells. Acinar cells express and secrete digestive enzymes to the digestive system, which assist in food digestion processes. These enzymes include lipases, proteases, amylases, ribonucleases, endonucleases, phospholipase A2, lysophospholipase, and cholesterol esterase [5]. The ductal cells compose the duct, which is the infrastructure by which the digestive enzymes reach the duodenum, and also secrete carbonate salts that increase the pH of the chyme that arrives from the stomach.

The pancreatic endocrine cells secrete different hormones. Some play an important role in glucose homeostasis. The endocrine cells are located in spherical clusters called islets of Langerhans. These clusters are composed of α, β, δ, ε, and PP cells [109]. α-Cells secrete glucagon, β-cells secrete insulin, δ-cells secrete somatostatin, ε-cells secrete ghrelin, and PP cells secrete pancreatic polypeptide (PP) [18].

α-Cells account for about 15% of the total endocrine cell mass; these cells produce and secrete glucagon when blood glucose levels are low. Glucagon secretion elevates blood glucose levels, which encourage the conversion of glycogen storage in the liver into glucose. In addition, glucagon activates fat conversion into free fatty acids and their subsequent arrival to the liver from adipose tissues [9]. Thus, glucagon is a crucial hormone in starving conditions.

The β-cell population is considered as the main endocrine cell type, and accounts for about 80% of the total endocrine cell population. When blood glucose levels are high, β-cells secrete insulin, which allows glucose entry into the cells of tissues that upregulate GLUT4 [2]. Thus, insulin is a crucial hormone in the fed state. β-Cells secrete another hormone termed amylin. Amylin is secreted together with insulin, in response to high levels of glucose. When amylin is secreted, glucagon secretion is inhibited, and the satiety feeling is stimulated [8].

δ-Cells constitute up to 10% of the total islet volume. These cells secrete the hormone somatostatin, which is known as the negative regulator hormone of the other endocrine hormones insulin and glucagon, and thus serves as a regulator of regulators of glucose metabolism in the body [110]. Somatostatin has two different isoforms, one is 28 amino acids long, and the other is 14 amino acids long. Furthermore, somatostatin plays an important role in neurotransmission systems and the immune system, in which it has an inhibitory effect on T-lymphocytes, and inhibits the secretion of different interleukins and cytokines [111–113]. There are five subtypes of GPCR somatostatin (SST) receptors (SSTR), named SSTR1–SSTR5 [114–116]. Somatostatin can bind and activate all five.

The ε-cells are another small endocrine cell population, constituting around 1% of the total pancreatic endocrine cells. ε-Cells express and secrete the hormone ghrelin during fasting. Ghrelin is a neuropeptide hormone, which acts as an appetite regulator on the central nervous system and induces hunger. In conditions of satiation, ghrelin secretion is inhibited [89]. Ghrelin is 28 amino acids long. Des-acyl ghrelin and *n*-octanoyl-modified ghrelin are the two isoforms of this hormone. Ghrelin expression is increased during fasting and its secretion stimulates appetite. Ghrelin also induces the secretion of different growth hormones [117]. The GPCR ghrelin receptor is known as the growth hormone secretagogue receptor. It is mainly expressed in the central nervous system and peripheral tissues.

Another endocrine cell population is the PP cells. This small cell population constitutes about 5% of the total islet cells. PP cells produce the inhibitor hormone PP. Its secretion is stimulated by food, physical activity, and fasting [118]. The PP hormone is 36 amino acids long. It plays a role in regulating energy hemostasis. The hormone is released during fasting [119]. The PP hormone stimulates gastric chief cells in addition to its inhibitory activity on pancreatic acinar and ductal cells, which secrete digestive enzymes and carbonate salts (CHO_3^-), respectively, and inhibits bile secretion [120]. Recently, perhaps as the term gamma-cells (γ) has been ignored [121], and led to the creation of a meaningless gap between β and δ, PP cells were renamed γ-cells by some opinion leaders [122]. PP hormone has not been sufficiently studied, possibly because no diseases discovered, as yet, are associated with PP hormone and its signaling. Still, PP's function in nutrient metabolism is not sufficiently understood.

23.7 Target tissues, bioconversion, and storage

23.7.1 Liver

When insulin is secreted, the liver and muscles convert glucose into glycogen. In addition to activating the glycogenesis process, the liver also converts glucose into fat in the lipogenesis pathway [123]. The fat is transported to the adipose tissues by very-low-density lipoprotein particles [124,125].

When blood glucose levels decrease, α-cells start to secrete glucagon. The liver responds to glucagon signaling by activating the glycogenolysis and gluconeogenesis pathways [126–129]. Glycogenolysis is the breakdown of liver-stored glycogen back to glucose, and gluconeogenesis is the synthesis of glucose from different precursors such as free fatty acids and glycerol that arrive from the adipose tissue [127–129].

23.7.2 Adipose tissues

In the fed state, adipose tissues can activate lipogenesis with the presence of glucose under insulin regulation [130]. In the fasting state, glucagon signaling will elicit the secretion of free fatty acids, glycerol, and high-density lipoproteins to the bloodstream, which are then directed to the liver to take part in the gluconeogenesis pathway described above [17,19,131–134].

23.8 Secretion in the kidney

The nephron is the functional unit of the kidney. When blood reaches the glomerulus all the smaller particles are filtered out to create the primary urine [135]. At that stage, reabsorption processes are activated to prevent the loss of essential nutrients and substances. All of the glucose reabsorption processes are mediated by SGLT1 and SGLT2, found in segment 1 of the proximal tube. About 90% are reabsorbed by SGLT2 while the remaining 10% are reabsorbed by SGLT1 [136]. While SGLT1 is expressed in several tissues (such as the small intestine, see above), SGLT2 is almost exclusively expressed in the kidney [137]. As a result, SGLT2 is an ideal drug target to lower blood glucose levels via urination [138,139].

23.9 Diabetes

Diabetes is a heterogeneous metabolic disease caused by environmental or genetic factors. Diabetes is defined by elevated blood glucose levels. In all cases, this disease is a result of β-cell and/or insulin-sensing functional deficiency. There are several types of diabetes; T1D is an autoimmune disease, in which β-cells are destroyed by the immune system. This results in β-cell loss over time, and as a result, to lose insulin secretion [140,141]. Latent autoimmune diabetes of adults is another form of autoimmune diabetes, which appears in older age. T2D is largely divided into two causes. One is insulin resistance, in which the body does not respond to the normal insulin levels [142]. The second cause of T2D is obesity. To cope with the bodily demand for effective insulin concentrations, the β-cells need to increase insulin production and secretion. The additional consumption of sugars and fat that traditionally go hand-in-hand with obesity and will add to the load on β-cells, lead to their dysfunction and eventually to their distraction. Maturity-onset diabetes of the young is a rare type of diabetes, which is a collection of hereditary manifestations of genetic mutations. Surprisingly, most of these mutations are localized to auxiliary pathways of diabetes, and in fact less than 1% are mapped to insulin or its receptor genes [143,144].

23.10 Insulin and insulin mimetics

23.10.1 Mammalian insulin

Human insulin is very similar to other mammalian insulins, it differs in one amino acid from porcine, and in three amino acids from bovine, both porcine and bovine insulins have been used (and in the case of porcine, it still is, as in the case of Hypurin by Wockhardt) for diabetic treatment and have an effective clinical outcome on humans.

23.10.2 Recombinant human insulin

Recombinant insulin manufacturing using single-cell organisms is another effective and economic strategy to produce high quantities of human insulin. Humulin by Eli Lilly and Co. (produced in *Escherichia coli*) was the first application for commercial recombinant insulin [145]. Today, the vast majority of recombinant insulin is manufactured in yeast [146].

23.10.3 Synthetic insulin

The main challenge in synthetic insulin manufacturing is that the insulin is composed of two individual chains—A and B—which are linked together with disulfide bonds [109,147,148]. The combination of the two chains results in a mixture of different dipeptides, from which only one is the "true" insulin with appropriate disulfide bonds. In the 1960s three individual research groups succeeded in producing synthetic insulin, albeit in very low yields [149–151].

Later, improved biochemical approaches to encourage the correct A- and B-chain recombination allowed for the appropriate formation of the disulfide bonds. The resultant chemically produced native insulin was manufactured in sufficient quantities. These approaches confirmed the importance of the native cysteine disulfide link formation to the hormone's biological functions [152,153]. Still, for production purposes, the yields remained very low.

Crystal structure methods indicated that the A-chain N-terminus in the insulin structure are positioned closely to the B-chain C-terminus [154]. This finding had since encouraged the production of different single-chain insulin products. However, this now single-chain long peptide has imposed additional yield complexity that was only overcome later [155]. Solid-phase peptide synthesis enabled the formation of a B2Aib desB30 human mini-proinsulin [156]. However, the challenge of an approximately 60-amino-acid peptide was too expensive for commercial use. Only the development of native chemical ligation has enabled an efficient strategy to overcome efficient proinsulin synthesis [157]. In 2009, the chemical synthesis of mature insulin was achieved, by combining chemical ligation of the A- and B-chains of insulin that included a temporal chemical tether sequence, subsequently proteolytically removed [155]. This important achievement has paved the way for the introduction of natural and nonnatural mutations for further biochemical and chemical researches. Another approach was the insertion of selenocysteine, which can be fused into proteins both by natural translation and manipulating biochemical pathways of certain organisms or by means of native chemical linkers [158]. An important significant milestone in that direction was gained by the synthesis of insulin with two modifications on both the A- and B-chains in position 7 [159–161]. These changes allowed the creation of a 49-amino-acid peptide named DesDi. This modification and its derivatives could be efficiently bio-reacted in the HEK 293 cell line and had improved half-life and stability due to the replacement of cysteine with selenocysteine [147,162]. The correct in vitro folding of the insulin A- and B-chains combined with efficient yields were agreed to be unachievable. Surprisingly, correct disulfide linkers and proper insulin folding were achieved with unmodified and unprotected insulin A- and B-chains, using native chain assembly in approximately a 40% yield [148].

23.10.4 Short-acting and long-acting insulin as drugs

Insulin is stored as hexamers in secretion vesicles of β-cells as the inactivated form [163,164]. The hexamer structure is held in place by Thr30 and Lys29 found in the B-chain (Fig. 23.6A) [163,164]. The conversion of the insulin hexamer form to monomer constitutes an aspect of insulin formulation and drug design. Furthermore, by changing the N-terminal of the B-chain, the balance between the inactive hexamer and the monomer forms can be altered [164], and thus can modulate the available signaling monomer effective concentrations. The use of monomer insulin units is an optimal strategy in hyperglycemic events. To achieve that, pharmaceutical companies created specific mutations that, on the one hand, had only slightly affected insulin binding to its receptor and downstream signaling and, on the other hand, destabilize the hexamer structure. Insulin Lispro switches between LysB29 and ProB28 of the natural insulin on the B-chain. Another form of short-acting insulin termed insulin Glulisine, replaces AspB3 with lysine and LysB29 with glutamate. Insulin Aspart replaces the single ProB28 with aspartate (Fig. 23.6B–D, respectively).

In addition to fighting hyperglycemic events, it is essential to attain a basal and steady supply of insulin and to minimize the number of insulin injections throughout the day, even at the expense of larger insulin quantities per injection, in order to diminish diabetic patient discomfort. To achieve that, it is important to stabilize the hexamer structure and to decrease the dissociation to monomers. Insulin Glargine has replaced AspA21 on the A-chain with glycine, and in addition, added two positively charged arginine residues in the C-terminus of the B-chain (Fig. 23.6E), the addition of the two arginine residues also led to a change in the isoelectric point of insulin, resulting in slower insulin absorption into the bloodstream [163]. The formulation of the degludec and detemir insulin analogs used a different strategy. Instead of stabilizing the hexamer structure, there was a removal of ThrB30 and a supplement of a large hydrophobic residue on the B-chain to strongly bind them to albumin, an abundant protein carrier in the blood (Fig. 23.6F).

Another strategy, as demonstrated with the analog insulin peglispro, was to substantially increase the mass of the insulin molecule, which results in a slower subcutaneous absorption rate, as well as a substantially reduced glomerular filtration rate. These two traits increased the half-life of the insulin

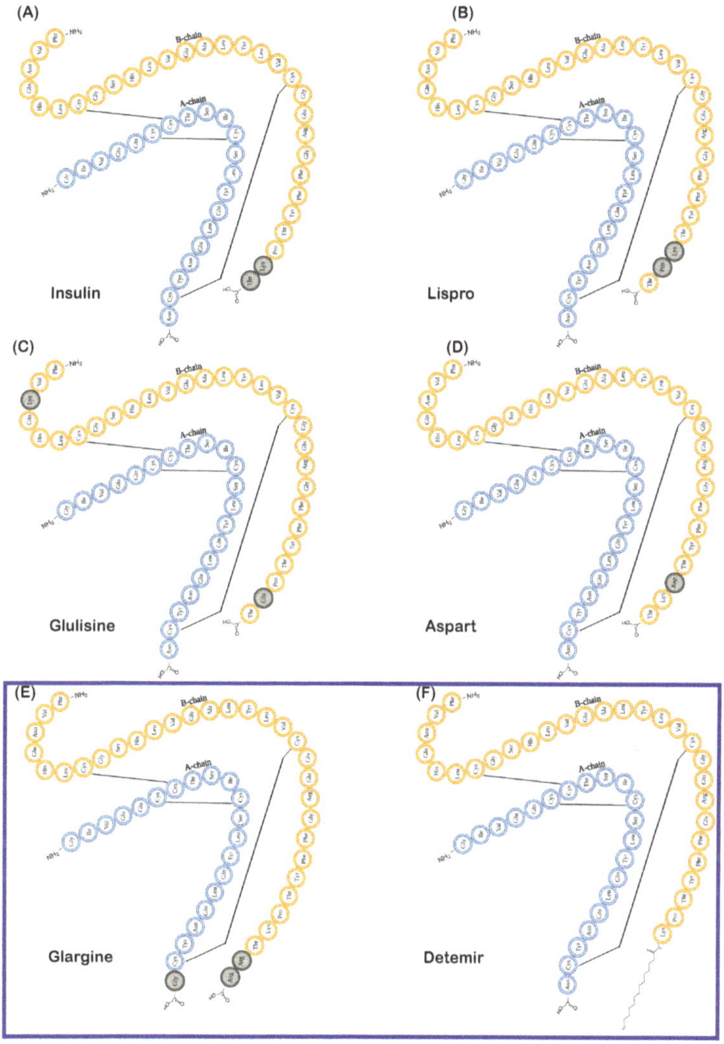

FIGURE 23.6 Insulin and its marketed analogs. (A) Insulin. Lys29 and Thr30 stabilize the insulin hexamer structure and therefore are highlighted in *gray*. Several strategies were employed to generate fast-acting analogs. (B) Insulin Lispro. Pro28 and Lys29 switched positions on the peptide sequence (in *gray*). (C) Insulin Glulisine. Lys3 and Glu29 are highlighted in *gray*. (D) Insulin Aspart. Asp28 is highlighted in *gray*. To generate slow-acting analogs the following molecules were generated (in the *blue box*). (E) Insulin Glargine. Gly21 is introduced at the C-terminus of the A-chain and two arginine modifications were made at the C-terminus of the B-chain. (F) Insulin Detemir. In insulin degludec the residue on LysB29 is different.

analog. The production of peglispro insulin was achieved by covalently attaching a 2000 Da polyethylene glycol (PEG) to the LysB28 side-chain of the analog. Thus, resulting in an enlarged hydrodynamic size of the analog due to the PEG conjugate [165]. Peglispro production was eventually discontinued due to safety issues.

23.11 Synthetic control of blood glucose levels

23.11.1 GLP-1 analogs

In early research, valinepyrrolidide, a DPP-4i, was successful in reducing plasma DPP-4 activity to inhibit infused GLP-1 degradation in pigs. Moreover, this observation was followed by an intense insulin response, proving that DPP-4 mediated degradation of GLP-1, playing an important role in reducing its insulinotropic effect [166]. Moreover, DPP-4i led to a significant blood glucose decrease in murine models of T2D (the obese Zucker rat) [167]. Furthermore, it was shown that isoleucine thiazolidide inhibited DPP-4 and induced insulin secretion while improving glucose tolerance after an oral glucose tolerance test [167]. On the other hand, no effect was detected when glucose was not administered. DPP-4 inhibition was proven to have an antihyperglycemic effect by preventing GLP-1 degradation [168]. In addition, DPP-4 inhibition did not have increased risk of hypoglycemia, despite the improvement in glycemic control [169–171]. These studies demonstrated that inhibition of DPP-4 activity could be exploited as a therapeutic approach for T2D. These studies were followed by phase 2 and 3 clinical trials introducing several DPP-4i, all showing that DPP-4 inhibition is safe and well-tolerated to achieve glycemic control in T2D patients.

The GLP-1 analog, extendin-4, is a natural peptide produced and secreted from the saliva of the lizard *Heloderma suspectum*, or the Gila monster (Fig. 23.7A). The sequence of extendin-4 is similar in 53% to the GLP-1 sequence (Fig. 23.5A). Today, there are at least 11 different GLP-1 agonists that have been developed and approved for therapy of T2D [183,184]. These medications include Lixisenatide (Fig. 23.7B), Liraglutide (Fig. 23.7C), Semaglutide (Fig. 23.7D), Albiglutide (Fig. 23.7E), Dulaglutide (Fig. 23.7F), and Detemir (Fig. 23.6F).

23.11.2 SGLT2 inhibitors reduce hyperglycemia and treat several types of kidney disease

SGLT2 is responsible for the reabsorption of glucose back to the bloodstream from the primary urine in the kidneys. Some approaches to lower blood glucose levels in cardiovascular and diabetic patients use SGLT2 inhibitors (SGLT2i). Different SGLT2i have been developed such as dapagliflozin, canagliflozin, empagliflozin, and ertugliflozin. And others are already in

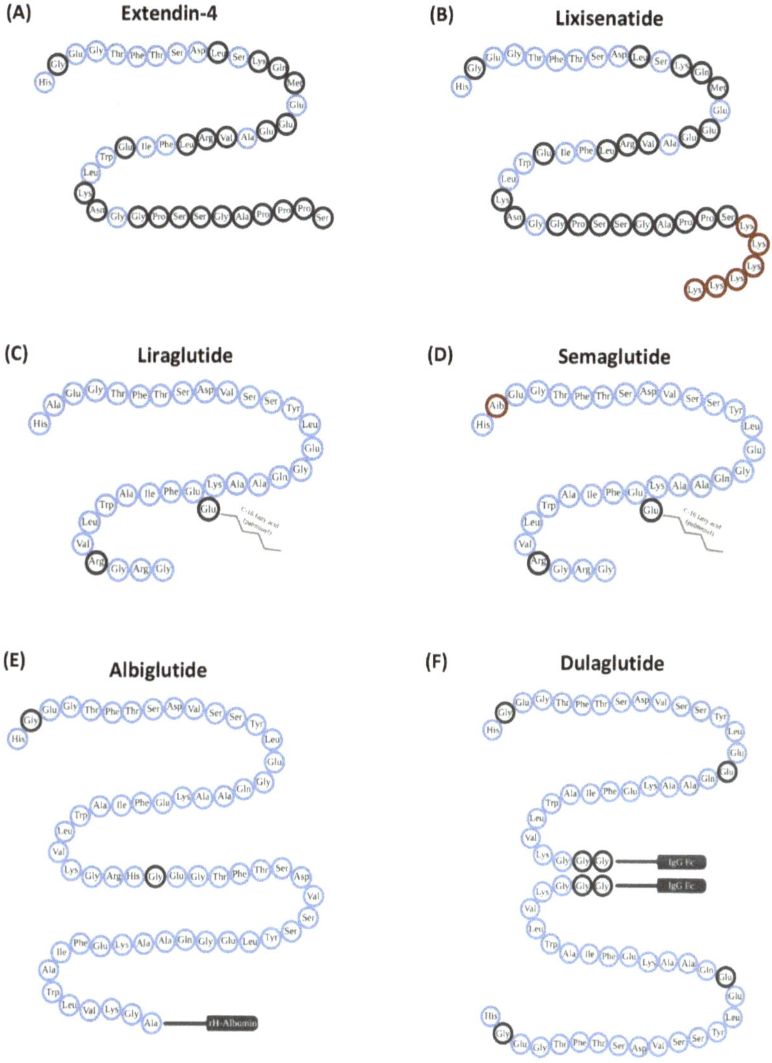

FIGURE 23.7 GLP-1 agonists. (A) Extendin-4. The amino acids that are identical to GLP-1 are in *blue*, and the amino acids that differ from GLP-1 are in *black*. (B) Lixisenatide. The engineered addition of the six lysine residues is in *red*. (C) Liraglutide. The addition of a C16 fatty acid linked to glutamate is covalently bound to Lys23, which is in *black*, as well as the newly introduced Arg28. (D) Semaglutide. The subsequent modification from Liraglutide, α-aminoisobutyric acid on position 2 is in *red*. (E) Albiglutide. The human GLP-1 was modified with Gly2 (*black*) and the hormone was fused to human albumin. (F) Dulaglutide. Two GLP-1 human analogs were bound to an antibody FC, while Efpeglenatide is a modification of extendin-4 (Efpeglenatide is not shown).

their clinical trial stages [138]. SGLT2i can reduce HbA1c levels from baseline in T2D patients. In addition, SGLT2i which was coadministered with a GLP-1 analog was able to significantly reduce HbA1c in nondiabetic obese patients [172,173].

The SGLT2i glucose-lowering capacity is dependent upon blood glucose levels [139], and therefore adverse hypoglycemic incidences are reduced. SGLT2i also lower body weight [174–177], and this effect is dose-dependent [178–181]. However, it was expected that SGLT2i administration would have a greater impact on weight loss, as the energy secreted by glycosuria is of high caloric value. It was postulated that sustained SGLT2i administration stimulates an adaptive reaction, in which patients compensate with increased food intake [182]. Therefore, it was hypothesized that combining SGLT2i with additional drugs that act by a compensatory pathway may prove more effective [183].

In addition, SGLT2i can reduce cardiovascular events [184–186] and improve blood pressure measurements [187,188]. Despite their advantage to T2D and obese patients, for T1D SGLT2i administration is more problematic. SGLT2i lower blood glucose levels and therefore, a marked increase of diabetic ketoacidosis is observed when SGLT2i are prescribed to T1D patients [138].

SGLT1, in contrast to SGLT2, is not predominantly expressed in the kidney and exhibits a broad tissue distribution. Therefore, SGLT1 inhibition suffers from significant side effects and as a result was not chosen as a specific drug candidate. However, recently more and more drugs that act as SGLT1/2 coinhibitors have been developed and approved for clinical trials [102].

23.12 Summary

In this chapter, we described the processes that act upon carbohydrates in general and glucose in particular from consumption or secretion. We discussed the importance of maintaining balanced blood glucose levels, the related mechanisms, and what happens when these mechanisms fail, leading to diabetes mellitus. Specific attention was devoted to peptidomimetic GLP-1 agonists and SGLT2i that hold promise for T2D treatment. For T1D, insulin remains the standard of care. New insulin analogs will allow injecting fewer doses per day and smaller volumes, while still maintaining the required insulin levels.

References

[1] Bell GI, Pictet RL, Rutter WJ, Cordell B, Tischer E, Goodman HM. Sequence of the human insulin gene. Nature. 1980;284(5751):26–32.

[2] Weiss M, Steiner DF, Philipson LH. Insulin biosynthesis, secretion, structure, and structure-activity relationships. In: DeGroot LJ, Chrousos G, Dungan K, Feingold KR, Grossman A, Hershman JM, et al., editors. Endotext. South Dartmouth, MA: MDText.com, Inc.; 2000.

[3] Slack JM. Developmental biology of the pancreas. Development. 1995;121(6):1569–80.

[4] Da Silva Xavier G. The cells of the islets of langerhans. J Clin Med 2018;7(3) Mar 12.
[5] Liu M, Lara-Lemus R, Shan S, Wright J, Haataja L, Barbetti F, et al. Impaired cleavage of preproinsulin signal peptide linked to autosomal-dominant diabetes. Diabetes. 2012;61 (4):828–37.
[6] Reichert M, Rustgi AK. Pancreatic ductal cells in development, regeneration, and neoplasia. J Clin Invest 2011;121(12):4572–8.
[7] Steiner DJ, Kim A, Miller K, Hara M. Pancreatic islet plasticity: interspecies comparison of islet architecture and composition. Islets. 2010;2(3):135–45.
[8] Zhang X-X, Pan Y-H, Huang Y-M, Zhao H-L. Neuroendocrine hormone amylin in diabetes. World J Diabetes 2016;7(9):189–97.
[9] Heppner KM, Habegger KM, Day J, Pfluger PT, Perez-Tilve D, Ward B, et al. Glucagon regulation of energy metabolism. Physiol Behav 2010;100(5):545–8.
[10] Svendsen B, Larsen O, Gabe MBN, Christiansen CB, Rosenkilde MM, Drucker DJ, et al. Insulin secretion depends on intra-islet glucagon signaling. Cell Rep 2018;25(5):1127–34 e2.
[11] Banarer S, McGregor VP, Cryer PE. Intraislet hyperinsulinemia prevents the glucagon response to hypoglycemia despite an intact autonomic response. Diabetes. 2002;51 (4):958–65.
[12] Hope KM, Tran POT, Zhou H, Oseid E, Leroy E, Robertson RP. Regulation of alpha-cell function by the beta-cell in isolated human and rat islets deprived of glucose: the "switch-off" hypothesis. Diabetes. 2004;53(6):1488–95.
[13] Meier JJ, Kjems LL, Veldhuis JD, Lefèbvre P, Butler PC. Postprandial suppression of glucagon secretion depends on intact pulsatile insulin secretion: further evidence for the intraislet insulin hypothesis. Diabetes. 2006;55(4):1051–6.
[14] Gromada J, Franklin I, Wollheim CB. Alpha-cells of the endocrine pancreas: 35 years of research but the enigma remains. Endocr Rev 2007;28(1):84–116.
[15] Lonovics J, Devitt P, Watson LC, Rayford PL, Thompson JC. Pancreatic polypeptide. A review. Arch Surg 1981;116(10):1256–64.
[16] De Pagter-Holthuizen P, Höppener JW, Jansen M, Geurts van Kessel AH, van Ommen GJ, Sussenbach JS. Chromosomal localization and preliminary characterization of the human gene encoding insulin-like growth factor II. Hum Genet 1985;69(2):170–3.
[17] Jansen M, van Schaik FM, Ricker AT, Bullock B, Woods DE, Gabbay KH, et al. Sequence of cDNA encoding human insulin-like growth factor I precursor. Nature. 1983;306(5943):609–11.
[18] Rinderknecht E, Humbel RE. The amino acid sequence of human insulin-like growth factor I and its structural homology with proinsulin. J Biol Chem 1978;253(8):2769–76.
[19] Qiu Q, Jiang J-Y, Bell M, Tsang BK, Gruslin A. Activation of endoproteolytic processing of insulin-like growth factor-II in fetal, early postnatal, and pregnant rats and persistence of circulating levels in postnatal life. Endocrinology. 2007;148(10):4803–11.
[20] Jiang G, Zhang BB. Glucagon and regulation of glucose metabolism. Am J Physiol Endocrinol Metab 2003;284(4):E671–8.
[21] Perry RJ, Zhang D, Guerra MT, Brill AL, Goedeke L, Nasiri AR, et al. Glucagon stimulates gluconeogenesis by INSP3R1-mediated hepatic lipolysis. Nature. 2020;579 (7798):279–83.
[22] Arafat AM, Kaczmarek P, Skrzypski M, Pruszyńska-Oszmalek E, Kołodziejski P, Szczepankiewicz D, et al. Glucagon increases circulating fibroblast growth factor 21 independently of endogenous insulin levels: a novel mechanism of glucagon-stimulated lipolysis? Diabetologia. 2013;56(3):588–97.

[23] White JW, Saunders GF. Structure of the human glucagon gene. Nucleic Acids Res 1986;14(12):4719–30.

[24] Pollock HG, Hamilton JW, Rouse JB, Ebner KE, Rawitch AB. Isolation of peptide hormones from the pancreas of the bullfrog (*Rana catesbeiana*). Amino acid sequences of pancreatic polypeptide, oxyntomodulin, and two glucagon-like peptides. J Biol Chem 1988;263(20):9746–51.

[25] Unger RH, Orci L. Glucagon and the A cell: physiology and pathophysiology (second of two parts). N Engl J Med 1981;304(26):1575–80.

[26] Svoboda M, Tastenoy M, Vertongen P, Robberecht P. Relative quantitative analysis of glucagon receptor mRNA in rat tissues. Mol Cell Endocrinol 1994;105(2):131–7.

[27] Iakoubov R, Izzo A, Yeung A, Whiteside CI, Brubaker PL. Protein kinase Czeta is required for oleic acid-induced secretion of glucagon-like peptide-1 by intestinal endocrine L cells. Endocrinology. 2007;148(3):1089–98.

[28] Hsieh J, Longuet C, Maida A, Bahrami J, Xu E, Baker CL, et al. Glucagon-like peptide-2 increases intestinal lipid absorption and chylomicron production via CD36. Gastroenterology. 2009;137(3):997–1005 1005.e1.

[29] Hsieh J, Longuet C, Baker CL, Qin B, Federico LM, Drucker DJ, et al. The glucagon-like peptide 1 receptor is essential for postprandial lipoprotein synthesis and secretion in hamsters and mice. Diabetologia. 2010;53(3):552–61.

[30] Meier JJ, Nauck MA. Incretins and the development of type 2 diabetes. Curr Diab Rep 2006;6(3):194–201.

[31] Baggio LL, Drucker DJ. Biology of incretins: GLP-1 and GIP. Gastroenterology. 2007;132(6):2131–57.

[32] Cheeseman CI. Upregulation of SGLT-1 transport activity in rat jejunum induced by GLP-2 infusion in vivo. Am J Physiol 1997;273(6):R1965–71.

[33] Au A, Gupta A, Schembri P, Cheeseman CI. Rapid insertion of GLUT2 into the rat jejunal brush-border membrane promoted by glucagon-like peptide 2. Biochem J 2002;367 (Pt 1):247–54.

[34] Pocai A. Action and therapeutic potential of oxyntomodulin. Mol Metab 2014;3 (3):241–51.

[35] Blache P, Kervran A, Bataille D. Oxyntomodulin and glicentin: brain-gut peptides in the rat. Endocrinology. 1988;123(6):2782–7.

[36] Thim L, Moody AJ. The primary structure of porcine glicentin (proglucagon). Regul Pept 1981;2(2):139–50.

[37] Landgraf D, Tsang AH, Leliavski A, Koch CE, Barclay JL, Drucker DJ, et al. Oxyntomodulin regulates resetting of the liver circadian clock by food. elife. 2015;4: e06253.

[38] Muppidi A, Zou H, Yang PY, Chao E, Sherwood L, Nunez V, et al. Design of potent and proteolytically stable oxyntomodulin analogs. ACS Chem Biol 2016;11(2):324–8.

[39] Perakakis N, Mantzoros CS. The role of glicentin and oxyntomodulin in human metabolism: new evidence and new directions. J Clin Endocrinol Metab 2020;105(8) Aug 1.

[40] Smeets PAM, Erkner A, de Graaf C. Cephalic phase responses and appetite. Nutr Rev 2010;68(11):643–55.

[41] Smith GP. Pavlov and appetite. Integr Physiol Behav Sci 1995;30(2):169–74.

[42] Teff K. Nutritional implications of the cephalic-phase reflexes: endocrine responses. Appetite. 2000;34(2):206–13.

[43] Power ML, Schulkin J. Anticipatory physiological regulation in feeding biology: cephalic phase responses. Appetite. 2008;50(2–3):194–206.

[44] Wooley SC, Wooley OW. Salivation to the sight and thought of food: a new measure of appetite. Psychosom Med 1973;35(2):136–42.

[45] Feldman M, Richardson CT. Role of thought, sight, smell, and taste of food in the cephalic phase of gastric acid secretion in humans. Gastroenterology. 1986;90(2):428–33.

[46] Rodin J. Insulin levels, hunger, and food intake: an example of feedback loops in body weight regulation. Health Psychol 1985;4(1):1–24.

[47] Pavlov I, Thompson WH. The work of digestive glands: lectures by Professor J. P. Pawlow. London, Philadelphia: Charles Griffin & Company, Limited; J.B. Lippincott Company:; 1910.

[48] Pavlov I. Conditioned reflexes: an investigation of the physiological activity of the cerebral cortex. Oxford, England: Oxford Univ. Press:; 1927.

[49] Berthoud HR, Bereiter DA, Trimble ER, Siegel EG, Jeanrenaud B. Cephalic phase, reflex insulin secretion neuroanatomical and physiological characterization. Diabetologia. 1981;20(Suppl 1):393–401.

[50] Christensen CM, Navazesh M. Anticipatory salivary flow to the sight of different foods. Appetite. 1984;5(4):307–15.

[51] Richardson CT, Feldman M. Salivary response to food in humans and its effect on gastric acid secretion. Am J Physiol 1986;250(1 Pt 1):G85–91.

[52] Janowitz HD, Hollander F, Orringer D, Levy MH, Winkelstein A, Kaufman R, et al. A quantitative study of the gastric secretory response to sham feeding in a human subject. Gastroenterology. 1950;16(1):104–16.

[53] Love JA, Yi E, Smith TG. Autonomic pathways regulating pancreatic exocrine secretion. Auton Neurosci 2007;133(1):19–34.

[54] Ohara I, Otsuka S, Yugari Y. Cephalic phase response of pancreatic exocrine secretion in conscious dogs. Am J Physiol 1988;254(3 Pt 1):G424–8.

[55] Strubbe JH, Steffens AB. Rapid insulin release after ingestion of a meal in the unanesthetized rat. Am J Physiol 1975;229(4):1019–22.

[56] Teff KL, Mattes RD, Engelman K. Cephalic phase insulin release in normal weight males: verification and reliability. Am J Physiol 1991;261(4 Pt 1):E430–6.

[57] Powley TL. The ventromedial hypothalamic syndrome, satiety, and a cephalic phase hypothesis. Psychol Rev 1977;84(1):89–126.

[58] Nederkoorn C, Smulders FT, Jansen A. Cephalic phase responses, craving and food intake in normal subjects. Appetite. 2000;35(1):45–55.

[59] Katschinski M, Dahmen G, Reinshagen M, Beglinger C, Koop H, Nustede R, et al. Cephalic stimulation of gastrointestinal secretory and motor responses in humans. Gastroenterology. 1992;103(2):383–91.

[60] McGregor IS, Lee AM. Changes in respiratory quotient elicited in rats by a conditioned stimulus predicting food. Physiol Behav 1998;63(2):227–32.

[61] LeBlanc J, Cabanac M. Cephalic postprandial thermogenesis in human subjects. Physiol Behav 1989;46(3):479–82.

[62] Veedfald S, Plamboeck A, Deacon CF, Hartmann B, Knop FK, Vilsbøll T, et al. Cephalic phase secretion of insulin and other enteropancreatic hormones in humans. Am J Physiol Gastrointest Liver Physiol 2016;310(1):G43–51.

[63] LeBlanc J. Nutritional implications of cephalic phase thermogenic responses. Appetite. 2000;34(2):214–16.

[64] Regazzi R, Rodriguez-Trejo A, Jacovetti C. Insulin secretion in health and disease: nutrients dictate the pace. Proc Nutr Soc 2016;75(1):19–29.

[65] Ahrén B. Autonomic regulation of islet hormone secretion–implications for health and disease. Diabetologia. 2000;43(4):393–410.

[66] Picard F, Naïmi N, Richard D, Deshaies Y. Response of adipose tissue lipoprotein lipase to the cephalic phase of insulin secretion. Diabetes. 1999;48(3):452–9.

[67] Mitrakou A, Kelley D, Mokan M, Veneman T, Pangburn T, Reilly J, et al. Role of reduced suppression of glucose production and diminished early insulin release in impaired glucose tolerance. N Engl J Med 1992;326(1):22–9.

[68] Calles-Escandon J, Robbins DC. Loss of early phase of insulin release in humans impairs glucose tolerance and blunts thermic effect of glucose. Diabetes. 1987;36(10):1167–72.

[69] Teff KL, Townsend RR. Early phase insulin infusion and muscarinic blockade in obese and lean subjects. Am J Physiol 1999;277(1):R198–208.

[70] Kraegen EW, Chisholm DJ, McNamara ME. Timing of insulin delivery with meals. Horm Metab Res 1981;13(7):365–7.

[71] Bruttomesso D, Pianta A, Mari A, Valerio A, Marescotti MC, Avogaro A, et al. Restoration of early rise in plasma insulin levels improves the glucose tolerance of type 2 diabetic patients. Diabetes. 1999;48(1):99–105.

[72] Schwartz TW, Stenquist B, Olbe L. Cephalic phase of pancreatic-polypeptide secretion studied by sham feeding in man. Scand J Gastroenterol 1979;14(3):313–20.

[73] Taylor IL, Feldman M. Effect of cephalic-vagal stimulation on insulin, gastric inhibitory polypeptide, and pancreatic polypeptide release in humans. J Clin Endocrinol Metab 1982;55(6):1114–17.

[74] Floyd JC, Fajans SS, Pek S, Chance RE. A newly recognized pancreatic polypeptide; plasma levels in health and disease. Recent Prog Horm Res 1976;33:519–70.

[75] Schwartz TW. Pancreatic polypeptide: a hormone under vagal control. Gastroenterology. 1983;85(6):1411–25.

[76] Schwartz TW. Pancreatic polypeptide: a unique model for vagal control of endocrine systems. J Auton Nerv Syst 1983;9(1):99–111.

[77] Mcintyre N, Holdsworth CD, Turner DS. New interpretation of oral glucose tolerance. Lancet 1964;2(7349):20–1.

[78] McIntyre N, Holdsworth CD, Turner DS. Intestinal factors in the control of insulin secretion. J Clin Endocrinol Metab 1965;25(10):1317–24.

[79] Edie ES. Further observations on the treatment of diabetes mellitus by acid extract of duodenal mucous membrane. Biochem J 1906;1(8–9):446–54.

[80] Baskin DG. A historical perspective on the identification of cell types in pancreatic islets of langerhans by staining and histochemical techniques. J Histochem Cytochem 2015;63(8):543–58.

[81] Brown JC, Dryburgh JR. A gastric inhibitory polypeptide. II. The complete amino acid sequence. Can J Biochem 1971;49(8):867–72.

[82] Brown JC. A gastric inhibitory polypeptide. I. The amino acid composition and the tryptic peptides. Can J Biochem 1971;49(2):255–61.

[83] Kieffer TJ. Gastro-intestinal hormones GIP and GLP-1. Ann Endocrinol (Paris) 2004;65 (1):13–21.

[84] Nauck MA, Niedereichholz U, Ettler R, Holst JJ, Orskov C, Ritzel R, et al. Glucagon-like peptide 1 inhibition of gastric emptying outweighs its insulinotropic effects in healthy humans. Am J Physiol 1997;273(5):E981–8.

[85] Fujita Y, Wideman RD, Asadi A, Yang GK, Baker R, Webber T, et al. Glucose-dependent insulinotropic polypeptide is expressed in pancreatic islet alpha-cells and promotes insulin secretion. Gastroenterology. 2010;138(5):1966–75.

[86] Buhren BA, Gasis M, Thorens B, Müller HW, Bosse F. Glucose-dependent insulinotropic polypeptide (GIP) and its receptor (GIPR): cellular localization, lesion-affected expression, and impaired regenerative axonal growth. J Neurosci Res 2009;87(8):1858–70.

[87] Moffett RC, Naughton V. Emerging role of GIP and related gut hormones in fertility and PCOS. Peptides. 2020;125:170233.

[88] Brereton MF, Vergari E, Zhang Q, Clark A. Alpha-, Delta- and PP-cells: are they the architectural cornerstones of islet structure and co-ordination? J Histochem Cytochem 2015;63(8):575–91.

[89] Andralojc KM, Mercalli A, Nowak KW, Albarello L, Calcagno R, Luzi L, et al. Ghrelin-producing epsilon cells in the developing and adult human pancreas. Diabetologia. 2009;52(3):486–93.

[90] Zhuo F, Elizabeth RG, Dongmin L. Regulation of insulin synthesis and secretion and pancreatic Beta-cell dysfunction in diabetes. Curr Diabetes Rev 2013;9(1):25–53.

[91] Lund PK, Goodman RH, Dee PC, Habener JF. Pancreatic preproglucagon cDNA contains two glucagon-related coding sequences arranged in tandem. Proc Natl Acad Sci USA 1982;79(2):345–9.

[92] Lund PK, Goodman RH, Montminy MR, Dee PC, Habener JF. Anglerfish islet preproglucagon II. Nucleotide and corresponding amino acid sequence of the cDNA. J Biol Chem 1983;258(5):3280–4.

[93] Deacon CF, Pridal L, Klarskov L, Olesen M, Holst JJ. Glucagon-like peptide 1 undergoes differential tissue-specific metabolism in the anesthetized pig. Am J Physiol 1996;271(3 Pt 1):E458–64.

[94] Van Dijk G, Thiele TE. Glucagon-like peptide-1 (7–36) amide: a central regulator of satiety and interoceptive stress. Neuropeptides. 1999;33(5):406–14.

[95] Hansen L, Hartmann B, Bisgaard T, Mineo H, Jørgensen PN, Holst JJ. Somatostatin restrains the secretion of glucagon-like peptide-1 and -2 from isolated perfused porcine ileum. Am J Physiol Endocrinol Metab 2000;278(6):E1010–18.

[96] Matikainen N, Mänttäri S, Schweizer A, Ulvestad A, Mills D, Dunning BE, et al. Vildagliptin therapy reduces postprandial intestinal triglyceride-rich lipoprotein particles in patients with type 2 diabetes. Diabetologia. 2006;49(9):2049–57.

[97] Mulvihill EE, Drucker DJ. Pharmacology, physiology, and mechanisms of action of dipeptidyl peptidase-4 inhibitors. Endocr Rev 2014;35(6):992–1019.

[98] Mroz PA, Finan B, Gelfanov V, Yang B, Tschöp MH, DiMarchi RD, et al. Optimized GIP analogs promote body weight lowering in mice through GIPR agonism not antagonism. Mol Metab 2019;20:51–62.

[99] Hinke SA, Gelling RW, Pederson RA, Manhart S, Nian C, Demuth H-U, et al. Dipeptidyl peptidase IV-resistant [D-Ala(2)]glucose-dependent insulinotropic polypeptide (GIP) improves glucose tolerance in normal and obese diabetic rats. Diabetes. 2002;51 (3):652–61.

[100] Vienberg SG, Bouman SD, Sørensen H, Stidsen CE, Kjeldsen T, Glendorf T, et al. Receptor-isoform-selective insulin analogues give tissue-preferential effects. Biochem J 2011;440(3):301–8.

[101] Song P, Onishi A, Koepsell H, Vallon V. Sodium glucose cotransporter SGLT1 as a therapeutic target in diabetes mellitus. Expert Opin Ther Targets 2016;20(9):1109–25.

[102] Tsimihodimos V, Filippas-Ntekouan S, Elisaf M. SGLT1 inhibition: Pros and cons. Eur J Pharmacol 2018;838:153–6.

[103] Olson AL, Pessin JE. Structure, function, and regulation of the mammalian facilitative glucose transporter gene family. Annu Rev Nutr 1996;16:235–56.

[104] Nishimura H, Pallardo FV, Seidner GA, Vannucci S, Simpson IA, Birnbaum MJ. Kinetics of GLUT1 and GLUT4 glucose transporters expressed in Xenopus oocytes. J Biol Chem 1993;268(12):8514–20.

[105] Ghezzi C, Loo DDF, Wright EM. Physiology of renal glucose handling via SGLT1, SGLT2 and GLUT2. Diabetologia. 2018;61(10):2087–97.

[106] Leino RL, Gerhart DZ, van Bueren AM, McCall AL, Drewes LR. Ultrastructural localization of GLUT 1 and GLUT 3 glucose transporters in rat brain. J Neurosci Res 1997;49(5):617–26.

[107] James DE, Brown R, Navarro J, Pilch PF. Insulin-regulatable tissues express a unique insulin-sensitive glucose transport protein. Nature. 1988;333(6169):183–5.

[108] Huang S, Czech MP. The GLUT4 glucose transporter. Cell Metab 2007;5(4):237–52.

[109] Musiol H-J, Moroder L. Two-chain insulin from a single-chain branched depsipeptide precursor: the end of a long journey. Angew Chem Int Ed Engl 2010;49(42):7624–6.

[110] Strowski MZ, Parmar RM, Blake AD, Schaeffer JM. Somatostatin inhibits insulin and glucagon secretion via two receptors subtypes: an in vitro study of pancreatic islets from somatostatin receptor 2 knockout mice. Endocrinology. 2000;141(1):111–17.

[111] Pintér E, Helyes Z, Szolcsányi J. Inhibitory effect of somatostatin on inflammation and nociception. Pharmacol Ther 2006;112(2):440–56.

[112] Elekes K, Helyes Z, Kereskai L, Sándor K, Pintér E, Pozsgai G, et al. Inhibitory effects of synthetic somatostatin receptor subtype 4 agonists on acute and chronic airway inflammation and hyperreactivity in the mouse. Eur J Pharmacol 2008;578(2–3):313–22.

[113] Helyes Z, Pintér E, Németh J, Kéri G, Thán M, Oroszi G, et al. Anti-inflammatory effect of synthetic somatostatin analogues in the rat. Br J Pharmacol 2001;134(7):1571–9.

[114] Patel YC. Somatostatin and its receptor family. Front Neuroendocrinol 1999;20(3):157–98.

[115] Shen LP, Pictet RL, Rutter WJ. Human somatostatin I: sequence of the cDNA. Proc Natl Acad Sci U S A 1982;79(15):4575–9.

[116] Shen LP, Rutter WJ. Sequence of the human somatostatin I gene. Science. 1984;224(4645):168–71.

[117] Kojima M, Hosoda H, Date Y, Nakazato M, Matsuo H, Kangawa K. Ghrelin is a growth-hormone-releasing acylated peptide from stomach. Nature. 1999;402(6762):656–60.

[118] Das R, Dobens LL. Conservation of gene and tissue networks regulating insulin signalling in flies and vertebrates. Biochem Soc Trans 2015;43(5):1057–62.

[119] Boel E, Schwartz TW, Norris KE, Fiil NP. A cDNA encoding a small common precursor for human pancreatic polypeptide and pancreatic icosapeptide. EMBO J 1984;3(4):909–12.

[120] Schwartz TS, Bronikowski AM. Evolution and function of the insulin and insulin-like signaling network in ectothermic reptiles: some answers and more questions. Integr Comp Biol 2016;56(2):171–84.

[121] Kadakia R, Ma M, Josefson JL. Neonatal adiposity increases with rising cord blood IGF-1 levels. Clin Endocrinol (Oxf) 2016;85(1):70–5.

[122] Höppener JW, de Pagter-Holthuizen P, Geurts van Kessel AH, Jansen M, Kittur SD, Antonarakis SE, et al. The human gene encoding insulin-like growth factor I is located on chromosome 12. Hum Genet 1985;69(2):157–60.

[123] Glimcher LH, Lee A-H. From sugar to fat: how the transcription factor XBP1 regulates hepatic lipogenesis. Ann N Y Acad Sci 2009;1173(Suppl 1):E2–9.

[124] Rui L. Energy metabolism in the liver. Compr Physiol 2014;4(1):177–97.

[125] Gibbons GF, Wiggins D, Brown AM, Hebbachi AM. Synthesis and function of hepatic very-low-density lipoprotein. Biochem Soc Trans 2004;32(Pt 1):59–64.

[126] Nordlie RC, Foster JD, Lange AJ. Regulation of glucose production by the liver. Annu Rev Nutr 1999;19:379–406.

[127] Kehlenbrink S, Tonelli J, Koppaka S, Chandramouli V, Hawkins M, Kishore P. Inhibiting gluconeogenesis prevents fatty acid-induced increases in endogenous glucose production. Am J Physiol Endocrinol Metab 2009;297(1):E165–73.

[128] Bongaerts GPA, Wagener DJT. Increased hepatic gluconeogenesis: the secret of Lance Armstrong's success. Med Hypotheses 2007;68(1):9–11.

[129] Ezaki J, Matsumoto N, Takeda-Ezaki M, Komatsu M, Takahashi K, Hiraoka Y, et al. Liver autophagy contributes to the maintenance of blood glucose and amino acid levels. Autophagy. 2011;7(7):727–36.

[130] Scherer T, O'Hare J, Diggs-Andrews K, Schweiger M, Cheng B, Lindtner C, et al. Brain insulin controls adipose tissue lipolysis and lipogenesis. Cell Metab 2011;13(2):183–94.

[131] Engström W, Shokrai A, Otte K, Granérus M, Gessbo A, Bierke P, et al. Transcriptional regulation and biological significance of the insulin like growth factor II gene. Cell Prolif 1998;31(5–6):173–89.

[132] Belfiore A, Frasca F, Pandini G, Sciacca L, Vigneri R. Insulin receptor isoforms and insulin receptor/insulin-like growth factor receptor hybrids in physiology and disease. Endocr Rev 2009;30(6):586–623.

[133] Rinderknecht E, Humbel RE. Primary structure of human insulin-like growth factor II. FEBS Lett 1978;89(2):283–6.

[134] Sciacca L, Le Moli R, Vigneri R. Insulin analogs and cancer. Front Endocrinol (Lausanne) 2012;3:21 Feb 10.

[135] Ichimura K, Sakai T. Evolutionary morphology of podocytes and primary urine-producing apparatus. Anat Sci Int 2017;92(2):161–72.

[136] Andersen M, Nørgaard-Pedersen D, Brandt J, Pettersson I, Slaaby R. IGF1 and IGF2 specificities to the two insulin receptor isoforms are determined by insulin receptor amino acid 718. PLoS ONE 2017;12(6):e0178885.

[137] Mather A, Pollock C. Glucose handling by the kidney. Kidney Int Suppl 2011;120:S1–6.

[138] Pereira MJ, Eriksson JW. Emerging Role of SGLT-2 Inhibitors for the Treatment of Obesity. Drugs. 2019;79(3):219–30.

[139] Monami M, Nardini C, Mannucci E. Efficacy and safety of sodium glucose co-transport-2 inhibitors in type 2 diabetes: a *meta*-analysis of randomized clinical trials. Diabetes Obes Metab 2014;16(5):457–66.

[140] Ziegler A-G, Nepom GT. Prediction and pathogenesis in type 1 diabetes. Immunity. 2010;32(4):468–78.

[141] Wyatt R, Williams AJK. Islet autoantibody analysis: radioimmunoassays. Methods Mol Biol 2016;1433:57–83.

[142] Isganaitis E, Lustig RH. Fast food, central nervous system insulin resistance, and obesity. Arterioscler Thromb Vasc Biol 2005;25(12):2451–62.

[143] Fajans SS, Bell GI, Polonsky KS. Molecular mechanisms and clinical pathophysiology of maturity-onset diabetes of the young. N Engl J Med 2001;345(13):971–80.

[144] Kleinberger JW, Pollin TI. Undiagnosed MODY: time for action. Curr Diab Rep 2015;15(12):110.

[145] Recombinant DNA technology in the synthesis of human insulin [Internet]. <http://www.littletree.com.au/dna.htm>; [accessed 23.09.19].

[146] Kjeldsen T. Yeast secretory expression of insulin precursors. Appl Microbiol Biotechnol 2000;54(3):277–86.

[147] Weil-Ktorza O, Rege N, Lansky S, Shalev DE, Shoham G, Weiss MA, et al. Substitution of an internal disulfide bridge with a diselenide enhances both foldability and stability of human insulin. Chem - A Eur J 2019;25(36):8513–21.

[148] Arai K, Takei T, Shinozaki R, Noguchi M, Fujisawa S, Katayama H, et al. Characterization and optimization of two-chain folding pathways of insulin via native chain assembly. Commun Chem 2018;1(1):26.

[149] Meienhofer J, Schnabel E, Bremer H, Brinkhoff O, Zabel R, Sroka W, et al. [synthesis of insulin chains and their combination to insulin-active preparations]. Z Naturforsch B 1963;18:1120–1.

[150] Katsoyannis PG, Fukuda K, Tometsko A, Suzuki K, Tilak M. Insulin peptides. X. The synthesis of the B-chain of insulin and its combination with natural or synthetis A-chain to generate insulin activity. J Am Chem Soc 1964;86(5):930–2.

[151] Kung YT, Du YC, Huang WT, Chen CC, Ke LT. Total synthesis of crystalline bovine insulin. Sci Sin 1965;14(11):1710–16.

[152] Akaji K, Fujino K, Tatsumi T, Kiso Y. Total synthesis of human insulin by regioselective disulfide formation using the silyl chloride-sulfoxide method. J Am Chem Soc 1993;115(24):11384–92.

[153] Sieber P, Kamber B, Hartmann A, Jöhl A, Riniker B, Rittel W. Total synthesis of human insulin under directed formation of the disulfide bonds. Helv Chim Acta 1974;57 (8):2617–21.

[154] Adams MJ, Blundell TL, Dodson EJ, Dodson GG, Vijayan M, Baker EN, et al. Structure of rhombohedral 2 zinc insulin crystals. Nature. 1969;224(5218):491–5.

[155] Sohma Y, Kent SBH. Biomimetic synthesis of lispro insulin via a chemically synthesized "mini-proinsulin" prepared by oxime-forming ligation. J Am Chem Soc 2009;131 (44):16313–18.

[156] Tofteng AP, Jensen KJ, Schäffer L, Hoeg-Jensen T. Total synthesis of desB30 insulin analogues by biomimetic folding of single-chain precursors. Chembiochem. 2008;9 (18):2989–96.

[157] Dawson PE, Muir TW, Clark-Lewis I, Kent SB. Synthesis of proteins by native chemical ligation. Science. 1994;266(5186):776–9.

[158] Metanis N, Hilvert D. Natural and synthetic selenoproteins. Curr Opin Chem Biol 2014;22:27–34.

[159] Arai K, Takei T, Okumura M, Watanabe S, Amagai Y, Asahina Y, et al. Preparation of selenoinsulin as a long-lasting insulin analogue. Angew Chem Int Ed Engl 2017;56 (20):5522–6.

[160] Mousa R, Notis Dardashti R, Metanis N. Selenium and selenocysteine in protein chemistry. Angew Chem Int Ed Engl 2017;56(50):15818–27.

[161] Dhayalan B, Chen Y-S, Phillips NB, Swain M, Rege NK, Mirsalehi A, et al. Reassessment of an innovative insulin analogue excludes protracted action yet highlights the distinction between external and internal diselenide bridges. Chem - A Eur J 2020;26 (21):4695–700.

[162] Zaykov AN, Mayer JP, Gelfanov VM, DiMarchi RD. Chemical synthesis of insulin analogs through a novel precursor. ACS Chem Biol 2014;9(3):683–91.

[163] Hilgenfeld R, Seipke G, Berchtold H, Owens DR. The evolution of insulin glargine and its continuing contribution to diabetes care. Drugs. 2014;74(8):911–27.

[164] Brange J, Ribel U, Hansen JF, Dodson G, Hansen MT, Havelund S, et al. Monomeric insulins obtained by protein engineering and their medical implications. Nature. 1988;333(6174):679–82.

[165] Zaykov AN, Mayer JP, DiMarchi RD. Pursuit of a perfect insulin. Nat Rev Drug Discov 2016;15(6):425–39.

[166] Deacon CF, Hughes TE, Holst JJ. Dipeptidyl peptidase IV inhibition potentiates the insulinotropic effect of glucagon-like peptide 1 in the anesthetized pig. Diabetes. 1998;47(5):764–9.

[167] Pederson RA, White HA, Schlenzig D, Pauly RP, McIntosh CH, Demuth HU. Improved glucose tolerance in Zucker fatty rats by oral administration of the dipeptidyl peptidase IV inhibitor isoleucine thiazolidide. Diabetes. 1998;47(8):1253–8.

[168] Deacon CF, Nauck MA, Toft-Nielsen M, Pridal L, Willms B, Holst JJ. Both subcutaneously and intravenously administered glucagon-like peptide I are rapidly degraded from the NH2-terminus in type II diabetic patients and in healthy subjects. Diabetes. 1995;44(9):1126–31.

[169] Ahrén B, Simonsson E, Larsson H, Landin-Olsson M, Torgeirsson H, Jansson P-A, et al. Inhibition of dipeptidyl peptidase IV improves metabolic control over a 4-week study period in type 2 diabetes. Diabetes Care 2002;25(5):869–75.

[170] Ahrén B, Landin-Olsson M, Jansson P-A, Svensson M, Holmes D, Schweizer A. Inhibition of dipeptidyl peptidase-4 reduces glycemia, sustains insulin levels, and reduces glucagon levels in type 2 diabetes. J Clin Endocrinol Metab 2004;89(5):2078–84.

[171] Ahrén B, Gomis R, Standl E, Mills D, Schweizer A. Twelve- and 52-week efficacy of the dipeptidyl peptidase IV inhibitor LAF237 in metformin-treated patients with type 2 diabetes. Diabetes Care 2004;27(12):2874–80.

[172] Lundkvist P, Sjöström CD, Amini S, Pereira MJ, Johnsson E, Eriksson JW. Dapagliflozin once-daily and exenatide once-weekly dual therapy: a 24-week randomized, placebo-controlled, phase II study examining effects on body weight and prediabetes in obese adults without diabetes. Diabetes Obes Metab 2017;19(1):49–60.

[173] Lundkvist P, Pereira MJ, Katsogiannos P, Sjöström CD, Johnsson E, Eriksson JW. Dapagliflozin once daily plus exenatide once weekly in obese adults without diabetes: sustained reductions in body weight, glycaemia and blood pressure over 1 year. Diabetes Obes Metab 2017;19(9):1276–88.

[174] Zaccardi F, Webb DR, Htike ZZ, Youssef D, Khunti K, Davies MJ. Efficacy and safety of sodium-glucose co-transporter-2 inhibitors in type 2 diabetes mellitus: systematic review and network *meta*-analysis. Diabetes Obes Metab 2016;18(8):783–94.

[175] Liu X-Y, Zhang N, Chen R, Zhao J-G, Yu P. Efficacy and safety of sodium-glucose cotransporter 2 inhibitors in type 2 diabetes: a *meta*-analysis of randomized controlled trials for 1 to 2years. J Diabetes Complicat 2015;29(8):1295–303.

[176] Maruthur NM, Tseng E, Hutfless S, Wilson LM, Suarez-Cuervo C, Berger Z, et al. diabetes medications as monotherapy or metformin-based combination therapy for type 2 diabetes: a systematic review and *meta*-analysis. Ann Intern Med 2016;164(11):740–51.

[177] Mearns ES, Sobieraj DM, White CM, Saulsberry WJ, Kohn CG, Doleh Y, et al. Comparative efficacy and safety of antidiabetic drug regimens added to metformin monotherapy in patients with type 2 diabetes: a network *meta*-analysis. PLoS ONE 2015;10(4):e0125879.

[178] Bolinder J, Ljunggren Ö, Johansson L, Wilding J, Langkilde AM, Sjöström CD, et al. Dapagliflozin maintains glycaemic control while reducing weight and body fat mass over 2 years in patients with type 2 diabetes mellitus inadequately controlled on metformin. Diabetes Obes Metab 2014;16(2):159–69.

[179] Bailey CJ, Morales Villegas EC, Woo V, Tang W, Ptaszynska A, List JF. Efficacy and safety of dapagliflozin monotherapy in people with type 2 diabetes: a randomized double-blind placebo-controlled 102-week trial. Diabet Med 2015;32(4):531–41.

[180] Del Prato S, Nauck M, Durán-Garcia S, Maffei L, Rohwedder K, Theuerkauf A, et al. Long-term glycaemic response and tolerability of dapagliflozin vs a sulphonylurea as add-on therapy to metformin in patients with type 2 diabetes: 4-year data. Diabetes Obes Metab 2015;17(6):581–90.

[181] Cai X, Yang W, Gao X, Chen Y, Zhou L, Zhang S, et al. The association between the dosage of SGLT2 inhibitor and weight reduction in type 2 diabetes patients: a *meta*-analysis. Obes (Silver Spring) 2018;26(1):70–80.

[182] Ferrannini G, Hach T, Crowe S, Sanghvi A, Hall KD, Ferrannini E. Energy balance after sodium-glucose cotransporter 2 inhibition. Diabetes Care 2015;38(9):1730–5.

[183] Leibel RL, Rosenbaum M, Hirsch J. Changes in energy expenditure resulting from altered body weight. N Engl J Med 1995;332(10):621–8.

[184] Neal B, Perkovic V, Mahaffey KW, de Zeeuw D, Fulcher G, Erondu N, et al. Canagliflozin and cardiovascular and renal events in type 2 diabetes. N Engl J Med 2017;377(7):644–57.

[185] Zinman B, Wanner C, Lachin JM, Fitchett D, Bluhmki E, Hantel S, et al. Empagliflozin, cardiovascular outcomes, and mortality in type 2 diabetes. N Engl J Med 2015;373 (22):2117–28.

[186] Wiviott SD, Raz I, Bonaca MP, Mosenzon O, Kato ET, Cahn A, et al. Dapagliflozin and cardiovascular outcomes in type 2 diabetes. N Engl J Med 2019;380(4):347–57.

[187] Busch RS, Kane MP. Combination SGLT2 inhibitor and GLP-1 receptor agonist therapy: a complementary approach to the treatment of type 2 diabetes. Postgrad Med 2017;129 (7):686–97.

[188] Oliva RV, Bakris GL. Blood pressure effects of sodium-glucose co-transport 2 (SGLT2) inhibitors. J Am Soc Hypertens 2014;8(5):330–9.

Chapter 24

Peptide therapeutics in anesthesiology

Nana-Maria Wagner[1], Marvin N. Driessen[1], Vanessa O. Zambelli[2,3] and Eric R. Gross[3]

[1]*Anesthesiology, University Hospital Muenster, Münster, Germany,* [2]*Laboratory of Pain and Signaling, Butantan Institute, São Paulo, São Paulo, Brazil,* [3]*Department of Anesthesiology, Perioperative and Pain Medicine, School of Medicine, Stanford University, Stanford, CA, United States*

24.1 Introduction

Anesthesiologists administer many synthetic peptides and peptide-like drugs either in the operating room, the intensive care unit (ICU), the pain clinic, or the obstetric suite. With primarily nonoral routes of administration, in addition to a peptide's short half-life and rapid degradation, peptide drug profiles complement well the needs anesthesiologists require for regulating acute physiological processes. Furthermore, due to an anesthesiologist's expertise regarding pharmacokinetics and pharmacodynamics, along with most peptides generally being delivered parenterally, the number of peptide-derived therapeutics an anesthesiologist may face are likely to steadily increase as new therapeutics are being developed.

Peptides are typically used clinically as replacement therapies to regulate physiological processes. While a patient has surgery, the anesthesiologist must control a multitude of variables that affect a person's physiology including fluid balance, glucose homeostasis, hemodynamics, ventilation, and body temperature. For anesthesiologists, the attraction of using peptides is the benefit of having a rapidly acting drug with a precise target and low side effect profile. Further, intravenous administration allows for rapid drug titration which is important for the dynamically changing environment of maintaining homeostasis during and after surgery.

Therefore, the purpose of this chapter is to review the indications and use of peptides presently being administered by anesthesiologists. Further, we will discuss potential drug-drug interactions that may occur with peptides and potential future applications for peptide therapeutics in anesthesiology.

Peptide and Peptidomimetic Therapeutics. DOI: https://doi.org/10.1016/B978-0-12-820141-1.00028-5

24.2 Peptide use in anesthesiology practice

Human physiologic functions are commonly mediated via endogenous peptides, including the analgesia occurring from endogenous opioid peptides (enkephalin, endorphin, and dynorphin). Peptides are formed by species of all animal kingdoms, with peptide isolates from nonhuman species providing the basis for some agents given clinically. Peptides can be evolutionary conserved, such as insulin, while other peptides are only within a specific class, such as peptides in snake and spider venom [1–3]. Peptides achieve their characteristics by their amino acid composition, ordering, spatial conformation, and synthetic modulations. In the following sections, we discuss the characteristics of several peptides used within anesthesiology practice and their indications of use.

24.2.1 Insulin

Insulin is used by anesthesiologists to control blood glucose levels during and after surgery. Intraoperatively, stress responses to the surgical procedure result in an increase in glucose levels in diabetic and also nondiabetic patients, making insulin one of the most often used peptide drugs in anesthesiology.

Insulin is primarily administered intravenously, either as a bolus and/or by continuous infusion. Insulin is a 51 amino acid peptide hormone well conserved amongst mammals. The peptide hormone consists of α and β chains joined by two disulfide bonds (Fig. 24.1). Insulin forms hexamers around a

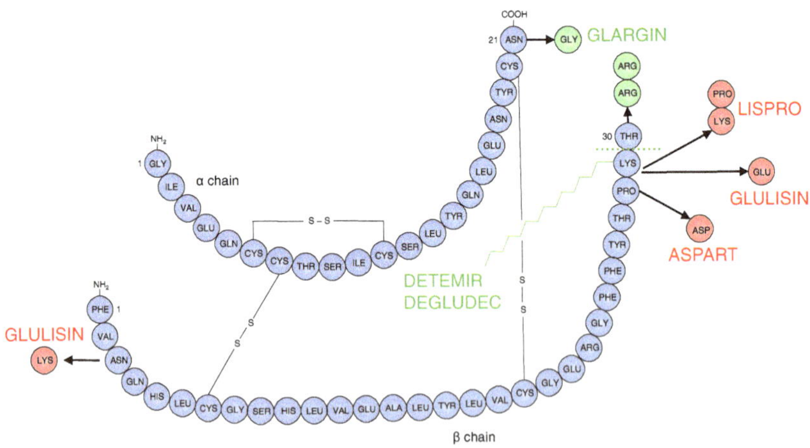

FIGURE 24.1 Insulin and modifications made to the insulin that results in different insulin formulations. Amino acid substitutions labeled in red are common for rapid-acting insulin formulations with a short half-life. Changes to the native sequence in green extend half-life by leading to slower dissociation and release into the bloodstream.

zinc molecule and can be administered intravenously, intramuscular, or subcutaneously. To enter the bloodstream, insulin must dissociate into dimers and monomers [4]. Insulin acts via binding to the insulin receptor, a tyrosine kinase that also becomes activated by insulin-like growth factor (IGF) I and II. After ligand binding and autophosphorylation of tyrosine residues, activation of the second messengers phosphoinositide 3-kinase (PI3K) and protein kinase B (PKB, also known as Akt) result in translocation of the glucose transporter (GLUT4) to the cell membrane. This in turn regulates cellular glucose uptake and glucose clearance from the plasma.

Human insulin is now synthetically manufactured, but prior to 2005, bovine or porcine insulin, which has close sequence homology to humans, was manufactured to treat human diabetics. When delivered subcutaneously in diabetics for control of postprandial and fasting glucose, insulin is manufactured in different formulations to extend its timespan of action. This is achieved by single amino acid substitutions within the peptide sequence (Fig. 24.1). For example, modification of insulin by switching the amino acid proline to lysine on the β chain C-terminus alters the peptide and blocks insulin from forming dimers and hexamers. First called LYSPRO, lispro insulin was named after the amino acid substitution of lysine to proline, allowing lispro insulin to be more rapidly released into the bloodstream. These modifications result in different rates of dissociation into monomers that then enter the bloodstream.

Uptake into the bloodstream can further be influenced by additives such as ethylenediamine tetraacetic acid (EDTA), magnesium, chaperones, or niacinamide (vitamin B3) that facilitate resorption. With fast-acting variants, postprandial hyperglycemia can be ameliorated. By the addition of the zinc neutral protamine hagedorn, insulin is maintained in hexamers and the formulation turns into a suspension that is delivered from the subcutaneous depot over several hours. By the addition of a fatty acid (Levemir/Detemir/Degludec), the insulin effects can be extended by binding to albumin within the bloodstream. This modification causes a slow dissociation of the insulin from albumin that is available for insulin receptor activation (Fig. 24.1).

In addition to its glucose-lowering effects, insulin has a variety of other actions that can be used intraoperatively by anesthesiologists. For example, insulin lowers plasma potassium levels and continuous insulin infusion can be used to treat hyperkalemia, such as in patients with acute intraoperative kidney failure. To maintain glucose homeostasis, glucose is coadministered intravenously as 2 g glucose/1 Unit of insulin. 1 Unit of insulin thereby lowers potassium levels by shifting approximately 1 mmol intracellularly.

Intraoperatively occurring high levels of glucose are associated with increased morbidity and mortality [5,6], making insulin an important drug used by anesthesiologists to maintain glucose homeostasis during surgery. Biochemical modifications of insulin have been successfully developed to tailor insulin pharmacodynamics to the individual diabetic patient's needs.

Additional physiological actions of insulin such as its potassium lowering effects render insulin an important tool for anesthesiologists.

24.2.2 Protamine

Protamine is used to antagonize the anticoagulative properties of heparin. During specific surgical procedures, such as when patients undergo cardiopulmonary bypass during cardiac surgery, very high amounts of heparin are applied to prevent activation of the coagulation system and clotting from contact activation in the circuit. These effects of heparin require rapid reversal after the patient is disconnected from cardiopulmonary bypass to prevent bleeding complications and is antagonized using protamine.

Protamines are small, nuclear, basic, arginine-rich, positively charged proteins that rapidly bind to the negatively charged heparin molecules (Fig. 24.2). Protamines consist of approximately 32 amino acids that share structural similarities with histones. Protamines can be found enriched in the sperm head where they are involved in compact folding and stabilization of DNA. They were first identified in salmon sperm in the 1930s [7] and are now produced by recombinant biotechnology.

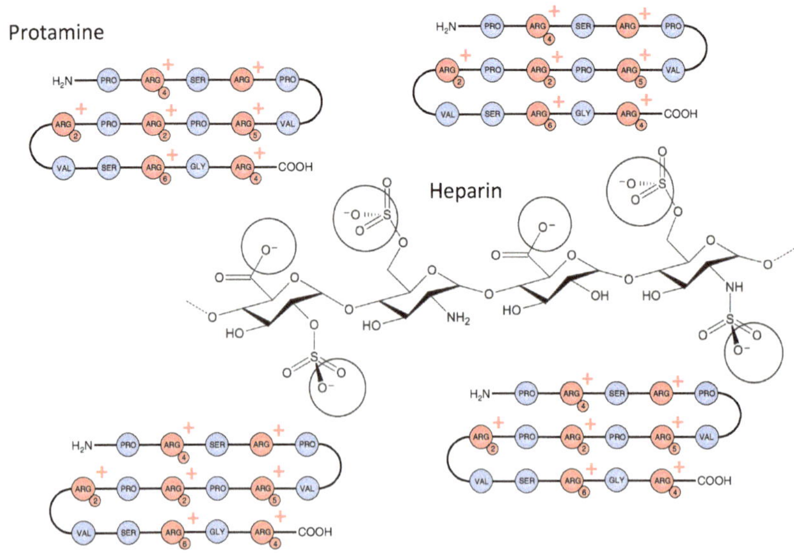

FIGURE 24.2 Neutralization of heparin by protamine. The arginine-rich positively charged protamine forms complexes with negatively charged heparin and antagonizes the anticoagulative properties of heparin.

In the first half of the 20th century, protamine was primarily used in insulin preparations such as Hagedorn insulin to prolong the hypoglycemic actions of insulin (see above). At the beginning of the 20th century, the hemostatic effects of protamine were first described from the discovery of its neutralizing effects on the anticoagulant properties of heparin (Fig. 24.2). Protamine sequences vary in peptide structure, with the ability for protamine to neutralize heparin through electrostatic binding between the cationic arginine groups of protamine and the anionic heparin in a ratio of 1:1 [8]. The resulting protamine-heparin salt aggregates are clearly visible as white suspension and are formed within seconds. The aggregates can also be associated with an immunogenic response by inducing antibody formation associated with thrombocytopenia. Protamine can also reduce platelet aggregation by reducing thrombin generation, the release of intracellularly stored adenosine diphosphate (ADP) and platelet factor 4 (PF4), which exerts an overall anticoagulant side-effect [9].

Due to the use of protamine in most patients undergoing cardiac surgery or cardiac ablation procedures, protamine is applied to an estimated more than 2 million patients undergoing surgery each year. Protamine has a variety of side effects that are associated with immunological and inflammatory alterations. Protamine activates the complement system through the classic pathway, induces IgG and IgE antibody formation and can lead to severe anaphylactic reactions in approximately 0.28%–6% of patients [10,11]. Patients allergic to the fish protein, diabetics using insulin preparations containing protamine, and vasectomized men are at particular risk for adverse allergic reactions in response to protamine exposure. Bivalirudin may be used as an alternative anticoagulant agent instead of heparin in this patient population [12].

A protamine infusion is also associated with a transient drop in blood pressure and an increase in pulmonary arterial pressure. This is caused by an increase in thromboxane release and occurs in approximately 1%–2% of patients during cardiac surgery after taking them off cardiopulmonary bypass. In isolated endothelial cells, protamine causes an increased mechanical stretch-induced intracellular endothelial calcium release and results in nitric oxide production that likely mediates these clinically observed phenomena [9]. Protamine is a peptide drug essential to surgical and anesthetic management of patients undergoing cardiac surgery with or without cardiopulmonary bypass, as it antagonizes the intraoperatively applied high amounts of heparin. Studies suggest that protamine associated side-effects are a matter of dosing. However, there is currently no evidence from larger clinical trials that protamine applied in a ratio of less than 1:1 to heparin reduces the associated side effects [9].

24.2.3 Bivalirudin and desirudin

In 1%–5% of patients, the use of heparin as an anticoagulant causes a heparin-induced drop in platelets and an increased pro-coagulant state with the risk of thrombosis development [13]. In these patients, a substitute for heparin in cardiac or vascular surgery is required.

In these patients, bivalirudin is used. Bivalirudin is derived from hirudin, a mixture of 65 amino acid peptides isolated from *Hirudo medicinalis* which act as direct thrombin inhibitors. The peptides directly bind thrombin, have a short half-life of 30 min and inhibit thrombin by reversibly binding to the catalytic and the anion-binding site of both circulating and clot-bound thrombin. Thrombin then cleaves bivalirudin at the third and fourth amino acids of the arginine-proline bond (Fig. 24.3). Bivalirudin is a 20-amino acid partial sequence of hirudin, is administered intravenously and is cleared from the plasma by a combination of renal elimination and proteolytic cleavage. In patients with renal impairment, clearance is reduced to 20% and 80% in those depending on dialysis. There is no agent reversing the effects of bivalirudin, but it is hemodialyzable, achieving clearance of approximately 25%.

Due to its thrombin inhibiting action, bivalirudin in addition to clopidogrel is indicated as an anticoagulant in adult patients undergoing percutaneous coronary intervention (PCI) when suffering from acute or subacute myocardial infarction. In a randomized controlled trial enrolling 4570 patients undergoing PCI, bivalirudin did not reduce the incidence of death or myocardial infarction but reduced the incidence of major bleeding [14]. While a *meta*-analysis identified an increased risk for thrombosis and myocardial ischemia, reduced risk of bleeding was confirmed [15]. Bivalirudin is administered as a bolus of 0.75 mg/kg body weight intravenously followed by a continuous infusion rate of 1.75 mg/kg body weight/hour intraoperatively or during the procedure requiring heparinization. Bivalirudin may be

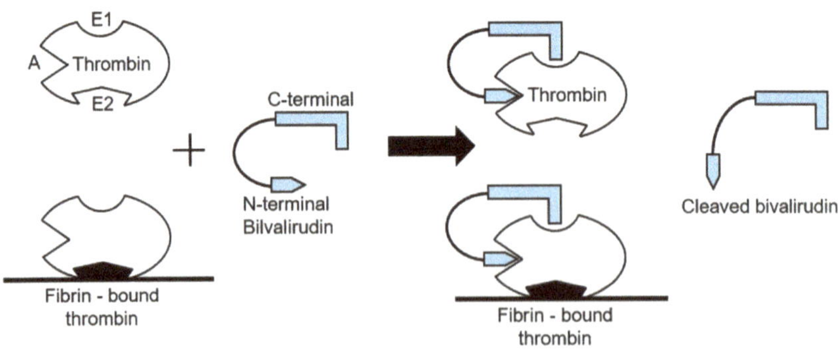

FIGURE 24.3 Interaction of bivalirudin with thrombin. Bivalirudin reversibly binds both clot-bound and free thrombin and is cleaved by thrombin.

continued postintervention for 4–12 h using a reduced infusion dose of 0.25 mg/kg/h as clinically necessary. The dose is reduced in patients with renal impairment.

In addition to its use in PCI, bivalirudin is most commonly used in patients requiring anticoagulation with heparin-induced thrombocytopenia (HIT) or heparin-induced thrombocytopenia and thrombosis syndrome (HITTS). An initial study identified hirudin as an alternative method of anticoagulation for patients with heparin-induced thrombocytopenia (HIT) requiring coronary artery bypass [16]. Subsequently, both lepirudin and bivalirudin use were reported during cardiac bypass with success [17,18]. However, bivalirudin use is more frequently reported and studied in the literature [19–21].

Desirudin is the synthetic derivative of hirudin and is only approved for deep vein thrombosis prevention in patients with HIT and is used postoperatively with success in patients undergoing cardiac or thoracic surgery [22]. Desirudin is licensed for the prevention of deep vein thrombosis in patients that have undergone elective hip replacement surgery. In this clinical context, it is administered subcutaneously at a dose of 15 mg every 12 hs and the dose must be reduced in patients with renal impairment to 1.7 mg every 12 h. In 10% of patients receiving desirudin after hip surgery, the development of antihirudin antibodies was observed that had no effect on the incidence of deep vein thrombosis, pulmonary embolism, allergic reactions or hemorrhage [23].

Taken together, bivalirudin is a clinically important alternative to heparin in patients where heparin is contraindicated due to heparin-induced side effects such as thrombocytopenia. Although some studies suggest that bivalirudin is associated with reduced complications due to major bleeding, recent *meta*-analysis regarding the use of bivalirudin compared to heparin suggested potentially less effective anticoagulation properties of bivalirudin compared to heparin [24]. Thus, bivalirudin will likely remain a second-line agent for anticoagulation in clinical use.

24.2.4 Vasopressin

Vasopressin is an important peptide drug delivered intravenously in the operating room by anesthesiologists to promote vasoconstriction. Although catecholamines such as norepinephrine are first-line agents to increase peripheral resistance in the vascular system and maintain blood pressure, some patients exhibit "resistance" to catecholamines. In these patients, vasopressin is used routinely by anesthesiologists.

Vasopressin is a nonapeptide synthetic modification of the endogenous antidiuretic hormone (ADH) synthesized in neurons of the hypothalamus (Fig. 24.4). The peptide's cysteines form a disulfide bond, and the

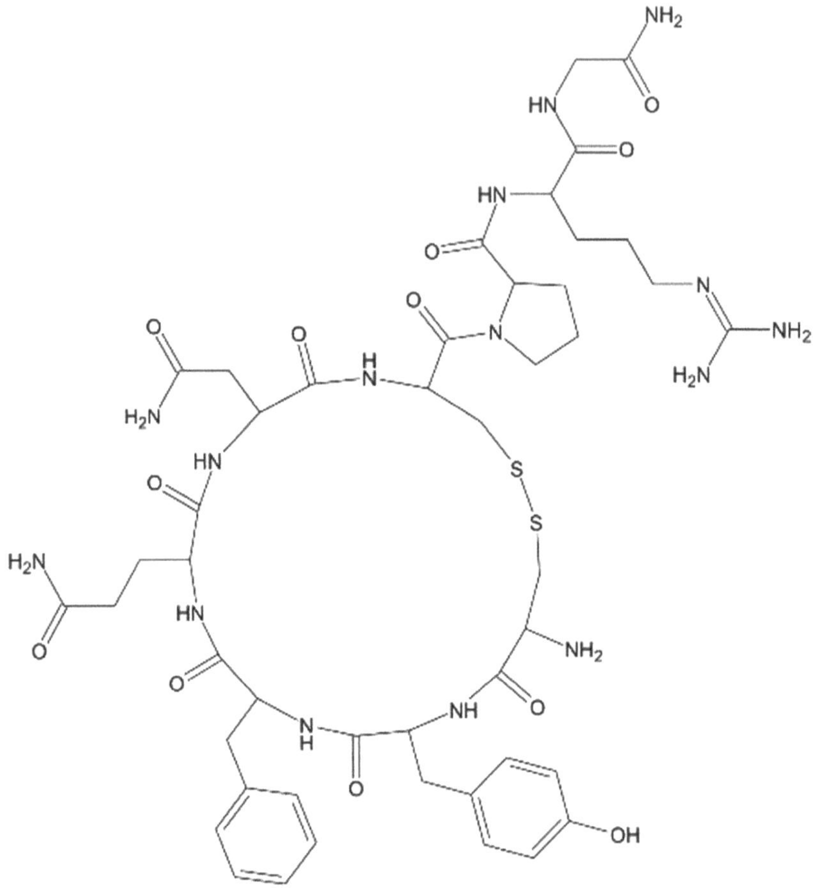

FIGURE 24.4 Structure of vasopressin.

C-terminus holds a primary amide. Vasopressin differs from ADH at the eighth amino acid modified to arginine or lysine, which improves peptide stability. Thus, vasopressin is named after the eighth amino acid, called arginine or lysine vasopressin. Vasopressin binds to the arginine-vasopressin receptor 1 A (AVPR1A) expressed in the liver, kidney and vasculature. Activation of AVPR1A by vasopressin and activation of phosphatidylinositol and calcium as second messengers leads to vasoconstriction, gluconeogenesis, platelet aggregation, and release of VIII and von Willebrand factor. Binding to AVPR2 results in an increase of cyclic adenosine monophosphate and insertion of aquaporin-2 channels in the basolateral membrane of cells in the collecting ducts of the kidney, which facilitates the reabsorption of water. Vasopressin is metabolized in a way similar to endogenous vasopressin and has a half-life of 24 min.

Vasopressin is primarily used in septic shock patients not responding to fluid resuscitation or infusion of catecholamines to treat hypotension. The Surviving Sepsis Campaign guidelines recommend the very early management of the sepsis focusing on the hour-1 bundle. This includes the use of vasopressin at 0.03 units/min as an add-on to norepinephrine with the intent of either raising the mean arterial pressure or decreasing the norepinephrine dosage [25]. There is still an ongoing discussion about the use of vasopressin in patients with cardiac arrest, as outcomes for these patients with respect to the incidence of return to spontaneous circulation or survival at hospital discharge have not been reproduced [26].

In cardiac surgery patients, vasopressin is used to treat catecholamine-refractory hypotension due to vasoplegia associated with preoperative use of angiotensin-converting-enzyme inhibitor use. In these patients, the use of vasopressin was associated with potentially reduced complications and no increase in mortality. Vasopressin is an important drug in clinical use, as it induces vasoconstriction by catecholamine-independent mechanisms [27,28]. Due to its short half-life, the dose applied intravenously by continuous infusion can be rapidly adjusted to the requirements for adequate blood pressure.

24.2.5 Desmopressin

Desmopressin is a modification of arginine vasopressin and is used in the operating room and within the ICU. In the operating room, desmopressin is used to optimize coagulation during surgery. In the ICU, it is primarily administered as a substitute for endogenous antidiuretic hormone (ADH). Desmopressin differs from vasopressin by deaminated first amino acid and the eighth amino acid being D-Arg instead of L-Arg. With these modifications, desmopressin causes little change in blood pressure and facilitates a greater increase in levels of factor VIII activity compared to vasopressin. In addition, these peptide modifications of desmopressin allow for slower peptide degradation, allowing administration via oral or nasal routes.

Desmopressin stimulates the release of von Willebrand factor from endothelial cells and is most commonly used in anesthesiology for the treatment of perioperative bleeding disorders. It is usually the first-line treatment for mild to moderate type I von Willebrand disease and mild hemophilia Type A (5%–50% of the normal concentration of clotting factors VIII and IX). Desmopressin is not recommended in severe disease or those with abnormal factor VIII, but it may be used in bleeding due to trauma or surgery and is frequently used to treat postoperative bleeding in patients undergoing cardiac surgery with impaired platelet function (0.3–0.4 μg/kg) [29]. Desmopressin is not effective in the treatment of hemophilia B (factor IX deficiency) or severe hemophilia A. In addition, desmopressin is used in the treatment of central diabetes insipidus (1–4 μg i.v.). The drug functions as a replacement for an endogenous antidiuretic hormone that is insufficient in quantity due to decreased or nonexistent secretion or production of ADH by the posterior pituitary or hypothalamus, respectively. Desmopressin is also used in the diagnostic workup for diabetes insipidus. In children, desmopressin is used to treat nocturnal enuresis (bedwetting).

Due to its multiple physiological actions, desmopressin is an important drug in anesthesiologic management. Despite its short half-life, desmopressin can induce beneficial changes to coagulation and fluid balance that persist over several hours. To serve as a depot, desmopressin can also be administered subcutaneously.

24.2.6 Octreotide

Octreotide is used in anesthesiologic practice for patients undergoing tumor removal of growth hormone-secreting tumors or neuroendocrine tumors that secrete vasoactive substances. Octreotide is an octapeptide mimetic of somatostatin that inhibits the release of hormones by binding to G-protein coupled receptors (GPCRs). Octreotide differs from somatostatin through forming a cyclic structure by a disulfide bond in addition to D-Tyr and D-Trp substitutions in the peptide sequence (Fig. 24.5). The octreotide modifications cause greater stability and a longer half-life of 1.7–1.9 h compared to the somatostatin half-life of 1–3 min [30]. Octreotide was first synthesized in 1979 by the chemist Wilfried Bauer and more potently inhibits the effects of growth hormone, glucagon and insulin compared to somatostatin. Octreotide can be administered either subcutaneously or intravenously with a peak concentration reached about 24 min. Once in the bloodstream, 65% of octreotide is bound to plasma protein, with the elimination half-life having 2 phases, 10 min and 80 min, due to a first-pass renal elimination of 32% [30,31]. The three indications for the treatment with octreotide are acromegaly, where endogenous growth hormone production is increased due to an endocrine tumor, severe diarrhea/flushing episodes associated with metastatic carcinoid tumors and profuse watery diarrhea associated with tumors secreting vasoactive intestinal peptide (VIP).

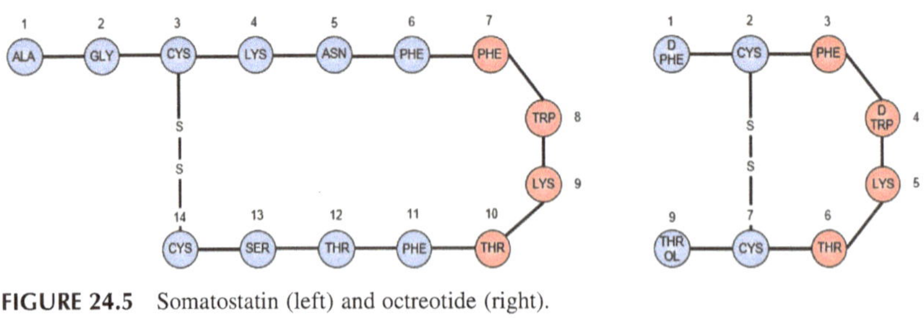

FIGURE 24.5 Somatostatin (left) and octreotide (right).

During anesthesia, octreotide is primarily used intravenously during carcinoid tumor resection to decrease the release of vasoactive substances, which was first reported in the literature over two decades ago [32]. When given as a bolus preoperatively and continued as an infusion perioperatively, octreotide decreases perioperative complications, including hypotension [33,34]. Octreotide can be administered safely in those with carcinoid syndrome requiring cardiac surgery or liver transplantation [35,36]. However, one case report suggested an intravenous octreotide bolus caused bradycardia and heart block [37]. The risk of sinus bradycardia is known to occur in 25% of acromegaly patients treated with octreotide [30]. Off-label uses of octreotide are also common, including decreasing splanchnic blood flow to assist in controlling hepatorenal syndrome prior to liver transplantation or intrahepatic portosystemic stent shunt (TIPS) [38,39]. In a randomized off-label study, octreotide administered peri- and postoperatively to patients undergoing major abdominal surgery reduced postoperative opioid requirements [40]. These opioid-sparing effects could be secondary to a reduction of CYP3A4, which is known to metabolize opioids [30].

Clinical use of octreotide is reserved for special indications mostly related to patients with neuroendocrinological tumors and excess endogenous production of growth hormone. In these patients, exogenously applied octreotide substitutes for a relative lack of endogenous somatostatin due to excess release of growth hormone.

24.2.7 Oxytocin

Oxytocin is used by anesthesiologists in the obstetric suite to control uterus contraction during labor and prevent hemorrhage. Oxytocin is a hormone and neuropeptide that is produced in the hypothalamus and released by the posterior pituitary. For clinical use, oxytocin is available as a synthetic version (Pitocin). The structure is similar to but differs from vasopressin by two amino acids. A pitocin derivative, carbetocin, is also available, which is a modification of the pitocin sequence to a methyltyrosine with deamination that increases peptide stability by reducing disulfidase and aminopeptidase cleavage [41]. These modifications extend the stability and prolong the half-life of carbetocin compared to oxytocin.

Oxytocin binds to oxytocin receptors highly expressed in the uterus shortly before birth where it induces contraction of the myocytes. Oxytocin receptors are also expressed in the mammary gland where myocyte contraction facilitates lactation. In contrast to oxytocin, carbetocin does not need to be refrigerated, giving it an advantage, especially in developing countries as a substitute for oxytocin. Reports consider carbetocin as effective as pitocin during cesarean section [42,43].

Clinically, oxytocin can be used to modulate uterus activity during labor. Oxytocin is licensed for use before delivery (antepartum) and after delivery (postpartum). Antepartum oxytocin has three main indications. The drug may be used for the induction of labor in patients with a medical indication for the initiation of labor. Additionally, oxytocin may be used to reinforce labor and as adjunctive therapy in the management of incomplete or inevitable abortion. Postpartum oxytocin is indicated to stimulate or strengthen uterine contractions to control postpartum bleeding or hemorrhage.

Given the high expression of oxytocin receptors on the maternal myometrium during labor, oxytocin is an important therapeutic tool used in anesthesiologic management of patients in the obstetric suite. Due to its short half-life, dosages applied can be titrated to the desired effect on the uterus peripartum. In the anesthesiologic management postpartum, oxytocin can help to prevent life-threatening hemorrhage by inducing sufficient uterus contraction.

24.2.8 Eptifibatide

Eptifibatide is clinically used to reduce platelet aggregation in patients at risk for thrombosis leading to obstruction of arterial perfusion. Eptifibatide is composed of six amino acids with a mercaptopropionyl residue located at the C-terminus creating a cyclic peptide via the bonded cysteine and mercaptopropionyl residues. The synthetic peptide was derived and modified from the venom of the *Sistrurus miliarius barbouri* rattlesnake, having a half-life of 2.5 h.

Eptifibatide is an antiplatelet drug targeting glycoprotein IIb/IIIa given during unstable angina, myocardial infarction, and in patients undergoing PCI [44–46]. The drug is administered intravenously and undergoes renal clearance by 50% without being metabolized. In comparison to abciximab, an inhibitory antibody fragment acting on the GPIIb/IIIa receptor, eptifibatide is less expensive and induces a reversible receptor occupation with platelet function being normalized 4–6 h after termination of infusion. When clinically compared to abciximab, eptifibatide was as effective as abciximab and not associated with increased adverse effects [47,48]. The adult loading dose is 180 μg/kg i.v. over 1 to 2 min immediately after diagnosis, followed by continuous intravenous infusion of 2 μg/kg/min until either hospital discharge or the initiation of coronary artery bypass grafting, or for up to 72 h. In patients with renal impairment, continuous infusion rates must be reduced. Patients weighing >120 kg should receive a maximum bolus of 22.6 mg followed by a maximum infusion rate of 15 mg/h. Patients with renal impairment evidenced by CrCl <50 mL/min should receive the standard 180 μg/kg loading dose followed by infusion at 1 μg/kg/m.

Eptifibatide is a highly selective drug that rapidly dissociates from its receptor after cessation of therapy. Studies are currently ongoing to further examine the safety and efficacy profile of eptifibatide in patients requiring PCI. So far, eptifibatide is not considered as a first-line treatment regime in this patient population.

24.2.9 Secretin

Secretin is used by anesthesiologists to support diagnosis and interventions in patients undergoing gastrointestinal procedures. Secretin is a natural endogenous peptide released from intestinal mucosa enterochromaffin cells. When binding to secretin receptors, secretin releases pancreatic fluid rich in bicarbonate and controls secretions into the duodenum. The peptide is composed of 27 amino acids and is a naturally occurring human peptide, forming an α-helix secondary structure between amino acids 5 and 13, with structural similarity to glucagon. Secretin is normally administered with a 0.2 mg test dose followed by 0.2 mcg/kg intravenous injection and has a 45-min half-life [49].

Secretin is clinically approved for endoscopic retrograde cholangiopancreatography (ERCP) to identify the ampulla of Vater and accessory papilla [50]. The administration also reduces the incidence of ERCP-induced pancreatitis [51]. However, anticholinergics may cause a decreased secretin response [52].

24.2.10 Nesiritide

Nesiritide is a synthetic β natriuretic peptide of 32 amino acids endogenously produced by the ventricular myocardium. Endogenous nesiritide is potentially useful as a biomarker to monitor patients in the ICU or undergoing vascular or cardiac surgery [53,54]. Nesiritide is also used to treat patients with acute decompensated heart failure by decreasing pulmonary capillary wedge pressure and dyspnea [55]. However, nesiritide use is limited in cardiac patients due to hypotension when administered in the ICU. Neseritide was reviewed extensively elsewhere [56].

24.2.11 Cyclosporine

Cyclosporine is administered in the operating room and the ICU as an immunosuppressive agent in patients subjected to transplantation. Cyclosporine is an 11 amino acid cyclic peptide originally synthesized from *Tolypocladium inflatum*. Cyclosporine is hydrophobic and thus suspended in an emulsion due to poor solubility in water, with a 3:1 equivalent oral to intravenous

[57]. Intramuscular delivery is considered a slow and less reliable route than intravenous delivery [58]. Oral administration is enabled due to cyclization and encasement in a gelatin capsule which slows degradation. Oral dosing is the preferred route of administration due to a higher incidence of anaphylactoid reactions when given intravenously [57]. Routine blood tests to determine therapeutic values are needed since chronic oral absorption is variable [59].

Cyclosporine is commonly given to patients as an immunosuppressive drug to prevent transplant rejection and is also approved for psoriasis treatment. Off-label use in cases of Steven Johnson's syndrome and toxic epidermal necrolysis are reported [60]. Cyclosporine is also potentially an effective adjunct treatment to dantrolene for malignant hyperthermia [61]. Chronic cyclosporine administration may also increase volatile anesthetic minimum alveolar concentration and perhaps prolong vecuronium neuromuscular blockade [62,63].

24.2.12 Ziconotide

Ziconotide is used by anesthesiologists to treat patients with chronic pain. Ziconotide is a synthetically generated 25 amino acid peptide originally derived from the marine snail ω-conotoxin of *Conus magus*. The structure is stabilized by three disulfide bonds and is hydrophobic secondary to the amino acid composition. Ziconotide targets N-type calcium channels, reducing the release of pronociceptive neurotransmitters.

Ziconotide is only given intrathecally secondary to hypotension with intravenous use and is rapidly degraded when given by routes other than intrathecal delivery [64]. The clearance of ziconotide appears dependent upon turnover of cerebral spinal fluid (CSF) approximating 4.5 h and volume of distribution approximating CSF volume [65]. The characteristics of the peptide and route of administration also limit the plasma concentration measurable after a 1 h infusion [66]. Naïve pump delivery systems also absorb a significant amount of ziconotide (17 mcg for a 20 mL pump) with the initial dosing reflecting this absorption [65]. Once a monolayer is formed on the pump, absorption is negligible [65].

Ziconotide is indicated for chronic pain refractory to traditional analgesic treatment with efficacy proven in double-blinded randomized controlled trials and an open trial [67–69]. Ziconotide is also suggested in case reports to be a potential therapeutic for complex regional pain syndrome and in treating trigeminal neuralgia [70,71]. Although ziconotide is formulated at a pH of 4.5, an initial study suggests combining morphine with ziconotide has no differences in ziconotide efficacy [72]. Short acute postoperative administration for pain control was also explored [73]. So far, however, clinical trials convincingly showing a benefit-to-risk profile of the drug as monotherapy or in combination that is superior to standard pain medication are lacking.

24.2.13 Ecallantide

Ecallantide is approved for treating hereditary angioedema caused by C1 esterase inhibitor deficiency. Ecallantide is a 60 amino acid peptide that decreases the ability of kallikrein to generate bradykinin. The peptide selectively and reversibly inhibits kallikrein by targeting the Kunitz domain, which acts as a synthetic enzyme inhibitor similar to C1 esterase, inhibiting the conversion of kininogen to bradykinin. Hereditary angioedema (HAE) is a rare genetic disorder (1:50,000) [74] caused by autosomal dominant trait mutations of the C1-esterase-inhibitor (C1-INH). HAE is characterized by low levels of C1-INH activity and low levels of C4. In HAE, normal regulation of plasma kallikrein activity and the classical complement cascade is not present. During attacks, the unregulated activity of plasma kallikrein results in excessive bradykinin generation. Bradykinin is a vasodilator that is thought by some to be responsible for the characteristic HAE symptoms of localized swelling, inflammation, and pain [75]. Following the administration of a single 30 mg subcutaneous dose of ecallantide to healthy subjects, a mean maximum plasma concentration of 586 ± 106 ng/mL was observed approximately 2–3 h postdose. Ecallantide undergoes renal elimination with a mean elimination half-life of 2.0 ± 0.5 h and is licensed in the US for the treatment of hereditary angioedema, but not in Europe. In a double-blinded placebo-controlled trial involving hereditary angioedema, ecallantide improved patient-reported treatment outcome scores and mean symptom complex severity scores at 4 h after treatment [76]. In angioedema induced by angiotensin-converting enzyme inhibitors (ACE-I), ecallantide was not beneficial and should be avoided [77,78]. As possible side effects, ecallantide can cause anaphylaxis, pruritus, rash, and urticaria within one hour of administration [79].

Prior to the development of ecallantide, aprotinin was the first Kunitz serine protease inhibitor used in clinical practice and was regarded as a nonspecific serine protease inhibitor used to reduce fibrinolysis during cardiac surgery. However, aprotinin is now permanently withdrawn from use due to increased mortality [80]. Since ecallantide is regarded as a more specific serine protease inhibitor than aprotinin since it only targets kallikrein, ecallantide was tested to determine if the peptide reduces postoperative bleeding from coronary artery bypass grafting surgery. However, a phase 2 prospective randomized double-blinded trial was stopped early when a statistically significant 30-day mortality was reported for ecallantide (12%) compared to tranexamic acid (4%) treated patients [81].

A marketing authorization application filed with the European Medicine Agencies was withdrawn in 2011. Although further marketing approvals for Australia, the Middle East and Latin America were planned, ecallantide is currently only approved in the US.

24.3 Peptide-like agents

Peptide-like agents do not clearly fit the definition of a peptide. However, these compounds possess many characteristics similar to peptide agents and include amino acid analaogs and peptide-like drugs, which are discussed in this section.

24.3.1 Amino acid analogs, aminocaproic acid and tranexamic acid

Aminocaproic and tranexamic acid are commonly given during cardiac bypass surgery to reduce fibrinolysis and are analogs of the single amino acid lysine. Both agents achieve their effect by binding to the lysine binding sites of enzymes. Specifically, these lysine analogs modulate the lysine binding sites on plasminogen, preventing plasminogen conversion to plasmin and plasmin-induced degradation of fibrin. These agents were recently reviewed [31,82].

24.3.2 Peptide-like drugs

Bacteria, plants and fungi can synthesize peptide-like substances independent of mRNA and ribosomes in a modular fashion, known as nonribosomal synthesis. With nonribosomal synthesis, nontraditional nonamino acid insertions can occur. Many antibiotics are generated in this fashion and are considered peptide-like therapeutics.

For example, penicillin and cephalosporin contain a beta-lactam backbone, formed by the covalent bonding of the amino acids cysteine and valine. The modified amino acid side chains cause the variability in antibiotic specificity to bacteria. Cephalosporins differ from the typical penicillin structure by having a six-member instead of a five-member ring within the beta-lactam component. Anaphylaxis in humans to penicillin and the cross-reactivity to cephalosporins is due to the beta-lactam ring binding to serum proteins. Other antibiotics, such as vancomycin, daptomycin, bacitracin and polymixin, and the anticancer agent bleomycin, are also peptide-like therapeutics isolated from nonhuman species.

24.4 Properties to consider when administering peptides

Peptides continue to enter anesthetic practice for use intraoperatively and within subspecialty disciplines. A few considerations relevant to anesthetic practice must be made for all peptides given.

Peptides are generally free of drug interactions; however, positively charged peptides when given concomitantly with a negatively charged agent, such as heparin, are neutralized. When heparin is given during an infusion, which commonly occurs intraoperatively or in the ICU, there is a potential for the therapeutic efficacy of a positively charged peptide to be limited. This also includes agents which use a carrier for intracellular delivery, such as TAT (GRKKRRQRRR), as they may potentially be neutralized by the heparin and not delivered intracellularly. Therefore, it is important to be aware of this potential clinical drug–drug interaction that may occur for peptide therapeutics that are highly positively charged.

Since a peptide is a biological compound, there is also a potential risk of immunogenicity which must be considered particularly after repeated dosing [83]. In addition, as peptides are biological compounds, there is also a potential for formulation instability, and in turn, the peptide may denature as particulate or precipitate. Aggregation may also occur, leading to precipitation or decreased activity. Hydrophobic peptides can also potentially form liquid crystals in aqueous solutions and require the addition of an organic solvent to the formulation, such as propylene glycol, to ameliorate liquid crystal formation [84,85]. Peptides such as eptifibatide, desmopressin, and ziconotide are also placed in acidic environments (pH 4–5.5) to maintain peptide stability [65,86]. Coadministration of peptides with basic drugs, such as opioids, also may potentially denature and decrease peptide stability.

The intravenous administration route for peptides typically results in rapid onset of effects. However, peptides exposed to surfaces such as plastic tubing may be adsorbed or adhere to its surface since the nonpolar surfaces can expose hydrophobic sequences within the peptides [59,87]. Similarly, when peptides such as ziconotide are administered intrathecally by a pump, surface adsorption occurs until a peptide monolayer forms on the pump surface, preventing further significant adsorption [87]. Therefore, special consideration is needed for the components of the peptide delivery system.

Although there are potential disadvantages as described above for developing peptide therapeutics, an important advantage of peptide therapeutics is that metabolism and elimination are independent of liver metabolism or liver function, thus reducing the risk of drug–drug interactions. Thus, peptides are also particularly useful in patients with liver dysfunction. The primary elimination route of peptides is glomerular filtration, involving either unchanged peptides or degraded peptide products. In general, filtration is determined by the glomerular sieving coefficient, which is the peptide average concentration in Bowman's space divided by the afferent plasma concentration [88]. First pass elimination depends upon the peptide size and charge. As much as 20% of a peptide with a weight less than 30 kDa can be cleared within the first pass [59,87]. Positively charged peptides are also more easily filtered at the negatively charged glomerulus than negatively charged peptides since elimination occurs through passive diffusion. However, with renal filtration compromised, peptide elimination is potentially reduced.

24.5 Summary and future directions

With the technological advances over the past two decades, new and modified synthetic peptides can significantly influence anesthetic practice. This technology can improve anesthesiology practice by the development of peptide therapeutics with minimal side effects, drug–drug interactions and rapid degradation. Generating novel or improved peptides for use in anesthetic practice may rely on synthetically modifying endogenous peptides, designing peptides based upon a portion of a protein sequence to enhance or disrupt protein-protein interactions, and discovering novel peptides.

For example, targeted changes in endogenous or synthetic peptide sequences may improve the therapeutic potential and reduce the unwanted side-effects of some peptides presently used. Slight amino acid changes to vasopressin subsequently eliminated blood pressure effects and produced desmopressin, which is used for unique indications compared to vasopressin. Perhaps by slightly modifying an existing amino acid sequence, peptides such as nesiritide or ziconotide may be conferred improved therapeutic potential and reduced hypotensive effects, which currently limit their efficacy and routes of delivery [89] (Table 24.1).

Designing peptides targeted to modify enzymes or proteins through allosteric effects can have an almost instantaneous effect and a short half-life, which may be used to alter anesthetic drug kinetics or target receptors to modify drug function. Peptides could be designed to interact with plasma proteins at a receptor on the cell surface or could target intracellular protein-protein interactions, displacing or enhancing protein binding to a specific protein partner. This is the premise for a family of peptide inhibitors and activators targeting selective protein kinase C (PKC) isozymes by disrupting a specific PKC protein isozyme interaction with a binding partner [90,91]. Many proteins targeting specific cellular protein interactions are presently being tested in clinical trials (Table 24.2).

Additional natural peptides and peptide-like compounds, such as those derived from spiders and snakes, are also potential future therapeutics, particularly in pain management, serving a unique need to provide nonaddicting pain therapeutics [104]. In addition, the large molecular weight and charge of insulin limits maternal insulin transport to a fetus during pregnancy. Perhaps this method could be exploited to develop natural peptide compounds that do not cross the placenta for pain management of pregnant patients.

TABLE 24.1 Overview of peptides used in anesthesiology.

Therapeutic agent	Indication	Dosage	Half-life	Dose reduction due to renal impairment
Insulin	Diabetes type I & II Hyperglycemia	Depends on Formulation	Depends on formulation	None
Protamine (Protamin Ipex)	Antagonize heparin	1 I.U. protamine antagonizes 1 I.U. heparin	7 min	None
Vasopressin (EmpressinVasostrict)	Severe hypotension	0.01–0.03 I.E./min	10 min	None
Desmopressin	1) Diabetes insipidus centralis 2) Enuresis nocturna 3) Hemophilia A 4) Von Willebrand disease	1) 0.2–1.2 mg/day 2) 0.2–0.4 mg/day 3) 0.3–0.4 μg/kg body weight	2.6 h	Contraindicated in moderate renal impairment
Octreotide (Benda-Treotid, Sandostatin)	1) Acromegaly 2) Treatment of tumors of the gastroenteropancreatic endocrine system	1) 0.05–0.1 mg every 8–12 h dosing up to 1.5 mg/day 2) 0.05 mg/12 hs up to 0.1–0.2 mg	1.8 h	GFR < 10 mL/min
Oxytocin (Pitocin)	1) Induction of labor in patients with a medical indication. 2) Management of incomplete or inevitable abortion.	1) Induction of labor 0.5–1 mU/min 2) Postpartum hemorrhage 3–5 I.E. diluted in saline	4–10 min	None

(*Continued*)

TABLE 24.1 (Continued)

Therapeutic agent	Indication	Dosage	Half-life	Dose reduction due to renal impairment
	3) Strengthen uterine contractions to control postpartum bleeding or hemorrhage.	solution, slow infusion 3) Subsequent continuous infusion of 10–40 I.E. oxytocin diluted in 500–1000 mL of saline		
Ecallantide (Kallikrein)	Hereditary angioedema for persons over 16 years of age	30 mg s.c.	1.5–2.5 h	None
Bilvirudin (Angiox, Angiomax)	1. Anticoagulant in adult patients undergoing percutaneous coronary intervention (PCI), including patients with ST-segment elevation myocardial infarction (STEMI) undergoing primary PCI. 2. Anticoagulant in adult patients with unstable angina/non-ST segment elevation myocardial infarction (UA/NSTEMI) planned for urgent or early intervention. 3. Anticoagulant in patients with heparin-induced thrombocytopenia (HIT) or heparin-induced thrombocytopenia and thrombosis syndrome (HITTS), undergoing PCI.	1) Bolus of 0.75 mg/kg body weight intravenously 2) Continuous infusion of 1.75 mg/kg body weight/h	22 min to 3 hours depending on renal function	No bolus reduction required Continuous infusion in renal impairment: Moderate: 1.75* Severe: 1* Dialyzed: 0.25* *mg/kg/h

Desirudin (Revasc)	Prophylactic agent for deep vein thrombosis prevention in patients undergoing elective hip replacement surgery.	15 mg every 12 h		Renal impairment: Moderate: 5 mg every 12 h Severe: 1.7 mg every 12 h
Eptifibatide (Integrilin)	1. Treatment of acute coronary syndrome (ACS) managed medically or with percutaneous coronary intervention (PCI) 2. Treatment of patients undergoing PCI (including intracoronary stenting)	1) Bolus of 180 μg/kg i.v. over 1 to 2 min 2) Followed by continuous i.v.-infusion of 2 μg/kg per min	2.5 h	Moderate or severe: initial bolus is not altered. The subsequent continuous infusion is reduced to 1 μg/kg/min
Ziconotide (Prialt)	Severe chronic pain with the need for intrathecal analgesia	Start at 2.4 μg/day and titrate up to 21.6 μg/day. Best efficiency at 9.6 μg/day.	4.5 h in CSF	None

TABLE 24.2 Overview of peptides in clinical trials.

Peptide	Use	Molecular target
Lucinactant	Newborn ARDS	Protein B mimetic [92]
Linaclotide	IBS, Abdominal Pain, Colorectal Cancer	Guanylate cyclase-C [93]
Neurotensin	Pain Management	Neurotensin receptor [94]
CGX 1160	Pain Management	Neurotensin receptor [95]
CR845	Pain and Pruritis Management	Kappa opioid receptor [95]
Bendavia (MTP-131/Elampretide/SS-31)	Myocardial Infarction	Mitochondrial cardiolipin [96]
NA-1	Stroke, Traumatic Brain Injuries and Subarachnoid Hemorrhage	Postsynaptic density protein 95 (synapse-associated protein 90) [97]
XG-102	Ocular Inflammation, Acoustic Trauma	c-Jun kinase [98]
Apelin	Type 2 Diabetes, Chronic Kidney Disease	APJ receptor [98]
Cenderitide	Heart Failure	Natriuretic peptide receptors (guanylyl cyclase A and B) [99]
AB-103	Septic Shock, Necrotizing Soft Tissue Infections	CD28 [100]
Vasoactive Intestinal Peptide (VIP)	Septic shock, Migraine, COPD	VIP receptor [101]
PXL01	Postsurgical Wound Healing	Human lactoferrin derivative [102]
SOR-C13	Cancer	TRPV6 calcium channel [103]

Acknowledgments

NMW is funded by grants from the Deutsche Forschungsgemeinschaft (DFG, German Research Foundation) (WA3786/2-1 and WA3786/3−1). VOZ is supported by Fundação de Amparo à Pesquisa do Estado de São Paulo (FAPESP) 2017/16071-5 and 2018/19332-7 and Stanford Anesthesiology Perioperative and Pain Medicine FIDL 2020. ERG is supported by National Institutes of Health, NIGMS 119522.

References

[1] Zambelli VO, de Oliveira Fernandes AC, Gutierrez VP, Ferreira JCB, Parada CA, Mochly-Rosen D, et al. Peripheral sensitization increases opioid receptor expression and activation by crotalphine in rats. PLoS One 2014;9:e90576.

[2] Bohlen CJ, Priel A, Zhou S, King D, Siemens J, Julius D. A bivalent tarantula toxin activates the capsaicin receptor, TRPV1, by targeting the outer pore domain. Cell 2010;141:834–45.

[3] Conlon JM. Evolution of the insulin molecule: insights into structure-activity and phylogenetic relationships. Peptides 2001;22:1183–93.

[4] Mayer JP, Zhang F, DiMarchi RD. Insulin structure and function. Peptide Sci Original Res Biomol 2007;88:687–713.

[5] Lazar HL, Chipkin SR, Fitzgerald CA, Bao Y, Cabral H, Apstein CS. Tight glycemic control in diabetic coronary artery bypass graft patients improves perioperative outcomes and decreases recurrent ischemic events. Circulation 2004;109:1497–502.

[6] Ouattara A, Lecomte P, Le Manach Y, Landi M, Jacqueminet S, Platonov I, et al. Poor intraoperative blood glucose control is associated with a worsened hospital outcome after cardiac surgery in diabetic patients. Anesthesiology 2005;103:687–94.

[7] Chargaff EO. K.B. Studies on the chemistry of blood coagulation. VI. Studies on the action of heparin and other anticoagulants. The influence of protamine on the anticoagulant effect in vivo. J Biol Chem 1937;122:153.

[8] Ando T, Watanabe S. A new method for fractionation of protamines and the amino acid sequences of salmine and three components of iridine. Int J Protein Res 1969;1:221–4.

[9] Boer C, Meesters M, Veerhoek D, Vonk A. Anticoagulant and side-effects of protamine in cardiac surgery: a narrative review. Br J Anaesth 2018;120:914–27.

[10] Campbell FW, Goldstein MF, Atkins PC. Management of the patient with protamine hypersensitivity for cardiac surgery. Anesthesiol J Am Soc Anesthesiol 1984;61:761–3.

[11] Welsby IJ, Monroe DM, Lawson JH, Hoffmann M. Recombinant activated factor VII and the anaesthetist. Anaesthesia. 2005;60:1203–12.

[12] Walker W, Reid K, Hider C, Davidson I, Boulton F, Yap P. Successful cardiopulmonary bypass in diabetics with anaphylactoid reactions to protamine. Heart. 1984;52:112–14.

[13] Ahmed I, Majeed A, Powell R. Heparin induced thrombocytopenia: diagnosis and management update. Postgrad Med J 2007;83:575–82.

[14] Kastrati A, Neumann F-J, Mehilli J, Byrne RA, Iijima R, Büttner HJ, et al. Bivalirudin vs unfractionated heparin during percutaneous coronary intervention. N Engl J Med 2008;359:688–96.

[15] Cavender MA, Sabatine MS. Bivalirudin vs heparin in patients planned for percutaneous coronary intervention: a meta-analysis of randomised controlled trials. Lancet 2014;384:599–606.

[16] Koster A, Hansen R, Kuppe H, Hetzer R, Crystal GJ, Mertzlufft F. Recombinant hirudin as an alternative for anticoagulation during cardiopulmonary bypass in patients with heparin-induced thrombocytopenia type II: a 1-year experience in 57 patients. J Cardiothorac Vasc Anesth 2000;14:243–8.

[17] Koster A, Chew D, Grundel M, Bauer M, Kuppe H, Spiess BD. Bivalirudin monitored with the ecarin clotting time for anticoagulation during cardiopulmonary bypass. Anesth Analg 2003;96:383–6 table of contents.

[18] Longrois D, de Maistre E, Bischoff N, Dopff C, Meistelman C, Angioi M, et al. Recombinant hirudin anticoagulation for aortic valve replacement in heparin-induced thrombocytopenia. Can J Anaesth 2000;47:255–60.

[19] Koster A, Dyke CM, Aldea G, Smedira NG, McCarthy 2nd HL, Aronson S, et al. Bivalirudin during cardiopulmonary bypass in patients with previous or acute heparin-induced thrombocytopenia and heparin antibodies: results of the CHOOSE-ON trial. Ann Thorac Surg 2007;83:572–7.

[20] Koster A, Spiess B, Jurmann M, Dyke CM, Smedira NG, Aronson S, et al. Bivalirudin provides rapid, effective, and reliable anticoagulation during off-pump coronary revascularization: results of the "EVOLUTION OFF" trial. Anesth Analg 2006;103:540–4.

[21] Koster A, Buz S, Krabatsch T, Yeter R, Hetzer R. Bivalirudin anticoagulation during cardiac surgery: a single-center experience in 141 patients. Perfusion. 2009;24:7–11.

[22] Avidan MS, Smith JR, Skrupky LP, Hill L, Jacobsohn E, Burnside B, et al. The occurrence of antibodies to heparin-platelet factor 4 in cardiac and thoracic surgical patients receiving desirudin or heparin for postoperative venous thrombosis prophylaxis. Thromb Res 2011;.

[23] Greinacher A, Eichler P, Albrecht D, Strobel U, Pötzsch B, Eriksson BI. Antihirudin antibodies following low-dose subcutaneous treatment with desirudin for thrombosis prophylaxis after hip-replacement surgery: incidence and clinical relevance. Blood. 2003;101:2617–19.

[24] Hamdi A, Hafeez A, Dali AF, Nuri M, Huda T, Saleh MS, et al. Safety and effectiveness of bivalirudin in patients undergoing percutaneous coronary intervention: a systematic review and meta-analysis. Front Pharmacol 2017;8:410.

[25] Dellinger R, Levy MM, Rhodes A, Annane D, Gerlach H, Opal SM, et al. Surviving sepsis campaign: international guidelines for management of severe sepsis and septic shock. Crit Care Med 2013;41:580–637.

[26] Aung K, Htay T. Vasopressin for cardiac arrest: a systematic review and meta-analysis. Arch Intern Med 2005;165:17–24.

[27] Hedman KF, Mann CL, Spulecki C, Castner J. Low-dose vasopressin and analogues to treat intraoperative refractory hypotension in patients prescribed angiotensin-converting enzyme inhibitors undergoing general anesthesia: a systematic review. Am Assoc Nurse Anesthesits 2016;84:413–19.

[28] Dünser MW, Bouvet O, Knotzer H, Arulkumaran N, Hajjar LA, Ulmer H, et al. Vasopressin in cardiac surgery: A meta-analysis of randomized controlled trials. J Cardiothorac Vasc Anesthesia 2018;32:2225–32.

[29] Pagano D, Milojevic M, Meesters MI, Benedetto U, Bolliger D, von Heymann C, et al. EACTS/EACTA Guidelines on patient blood management for adult cardiac surgery. Eur J Cardio-Thoracic Surg 2017;2017(53):79–111.

[30] Medicines TP. Octreotide product insert. 2007.

[31] Evers ASM, M., Kharasch ED. Anesthetic pharmacology: basic principles and clinical practice. 2nd ed. Cambridge Press; 2011.

[32] Kvols LK, Martin JK, Marsh HM, Moertel CG. Rapid reversal of carcinoid crisis with a somatostatin analogue. N Engl J Med 1985;313:1229–30.

[33] Kinney MA, Warner ME, Nagorney DM, Rubin J, Schroeder DR, Maxson PM, et al. Perianaesthetic risks and outcomes of abdominal surgery for metastatic carcinoid tumours. Br J Anaesth 2001;87:447–52.

[34] Fouché M, Bouffard Y, Legoff M-C, Prothet J, Malavieille F, Sagnard P, et al. Intraoperative carcinoid syndrome during small-bowel neuroendocrine tumour surgery. Endocr Connect 2018;1.

[35] Weingarten TN, Abel MD, Connolly HM, Schroeder DR, Schaff HV. Intraoperative management of patients with carcinoid heart disease having valvular surgery: a review of one hundred consecutive cases. Anesth Analg 2007;105:1192–9.

[36] Claure RE, Drover DD, Haddow GR, Esquivel CO, Angst MS. Orthotopic liver transplantation for carcinoid tumour metastatic to the liver: anesthetic management. Can J Anaesth 2000;47:334–7.

[37] Dilger JA, Rho EH, Que FG, Sprung J. Octreotide-induced bradycardia and heart block during surgical resection of a carcinoid tumor. Anesth Analg 2004;98:318–20.

[38] Angeli P, Volpin R, Gerunda G, Craighero R, Roner P, Merenda R, et al. Reversal of type 1 hepatorenal syndrome with the administration of midodrine and octreotide. Hepatology. 1999;29:1690–7.

[39] Wong F, Pantea L, Sniderman K. Midodrine, octreotide, albumin, and TIPS in selected patients with cirrhosis and type 1 hepatorenal syndrome. Hepatology. 2004;40:55–64.

[40] Dahaba AA, Mueller G, Mattiassich G, Rumpold-Seitlinger G, Bornemann H, Rehak PH, et al. Effect of somatostatin analogue octreotide on pain relief after major abdominal surgery. Eur J Pain 2009;13:861–4.

[41] Dyer RA, Butwick AJ, Carvalho B. Oxytocin for labour and caesarean delivery: implications for the anaesthesiologist. Curr Opin Anaesthesiol 2011;24:255–61.

[42] Attilakos G, Psaroudakis D, Ash J, Buchanan R, Winter C, Donald F, et al. Carbetocin vs oxytocin for the prevention of postpartum haemorrhage following caesarean section: the results of a double-blind randomised trial. BJOG: An Int J Obstet Gynaecol 2010;117:929–36.

[43] Su L, Rauff M, Chan Y, Mohamad Suphan N, Lau T, Biswas A, et al. Carbetocin vs syntometrine for the third stage of labour following vaginal delivery—a double-blind randomised controlled trial. BJOG: An Int J Obstet & Gynaecol 2009;116:1461–6.

[44] Randomised placebo-controlled trial of effect of eptifibatide on complications of percutaneous coronary intervention: IMPACT-II. Integrilin to minimise platelet aggregation and coronary thrombosis-II. Lancet. 1997;349:1422–8.

[45] Inhibition of platelet glycoprotein IIb/IIIa with eptifibatide in patients with acute coronary syndromes. The PURSUIT trial investigators. Platelet glycoprotein IIb/IIIa in unstable angina: receptor suppression using integrilin therapy. N Engl J Med. 1998;339:436–43.

[46] Novel dosing regimen of eptifibatide in planned coronary stent implantation (ESPRIT): a randomised, placebo-controlled trial. Lancet. 2000;356:2037–44.

[47] Zeymer U, Margenet A, Haude M, Bode C, Lablanche J-M, Heuer H, et al. Randomized comparison of eptifibatide vs abciximab in primary percutaneous coronary intervention in patients with acute ST-segment elevation myocardial infarction: results of the EVA-AMI Trial. J Am Coll Cardiol 2010;56:463–9.

[48] Åkerblom A, James SK, Koutouzis M, Lagerqvist B, Stenestrand U, Svennblad B, et al. Eptifibatide is noninferior to abciximab in primary percutaneous coronary intervention: results from the SCAAR (Swedish Coronary Angiography and Angioplasty Registry). J Am Coll Cardiol 2010;56:470–5.

[49] Incorporated C. Secretin product insert. 2006.

[50] Devereaux BM, Fein S, Purich E, Trout JR, Lehman GA, Fogel EL, et al. A new synthetic porcine secretin for facilitation of cannulation of the dorsal pancreatic duct at ERCP in patients with pancreas divisum: a multicenter, randomized, double-blind comparative study. Gastrointest Endosc 2003;57:643–7.

[51] Jowell PS, Branch MS, Fein SH, Purich ED, Kilaru R, Robuck G, et al. Intravenous synthetic secretin reduces the incidence of pancreatitis induced by endoscopic retrograde cholangiopancreatography. Pancreas 2011;40:533–9.

[52] You CH, Chey WY. Atropine abolishes the potentiation effect of secretin and cholecystokinin-octapeptide on exocrine pancreatic secretion in humans. Pancreas 1988;3:99–103.

[53] Biccard BM, Naidoo P. The role of brain natriuretic peptide in prognostication and reclassification of risk in patients undergoing vascular surgery. Anaesthesia 2011;66:379–85.

[54] Mitchell J, Webb ST. Is brain natriuretic peptide a marker for adverse postoperative outcomes in patients undergoing cardiac surgery? Interact Cardiovasc Thorac Surg 2011;12:467–72.

[55] Intravenous nesiritide vs nitroglycerin for treatment of decompensated congestive heart failure: a randomized controlled trial. JAMA. 2002;287:1531–40.

[56] Keating GM, Goa KL. Nesiritide: a review of its use in acute decompensated heart failure. Drugs. 2003;63:47–70.

[57] Kahan BD. Individualization of cyclosporine therapy using pharmacokinetic and pharmacodynamic parameters. Transplantation. 1985;40:457–76.

[58] Keown PA, Stiller CR, Ulan RA, Sinclair NR, Wall WJ, Carruthers G, et al. Immunological and pharmacological monitoring in the clinical use of cyclosporin A. Lancet. 1981;1:686–9.

[59] Banga AK. Therapeutic peptides and proteins. Boca Raton:: CRC Press;; 2006.

[60] Reese D, Henning JS, Rockers K, Ladd D, Gilson R. Cyclosporine for SJS/TEN: a case series and review of the literature. Cutis. 2011;87:24–9.

[61] Ducreux S, Zorzato F, Muller C, Sewry C, Muntoni F, Quinlivan R, et al. Effect of ryanodine receptor mutations on interleukin-6 release and intracellular calcium homeostasis in human myotubes from malignant hyperthermia-susceptible individuals and patients affected by central core disease. J Biol Chem 2004;279:43838–46.

[62] Niemann CU, Stabernack C, Serkova N, Jacobsen W, Christians U, Eger 2nd. EI. Cyclosporine can increase isoflurane MAC. Anesth Analg 2002;95:930–4 table of contents.

[63] Wood GG. Cyclosporine-vecuronium interaction. Can J Anaesth 1989;36:358.

[64] McGuire D, Bowersox S, Fellmann JD, Luther RR. Sympatholysis after neuron-specific, N-type, voltage-sensitive calcium channel blockade: first demonstration of N-channel function in humans. J Cardiovasc Pharmacol 1997;30:400–3.

[65] Pharmaceuticals E. Ziconotide product insert. 2007.

[66] Wermeling D, Drass M, Ellis D, Mayo M, McGuire D, O'Connell D, et al. Pharmacokinetics and pharmacodynamics of intrathecal ziconotide in chronic pain patients. J Clin Pharmacol 2003;43:624–36.

[67] Wallace MS, Rauck R, Fisher R, Charapata SG, Ellis D, Dissanayake S. Intrathecal ziconotide for severe chronic pain: safety and tolerability results of an open-label, long-term trial. Anesth Analg 2008;106:628–37 table of contents.

[68] Rauck RL, Wallace MS, Leong MS, Minehart M, Webster LR, Charapata SG, et al. A randomized, double-blind, placebo-controlled study of intrathecal ziconotide in adults with severe chronic pain. J Pain Symptom Manage 2006;31:393–406.

[69] Staats PS, Yearwood T, Charapata SG, Presley RW, Wallace MS, Byas-Smith M, et al. Intrathecal ziconotide in the treatment of refractory pain in patients with cancer or AIDS: a randomized controlled trial. JAMA. 2004;291:63–70.

[70] Kapural L, Lokey K, Leong MS, Fiekowsky S, Stanton-Hicks M, Sapienza-Crawford AJ, et al. Intrathecal ziconotide for complex regional pain syndrome: seven case reports. Pain Pract 2009;9:296–303.

[71] Michiels WB, McGlthlen GL, Platt BJ, Grigsby EJ. Trigeminal neuralgia relief with intrathecal ziconotide. Clin J Pain 2011;27:352–4.

[72] Webster LR, Fakata KL, Charapata S, Fisher R, MineHart M. Open-label, multicenter study of combined intrathecal morphine and ziconotide: addition of morphine in patients receiving ziconotide for severe chronic pain. Pain Med 2008;9:282–90.

[73] Atanassoff PG, Hartmannsgruber MW, Thrasher J, Wermeling D, Longton W, Gaeta R, et al. Ziconotide, a new N-type calcium channel blocker, administered intrathecally for acute postoperative pain. Reg Anesth Pain Med 2000;25:274–8.

[74] Bygum A. Hereditary angio-oedema in Denmark: a nationwide survey. Br J Dermatology 2009;161:1153–8.

[75] Zanichelli A, Arcoleo F, Barca MP, Borrelli P, Bova M, Cancian M, et al. A nationwide survey of hereditary angioedema due to C1 inhibitor deficiency in Italy. Orphanet J Rare Dis 2015;10:11.

[76] Cicardi M, Levy RJ, McNeil DL, Li HH, Sheffer AL, Campion M, et al. Ecallantide for the treatment of acute attacks in hereditary angioedema. N Engl J Med 2010;363:523–31.

[77] Scalese MJ, Reinaker TS. Pharmacologic management of angioedema induced by angiotensin-converting enzyme inhibitors. Am J Health-System Pharm 2016;73:873–9.

[78] Lewis LM, Graffeo C, Crosley P, Klausner HA, Clark CL, Frank A, et al. Ecallantide for the acute treatment of angiotensin-converting enzyme inhibitor–induced angioedema: a multicenter, randomized, controlled trial. Ann Emerg Med 2015;65:204–13.

[79] Zuraw B, Yasothan U, Kirkpatrick PE. Nat Rev Drug Discov 2010;9:189–90.

[80] Mangano DT, Tudor IC, Dietzel C. The risk associated with aprotinin in cardiac surgery. N Engl J Med 2006;354:353–65.

[81] Bokesch PM, Szabo G, Wojdyga R, Grocott HP, Smith PK, David Mazer C, et al. A phase 2 prospective, randomized, double-blind trial comparing the effects of tranexamic acid with ecallantide on blood loss from high-risk cardiac surgery with cardiopulmonary bypass (CONSERV-2 Trial). J Thorac Cardiovasc Surg 2011;.

[82] Ide M, Bolliger D, Taketomi T, Tanaka KA. Lessons from the aprotinin saga: current perspective on antifibrinolytic therapy in cardiac surgery. J Anesth 2010;24:96–106.

[83] Greinacher A, Lubenow N, Eichler P. Anaphylactic and anaphylactoid reactions associated with lepirudin in patients with heparin-induced thrombocytopenia. Circulation. 2003;108:2062–5.

[84] Powell MF, Fleitman J, Sanders LM, Si VC. Peptide liquid crystals: inverse correlation of kinetic-formation and thermodynamic stability in aqueous solution. Pharm Res 1994;11:1352–4.

[85] Powell MF, Sanders LM, Rogerson A, Si V. Parenteral peptide formulations: chemical and physical properties of native luteinizing hormone-releasing hormone (LHRH) and hydrophobic analogues in aqueous solution. Pharm Res 1991;8:1258–63.

[86] Powell MF. Peptide stability in aqueous parenteral formulations. In: Cleland, R. JLL, editor. Formulation and delivery of proteins and peptides. Washington, DC: American Chemical Society; 1994. p. 100–17.

[87] Ghosh TKJ, Bhaskara R. Theory and practice of contemporary pharmaceutics 2004;.

[88] Deen WM, Lazzara MJ, Myers BD. Structural determinants of glomerular permeability. Am J Physiol Ren Physiol 2001;281:F579–96.

[89] Lisy O, Huntley BK, McCormick DJ, Kurlansky PA, Burnett Jr. JC. Design, synthesis, and actions of a novel chimeric natriuretic peptide: CD-NP. J Am Coll Cardiol 2008;52:60–8.

[90] Gray MO, Karliner JS, Mochly-Rosen D. A selective epsilon-protein kinase C antagonist inhibits protection of cardiac myocytes from hypoxia-induced cell death. J Biol Chem 1997;272:30945–51.

[91] Johnson JA, Gray MO, Chen CH, Mochly-Rosen D. A protein kinase C translocation inhibitor as an isozyme-selective antagonist of cardiac function. J Biol Chem 1996;271:24962–6.

[92] Piehl E, Fernandez-Bustamante A. Lucinactant for the treatment of respiratory distress syndrome in neonates. Drugs today (Barcelona, Spain: 1998) 2012;48:587.

[93] Nee JW, Johnston JM, Shea EP, Walls CE, Tripp K, Shiff S, et al. Safety and tolerability of linaclotide for the treatment of chronic idiopathic constipation and irritable bowel syndrome with constipation: pooled Phase 3. Anal Expert Rev Gastroenterol Hepatol 2019;13:397–406.

[94] Sang CN, Barnabe KJ, Kern SE. Phase IA clinical trial evaluating the tolerability, pharmacokinetics, and analgesic efficacy of an intrathecally administered neurotensin A analogue in central neuropathic pain following spinal cord injury. Clpharmacol Drug Dev 2016;5:250–8.

[95] Fishbane S, Mathur V, Germain MJ, Shirazian S, Bhaduri S, Munera C, et al. Randomized controlled trial of difelikefalin for chronic pruritus in hemodialysis patients. Kidney Int Rep 2020;.

[96] Karaa A, Haas R, Goldstein A, Vockley J, Weaver WD, Cohen BH. Randomized dose-escalation trial of elamipretide in adults with primary mitochondrial myopathy. Neurology. 2018;90:e1212–21.

[97] Ballarin B, Tymianski M. Discovery and development of NA-1 for the treatment of acute ischemic stroke. Acta Pharmacologica Sin 2018;39:661–8.

[98] Beydoun T, Deloche C, Perino J, Kirwan B-A, Combette J-M, Behar-Cohen F. Subconjunctival injection of XG-102, a JNK inhibitor peptide, in patients with intraocular inflammation: a safety and tolerability study. J Ocul PharmacolTherap 2015;31:93–9.

[99] Kawakami R, Lee CY, Scott C, Bailey KR, Schirger JA, Chen HH, et al. A human study to evaluate safety, tolerability, and cyclic GMP activating properties of cenderitide in subjects with stable chronic heart failure. Clin Pharmacol Therap 2018;104:546–52.

[100] Bulger EM, Maier RV, Sperry J, Joshi M, Henry S, Moore FA, et al. A novel drug for treatment of necrotizing soft-tissue infections: a randomized clinical trial. JAMA Surg 2014;149:528–36.

[101] Hansen J, Sitarz J, Birk S, Rahmann A, Oturai P, Fahrenkrug J, et al. Vasoactive intestinal polypeptide evokes only a minimal headache in healthy volunteers. Cephalalgia. 2006;26:992–1003.

[102] Wiig ME, Dahlin LB, Friden J, Hagberg L, Larsen SE, Wiklund K, et al. PXL01 in sodium hyaluronate for improvement of hand recovery after flexor tendon repair surgery: randomized controlled trial. PLoS One 2014;9:e110735.

[103] Fu S, Hirte H, Welch S, Ilenchuk T, Lutes T, Rice C, et al. First-in-human phase I study of SOR-C13, a TRPV6 calcium channel inhibitor, in patients with advanced solid tumors. Investig N drugs 2017;35:324–33.

[104] Konno K, Picolo G, Gutierrez VP, Brigatte P, Zambelli VO, Camargo AC, et al. Crotalphine, a novel potent analgesic peptide from the venom of the South American rattlesnake Crotalus durissus terrificus. Peptides 2008;29:1293–304.

Chapter 25

Cardiovascular-derived therapeutic peptidomimetics in cardiovascular disease

Adi Litmanovich[1,2], Gassan Moady[3], Mussa Saad[1,4], Yair Blumberg[1], Offir Ertracht[5] and Nir Qvit[1]

[1]*The Azrieli Faculty of Medicine in the Galilee, Bar-Ilan University, Safed, Israel,* [2]*Galilee Medical Center, Nahariya, Israel,* [3]*Cardiology Unit, Galilee Medical Center, Nahariya, Israel,* [4]*Cardiology Department, Ziv Medical Center, Safed, Israel,* [5]*The Cardiovascular Research Laboratory, Research Institute, Galilee Medical Center, Nahariya, Israel*

25.1 Introduction

Peptides are naturally occurring molecules that regulate numerous physiological processes in the body, mainly by binding to specific receptors and inducing a series of signals, neurotransmissions, and/or release of growth factors. Numerous naturally occurring peptides produced by the cardiovascular system vary in structure and function and play vital roles within the same system. In this chapter, we review these peptides produced by the cardiovascular system and their potential as therapeutic leads or approved therapies for cardiovascular disease (CVD).

25.2 Adrenomedullin

25.2.1 Background

Adrenomedullin (ADM) is a 52-amino-acid peptide hormone, belonging to the calcitonin gene peptide superfamily, that was discovered in 1993 by Kitamura and colleagues [1]. Both its ring structure, which forms via a disulfide bond, and its C-terminal amide, are essential for binding to ADM receptors. It imparts a significant vasodilatory effect and is involved in a multitude of other biological processes, including antiinflammatory reactions, stabilization of endothelial barrier function, and regulation of vascular tone. The group of receptor activity-modifying proteins (RAMP) is essential for the biological activity of ADM [2,3] (Fig. 25.1).

Peptide and Peptidomimetic Therapeutics. DOI: https://doi.org/10.1016/B978-0-12-820141-1.00011-X

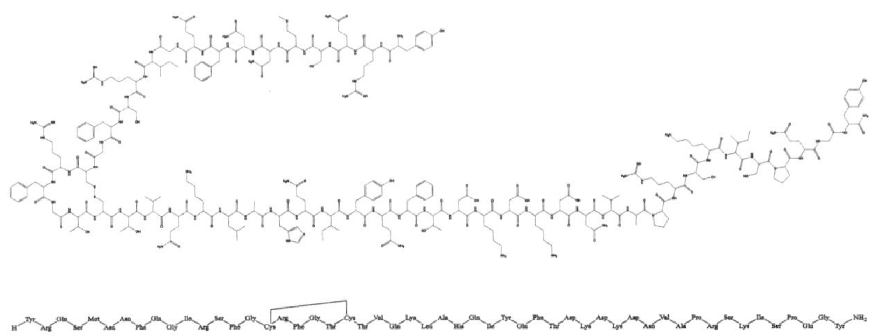

H-Tyr-Arg-Gln-Ser-Met-Asn-Asn-Phe-Gln-Gly-Ile-Arg-Ser-Phe-Gly-Cys(1)-Arg-Phe-Gly-Thr-Cys(1)-Thr-Val-Gln-Lys-Leu-Ala-His-Gln-Ile-Tyr-Gln-Phe-Thr-Asp-Lys-Asp-Lys-Asp-Asn-Val-Ala-Pro-Arg-Ser-Lys-Ile-Ser-Pro-Gln-Gly-Tyr-NH_2

FIGURE 25.1 Chemical structure and amino acid sequence of adrenomedullin.

25.2.2 Genetics and translation

ADM is a protein-coding gene, comprised of four exons and three introns, localized at a single locus on Chromosome 11. Cleavage of the signal peptide of the 185-amino acids preprohormone gives rise to a 164-amino acid peptide called pro-ADM. The pro-ADM has three vasoactive peptides, ADM, the amino-terminal peptide of pro-ADM and adrenotensin. Pro-ADM, is a 20-residue hypotensive peptide termed "pro-ADM N-terminal 20 peptide" (PAMP). Both ADM and PAMP show hypotensive effects in rats, but each act via a different hypotensive mechanism. Plasma ADM and PAMP concentrations are significantly increased in various CVDs, including hypertension and congestive heart failure [4]. The mid-regional proadrenomedullin (Mr-pro-ADM) is a region without known activity [5].

25.2.3 Distribution

While ADM was first discovered in pheochromocytoma originating from the adrenal medulla (hence the name "ADM"), further investigation showed that it is synthesized by many other tissues/cells, especially endothelial and vascular smooth muscle cells (VSMCs), and due to its small size (6 kDa), it diffuses freely between the blood and interstitium. ADM receptors and binding sites are ubiquitous in the body, with cardiovascular and lung tissues having the highest density of these binding sites [6].

ADM synthesis and secretion are stimulated by angiotensin II (ATII), norepinephrine, endothelin (ET), bradykinin, volume overload, and shear stress. Circulating ADM has a half-life of approximately 22 min and is rapidly degraded from its N-terminus by proteases [7]. ADM is primarily metabolized by a neutral endopeptidase (NEP), also known as neprilysin [8].

Sacubtril, a neprilysin inhibitor that inhibits the breakdown of ADM and several other peptide hormones, is a drug associated with reduced mortality and morbidity in heart failure patients [9,10]. ADM is also cleared via internalization and degradation after it binds its receptors [11,12].

25.2.4 Function

The most recognized function of ADM is vasodilatation, which is triggered via stimulation of medial adenylate cyclase [13] and endothelial nitric oxide (NO) production [14], thereby lowering blood pressure while increasing blood flow.

ADM also inhibits the renin–ATII–aldosterone system (RAAS) [15–18]. It consistently attenuates the pressor effects of ATII (but not norepinephrine) and protects against cardiac hypertrophy and renal damage induced by ATII. Furthermore, it is a positive inotrope, diuretic, natriuretic, bronchodilator, and inhibitor of insulin secretion [19].

Even low doses induce vasodilatation, indicating that the plasma levels of ADM under conditions such as heart failure are in the range that directly affects vascular tone [20]. In clinical studies, ADM infusion significantly reduced mean arterial pressure, pulmonary arterial pressure and systemic and pulmonary vascular resistance (PVR) and increased cardiac output, while some studies showed no change in heart rate [21], others demonstrated an increase in heart rate [22]. Moreover, intravascular ADM is thought to improve vascular integrity and decrease vascular permeability through the ADM-RAMP-2 system, which regulates cortical actin formation in vascular endothelial cells and barrier integrity through upregulation of cyclic adenosine monophosphate (cAMP), activation of Rac1 and suppression of oxidative stress [23–25]. Indeed, disruption of the ADM system resulted in vascular leakage and systemic and pulmonary edema as demonstrated in several studies [26–28].

In the heart, ADM exerts potent cardiovascular effects [29], as shown by its compensatory role in heart failure, achieved by offsetting the deleterious effects of excessive peripheral vasoconstriction. Its effects included a reduction in myocardial infarct size, cardiac myocyte apoptosis, left ventricular remodeling (in animals), aldosterone levels (animals and humans) and improvement of hemodynamics (in both humans and animals) and survival (in animals).

Circulating concentrations of ADM are elevated in CVD and heart failure [30] in proportion to the severity of cardiac and hemodynamic impairment. In addition, there is a strong negative correlation between higher levels of ADM and mortality in heart failure patients [31].

25.2.5 Mechanism of action

ADM acts through a 395-amino acid G protein-like-coupled receptor, belonging to the calcitonin receptor-like receptor (CRLR) family. The specificity of CRLR for its ligands is regulated by a group of three RAMPs,

RAMP1, RAMP-2, and RAMP-3. ADM exerts its biological effects by activation of the CRLR as a complex with one of the RAMPs, which alter the ligand affinity.

Receptor activation in VSMCs increases intracellular cAMP and calcium-activated potassium channel activity, leading to prolonged, dose-dependent, generalized vasodilation and hypotension, accompanied by increases in heart rate and cardiac output. The vasodilating effect of ADM is also stimulated by calcium-dependent NO synthesis in endothelial cells. A third pathway triggered by ADM involves the calcium-activated potassium channel, which mediates its vascular muscle relaxant effect [2,32,33].

The positive inotropic effect of ADM appears to be mediated by another mechanism, independent of cyclic AMP, likely via activation of protein kinase, and Ca^{2+} influx through L-type Ca^{2+} channels [34].

25.2.6 Therapeutics

Recent studies have shown that ADM induces beneficial hemodynamic, hormonal and myocardial changes in both animal experimental models [35–42] and patients with heart failure [21,22,43,44].

ADM has a very short half-life and easily adheres to surfaces, limiting its clinical applicability. The nonneutralizing anti-ADM antibody adrecizumab that binds the N-terminal epitope of ADM, leaving the C-terminal side free to bind to its receptor, provides a new alternative therapy, by increasing plasma concentrations of ADM, likely via drainage of ADM from the interstitium into the circulation. In animal models, adrecizumab reduced inflammation attenuated vascular leakage, and improved hemodynamics, kidney function, and survival [45]. Both animal and phase I human studies demonstrated its excellent safety profile [46]. Adrenomed AG has reported positive results from its proof-of-concept Phase II AdrenOSS-2 clinical trial, in which adrecizumab was administered to patients suffering from early septic shock [47]. Its phase II study on the effects of adrecizumab in patients with acute decompensated heart failure with elevated ADM is currently in progress [48].

A new diagnostic immunoassay has been developed to specifically measure the levels of biologically active ADM (bio-ADM) in plasma, which have been associated with clinically relevant 30-day outcomes [49]. Bio-ADM is a potential marker of cardiac congestion status both at admission and during/after hospitalization and might be useful to guide decongestive therapies [50].

25.3 Angiotensin II

25.3.1 Background

ATII is a naturally occurring octapeptide hormone component of the RAAS and is a potent vasoconstrictor [51]. It plays central roles in cardiovascular, neurologic, and renal physiology, including maintenance of blood pressure, renal blood flow, glomerular filtration rate (GFR), and electrolyte and free water homeostasis [52]. ATII has also been implicated as a contributor to the pathophysiology of certain conditions such as hypertension, cardiovascular and renal damage. Its effects on myocardial structure, particularly adverse remodeling in the setting of congestive heart failure, are mediated through a variety of receptors and mechanisms that are well described [53] (Fig. 25.2).

25.3.2 Genetics and translation

ATII encodes a protein called angiotensinogen, the sole precursor of all AT peptides. It is located on the long (q) arm of chromosome 1 at position 42.2 (genetics home reference, NIH, http://ghr.nlm.nih.gov/). ATII abundance is regulated at the transcriptional level through hormonal and cell-type-specific regulators. Preangiotensinogen, or angiotensinogen precursor, is primarily synthesized by the liver and circulates in plasma, where it is cleaved by renin in response to lowered blood pressure [54,55]. Renin enzymatically cleaves two amino acids from the prohormone, forming ATI, which is later converted to ATII through the removal of two C-terminal residues by the angiotensin-converting enzyme (ACE), primarily within the lung (but also in endothelial cells, kidney epithelial cells, heart, and brain). Aminopeptidases A and N cleave ATII to form ATIII and ATIII to form ATIV, respectively.

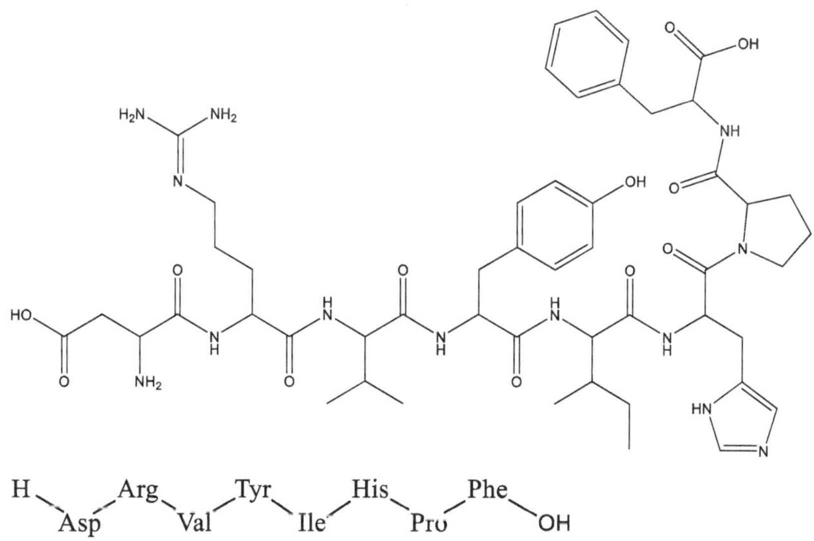

H-Asp-Arg-Val-Tyr-Ile-His-Pro-Phe-OH

FIGURE 25.2 Chemical structure and amino acid sequence of angiotensin II.

AT 1–7 (Ang1–7) is formed from ATII by ACE2, prolyl endopeptidase, or prolyl carboxypeptidase [55–57].

25.3.3 Distribution

ATII has a half-life in the circulation of around 30 s; however, in the tissues, the half-life is extended to 15–30 min. Ang II regulates fluid and electrolytes and maintains hemodynamic stability by binding to specific AT receptors on the cell membrane [58]

25.3.4 Function

ATII is a potent vasopressor, acting on vascular endothelial receptors. Studies in the early 1960s documented the following rapid physiologic changes upon ATII intravenous administration: increased systemic blood pressure, increased pulmonary arterial and wedge pressures (without change in ventilatory function), reduced heart rate and cardiac output, decreased peripheral blood flow, decreased renal blood flow and GFR, reduced-sodium and water excretion, and increased plasma aldosterone [59–65]. Its hypertensive effect was reported in the majority of studies [66]. Several other studies demonstrated that intravenous infusion of ATII resulted in a significant increase in pulmonary mean arterial pressure and PVR, and a reduction in heart rate and cardiac index (CI), but with no change in peripheral pulse pressure [62,67–70]. The lack of increase in peripheral pulse pressure is likely to be due to the reduction in CI during infusion of ATII, probably due to baroreceptor stimulation and reflex cardioinhibition. ATII also increases aortic and systemic arterial stiffness.

ATII regulates cardiac contractility, cell communication, and impulse propagation. It is involved in cardiac remodeling, growth, and apoptosis. It stimulates both hypertrophy and hyperplasia of cardiac myocytes as well as VSMCs by activating proto-oncogenes; a close correlation has been reported in experimental animals between left ventricular hypertrophy (LVH) and myocardial levels of ATII [71]. It stimulates cardiac fibroblast proliferation, which increases extracellular matrix deposition, resulting in cardiac fibrosis [72]. ATII promotes myocardial damage by affecting normal oxygen metabolism, and increases the activity of the NADH and NADPH oxidases, resulting in production of large amounts of superoxide anions [73]. The arrhythmogenic potential of ATII has also been demonstrated [74]. Currently evolving evidence suggests that ATII participates in the initiation and propagation of abdominal aortic aneurysm [75] and promotes the onset and progression of vascular senescence [76].

25.3.5 Mechanism of action

ATII acts as an endocrine, autocrine/paracrine, and intracrine hormone. Its biologic effects are mediated through its interaction with one of its receptors, which belong to the seven-transmembrane, G protein–coupled receptor (GPCR) family [77,78]. AT1 receptors are predominantly coupled to Gq/11 and signal through phospholipases A, C, D, inositol phosphates, calcium channels, and a variety of serine/threonine and tyrosine kinases. Many AT1-induced growth responses are mediated by the transactivation of growth factor receptors [79]. All classic physiological effects of ATII, such as vasoconstriction, aldosterone and vasopressin release, sodium and water retention and sympathetic facilitation, are mediated by the AT1 receptor. The AT1 receptor also mediates ATII effects on cell proliferation, LVH, nephrosclerosis, vascular media hypertrophy, endothelial dysfunction, neointima formation and processes leading to athero-thrombosis [78,79]. ATII receptors predominate in fetal life [80,81] and are much less abundant in adult tissues but are upregulated in certain pathological conditions. Its signaling pathways involve the activity of serine and tyrosine phosphatases, phospholipase A2, NO, and cyclic guanosine monophosphate. The ATII receptor counteracts several of the growth responses initiated by the AT1 and growth factor receptors and contributes to the regulation of blood pressure and renal function [78]. The AT4 receptor specifically binds ATIV (Ang 3–8) and is located in the brain and kidney. Its signaling mechanisms are unknown, but it influences local blood flow and is associated with cognitive processes and sensory and motor functions.

ATII increases blood pressure by stimulating the Gq protein in VSMCs, which, in turn, stimulates Ca^{2+}/calmodulin-dependent phosphorylation of myosin, which ultimately leads to contraction. The binding of ATII to the AT1 receptor in the zona glomerulosa stimulates aldosterone production via the Ca^{2+}/calmodulin kinase, mitogen-activated protein kinase (MAPK), and cAMP cascade [82]. ATII was shown to enhance the activity of the transcription factor nuclear factor–κB, the expression of proinflammatory molecules, such as the vascular cell adhesion molecule, and the release of cytokines, such as interleukin-6 [83].

25.3.6 Therapeutics

Since 1941, ATII has been administered to humans in studies of vascular resistance and hypertension [57]. Intravenous ATII has also been administered to patients with a broad range of medical conditions, including cardiovascular, renal, hepatic, and pulmonary diseases, endocrine and metabolic disorders, and traumatic injuries and shock [84].

Intravenously administered (IV) ATII safely restored mean arterial pressure in patients with catecholamine-refractory hypotension following ACE inhibitor overdose [85–87]. A number of case reports and studies demonstrated the successful use of ATII in the treatment of vasopressor-resistant hypotensive septic shock [88–92]. ATII also effectively increased blood pressure in patients with vasodilatory shock who did not respond to high doses of conventional vasopressors [93]; the data from this single study supported approval of ATII by the Food and Drug Administration (FDA) for marketing in the USA. IV ATII has been shown to exacerbate only two underlying diseases, asthma [94] and congestive heart failure [95]. ACE inhibitors, ATII receptor inhibitors, and other modulators of the RAAS are currently very commonly used to treat CVD. Their efficacy is well established [84].

25.4 Endothelin

25.4.1 Background

Between 1985 and 1988, several publications showed that porcine endothelial cell-primed medium may induce slow-developing and long-lasting contraction of coronary arteries [96–99]. This effect was later attributed to a 21-amino-acid-long peptide termed endothelin. Later work isolated three human isoforms, which were designated ET-1, ET-2, and ET-3. Afterwards, two receptors were also characterized, ET_A and ET_B. Natural isoforms of ETs were found in other species, including pigs and rats. Interestingly, sarafotoxins (S6a-S6d), four highly toxic isoforms of ET were found in the venom of *Atractaspis engaddensis*, a snake endemic to the Middle East [100] (Fig. 25.3).

25.4.2 Genetics and translation

The three human ETs are encoded by distinct genes, but show highly conserved amino acid residues, with 77%–82% identity. This led to the assumption that they originated from a common ancestral gene under strong pressure to preserve mature ET sequences [101]. The ET translation process yields a 212-amino acid sequence, which subsequently undergoes two cleavages by the endopeptidase furin, to produce the big-ET which is composed of 38 amino acid residues. The specific ET-converting enzyme cleaves the big-ET to form active ET-1 via rare proteolytic processing between Trp^{73} and Val^{74}. The two Cys–Cys intramolecular disulfide bonds (between the first and the 15th and between the third and the 11th Cys amino acid of the mature peptide) are a distinct feature of the mature ET though they can be found in several toxins as well [99].

25.4.3 Distribution

Although vascular endothelial cells are the major source of ET-1, genes encoding the 3 ET isopeptides are expressed in a wide variety of cell types, including cardiac myocytes, VSMCs, pituitary cells, macrophages, and mast

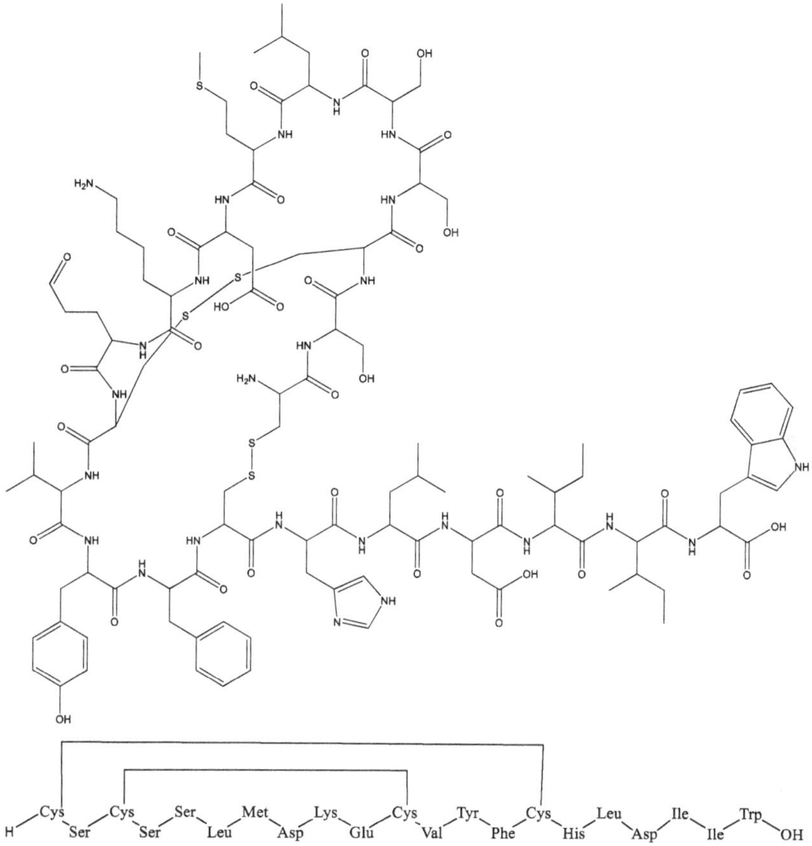

H-Cys(1)-Ser-Cys(2)-Ser-Ser-Leu-Met-Asp-Lys-Glu-Cys(2)-Val-Tyr-Phe-Cys(1)-His-Leu-Asp-Ile-Ile-Trp-OH

FIGURE 25.3 Chemical structure and amino acid sequence of endothelin 1.

cells, suggesting that the peptides participate independently in complex regulatory mechanisms in various organs.

Plasma concentrations of ET-1 in healthy subjects fall in the range of 1.0–2.0 pg/mL, which is below the threshold concentration triggering vasoconstriction and therefore more likely a locally acting (autocrine or paracrine) rather than a circulating hormone [102–104]. ET-1 has a blood half-life of only several minutes (~5), mainly due to its rapid uptake by the lungs, kidneys, spleen, and liver cells [105]. The lungs, via high levels of ET_B receptor expression, uptake approximately 60% of the circulating ET-1. Despite its rapid clearance from the circulation, the blood pressure-elevating effect of ET-1 persists for an extremely long period; stabilization of the

ET-1-ET_A receptor complex may contribute to this long-lasting effect, but the exact reason for this phenomenon remains unknown.

25.4.4 Function

The main effect of ET is mediated through the ET_A receptor and results in constriction of vascular, bronchial and uterine smooth muscle cells, as well as induction of aldosterone secretion. Additional ET_A mediated effects attributed to ETs include positive inotropic and chronotropic cardiac effects, coronary vessel, airway, gastrointestinal tract, bladder, afferent and efferent arteriole and uterus contraction, and adrenocorticotropic hormone (ACTH) and prolactin release. The less prominent effects of ET, which are mediated by the ET_B receptor, include endothelium-dependent vascular relaxation, which is attenuated in the absence of functional endothelial layer (e.g., atherosclerosis), and astrocyte proliferation. Other ET_B-mediated effects include venous contraction, endothelium-derived relaxing factor release and enhanced renal Na^+ excursion [106].

25.4.5 Mechanism of action

Acute ET effects are mediated by the ET_A and ET_B receptors which are coupled to various G proteins, such as Gq, Gs, Gi, or Go, which trigger short-term effects, that is, vasoconstriction. Its long-term effects are exerted via potent mitogenic actions on VSMCs and cardiac myocytes, which can lead to adverse vascular and cardiac hypertrophy [107–109]. Specifically, these effects are regulated through intracellular pathways involving raf-1, MAPK, MAPK kinase, and S6 kinase II [107,110,111].

25.4.6 Therapeutics

ET receptor antagonists were studied as early as 4 years after ET discovery and the establishment of its role as a vasoconstrictor [112]. The first tested ET receptor antagonist, bosentan, was FDA-approved in early 2000, to treat pulmonary arterial hypertension (PAH) [113,114]. Apart from PAH management, ET receptor antagonists proved effective in the treatment of renal diseases in nondiabetic patients but proved less effective for other tested indications, despite promising preclinical data [115].

To date, bosentan, ambrisentan, and macitentan are the only clinically available ET_A/ET_B nonspecific antagonists. As ET-1 is thought to contribute to the development of pathogenic processes including atherosclerosis, hypertension, angina pectoris, myocardial infarction, and heart failure, the search for additional antagonists continues.

25.5 Bradykinin

25.5.1 Background

Bradykinin was originally isolated from the venom of the Central American snake *Bothrops jararaca* by Rocha e Silva and colleagues and was identified as a powerful hypotensive compound [116]. It was later discovered to be a part of the kallikrein-kinin system, a pathway discovered by Frey and Kraut, that plays a role in several physiological processes, including inflammation, blood pressure control, coagulation, and pain sensation [117].

Bradykinin is a nine-amino acid-long peptide with a short half-life ($\sim$15 s) in circulation, due to its rapid inactivation by kininases, including metalloproteinases, NEP and ACE (also known as kininase II) (Fig. 25.4).

25.5.2 Genetics and translation

Bradykinin is produced from the posttranslational cleavage of kininogen. The human kininogen gene is located on chromosome 3q27 and has 11 exons, which encode two alternatively spliced kininogen isoforms that differ in structure, size, and susceptibility to cleavage by plasma and tissue kallikrein serine proteases. Human high-molecular-weight kininogen (HMWK) is 626-amino-acid-long, whereas low-molecular-weight kininogen (LMWK) is composed of 409 amino acids [118].

In humans, bradykinin can be produced by two distinct mechanisms that involve cleavage by kallikreins [116]. In the plasma, HMWK is cleaved by kallikrein to produce bradykinin. The serine protease kallikreins are first produced as inactive precursors called prekallikreins in the plasma and prokallikreins in tissues. The production of bradykinin in the plasma is also mediated by factor XII (Hageman factor), which is involved in the clotting cascade. Bradykinin is therefore the end-product of the contact activation system. The complex that includes factor XII (FXII)-plasma prekallikrein and HMWK is linked to the intrinsic coagulation system via factor XI (FXI), which is cleaved by factor XII (FXII), leading to the initiation of the coagulation cascade [117,119]. In tissues, cleavage of HMWK and LMWK by tissue kallikrein produces kallidin, which is essentially bradykinin carrying an additional lysine at its N-terminus. Kallidin can be transformed to bradykinin by an aminopeptidase.

25.5.3 Distribution

The liver is the main site of kininogen-1 (*KNG1*) gene expression. HMWK mRNA can also be found in the kidney, pancreas, placenta, heart, and colon, whereas LMWK mRNA is found in the kidney, brain, placenta, testis, pancreas, thymus, heart, spleen, lung, colon, and small intestine.

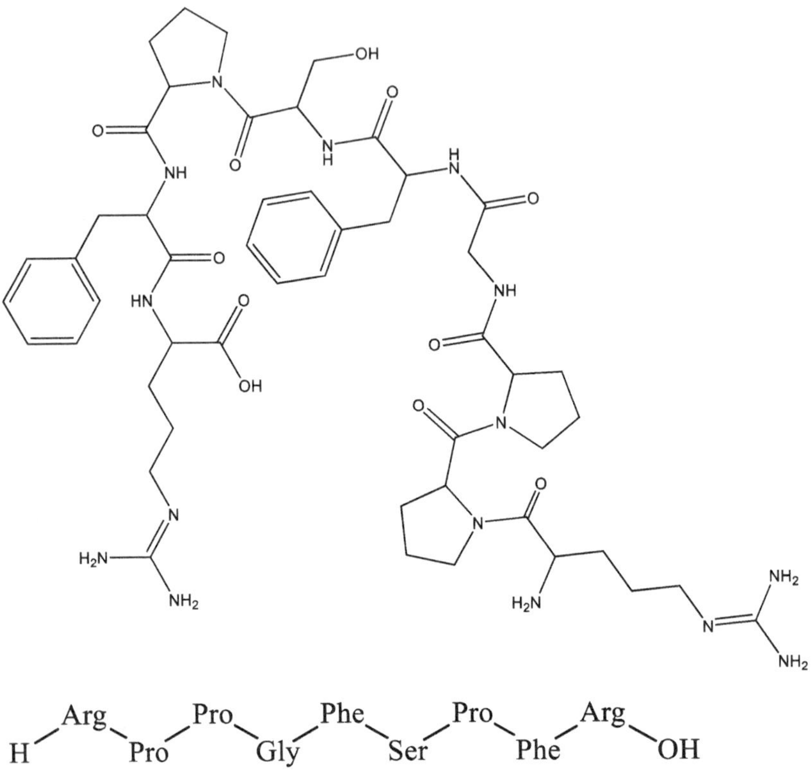

H-Arg-Pro-Pro-Gly-Phe-Ser-Pro-Phe-Arg-OH

FIGURE 25.4 Chemical structure and amino acid sequence of bradykinin.

Bradykinin is present in almost all tissues, with animal studies showing that bradykinin levels are higher in tissues (as high as 100–300 fmol/g tissue) than in the circulation (generally <3 fmol/mL). Within the circulation, bradykinin levels are higher in veins than in arteries [120,121].

25.5.4 Function

Bradykinin acts through B1 and B2, two GPCRs that are coexpressed in many cell types and tissues. In blood vessels, it is predominantly localized in the endothelium. In small arterioles, however, it is found in the smooth muscle layer [122]. While B2 is constitutively expressed and predominant in different tissues, B1 receptor expression is associated with pathological conditions such as inflammation and cellular injury. Activation of the B1 receptor by bradykinin in blood vessels causes vasodilatation, and activation of the B2 receptor in the cardiovascular system stimulates the release of NO from the endothelium, rendering it a potent vasodilator. Moreover, it acts as an antiarrhythmic agent in the heart and as an antithrombotic in the vasculature. Finally, it can improve myocardial oxygen demand in heart failure by attenuating endothelial dysfunction. Since bradykinin is a product of the contact system of endothelial and neutrophil cell membranes, it can, in some cases, lead to the development of angioedema [123]. In the kidney, it can modulate renal function by affecting both natriuresis and glomerular filtration. During inflammation, it can lead to leukocyte recruitment, initiation of the inflammatory responses and induction of pain [120,121].

25.5.5 Mechanism of action

B1 and B2 receptor activation by bradykinin largely trigger the same cellular response. First, phospholipase C (PLC) is activated and calcium is mobilized into the cells. Then, endothelium-dependent relaxation factor is released, which activates protein kinase C (PKC) and cytosolic phospholipase A2 (PLA2), resulting in the release of prostaglandins. Endothelial NO synthase (eNOS) is also stimulated by B2 activation, leading to a rise in NO levels, stimulation of guanylyl cyclase, and an increase in cyclic guanosine monophosphate (cGMP).

25.5.6 Therapeutics

ACE hydrolyzes a carboxy-terminal dipeptide from bradykinin, which leads to its inactivation [123]. Therefore ACE inhibitors increase the amount of bradykinin in the circulation and thus help to maintain a normal blood pressure. However, increased bradykinin levels also trigger a dry cough as it accumulates in the lungs, considered to be one of the main side effects of ACE inhibitors. Inhibition of bradykinin degradation by ACE inhibitors results in plasma extravasation into the submucosal tissue leading to angioedema [124].

Bradykinin B2 receptor activation has also been implicated in cardioprotective mechanisms. The best-characterized bradykinin B2 receptor agonists are JMV-1116 and RMP-7 [125]. Their mechanism of action involves resistance to enzymatic degradation by kininases.

Several bradykinin receptor antagonists also exist. For example, Des-Arg9-(leu8)-bradykinin, which lacks the Arg9 residue, was the first B1 receptor antagonist described in rodents [125]. Icatibant is a synthetic peptidomimetic selective bradykinin B2 receptor antagonist, which does not interact with bradykinin B1 receptors or other peptide receptors [123], has been approved for use in hereditary angioedema and is under investigation for a number of other conditions in which bradykinin is thought to play a significant role. Other bradykinin B2 receptor antagonists include

FR-167344, FR-173657 and Bradyzide, which are potent, selective and orally administered. These antagonists may enhance our understanding of the pathological states that involve the hyperactivity of kinins [126].

25.6 Natriuretic peptides

25.6.1 Background

Natriuretic peptides (NPs) are a family of hormones that are secreted by cardiac tissue as a response to pressure and volume overload and mediate natriuretic (sodium excretion) and vasodilator pathways. There are six known members of the NP family: atrial NP (ANP), brain NP (BNP), C-type NP (CNP), dendroaspis NP (DNP), ventricular NP (VNP), and urodilatin. To date, three NP receptors (NPRs) are known: NPR-A, NPR-B, and NPR-C, all of which share a large extracellular ligand-binding domain and a single membrane-spanning region. NPR-A functions as the receptor for ANP and BNP and is expressed in kidney, lung, adipose, adrenal, brain, heart, testis, and vascular smooth muscle tissue. NPR-B is the main receptor for CNP and is expressed in bone, brain, fibroblasts, heart, kidney, liver, lung, uterine, and vascular smooth muscle tissue. NPR-C is the most expressed receptor in this family and is found in adrenal, brain, heart, kidney, mesentery, and vascular smooth muscle tissue and binds NPs in the following affinity: ANP > CNP > BNP. Through binding these receptors, NPs mediate several physiologic actions, including blood pressure homeostasis, cardiac remodeling and fibrosis, and bone growth. NPs are inactivated by NPR-C, NEP, dipeptidyl peptidase-4, and insulin-degrading enzymes. These peptides have strengthened their position in the CVD arena during the last years, and a number of recombinant forms have been investigated for therapeutic purposes.

25.6.2 Atrial natriuretic peptide

ANP is the first diuretic and vasodilatative hormone isolated from cardiac tissue.

Genetics and translation: ANP is a 28-amino-acid peptide encoded by the natriuretic peptide precursor A (*NPPA*) gene and synthesized as an inactive pre-pro-hormone which is then cleaved into pro-ANP, the major form of ANP stored in atrial granules [127–129]. In response to an atrial wall stretching, pro-ANP is converted to its active form C-terminal-ANP by the serine protease Corin [130]. In healthy subjects, ANP is mainly produced by atria; however, in the failing heart, the ventricles become the main producer of ANP.

Distribution: The active form of ANP enters the circulation and acts by binding its three cell surface receptors. Plasma levels of ANP are relatively low, with plasma half-life being approximately 2–4 min [131]. As with other NPs, ANP is degraded into inactive forms by the endopeptidase neprilysin and through binding NPR-C [132].

Functions: By activating transmembrane guanylyl cyclase, ANP induces an elevation in cyclic guanosine monophosphate (c-GMP) levels, and by doing this, it causes smooth muscle relaxation in arterioles [132]. In addition, ANP exerts natriuretic effects by enhancing renal sodium excretion by acting on the loop of Henle, suppressing renin secretion from juxtaglomerular cells, and inhibiting aldosterone synthesis in the zona glomerulosa. It also promotes lipolysis and modulates the innate immune system [133]. In cardiac tissue, ANP inhibits cardiac hypertrophy and fibrosis by inducing myocyte apoptosis and inhibiting fibroblast migration and proliferation [133–136].

Therapeutics: Due to its favorable hemodynamic effects, human ANP has been investigated for the treatment of acute heart failure [137]. In 1995, the Ministry of Health and Welfare of Japan approved a recombinant ANP (Carperitide) for intravenous administration to patients with acute decompensated heart failure (Fig. 25.5).

25.6.3 Carperitide

Carperitide is a recombinant α-human ANP that exhibits diuretic, natriuretic and vasodilatory effects [138].

Function: Carperitide acts by inhibiting ET-1 secretion stimulated by ATII and thrombin [139]. Its hemodynamic effects in patients with heart failure may be mediated by inhibition of the RAAS. Its beneficial clinical impact on patients with acute myocardial infarction is achieved by increased coronary blood flow, which subsequently reduces the myocardial infarct size [140].

Therapeutics: As mentioned above, carperitide has been approved in Japan to treat acute heart failure due to its cardio-protective and hemodynamic features. One retrospective study showed that carperitide treatment of patients with moderate to severe pulmonary congestion was associated with

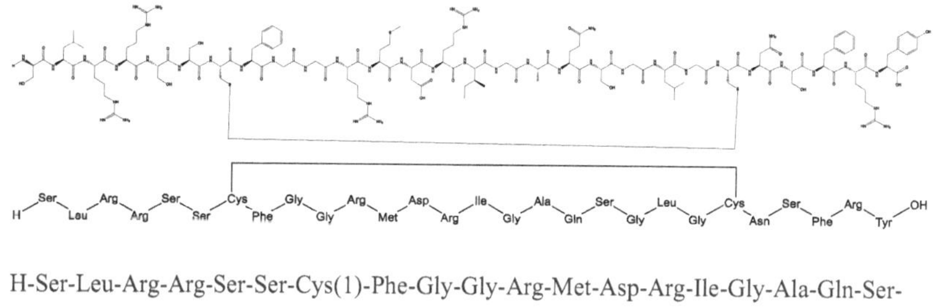

H-Ser-Leu-Arg-Arg-Ser-Ser-Cys(1)-Phe-Gly-Gly-Arg-Met-Asp-Arg-Ile-Gly-Ala-Gln-Ser-Gly-Leu-Gly-Cys(1)-Asn-Ser-Phe-Arg-Tyr-OH

FIGURE 25.5 Chemical structure and amino acid sequence of atrial natriuretic peptide.

more effective decongestion in the short-term and better long-term outcomes [141]. However, in a large retrospective matched-cohort study, administration of carperitide for early inpatient care of acute heart failure was associated with worse outcomes when compared to nitrate use [142].

25.6.4 B-type natriuretic peptide

BNP was initially extracted from the porcine brain [143] and is now recognized for its roles in cardiac tissue.

Genetics and translation: BNP is encoded by the natriuretic peptide precursor B (*NPPB*) gene, located on chromosome 1 [144]. BNP is synthesized as a 134-amino acid pre-pro-BNP, which is subsequently modified to pro-BNP and then further cleaved by corin and furin to yield its active form and the inactive N-terminal pro-BNP (NT-pro-BNP).

Distribution and function: In healthy individuals, BNP is secreted from the atria, where it is stored in granules; however, in the failing heart it is transcribed and released mainly from the ventricles [145]. In healthy humans, plasma BNP concentrations are very low, with a plasma half-life of approximately 20 min [146]. After degradation by neprilysin [147] and NPR-C, BNP is cleared from the circulation. BNP acts on several targets and mediates physiological processes by binding NPR-A [148]. In the vasculature, it activates c-GMP and inhibits the effect of catecholamine, resulting in vasodilation. In the kidneys, BNP increases GFR, decreases sodium reabsorption (natriuresis), and inhibits RAAS [149]. In cardiac tissue, BNP participates in the complex pathways of cardiac hypertrophy and fibrosis.

Therapeutics: Over the last few years, BNP and the inactive NT-pro-BNP form have been increasingly viewed as accurate heart failure diagnostic and prognostic markers [150,151]. Their diagnostic value is being exploited to enhance triage and management of patients with a medical history of heart failure and chronic obstructive pulmonary disease. High BNP levels also have prognostic implications in patients with heart failure, as it suggests more progressive disease and a higher risk for recurrent hospitalization [152]. BNP was also suggested as a prognostic marker for the acute coronary syndrome, pulmonary hypertension, valvular disease, and pulmonary embolism [153]. In patients with severe aortic valve stenosis, high levels of BNP may indicate the need for surgery, even in asymptomatic cases. LCZ696, which combines valsartan (AT receptor antagonist) and sacubitril (neprilysin inhibitor), was demonstrated to effectively reduce mortality and heart failure-related hospitalization in patients with symptomatic heart failure with reduced ejection fraction [9]. These effects are partially related to the elevation of BNP levels that is achieved by neprilysin inhibition. Although this drug was associated with improved New York Heart Association functional class and decreased NT-pro-BNP levels in patients with heart failure with preserved ejection fraction, it did not reduce mortality in this subset of

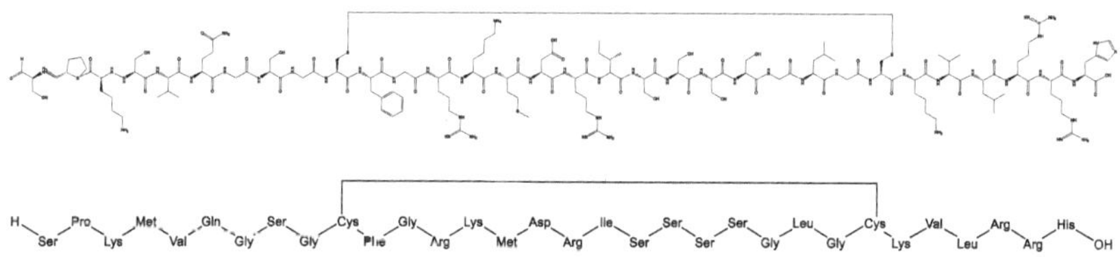

H-Ser-Pro-Lys-Met-Val-Gln-Gly-Ser-Gly-Cys(1)-Phe-Gly-Arg-Lys-Met-Asp-Arg-Ile-Ser-Ser-Ser-Ser-Gly-Leu-Gly-Cys(1)-Lys-Val-Leu-Arg-Arg-His-OH

FIGURE 25.6 Chemical structure and amino acid sequence of brain natriuretic peptide.

patients [154]. Elevated BNP and NT-pro-BNP levels are not specific for heart failure; such elevations (albeit, usually to a lesser extent) may be found in patients with renal failure, atrial fibrillation, sepsis, myocarditis, and/or of advanced age [153] (Fig. 25.6).

25.6.5 Nesiretide

Nesiritide is the recombinant form of BNP that was designed for the treatment of acute heart failure and was approved in 2001 after proving effective in decreasing pulmonary capillary wedge pressure and providing significant relief of dyspnea [155].

Therapeutics: Nesiretide acts by binding to receptors in the vasculature and kidney and stimulating cyclic guanosine monophosphate, leading to rapid and balanced vasodilation with a subsequent decrease in right and left ventricular filling pressure and increased natriuresis in the kidney. After being cleared for relief of dyspnea, nesiritide was tested in the Acute Study of Clinical Effectiveness of Nesiritide in Decompensated Heart Failure (ACSEND-HF) trial, which showed no effect on mortality or heart failure-related hospitalization rates [154]. The most common side effect of intravenous nesiritide was hypotension. Based on the ASCEND-HF trial, nesiretide is not routinely recommended for use in acute heart failure patients.

25.6.6 C-type natriuretic peptide

CNP was initially purified from porcine brain tissue in 1990 [156]. CNP drives a variety of cardiovascular protective effects via NPR-B and NPR-C.

Genetics and translation: CNP is encoded by natriuretic peptide precursor C (*NPPC*), a gene located on chromosome 2 and is highly expressed in the brain, chondrocytes, and endothelial cells. After removal of the signal peptide, pre-pro-CNP is converted to pro-CNP, which is then cleaved into the 53-amino-acid biologically active peptide (CNP-53), the major active form at the tissue level. However, in the systemic circulation, a shorter 22-amino-acid form (CNP-22) dominates.

Mechanism of action: In cultured cells, CNP-53 and CNP-22 showed equal affinity to NPR-B [157]. CNP-22 is rapidly inactivated by NEP and cleared by NPR-C, resulting in its low half-life of 2.6 min [158]. As CNP is found in low concentrations in the circulatory system and mainly acts via NPR-B, it does not possess the potent diuretic effects of ANP and BNP. While mean CNP concentrations are very low in healthy individuals, they are elevated in patients with congestive heart failure, albeit, to a much lower extent than ANP and BNP [159,160].

Function: CNP is secreted from endothelial cells, cardiac myocytes, and fibroblasts in response to endothelial sheer stress, and release of growth factors [161]. Knockout mice with underexpression of CNP were dwarfs due to impaired endochondral ossification and had reduced life spans, while overexpression of CNP resulted in increased growth of chondrocytes and prolonged survival [162]. In humans, CNP plays a central role in long bone growth and in regulating vascular tone and blood pressure by acting on the endothelium, and mediates several cardiovascular processes, such as inflammation, atherosclerosis, myocardial contractility and hypertrophy, and cardiac fibrosis [163].

Therapeutics: The CNP analog vasoritide has potent endochondral ossification effects and can safely increase the annualized growth velocity in children with achondroplasia, a genetic disorder characterized by short stature [164] (Fig. 25.7).

25.6.7 Dendroaspis natriuretic peptide

DNP consists of 38 amino acids and was originally isolated from green mamba venom snake [165].

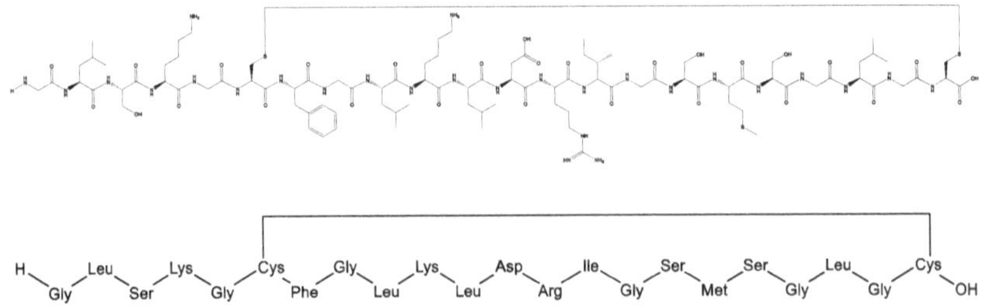

H-Gly-Leu-Ser-Lys-Gly-Cys(1)-Phe-Gly-Leu-Lys-Leu-Asp-Arg-Ile-Gly-Ser-Met-Ser-Gly-Leu-Gly-Cys(1)-OH

FIGURE 25.7 Chemical structure and amino acid sequence of C-type natriuretic peptide.

Distribution: Although DNP has not been purified from human blood, its immunoreactivity was demonstrated in myocardial atria and human plasma. DNP is structurally similar to the other members of the NP family but less susceptible to enzymatic degradation.

Function: DNP effects on the cardiovascular system have not been fully clarified yet. Similar to other NPs, plasma levels of DNP seems to increase in patients with congestive heart failure [165].

Therapeutics: In animal models, intravenously administered synthetic DNP showed natriuretic and diuretic properties, with subsequent beneficial cardiovascular and renal effects [166] (Fig. 25.8).

25.6.8 Ventricular natriuretic peptide

VNP is a cardiac hormone found in chondrostei and primitive teleostei, such as sturgeon, bichir, eel, and trout. VNP has a long C-terminal tail sequence extending from an intramolecular ring, which is distinct in molecular structure from other NPs. It is released in response to acute salt load in eels or chronic volume overload in trout [167]. VNP does not exist in mammals (Fig. 25.9).

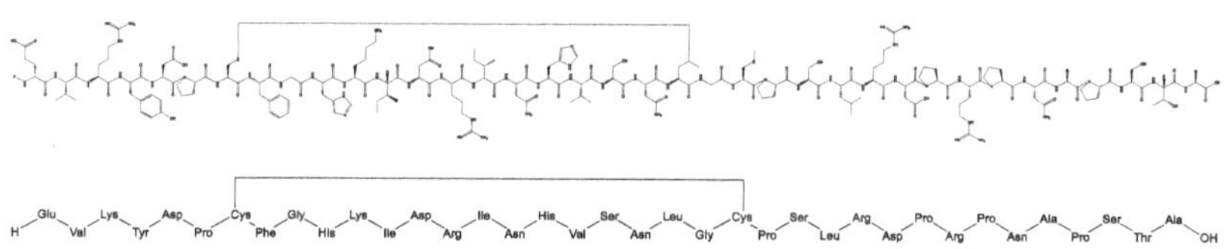

H-Glu-Val-Lys-Tyr-Asp-Pro-Cys(1)-Phe-Gly-His-Lys-Ile-Asp-Arg-Ile-Asn-His-Val-Ser-Asn-Leu-Gly-Cys(1)-Pro-Ser-Leu-Arg-Asp-Pro-Arg-Pro-Asn-Ala-Pro-Ser-Thr-Ser-Ala-OH

FIGURE 25.8 Chemical structure and amino acid sequence of dendroaspis natriuretic peptide.

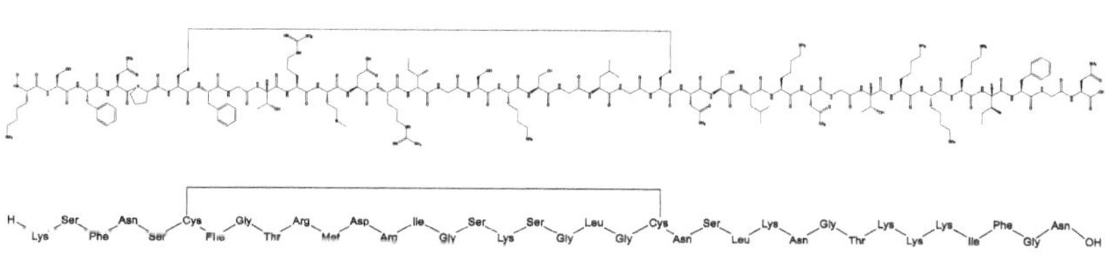

H-Lys-Ser-Phe-Asn-Ser-Cys(1)-Phe-Gly-Thr-Arg-Met-Asp-Arg-Ile-Gly-Ser-Lys-Ser-Gly-Leu-Gly-Cys(1)-Asn-Ser-Leu-Lys-Asn-Gly-Thr-Lys-Lys-Lys-Ile-Phe-Gly-Asn-OH

FIGURE 25.9 Chemical structure and amino acid sequence of ventricular natriuretic peptide.

25.6.9 Urodilatin

Urodilatin is a member of the NP family and consists of 32 amino acids, first isolated from human urine in 1988 by Schulz-Knappe et al. [168]. While Pro-ANP produces ANP in cardiac tissue, in the kidneys it is processed by an alternative pathway resulting in the formation of urodilatin [168]. In the kidneys, urodilatin is synthesized in the distal and collecting tubular cells and is secreted to the urine to regulate sodium and water homeostasis. It has a longer half-life and is more potent than ANP, due to its resistance to degradation [169].

Therapeutics: A synthetic urodilatin (ularitide) was found to decrease capillary wedge pressure and cardiac output without decreasing blood pressure. Despite its favorable physiological effects in acute heart failure patients, it did not reduce cardiovascular mortality [170] (Fig. 25.10).

25.6.10 Vasonatrin

Vasonatrin is a synthetic NP, comprised of the 22 amino acids of CNP with the 5-amino acid carboxyl terminus of ANP [171], offering both the venodilation property of CNP and the natriuretic property of ANP. Vasonatrin binds NPR-A and NPR-B, with a preference for the latter. In one preclinical study, vasonatrin was shown to protect the diabetic heart against ischemia-reperfusion injury by inhibiting endoplasmic reticulum stress via the cGMP-protein kinase G (PKG) signaling pathway, indicating that vasonatrin may have potential therapeutic value in patients with diabetes and ischemic heart disease [172]. In another model, vasonatrin reduced the cytotoxicity of *N*-methyl-4-phennylpyridinium by increasing dopaminergic neuron axon counts and lengths and was therefore postulated as a potential therapy for Parkinson's disease [173]. Vasonatrin was also shown to increase blood perfusion, decrease inflammatory mediators, and increase survival in rats [174].

25.6.11 Cenderitide

Cenderitide is a novel synthetic NP consisting of 37 amino acids, developed by the Mayo Clinic as a potential treatment for heart failure [175].

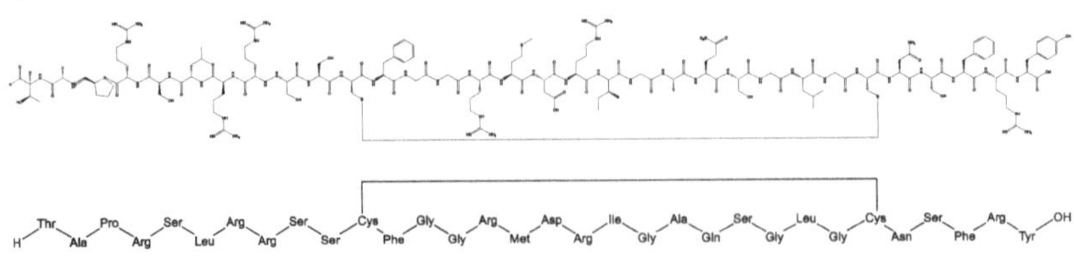

H-Thr-Ala-Pro-Arg-Ser-Leu-Arg-Arg-Ser-Ser-Cys(1)-Phe-Gly-Gly-Arg-Met-Asp-Arg-Ile-Gly-Ala-Gln-Ser-Gly-Leu-Gly-Cys(1)-Asn-Ser-Phe-Arg-Tyr-OH

FIGURE 25.10 Chemical structure and amino acid sequence of urodilatin.

Genetics and mechanism: Cenderitide is a chimeric peptide consisting of the 15-amino-acid C-terminus of DNP fused to the full CNP. The rationale for this combination is to activate two NPRs, that is, guanylyl cyclase A (the target of DNP) and guanylyl cyclase B (the target of CNP), to inhibit cardiac fibrosis without inducing significant hypotension [175].

Therapeutics: Dickey et al. demonstrated that cenderitide is more resistant to degradation by NEP than native ANP, BNP, and CNP [176]. Owing to its antiproliferative effects, antifibrotic properties, and its role in inhibiting the RAAS, cenderitide is currently being tested in clinical studies involving patients with heart failure with reduced ejection fraction (NCT02359227). The natriuretic effect of cenderitide was achieved without significant change in creatinine clearance. In one study, infusion of cenderitide to patients with heart failure was associated with increased plasma c-GMP and urinary c-GMP excretion and preservation of kidney function, without significant hypotension [177]. A patch-like device is also being tested in experimental cardiovascular models [178].

25.6.12 CU-NP

CU-NP is a NP that was synthesized by combining the ring of CNP with the N- and C-termini of urodilatin [179]. In animal models, CU-NP was demonstrated to increase renal blood flow, decrease capillary wedge pressure, and increase sodium and water excretion by activating NPR-B and NPR-A, without inducing significant systemic hypotension [180]. CU-NP also imparts an antihypertrophic effect via sodium-hydrogen exchanger-1 inhibition, indicating that this chimeric peptide may be used as a drug to prevent cardiac hypertrophy and failure. These effects are likely partially mediated by ATII and aldosterone suppression.

25.6.13 M-ANP

M-ANP (ZD100) was discovered from a familial frameshift mutation of the ANP gene, *NPPA*. M-ANP has the same 28 amino acids of the ANP core, along with a 12-amino-acid C-terminus, resulting in a 40 amino acids peptide. It acts like ANP as a specific guanylyl cyclase activator, is resistant to degradation by neprilysin, rendering it more potent and with a longer half-life than ANP [181]. Physiologic effects of M-ANP include lowered blood pressure through vasodilation, suppressed aldosterone action, inhibited sympathetic activity and increased renal blood flow. In a canine model of heart failure in the setting of hypertension, M-ANP was more potent in preserving renal blood flow and suppressing aldosterone than nitroglycerine. The above-mentioned actions position M-ANP as an attractive potential drug for hypertension. In a phase I, first-in-human trial, ZD100 injection to stable hypertensive patients resulted in both systolic and diastolic blood pressure reduction, with no serious adverse effects [182].

25.6.14 ANX042

ANX042 is a truncated peptide form of an alternatively spliced BNP, consisting of 26 amino acids from BNP and 16 amino acids from the C-terminus of the alternative spliced transcript of BNP. ANX042 does not induce cGMP activation in smooth muscles and endothelial cells and therefore has no effect on blood pressure [182]. In contrast, it can induce more cGMP production in renal mesangial cells than BNP, and in animal models with heart failure, infusion of ANX042 induced natriuresis and diuresis without affecting blood pressure [182]. In healthy volunteers, ANX-042 was safe and led to activation of cGMP at doses that did not result in symptomatic reduction of blood pressure. Based on these findings, ANX042 may play a role in the treatment of patients with acute heart failure accompanied by renal dysfunction (cardiorenal syndrome).

25.6.15 Lebetins

Lebetins are NPs that were extracted from *Macrovipera lebetina* snake venom and constitute a new class of inhibitors of platelet aggregation mediated by thrombin, platelet aggregating factor, and collagen [183]. They also inhibit fibrinogen-induced aggregation of alpha-chymotrypsin-treated platelets and were shown to inhibit collagen-induced thrombocytopenia in rats. There are two groups of lebetin peptides, lebetin 1 and lebetin 2, both of which improved contractile recovery after ischemia-reperfusion injury, while only lebetin 2 increased coronary flow and reduced severe cardiac dysfunction after ischemia. In animal models, lebetin 2 treatment exhibited cardioprotective properties in acute ischemia through stimulation of NPRs [184]. Therefore lebetins may prove to be a potent antithrombotic therapy for ischemic heart disease.

25.7 Urotensin II

25.7.1 Background

Urotensin II (UTII) is a cyclic peptide initially isolated in the 1960s from a goby fish indigenous to the Californian coastline [185] (Fig. 25.11).

25.7.2 Genetics and translation

UTII is encoded by the *UTII* gene located on chromosome 1. Its length varies across species and ranges from 11 amino acids in humans to 17 amino

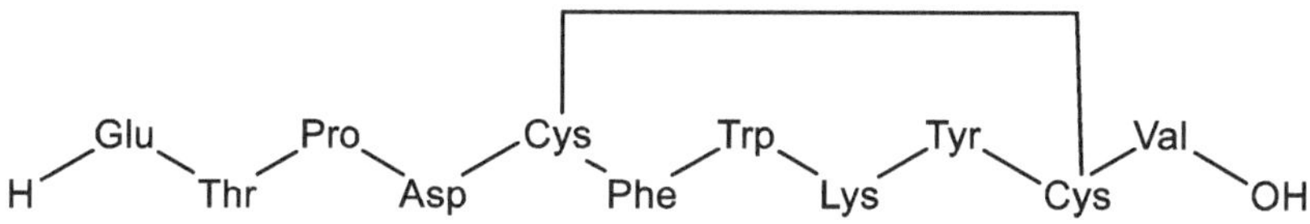

H-Glu-Thr-Pro-Asp-Cys(1)-Phe-Trp-Lys-Tyr-Cys(1)-Val-OH

FIGURE 25.11 Chemical structure and amino acid sequence of urotensin II.

acids in mice [186]. A UTII paralog called UTII-related peptide also exists in tetrapods and teleosts.

25.7.3 Mechanism of action

UTII and urotensin-related peptides constitute the urotensin system that activates the urotensin receptor. UTII binds to the Gq protein UTII receptor, a GPCR that is expressed in the brain, vasculature, cardiac tissue, renal cortex, and skeletal muscle. Unlike UII-related peptides, UTII has an acidic amino acid (glutamic or aspartic) that precedes the core sequence, and, while the amino acid is not necessary for the activation of the receptor, its conservation in different species suggests a potential biological function [187]. After binding the UTII receptor, protein kinase C is activated, which, in turn, activates phospholipase C and finally leads to an increase in intracellular calcium concentrations [188].

25.7.4 Functions

UTII is considered the most powerful mammalian vasoconstrictor known to date, with vasoconstrictive potency one order of magnitude greater than that of ET-1. It plays a central role in controlling blood pressure, the pathophysiology of renal diseases, insulin resistance, and the pathogenesis of atherosclerosis [186]. Intracerebroventricular injection of UTII induces blood pressure and heart rate elevation in rats increases ACTH level and reduces insulin secretion [189]. UTII has also been linked to the pathogenesis of cardiac hypertrophy and fibrosis in heart failure [186] and is elevated in the plasma of subjects with heart failure, hypertension and diabetes mellitus.

25.7.5 Therapeutics

Despite promising effects of UTII antagonists in heart failure, hypertension, chronic kidney disease, atherosclerosis, and metabolic syndrome animal models, its efficacy in human clinical trials is still limited. The new oral long-acting competitive UTII receptor antagonist (SAR101099) was shown to reduce proteinuria and preserve contractile reserve after myocardial infarction in animal models, and to be safe at all tested doses in clinical phase I trials [190]. The effectiveness of this new UTII antagonist in treating chronic kidney is currently under investigation.

25.8 Conclusions

CVD is the leading cause of death globally, accounting for more than 17.9 million deaths per year in 2015. Despite considerable progress in the area of treatments for CVDs, significant challenges remain in both the prevention and design of more efficient therapies. These limitations have prompted extensive research on the use of naturally occurring therapeutic peptides as well as rationally designed analogs for treating CVD, which demonstrate increasingly development promise and clinical utility.

Funding

This research was supported by a grant from the Ministry of Health (MoH), Jerusalem, Israel, and a grant from the Israel Science Foundation (ISF), Israel, both to N.Q.

References

[1] Kitamura K, Kangawa K, Kawamoto M, et al. Adrenomedullin: a novel hypotensive peptide isolated from human pheochromocytoma. Biochem Biophys Res Commun 1993;192 (2):553–60. Available from: https://doi.org/10.1006/bbrc.1993.1451.

[2] Brain SD, Grant AD. Vascular actions of calcitonin gene-related peptide and adrenomedullin. Physiol Rev 2004;84(3):903–34. Available from: https://doi.org/10.1152/physrev.00037.2003.

[3] Autelitano DJ, Ridings R. Adrenomedullin signalling in cardiomyocytes is dependent upon CRLR and RAMP2 expression. Peptides 2001;22(11):1851–7. Available from: https://doi.org/10.1016/s0196-9781(01)00536-8.

[4] Kitamura K, Eto T. Nihon Rinsho 1997;55(8):1963–70.

[5] Ishimitsu T, Kojima M, Kangawa K, et al. Genomic structure of human adrenomedullin gene. Biochem Biophys Res Commun 1994;203(1):631–9. Available from: https://doi.org/10.1006/bbrc.1994.2229.

[6] Owji AA, Smith DM, Coppock HA, et al. An abundant and specific binding site for the novel vasodilator adrenomedullin in the rat. Endocrinology 1995;136(5):2127–34. Available from: https://doi.org/10.1210/endo.136.5.7720662.

[7] Meeran K, O'Shea D, Upton PD, et al. Circulating adrenomedullin does not regulate systemic blood pressure but increases plasma prolactin after intravenous infusion in humans: a pharmacokinetic study. J Clin Endocrinol Metab 1997;82(1):95–100. Available from: https://doi.org/10.1210/jcem.82.1.3656.

[8] Lisy O, Jougasaki M, Schirger JA, Chen HH, Barclay PT, Burnett Jr. JC. Neutral endopeptidase inhibition potentiates the natriuretic actions of adrenomedullin. Am J Physiol 1998;275(3):F410–14. Available from: https://doi.org/10.1152/ajprenal.1998.275.3.F410.

[9] McMurray JJ, Packer M, Desai AS, et al. Angiotensin-neprilysin inhibition vs enalapril in heart failure. N Engl J Med 2014;371(11):993–1004. Available from: https://doi.org/10.1056/NEJMoa1409077.

[10] Ponikowski P, Voors AA, Anker SD, et al. 2016 ESC Guidelines for the diagnosis and treatment of acute and chronic heart failure: The Task Force for the diagnosis and treatment of acute and chronic heart failure of the European Society of Cardiology (ESC) developed with the special contribution of the Heart Failure Association (HFA) of the ESC [published correction appears in Eur Heart J 2016]Eur Heart J 2016;37 (27):2129–200. Available from: https://doi.org/10.1093/eurheartj/ehw128.

[11] Schönauer R, Kaiser A, Holze C, et al. Fluorescently labeled adrenomedullin allows real-time monitoring of adrenomedullin receptor trafficking in living cells. J Pept Sci 2015;21 (12):905–12. Available from: https://doi.org/10.1002/psc.2833.

[12] Lewis LK, Smith MW, Brennan SO, Yandle TG, Richards AM, Nicholls MG. Degradation of human adrenomedullin (1–52) by plasma membrane enzymes and identification of metabolites. Peptides 1997;18(5):733–9. Available from: https://doi.org/10.1016/s0196-9781(97)00005-3.

[13] Ishizaka Y, Ishizaka Y, Tanaka M, et al. Adrenomedullin stimulates cyclic AMP formation in rat vascular smooth muscle cells. Biochem Biophys Res Commun 1994;200 (1):642–6. Available from: https://doi.org/10.1006/bbrc.1994.1496.

[14] Shimekake Y, Nagata K, Ohta S, et al. Adrenomedullin stimulates two signal transduction pathways, cAMP accumulation and Ca2 + mobilization, in bovine aortic endothelial cells. J Biol Chem 1995;270(9):4412–17. Available from: https://doi.org/10.1074/jbc.270.9.4412.

[15] Voors AA, Kremer D, Geven C, et al. Adrenomedullin in heart failure: pathophysiology and therapeutic application. Eur J Heart Fail 2019;21(2):163–71. Available from: https://doi.org/10.1002/ejhf.1366.

[16] Sugo S, Minamino N, Kangawa K, et al. Endothelial cells actively synthesize and secrete adrenomedullin [published correction appears in Biochem Biophys Res Commun 1994;203(2):1363]Biochem Biophys Res Commun 1994;201(3):1160–6. Available from: https://doi.org/10.1006/bbrc.1994.1827.

[17] Schönauer R, Els-Heindl S, Beck-Sickinger AG. Adrenomedullin—new perspectives of a potent peptide hormone. J Pept Sci 2017;23(7–8):472–85. Available from: https://doi.org/10.1002/psc.2953.

[18] Charles CJ, Lainchbury JG, Nicholls MG, Rademaker MT, Richards AM, Troughton RW. Adrenomedullin and the renin-angiotensin-aldosterone system. Regul Pept 2003;112 (1–3):41–9. Available from: https://doi.org/10.1016/s0167-0115(03)00021-1.

[19] Kitamura K, Kangawa K, Eto T. Adrenomedullin and PAMP: discovery, structures, and cardiovascular functions. Microsc Res Tech 2002;57(1):3–13. Available from: https://doi.org/10.1002/jemt.10052.

[20] Cockcroft JR, Noon JP, Gardner-Medwin J, Bennett T. Haemodynamic effects of adrenomedullin in human resistance and capacitance vessels. Br J Clin Pharmacol 1997;44 (1):57–60. Available from: https://doi.org/10.1046/j.1365-2125.1997.00622.x.

[21] Nishikimi T, Karasawa T, Inaba C, et al. Effects of long-term intravenous administration of adrenomedullin (AM) plus hANP therapy in acute decompensated heart failure: a pilot study. Circ J 2009;73(5):892–8. Available from: https://doi.org/10.1253/circj.cj-08-0487.

[22] Nagaya N, Satoh T, Nishikimi T, et al. Hemodynamic, renal, and hormonal effects of adrenomedullin infusion in patients with congestive heart failure. Circulation 2000;101 (5):498–503. Available from: https://doi.org/10.1161/01.cir.101.5.498.

[23] Tanaka M, Koyama T, Sakurai T, et al. The endothelial adrenomedullin-RAMP2 system regulates vascular integrity and suppresses tumour metastasis. Cardiovasc Res 2016;111 (4):398–409. Available from: https://doi.org/10.1093/cvr/cvw166.

[24] Koyama T, Sakurai T, Kamiyoshi A, Ichikawa-Shindo Y, Kawate H, Shindo T. Adrenomedullin-RAMP2 system in vascular endothelial cells. J Atheroscler Thromb 2015;22(7):647–53. Available from: https://doi.org/10.5551/jat.29967.

[25] Koyama T, Ochoa-Callejero L, Sakurai T, et al. Vascular endothelial adrenomedullin-RAMP2 system is essential for vascular integrity and organ homeostasis. Circulation 2013;127 (7):842–53. Available from: https://doi.org/10.1161/CIRCULATIONAHA.112.000756.

[26] Ochoa-Callejero L, Pozo-Rodrigálvarez A, Martínez-Murillo R, Martínez A. Lack of adrenomedullin in mouse endothelial cells results in defective angiogenesis, enhanced vascular permeability, less metastasis, and more brain damage [published correction appears in Sci Rep 2020;10(1):6283]Sci Rep 2016;6:33495. Available from: https://doi.org/10.1038/srep33495.

[27] Hippenstiel S, Witzenrath M, Schmeck B, et al. Adrenomedullin reduces endothelial hyperpermeability. Circ Res 2002;91(7):618–25. Available from: https://doi.org/10.1161/01.res.0000036603.61868.f9.

[28] Geven C, Kox M, Pickkers P. Adrenomedullin and adrenomedullin-targeted therapy as treatment strategies relevant for sepsis. Front Immunol 2018;9:292. Available from: https://doi.org/10.3389/fimmu.2018.00292.

[29] Nishikimi T, Saito Y, Kitamura K, et al. Increased plasma levels of adrenomedullin in patients with heart failure. J Am Coll Cardiol 1995;26(6):1424–31. Available from: https://doi.org/10.1016/0735-1097(95)00338-X.

[30] Jougasaki M, Wei CM, McKinley LJ, Burnett Jr. JC. Elevation of circulating and ventricular adrenomedullin in human congestive heart failure. Circulation 1995;92(3):286–9. Available from: https://doi.org/10.1161/01.cir.92.3.286.

[31] Maisel A, Mueller C, Nowak R, et al. Mid-region pro-hormone markers for diagnosis and prognosis in acute dyspnea: results from the BACH (biomarkers in acute heart failure) trial. J Am Coll Cardiol 2010;55(19):2062–76. Available from: https://doi.org/10.1016/j.jacc.2010.02.025.

[32] Valenzuela-Sánchez F, Valenzuela-Méndez B, Rodríguez-Gutiérrez JF, Estella-García Á, González-García MÁ. New role of biomarkers: mid-regional pro-adrenomedullin, the biomarker of organ failure. Ann Transl Med 2016;4(17):329. Available from: https://doi.org/10.21037/atm.2016.08.65.

[33] Wimalawansa SJ. Amylin, calcitonin gene-related peptide, calcitonin, and adrenomedullin: a peptide superfamily. Crit Rev Neurobiol 1997;11(2-3):167–239. Available from: https://doi.org/10.1615/critrevneurobiol.v11.i2-3.40.

[34] Szokodi I, Kinnunen P, Tavi P, Weckström M, Tóth M, Ruskoaho H. Evidence for cAMP-independent mechanisms mediating the effects of adrenomedullin, a new inotropic peptide. Circulation 1998;97(11):1062–70. Available from: https://doi.org/10.1161/01.cir.97.11.1062.

[35] Nakamura R, Kato J, Kitamura K, et al. Beneficial effects of adrenomedullin on left ventricular remodeling after myocardial infarction in rats. Cardiovasc Res 2002;56 (3):373–80. Available from: https://doi.org/10.1016/s0008-6363(02)00594-1.

[36] Okumura H, Nagaya N, Kangawa K. Adrenomedullin infusion during ischemia/reperfusion attenuates left ventricular remodeling and myocardial fibrosis in rats. Hypertens Res 2003;26(Suppl):S99–104. Available from: https://doi.org/10.1291/hypres.26.s99.

[37] Niu P, Shindo T, Iwata H, et al. Accelerated cardiac hypertrophy and renal damage induced by angiotensin II in adrenomedullin knockout mice. Hypertens Res 2003;26 (9):731–6. Available from: https://doi.org/10.1291/hypres.26.731.

[38] Nishikimi T, Yoshihara F, Horinaka S, et al. Chronic administration of adrenomedullin attenuates transition from left ventricular hypertrophy to heart failure in rats. Hypertension 2003;42 (5):1034–41. Available from: https://doi.org/10.1161/01.HYP.0000097604.64716.D2.

[39] Okumura H, Nagaya N, Itoh T, et al. Adrenomedullin infusion attenuates myocardial ischemia/reperfusion injury through the phosphatidylinositol 3-kinase/Akt-dependent pathway. Circulation 2004;109(2):242–8. Available from: https://doi.org/10.1161/01.CIR.0000109214.30211.7C.

[40] Nakamura R, Kato J, Kitamura K, et al. Adrenomedullin administration immediately after myocardial infarction ameliorates progression of heart failure in rats. Circulation 2004;110 (4):426–31. Available from: https://doi.org/10.1161/01.CIR.0000136085.34185.83.

[41] Niu P, Shindo T, Iwata H, et al. Protective effects of endogenous adrenomedullin on cardiac hypertrophy, fibrosis, and renal damage. Circulation 2004;109(14):1789–94. Available from: https://doi.org/10.1161/01.CIR.0000118466.47982.CC.

[42] Looi YH, Kane KA, McPhaden AR, Wainwright CL. Adrenomedullin acts via nitric oxide and peroxynitrite to protect against myocardial ischaemia-induced arrhythmias in anaesthetized rats. Br J Pharmacol 2006;148(5):599–609. Available from: https://doi.org/10.1038/sj.bjp.0706771.

[43] Nakamura M, Yoshida H, Makita S, Arakawa N, Niinuma H, Hiramori K. Potent and long-lasting vasodilatory effects of adrenomedullin in humans. Comparisons between normal subjects and patients with chronic heart failure. Circulation 1997;95(5):1214–21. Available from: https://doi.org/10.1161/01.cir.95.5.1214.

[44] Kataoka Y, Miyazaki S, Yasuda S, et al. The first clinical pilot study of intravenous adrenomedullin administration in patients with acute myocardial infarction. J Cardiovasc Pharmacol 2010;56(4):413–19. Available from: https://doi.org/10.1097/FJC.0b013e3181f15b45.

[45] Geven C, Bergmann A, Kox M, Pickkers P. Vascular effects of adrenomedullin and the anti-adrenomedullin antibody adrecizumab in sepsis. Shock 2018;50(2):132–40. Available from: https://doi.org/10.1097/SHK.0000000000001103.

[46] Geven C, van Lier D, Blet A, et al. Safety, tolerability and pharmacokinetics/pharmacodynamics of the adrenomedullin antibody adrecizumab in a first-in-human study and during experimental human endotoxaemia in healthy subjects. Br J Clin Pharmacol 2018;84 (9):2129–41. Available from: https://doi.org/10.1111/bcp.13655.

[47] Geven C, Blet A, Kox M, et al. A double-blind, placebo-controlled, randomised, multicentre, proof-of-concept and dose-finding phase II clinical trial to investigate the safety, tolerability and efficacy of adrecizumab in patients with septic shock and elevated adrenomedullin concentration (AdrenOSS-2). BMJ Open 2019;9(2):e024475. Available from: https://doi.org/10.1136/bmjopen-2018-024475.

[48] Adrecizumab dose escalation safety and tolerability evaluation (ADESTE), NCT04252937.

[49] Kremer D, Ter Maaten JM, Voors AA. Bio-adrenomedullin as a potential quick, reliable, and objective marker of congestion in heart failure. Eur J Heart Fail 2018;20(9):1363–5. Available from: https://doi.org/10.1002/ejhf.1245.

[50] Tolppanen H, Rivas-Lasarte M, Lassus J, et al. Adrenomedullin: a marker of impaired hemodynamics, organ dysfunction, and poor prognosis in cardiogenic shock. Ann Intensive Care 2017;7(1):6. Available from: https://doi.org/10.1186/s13613-016-0229-2.

[51] Basso N, Terragno NA. History about the discovery of the renin-angiotensin system. Hypertension 2001;38(6):1246–9. Available from: https://doi.org/10.1161/hy1201.101214.

[52] Struthers AD, MacDonald TM. Review of aldosterone- and angiotensin II-induced target organ damage and prevention. Cardiovasc Res 2004;61(4):663–70. Available from: https://doi.org/10.1016/j.cardiores.2003.11.037.

[53] Weber KT. Extracellular matrix remodeling in heart failure: a role for de novo angiotensin II generation. Circulation 1997;96(11):4065–82. Available from: https://doi.org/10.1161/01.cir.96.11.4065.

[54] Gaillard-Sanchez I, Mattei MG, Clauser E, Corvol P. Assignment by in situ hybridization of the angiotensinogen gene to chromosome band 1q4, the same region as the human renin gene. Hum Genet 1990;84(4):341–3. Available from: https://doi.org/10.1007/BF00196230.

[55] Gaillard I, Clauser E, Corvol P. Structure of human angiotensinogen gene. DNA 1989;8(2):87–99. Available from: https://doi.org/10.1089/dna.1.1989.8.87.

[56] Yang HY, Erdös EG, Levin Y. Characterization of a dipeptide hydrolase (kininase II: angiotensin I converting enzyme). J Pharmacol Exp Ther 1971;177(1):291–300.

[57] Ferrario CM, Iyer SN. Angiotensin-(1-7): a bioactive fragment of the renin-angiotensin system. Regul Pept 1998;78(1-3):13–18. Available from: https://doi.org/10.1016/s0167-0115(98)00134-7.

[58] Paul M, Poyan Mehr A, Kreutz R. Physiology of local renin-angiotensin systems. Physiol Rev 2006;86(3):747–803. Available from: https://doi.org/10.1152/physrev.00036.2005.

[59] Bradley SE, Parker B. The hemodynamic effects of angiotonin in normal man. J Clin Invest 1941;20(6):715–19. Available from: https://doi.org/10.1172/JCI101265.

[60] Johnson WP, Bruce RA. Hemodynamic and metabolic effects of angiotensin II during rest and exercise in normal healthy subjects. Am Heart J 1962;63:212–18. Available from: https://doi.org/10.1016/0002-8703(62)90197-7.

[61] Radice M, Folli G, Giani P, Tavecchia A, Mariotti G. Latent myocardial contractile impairment in patients with angina pectoris. Acta Cardiol 1975;30(5):333–41.

[62] Ahmed SS, Levinson GE, Weisse AB, Regan TJ. The effect of angiotensin on myocardial contractility. J Clin Pharmacol 1975;15(4 Pt 1):276–85. Available from: https://doi.org/10.1002/j.1552-4604.1975.tb01449.x.

[63] Brod J, Hejl Z, Hornych A, Jirka J, Slechta V, Buriánová B. Comparison of haemodynamic effects of equipressor doses of intravenous angiotensin and noradrenaline in man. Clin Sci 1969;36(2):161–72.

[64] Cachovan M, Brod J, Bahlmann J, Sippel R, Celsen B, Hundeshagen H. The effect of intravenous angiotensin II on the peripheral circulation with particular reference to its bearing on general haemodynamics. Clin Sci Mol Med Suppl 1976;3:219s–21s. Available from: https://doi.org/10.1042/cs051219s.

[65] Cargill RI, Coutie WJ, Lipworth BJ. The effects of angiotensin II on circulating levels of natriuretic peptides. Br J Clin Pharmacol 1994;38(2):139–42. Available from: https://doi.org/10.1111/j.1365-2125.1994.tb04337.x.

[66] Yu PN, Luria MN, Finlayson JK, Stanfield CA, Constantine H, Flatley FJ. The effects of angiotensin on pulmonary circulation and ventricular function. Circulation 1961;24:1326–37. Available from: https://doi.org/10.1161/01.cir.24.6.1326.

[67] Wilkinson IB, MacCallum H, Hupperetz PC, van Thoor CJ, Cockcroft JR, Webb DJ. Changes in the derived central pressure waveform and pulse pressure in response to

angiotensin II and noradrenaline in man. J Physiol 2001;530(Pt 3):541–50. Available from: https://doi.org/10.1111/j.1469-7793.2001.0541k.x.
[68] Ramsay M, Broughton Pipkin F, Rubin P. Comparative study of pressor and heart rate responses to angiotensin II and noradrenaline in pregnant and non-pregnant women. Clin Sci (Lond) 1992;82(2):157–62. Available from: https://doi.org/10.1042/cs0820157.
[69] Brown MA, Broughton Pipkin F, Symonds EM. The effects of intravenous angiotensin II upon blood pressure and sodium and urate excretion in human pregnancy. J Hypertens 1988;6(6):457–64. Available from: https://doi.org/10.1097/00004872-198806000-00005.
[70] Motwani JG, Struthers AD. Dose-response study of the redistribution of intravascular volume by angiotensin II in man. Clin Sci (Lond) 1992;82(4):397–405. Available from: https://doi.org/10.1042/cs0820397.
[71] Nagano M, Higaki J, Mikami H, et al. Converting enzyme inhibitors regressed cardiac hypertrophy and reduced tissue angiotensin II in spontaneously hypertensive rats. J Hypertens 1991;9(7):595–9. Available from: https://doi.org/10.1097/00004872-199107000-00003.
[72] Hsueh WA, Do YS, Anderson PW, Law RE. Angiotensin II in cell growth and matrix production. Adv Exp Med Biol 1995;377:217–23. Available from: https://doi.org/10.1007/978-1-4899-0952-7_12.
[73] Griendling KK, Minieri CA, Ollerenshaw JD, Alexander RW. Angiotensin II stimulates NADH and NADPH oxidase activity in cultured vascular smooth muscle cells. Circ Res 1994;74(6):1141–8. Available from: https://doi.org/10.1161/01.res.74.6.1141.
[74] De Mello WC, Crespo MJ. Cardiac refractoriness in rats is reduced by angiotensin II. J Cardiovasc Pharmacol 1995;25(1):51–6. Available from: https://doi.org/10.1097/00005344-199501000-00009.
[75] Daugherty A, Cassis L. Angiotensin II and abdominal aortic aneurysms. Curr Hypertens Rep 2004;6(6):442–6. Available from: https://doi.org/10.1007/s11906-004-0038-0.
[76] Min LJ, Mogi M, Iwai M, Horiuchi M. Signaling mechanisms of angiotensin II in regulating vascular senescence. Ageing Res Rev 2009;8(2):113–21. Available from: https://doi.org/10.1016/j.arr.2008.12.002.
[77] Bernstein KE, Berk BC. The biology of angiotensin II receptors. Am J Kidney Dis 1993;22(5):745–54. Available from: https://doi.org/10.1016/s0272-6386(12)80441-0.
[78] Kaschina E, Unger T. Angiotensin AT1/AT2 receptors: regulation, signalling and function. Blood Press 2003;12(2):70–88. Available from: https://doi.org/10.1080/08037050310001057.
[79] de Gasparo M, Catt KJ, Inagami T, Wright JW, Unger T. International union of pharmacology. XXIII. The angiotensin II receptors. Pharmacol Rev 2000;52(3):415–72.
[80] Sechi LA, Grady EF, Griffin CA, Kalinyak JE, Schambelan M. Distribution of angiotensin II receptor subtypes in rat and human kidney. Am J Physiol 1992;262(2 Pt 2):F236–40. Available from: https://doi.org/10.1152/ajprenal.1992.262.2.F236.
[81] Yoo KH, Thornhill BA, Wolstenholme JT, Chevalier RL. Tissue-specific regulation of growth factors and clusterin by angiotensin II. Am J Hypertens 1998;11(6 Pt 1):715–22. Available from: https://doi.org/10.1016/s0895-7061(98)00018-1.
[82] Hattangady NG, Olala LO, Bollag WB, Rainey WE. Acute and chronic regulation of aldosterone production. Mol Cell Endocrinol 2012;350(2):151–62. Available from: https://doi.org/10.1016/j.mce.2011.07.034.
[83] Kranzhöfer R, Schmidt J, Pfeiffer CA, Hagl S, Libby P, Kübler W. Angiotensin induces inflammatory activation of human vascular smooth muscle cells. Arterioscler Thromb Vasc Biol 1999;19(7):1623–9. Available from: https://doi.org/10.1161/01.atv.19.7.1623.

[84] Busse LW, Wang XS, Chalikonda DM, et al. Clinical experience with IV angiotensin II administration: a systematic review of safety. Crit Care Med 2017;45(8):1285–94. Available from: https://doi.org/10.1097/CCM.0000000000002441.

[85] Jackson T, Corke C, Agar J. Enalapril overdose treated with angiotensin infusion. Lancet 1993;341(8846):703. Available from: https://doi.org/10.1016/0140-6736(93)90479-z.

[86] Newby DE, Lee MR, Gray AJ, Boon NA. Enalapril overdose and the corrective effect of intravenous angiotensin II. Br J Clin Pharmacol 1995;40(1):103–4. Available from: https://doi.org/10.1111/j.1365-2125.1995.tb04546.x.

[87] Trilli LE, Johnson KA. Lisinopril overdose and management with intravenous angiotensin II. Ann Pharmacother 1994;28(10):1165–8. Available from: https://doi.org/10.1177/106002809402801006.

[88] Ryding J, Heslet L, Hartvig T, Jønsson V. Reversal of 'refractory septic shock' by infusion of amrinone and angiotensin II in an anthracycline-treated patient. Chest 1995;107(1):201–3. Available from: https://doi.org/10.1378/chest.107.1.201.

[89] Whiteley SM, Dade JP. Treatment of hypotension in septic shock. Lancet 1996;347(9001):622.

[90] Yunge M, Petros A. Angiotensin for septic shock unresponsive to noradrenaline. Arch Dis Child 2000;82(5):388–9. Available from: https://doi.org/10.1136/adc.82.5.388.

[91] Thomas VL, Nielsen MS. Administration of angiotensin II in refractory septic shock. Crit Care Med 1991;19(8):1084–6. Available from: https://doi.org/10.1097/00003246-199108000-00020.

[92] Wray GM, Coakley JH. Severe septic shock unresponsive to noradrenaline. Lancet 1995;346(8990):1604. Available from: https://doi.org/10.1016/s0140-6736(95)91933-3.

[93] Chawla LS, Busse L, Brasha-Mitchell E, et al. Intravenous angiotensin II for the treatment of high-output shock (ATHOS trial): a pilot study. Crit Care 2014;18(5):534. Available from: https://doi.org/10.1186/s13054-014-0534-9.

[94] Millar EA, Angus RM, Hulks G, Morton JJ, Connell JM, Thomson NC. Activity of the renin-angiotensin system in acute severe asthma and the effect of angiotensin II on lung function. Thorax 1994;49(5):492–5. Available from: https://doi.org/10.1136/thx.49.5.492.

[95] Cohn JN, Luria MH. Studies in clinical shock and hypotension. II. Hemodynamic effects of norepinephrine and angiotensin. J Clin Invest 1965;44(9):1494–504. Available from: https://doi.org/10.1172/JCI105256.

[96] Hickey KA, Rubanyi G, Paul RJ, Highsmith RF. Characterization of a coronary vasoconstrictor produced by cultured endothelial cells. Am J Physiol 1985;248(5 Pt 1):C550–6. Available from: https://doi.org/10.1152/ajpcell.1985.248.5.C550.

[97] Gillespie MN, Owasoyo JO, McMurtry IF, O'Brien RF. Sustained coronary vasoconstriction provoked by a peptidergic substance released from endothelial cells in culture. J Pharmacol Exp Ther 1986;236(2):339–43.

[98] O'Brien RF, Robbins RJ, McMurtry IF. Endothelial cells in culture produce a vasoconstrictor substance. J Cell Physiol 1987;132(2):263–70. Available from: https://doi.org/10.1002/jcp.1041320210.

[99] Yanagisawa M, Inoue A, Ishikawa T, et al. Primary structure, synthesis, and biological activity of rat endothelin, an endothelium-derived vasoconstrictor peptide. Proc Natl Acad Sci USA 1988;85(18):6964–7. Available from: https://doi.org/10.1073/pnas.85.18.6964.

[100] Takasaki C, Tamiya N, Bdolah A, Wollberg Z, Kochva E. Sarafotoxins S6: several isotoxins from Atractaspis engaddensis (burrowing asp) venom that affect the heart.

Toxicon 1988;26(6):543–8. Available from: https://doi.org/10.1016/0041-0101(88)90234-6.

[101] Inoue A, Yanagisawa M, Kimura S, et al. The human endothelin family: three structurally and pharmacologically distinct isopeptides predicted by three separate genes. Proc Natl Acad Sci USA 1989;86(8):2863–7. Available from: https://doi.org/10.1073/pnas.86.8.2863.

[102] Maeda S, Miyauchi T, Goto K, Matsuda M. Differences in the change in the time course of plasma endothelin-1 and endothelin-3 levels after exercise in humans. The response to exercise of endothelin-3 is more rapid than that of endothelin-1. Life Sci 1997;61(4):419–25. Available from: https://doi.org/10.1016/s0024-3205(97)00399-8.

[103] Maeda S, Miyauchi T, Kakiyama T, et al. Effects of exercise training of 8 weeks and detraining on plasma levels of endothelium-derived factors, endothelin-1 and nitric oxide, in healthy young humans. Life Sci 2001;69(9):1005–16. Available from: https://doi.org/10.1016/s0024-3205(01)01192-4.

[104] Maeda S, Tanabe T, Miyauchi T, et al. Aerobic exercise training reduces plasma endothelin-1 concentration in older women. J Appl Physiol (1985) 2003;95(1):336–41. Available from: https://doi.org/10.1152/japplphysiol.01016.2002.

[105] Kawanabe Y, Nauli SM. Endothelin. Cell Mol Life Sci 2011;68(2):195–203. Available from: https://doi.org/10.1007/s00018-010-0518-0.

[106] Miyauchi T. Endothelins. In: Kastin A, editor. Handbook of biologically active peptides. 2nd ed. Academic press; 2013. p. 1402–7.

[107] Miyauchi T, Masaki T. Pathophysiology of endothelin in the cardiovascular system. Annu Rev Physiol 1999;61:391–415. Available from: https://doi.org/10.1146/annurev.physiol.61.1.391.

[108] Shimojo N, Jesmin S, Zaedi S, et al. Eicosapentaenoic acid prevents endothelin-1-induced cardiomyocyte hypertrophy in vitro through the suppression of TGF-beta 1 and phosphorylated JNK. Am J Physiol Heart Circ Physiol 2006;291(2):H835–45. Available from: https://doi.org/10.1152/ajpheart.01365.2005.

[109] Shimojo N, Jesmin S, Zaedi S, et al. Contributory role of VEGF overexpression in endothelin-1-induced cardiomyocyte hypertrophy. Am J Physiol Heart Circ Physiol 2007;293(1):H474–81. Available from: https://doi.org/10.1152/ajpheart.00922.2006.

[110] Goto K, Hama H, Kasuya Y. Molecular pharmacology and pathophysiological significance of endothelin. Jpn J Pharmacol 1996;72(4):261–90. Available from: https://doi.org/10.1254/jjp.72.261.

[111] Goto K. Basic and therapeutic relevance of endothelin-mediated regulation. Biol Pharm Bull 2001;24(11):1219–30. Available from: https://doi.org/10.1248/bpb.24.1219.

[112] Barton M, Yanagisawa M. Endothelin: 20 years from discovery to therapy. Can J Physiol Pharmacol 2008;86(8):485–98. Available from: https://doi.org/10.1139/Y08-059.

[113] Channick RN, Simonneau G, Sitbon O, et al. Effects of the dual endothelin-receptor antagonist bosentan in patients with pulmonary hypertension: a randomised placebo-controlled study. Lancet 2001;358(9288):1119–23. Available from: https://doi.org/10.1016/S0140 6736(01)06250-X.

[114] Khimji AK, Rockey DC. Endothelin–biology and disease. Cell Signal 2010;22(11):1615–25. Available from: https://doi.org/10.1016/j.cellsig.2010.05.002.

[115] Enevoldsen FC, Sahana J, Wehland M, Grimm D, Infanger M, Krüger M. Endothelin receptor antagonists: status quo and future perspectives for targeted therapy. J Clin Med 2020;9(3):824. Available from: https://doi.org/10.3390/jcm9030824.

[116] Rocha E, Silva M, Beraldo WT, Rosenfeld G. Bradykinin, a hypotensive and smooth muscle stimulating factor released from plasma globulin by snake venoms and by trypsin. Am J Physiol 1949;156(2):261–73. Available from: https://doi.org/10.1152/ajplegacy.1949.156.2.261.

[117] Sainz IM, Pixley RA, Colman RW. Fifty years of research on the plasma kallikrein-kinin system: from protein structure and function to cell biology and in-vivo pathophysiology. Thromb Haemost 2007;98(1):77–83.

[118] Takagaki Y, Kitamura N, Nakanishi S. Cloning and sequence analysis of cDNAs for human high molecular weight and low molecular weight prekininogens. Primary structures of two human prekininogens. J Biol Chem 1985;260(14):8601–9.

[119] Hofman Z, de Maat S, Hack CE, Maas C. Bradykinin: inflammatory product of the coagulation system. Clin Rev Allergy Immunol 2016;51(2):152–61. Available from: https://doi.org/10.1007/s12016-016-8540-0.

[120] Campbell DJ. The kallikrein-kinin system in humans. Clin Exp Pharmacol Physiol 2001;28(12):1060–5. Available from: https://doi.org/10.1046/j.1440-1681.2001.03564.x.

[121] Campbell DJ. Bradykinin peptides. Handbook of biologically active peptides. 2nd ed. Elsevier Inc; 2013. p. 1386–93.

[122] Figueroa CD, Marchant A, Novoa U, et al. Differential distribution of bradykinin B(2) receptors in the rat and human cardiovascular system. Hypertension 2001;37(1):110–20. Available from: https://doi.org/10.1161/01.hyp.37.1.110.

[123] Marceau F, Regoli D. Bradykinin receptor ligands: therapeutic perspectives. Nat Rev Drug Discov 2004;3(10):845–52. Available from: https://doi.org/10.1038/nrd1522.

[124] Jancso G, Jaberansari M, Gasz B, Szanto Z, Cserepes B, Röth E. Bradykinin and angiotensin-converting enzyme inhibition in cardioprotection. Exp Clin Cardiol 2004;9(1):21–5.

[125] Marceau F, Bachelard H, Bouthillier J, et al. Bradykinin receptors: agonists, antagonists, expression, signaling, and adaptation to sustained stimulation [published online ahead of print]Int Immunopharmacol 2020;82:106305. Available from: https://doi.org/10.1016/j.intimp.2020.106305.

[126] Aramori I, Zenkoh J, Morikawa N, et al. Novel subtype-selective nonpeptide bradykinin receptor antagonists FR167344 and FR173657. Mol Pharmacol 1997;51(2):171–6. Available from: https://doi.org/10.1124/mol.51.2.171.

[127] Man J, Barnett P, Christoffels VM. Structure and function of the Nppa-Nppb cluster locus during heart development and disease. Cell Mol Life Sci 2018;75(8):1435–44. Available from: https://doi.org/10.1007/s00018-017-2737-0.

[128] McGrath MF, de Bold AJ. Determinants of natriuretic peptide gene expression. Peptides 2005;26(6):933–43. Available from: https://doi.org/10.1016/j.peptides.2004.12.022.

[129] Garbers DL, Chrisman TD, Wiegn P, et al. Membrane guanylyl cyclase receptors: an update. Trends Endocrinol Metab 2006;17(6):251–8. Available from: https://doi.org/10.1016/j.tem.2006.06.006.

[130] Cannone V, Cabassi A, Volpi R, Burnett Jr. JC. Atrial natriuretic peptide: a molecular target of novel therapeutic approaches to cardio-metabolic disease. Int J Mol Sci 2019;20(13):3265. Available from: https://doi.org/10.3390/ijms20133265.

[131] Benomar K, Espiard S, Loyer C, Jannin A, Vantyghem MC. Hormones natriurétiques et syndrome métabolique: mise au point [atrial natriuretic hormones and metabolic syndrome: recent advances]. Presse Med 2018;47(2):116–24. Available from: https://doi.org/10.1016/j.lpm.2017.12.002.

[132] Volpe M, Carnovali M, Mastromarino V. The natriuretic peptides system in the pathophysiology of heart failure: from molecular basis to treatment. Clin Sci (Lond) 2016;130(2):57–77. Available from: https://doi.org/10.1042/CS20150469.

[133] Wang D, Oparil S, Feng JA, et al. Effects of pressure overload on extracellular matrix expression in the heart of the atrial natriuretic peptide-null mouse. Hypertension 2003;42(1):88–95. Available from: https://doi.org/10.1161/01.HYP.0000074905.22908.A6.

[134] Goodfriend TL, Calhoun DA. Resistant hypertension, obesity, sleep apnea, and aldosterone: theory and therapy. Hypertension 2004;43(3):518–24. Available from: https://doi.org/10.1161/01.HYP.0000116223.97436.e5.

[135] Song W, Wang H, Wu Q. Atrial natriuretic peptide in cardiovascular biology and disease (NPPA). Gene 2015;569(1):1–6. Available from: https://doi.org/10.1016/j.gene.2015.06.029.

[136] Hayashi D, Kudoh S, Shiojima I, et al. Atrial natriuretic peptide inhibits cardiomyocyte hypertrophy through mitogen-activated protein kinase phosphatase-1. Biochem Biophys Res Commun 2004;322(1):310–19. Available from: https://doi.org/10.1016/j.bbrc.2004.07.119.

[137] Kobayashi D, Yamaguchi N, Takahashi O, Deshpande GA, Fukui T. Human atrial natriuretic peptide treatment for acute heart failure: a systematic review of efficacy and mortality. Can J Cardiol 2012;28(1):102–9. Available from: https://doi.org/10.1016/j.cjca.2011.04.011.

[138] Mitaka C, Kudo T, Haraguchi G, Tomita M. Cardiovascular and renal effects of carperitide and nesiritide in cardiovascular surgery patients: a systematic review and meta-analysis. Crit Care 2011;15(5):R258. Available from: https://doi.org/10.1186/cc10519.

[139] Kohno M, Yasunari K, Yokokawa K, Murakawa K, Horio T, Takeda T. Inhibition by atrial and brain natriuretic peptides of endothelin-1 secretion after stimulation with angiotensin II and thrombin of cultured human endothelial cells. J Clin Invest 1991;87(6):1999–2004. Available from: https://doi.org/10.1172/JCI115228.

[140] Kitakaze M, Asakura M, Kim J, et al. Human atrial natriuretic peptide and nicorandil as adjuncts to reperfusion treatment for acute myocardial infarction (J-WIND): two randomised trials [published correction appears in Lancet 2008;370(9605):2102]Lancet 2007;370(9597):1483–93. Available from: https://doi.org/10.1016/S0140-6736(07)61634-1.

[141] Kawase Y, Hata R, Tada T, Katoh H, Kadota K. Effects of carperitide on degree of pulmonary congestion in treatment of acute heart failure. Circ J 2018;82(8):2079–88. Available from: https://doi.org/10.1253/circj.CJ-18-0057.

[142] Nagai T, Iwakami N, Nakai M, et al. Effect of intravenous carperitide vs nitrates as first-line vasodilators on in-hospital outcomes in hospitalized patients with acute heart failure: Insight from a nationwide claim-based database. Int J Cardiol 2019;280:104–9. Available from: https://doi.org/10.1016/j.ijcard.2019.01.049.

[143] Sudoh T, Kangawa K, Minamino N, Matsuo H. A new natriuretic peptide in porcine brain. Nature 1988;332(6159):78–81. Available from: https://doi.org/10.1038/332078a0.

[144] Yoshimura M, Yasue H, Ogawa H. Pathophysiological significance and clinical application of ANP and BNP in patients with heart failure. Can J Physiol Pharmacol 2001;79(8):730–5.

[145] Hall C. Essential biochemistry and physiology of (NT-pro)BNP. Eur J Heart Fail 2004;6(3):257–60. Available from: https://doi.org/10.1016/j.ejheart.2003.12.015.

[146] Mukoyama M, Nakao K, Hosoda K, et al. Brain natriuretic peptide as a novel cardiac hormone in humans. Evidence for an exquisite dual natriuretic peptide system, atrial natriuretic peptide and brain natriuretic peptide. J Clin Invest 1991;87(4):1402–12. Available from: https://doi.org/10.1172/JCI115146.

[147] Pankow K, Wang Y, Gembardt F, et al. Successive action of meprin A and neprilysin catabolizes B-type natriuretic peptide. Circ Res 2007;101(9):875–82. Available from: https://doi.org/10.1161/CIRCRESAHA.107.153585.

[148] Nakagawa Y, Nishikimi T, Kuwahara K. Atrial and brain natriuretic peptides: Hormones secreted from the heart. Peptides 2019;111:18–25. Available from: https://doi.org/10.1016/j.peptides.2018.05.012.

[149] Maric C, Zheng W, Walther T. Interactions between angiotensin ll and atrial natriuretic peptide in renomedullary interstitial cells: the role of neutral endopeptidase. Nephron Physiol 2006;103(3):p149–56. Available from: https://doi.org/10.1159/000092457.

[150] Maisel AS, Krishnaswamy P, Nowak RM, et al. Rapid measurement of B-type natriuretic peptide in the emergency diagnosis of heart failure. N Engl J Med 2002;347(3):161–7. Available from: https://doi.org/10.1056/NEJMoa020233.

[151] Januzzi Jr JL, Camargo CA, Anwaruddin S, et al. The N-terminal Pro-BNP investigation of dyspnea in the emergency department (PRIDE) study. Am J Cardiol 2005;95 (8):948–54. Available from: https://doi.org/10.1016/j.amjcard.2004.12.032.

[152] Santaguida PL, Don-Wauchope AC, Oremus M, et al. BNP and NT-proBNP as prognostic markers in persons with acute decompensated heart failure: a systematic review. Heart Fail Rev 2014;19(4):453–70. Available from: https://doi.org/10.1007/s10741-014-9442-y.

[153] Wang TJ, Larson MG, Levy D, et al. Plasma natriuretic peptide levels and the risk of cardiovascular events and death. N Engl J Med 2004;350(7):655–63. Available from: https://doi.org/10.1056/NEJMoa031994.

[154] Solomon SD, McMurray JJV, Anand IS, et al. Angiotensin-neprilysin inhibition in heart failure with preserved ejection fraction. N Engl J Med 2019;381(17):1609–20. Available from: https://doi.org/10.1056/NEJMoa1908655.

[155] Colucci WS, Elkayam U, Horton DP, et al. Intravenous nesiritide, a natriuretic peptide, in the treatment of decompensated congestive heart failure. Nesiritide Study Group [published correction appears in N Engl J Med 2000;343(20):1504 and N Engl J Med 2000;343(12):896]N Engl J Med 2000;343(4):246–53. Available from: https://doi.org/10.1056/NEJM200007273430403.

[156] Sudoh T, Minamino N, Kangawa K, Matsuo H. C-type natriuretic peptide (CNP): a new member of natriuretic peptide family identified in porcine brain. Biochem Biophys Res Commun 1990;168(2):863–70. Available from: https://doi.org/10.1016/0006-291x(90)92401-k.

[157] Wu C, Wu F, Pan J, Morser J, Wu Q. Furin-mediated processing of Pro-C-type natriuretic peptide. J Biol Chem 2003;278(28):25847–52. Available from: https://doi.org/10.1074/jbc.M301223200.

[158] Yeung VT, Ho SK, Nicholls MG, Cockram CS. Binding of CNP-22 and CNP-53 to cultured mouse astrocytes and effects on cyclic GMP. Peptides 1996;17(1):101–6. Available from: https://doi.org/10.1016/0196-9781(95)02099-3.

[159] Charles CJ, Prickett TC, Espiner EA, Rademaker MT, Richards AM, Yandle TG. Regional sampling and the effects of experimental heart failure in sheep: differential responses in A, B and C-type natriuretic peptides. Peptides 2006;27(1):62–8. Available from: https://doi.org/10.1016/j.peptides.2005.06.019.

[160] Del Ry S, Passino C, Maltinti M, Emdin M, Giannessi D. C-type natriuretic peptide plasma levels increase in patients with chronic heart failure as a function of clinical severity. Eur J Heart Fail 2005;7(7):1145–8. Available from: https://doi.org/10.1016/j.ejheart.2004.12.009.

[161] Chun TH, Itoh H, Ogawa Y, et al. Shear stress augments expression of C-type natriuretic peptide and adrenomedullin. Hypertension 1997;29(6):1296–302. Available from: https://doi.org/10.1161/01.hyp.29.6.1296.

[162] Chusho H, Tamura N, Ogawa Y, et al. Dwarfism and early death in mice lacking C-type natriuretic peptide. Proc Natl Acad Sci USA 2001;98(7):4016–21. Available from: https://doi.org/10.1073/pnas.071389098.

[163] Moyes AJ, Hobbs AJ. C-type natriuretic peptide: a multifaceted paracrine regulator in the heart and vasculature. Int J Mol Sci 2019;20(9):2281. Available from: https://doi.org/10.3390/ijms20092281.

[164] Savarirayan R, Irving M, Bacino CA, et al. C-type natriuretic peptide analogue therapy in children with achondroplasia. N Engl J Med 2019;381(1):25–35. Available from: https://doi.org/10.1056/NEJMoa1813446.

[165] Schweitz H, Vigne P, Moinier D, Frelin C, Lazdunski M. A new member of the natriuretic peptide family is present in the venom of the green mamba (*Dendroaspis angusticeps*). J Biol Chem 1992;267(20):13928–32.

[166] Lisy O, Lainchbury JG, Leskinen H, Burnett Jr. JC. Therapeutic actions of a new synthetic vasoactive and natriuretic peptide, dendroaspis natriuretic peptide, in experimental severe congestive heart failure. Hypertension 2001;37(4):1089–94. Available from: https://doi.org/10.1161/01.hyp.37.4.1089.

[167] Handbook of hormones comparative endocrinology for basic and clinical research; 2016. p. 28–29, e2C-1.

[168] Schulz-Knappe P, Forssmann K, Herbst F, Hock D, Pipkorn R, Forssmann WG. Isolation and structural analysis of "urodilatin," a new peptide of the cardiodilatin-(ANP)-family, extracted from human urine. Klin Wochenschr 1988;66(17):752–9. Available from: https://doi.org/10.1007/BF01726570.

[169] Hirsch JR, Meyer M, Forssmann WG. ANP and urodilatin: who is who in the kidney. Eur J Med Res 2006;11(10):447–54.

[170] Packer M, O'Connor C, McMurray JJV, et al. Effect of ularitide on cardiovascular mortality in acute heart failure. N Engl J Med 2017;376(20):1956–64. Available from: https://doi.org/10.1056/NEJMoa1601895.

[171] Wei CM, Kim CH, Miller VM, Burnett Jr. JC. Vasonatrin peptide: a unique synthetic natriuretic and vasorelaxing peptide. J Clin Invest 1993;92(4):2048–52. Available from: https://doi.org/10.1172/JCI116800.

[172] Shi Z, Fu F, Yu L, et al. Vasonatrin peptide attenuates myocardial ischemia-reperfusion injury in diabetic rats and underlying mechanisms. Am J Physiol Heart Circ Physiol 2015;308(4):H281–90. Available from: https://doi.org/10.1152/ajpheart.00666.2014.

[173] Zhao HK, Chen BY, Chang R, et al. Vasonatrin peptide, a novel protector of dopaminergic neurons against the injuries induced by n-methyl-4-phenylpyridiniums. Peptides 2013;49:117–22. Available from: https://doi.org/10.1016/j.peptides.2013.09.005.

[174] Wang SP, Lan ZY, Xia W, et al. The effects of vasonatrin peptide on random pattern skin flap survival. Ann Plast Surg 2014;72(1):94–9. Available from: https://doi.org/10.1097/SAP.0b013e318255a3eb.

[175] Ichiki T, Dzhoyashvili N, Burnett Jr. JC. Natriuretic peptide based therapeutics for heart failure: cenderitide: a novel first-in-class designer natriuretic peptide. Int J Cardiol 2019;281:166–71. Available from: https://doi.org/10.1016/j.ijcard.2018.06.002.

[176] Dickey DM, Potter LR. Dendroaspis natriuretic peptide and the designer natriuretic peptide, CD-NP, are resistant to proteolytic inactivation. J Mol Cell Cardiol 2011;51(1):67–71. Available from: https://doi.org/10.1016/j.yjmcc.2011.03.013.
[177] Kawakami R, Lee CYW, Scott C, et al. A human study to evaluate safety, tolerability, and cyclic GMP activating properties of cenderitide in subjects with stable chronic heart failure. Clin Pharmacol Ther 2018;104(3):546–52. Available from: https://doi.org/10.1002/cpt.974.
[178] Ng XW, Huang Y, Chen HH, Burnett Jr JC, Boey FY, Venkatraman SS. Cenderitide-eluting film for potential cardiac patch applications. PLoS One 2013;8(7):e68346. Available from: https://doi.org/10.1371/journal.pone.0068346.
[179] Zakeri R, Burnett JC. Designer natriuretic peptides: a vision for the future of heart failure therapeutics. Can J Physiol Pharmacol 2011;89(8):593–601. Available from: https://doi.org/10.1139/y11-048.
[180] From AH. Do engineered natriuretic peptides have greater therapeutic potential than do native peptides? Cardiovasc Res 2010;88(3):391–2. Available from: https://doi.org/10.1093/cvr/cvq319.
[181] McKie PM, Ichiki T, Burnett Jr. JC. M-atrial natriuretic peptide: a novel antihypertensive protein therapy. Curr Hypertens Rep 2012;14(1):62–9. Available from: https://doi.org/10.1007/s11906-011-0244-5.
[182] Meems LMG, Burnett Jr. JC. Innovative therapeutics: designer natriuretic peptides. JACC Basic Transl Sci 2016;1(7):557–67. Available from: https://doi.org/10.1016/j.jacbts.2016.10.001.
[183] Barbouche R, Marrakchi N, Mansuelle P, et al. Novel anti-platelet aggregation polypeptides from *Vipera lebetina* venom: isolation and characterization. FEBS Lett 1996;392(1):6–10. Available from: https://doi.org/10.1016/0014-5793(96)00774-0.
[184] Tourki B, Matéo P, Morand J, et al. Lebetin 2, a snake venom-derived natriuretic peptide, attenuates acute myocardial ischemic injury through the modulation of mitochondrial permeability transition pore at the time of reperfusion. PLoS One 2016;11(9):e0162632. Available from: https://doi.org/10.1371/journal.pone.0162632.
[185] Bern HA, Lederis K. A reference preparation for the study of active substances in the caudal neurosecretory system of teleosts. J Endocrinol 1969;45(1):xi–xii.
[186] Vaudry H, Leprince J, Chatenet D, et al. International union of basic and clinical pharmacology. XCII. Urotensin II, urotensin II-related peptide, and their receptor: from structure to function. Pharmacol Rev 2015;67(1):214–58. Available from: https://doi.org/10.1124/pr.114.009480.
[187] Nothacker HP, Clark S. From heart to mind. The urotensin II system and its evolving neurophysiological role. FEBS J 2005;272(22):5694–702. Available from: https://doi.org/10.1111/j.1742-4658.2005.04983.x.
[188] Gibson A, Conyers S, Bern HA. The influence of urotensin II on calcium flux in rat aorta. J Pharm Pharmacol 1988;40(12):893–5. Available from: https://doi.org/10.1111/j.2042-7158.1988.tb06298.x.
[189] Watson AM, Lambert GW, Smith KJ, May CN. Urotensin II acts centrally to increase epinephrine and ACTH release and cause potent inotropic and chronotropic actions. Hypertension 2003;42(3):373–9. Available from: https://doi.org/10.1161/01.
[190] Ozoux ML, Briand V, Pelat M, et al. Potential therapeutic value of urotensin II receptor antagonist in chronic kidney disease and associated comorbidities. J Pharmacol Exp Ther 2020;374(1):24–37. Available from: https://doi.org/10.1124/jpet.120.265496.

Chapter 26

Noncardiovascular-derived therapeutic peptidomimetics in cardiovascular disease

Adi Litmanovich[1,2], Offir Ertracht[3], Yair Blumberg[1], Mussa Saad[1,4], Gassan Moady[5] and Nir Qvit[1]

[1]*The Azrieli Faculty of Medicine in the Galilee, Bar-Ilan University, Safed, Israel,* [2]*Galilee Medical Center, Nahariya, Israel,* [3]*The Cardiovascular Research Laboratory, Research Institute, Galilee Medical Center, Nahariya, Israel,* [4]*Cardiology Department, Ziv Medical Center, Safed, Israel,* [5]*Cardiology Unit, Galilee Medical Center, Nahariya, Israel*

26.1 Introduction

Peptides constitute one of the most promising platforms for drug development due to their biocompatibility, chemical diversity, and resemblance to proteins. More than 100 peptides with functions in the central and peripheral nervous systems, immunological processes, cardiovascular system (CVS), and intestines are known to function as neurotransmitters, neuromodulators, and hormones via receptor-mediated signal transduction. Increasing understanding of bioactive peptides led to growing interest in pharmacology and medical application of these compounds. Further, recent work has also demonstrated a variety of cardiovascular modulating properties ascribed to peptides generally considered to function primary outside the CVS. Herein we discuss the therapeutic potential and application of these peptides that primarily affect organ systems other than the CVS as therapeutic leads and treatments for cardiovascular disease (CVD).

26.2 Oxytocin and vasopressin

26.2.1 Background

Oxytocin and vasopressin are neurohormones produced in the hypothalamus and transmitted to the posterior lobe of the pituitary gland, also called the neurohypophysis. They control many biological processes, including water reabsorption in the kidneys, uterine contractions, milk ejection and sexual response [1]. Their

Peptide and Peptidomimetic Therapeutics. DOI: https://doi.org/10.1016/B978-0-12-820141-1.00013-3

hormonal activities and origin of secretion were first discovered between the years 1895 and 1915. Their structure was later elucidated by Du Vigneud in 1953 and with the development of radioactive labeling of hormones in the 1950s, their receptors were also identified [2]. Elucidation of the structure of these two peptides was a substantial push to Du Vigneud's Nobel Prize candidacy, which he was awarded two years later, in 1955 [3]. Since their discovery in 1953, a wealth of knowledge has accumulated regarding their activities. Their effects on the cardiovascular (CV) system have recently gained much attention, with the development of new CV therapeutics.

26.2.2 Genetics and translation

Both oxytocin and vasopressin are nine amino acid (AA) neuropeptides, resulting from ancient gene duplication. They are oriented in opposing transcriptional directions on adjacent loci on chromosome 20p13, with the region between them containing important enhancer elements [4]. The two mature peptides differ in only two out of their nine amino acids, and both have a disulfide ring at one end (Fig. 26.1). They are both posttranslationally modified from larger precursors termed pre-pro-oxyphysin and pre-pro-vasophysin, respectively, by cleavage of the N-terminal signal peptide. The prohormones are then proteolytically cleaved to their mature forms—oxytocin and vasopressin, together with their associated cosecreted peptides—neurophysin I associated with vasopressin and neurophysin II with oxytocin [5].

26.2.3 Distribution

Oxytocin and vasopressin are produced as preprohormones in the magnocellular neurons of the hypothalamus, from where they are transferred to the neurohypophysis directly via the hypothalamic-hypophyseal tract, a bundle of nerve processes with cell bodies in the hypothalamus that release their neurosecretions directly in the posterior hypophysis. They then undergo maturation and are stored in Herring bodies in the posterior pituitary, together with the associated neurophysins I and II. From there, the two hormones, along with the neurophysins, are released into the circulation in response to electrical stimuli received in the hypothalamus [5]. Native vasopressin in the circulation is unstable and rapidly cleared with a half-life of 5–15 min [6]. The native oxytocin plasma half-life is assessed to be 1–5 min but varies greatly between pregnant and nonpregnant women, as well during the different phases of the menstrual cycle [7,8].

26.2.4 Function

Oxytocin and vasopressin share some common properties, due to their similar structure; however, their functions at physiological levels vary greatly.

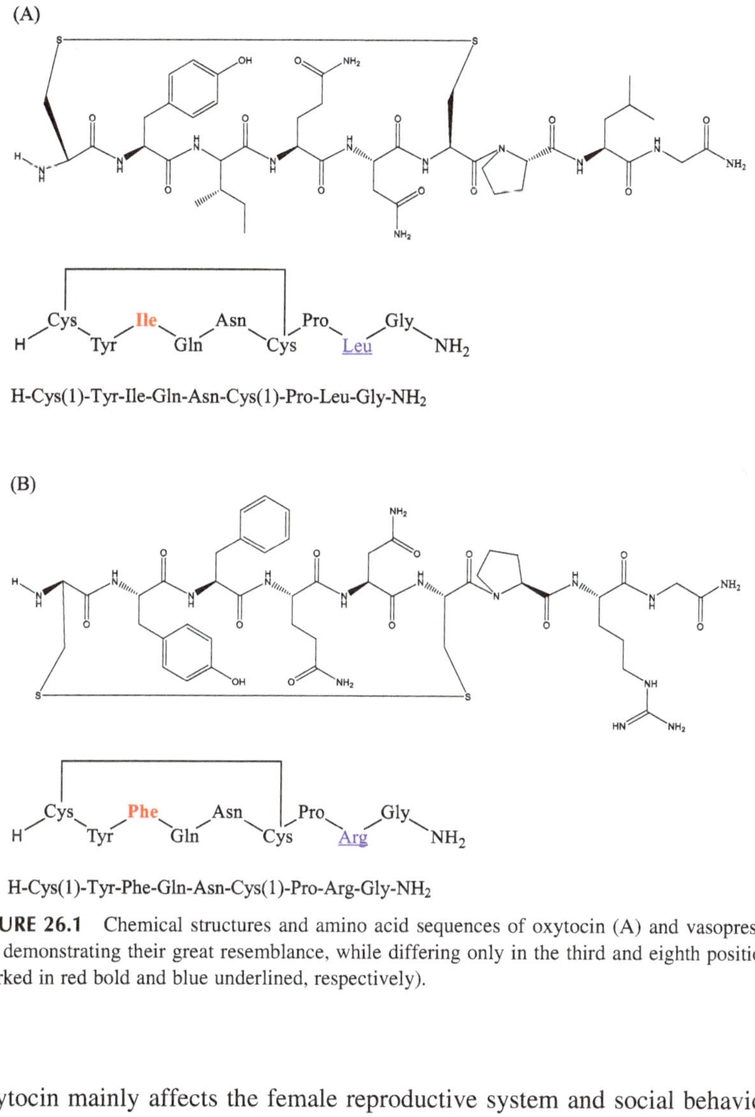

FIGURE 26.1 Chemical structures and amino acid sequences of oxytocin (A) and vasopressin (B), demonstrating their great resemblance, while differing only in the third and eighth positions (marked in red bold and blue underlined, respectively).

Oxytocin mainly affects the female reproductive system and social behavior. Oxytocin receptors are found in the uterus and ovaries of both pregnant and nonpregnant women, and oxytocin is in fact thought to be synthesized within the uterine tissue. Among its main roles are increasing peristaltic activity, basal tone and contractions of the uterus, and induction of lactation. In the male reproductive system, oxytocin is thought to be involved in spermatogenesis and to promote ejaculations. In addition, it is thought to influence carbohydrate metabolism, kidney function and the brain itself [9,10]. Vasopressin acts mainly on the collecting ducts of the kidney, where it increases the permeability of the collecting duct via translocation of aquaporins to its luminal membrane, thus enabling water reabsorption and the concentration of urine, together with a decrease in plasma osmolality. In addition, it has a paracrine effect on the anterior pituitary, which stimulates the secretion of corticotropin from corticotrophs [10].

Both oxytocin and vasopressin are thought to influence the CVS, both centrally and through direct effects on the kidneys, vasculature and the heart itself. Certain vasopressin receptors found in blood vessels grant it powerful vasoconstrictive properties, thereby increasing peripheral resistance. In some vascular beds, however, vasopressin may cause vasodilation, likely due to nonspecific activation of oxytocin receptors. Other types of receptors found in the collecting ducts of the kidneys mediate its effect on water balance regulation and on the normalization of blood osmolality and volume through water reabsorption [11,12]. The existence of direct effects of vasopressin on the heart is controversial and probably dose-dependent.

Oxytocin synthesized in the heart is thought to have cardioprotective properties. However, most of the data demonstrating such effects have been gathered from animal studies; human data are still lacking [11]. Centrally released stress-induced oxytocin and exogenous intracerebroventricular administration of oxytocin ameliorated ischemia/reperfusion (I/R) injury in rat hearts [13], inhibited myocardial injury in rats [14,15] and had cardioprotective effects in I/R and postinfraction rabbits [16,17]. In addition, oxytocin is suspected to have a protective role against coronary artery disease (CAD), as oxytocin receptor (OTR) signaling was found to be suppressed in CAD [18]. Finally, it may nonselectively bind to vasopressin receptors and induce weak vasoconstriction in peripheral arteries.

26.2.5 Mechanism of action

Vasopressin binds to three very specific membrane G protein-coupled receptors (GPCRs). Type 1 receptors (V1aR) are mainly distributed in vascular smooth muscle cells (VSMCs) and are coupled to G_q/11 protein, phospholipases C, D, and A2 and calcium influx. Type 2 receptors (V2R) are coupled to cyclic 3′-5′ -adenosine monophosphate (cAMP) production and are specifically located in the distal tubules of the kidney. Finally, a unique type 1 receptor, V1bR, is found in the anterior pituitary, the limbic system and the pancreas. Oxytocin binds to OTRs, coupled to G_q/11 protein, and vasopressin receptor V1aR [11]. Due to their similar structures, some cross-reactivity exists between vasopressin, oxytocin and their respective receptors. The amount of crosstalk between endogenous or exogenous oxytocin and vasopressin, as well as the influence of the activation of their canonical receptors, are not yet fully understood; however, they are thought to be of greater importance at supraphysiological levels, such as when administered exogenously [19]. A better understanding of this crosstalk may lead to improved pharmacological outcomes.

26.2.6 Therapeutics

Recent discoveries regarding the roles of oxytocin and vasopressin in cardiovascular homeostasis have advanced the development of new drugs for the treatment of different CV diseases. Due to the critical influence of vasopressin on osmolarity regulation and CV homeostasis, both specific agonists and antagonists have emerged for the treatment of CV-related diseases.

Conivaptan was the first vasopressin antagonist to be discovered and marketed, with a 10:1 selectivity for V1aR, and was shown to induce acute hemodynamic improvements with aquaretic effects without causing hyponatremia in both animal and human studies. However, it suffered from low bioavailability and lacked a long-term effect [12]. Tolvaptan, an oral V2R antagonist, was assessed in a large-scale clinical trial called EVEREST (NCT00071331), in patients suffering from chronic heart failure (HF). Patients with worsening symptoms of congestive HF and left ventricular (LV) ejection fraction (LVEF) $\leq 40\%$ tolerated tolvaptan and had favorable outcomes in weight loss, edema score and general HF symptoms during hospitalization. However, in the long-term, tolvaptan was neither beneficial in improving morbidity or mortality nor was it harmful [12,20]. In contrast, vasopressin agonists, terlipressin and the more recently developed selepressin, and vasopressin analogs have been associated with some hemodynamic improvement in states of refractory septic shock and vasoplegic shock after cardiac surgery [21,22].

There are currently no oxytocin-based drugs, analogs, agonists, or antagonists in regular clinical use for treating CV diseases, and the existing drugs are mainly used either to induce or delay labor. Intranasal oxytocin delivery was found to have an antiarrhythmic effect in humans [23] as well as a cardioprotective effect in CAD patients [24]. However, at high doses, oxytocin drugs may cause adverse CV effects, which has limited their development [25–27].

26.3 Calcitonin gene-related protein

26.3.1 Background

Calcitonin gene-related protein (CGRP) was thoroughly described in 1982 [28]. Interestingly, its structure was deciphered as a product of alternative splicing process [28]. Specifically, this 37 AA peptide was used as an example for the post-mRNA transcription splicing process, which translates a single mRNA into two distinct peptides, the first being calcitonin, and the second being CGRP [28]. While calcitonin mRNA was shown to be largely confined to the thyroid tissue, CGRP mRNA is considered to predominate throughout the sensory nervous system [28] (Fig. 26.2).

(A)

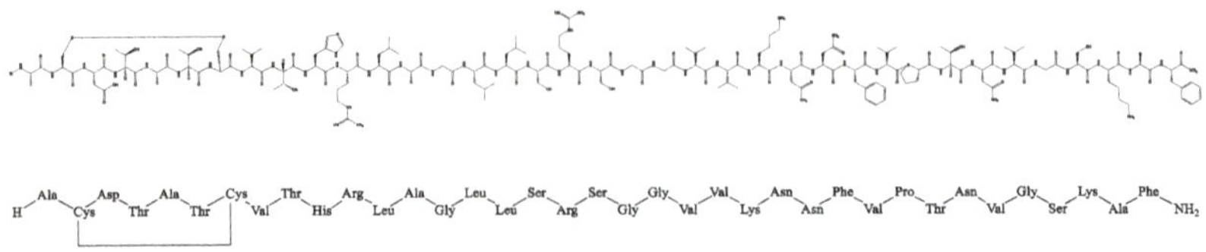

H-Ala-Cys(1)-Asp-Thr-Ala-Thr-Cys(1)-Val-Thr-His-Arg-Leu-Ala-Gly-Leu-Leu-Ser-Arg-Ser-Gly-Gly-Val-Val-Lys-Asn-Asn-Phe-Val-Pro-Thr-Asn-Val-Gly-Ser-Lys-Ala-Phe-NH_2

(B)

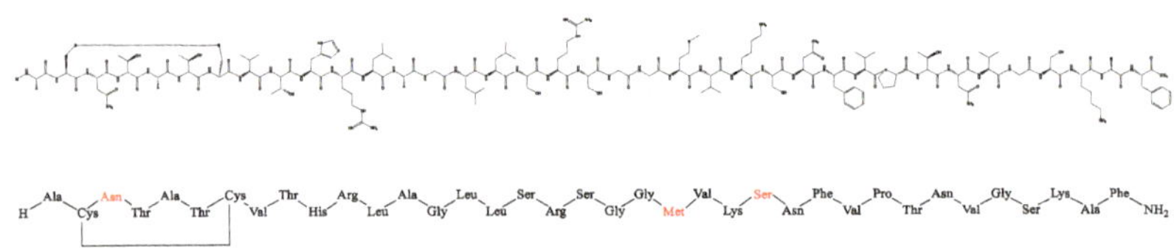

H-Ala-Cys(1)-Asn-Thr-Ala-Thr-Cys(1)-Val-Thr-His-Arg-Leu-Ala-Gly-Leu-Leu-Ser-Arg-Ser-Gly-Gly-Met-Val-Lys-Ser-Asn-Phe-Val-Pro-Thr-Asn-Val-Gly-Ser-Lys-Ala-Phe-NH_2

FIGURE 26.2 Chemical structures and amino acid sequences of α-calcitonin gene-related protein (A) and β-calcitonin gene-related protein (B), demonstrating their great resemblance, with differences in three amino acids only (marked in red underlined).

26.3.2 Genetics and translation

Calcitonin mRNA comprises exons 1, 2, 3, and 4 of the calcitonin gene, whereas CGRP mRNA comprises exons 1, 2, 3, 5, and 6. Two distinct isoforms of CGRP, termed αCGRP and βCGRP, have been identified. Interestingly, αCGRP is formed from the alternative splicing of mRNA transcribed from the calcitonin gene itself, which is located on chromosome 11. The β isoform is transcribed from its own separate gene which is also located on chromosome 11, that is, not through the alternative splicing process [29]. Both isoforms are structurally similar, with the β isoform differing from the alpha isoform by only 3 amino acid residues in humans and single amino acid in rodents. Both isoforms display comparable biological activity and primarily differ in their tissue distribution.

26.3.3 Distribution

CGRP is primarily located in unmyelinated, small-diameter sensory C and thinly myelinated A delta (Aδ) fibers, which lay adjacent to the vasculature, and primarily on the arterial side of the circulatory system. αCGRP is most commonly found within peripheral sensory neurons, whereas βCGRP seems to be more restricted to nerves of the enteric plexuses.

CGRP levels in the plasma are in the pico-molar range, which is the result of overspill from its site of release, that is, the vascular nerve endings [30]. Other studies have shown CGRP presence in progenitor and mature endothelial cells and human lymphocytes [31]. CGRP half-life is approximately 18 min [32]. While its elimination pathway is not fully understood, endogenous cleavage sites within the primary sequence of the peptide have been found. Thus, it is presumed that metalloproteases, insulin-degrading enzymes and endothelin converting enzymes may degrade CGRP. Alternatively, it has also been suggested that CGRP undergoes re-uptake into sympathetic nerves [33].

26.3.4 Function and pathological implications

CGRP is a potent vasodilator. Brain et al. showed that subcutaneous CGRP administration induced skin erythema, due to an increase in local blood flow [34]. Intravenous (IV) administration caused hypotension in rodents, activation of the sympathetic system and subsequent positive cardiac inotropic and chronotropic effects [30]. At the same time, several other studies failed to show CGRP as a potent systemic hemodynamic effector; however, since it acts at the microcirculation level, it may have effects on neovascularization and angiogenesis [30,35,36].

One of the most important effects of CGRP is its ability to relax cerebral arteries, which leads to the development of migraine cluster headaches and neuralgia-sourced headaches. Studies have shown that CGRP concentrations are high in the extracerebral tissue of migraine patients as compared to normal controls. This finding is in accordance with the theory that migraines arise following the relaxation of extracerebral blood vessels which supply blood to the dura mater.

26.3.5 Mechanism of action

CGRP is released from nerve endings through exocytosis. Once released, its main effect is local. Two distinct CGRP receptors are known, both of which belong to the GPCRs family, and are termed "calcitonin-like receptor" (CLR) [37]. CLR is activated after it binds CGRP and the receptor-activity modifying protein 1 (RAMP1) and subsequently dimerizes. Receptor activation is associated with increased cAMP levels and activation of protein kinase A (PKA), which is followed by the opening of the ATP-sensitive potassium and L-type calcium channels. Additionally, the complex through the cAMP response element-binding (CREB) protein affects transcription of several genes which are responsible for bone homeostasis (Jdp2) and angiogenesis (VEGF, CD31) [38]. Termination of its activity is dependent upon desensitization by the agonist and receptor internalization [39].

CLR and RAMP1 are present in various cells and tissues, including VSMC, endothelial cells, cardiac myocytes, some lymphocytes, and perivascular neuron bodies. Responses can differ by tissue and cell type; for example, VSMCs can develop hyperpolarization and relaxation of their tone, thus inducing hypotension, while in endothelial cells, CGRP activates endothelial nitric oxide (NO) synthase and NO release, which further enhances VSMC relaxation and subsequent vasodilatation [30].

26.3.6 Therapeutics

Studies assessed CGRP's cardioprotective effect against ischemia, and while some showed its efficacy against ischemia and postinfarction apoptosis and in improving cardiac physiology; however, to date, it is still not in clinical use [40]. CGRP inhibitors and receptor monoclonal antibodies showed significant antimigraine efficacy. Extensive research led to the development of two CGRP inhibitors, that is, the IV administered olcegeplant and the per os (PO) telcagepant [41]. Additionally, CGRP receptor antibodies, for example, erenumab, eptinezumab, galcanezumab, and fremanezumab, have been approved for clinical use [42].

Raynaud's syndrome is a peripheral nerve phenomenon, presenting as inappropriate vasoconstriction and pain of the cutaneous microcirculation. Raynaud's patients suffer from high vasopressor, for example, endothelin, levels and from low levels of vasodilators, such as CGRP. CGRP infusion improved their symptoms; thus, CGRP or a CGRP agonist may be beneficial for this patient population [43]. Local involvement of CGRP in hypertension was also suggested by various in vivo studies. Specifically, CGRP may attenuate endothelial progenitor cells, endothelial cells and VSMCs long-term deleterious effects of hypertension, such as vascular remodeling and endothelial dysfunction [31,44].

26.4 Amylin

26.4.1 Background

Human amylin was probably first identified early as 1901 when it was described as hyaline deposits found in the pancreatic islets of patients with type 2 diabetes [45]. It was later independently rediscovered in the pancreas by two research groups, Copper et al. and Westermark et al., who named it diabetes-associated peptide or amyloid polypeptide (IAPP), respectively [46,47]. Both groups identified the 37 AA amylin peptide, with about 46% identity to CGRP, a weaker identity to the A chain of insulin, and some similarity to insulin [45]. Both amylin and CGRP were described as potent inhibitors of insulin-stimulated glycogen synthesis in skeletal muscle in vitro [46].

Due to its structure and some of its functions, amylin has been linked to CGRP; both have been found to bear some resemblance to the insulin α-chain and are thus collectively related to the amylin superfamily. The members of this family, together with insulin, insulin growth factors (IGFs) and relaxins, form the insulin-related family [46] (Fig. 26.3).

26.4.2 Genetics and translation

The amylin gene is located on chromosome 12, as is the gene encoding IGF-1; genes of other family members are located on the short arm of chromosome 11 [48–53]. Thus, it is believed that the two chromosomes developed from a single ancestor [54]. CGRP and amylin genes are very similar, suggesting a common ancestor gene for both, as well. Interestingly, the CGRP-2 gene lies in proximity to the homologous region of the amylin gene on chromosome 11 [55].

The amylin gene is composed of three exons, of which the last two encode the full prepropeptide. The human 89 AA prepropeptide undergoes initial cleavage at the endoplasmic reticulum, yielding the 67 AA proamylin, while its final cleavage to 37 AA amylin, posttranslational fold, disulfide (S-S) bridge formation, and C-terminus amidation all occur in the secretory vesicles [53,56]. Proamylin cleavage involves prohormone convertase 2 (PC2) and prohormone convertase 1/3 (PC1/3) and carboxypeptidase E (CPE).

26.4.3 Distribution

Amylin is synthesized at high levels in islet β-cells and secreted from this source [46,47,53]. Out of nine normal tissues examined, amylin mRNA was only detectable in islets, suggesting that amylin, like insulin, is specifically produced in the β-cells of the islets of Langerhans [57]. Additional minor

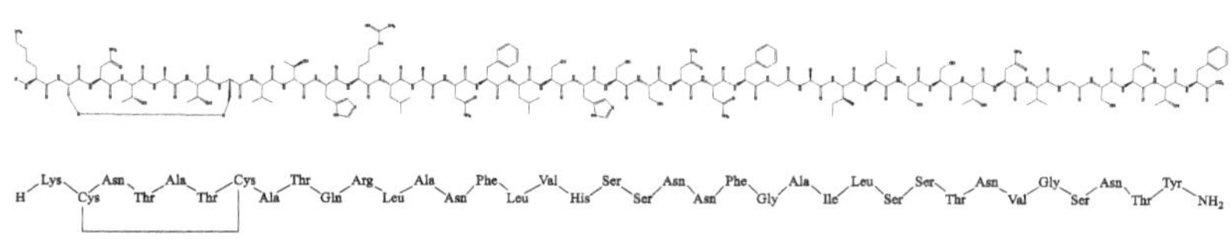

H-Lys-Cys-Asn-Thr-Ala-Thr-Cys-Ala-Thr-Gln-Arg-Leu-Ala-Asn-Phe-Leu-Val-His-Ser-Ser-

Asn-Asn-Phe-Gly-Ala-Ile-Leu-Ser-Ser-Thr-Asn-Val-Gly-Ser-Asn-Thr-Tyr-NH_2

FIGURE 26.3 Chemical structure and amino acid sequence of amylin.

amylin sources include neuroendocrine cells of the mammalian stomach, the dorsal root ganglion [58], and the developing kidney [59].

Amylin secretion tracks insulin secretion, with insulin:amylin secretion ratios of 15:1. Amylin is secreted following the same signals that induce insulin secretion, that is, glucose, arginine, and fatty acids. Its plasma level is in the pM scale of order. Amylin is metabolized rather quickly following its secretion (estimated $t_{1/2}$ of 13 min) in the kidneys by proteolytic degradation [60].

26.4.4 Function and pathological implications

Amylin regulates glucose homeostasis by inhibiting gastric emptying, inhibiting the release of the counter-regulatory hormone glucagon and inducing meal-ending satiety [60]. Physiologically, amylin functions as a glucoregulatory and satiety-inducing hormone, which is protective against postprandial spikes in blood glucose and overeating [60,61]. The main site of amylin action is the brain; after a meal, the pancreas-secreted amylin circulates in the blood to activate specific receptors in the brainstem. This results in the suppression of glucagon release from the pancreas, a reduction in food intake, and gastric emptying [62].

In type 2 diabetes mellitus (T2DM), amylin was associated with the downregulation of insulin-stimulated glycogen synthesis and glucose uptake in isolated, stripped soleus rat muscle [63,64]. Additionally, it was suggested that amylin participates in an amyloid deposition in the T2DM pancreas in humans, dogs and macaque, but not in cats or rats [47,65]. Even though the distinct process remains unknown, structure–function and comparative physiology data indicate amylin participation in the process.

26.4.5 Mechanism of action

Amylin action is mediated by two B-type G protein-coupled receptors (B GPCRs). Upon amylin association to B GPCRs, the receptor heterodimerizes with one of three RAMPs, to form the active complex. Two types of B GPCRs and three types of RAMPs dictate functionally six (2×3) different amylin receptors [62]. The amylin-GPCR-RAMP complex activates an intracellular Gs-coupled pathway, which results in increased intracellular cAMP levels. Alternatively, activation of the Gq pathway results in acute phosphorylation of extracellular signal-regulated kinase 1 and 2 (ERK1/2). Further studies will be necessary to characterize the signals triggered by amylin [60].

26.4.6 Therapeutics

Amylin, its analog pramlinitide, and several antagonists have been studied as potential therapeutic agents [60]. In preclinical and clinical studies, amylin was shown to downregulate insulin, preserve glucagonostatis, reduce diabetes-associated gastric hypermotility, enhance emptying, and reduce meal size.

Researchers have studied the effects of amylin on the CVS. In several preclinical studies, amylin exhibited a hypotensive and negative inotropic effect [66]. However, these effects were found to be the result of the structural resemblance of amylin to CGRP. Since physiological plasma amylin concentrations are approximately three magnitudes of order lower than the level needed to activate CGRP receptors, they concluded that amylin has a very small effect on this system [60].

The clinical effects of amylin and its agonists on diabetes, food intake, and weight loss specifically in morbidly obese patients have also been studied extensively. Several clinical studies indicated that pramlintide, the most common modified human amylin that lacks amyloidogenic properties, in combination with insulin provides a more physiologically balanced therapeutic effect in diabetes patients [60]. Subsequently, pramlintide, was approved for clinical treatment of diabetes patients [60].

Several clinical studies demonstrated that pramlintide treatment leads to meal size reduction, increased satiety scores, and reduced total food intake. Collectively, these studies suggest that obese subjects treated with pramlintide alone or as an adjunct to lifestyle intervention can achieve and maintain weight loss [60].

Additional studies administered oral pramlintide combined with two small-molecule anorexigenic agents (phentermine or sibutramineoral) to morbidly obese patients in efforts to improve treatment efficacy and prolong the dietary effect. Nevertheless, due to several adverse cardiovascular effects, each of these treatments was taken off the market [60].

26.5 Vasoactive intestinal peptide

26.5.1 Background

Vasoactive intestinal peptide (VIP) was first isolated from the porcine duodenum in 1970 by Sami I. Said and Viktor Mutt [67]. Soon after, it was identified in a wide variety of tissues, including the brain, heart and central and peripheral nervous systems, where it acts both as a neurotransmitter and a hormone and is thus categorized today as a neuropeptide [68]. VIP has been found to affect many CV processes and holds immense potential as a novel therapeutic agent for CVDs (Fig. 26.4).

26.5.2 Genetics and translation

VIP is a member of a peptide family which includes, among others, glucagon, secretin, and growth hormone-releasing factors. The VIP gene consists

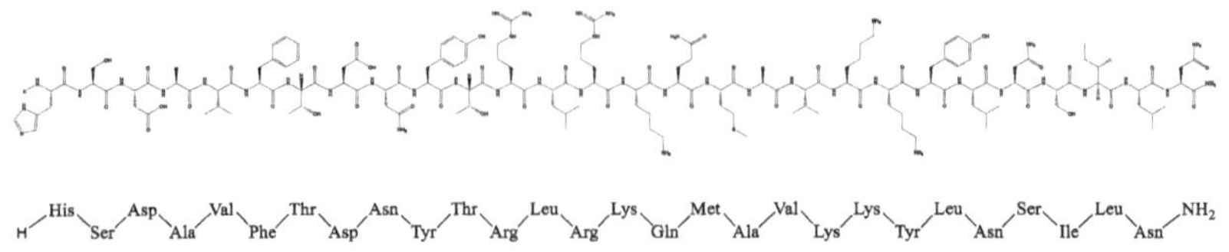

H-His-Ser-Asp-Ala-Val-Phe-Thr-Asp-Asn-Tyr-Thr-Arg-Leu-Arg-Lys-Gln-Met-Ala-Val-Lys-Lys-Tyr-Leu-Asn-Ser-Ile-Leu-Asn-NH_2

FIGURE 26.4 Chemical structure and amino acid sequence of vasoactive intestinal peptide.

of seven exons, encoding seven different functional domains in the VIP mRNA [68]. It is located on chromosome 6q24, while the genes encoding its two high-affinity receptors, vasoactive intestinal peptide receptor 1 (VPAC1) and VPAC2, are located on chromosomes 3p22 and 7q36.3, respectively. The mature form of VIP, consisting of 28 amino acids, is proteolytically cleaved from the 170 AA pre-pro-VIP, together with other cleavage products bearing little to no biological activity [69]. Due to its metabolism in the lungs, liver and possibly the kidneys, VIP is cleared rapidly from the circulation with a half-life of approximately one minute [70].

26.5.3 Distribution

VIP transcripts are primarily found in neurons or neuron-related cells in different areas of the central and peripheral nervous systems [71]. VIP was found to innervate the thymus, the spleen and lymph nodes among other tissues [72,73]. In the heart, VIP is secreted, together with acetylcholine, from postganglionic vagal VIP-containing neurons which are mainly found throughout the atria and near nodal cells [74]. VIP exerts its effects by binding to VPAC1 and VPAC2 with equally high affinity, or to peptide receptor 1 (PAC1), with much lower affinity. These receptors are located in numerous areas of the central nervous system [75]. Recently, VIP and its VPAC2 receptor were found to be highly expressed in the hypothalamic suprachiasmatic nucleus as well [76].

26.5.4 Function

The biological effects of infused VIP were first investigated in anesthetized dogs, where it was found to increase blood flow and total cardiac output (29% increase in stroke volume and 8% increase in heart rate), and to decrease the mean systemic arterial blood pressure [67]. In addition, it augmented respiratory minute volume by increasing both frequency and tidal volume and caused hyperglycemia soon after injection. Due to its inactivation in the liver, VIP was first thought to mainly affect the gastrointestinal tract (GIT) [67]. However, in the fifty years since its discovery, VIP was found to have numerous physiological functions, owing to the wide distribution of its high-affinity receptors throughout the body. It acts as a neurotransmitter, neuromodulator, and secretagogue, driving a myriad of biological processes. VIP affects secretion and motility in the GIT [77] and is suggested to play multiple roles in the respiratory system, mostly in pulmonary vascular relaxation and bronchodilation [78] and the immune system [79,80]. Its powerful systemic vasodilatory properties impact many organs, including the brain and the central nervous system [81]. In addition, in recent years, VIP was found to play an important role in regulating circadian rhythms, including indirectly affecting heart rate [76,82].

The effects of VIP on the CV system are vast and have been thoroughly studied in both animals and humans. VIP strongly vasodilates coronary vessels, with a greater effect on arteries than on veins, acts as a primary positive inotrope and reduces systemic mean arterial pressure. Due to the presence of VIP-ergic fibers near the sinus and atrioventricular nodes, VIP directly increases heart rate, thus the opposing chronotropic effects of acetylcholine [83]. In addition, VIP infusion has been suggested to reverse existing myocardial fibrosis in rats [84], ameliorate myocarditis and atherosclerosis in mice [85], and play a protective role in cardiomyopathy, HF and pulmonary arterial hypertension (PAH) [86,87].

26.5.5 Mechanism of action

VIP mainly functions through receptor-mediated systems, which activate signal transduction pathways. The VIP receptors are coupled to G-proteins, which activate adenylate cyclase, which in turn catalyzes cyclic adenosine monophosphate (cAMP) synthesis [88]. Downstream signaling then varies and includes the activation of different signaling pathways, such as PKA and mitogen-activated protein kinases (MAPK) signaling, exchange proteins directly activated by cAMP (EPAC), and the Ras-extracellular regulated kinases (ERK)-phosphoinositide 3-kinase (PI3K) pathway [89–91]. In addition, but to a lesser extent, VIP can act through increasing Ca^{2+} levels either through $G_{i/o}$ or G_q proteins [90] or through the phospholipase C (PLC)-inositol phosphate (IP3) pathway [92]. Ca^{2+} ions together with diacylglycerol (DAG) then activate protein kinase C (PKC), which then phosphorylates other kinases [93].

26.5.6 Therapeutics

Due to its multiple effects on the CV system, VIP is an attractive candidate novel therapeutic for CVDs, including HF, myocarditis, myocardial fibrosis, atherosclerosis, and pulmonary hypertension. However, its rapid degradation, short circulatory half-life and wide distribution of its receptors pose challenges in implementing its regular use in the clinic. The development of novel, improved delivery systems that may inhibit VIP degradation or enhance its selectivity will be necessary to benefit from its potential as a therapeutic agent.

Aviptadil is a synthetic VIP analog approved for the treatment of erectile dysfunction. When administered as a single inhaled dose to 20 patients with PAH during right heart catheterization, aviptadil was well tolerated with a modest, short-lived effect on oxygenation, pulmonary vasodilatation and right heart strain [94]. In a study performed in a rat model of PAH, aviptadil was given as a solo agent or in combination with bosentan, a dual endothelin receptor antagonist. Aviptadil alone was found to prevent and reverse PAH more efficiently than bosentan, but less efficiently compared with the combination of the two agents, likely due to a synergistic effect [95]. CPT-SSM-VIP (sterically stabilized micelles loaded with camptothecin and surface-modified with VIP) is a biocompatible, biodegradable, targeted phospholipid-based nanocarrier. When injected subcutaneously into mice with collagen-induced arthritis (CIA), it was found to abrogate joint inflammation without systemic toxicity [96]. Although only tested in a rheumatoid arthritis model, CPT-SSM-VIP may be beneficial in CVD as well. Other stable analogs of VIP have been developed and tested mainly in animal studies, but so far none have been approved for clinical use [97–99]. Lastly, VIP gene therapy has shown encouraging results in mouse models for CIA [100], experimental autoimmune encephalomyelitis and cecal ligation and puncture sepsis [101].

26.6 Glucagon-like peptide-1

26.6.1 Background

Glucagon-like peptide-1 (GLP-1) is a member of the incretin system which stimulates the release of insulin from pancreatic β cells upon elevation of glucose concentrations. [102]. First described in the 1960s, this effect was termed "the incretin effect" and was attributed to GLP-1 and glucose-dependent insulinotropic polypeptide (GIP) [103]. In the past years, GLP-1 analogs have been used for the management of type 2 diabetes (T2D) due to their incretin effect, but growing evidence suggests that GLP-1 may also play an important cardioprotective role, as CV events are a major cause of morbidity among diabetic patients [104] (Fig. 26.5).

26.6.2 Genetics and translation

Proglucagon is expressed in pancreatic α-cells and intestinal endocrine L cells and then trafficked to various tissues, where it is processed to glucagon, GLP-1, GLP-2, glicentin or oxyntomodulin, depending on the specific PCs expressed in the tissue. GLP-1 is specifically cleaved from proglucagon in

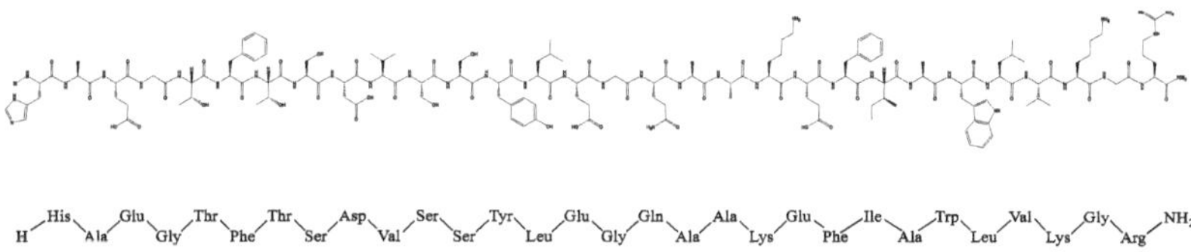

H-His-Ala-Glu-Gly-Thr-Phe-Thr-Ser-Asp-Val-Ser-Ser-Tyr-Leu-Glu-Gly-Gln-Ala-Ala-Lys-Glu-Phe-Ile-Ala-Trp-Leu-Val-Lys-Gly-Arg-NH_2

FIGURE 26.5 Chemical structure and amino acid sequence of glucagon-like peptide-1.

the gut by PC1 into a 30 AA peptide, after which it undergoes additional alterations at both its amino- and carboxyl-terminus. Upon glucose intake and neuroendocrine input, it is stored and released from secretory granules within L cells to the circulation [105].

26.6.3 Distribution

GLP-1 is secreted from intestinal L cells within minutes of food ingestion. GLP-1 receptor (GLP-1R) is a Gs protein-coupled receptor expressed in many tissues, with the highest expression in the lungs and pancreas and lower expression in the stomach, intestine, kidney, heart, and brain [106]. The circulation time of GLP-1 is very brief due to its extremely short (2 min) half-life and its rapid renal clearance [107]. GLP-1 concentrations were found to be highest in the ileum and throughout the small intestine. During fasting, GLP-1 plasma concentrations are low, but still measurable, and rise significantly postprandially [108].

26.6.4 Function

GLP-1 is an intestinal hormone of great importance in glucose metabolism, manifested by stimulation of insulin secretion and inhibition of glucagon release. In addition, it delays gastric emptying, thus suppressing appetite, decreases absorption of dietary nutrients and improves lipid metabolism [109]. GLP-1 affects the CVS as well, albeit, to a lesser extent. It was previously shown in rats and calves that the addition of exogenous IV GLP-1 increases arterial systolic and diastolic blood pressure and heart rate in a dose-dependent matter, thus exhibiting positive inotropic and chronotropic activity [110–112]. In addition, mice lacking the GLP-1 receptor (GLP-1$R^{-/-}$ mice) had lower age-dependent heart rates, elevated LV end-diastolic (LVED) pressure and increased ventricular rate [113]. In a more recent study performed in an angiotensin II–induced arterial hypertension mouse model, the GLP-1 analog liraglutide was found to reduce blood pressure and to provide endothelial protection by reducing vascular inflammation and oxidative stress [114]. Another study in rodents reported on the protective cardiac role from myocardial infarction (MI) upon GLP-1 administration together with valine pyrrolidide, an inhibitor of its breakdown, during an infarction challenge in the in vitro rat heart and an in vivo setting [115]. In humans, incretin-based therapy is known to lower systolic blood pressure [116]. GLP-1 administration was also found to decrease aldosterone levels in healthy volunteers, suggesting its influence on the renin-angiotensin-aldosterone system (RAAS) [117]. Growing evidence suggests that GLP-1 may also protect against endothelial dysfunction, dyslipidemia, atherosclerosis, and HF [108,118].

26.6.5 Mechanism of action

The main physiological effect of GLP-1 is its glucose-dependent insulinotropic activity triggered by the interaction between GLP-1 and its receptor located in β cells in the pancreas. However, GLP-1 has many other functions, including inhibition of glucagon secretion, gastrointestinal secretion and motility, and cardiovascular effects [108]. The mechanism of action of GLP-1 in the heart has not yet been fully elucidated; however, experimental and clinical observations, together with the discovery of GLP-1 receptors in the heart [108], suggest its great clinical potential in managing CVD.

26.6.6 Therapeutics

In T2D, the incretin effect is severely reduced. Currently, GLP-1 receptor agonists (GLP-1Ras) are used mainly for glycemic control in diabetes patients, where pharmacological dosages of GLP-1Ras restore insulin secretion to levels approaching normal and inhibit glucagon secretion, thereby normalizing fasting plasma glucose (FPG) levels in the short term [119,120]. GLP-1 may have additional benefits, such as promotion of weight loss and reduced cardiovascular risk [102]. There are currently many GLP-1 analogs approved for the treatment of diabetes including exenatide, liraglutide, exenatide ER (extended-release), albiglutide, lixisenatide, dulaglutide, semaglutide, efpeglenatide, itca 650 [121], and many more in preclinical and early phases of development [122]. To date, GLP-1R agonists are not yet widely used by cardiologists, but by October 2019 a total of seven cardiovascular trials were conducted covering together a total of over 56,000 T2D patients, demonstrating that liraglutide, semaglutide, albiglutide, lixisenatide, exenatide, and dulaglutide impart favorable cardiovascular effects, including a decrease in atherosclerosis-mediated events, a decrease in fatal/nonfatal stroke events, and reduction in all-cause mortality, CV-related deaths, and HF-associated hospitalizations [123–130]. Additionally, in patients suffering from acute MI, administration of a GLP-1 receptor agonist was shown to reduce infarct size and improve LV ejection fraction (LVEF) [131]. Due to this growing body of evidence, current guidelines recommend the addition of GLP-1RA to treatment protocols for T2D patients with a high risk of or established atherosclerotic CVD [104].

26.7 Ghrelin

26.7.1 Background

Ghrelin is a 28 AA peptide produced by the stomach and is the natural growth hormone secretagogue (GHS) receptor (GHSR) ligand. Two major molecular forms of ghrelin are known: des-acyl ghrelin (DAG) and acyl-modified ghrelin. Octanoylated ghrelin is generated by the octanoylation of serine in position 3 [132] (Fig. 26.6).

26.7.2 Genetics and translation

Ghrelin, growth hormone secretagogue receptor ligand (GHRL), is a protein-coding gene, located at the short (p) arm of chromosome 3, at position 25.3. It encodes the ghrelin-obestatin preproprotein, which is cleaved to yield ghrelin and obestatin. The human ghrelin gene comprises five exons, with the first exon encoding the 5′-untranslated region. There are two different transcriptional initiation sites in the ghrelin gene, resulting in two distinct mRNA transcripts (transcript-A and transcript-B). Transcript-A, a splicing product from exon 2 to exon 4, is the main form of human ghrelin mRNA in vivo. This mRNA is translated into the 117 AA ghrelin precursor (pre-proghrelin). Its subsequent protease cleavage and acyl-modification results in the production of the 28 AA active acyl-modified ghrelin peptide [133].

26.7.3 Distribution

N-octanoyl-modified and des-acyl ghrelin are the two major forms of ghrelin found in tissues and plasma. The gastric fundus is the most abundant source of ghrelin, although lower amounts of ghrelin are found in the intestine,

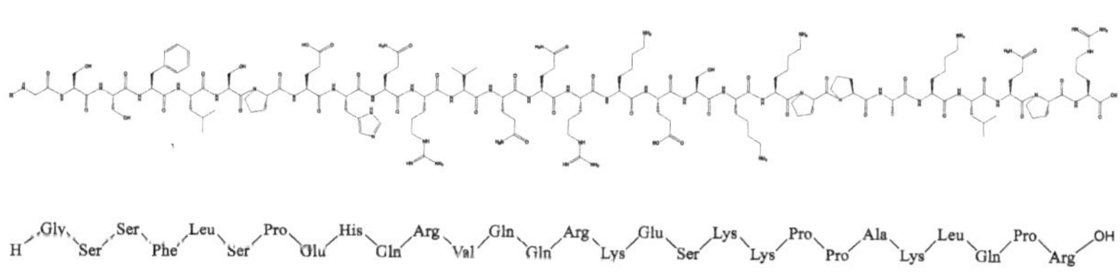

H-Gly-Ser-Ser-Phe-Leu-Ser-Pro-Glu-His-Gln-Arg-Val-Gln-Gln-Arg-Lys-Glu-Ser-Lys-Lys-Pro-Pro-Ala-Lys-Leu-Gln-Pro-Arg-OH

FIGURE 26.6 Chemical structure and amino acid sequence of ghrelin.

pancreas, pituitary, kidney, placenta, and cardiomyocytes [134]. Ghrelin is produced by distinctive endocrine cells known as P/D1 cells [133,135,136]. Plasma ghrelin concentrations are increased during periods of fasting or under conditions associated with negative energy balance, such as starvation or anorexia. In contrast, ghrelin levels are low after eating and in obesity [137,138].

26.7.4 Function

Ghrelin has a variety of physiological functions; it modulates feeding behavior, adiposity, and glucose- and energy-homeostasis, which are primarily mediated through its strong GH-releasing activity. In addition, it plays crucial roles in cardioprotection, muscle atrophy, bone metabolism [137,138], and fluid intake reduction [139].

The effects of ghrelin in the CV system are mediated not only via its growth hormone-releasing effect but also by its direct effects on the heart, for example, significantly decreased mean arterial pressure ($\sim$12 mm Hg) without a significant change in heart rate ($\sim$4 beats/min), as well as significantly increased cardiac index (+16%) and stroke volume index [140]. Recent studies demonstrated that ghrelin and GHSR exert some CV activities and protective effects, such as attenuation of inflammation, apoptosis inhibition, increased contractility, protection against HF, and vasodilatation [141–146]. Ghrelin was found to be a potent, endothelium-independent vasodilator, effectively reversing endothelin-1 (ET-1)-mediated constriction in vitro. The vasodilatation observed in response to ghrelin was similar in potency and maximum response to that induced by adrenomedullin (ADM) [147,148]. In an animal study ghrelin improved the survival prognosis of MI by reducing sympathetic nerve activity, thereby improving the function of the heart [141]. In addition, ghrelin directly protected the myocardium from apoptosis by increasing the Bcl-2/Bax ratio and partly through GHSR1a and GHSR1b [144].

The ghrelin-GHSR signaling pathway is also involved in the regulation of vascular tone, preventing diabetic complications by upregulation of osteoprotegerin (OPG) expression, downregulation of RANKL expression, activation of PI3K/Akt and inhibiting of reactive oxygen species (ROS) generation, and reversal of endothelial dysfunction in patients with metabolic syndrome [149–152].

26.7.5 Mechanism of action

Ghrelin acts through the ghrelin receptor, which is abundantly expressed in the hypothalamus [137], including the aorta, left ventricle, and left atria. GHSR, is a typical GPCR, with seven transmembrane domains, and stimulates phospholipase C, resulting in an increase in inositol 1,4,5-trisphosphate

(IP3) and intracellular Ca^{2+} levels [138]. As mentioned above, the effects of ghrelin in the CV system are mediated not only via its growth hormone-releasing effect but also via its direct effects on the heart.

26.7.6 Therapeutics

Several studies in rat models demonstrated that ghrelin administration induced cardiac and vascular protective changes, improved cardiac function and hemodynamics in HF after MI [145], promoted angiogenesis, reduced inflammatory responses, apoptosis, and oxidative stress, improved cardiac function in cardiopulmonary bypass [139], and inhibited neural remodeling and sympathetic hyperinnervation [142]. Overall, ghrelin may reduce cardiac sympathetic nerve activity, inflammation and oxidative stress in the heart, and induce angiogenesis [153–155]. In rats after cardiac arrest, ghrelin protected against cerebral mitochondria dysfunction [156]. Only a few clinical trials have been conducted with ghrelin, but none related to CV indications.

26.8 Salusins

26.8.1 Background

Salusins, a name stemming from the Latin word for health, were the first peptides discovered using a reverse pharmacology method that screens a cDNA library for sequences that were predicted to form active peptides of a standard known size of 90–250 AA and with specific cleavage sites [157]. Ota et al. searched for genes in a human cDNA library encoding for secreted or membrane proteins with possible physiological activity. The genes that matched the predictions were further screened for a biological response. The active peptides were then further explored for their specific bioactivity [158]. The actual presence of salusins (α and β) in body fluids was later established using radioimmunoassay and liquid chromatography techniques [159–161] (Fig. 26.7).

26.8.2 Genetics and translation

Salusins are derived from an alternatively spliced product of torsin family 2 member A (TOR2a). TOR2a is located at 9q34.11, comprised of 5 exons, and encodes a 321 AA protein [162,163]. Its splice variant, preprosalusin (PSEC0218, HEMBA 1005096, AK075520), contains a frameshift resulting from the deletion of exon 4, which deduces it to a 242 AA protcin with an alternate C-terminus. Preprosalusin, the precursor of salusins, contains a 26 AA signal peptide at the N-terminus, which is cleaved to yield the 216 AA prosalusin. Prosaluin is further cleaved to produce either salusin-α (28 AA) or salusin-ß (20 AA), depending on the cell type [164–166] (Fig. 26.8).

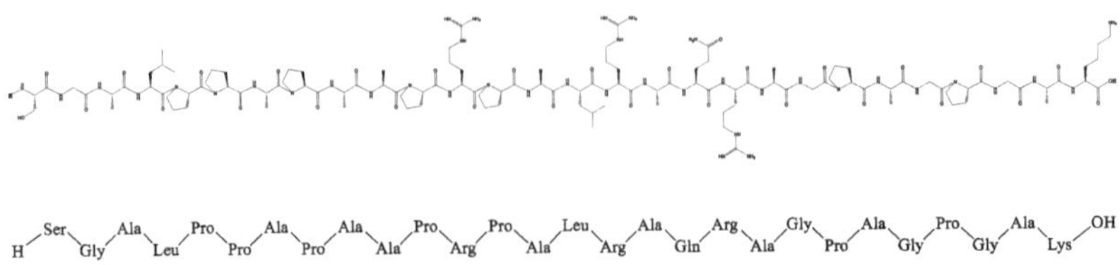

H-Ser-Gly-Ala-Leu-Pro-Pro-Ala-Pro-Ala-Ala-Pro-Arg-Pro-Ala-Leu-Arg-Ala-Gln-Arg-Ala-Gly-Pro-Ala-Gly-Pro-Gly-Ala-Lys-OH

FIGURE 26.7 Chemical structure and amino acid sequence of salusin-α.

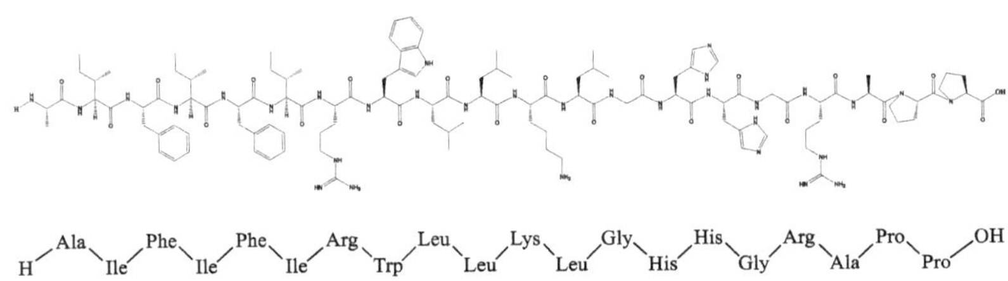

H-Ala-Ile-Phe-Ile-Phe-Ile-Arg-Trp-Leu-Leu-Lys-Leu-Gly-His-His-Gly-Arg-Ala-Pro-Pro-OH

FIGURE 26.8 Chemical structure and amino acid sequence of salusin-β.

26.8.3 Distribution

While TOR2a is expressed almost ubiquitously throughout the body tissues, preprosalusin mRNA is abundant in spleen, peripheral leukocytes, lung, kidneys, brain, and endocrine glands, such as pancreas, testis, ovary, and thymus, while its expression level is low in cardiac and skeletal muscle [159].

26.8.4 Function and pathological implications

Intravenously delivered salusin-β caused rapid, temporary, and profound decreases in blood pressure and heart rate in rats by triggering a parasympathetic response [159,167]. Concomitantly, it induced a negative inotropic effect via a direct myotropic effect [164], likely through inhibition of L-type Ca^{2+} channels [168]. In the cardiac muscle itself, both saluslins promote cardiomyocyte hypertrophy, and proliferation of smooth muscle cells and fibroblasts, as well as antiapoptotic effects.

Salusin-α inhibits acyl-coenzyme A:cholesterol acyltransferase-1 expression, the turnover of macrophages to foam cells and reduces serum cholesterol levels. It was therefore assumed to provide protection against atherosclerosis. In contrast, salusin-β had an opposite effect on acyl-coenzyme A:cholesterol acyltransferase-1, and may therefore be considered atherogenic [169].

Low salusin-α serum levels were associated with various CVDs, such as coronary artery lesions, hypertension, LV diastolic dysfunction, and common carotid intima-media thickening [170–173]. In renal failure, salusin-α level is low; however, it is unclear whether the clinical condition induces this reduction or vice versa [174]. Of note, all these correlations are probably true only in the absence of active inflammation, which may obscure the correlation [173]. On the other hand, salusin-β was found to induce hypotension by activating the parasympathetic nervous system. Salusin-β, which is synthesized in the neuroendocrine system, is released to the circulation through the pituitary and also signals the adjacent neurohypophyseal cells to stimulate the release of vasopressin and oxytocin. To date, salusin-β serum levels cannot be determined, due to the low sensitivity of the radioimmunoassay.

26.8.5 Mechanism of action

No direct receptor for salusins is currently known. The mechanism of action of salusins likely involves increases in cellular Ca^{2+}. Salusin-β induces overexpression of vasopressin and oxytocin in the pituitary gland, a process that involves increases in intracellular free Ca^{2+}, achieved by its influx through voltage-dependent Ca^{2+} channels. The hypertrophic effect of salusins on neonatal rat ventricular myocytes and the proliferation of smooth muscle cells and fibroblasts are also associated with an increase in intracellular calcium. Nevertheless, the variety of membrane and intracellular proteins involved in increased intracellular Ca^{2+} levels are large, and the exact mechanism of action remains unknown.

26.8.6 Therapeutics

Recently, several preclinical studies demonstrated the therapeutic effects of salusins [175–177]; exogenous administration of salusin-α to diabetic rats eliminated the typical characteristics of diabetic nephropathy [175]. Tang et al. suggested that salusin-α may reduce hepatic steatosis and atherosclerosis via pleiotropic effects, ameliorate lipid profiles and regulate some key molecules involved in lipid metabolism in the liver. Moreover, it bears antioxidative and antiinflammatory capacities [176,177].

Chang et al. found that the MrgprA1 receptor mediates the neuroprotective effects of salusin-β on the midbrain dopamine neuron [178]. Further, using a specific antigen, salusin-β was blocked in the central nervous system. This attenuated hypertension reduced circulating norepinephrine levels and improved cardiac hypertrophy and function in hypertensive rats [179]. It also reduced proinflammatory cytokine's nuclear factor-kappa B (NF-κB) activity, and ROS levels, and altered renin-angiotensin-aldosterone system components in the paraventricular nucleus [179]. Additionally, endogenous salusin-β suppresses angiogenesis, while its inhibition promotes post-MI angiogenesis, thereby preventing cardiac remodeling following I/R injury [180].

In addition to extensive preclinical data, clinical studies of the correlation between salusins, atherosclerosis and hypertension found a negative correlation between these disease processes and salusin α, while a positive with salusin β [166,181–184].

In another study, concomitant administration of salusins α and β reduced caspase-3 expression, gastric mucosal damage, and protected against renal I/R damage. Thus, these two hormones could also serve as gastro- or renoprotective agents [185,186].

26.9 Apelin/APJ system

26.9.1 Background

Apelin is a peptide first isolated and characterized in 1998 as a ligand for an orphan GPCR named the apelin receptor (APJ) [187]. It carries several physiological functions, depending on the tissue expressing it, including regulation of blood pressure, endocrine stress response, cardiac contractility, angiogenesis, and energy metabolism [188]. APJ was identified in 1993 by homologous cloning in the Human Genome Project and was initially considered an orphan receptor. Recently, another peptide named elabela was also shown to function as an APJ receptor ligand [189]. The APJ sequence bears strong homology to the angiotensin II type 1 receptor (AT1R) (54% in transmembrane domains and 30% for the entire sequence) (Fig. 26.9).

26.9.2 Genetics and translation

The human apelin gene is located on chromosome Xq25–26.1 and consists of three exons. The gene shows great sequence homology across all mammalian species examined. The apelin preproprotein contains 77 AAs, which can be enzymatically hydrolyzed from its C-terminus into more biologically active fragments, for example, apelin-36, -17, -16, -13 and -12. The two most active apelin peptides are apelin-36 and apelin-13. Apelin-13 can be posttranslationally modified to form pyroglutaminated apelin-13 ([Pyr1]-apelin-13), a much more stable form of apelin-13, due to its resistance to degradation by exopeptidases [190].

26.9.3 Distribution

Apelin receptor is distributed extensively in the central nervous system and peripheral tissues. In addition, apelin is present in the heart, mainly in vascular and endocardial endothelium. It is present in normal human plasma, with

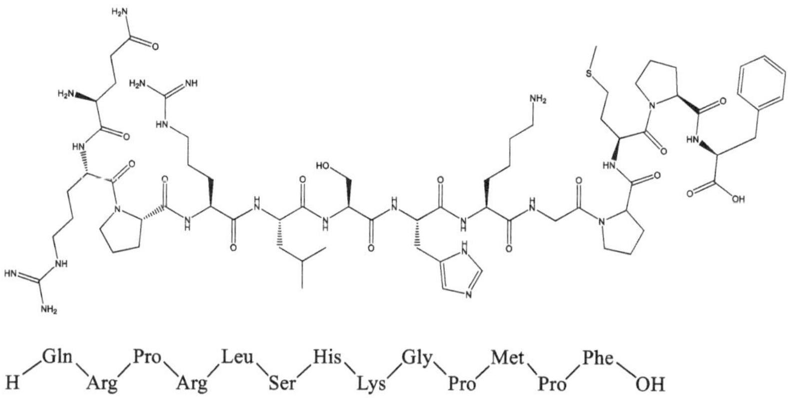

H-Gln-Arg-Pro-Arg-Leu-Ser-His-Lys-Gly-Pro-Met-Pro-Phe-OH

FIGURE 26.9 Chemical structure and amino acid sequence of apelin-13.

a brief half-life of <5 min. The plasma concentration of immunoreactive apelin is approximately 0.5–2 ng/mL [189].

26.9.4 Mechanism of action

Several signaling pathways have been suggested to be activated after apelin binds its receptor on vascular endothelial cells (VEC) and VSMC. Apelin binding to VEC can activate two subfamilies of Gα subunit, Gαi/o and Gαq, thus promoting relaxation of smooth muscle cells through PI3K activation and the release of NO and prostacyclin. Gαi/o coupling with APJ promotes Akt (protein kinase B, or PKB) and extracellular signal-regulated kinase (ERK) activation and inactivation of PKA, leading to a decrease in cyclic adenosine monophosphate (cAMP) production. In the absence of endothelium, apelin can directly bind its receptor on smooth muscle cells and lead to constriction, through Gαq coupling with APJ and the activation of the phospholipase C (PLC) protein kinase C β (PKCβ) pathway. The downstream effect of this pathway includes an increase in Na^+-H^+ exchanger (NHE) and Na^+-Ca^{2+} exchanger (NCX) expression, which then leads to increased Ca^{2+} sensitivity of the filaments. APJ has also been reported to act as a mechanosensor for stretch, which transduces signaling through a β-arrestin-dependent and G-protein-independent pathway [189].

26.9.5 Function

Apelin/APJ was shown to induce peripheral and coronary vasodilation, lower arterial blood pressure, decrease cardiac preload and afterload and increase cardiac output through a NO-dependent mechanism. Apelin has a strong inotropic effect, mediated by PKC. The APJ system counter-regulates the renin-angiotensin-aldosterone system by antagonizing the activity of angiotensin II, due to its high sequence similarity to angiotensin receptor 1 [189].

26.9.6 Pathophysiology

The apelin/APJ system plays a significant role in cardiac hypertrophy. Rats devoid of the apelin/APJ system were protected from cardiac hypertrophy [191]. Apelin reduces fibrosis in the heart by counteracting angiotensin II activity and decreases activator inhibitor type-1 production [192]. In the early stages of HF, apelin levels rise in the plasma and in the LV, but decrease significantly in severe, end-stage HF, regardless of etiology or LV ejection fraction [193]. In atherosclerosis, there is a discrepancy regarding the effect of apelin/APJ. In some studies on apelin and apolipoprotein E (a protein involved in fat metabolism) knockout mice, a significant atherosclerotic lesion area appeared, while other studies done on double knockout mice showed a smaller atherosclerotic lesion area compared to control animals. Meanwhile, apelin has a protective effect against I/R injury through the activation of PI3Ks and ERK in the salvage kinase (RISK) pathway [194].

26.9.7 Therapeutics

There are currently no clinically used drugs that target the apelin/APJ system, although a few potential therapeutic agents are being studied. The apelin/APJ and elabela/APJ systems may represent as drug targets for the treatment of CVD and other diseases, and drugs that are developed, both agonists and antagonists of either system, may have significant therapeutic potential [195].

26.10 Urocortin

26.10.1 Background

Urocortin is a member of the corticotrophin-releasing factor family that was extracted from the mammalian brain in 1981 [192] (Fig. 26.10).

26.10.2 Genetics and translation

Urocortin is a 40 AA peptide that is encoded by the UCN gene, which contains two exons and encodes a member of the corticotropin-releasing factor/urotensin I family. Urocortins 1, 2 and 3 genes are located in chromosomes 2, 3, and 10, respectively [196]. The encoded preproprotein is further processed to generate the mature peptide, a ligand of both the corticotropin-releasing factor receptor 1 and the corticotropin-releasing factor receptor 2.

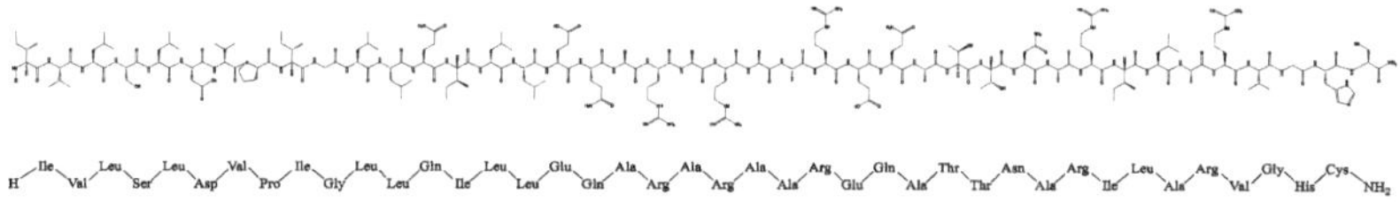

H-Ile-Val-Leu-Ser-Leu-Asp-Val-Pro-Ile-Gly-Leu-Leu-Gln-Ile-Leu-Leu-Glu-Gln-Ala-Arg-Ala-Arg-Ala-Ala-Arg-Glu-Gln-Ala-Thr-Thr-Asn-Ala-Arg-Ile-Leu-Ala-Arg-Val-Gly-His-Cys-NH_2

FIGURE 26.10 Chemical structure and amino acid sequence of urocortin.

26.10.3 Distribution

Similar to the corticotrophin-releasing factor, urocortin is expressed mainly in the central and peripheral nervous systems, spleen, thymus, synovial epithelial cells, inflammatory cells, and gastric epithelial cells, but is also expressed in cardiac tissue [197].

26.10.4 Mechanism of action

Urocortin exhibits its effects by binding CRF-1 and CRF-2 receptors, and binds CFR-2 (both α and β isoforms) with a higher affinity as compared to corticotrophin-releasing hormone [198].

26.10.5 Function

In the heart, urocortin exhibits unique cardioprotective properties and may be beneficial in preventing I/R injury when binding CRF-2beta receptors [199]. Bale et al. found that IV injection of urocortin 2 resulted in positive inotropic effects and enhanced diastolic function in rats [200]. In healthy volunteers, infusion of urocortin 2 dose-dependently resulted in increased LV ejection fraction and cardiac output, decreased systemic vascular resistance, activated plasma renin-angiotensin II, and norepinephrine [201]. In stable HF patients, urocortin 2 induced vasodilation and increased sympathetic nerve activity in the lower limbs [202].

26.10.6 Therapeutics

The main interest of urocortin 2 in the cardiovascular arena is in the field of HF. Rademaker et al. demonstrated that the addition of urocortin 2 to ACE inhibitor therapy in HF models improved cardiac and renal hemodynamics [203]. In patients with HF with reduced ejection fraction, IV human urocortin 2 improved cardiac function and reduced blood pressure [204]. In a double-blind placebo-controlled trial in patients with acute HF, Chan et al. demonstrated that urocortin 2 infusion induced a hypotensive effect, as well as an increase in cardiac output [205]. In the nervous system, urocortin participates in regulating anxiety, memory, learning, body temperature, suppressing appetite, delaying gastric emptying, and increasing locomotor activity mediated by plasma corticotrophin levels [206]. Its neurological actions may make it a potential neuroprotective agent against oxidative and excitotoxic insults [207].

26.11 Conclusion

The increasing need for new therapeutic approaches and treatments for CVD combined with the promise of peptides and peptidomimetics as active pharmaceutical agents for effective drug formulation is contributing to their extensive development, including new synthetic strategies, along with alternative routes of administration. While peptides are generally considered to be poor drug candidates because of their low oral bioavailability and propensity to be rapidly metabolized, these limitations can be overcome by modifying existing peptide sequences to become peptidomimetics, which provide the advantages of increased bioavailability, biostability, bioefficiency, and bioselectivity. These tools are allowing the application of peptides and peptide leads previously considered to be extra-cardiovascular to modulation of the CVS. Therefore we predict that these molecules will advance novel treatments for CVDs.

Funding

This research was supported by a grant from the Ministry of Health (MoH), Jerusalem, Israel, and a grant from the Israel Science Foundation (ISF), Israel, both to N.Q.

References

[1] Sanchez Jimenez JG, De Jesus O. Hypothalamic dysfunction July 15 StatPearls. Treasure Island (FL): StatPearls Publishing; 2020.

[2] Morris JF. VP/OT. Handbook of biologically active peptides. Academic Press; 2013. p. 975–81.

[3] Ragnarsson U. The nobel trail of Vincent du Vigneaud. J Pept Sci 2007;13(7):431–3. Available from: https://doi.org/10.1002/psc.864.

[4] Caldwell HK. Oxytocin and vasopressin: powerful regulators of social behavior. Neuroscientist. 2017;23(5):517–28. Available from: https://doi.org/10.1177/1073858417708284.

[5] Koeppen BM, Stanton BA. Berne and levy physiology e-book. Elsevier Health Sciences; 2017.

[6] Gutkowska J, Jankowski M. VP/OT. Handbook of biologically active peptides. Academic Press; 2013. p. 1450–7.

[7] MacLean EL, Wilson SR, Martin WL, Davis JM, Nazarloo HP, Carter CS. Challenges for measuring oxytocin: the blind men and the elephant? Psychoneuroendocrinology 2019;107:225–31. Available from: https://doi.org/10.1016/j.psyneuen.2019.05.018.

[8] Rydén G, Sjöholm I. Half-life of oxytocin in blood of pregnant and non-pregnant women. Acta Endocrinol (Copenh) 1969;61(3):425–31.

[9] Lippert TH, Mueck AO, Seeger H, Pfaff A. Effects of oxytocin outside pregnancy. Horm Res 2003;60(6):262–71. Available from: https://doi.org/10.1159/000074243.

[10] Amar AP, Weiss MH. Pituitary anatomy and physiology. Neurosurg Clin N Am 2003;14 (1). Available from: https://doi.org/10.1016/s1042-3680(02)00017-7 11-v.

[11] Japundžić-Žigon N, Lozić M, Šarenac O, Murphy D. Vasopressin & oxytocin in control of the cardiovascular system: an updated review. Curr Neuropharmacol 2020;18 (1):14–33. Available from: https://doi.org/10.2174/1570159X17666190717150501.

[12] Sanghi P, Uretsky BF, Schwarz ER. Vasopressin antagonism: a future treatment option in heart failure. Eur Heart J 2005;26(6):538–43. Available from: https://doi.org/10.1093/eurheartj/ehi145.

[13] Moghimian M, Faghihi M, Karimian SM, Imani A, Houshmand F, Azizi Y. Role of central oxytocin in stress-induced cardioprotection in ischemic-reperfused heart model. J Cardiol 2013;61(1):79–86. Available from: https://doi.org/10.1016/j.jjcc.2012.08.021.

[14] Moghimian M, Faghihi M, Karimian SM, Imani A. The effect of acute stress exposure on ischemia and reperfusion injury in rat heart: role of oxytocin. Stress. 2012;15(4):385–92. Available from: https://doi.org/10.3109/10253890.2011.630436.

[15] Polshekan M, Jamialahmadi K, Khori V, et al. RISK pathway is involved in oxytocin postconditioning in isolated rat heart. Peptides. 2016;86:55–62. Available from: https://doi.org/10.1016/j.peptides.2016.10.001.

[16] Faghihi M, Alizadeh AM, Khori V, Latifpour M, Khodayari S. The role of nitric oxide, reactive oxygen species, and protein kinase C in oxytocin-induced cardioprotection in ischemic rat heart. Peptides. 2012;37(2):314–19. Available from: https://doi.org/10.1016/j.peptides.2012.08.001.

[17] Kobayashi H, Yasuda S, Bao N, et al. Postinfarct treatment with oxytocin improves cardiac function and remodeling via activating cell-survival signals and angiogenesis. J Cardiovasc Pharmacol 2009;54(6):510–19. Available from: https://doi.org/10.1097/FJC.0b013e3181bfac02.

[18] Wang P, Wang SC, Yang H, et al. Therapeutic potential of oxytocin in atherosclerotic cardiovascular disease: mechanisms and signaling pathways. Front Neurosci 2019;13:454. Available from: https://doi.org/10.3389/fnins.2019.00454 Published 2019 May 21.

[19] Song Z, Albers HE. Cross-talk among oxytocin and arginine-vasopressin receptors: relevance for basic and clinical studies of the brain and periphery. Front Neuroendocrinol 2018;51:14–24. Available from: https://doi.org/10.1016/j.yfrne.2017.10.004.

[20] Greenberg A. Tolvaptan, an oral vasopressin V2 receptor antagonist for heart failure? Am J Kidney Dis 2007;50(6):904–7. Available from: https://doi.org/10.1053/j.ajkd.2007.10.006.

[21] Holmes CL, Landry DW, Granton JT. Science review: vasopressin and the cardiovascular system part 2—clinical physiology. Crit Care 2004;8(1):15–23. Available from: https://doi.org/10.1186/cc2338.

[22] Demiselle J, Fage N, Radermacher P, Asfar P. Vasopressin and its analogues in shock states: a review. Ann Intensive Care 2020;10(1):9. Available from: https://doi.org/10.1186/s13613-020-0628-2 Published 2020 Jan 22.

[23] Jain V, Marbach J, Kimbro S, et al. Benefits of oxytocin administration in obstructive sleep apnea. Am J Physiol Lung Cell Mol Physiol 2017;313(5):L825–33. Available from: https://doi.org/10.1152/ajplung.00206.2017.

[24] Zhang H, Wu C, Chen Q, et al. Treatment of obesity and diabetes using oxytocin or analogs in patients and mouse models. PLoS One 2013;8(5):e61477. Available from: https://doi.org/10.1371/journal.pone.0061477 Published 2013 May 20.

[25] Johnstone M. The cardiovascular effects of oxytocic drugs. Br J Anaesth 1972;44(8):826–34. Available from: https://doi.org/10.1093/bja/44.8.826.

[26] Weissman A, Tobia RS, Burke YZ, Maxymovski O, Drugan A. The effects of oxytocin and atosiban on the modulation of heart rate in pregnant women. J Matern Fetal Neonatal Med 2017;30(3):329–33. Available from: https://doi.org/10.3109/14767058.2016.1172564.

[27] Rabow S, Hjorth U, Schönbeck S, Olofsson P. Effects of oxytocin and anaesthesia on vascular tone in pregnant women: a randomised double-blind placebo-controlled study using non-invasive pulse wave analysis. BMC Pregnancy Childbirth 2018;18(1):453. Available from: https://doi.org/10.1186/s12884-018-2029-1 Published 2018 Nov 22.

[28] Amara SG, Jonas V, Rosenfeld MG, Ong ES, Evans RM. Alternative RNA processing in calcitonin gene expression generates mRNAs encoding different polypeptide products. Nature. 1982;298(5871):240–4. Available from: https://doi.org/10.1038/298240a0.

[29] Alevizaki M, Shiraishi A, Rassool FV, Ferrier GJ, MacIntyre I, Legon S. The calcitonin-like sequence of the beta CGRP gene. FEBS Lett 1986;206(1):47–52. Available from: https://doi.org/10.1016/0014-5793(86)81338-2.

[30] Brain SD, Grant AD. Vascular actions of calcitonin gene-related peptide and adrenomedullin. Physiol Rev 2004;84(3):903–34. Available from: https://doi.org/10.1152/physrev.00037.2003.

[31] Fang L, Chen MF, Xiao ZL, et al. Calcitonin gene-related peptide released from endothelial progenitor cells inhibits the proliferation of rat vascular smooth muscle cells induced by angiotensin II. Mol Cell Biochem 2011;355(1–2):99–108. Available from: https://doi.org/10.1007/s11010-011-0843-0.

[32] McEwan JR, Benjamin N, Larkin S, Fuller RW, Dollery CT, MacIntyre I. Vasodilatation by calcitonin gene-related peptide and by substance P: a comparison of their effects on resistance and capacitance vessels of human forearms. Circulation. 1988;77(5):1072–80. Available from: https://doi.org/10.1161/01.cir.77.5.1072.

[33] Russell FA, King R, Smillie SJ, Kodji X, Brain SD. Calcitonin gene-related peptide: physiology and pathophysiology. Physiol Rev 2014;94(4):1099–142. Available from: https://doi.org/10.1152/physrev.00034.2013.

[34] Brain SD, Williams TJ, Tippins JR, Morris HR, MacIntyre I. Calcitonin gene-related peptide is a potent vasodilator. Nature. 1985;313(5997):54–6. Available from: https://doi.org/10.1038/313054a0.

[35] Olesen J, Diener HC, Husstedt IW, et al. Calcitonin gene-related peptide receptor antagonist BIBN 4096 BS for the acute treatment of migraine. N Engl J Med 2004;350(11):1104–10. Available from: https://doi.org/10.1056/NEJMoa030505.

[36] Smillie SJ, Brain SD. Calcitonin gene-related peptide (CGRP) and its role in hypertension. Neuropeptides 2011;45(2):93–104. Available from: https://doi.org/10.1016/j.npep.2010.12.002.

[37] Tominaga M, Tominaga T. Structure and function of TRPV1. Pflug Arch 2005;451(1):143–50. Available from: https://doi.org/10.1007/s00424-005-1457-8.

[38] Xu J, Wang J, Chen X, Li Y, Mi J, Qin L. The effects of calcitonin gene-related peptide on bone homeostasis and regeneration. Curr Osteoporos Rep 2020;18(6):621–32. Available from: https://doi.org/10.1007/s11914-020-00624-0.

[39] Miller PS, Barwell J, Poyner DR, Wigglesworth MJ, Garland SL, Donnelly D. Non-peptidic antagonists of the CGRP receptor, BIBN4096BS and MK-0974, interact with the calcitonin receptor-like receptor via methionine-42 and RAMP1 via tryptophan-74. Biochem Biophys Res Commun 2010;391(1):437–42. Available from: https://doi.org/10.1016/j.bbrc.2009.11.076.

[40] Huang R, Karve A, Shah I, et al. Deletion of the mouse alpha-calcitonin gene-related peptide gene increases the vulnerability of the heart to ischemia-reperfusion injury. Am J Physiol Heart Circ Physiol 2008;294(3):H1291–7. Available from: https://doi.org/10.1152/ajpheart.00749.2007.

[41] Ho TW, Edvinsson L, Goadsby PJ. CGRP and its receptors provide new insights into migraine pathophysiology. Nat Rev Neurol 2010;6(10):573–82. Available from: https://doi.org/10.1038/nrneurol.2010.127.

[42] Rashid A, Manghi A. Calcitonin gene-related peptide receptor July 15 StatPearls. Treasure Island (FL): StatPearls Publishing; 2020.

[43] Meens MJ, Mattheij NJ, Nelissen J, et al. Calcitonin gene-related peptide terminates long-lasting vasopressor responses to endothelin 1 in vivo. Hypertension. 2011;58(1):99–106. Available from: https://doi.org/10.1161/HYPERTENSIONAHA.110.169128.

[44] Zhou Z, Hu CP, Wang CJ, Li TT, Peng J, Li YJ. Calcitonin gene-related peptide inhibits angiotensin II-induced endothelial progenitor cells senescence through up-regulation of klotho expression. Atherosclerosis. 2010;213(1):92–101. Available from: https://doi.org/10.1016/j.atherosclerosis.2010.08.050.

[45] Opie EL. On the relation of chronic interstitial pancreatitis to the islands of langerhans and to diabetes melutus. J Exp Med 1901;5(4):397–428. Available from: https://doi.org/10.1084/jem.5.4.397.

[46] Cooper GJ, Willis AC, Clark A, Turner RC, Sim RB, Reid KB. Purification and characterization of a peptide from amyloid-rich pancreases of type 2 diabetic patients. Proc Natl Acad Sci U S A 1987;84(23):8628–32. Available from: https://doi.org/10.1073/pnas.84.23.8628.

[47] Westermark P, Wilander E, Westermark GT, Johnson KH. Islet amyloid polypeptide-like immunoreactivity in the islet B cells of type 2 (non-insulin-dependent) diabetic and non-diabetic individuals. Diabetologia. 1987;30(11):887–92. Available from: https://doi.org/10.1007/BF00274799.

[48] Staden R. The current status and portability of our sequence handling software. Nucleic Acids Res 1986;14(1):217–31. Available from: https://doi.org/10.1093/nar/14.1.217.

[49] Chou PY, Fasman GD. Empirical predictions of protein conformation. Annu Rev Biochem 1978;47:251–76. Available from: https://doi.org/10.1146/annurev.bi.47.070178.001343.

[50] Kyte J, Doolittle RF. A simple method for displaying the hydropathic character of a protein. J Mol Biol 1982;157(1):105–32. Available from: https://doi.org/10.1016/0022-2836(82)90515-0.

[51] Morris HR, Panico M, Etienne T, Tippins J, Girgis SI, MacIntyre I. Isolation and characterization of human calcitonin gene-related peptide. Nature. 1984;308(5961):746–8. Available from: https://doi.org/10.1038/308746a0.

[52] Steenbergh PH, Höppener JW, Zandberg J, Lips CJ, Jansz HS. A second human calcitonin/CGRP gene. FEBS Lett 1985;183(2):403–7. Available from: https://doi.org/10.1016/0014-5793(85)80820-6.

[53] Westermark P, Andersson A, Westermark GT. Islet amyloid polypeptide, islet amyloid, and diabetes mellitus. Physiol Rev 2011;91(3):795–826. Available from: https://doi.org/10.1152/physrev.00042.2009.

[54] Mulderry PK, Ghatei MA, Rodrigo J, et al. Calcitonin gene-related peptide in cardiovascular tissues of the rat. Neuroscience. 1985;14(3):947–54. Available from: https://doi.org/10.1016/0306-4522(85)90156-3.

[55] Schifter S, Williams ED, Craig RK, Hansen HH. Calcitonin gene-related peptide and calcitonin in medullary thyroid carcinoma. Clin Endocrinol (Oxf) 1986;25(6):703–10. Available from: https://doi.org/10.1111/j.1365-2265.1986.tb03626.x.

[56] Cooper GJS. Amylin and related proteins: physiology and pathophysiology , vol. 10, supplement 21 Handbook of physiology, the endocrine system, the endocrine pancreas and regulation of metabolism. American Physiological Society. Wiley Online library; 2001. p. 303–68.

[57] Cooper GJ, Day AJ, Willis AC, Roberts AN, Reid KB, Leighton B. Amylin and the amylin gene: structure, function and relationship to islet amyloid and to diabetes mellitus. Biochim Biophys Acta 1989;1014(3):247–58. Available from: https://doi.org/10.1016/0167-4889(89)90220-6.

[58] Gebre-Medhin S, Mulder H, Pekny M, et al. Increased insulin secretion and glucose tolerance in mice lacking islet amyloid polypeptide (amylin). Biochem Biophys Res Commun 1998;250(2):271–7. Available from: https://doi.org/10.1006/bbrc.1998.9308.

[59] Wookey PJ, Cooper ME. Amylin: physiological roles in the kidney and a hypothesis for its role in hypertension. Clin Exp Pharmacol Physiol 1998;25(9):653–60. Available from: https://doi.org/10.1111/j.1440-1681.1998.tb02272.x.

[60] Hay DL, Chen S, Lutz TA, Parkes DG, Roth JD. Amylin: pharmacology, physiology, and clinical potential. Pharmacol Rev 2015;67(3):564–600. Available from: https://doi.org/10.1124/pr.115.010629.

[61] Hinshaw L, Schiavon M, Dadlani V, et al. Effect of pramlintide on postprandial glucose fluxes in type 1 diabetes. J Clin Endocrinol Metab 2016;101(5):1954–62. Available from: https://doi.org/10.1210/jc.2015-3952.

[62] Bower RL, Hay DL. Amylin structure-function relationships and receptor pharmacology: implications for amylin mimetic drug development. Br J Pharmacol 2016;173(12):1883–98. Available from: https://doi.org/10.1111/bph.13496.

[63] Cooper GJ, Leighton B, Dimitriadis GD, et al. Amylin found in amyloid deposits in human type 2 diabetes mellitus may be a hormone that regulates glycogen metabolism in skeletal muscle. Proc Natl Acad Sci USA 1988;85(20):7763–6. Available from: https://doi.org/10.1073/pnas.85.20.7763.

[64] Leighton B, Cooper GJ. Pancreatic amylin and calcitonin gene-related peptide cause resistance to insulin in skeletal muscle in vitro. Nature. 1988;335(6191):632–5. Available from: https://doi.org/10.1038/335632a0.

[65] Johnson KH, O'Brien TD, Hayden DW, et al. Immunolocalization of islet amyloid polypeptide (IAPP) in pancreatic beta cells by means of peroxidase-antiperoxidase (PAP) and protein A-gold techniques. Am J Pathol 1988;130(1):1–8.

[66] Young A. Cardiovascular effects. Adv Pharmacol 2005;52:239–50. Available from: https://doi.org/10.1016/S1054-3589(05)52014-3.

[67] Said SI, Mutt V. Polypeptide with broad biological activity: isolation from small intestine. Science. 1970;169(3951):1217–18. Available from: https://doi.org/10.1126/science.169.3951.1217.

[68] Gozes I, Brenneman DE. VIP: molecular biology and neurobiological function. Mol Neurobiol 1989;3(4):201–36. Available from: https://doi.org/10.1007/BF02740606.

[69] Henning RJ. VIP. Handbook of biologically active peptides. Academic Press; 2013. p. 1443–9.

[70] Domschke S, Domschke W, Bloom SR, et al. Vasoactive intestinal peptide in man: pharmacokinetics, metabolic and circulatory effects. Gut. 1978;19(11):1049–53. Available from: https://doi.org/10.1136/gut.19.11.1049.

[71] Said SI, Rosenberg RN. Vasoactive intestinal polypeptide: abundant immunoreactivity in neural cell lines and normal nervous tissue. Science. 1976;192(4242):907–8. Available from: https://doi.org/10.1126/science.1273576.

[72] Felten DL, Felten SY, Carlson SL, Olschowka JA, Livnat S. Noradrenergic and peptidergic innervation of lymphoid tissue. J Immunol 1985;135(2 Suppl):755s–65s.

[73] Bellinger DL, Lorton D, Brouxhon S, Felten S, Felten DL. The significance of vasoactive intestinal polypeptide (VIP) in immunomodulation. Adv Neuroimmunol 1996;6(1):5–27. Available from: https://doi.org/10.1016/s0960-5428(96)00008-3.

[74] Tiaho F, Nerbonne JM. VIP and secretin augment cardiac L-type calcium channel currents in isolated adult rat ventricular myocytes. Pflug Arch 1996;432(5):821–30. Available from: https://doi.org/10.1007/s004240050204.

[75] Staun-Olsen P, Ottesen B, Bartels PD, Nielsen MH, Gammeltoft S, Fahrenkrug J. Receptors for vasoactive intestinal polypeptide on isolated synaptosomes from rat cerebral cortex. Heterogeneity of binding and desensitization of receptors. J Neurochem 1982;39(5):1242–51. Available from: https://doi.org/10.1111/j.1471-4159.1982.tb12561.x.

[76] Hannibal J, Hsiung HM, Fahrenkrug J. Temporal phasing of locomotor activity, heart rate rhythmicity, and core body temperature is disrupted in VIP receptor 2-deficient mice. Am J Physiol Regul Integr Comp Physiol 2011;300(3):R519–30. Available from: https://doi.org/10.1152/ajpregu.00599.2010.

[77] Iwasaki M, Akiba Y, Kaunitz JD. Recent advances in vasoactive intestinal peptide physiology and pathophysiology: focus on the gastrointestinal system. F1000Res. 2019;8:F1000 Faculty Rev-1629. Published 2019 Sep 12. Available from: https://doi.org/10.12688/f1000research.18039.1.

[78] Said SI, Hamidi SA, Dickman KG, et al. Moderate pulmonary arterial hypertension in male mice lacking the vasoactive intestinal peptide gene. Circulation. 2007;115(10):1260–8. Available from: https://doi.org/10.1161/CIRCULATIONAHA.106.681718.

[79] Gomariz RP, Juarranz Y, Abad C, Arranz A, Leceta J, Martinez C. VIP-PACAP system in immunity: new insights for multitarget therapy. Ann N Y Acad Sci 2006;1070:51–74. Available from: https://doi.org/10.1196/annals.1317.031.

[80] Delgado M, Pozo D, Ganea D. The significance of vasoactive intestinal peptide in immunomodulation. Pharmacol Rev 2004;56(2):249–90. Available from: https://doi.org/10.1124/pr.56.2.7.

[81] Morell M, Souza-Moreira L, González-Rey E. VIP in neurological diseases: more than a neuropeptide. Endocr Metab Immune Disord Drug Targets 2012;12(4):323–32. Available from: https://doi.org/10.2174/187153012803832549.

[82] Todd WD, Venner A, Anaclet C, et al. Suprachiasmatic VIP neurons are required for normal circadian rhythmicity and comprised of molecularly distinct subpopulations. Nat Commun 2020;11(1):4410. Available from: https://doi.org/10.1038/s41467-020-17197-2 Published 2020 Sep 2.

[83] Henning RJ, Sawmiller DR. Vasoactive intestinal peptide: cardiovascular effects. Cardiovasc Res 2001;49(1):27–37. Available from: https://doi.org/10.1016/s0008-6363(00)00229-7.

[84] Duggan KA, Hodge G, Chen J, Hunter T. Vasoactive intestinal peptide infusion reverses existing myocardial fibrosis in the rat. Eur J Pharmacol 2019;862:172629. Available from: https://doi.org/10.1016/j.ejphar.2019.172629.

[85] Benitez R, Delgado-Maroto V, Caro M, et al. Vasoactive intestinal peptide ameliorates acute myocarditis and atherosclerosis by regulating inflammatory and autoimmune responses. J Immunol 2018;200(11):3697–710. Available from: https://doi.org/10.4049/jimmunol.1800122.

[86] Szema AM, Hamidi SA. Gene deletion of VIP leads to increased mortality associated with progressive right ventricular hypertrophy. J Cardiovasc Dis 2014;2(3):131–6.

[87] Szema AM, Hamidi SA, Smith SD, Benveniste H. VIP gene deletion in mice causes cardiomyopathy associated with upregulation of heart failure genes. PLoS One 2013;8(5):e61449. Available from: https://doi.org/10.1371/journal.pone.0061449 Published 2013 May 20.

[88] Dickson L, Finlayson K. VPAC and PAC receptors: from ligands to function. Pharmacol Ther 2009;121(3):294–316. Available from: https://doi.org/10.1016/j.pharmthera.2008.11.006.

[89] Gutiérrez-Cañas I, Juarranz MG, Collado B, et al. Vasoactive intestinal peptide induces neuroendocrine differentiation in the LNCaP prostate cancer cell line through PKA, ERK, and PI3K. Prostate. 2005;63(1):44–55. Available from: https://doi.org/10.1002/pros.20173.

[90] Crespo P, Xu N, Simonds WF, Gutkind JS. Ras-dependent activation of MAP kinase pathway mediated by G-protein beta gamma subunits. Nature. 1994;369(6479):418–20. Available from: https://doi.org/10.1038/369418a0.

[91] Maier U, Babich A, Nürnberg B. Roles of non-catalytic subunits in gbetagamma-induced activation of class I phosphoinositide 3-kinase isoforms beta and gamma. J Biol Chem 1999;274(41):29311–17. Available from: https://doi.org/10.1074/jbc.274.41.29311.

[92] MacKenzie CJ, Lutz EM, Johnson MS, Robertson DN, Holland PJ, Mitchell R. Mechanisms of phospholipase C activation by the vasoactive intestinal polypeptide/pituitary adenylate cyclase-activating polypeptide type 2 receptor. Endocrinology. 2001;142(3):1209–17. Available from: https://doi.org/10.1210/endo.142.3.8013.

[93] McCulloch DA, Lutz EM, Johnson MS, MacKenzie CJ, Mitchell R. Differential activation of phospholipase D by VPAC and PAC1 receptors. Ann N Y Acad Sci 2000;921:175–85. Available from: https://doi.org/10.1111/j.1749-6632.2000.tb06964.x.

[94] Leuchte HH, Baezner C, Baumgartner RA, et al. Inhalation of vasoactive intestinal peptide in pulmonary hypertension. Eur Respir J 2008;32(5):1289–94. Available from: https://doi.org/10.1183/09031936.00050008.

[95] Hamidi SA, Lin RZ, Szema AM, Lyubsky S, Jiang YP, Said SI. VIP and endothelin receptor antagonist: an effective combination against experimental pulmonary arterial hypertension. Respir Res 2011;12(1):141. Available from: https://doi.org/10.1186/1465-9921-12-141 Published 2011 Oct 26.

[96] Koo OM, Rubinstein I, Onyüksel H. Actively targeted low-dose camptothecin as a safe, long-acting, disease-modifying nanomedicine for rheumatoid arthritis. Pharm Res 2011;28(4):776–87. Available from: https://doi.org/10.1007/s11095-010-0330-4.

[97] Bolin DR, Cottrell J, Garippa R, et al. Comparison of cyclic and linear analogs of vasoactive intestinal peptide. Drug Des Discov 1996;13(3-4):107–14.

[98] Schmidt DT, Rühlmann E, Waldeck B, et al. The effect of the vasoactive intestinal polypeptide agonist Ro 25-1553 on induced tone in isolated human airways and pulmonary artery. Naunyn Schmiedebergs Arch Pharmacol 2001;364(4):314–20. Available from: https://doi.org/10.1007/s002100100458.

[99] Takeuchi S, Kawanai T, Yamauchi R, et al. Activation of the VPAC2 receptor impairs axon outgrowth and decreases dendritic arborization in mouse cortical neurons by a PKA-

dependent mechanism. Front Neurosci 2020;14:521. Available from: https://doi.org/10.3389/fnins.2020.00521 Published 2020 Jun 4.
[100] Delgado M, Toscano MG, Benabdellah K, et al. In vivo delivery of lentiviral vectors expressing vasoactive intestinal peptide complementary DNA as gene therapy for collagen-induced arthritis. Arthritis Rheum 2008;58(4):1026–37. Available from: https://doi.org/10.1002/art.23283.
[101] Toscano MG, Delgado M, Kong W, Martin F, Skarica M, Ganea D. Dendritic cells transduced with lentiviral vectors expressing VIP differentiate into VIP-secreting tolerogenic-like DCs. Mol Ther 2010;18(5):1035–45. Available from: https://doi.org/10.1038/mt.2009.293.
[102] Nauck M. Incretin therapies: highlighting common features and differences in the modes of action of glucagon-like peptide-1 receptor agonists and dipeptidyl peptidase-4 inhibitors. Diabetes Obes Metab 2016;18(3):203–16.
[103] D'Alessio D. Is GLP-1 a hormone: whether and when? J Diabetes Investig 2016;7(Suppl 1):50–5. Available from: https://doi.org/10.1111/jdi.12466 Suppl 1.
[104] Honigberg MC, Chang LS, McGuire DK, Plutzky J, Aroda VR, Vaduganathan M. Use of glucagon-like peptide-1 receptor agonists in patients with type 2 diabetes and cardiovascular disease: a review [published online ahead of print, 2020 Jun 17]JAMA Cardiol 2020;. Available from: https://doi.org/10.1001/jamacardio.2020.1966.
[105] Rowzee AM, Cawley NX, Chiorini JA, Di Pasquale G. Glucagon-like peptide-1 gene therapy. Exp Diabetes Res 2011;2011:601047. Available from: https://doi.org/10.1155/2011/601047.
[106] Pyke C, Heller RS, Kirk RK, et al. GLP-1 receptor localization in monkey and human tissue: novel distribution revealed with extensively validated monoclonal antibody. Endocrinology. 2014;155(4):1280–90. Available from: https://doi.org/10.1210/en.2013-1934.
[107] Cheang JY, Moyle PM. Glucagon-like peptide-1 (glp-1)-based therapeutics: current status and future opportunities beyond type 2 diabetes. ChemMedChem. 2018;13(7):662–71. Available from: https://doi.org/10.1002/cmdc.201700781.
[108] Holst JJ. The physiology of glucagon-like peptide 1. Physiol Rev 2007;87(4):1409–39. Available from: https://doi.org/10.1152/physrev.00034.2006.
[109] Yaribeygi H, Maleki M, Sathyapalan T, Jamialahmadi T, Sahebkar A. Incretin-based therapies and renin-angiotensin system: looking for new therapeutic potentials in the diabetic milieu. Life Sci 2020;256:117916. Available from: https://doi.org/10.1016/j.lfs.2020.117916.
[110] Barragán JM, Rodríguez RE, Blázquez E. Changes in arterial blood pressure and heart rate induced by glucagon-like peptide-1-(7-36) amide in rats. Am J Physiol 1994;266(3 Pt 1):E459–66. Available from: https://doi.org/10.1152/ajpendo.1994.266.3.E459.
[111] Edwards CM, Edwards AV, Bloom SR. Cardiovascular and pancreatic endocrine responses to glucagon-like peptide-1(7-36) amide in the conscious calf. Exp Physiol 1997;82(4):709–16. Available from: https://doi.org/10.1113/expphysiol.1997.sp004059.
[112] Yamamoto H, Lee CE, Marcus JN, et al. Glucagon-like peptide-1 receptor stimulation increases blood pressure and heart rate and activates autonomic regulatory neurons. J Clin Invest 2002;110(1):43–52. Available from: https://doi.org/10.1172/JCI15595.
[113] Gros R, You X, Baggio LL, et al. Cardiac function in mice lacking the glucagon-like peptide-1 receptor. Endocrinology. 2003;144(6):2242–52. Available from: https://doi.org/10.1210/en.2003-0007.

[114] Helmstädter J, Frenis K, Filippou K, et al. Endothelial GLP-1 (glucagon-like peptide-1) receptor mediates cardiovascular protection by liraglutide in mice with experimental arterial hypertension. Arterioscler Thromb Vasc Biol 2020;40(1):145–58. Available from: https://doi.org/10.1161/atv.0000615456.97862.30.

[115] Bose AK, Mocanu MM, Carr RD, Yellon DM. Myocardial ischaemia-reperfusion injury is attenuated by intact glucagon like peptide-1 (GLP-1) in the in vitro rat heart and may involve the p70s6K pathway. Cardiovasc Drugs Ther 2007;21(4):253–6. Available from: https://doi.org/10.1007/s10557-007-6030-6.

[116] Okerson T, Yan P, Stonehouse A, Brodows R. Effects of exenatide on systolic blood pressure in subjects with type 2 diabetes. Am J Hypertens 2010;23(3):334–9. Available from: https://doi.org/10.1038/ajh.2009.245.

[117] Baretić M, Kušec V, Pavlić-Renar I. Glucagon-like peptide-1 infusion suppresses aldosterone levels in healthy normal-weight individuals: double-blind, placebo-controlled crossover study. Diabetes Ther 2018;9:2315–24. Available from: https://doi.org/10.1007/s13300-018-0517-y.

[118] Del Olmo-Garcia MI, Merino-Torres JF. GLP-1 receptor agonists and cardiovascular disease in patients with type 2 diabetes. J Diabetes Res 2018;2018:4020492. Available from: https://doi.org/10.1155/2018/4020492 Published 2018 Apr 2.

[119] Hojberg PV, Vilsboll T, Rabol R, et al. Four weeks of near-normalisation of blood glucose improves the insulin response to glucagon-like peptide-1 and glucose-dependent insulinotropic polypeptide in patients with type 2 diabetes. Diabetologia 2009;52:199–207.

[120] Nauck MA, Heimesaat MM, Orskov C, Holst JJ, Ebert R, Creutzfeldt W. Preserved incretin activity of glucagon-like peptide 1 [7-36 amide] but not of synthetic human gastric inhibitory polypeptide in patients with type-2 diabetes mellitus. J Clin Invest 1993;91(1):301–7.

[121] Sharma D, Verma S, Vaidya S, Kalia K, Tiwari V. Recent updates on GLP-1 agonists: Current advancements & challenges. Biomed Pharmacother 2018;108:952–62. Available from: https://doi.org/10.1016/j.biopha.2018.08.088.

[122] He J, Chu Y. Small-molecule GLP-1 secretagogs: challenges and recent advances [published online ahead of print, 2020 Aug 21]. Drug Discov Today 2020;S1359-6446(20). Available from: https://doi.org/10.1016/j.drudis.2020.08.003 30308-1.

[123] Kristensen SL, Rorth R, Jhund PS, Docherty KF, Sattar N, Preiss D, et al. Cardiovascular, mortality, and kidney outcomes with GLP-1 receptor agonists in patients with type 2 diabetes: a systematic review and meta-analysis of cardiovascular outcome trials. Lancet Diabetes Endocrinol 2019;7(10):776–85. Available from: https://doi.org/10.1016/S2213-11218587(19)30249-9.

[124] Marso SP, Bain SC, Consoli A, et al. Semaglutide and cardiovascular outcomes in patients with type 2 diabetes. N Engl J Med 2016;375(19):1834–44. Available from: https://doi.org/10.1056/NEJMoa1607141.

[125] Marso SP, Daniels GH, Brown-Frandsen K, et al. Liraglutide and cardiovascular outcomes in type 2 diabetes. N Engl J Med 2016;375(4):311–22. Available from: https://doi.org/10.1056/NEJMoa1603827.

[126] Husain M, Birkenfeld AL, Donsmark M, et al. Oral semaglutide and cardiovascular outcomes in patients with type 2 diabetes. N Engl J Med 2019;381(9):841–51. Available from: https://doi.org/10.1056/NEJMoa1901118.

[127] Hernandez AF, Green JB, Janmohamed S, et al. Albiglutide and cardiovascular outcomes in patients with type 2 diabetes and cardiovascular disease (Harmony Outcomes): a

double-blind, randomised placebo-controlled trial. Lancet. 2018;392(10157):1519–29. Available from: https://doi.org/10.1016/S0140-6736(18)32261-X.

[128] Gerstein HC, Colhoun HM, Dagenais GR, et al. Dulaglutide and cardiovascular outcomes in type 2 diabetes (REWIND): a double-blind, randomised placebo-controlled trial. Lancet. 2019;394(10193):121–30. Available from: https://doi.org/10.1016/S0140-6736(19)31149-3.

[129] Pfeffer MA, Claggett B, Diaz R, et al. Lixisenatide in patients with type 2 diabetes and acute coronary syndrome. N Engl J Med 2015;373(23):2247–57. Available from: https://doi.org/10.1056/NEJMoa1509225.

[130] Holman RR, Bethel MA, Mentz RJ, et al. Effects of once-weekly exenatide on cardiovascular outcomes in type 2 diabetes. N Engl J Med 2017;377(13):1228–39. Available from: https://doi.org/10.1056/NEJMoa1612917.

[131] Huang M, Wei R, Wang Y, et al. Protective effect of glucagon-like peptide-1 agents on reperfusion injury for acute myocardial infarction: a meta-analysis of randomized controlled trials. Ann Med 2017;49(7):552–61. Available from: https://doi.org/10.1080/07853890.2017.1306653.

[132] Bohórquez DV, Liddle RA. Gastrointestinal hormones and neurotransmitters. In: Feldman M, Friedman LS, Brandt LJ, editors. Sleisenger and Fordtran's gastrointestinal and liver disease e-book: pathophysiology, diagnosis, management. Amsterdam, The Netherlands: Elsevier Health Sciences; 2016. p. 36–56.

[133] Kojima M, Kangawa K. Ghrelin: structure and function. Physiol Rev 2005;85(2):495–522. Available from: https://doi.org/10.1152/physrev.00012.2004.

[134] Iglesias MJ, Piñeiro R, Blanco M, et al. Growth hormone releasing peptide (ghrelin) is synthesized and secreted by cardiomyocytes. Cardiovasc Res 2004;62(3):481–8. Available from: https://doi.org/10.1016/j.cardiores.2004.01.024.

[135] Date Y, Kojima M, Hosoda H, et al. Ghrelin, a novel growth hormone-releasing acylated peptide, is synthesized in a distinct endocrine cell type in the gastrointestinal tracts of rats and humans. Endocrinology. 2000;141(11):4255–61. Available from: https://doi.org/10.1210/endo.141.11.7757.

[136] Hosoda H, Kojima M, Matsuo H, Kangawa K. Ghrelin and des-acyl ghrelin: two major forms of rat ghrelin peptide in gastrointestinal tissue. Biochem Biophys Res Commun 2000;279(3):909–13. Available from: https://doi.org/10.1006/bbrc.2000.4039.

[137] Pradhan G, Samson SL, Sun Y. Ghrelin: much more than a hunger hormone. Curr Opin Clin Nutr Metab Care 2013;16(6):619–24. Available from: https://doi.org/10.1097/MCO.0b013e328365b9be.

[138] Sato T, Nakamura Y, Shiimura Y, Ohgusu H, Kangawa K, Kojima M. Structure, regulation and function of ghrelin. J Biochem 2012;151(2):119–28. Available from: https://doi.org/10.1093/jb/mvr134.

[139] Plyler KS, Daniels D. Fourth ventricle injection of ghrelin decreases angiotensin II-induced fluid intake and neuronal activation in the paraventricular nucleus of the hypothalamus. Physiol Behav 2017;178:35–42. Available from: https://doi.org/10.1016/j.physbeh.2016.11.018.

[140] Nagaya N, Kojima M, Uematsu M, et al. Hemodynamic and hormonal effects of human ghrelin in healthy volunteers. Am J Physiol Regul Integr Comp Physiol 2001;280(5):R1483–7. Available from: https://doi.org/10.1152/ajpregu.2001.280.5.R1483.

[141] Mao Y, Tokudome T, Otani K, Kishimoto I, Miyazato M, Kangawa K. Excessive sympathoactivation and deteriorated heart function after myocardial infarction in male ghrelin knockout mice. Endocrinology. 2013;154(5):1854–63. Available from: https://doi.org/10.1210/en.2012-2132.

[142] Cao Y, Tang J, Yang T, et al. Cardioprotective effect of ghrelin in cardiopulmonary bypass involves a reduction in inflammatory response. PLoS One 2013;8(1):e55021. Available from: https://doi.org/10.1371/journal.pone.0055021.

[143] Yuan MJ, He-Huang HH, Li-Quan H-J, Huang CX. Myocardial angiogenesis after chronic ghrelin treatment in a rat myocardial infarction model. Regul Pept 2012;179(1-3):39–42. Available from: https://doi.org/10.1016/j.regpep.2012.08.013.

[144] Yang C, Yang P. GW29-e1601 Ghrelin partially protects against heart failure after myocardial infarction by increasing the Bcl-2/Bax ratio in a GHS-R1a-dependent manner. 2018.

[145] Yang C, Yang P. GW29-e1603 The effect of activin A-follistatin system imbalance on myocardial fibrosis of heart failure and the regulation of ghrelin. 2018.

[146] Yuan MJ, Huang CX, Tang YH, et al. A novel peptide ghrelin inhibits neural remodeling after myocardial infarction in rats. Eur J Pharmacol 2009;618(1-3):52–7. Available from: https://doi.org/10.1016/j.ejphar.2009.07.015.

[147] Wiley KE, Davenport AP. Comparison of vasodilators in human internal mammary artery: ghrelin is a potent physiological antagonist of endothelin-1. Br J Pharmacol 2002;136(8):1146–52. Available from: https://doi.org/10.1038/sj.bjp.0704815.

[148] Dincă M, Dumitriu IL, Gurzu MB, et al. Ghrelin and Ang 1-7 have cumulative vasodilatory effects on pulmonary vessels. Rev Med Chir Soc Med Nat Iasi 2010;114(3): 803–7.

[149] Suining Xu, Yan J, Wang Z. GW27-e0298 Octanoylated ghrelin attenuates vascular calcification in diabetes mellitus. J Am Coll Cardiol 2016;68(16S):C79.

[150] Xu S, Ye F, Li L, et al. Ghrelin attenuates vascular calcification in diabetic patients with amputation. Biomed Pharmacother 2017;91:1053–64. Available from: https://doi.org/10.1016/j.biopha.2017.05.031.

[151] Zhao H, Liu G, Wang Q, et al. Effect of ghrelin on human endothelial cells apoptosis induced by high glucose. Biochem Biophys Res Commun 2007;362(3):677–81. Available from: https://doi.org/10.1016/j.bbrc.2007.08.021.

[152] Tesauro M, Schinzari F, Iantorno M, et al. Ghrelin improves endothelial function in patients with metabolic syndrome. Circulation. 2005;112(19):2986–92. Available from: https://doi.org/10.1161/CIRCULATIONAHA.105.553883.

[153] Huang CX, Yuan MJ, Huang H, et al. Ghrelin inhibits post-infarct myocardial remodeling and improves cardiac function through anti-inflammation effect. Peptides. 2009;30 (12):2286–91. Available from: https://doi.org/10.1016/j.peptides.2009.09.004.

[154] Schwenke DO, Tokudome T, Kishimoto I, et al. Early ghrelin treatment after myocardial infarction prevents an increase in cardiac sympathetic tone and reduces mortality. Endocrinology. 2008;149(10):5172–6. Available from: https://doi.org/10.1210/en.2008-0472.

[155] Zhang G, Yin X, Qi Y, et al. Ghrelin and cardiovascular diseases. Curr Cardiol Rev 2010;6(1):62–70. Available from: https://doi.org/10.2174/157340310790231662.

[156] Xu H, Li Y, Liu R, et al. Protective effects of ghrelin on brain mitochondria after cardiac arrest and resuscitation. Neuropeptides. 2019;76:101936. Available from: https://doi.org/10.1016/j.npep.2019.05.007.

[157] Suzuki Y, Yoshitomo-Nakagawa K, Maruyama K, Suyama A, Sugano S. Construction and characterization of a full length-enriched and a 5'-end-enriched cDNA library. Gene. 1997;200(1-2):149–56. Available from: https://doi.org/10.1016/s0378-1119(97)00411-3.

[158] Southey BR, Rodriguez-Zas SL, Sweedler JV. Characterization of the prohormone complement in cattle using genomic libraries and cleavage prediction approaches. BMC

Genomics 2009;10:228. Available from: https://doi.org/10.1186/1471-2164-10-228 Published 2009 May 16.

[159] Shichiri M, Ishimaru S, Ota T, Nishikawa T, Isogai T, Hirata Y. Salusins: newly identified bioactive peptides with hemodynamic and mitogenic activities. Nat Med 2003;9(9):1166–72. Available from: https://doi.org/10.1038/nm913.

[160] Sato K, Koyama T, Tateno T, Hirata Y, Shichiri M. Presence of immunoreactive salusin-alpha in human serum and urine. Peptides. 2006;27(11):2561–6. Available from: https://doi.org/10.1016/j.peptides.2006.06.005.

[161] Sato K, Sato T, Susumu T, Koyama T, Shichiri M. Presence of immunoreactive salusin-beta in human plasma and urine. Regul Pept 2009;158(1-3):63–7. Available from: https://doi.org/10.1016/j.regpep.2009.07.017.

[162] Ozelius LJ, Hewett JW, Page CE, et al. The early-onset torsion dystonia gene (DYT1) encodes an ATP-binding protein. Nat Genet 1997;17(1):40–8. Available from: https://doi.org/10.1038/ng0997-40.

[163] Ozelius LJ, Page CE, Klein C, et al. The TOR1A (DYT1) gene family and its role in early onset torsion dystonia. Genomics. 1999;62(3):377–84. Available from: https://doi.org/10.1006/geno.1999.6039.

[164] Sato K, Koyama T, Shichiri M. Biosynthesis and secretion of salusin-alpha from human cells. Peptides. 2008;29(12):2203–7. Available from: https://doi.org/10.1016/j.peptides.2008.08.015.

[165] Nakayama C, Shichiri M, Sato K, Hirata Y. Expression of prosalusin in human neuroblastoma cells. Peptides. 2009;30(7):1362–7. Available from: https://doi.org/10.1016/j.peptides.2009.03.021.

[166] Matsumura K. Salusin and central regulation of blood pressure in hypertension. J Hypertens 2014;32(5):981–2. Available from: https://doi.org/10.1097/HJH.0000000000000156.

[167] Izumiyama H, Tanaka H, Egi K, Sunamori M, Hirata Y, Shichiri M. Synthetic salusins as cardiac depressors in rat. Hypertension. 2005;45(3):419–25. Available from: https://doi.org/10.1161/01.HYP.0000156496.15668.62.

[168] Shi JS, Li D, Li N, et al. Inhibition of L-type calcium currents by salusin-beta in rat cardiac ventricular myocytes. Peptides. 2010;31(6):1146–9. Available from: https://doi.org/10.1016/j.peptides.2010.03.016.

[169] Usta Atmaca H, Akbas F. Is salusin-alpha a new marker of cardiovascular disease risk in hypothyroidism? Acta Endocrinol (Buchar) 2017;13(1):53–9. Available from: https://doi.org/10.4183/aeb.2017.53.

[170] Watanabe T, Suguro T, Sato K, et al. Serum salusin-alpha levels are decreased and correlated negatively with carotid atherosclerosis in essential hypertensive patients. Hypertens Res 2008;31(3):463–8. Available from: https://doi.org/10.1291/hypres.31.463.

[171] Ti Y, Wang F, Wang ZH, et al. Associations of serum salusin-alpha levels with atherosclerosis and left ventricular diastolic dysfunction in essential hypertension. J Hum Hypertens 2012;26(10):603–9. Available from: https://doi.org/10.1038/jhh.2011.71.

[172] Watanabe T, Nishio K, Kanome T, et al. Impact of salusin-alpha and -beta on human macrophage foam cell formation and coronary atherosclerosis. Circulation. 2008;117(5):638–48. Available from: https://doi.org/10.1161/CIRCULATIONAHA.107.712539.

[173] Ozgen M, Koca SS, Dagli N, Balin M, Ustundag B, Isik A. Serum salusin-alpha level in rheumatoid arthritis. Regul Pept 2011;167(1):125–8. Available from: https://doi.org/10.1016/j.regpep.2010.12.003.

[174] Kimoto S, Sato K, Watanabe T, Suguro T, Koyama T, Shichiri M. Serum levels and urinary excretion of salusin-alpha in renal insufficiency. Regul Pept 2010;162(1-3):129−32. Available from: https://doi.org/10.1016/j.regpep.2010.03.009.

[175] Wang WJ, Jiang X, Gao CC, Chen ZW. Salusin-α mitigates diabetic nephropathy via inhibition of the Akt/mTORC1/p70S6K signaling pathway in diabetic rats [published online ahead of print, 2019 Oct 31]. Drug Chem Toxicol 2019;1−8. Available from: https://doi.org/10.1080/01480545.2019.1683572.

[176] Tang K, Wang F, Zeng Y, Chen X, Xu X. Salusin-α attenuates hepatic steatosis and atherosclerosis in high fat diet-fed low density lipoprotein receptor deficient mice. Eur J Pharmacol 2018;830:76−86. Available from: https://doi.org/10.1016/j.ejphar.2018.04.026.

[177] Esfahani M, Saidijam M, Goodarzi MT, Movahedian A, Najafi R. Salusin-α attenuates inflammatory responses in vascular endothelial cells. Biochem (Mosc) 2017;82(11):1314−23. Available from: https://doi.org/10.1134/S0006297917110098.

[178] Chang Y, Yoo J, Kim H, Park HJ, Jeon S, Kim J. Salusin-β mediate neuroprotective effects for Parkinson's disease. Biochem Biophys Res Commun 2018;503(3):1428−33. Available from: https://doi.org/10.1016/j.bbrc.2018.07.059.

[179] Li HB, Qin DN, Suo YP, et al. Blockade of salusin-β in hypothalamic paraventricular nucleus attenuates hypertension and cardiac hypertrophy in salt-induced hypertensive rats. J Cardiovasc Pharmacol 2015;66(4):323−31. Available from: https://doi.org/10.1097/FJC.0000000000000284.

[180] Masumura M, Watanabe R, Nagashima A, et al. Anti-salusin-β antibody enhances angiogenesis after myocardial ischemia reperfusion injury. Expert Opin Ther Targets 2013;17(9):1003−9. Available from: https://doi.org/10.1517/14728222.2013.819852.

[181] Wang Y, Wang S, Zhang J, et al. Salusin-β is superior to salusin-α as a marker for evaluating coronary atherosclerosis. J Int Med Res 2020;48(2). Available from: https://doi.org/10.1177/0300060520903868 300060520903868.

[182] Dervişoğlu P, Elmas B, Kösecik M, İşgüven ŞP, Büyükavcı M, Köroğlu M. Salusin-α levels are negatively correlated with diastolic blood pressure in children with obesity. Cardiol Young 2019;29(10):1225−9. Available from: https://doi.org/10.1017/S1047951119001173.

[183] Erden I, Uçak H, Demir B, et al. Serum salusin-α and salusin-β levels in patients with psoriasis. Eur J Dermatol 2015;25(4):352−3. Available from: https://doi.org/10.1684/ejd.2015.2575.

[184] Watanabe T, Sato K, Itoh F, Wakabayashi K, Shichiri M, Hirano T. Endogenous bioactive peptides as potential biomarkers for atherosclerotic coronary heart disease. Sens (Basel) 2012;12(4):4974−85. Available from: https://doi.org/10.3390/s120404974.

[185] Tanyeli A, Eraslan E, Polat E, Bal T. Protective effect of salusin-α and salusin-β against ethanol-induced gastric ulcer in rats. J Basic Clin Physiol Pharmacol 2017;28(6):623−30. Available from: https://doi.org/10.1515/jbcpp-2016-0100.

[186] Cakir M, Duzova H, Taslidere A, Orhan G, Ozyalin F. Protective effects of salusin-α and salusin-β on renal ischemia/reperfusion damage and their levels in ischemic acute renal failure [published correction appears in Biotech Histochem 2020 Oct;95(7):561] Biotech Histochem 2017;92(2):122−33. Available from: https://doi.org/10.1080/10520295.2017.1283056.

[187] Tatemoto K, Hosoya M, Habata Y, et al. Isolation and characterization of a novel endogenous peptide ligand for the human APJ receptor. Biochem Biophys Res Commun 1998;251(2):471−6. Available from: https://doi.org/10.1006/bbrc.1998.9489.

[188] Wysocka MB, Pietraszek-Gremplewicz K, Nowak D. The role of apelin in cardiovascular diseases, obesity and cancer. Front Physiol 2018;9:557. Available from: https://doi.org/10.3389/fphys.2018.00557 Published 2018 May 23.

[189] Kuba K, Sato T, Imai Y, Yamaguchi T. Apelin and Elabela/Toddler; double ligands for APJ/Apelin receptor in heart development, physiology, and pathology. Peptides. 2019;111:62–70. Available from: https://doi.org/10.1016/j.peptides.2018.04.011.

[190] Lee DK, Cheng R, Nguyen T, et al. Characterization of apelin, the ligand for the APJ receptor. J Neurochem 2000;74(1):34–41. Available from: https://doi.org/10.1046/j.1471-4159.2000.0740034.x.

[191] Scimia MC, Hurtado C, Ray S, et al. APJ acts as a dual receptor in cardiac hypertrophy. Nature. 2012;488(7411):394–8. Available from: https://doi.org/10.1038/nature11263.

[192] Vale W, Spiess J, Rivier C, Rivier J. Characterization of a 41-residue ovine hypothalamic peptide that stimulates secretion of corticotropin and beta-endorphin. Science. 1981;213(4514):1394–7. Available from: https://doi.org/10.1126/science.6267699.

[193] Goidescu CM, Vida-Simiti LA. The Apelin-APJ system in the evolution of heart failure. Clujul Med 2015;88(1):3–8. Available from: https://doi.org/10.15386/cjmed-380.

[194] Simpkin JC, Yellon DM, Davidson SM, Lim SY, Wynne AM, Smith CC. Apelin-13 and apelin-36 exhibit direct cardioprotective activity against ischemia-reperfusion injury. Basic Res Cardiol 2007;102(6):518–28. Available from: https://doi.org/10.1007/s00395-007-0671-2.

[195] Yang P, Maguire JJ, Davenport AP. Apelin, Elabela/Toddler, and biased agonists as novel therapeutic agents in the cardiovascular system. Trends Pharmacol Sci 2015;36(9):560–7. Available from: https://doi.org/10.1016/j.tips.2015.06.002.

[196] Donaldson CJ, Sutton SW, Perrin MH, et al. Cloning and characterization of human urocortin [published correction appears in Endocrinology 1996 Sep;137(9):3896] Endocrinology 1996;137(5):2167–70. Available from: https://doi.org/10.1210/endo.137.5.8612563.

[197] Gottowik J, Goetschy V, Henriot S, et al. Labelling of CRF1 and CRF2 receptors using the novel radioligand, [3H]-urocortin. Neuropharmacology. 1997;36(10):1439–46. Available from: https://doi.org/10.1016/s0028-3908(97)00098-1.

[198] Vaughan J, Donaldson C, Bittencourt J, et al. Urocortin, a mammalian neuropeptide related to fish urotensin I and to corticotropin-releasing factor. Nature. 1995;378(6554):287–92. Available from: https://doi.org/10.1038/378287a0.

[199] Brar BK, Jonassen AK, Stephanou A, et al. Urocortin protects against ischemic and reperfusion injury via a MAPK-dependent pathway. J Biol Chem 2000;275(12):8508–14. Available from: https://doi.org/10.1074/jbc.275.12.8508.

[200] Bale TL, Hoshijima M, Gu Y, et al. The cardiovascular physiologic actions of urocortin II: acute effects in murine heart failure. Proc Natl Acad Sci USA 2004;101(10):3697–702. Available from: https://doi.org/10.1073/pnas.0307324101.

[201] Davis ME, Pemberton CJ, Yandle TG, et al. Urocortin 2 infusion in healthy humans: hemodynamic, neurohormonal, and renal responses. J Am Coll Cardiol 2007;49(4):461–71. Available from: https://doi.org/10.1016/j.jacc.2006.09.035.

[202] Chan WYW, Charles CJ, Frampton CM, et al. Human muscle sympathetic nerve responses to urocortin-2 in health and stable heart failure. Clin Exp Pharmacol Physiol 2015;42(9):888–95. Available from: https://doi.org/10.1111/1440-1681.12449.

[203] Rademaker MT, Charles CJ, Nicholls MG, Richards AM. Urocortin 2 combined with angiotensin-converting enzyme inhibition in experimental heart failure. Clin Sci (Lond) 2008;114(10):635–42. Available from: https://doi.org/10.1042/CS20070364.

[204] Davis ME, Pemberton CJ, Yandle TG, et al. Urocortin 2 infusion in human heart failure. Eur Heart J 2007;28(21):2589–97. Available from: https://doi.org/10.1093/eurheartj/ehm340.

[205] Chan WY, Frampton CM, Crozier IG, Troughton RW, Richards AM. Urocortin-2 infusion in acute decompensated heart failure: findings from the UNICORN study (urocortin-2 in the treatment of acute heart failure as an adjunct over conventional therapy). JACC Heart Fail 2013;1(5):433–41. Available from: https://doi.org/10.1016/j.jchf.2013.07.003.

[206] Pan W, Kastin AJ. Urocortin and the brain. Prog Neurobiol 2008;84(2):148–56. Available from: https://doi.org/10.1016/j.pneurobio.2007.10.008.

[207] Pedersen WA, Wan R, Zhang P, Mattson MP. Urocortin, but not urocortin II, protects cultured hippocampal neurons from oxidative and excitotoxic cell death via corticotropin-releasing hormone receptor type I. J Neurosci 2002;22(2):404–12. Available from: https://doi.org/10.1523/JNEUROSCI.22-02-00404.2002.

Section 5

Drug development

Chapter 27

Clinical and preclinical data on therapeutic peptides

Subhasis Mahari[1], Deepshikha Shahdeo[1], Ivneet Banga[2], Samraggi Choudhury[1] and Sonu Gandhi[1]
[1]*DBT-National Institute of Animal Biotechnology (DBT-NIAB), Hyderabad, Telangana, India,*
[2]*Department of Bioengineering, University of Texas, Arlington, TX, United States*

27.1 Introduction: the evolution of peptide therapeutics

In recent times, peptides have been receiving significant attention as therapeutics due to their efficient targeting, chemical stability, and robust mode of action. Therapeutic peptides are short, linear sequences of amino acids mostly used in therapeutic applications. Development of therapeutics started parallel to the advancement of medicine and biochemistry. Peptides segregated from natural sources contributed to life saving drugs in the early 1900s [1]. Development of recombinant insulin for the treatment of diabetes was the first use of recombinant proteins in medical practice and dates back to the early 1980s [2]. Chemical synthesis of peptides led to the advancement in the development of therapeutic peptides in the 1950s with man-made oxytocin and vasopressin being used in clinical trials. Despite the fact that peptides are produced by natural and biosynthetic fermentation, chemical synthesis enables conjugation of other molecules or embodiment of nonnatural amino acids with the peptide [3]. They play an important role by acting as hormones, growth factors, neurotransmitters, and cell signaling pathways [4]. Their use as biologically active therapeutics has increased significantly over the past few decades as they possess inherent advantages such as specificity, sensitivity, and are clinically amenable by modification with chemical agents [2]. Many different peptide derivatives such as a simple dipeptide sugar complex like aspartame, which is being used as a low calorie sweetener, or a highly complex clinically employed hormone therapy such as calcitonin or oxytocin, have emerged as commercial products [5].

Therapeutic peptides have mostly been synthesized or produced by three methods [6]. Biological or bioactive peptides derived from animals, plants or humans (1) are isolated from natural peptide-based hormones or as

Peptide and Peptidomimetic Therapeutics. DOI: https://doi.org/10.1016/B978-0-12-820141-1.00007-8

subcomponents of larger proteins; (2) are produced using recombinant DNA technology; or (3) are discovered using chemical libraries. The production method used for a peptide is mostly governed by the size and the chemical structure of the peptide [7]. Various methods of peptide synthesis such as chemical pathway of production, recombinant DNA technology, and synthesis using transgenic animals or plants have evolved over time [8]. Chemical synthesis is the most widely used method of production, as most peptide therapeutics involve unnatural amino acids or the formation of (pseudo) peptide bonds, which can be easily produced using this pathway rather than employing recombinant DNA technology. Chemical synthesis allows the development of peptides ranging from 5 to 50 residues and produces one of the most pharmaceutically relevant molecules [7]. Peptides are mostly defined based upon their chemical method of synthesis, primarily, the solid-phase peptide synthesis [5,9]. Nowadays, development of therapeutic peptides by the pharmaceutical industries has become easily accessible due to the advancements in solid-phase peptide synthesis [9]. Solid-phase peptide synthesis allows for easy production of peptides in the early stages of preclinical research and the development of pharmaceutical agents using active peptides as targeting agents, or active pharmaceutical ingredients.

Solid-phase peptide synthesis involves three different routes of synthesis, namely, sequential, convergent, and chemical ligation synthesis. The process of sequential synthesis involves the consecutive incorporation of peptides until the desired sequence is attained. Convergent synthesis involves the same process such as sequential synthesis with the only difference being that peptides are split from the polymer backbone without the removal of the N/C-terminal or functional side chain protecting group which is then finally cross-linked via condensation in solution. This is also referred to as mixed phase synthesis [7]. Formation of long peptide and protein chains for formulation of any unique reaction product is called chemical ligation. This method uses coupling of unprotected fragments via selective chemical reactions [10]. The most commonly used form of chemical ligation for chemoselective reactions is native chemical ligation, which ensures proper assembly of long peptides from their associated peptide building blocks [11].

Peptides were proposed as a treatment method for cancer four decades ago by molecular biologists, as they are distinct pharmaceutical compounds that can be easily tuned and tailored to allow biodegradability and biocompatibility [12]. High molecular weight, presence of charged groups, and low hydrophobic character caused a challenge for absorption. Low bioavailability using oral uptake and short plasma life were limitations faced while using native peptides [1]. Other obstacles such as high polarity and molecular weight of peptides decreased intestinal permeability. This reduced the bioavailability as well as the half- life (<30 min) of the peptide [13]. Other delivery processes like intravenous or subcutaneous showed reduced efficacy due to conformational changes, dissociation of protein subunits and destruction of side-groups [14,15]. As oral delivery is patient friendly and noninvasive, researchers focus on developing embodiments based on oral mode of delivery for efficient administration of the peptides.

This chapter describes the evolution of peptide therapeutics with a brief introduction on synthesis, classification, and cellular uptake mechanisms. The later part of the chapter focuses on current peptides in the drug market, preclinical data involved, and the future aspects of using peptides as therapeutics.

27.2 Classification of peptides

27.2.1 Cell-penetrating peptides

The last two decades have seen the development of more than 100 therapeutic peptides (comprising 5–40 amino acid sequences) that can be successfully employed for the delivery of drugs in the plant, bacterial, or mammalian cells [16,17]. Classification of cell-penetrating peptides (CPPs) is often based on their origin or on the amino acid sequence. Mostly, CPPs are classified based upon the charge that they carry (mainly positively charged amino acids such as lysine and arginine within the physiological pH range) or the electrostatic interactions that are present with the cell surface negatively charged glycoproteins (prior to cellular internalization). Arginine has a guanidine head group that can easily form hydrogen bonds with sulfates and phosphates having negative charges, which allows easy cellular internalization. However, lysine does not possess guanidine as the head group and is unable to internalize within the cell and thereby cannot act as cargo for drug delivery.

Peptides are also classified as cationic, amphipathic and hydrophobic, based on their physical and chemical properties [18]. A cationic peptide was the first CPP derived from the HIV-1 protein TAT [19]. Nuclear localization sequence (NLS) is a type of CPP based upon lysine, arginine, or proline-rich motifs which can be transported into the nucleus via nuclear pore complex and can further be divided into monopartite and bipartite signals [20]. The amphipathic alpha-helix is a common structural motif that gets bound to membranes, as it contains a hydrophobic side at one end which attaches to the membrane while the other end to the NLS. Other amphipathic alpha-helices are derived from natural compounds like pVEC, ARF and BPrPr transmembrane domains, as well as N-terminal domains and motifs. Secondary amphipathic alpha CPPs have a patch of hydrophobic molecules on one face and cationic, anionic, or polar molecules present on another face, while amphipathic β-sheet peptides contain a hydrophobic and another hydrophilic stretch of amino acids exposed to the solvent [21]. There are also hydrophobic CPPs that either contains nonpolar residues or residues with hydrophobic motifs that help in the efficient uptake of molecules. Hydrophobic CPPs can further be divided into natural and chemically modified peptides, including pepducins and prenylated peptides. Natural peptides include anionic and cationic pentapeptides like PLMKE, VSSGK, and VPTLQ.

The cell transduction property of the CPP is determined primarily by the amino acid sequence of the CPP, mainly by the presence of arginine [16]. Low amphipathic CPPs include polyarginines, TAT and penetratin [22–24]. However, in certain high order amphipathic CPPs such as MAP, Pep-1, and Transportan, lysine residues are responsible for the charge contribution that allows cellular internalization due to low absorption affinity for the cell surface even though they do not have a guanidine head group [25,26]. CPPs can be used for targeting cancerous tissue and also as carriers for anticancer drugs, hormones, vaccines or radionuclides. Delivery of therapeutic CPP payloads can be self-assembled CPP mediated, nanoparticle-mediated, liposome associated, siRNA conjugated, or radionucleotide conjugated (Fig. 27.1) [18].

Usually, CPP mediated drug delivery is done by covalent conjugation via chemical or biological agents. Recently, Wu et al. demonstrated the targeting of tumor blood vessels using the SP5–52 peptide conjugated PEGylated distearoyl phosphatidyl ethanolamine (DSPE-PEG) liposome (Wu et al., 2007). Loading of the anticancer topoisomerase II inhibitor drug doxorubicin on

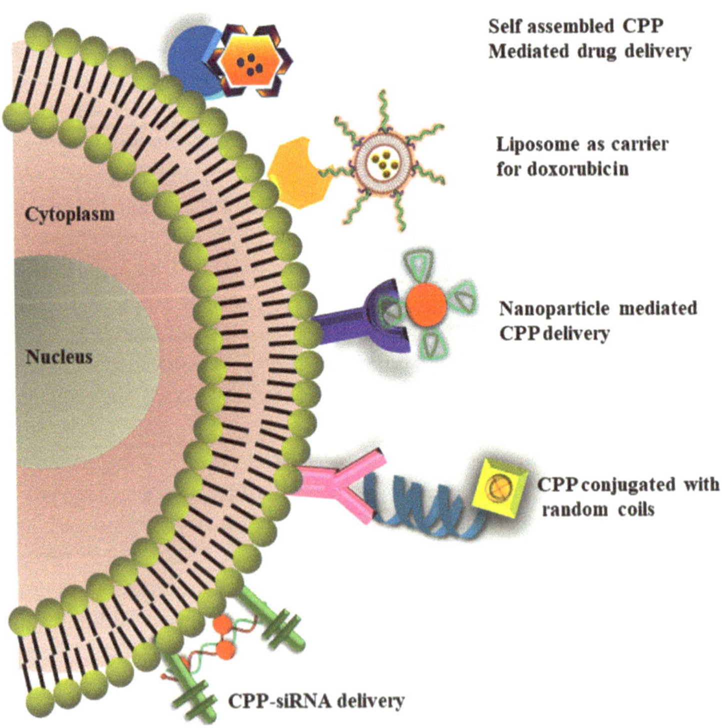

FIGURE 27.1 Cell-penetrating peptides acts as a carrier for the transportation of therapeutic agents, such as drugs, liposomes, nanoparticles, radionuclides, and siRNA.

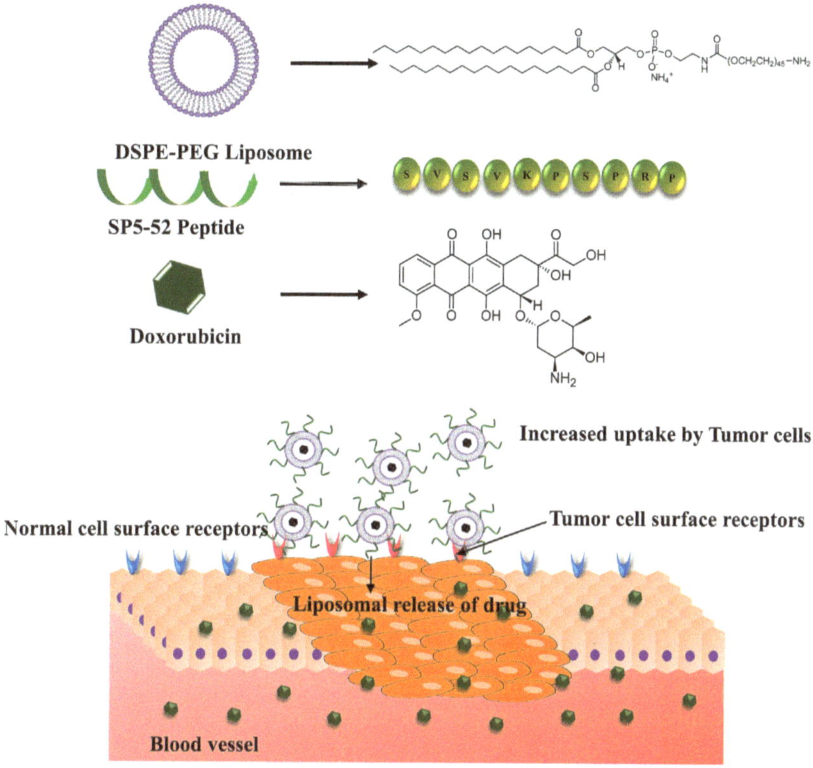

FIGURE 27.2 Drug delivery of doxorubicin using PEGylated distearoyl phosphatidyl ethanolamine (DSPE-PEG) liposome as a carrier. Doxorubicin was encapsulated with DSPE-PEG and further conjugated with SP5–52 peptide on its surface for efficient delivery at the tumor cell surface.

SP5–52-DSPE-PEG liposome particles decreased tumor blood vessel formation as shown in Fig. 27.2. Further, Stroh et al. encapsulated quantum dots with TAT-modified micelles, which were taken up by mouse endothelial cells in tumor endothelium [27]. CPP conjugated siRNA drug delivery was also considered a potential therapeutic agent because of its specificity in the suppression of genes that are over expressed in disease. Chiu et al. covalently conjugated TAT peptide to the 3′ end of an siRNA antisense strand using a noncleavable SMPB (succinimidyl 4-(p-maleimidophenyl) butyrate) crosslinker [28]. The siRNA-TAT conjugate was still able to induce targeted gene silencing with a 30% reduction in gene expression.

27.2.2 Cell-targeting peptides

Cell-targeting peptides (CTPs) are small peptides of 3–10 amino acids that are specific for certain cells or tissues [29]. CTPs have high specificity and affinity for particular cell lines or cell interactions [30]. Monoclonal antibodies have been generated for targeting different cell surface markers, but their large size, high production cost, and altered binding after conjugation are drawbacks that provide CTPs with a chance to establish in the market. CTPs are cost-effective, highly efficient, specific, and are easily prepared [31]. Peptides that specifically target cancerous cells can be characterized by two types of mechanisms: phage display and one bead one compound.

In phage display, the peptide is fused with the bacteriophage coat protein expressed on the surface of a virus or virion. This technology was first noted by George P. Smith in 1985 [32] as a display of foreign peptides on the surface of phage particles. This is done by splicing a gene consisting of the peptide into a gene containing the capsid structural protein [33]. The phage display technique is also called biopanning, as it aims to isolate target binding molecules. As shown in Fig. 27.3, phage display includes five steps for peptide selection. The first step includes "library construction and amplification" where the phage display library is constructed using cloning of different DNA combinations. The second step of phage display is "target capturing." Here, the target molecule is incubated with the phage library for binding. The third step constitutes "removing unbound and unspecific phages" by repeated washing. The fourth step includes "elution," where peptide bound targets are removed using a low pH buffer or by competitive elution. The fifth step is the "infection stage" where the eluted antigen-specific phages are infected with bacteria for amplification [34].

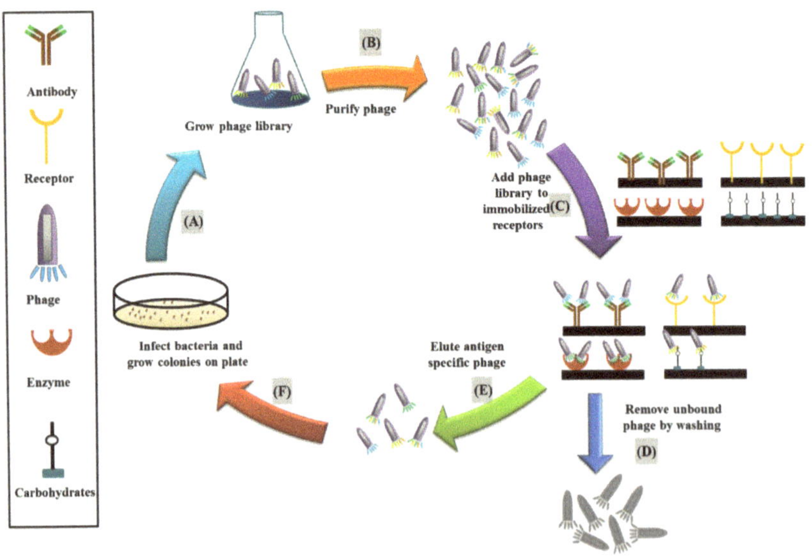

FIGURE 27.3 Application of biopanning with a random peptide library demonstrating phage display that can be used for targeting carbohydrates, antibodies, enzymes, and receptors.

In the one bead one compound approach, compound beads are made by the split-mix synthesis method that allows the display of a large number of the same peptide on a single bead [35]. Park et al. described a peptide ligand (Leu-Asp-Ile, Leu-Asp-Phe, or Leu-Asp-Val) that is specifically bound to malignant human lymphoid cells and not to normal human lymphocytes. Anti-α4 integrin antibodies could block the binding of peptide ligands to the cells indicating α4β1 integrin as the peptide target receptor [36]. In this method, the length of the peptide plays an important role, as peptide ligand binding is limited to a certain size. Usually, a peptide length of more than 15 amino acids is not considered feasible for peptide ligand binding. However, the affinity between peptides and their receptors can be increased via cyclization or multimerization [31].

27.2.3 Therapeutic and diagnostic agents

Therapeutic peptides include a family of proteins that have a common fold in which a single disulfide bond connects with two other disulfide bonds, forming a knot and conferring thermal, chemical, and proteolytic stability [37]. For example, knottins are cysteine knot mini proteins that evolved over time as a tool in therapeutics and diagnostics [38–40]. Knottins consist of loops with variable compositions of antiparallel β-strands that help in folding, structural maintenance, molecular interaction, and resisting amino acid mutations [39]. Knottins are present in fungi, plants, animals, and are responsible for protease inhibition, blockage of ion channels, and antimicrobial activity [41]. These knottins are being used in theranostics, as they have ideal biophysical and biochemical properties to be used in vivo. Their stability and small size made them amenable to recombinant expression. Moreover, multiple cell adhesion peptides, from the extracellular matrix and intercellular adhesion molecule proteins can be used for selective drug payload delivery, signal inhibition, and as diagnostic molecules [42].

A recent study documented that atrial natriuretic, a cardiac derived peptide hormone, was reported as a potential tool for colorectal cancer. Li et al. reported that a peptide called TCP-1 targeted blood vessels in tumor tissues [43]. Further, Wang et al. reported that F56 peptide conjugated nanoparticles loaded with vincristine (F56-VCR-NP) could target both primary tumor lesions and the neovasculature of lung metastases, leading to apoptosis of the neovasculature [44]. In addition, MPG peptides can be used to deliver oligonucleotides and plasmid DNA into cells. siRNA linked with the MPG peptide caused the temporary formation of transmembrane alpha-helices and beta sheets. The formation of these structures allowed insertion of the peptide complex into the membrane followed by translocation due to membrane potential as shown in Fig. 27.4. Several other peptides, such as contulakin-G and conantokin-G, used as neurotensin receptors are currently being studied

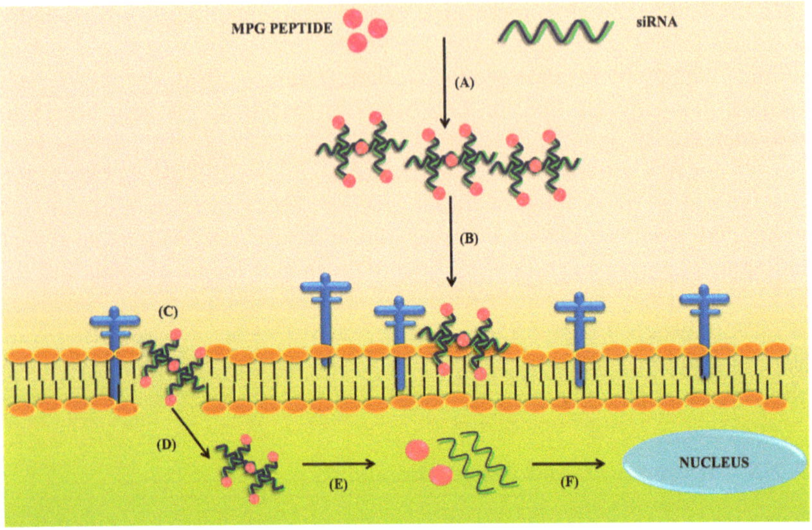

FIGURE 27.4 Delivery of siRNA peptide cargo. (A) MPG-siRNA forms a complex through electrostatic or hydrophobic interactions. (B) Interaction of the complex with cell surface via proteoglycans. (C) Direct penetration of the complex through the lipid bilayer of the cell. (D) The complex is released inside the cytoplasm. (E) Dissociation of the complex by proteases. (F) Targeted to the nucleus.

in phase I and II clinical trials. Application of peptides used in recent studies re described in Table 27.1.

Cellular uptake mechanisms for therapeutic peptides remain ambiguous in certain aspects. However, three distinct pathways have been identified for the internalization of peptides into the cell membrane. Choice of the pathway depends on three important factors, including the concentration of the therapeutic peptide, the amino acid sequence involved, and the lipid composition of the cellular membrane [2,56,57]. The uptake mechanism of the peptide also depends upon the charge, concentration, and polarity of the molecule. Generally, two methods of cellular uptake mechanism occur in cells: direct penetration, which does not require any energy, and endocytosis, which requires energy to bring fluids into the cell [18].

27.2.4 Direct cellular penetration

Direct cellular penetration of peptides via an energy-independent pathway involves uptake by the formation of structures, such as inverted micelle or pore formation. This can involve a membrane thinning model or carpet-like cellular uptake model for peptides. The first step is common in all energy-independent pathways, wherein the electrostatic interactions take place between the positively charged peptide head group and the negatively

TABLE 27.1 Cell-penetrating peptides (CPPs)and their application to therapeutics and imaging.

		Sequence	Applications	References
1 CPP				
1.1	MPG	GALFLGFLGAGS TMGAWSQPKKKRKV	Delivery of DNA, siRNA, plasmid DNA, and oligonucleotides	[45,46]
1.2	Pep-1	KETWWETWWTEWSQPKKKRKV	Peptides, proteins and PNA analogs delivered using Pep-1	[47]
1.3	Pep-2	KETWFETWFTEWSQPKKKRKV	Delivery of nucleic acids and peptide	[47]
1.4	Pep-3	KETWFETWFTEWSQPKKKRKV	Delivery of nucleic acids and peptide	[48]
1.5	CADY	AcGLWRALWRLLRSLWRLLWRA-Cya	Delivery of siRNA	[18]
1.6	Rath	TPWWRLWTKWHHKRRDLPRKPE	Binds to plasmid DNA, antibody and proteins	[49]
2 Peptides for targeting ligands				
2.1	LyP-1	CGNKRTRGC	Tumor hypoxia and tumor-induced lymphangiogenesis	[50]
2.2	SP5−52	SVSVGMKPSPRP	Inhibits angiogenesis	[51]
3 Target peptides				
3.1	NLS	PPKKKRKV	Transported to the nucleus	[52]
3.2		H2N-MMSFVSLLLVGILFWATEAEQLTKCEVFQ	Transported to the secretory pathway (plasma membrane of prokaryotes and endoplasmic reticulum of eukaryotes)	[53]
3.3		H2N-MLSRQSIRFPFKPATRTLCSSRYLL	Transported to the mitochondrial matrix	[54]
3.4	PTS 1	SKL-COOH	Transported to the peroxisome	[55]

charged cell membrane. This interaction leads to membrane destabilization and allows protein unfolding in order for the peptide to easily cross the lipid membrane layer. The direct penetration mechanism is usually adopted by peptides with high concentrations or amphipathic characteristics [58–60].

The inverted micelle model for peptide uptake follows the electrostatic interactions between the peptides and the membrane. However, this mode of cellular uptake is not suitable for peptides such as TAT, which is a cationic peptide, due to its hydrophobic nature [61–64]. Pore formation is another mode for cellular uptake where two different mechanisms are involved: the barrel stave model and the toroidal model [65]. In the first model, peptides that are helical collectively form a barrel-like structure, wherein the hydrophobic residues are nearer to the lipid bilayer and the central pore of the barrel is formed by the hydrophilic residues. In the toroidal pore model, the lipid bilayer modifies its structure in such a way that the peptides, as well as the lipid membrane collectively, form a pore. The latter model is also known as the two-stage models, as it shows a transition between inactive stage and active stages. In the inactive stage, peptide molecules are arranged parallel to the membrane lipid bilayer; when the peptide molecules are activated, they reorient into a perpendicular form, penetrating the membrane [66]. The third model of direct uptake is the carpet-like model [67] or the membrane thinning model [68], wherein the negatively charged phospholipid membrane interacts with positively charged peptides to cause carpeting and thinning of the outer cell membrane until transient pore formation occurs, allowing the peptides to translocate into the lipid bilayer.

27.2.5 Endocytosis

Endocytosis is a cellular ingestion process of taking substances into the cell. It primarily works by two mechanisms: phagocytosis, and pinocytosis. Cellular uptake of peptides by endocytosis is a natural process and requires energy to drive the translocation of peptides from the external environment into the cell [69] as shown in Fig. 27.5. Phagocytosis can be used as a useful mechanism for the removal of pathogens and cell debris. Phagocytosis was first reported by William Osler in 1876 and was named by Elie Metchnikoff in the early 1880s [70,71]. Phagocytosis is one of the most important innate immune defense systems. Pinocytosis, or fluid endocytosis, is a process where extracellular fluid is ingested into the cell by cell membrane invagination. Pinocytosis can be further divided into caveolin mediated, clathrin-mediated, micropinocytosis or clathrin, and caveolin independent endocytosis [72,73]. Phagocytosis can involve the uptake of larger particles such as bacteria and viruses, whereas pinocytosis involves the uptake of solutes. Type 1 phagocytosis involves membrane ruffling and extension of pseudopodia, and type 2 involves a sinking mechanism associated with a nascent phagosome [74]. Phagocytosis in immune cells such as neutrophils and macrophages is

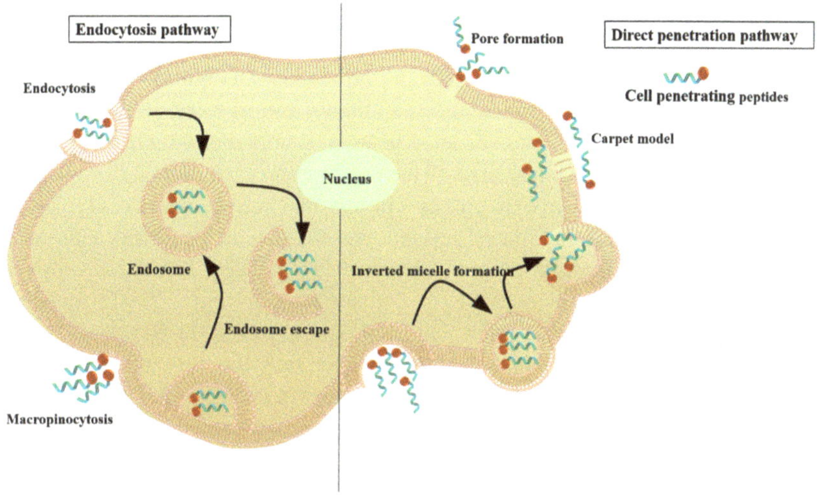

FIGURE 27.5 Uptake of cell-penetrating peptide (CPP) complexes by endocytosis or direct penetration pathways. The endocytosis pathway includes CPP peptide uptake via endocytosis and micropinocytosis. The direct penetration pathway involves localization of the CPP peptide complex into plasma membrane via pore formation, the carpet model, and inverted micelle formation.

facilitated by antibody recognition, leading to membrane blabbing around the target, called pseudopodia. This mechanism is followed by fusion with lysosomes, acidification of phagosome, and target degradation [75].

Endocytosis includes receptor-mediated endocytosis (clathrin-coated and caveolae-mediated), macropinocytosis, dynamin, and clathrin-independent endocytosis, receptor-mediated endocytosis involves clathrin and caveolin pits. These proteins are present in the intracellular cell membrane and are a prerequisite for receptor-mediated endocytosis. This method is useful in the regulation of signal transduction from the cell periphery to the nucleus [76]. This is also considered to be an important pathway through which viruses and other bioparticles can enter or leave cells. Caveolin mediated endocytosis forms vesicles that are smaller when compared to vesicles formed during clathrin-mediated endocytosis [77]. Caveolae are invaginations in the plasma membrane made of cholesterol and sphingolipids. The sensitivity of caveolin mediated endocytosis is based upon cholesterol depleting agents like filipin, nystatin and cyclodextrin [78]. Receptor-mediated endocytosis provides a suitable pathway for the delivery of nanomaterials into human or animal cells and has potential use as a gene delivery mechanism [79]. According to published studies, receptor-mediated endocytosis showed a greater uptake of particles less than 50 nm as compared to particles greater than 50 nm [80,81]. Recently, Matsuoka et al. demonstrated that muscarinic receptor stimulation in rat adrenal medullary and PC12 cells induced tyrosine phosphorylation of TWIK-related acid-sensitive K + 1 (TASK1) channels with clathrin-dependent endocytosis [82].

Macropinocytosis was discovered in the 1930s by Warren Lewis and refers to the engulfment of fluids and membranes into the cell [83]. The process involves invagination of the outer plasma membrane, which causes the formation of macropinosomes. Size of the micropinosomes ranges from 0.2 to 5 μm in diameter [84]. Cellular uptake via this mechanism requires dynamin [85]. Macropinocytosis facilitates antigen presentation in dendritic cells and modulation of the immune response more broadly [83]. Macropinocytosis is regulated by three processes: actin cytoskeleton-driven ruffle formation, closure of the ruffle into a vesicle, and dissociation of actin filaments from the vesicle. Macropinocytosis provides a route into the cell and acts as a delivery mechanism for pathogens and viruses. For example, *Salmonella enterica var. typhimurium* enhanced cell surface ruffling in epithelial cells and macrophages and entered phagosomes resembling macropinosomes [86,87]. Micropinocytosis also facilitates cell invasion by *Legionella pneumophila*, *Shigella flexneri*, and some protozoa [88–90].

27.3 Peptide drug market and preclinical data

The market for peptide drugs has been rising since its initial development. Around 7000 naturally occurring peptides have been identified that play a vital role in hormonal regulation, triggering intracellular channels and ion channel regulations [91]. The development of peptides started more than three decades ago in 1982, beginning with insulin given to patients intravenously. With time, there have been formulations of oral peptides, and recently in 2012, the first oral peptide Linzess was launched by the US and the EU. Considering data up to 2013, more than 100 therapeutic peptides have been approved by the Food and Drug Administration (FDA) [15].

The current worldwide annual market for therapeutic peptides is around $13 billion. A global industry analysis on peptide therapeutics predicted the compound annual growth rate of 9.1% from 2016 to 2024 and a hike up to US$70 billion in 2019. Several approved peptide drugs, such as liraglutide (Victoza) and glucagon-like peptide 1, garnered around $2 billion in sales per annum in 2018 [92]. The therapeutic peptides market is predicted to reach $43.26 billion by the year 2024. Unmet clinical needs relevant to the development of therapeutic peptides have not ceased since the inception of therapeutic peptide technology. Therapeutic peptides are being developed for the treatment of infections, inflammatory diseases, and countless other conditions.

Currently, peptide drugs are typically administered by the parenteral route or injection, but some may also be administered to patients orally, intranasal, or transdermally. Recent approaches in development for peptide drug delivery have included transbuccal routes and the use of nanoparticles. MidaSol therapeutics carried out peptide delivery using transbuccal administration of insulin-passivated gold nanoparticles [93]. The use of such alternative methods of administration could help improve numerous therapeutic applications [94]. Table 27.2 describes peptides of animal origins that have been approved as well as those in clinical trials.

27.3.1 Systemically absorbed oral peptides

27.3.1.1 Cyclosporin A

Sandimmune, the first oral form of cyclosporin A, contained a nonionic surfactant with long-chain mono, di and polyglycerides that helped increase bioavailability by 25%–30% [108]. In order to increase bioavailability and reduce other drawbacks, oil-based Sandimmune was formulated containing oil droplets of size around 150 nm [109]. Reducing the droplet size facilitates solubilization and uniform release of drugs rapidly [110]. Therapeutic effects include inhibition of the T-cell receptor signaling pathway via the formation of a cyclosporin A–cyclophilin complex that inhibits calcineurin. Cyclosporin A also inhibits nitric oxide synthesis induced by interlukin 1 alpha, lipopolysaccharides, and TNF-α, as well as the release of cytochrome C from mitochondria.

27.3.1.2 Desmopressin acetate (DDVAP)

DDVAP is a potent hydrophobic nanopeptide sold as Minirin and produced by Ferring Pharmaceuticals, which was approved by the FDA in 1978. DDVAP corresponds to arginine vasopressin with slight modifications, including deamination of the first amino acid and substitution of the eighth amino acid L-arginine. The drug is an agonist for the vasopressin V2 receptor in endothelia of blood vessels and kidneys and has been used for the treatment of diabetes insipidus, primary nocturnal enuresis, and connective tissue disorders but also possesses antidiuresis and antimetastasis properties [111,112]. DDVAP was also approved for the treatment of mild classical hemophilia and von Willebrand's disease [113] and is available as an intranasal solution, intravenous solution, oral tablet, and oral lyophilisate formulation [114].

27.3.1.3 Taltirelin

Taltirelin is an analog of thyrotropin-releasing hormone (TRH) that prevents the formation of proglutamyl peptide bonds from hydrolysis [115]. The TRH endocrine hormone acts through the hypothalamus, pituitary and thyroid axes. According to studies, TRH plays an important role in the release of dopamine in silico [116] and in vivo [117–119]. Subcutaneous injection of TRH sustained-release microparticles significantly improved motor functions of encephalitis-induced Parkinson's disease in rats [120]. The peptide

TABLE 27.2 Approved and select investigational therapeutic peptides with animal origins.

Compound name	Source of peptide	Current status	Date of clinical trials	Peptide target	References
Insulin	Isolated from canine and bovine pancreata	Replaced by Genetically engineered insulin	1920s	PTEN, Adipocytes	
Adenocorticotropic hormone (ACTH)	Isolated from bovine and porcine pituitary glands	Approved	1950s	Adrenal cortex	[95]
Elcatonin	Ultimobranchial body of the eel (*Anguilla japonica*)	Approved	1970s	Calcitonin receptor	[96]
Calcitonin	Isolated from salmon ultimobranchial gland	Approved	1971	Bone matrix, Kidney tubules	[97]
Eledoisin	Posterior salivary glands of two mollusks (*Eledone moschata, E. aldrovandi*)	Previously approved; current marketing status unclear	1970s	Neurokinin receptors	[98]
Enalapril	Venom of the pit viper (*Bothrops jararaca*)	Approved	1980s	Angiotensin-converting enzyme	[99]
Bivalirudin	Salivary glands of medicinal leeches (*Hirudo medicinalis*)	Approved	1990s	Thrombin	[100]
Cemadotin	Sea hare (*Dolabella auricularia*)	Discontinued at Phase 2	1990s	Microtubules	[101]
Exenatide	Saliva of the Gila monster (*Heloderma suspectum*)	Approved	1990s	GLP-1 receptor	[102]

Pexiganan	Skin of the African clawed frog (*Xenopus laevis*)	Discontinued at Phase 3	1990s	Bacterial cell membrane	[103]
Ziconotide	Cone snail (*Conus magus*)	Approved	1990s	N-type calcium channels	[104]
Chlorotoxin I-135 conjugate	Venom of the deathstalker scorpion (*Leiurus quinquestriatus*)	Discontinued at Phase 2	2000s	Chloride ion channels	[105]
Dalazatide	Toxin from a sea anemone (*Stoichactis helianthus*)	Phase 1	2010s	Kv1.3 channel	[106]
SOR-C13	Venom of the northern short-tailed shrew (*Blarina brevicauda*)	Phase 1	2010s	TRPV6	[107]

(taltirelin) increased stability in the blood and in the brain, leading to 10- to 100-fold increased CNS stimulatory action and increased half-life in comparison to TRH [115]. TRH can be given orally, and its analog taltirelin has been developed for administration via oral disintegrating tablets. Taltirelin has been used to treat spinocerebellar ataxia and other neurodegenerative disorders [121,122] and has demonstrated antiataxic [123], neuroprotective [122,124] and analgesic [121] properties. However, TRH is limited by a short half-life, poor lipophilicity, and a strong HPT axis stimulating effect that severely reduces its potential applications [125].

27.3.1.4 Reduced L-glutathione

Glutathione is a natural antioxidant present in almost every cell of the body, plays a role in the detoxification of drugs and xenobiotics, and is used as a supplement in the treatment of AIDS-associated cachexia. Reduced L-glutathione was previously given GRAS status by the FDA for use as a food additive [108]. Cachexon coated reduced glutathione, an orphan drug for AIDS-associated cachexia, and other oral glutathione formulations used in the treatment of cystic fibrosis were removed from the market due to complications [108]. To date, there is no FDA approved oral formulation of reduced glutathione, neither for cachexia nor for cystic fibrosis.

27.3.1.5 Linaclotide

Linaclotide has a molecular weight of 1526 Da and is antisecretory in nature [126]. The compound is a heterodetic cyclic peptide consisting of fourteen amino acids. Linaclotide was approved in 2012 for the treatment of irritable bowel syndrome with constipation. The principle mode of action involves the activation of guanylyl cyclase C by guanylin peptides that results in increased levels of cyclic guanosine monophosphate (cGMP). This leads to activation of the cGMP protein kinase and phosphorylation of downstream targets, including cystic fibrosis transmembrane conductance regulator (CFTR). Activation of the CFTR channel leads to increased secretion of chloride and bicarbonate ions into the intestinal lumen followed by binding of apically expressed receptors in the small intestine and colon. Currently, research exploring linaclotide effects in specific regions of the gastrointestinal tract is ongoing.

27.3.1.6 Vancomycin

Vacomycin is a tricyclic glycopeptide derived from *Streptococcus orientalis*. The compound is an injected antibiotic that was developed by Eli Lily Company in the 1950s for penicillin resistance in *Staphylococci* but was never the first-line drug because of its toxicity and undesirable early formulations [127]. The rate of absorption of vancomycin decreases to a great extent when taken orally; hence, the drug was preferred intravenously. Intravenous injection of vancomycin for treatment of *Staphylococcus aureus* resulted in inhibition of cell wall biosynthesis. Clinical uses of vancomycin approved by FDA included sepsis, skin and soft tissue infections, bone infections, and lower respiratory tract infections. Oral administration of vancomycin was only preferred in *Clostridium difficile*-associated diarrhea where vancomycin stayed in the intestine to stop the growth of causative bacteria. There are several side effects of vancomycin, including injection site reactions, a few cases of hearing loss, low blood pressure, and bone marrow suppression [128].

27.3.1.7 Tyrothricin

Tyrothricin is a commercially available peptide consumed orally. The compound was isolated from *Baccilus brevis* [129]. Dubos and Rollin Hotchkiss showed thyrothricin to be a mixture of two different antibiotics: gramicidin and tyrocidine. Tyrothricin can disrupt cell membranes of gram-positive bacteria including *Staphylococcus*, *Stroptococcus* and *Diplococcus pneumonia*. The drug is used as a topical antibiotic for skin and oropharyngeal infections and is administered orally to avoid toxicity [130].

27.4 The future of peptide therapeutics

Historically, therapeutic peptides usually express their excellence at the primary stage but do not get approved by the FDA due to disadvantages including short half-life, low bioavailability, poor solubility, and high-manufacturing cost. Researchers are working on approaches to reduce the disadvantages and enhance activity by using multifunctional peptides with more than one site of action. Even after several efforts by industries to selection of agonists and antagonists for commercial purposes, very few ligands have been made available for this purpose [131]. Currently, drug development programs include antimicrobial peptides with additional properties like immune activation or wound healing. Some of the established drugs like Byetta (exenatide), Bydureon (exenatide), Victoza (liraglutide), Lyxumia (lixisenatide), and Tanzeum (albiglutide) are on the market. Pharmaceutical industries are showing an interest in the treatment of type 2 diabetes and obesity using peptides [132]. Along with that, research is also being focused on drug delivery using nanoparticles, which increase bioavailability and stability of the peptides to a greater extent. Furthermore, chemical modification of peptides using linkers or chimeras is being practiced for the formulation of multifunctional peptides. Single peptides with more than one receptor specificity are under clinical trials [91,133]. Production of oral peptides on a large scale will improve the peptide market with convenient administration, targeted delivery and efficacy eventually improved with more work. Several challenges in oral peptides including enzymatic degradation will always be an obstacle. Hence, current research on chemically modified peptides, such as building hydrophobic faces, cyclization,

TABLE 27.3 Widely used therapeutic peptides approved since 1920.

Compound name	Year of approval	Target organ or field	Route of administration	Chemical basis	References
Corticotropin	1952	CNS	Subcutaneous	Native	
Vassopresin	1962	Urology	Intravenous	Analog	[134]
Oxytocin	1962	Infertility and OB-GYN	Intramuscular, intravenous, intranasal	Analog	[135]
Eledosin	1970s	Opthalmology	Topical	Native	[136]
Somatostatin	1970s	Hepatology	Intravenous	Native	
Felypressin	1970s	Pain	Subcutaneous	Analog	[137]
Calcitonin (Salmon)	1971	Bones and connective tissue	Intranasal, subcutaneous	Native	[138]
Ornipressin	1971	Hematology	Intravenous	Analog	[139]
Desmopressin	1972	Urology	Oral, intranasal	Analog	[140]
Telepressin	1978	Hepatology	Intravenous	Analog	[141]
Tetracosactide	1980	Endocrinology	Subcutaneous	Native	[142]
Elcatonin	1981	Bones and connective tissue	Intramuscular	Analog	[96]
Buserelin	1984	Urology	Intranasal, subcutaneous	Analog	[143]
Enalapril	1985	Cardiovascular	Oral	Analog	[144]
Thymopentin	1985	Allergy and Immunology	Subcutaneous, intramuscular	Analog	[145]

Triptorelin	1986	Urology	Intramuscular	Analog	[146]
Calcitonin (Human)	1986	Bones and connective tissue	Intramuscular, intranasal, subcutaneous	Native	[147]
Lisinopril	1987	Cardiovascular	Oral	Analog	
Octreotide	1988	Endocrinology	Subcutaneous, intramuscular, intravenous	Analog	
Gonadorelin	1989	Infertility and OB-GYN	Subcutaneous	Native	[148]
Glucagon	1989	Metabolic	Subcutaneous	Native	[149]
Nafarelin	1990	Infertility and OB-GYN	Intranasal	Analog	[150]
Histrelin	1991	Infertility and OB-GYN	Subcutaneous, intramuscular	Analog	[151]
Glatiramer acetate	1996	CNS	Subcutaneous	Heterologous	[152]
Bivalirudin	1999	Hematology	Intravenous	Analog	[100]
P-15	1999	Orthopedics	Implanted	Analog	[153]
Taltirelin	2000	CNS	Oral	Analog	[122]
Carbetocin	2001	Infertility and OB-GYN	Intravenous	Analog	[154]
Nesirtide	2001	Cardiovascular	Intravenous	Native	[155]
Teriparatide	2002	Bones and connective tissue	Subcutaneous	Analog	[156]
Enfuvirtide	2003	Antimicrobial, Antiviral	Subcutaneous	Analog	[157]
Ziconotide	2004	Pain	Intrathecal	Native	[104]

(*Continued*)

TABLE 27.3 (Continued)

Compound name	Year of approval	Target organ or field	Route of administration	Chemical basis	References
Pramlintide	2005	Metabolic	Subcutaneous	Analog	[158]
Exenatide	2005	Metabolic	Subcutaneous	Native	[102]
Degarelix	2008	Urology	Subcutaneous	Analog	[159]
Romiplostim	2008	Hematology	Subcutaneous	Heterologous	[160]
Icatibant	2008	Cardiovascular	Subcutaneous	Analog	[161]
Mifamurtide	2009	Oncology	Intravenous	Analog	[162]
Pasireotide	2012	Endocrinology	Subcutaneous	Analog	[163]
Carfilzomib	2012	Oncology	Intravenous	Analog	[164]
Linaclotide	2012	Gastroenterology	Oral	Analog	[165]
Teduglutide	2012	Gastroenterology	Subcutaneous	Analog	[166]
Albiglutide	2014	Metabolic	Subcutaneous	Analog	
Dulaglutide	2014	Metabolic	Subcutaneous	Analog	[11]
Afamelanotide	2014	Dermatology	Subcutaneous	Analog	[167]
Etelcalcetide	2016	Endocrinology	Intravenous	Heterologous	[168]
Abaloparatide	2017	Bones and connective tissue	Subcutaneous	Analog	[169]
Plecanatide	2017	Gastroenterology	Oral	Analog	[170]

N-methylation, and intramolecular hydrogen bonds is necessary and will further improve bioavailability and efficiency of therapeutic peptides.

27.5 Concluding remarks

An increase in the efficiency of peptide development has attracted both researchers and industries. Currently, more than 100 peptides have been approved and reached patients for the treatment of disease, many of which are widely used in the clinic (Table 27.3). Along with that, hundreds of peptides are under clinical and preclinical trials and have the potential to cure numerous disorders. The next step is will include the development of multifunctional CPPs and drug conjugates that will expand possible applications. The efficacy of existing compounds can also be enhanced by advanced modifications. We conclude that the emerging market of peptides will expand in the upcoming years and will play a vital role in novel therapeutics.

References

[1] Lau JL, Dunn MK. Therapeutic peptides: historical perspectives, current development trends, and future directions. Bioorg Med Chem 2018;26(10):2700–7. Available from: https://doi.org/10.1016/j.bmc.2017.06.052.

[2] Guidotti G, Brambilla L, Rossi D. Cell-penetrating peptides: from basic research to clinics. Trends Pharmacol Sci 2017;38(4):406–24. Available from: https://doi.org/10.1016/j.tips.2017.01.003.

[3] Shahdeo D, et al. Self-assembled chitosan polymer intercalating peptide functionalized gold nanoparticles as nanoprobe for efficient imaging of urokinase plasminogen activator receptor in cancer diagnostics. Carbohydrate Polymers 2021;266. Available from: https://doi.org/10.1016/j.carbpol.2021.118138.

[4] Barker JL. Physiological roles of peptides in the nervous system. In: Gainer H, editor. Peptides in neurobiology. Springer US; 1977. p. 295–343. Available from: https://doi.org/10.1007/978-1-4613-4130-7_11.

[5] Fields GB. Introduction to peptide synthesis. Curr Protoc Immunol 2002;47(1):9.1.1–9.1.9. Available from: https://doi.org/10.1002/0471142735.im0901s47.

[6] Latham PW. Therapeutic peptides revisited. Nat Biotechnol 1999;17(8):755–7. Available from: https://doi.org/10.1038/11686.

[7] Kesarwani V. Identification of unique peptides for SARS-CoV-2 diagnostics and vaccine development by an in silico proteomics approach. Front. Immunol. 2021. Available from: https://doi.org/10.3389/fimmu.2021.725240.

[8] Gomez JA, Chen J, Ngo J, Hajkova D, Yeh IJ, Gama V, et al. Cell-penetrating penta-peptides (CPP5s): measurement of cell entry and protein-transduction activity. Pharmaceuticals 2010;3(12):3594–613. Available from: https://doi.org/10.3390/ph3123594.

[9] Merrifield RB. Solid phase peptide synthesis. I. The synthesis of a tetrapeptide. J Am Chem Soc 1963;85(14):2149–54. Available from: https://doi.org/10.1021/ja00897a025.

[10] Shahdeo D, et al. Urokinase plasminogen activator receptor-mediated targeting of a stable nanocomplex coupled with specific peptides for imaging of cancer. Anal Chem . 2021;93(34):11868–77. Available from: https://doi.org/10.1021/acs.analchem.1c02697.

[11] Shahdeo D, et al. Polymeric biocompatible iron oxide nanoparticles labeled with peptides for imaging in ovarian cancer. Biosci Rep. 2022;42(2). Available from: https://doi.org/10.1042/BSR20212622.

[12] Gandhi S, Arami H, Krishnan KM. Detection of cancer-specific proteases using magnetic relaxation of peptide-conjugated nanoparticles in biological environment. Nano Lett 2016;16(6):3668–74. Available from: https://doi.org/10.1021/acs.nanolett.6b00867.

[13] Borchardt R T, Jeffrey A, Siahaan TJ, Gangwar S, Pauletti GM. Improvement of oral peptide bioavailability: peptidomimetics and prodrug strategies. Adv Drug Delivery Rev 1997;27(2–3):235–56.

[14] Zhao B, et al. Stabilizing a flexible interdomain hinge region harboring the SMB binding site drives uPAR into its closed conformation. J Mol Biol. 2015;427:1389–403. Available from: https://doi.org/10.1016/j.jmb.2015.01.022.

[15] Zhao B, et al. Mapping the topographic epitope landscape on the urokinase plasminogen activator receptor (uPAR) by surface plasmon resonance and X-ray crystallography. Data Brief. 2015;5:107–13. Available from: https://doi.org/10.1016/j.dib.2015.08.027.

[16] Lindgren M, Langel Ü. Classes and prediction of cell-penetrating peptides. In: Langel Ü, editor. Cell-penetrating peptides: methods and protocols. Humana Press; 2011. p. 3–19. Available from: https://doi.org/10.1007/978-1-60761-919-2_1.

[17] Hudecz F, Bánóczi Z, Csík G. Medium-sized peptides as built in carriers for biologically active compounds. Med Res Rev 2005;25(6):679–736. Available from: https://doi.org/10.1002/med.20034.

[18] Tripathi PP, Arami H, Banga I, Gupta J, Gandhi S. Cell penetrating peptides in preclinical and clinical cancer diagnosis and therapy. Oncotarget 2018;9(98):37252–67. Available from: https://doi.org/10.18632/oncotarget.26442.

[19] Green M, Ishino M, Loewenstein PM. Mutational analysis of HIV-1 Tat minimal domain peptides: Identification of trans-dominant mutants that suppress HIV-LTR-driven gene expression. Cell 1989;58(1):215–23. Available from: https://doi.org/10.1016/0092-8674(89)90417-0.

[20] Milletti F. Cell-penetrating peptides: classes, origin, and current landscape. Drug Discovery Today 2012;17(15–16):850–60. Available from: https://doi.org/10.1016/j.drudis.2012.03.002.

[21] Oehlke J, Krause E, Wiesner B, Beyermann M, Bienert M. Extensive cellular uptake into endothelial cells of an amphipathic beta-sheet forming peptide. FEBS Lett 1997;415(2):196–9. Available from: https://doi.org/10.1016/s0014-5793(97)01123-x.

[22] Derossi D, Calvet S, Trembleau A, Brunissen A, Chassaing G, Prochiantz A. Cell internalization of the third helix of the antennapedia homeodomain is receptor-independent. J Biol Chem 1996;271(30):18188–93. Available from: https://doi.org/10.1074/jbc.271.30.18188.

[23] Futaki S, Suzuki T, Ohashi W, Yagami T, Tanaka S, Ueda K, et al. Arginine-rich peptides: an abundant source of membrane-permeable peptides having potential as carriers for intracellular protein delivery. J Biol Chem 2001;276(8):5836–40. Available from: https://doi.org/10.1074/jbc.M007540200.

[24] Jeang K-T, Xiao H, Rich EA. Multifaceted activities of the HIV-1 transactivator of transcription, Tat. J Biol Chem 1999;274(41):28837–40. Available from: https://doi.org/10.1074/jbc.274.41.28837.

[25] Pooga M, Hällbrink M, Zorko M, Langel U. Cell penetration by transportan. FASEB J 1998;12(1):67–77. Available from: https://doi.org/10.1096/fasebj.12.1.67.

[26] Almarwani B, Phambu EN, Alexander C, Nguyen HAT, Phambu N, Sunda-Meya A. Vesicles mimicking normal and cancer cell membranes exhibit differential responses to

the cell-penetrating peptide Pep-1. Biochim Biophys Acta, Biomembr 2018;1860 (6):1394–402. Available from: https://doi.org/10.1016/j.bbamem.2018.03.022.

[27] Stroh M, Zimmer JP, Duda DG, Levchenko TS, Cohen KS, Brown EB, et al. Quantum dots spectrally distinguish multiple species within the tumor milieu in vivo. Nat Med 2005;11(6):678–82. Available from: https://doi.org/10.1038/nm1247.

[28] Chiu YL, Ali A, Chu CY, Cao H, Rana TM. Visualizing a correlation between siRNA localization, cellular uptake, and RNAi in living cells. Chem Biol 2004;11(8):1165–75. Available from: https://doi.org/10.1016/j.chembiol.2004.06.006.

[29] Mousavizadeh A, Jabbari A, Akrami M, Bardania H. Cell targeting peptides as smart ligands for targeting of therapeutic or diagnostic agents: a systematic review. Colloids Surf, B: Biointerfaces, 2017;158:507–17. Available from: https://doi.org/10.1016/j.colsurfb.2017.07.012.

[30] Tomitaka A, Arami H, Gandhi S, Krishnan KM. Lactoferrin conjugated iron oxide nanoparticles for targeting brain glioma cells in magnetic particle imaging. Nanoscale 2015;7 (40):16890–8. Available from: https://doi.org/10.1039/c5nr02831k.

[31] Vivès E, Schmidt J, Pèlegrin A. Cell-penetrating and cell-targeting peptides in drug delivery. Biochim Biophys Acta, Mol Cell Res 2008;1786(2). Available from: https://doi.org/10.1016/j.bbcan.2008.03.001ï.

[32] Wu CH, Liu IJ, Lu RM, Wu HC. Advancement and applications of peptide phage display technology in biomedical science. J Biomed Sci 2016;23(1). Available from: https://doi.org/10.1186/s12929-016-0223-x.

[33] Bratkovič T. Progress in phage display: evolution of the technique and its applications. Cell Mol Life Sci 2010;67(5):749–67. Available from: https://doi.org/10.1007/s00018-009-0192-2.

[34] Saw PE, Song EW. Phage display screening of therapeutic peptide for cancer targeting and therapy. Protein Cell 2019;10(11):787–807. Available from: https://doi.org/10.1007/s13238-019-0639-7.

[35] Lam KS, Salmon SE, Hersh EM, Hruby VJ, Kazmierskit WM, Knappt RJ. A new type of synthetic peptide library for identifying ligand-binding activity. Nature 1991;354 (6348):82–4. Available from: https://doi.org/10.1038/354082a0.

[36] Park SI, Renil M, Vikstrom B, Amro N, Song Lw, Xu Bl, et al. The use of one-bead one-compound combinatorial library method to identify peptide ligands for α4β1 integrin receptor in non-Hodgkin's lymphoma. Lett Peptide Sci 2001;8(3–5):171–8. Available from: https://doi.org/10.1023/A:1016297601361.

[37] Kolmar H. Alternative binding proteins: biological activity and therapeutic potential of cystine-knot miniproteins. FEBS J 2008;275(11):2684–90. Available from: https://doi.org/10.1111/j.1742-4658.2008.06440.x.

[38] Craik DJ, Clark RJ, Daly NL. Potential therapeutic applications of the cyclotides and related cystine knot mini-proteins. Exp Opin Investig Drugs 2007;16(5):595–604. Available from: https://doi.org/10.1517/13543784.16.5.595.

[39] Kolmar H. Biological diversity and therapeutic potential of natural and engineered cystine knot miniproteins. Curr Opin Pharmacol 2009;9(5):608–14. Available from: https://doi.org/10.1016/j.coph.2009.05.004.

[40] Kolmar H. Engineered cystine-knot miniproteins for diagnostic applications. Expert Rev Mol Diag 2010;10(3):361–8. Available from: https://doi.org/10.1586/erm.10.15.

[41] Gracy J, Le-Nguyen D, Gelly JC, Kaas Q, Heitz A, Chiche L. KNOTTIN: the knottin or inhibitor cystine knot scaffold in 2007. Nucleic Acids Res 2008;36(Suppl 1):D314. Available from: https://doi.org/10.1093/nar/gkm939.

[42] Moral MEG, Siahaan TJ. Conjugates of cell adhesion peptides for therapeutics and diagnostics against cancer and autoimmune diseases. Curr Top Med Chem 2018;17(32):3425–43. Available from: https://doi.org/10.2174/1568026618666180118154514.

[43] Li ZJ, Wu WKK, Ng SSM, Yu L, Li HT, Wong CCM, et al. A novel peptide specifically targeting the vasculature of orthotopic colorectal cancer for imaging detection and drug delivery. J Controlled Release 2010;148(3):292–302. Available from: https://doi.org/10.1016/j.jconrel.2010.09.015.

[44] Wang C, Zhao M, Liu YR, Luan X, Guan YY, Lu Q, et al. Suppression of colorectal cancer subcutaneous xenograft and experimental lung metastasis using nanoparticle-mediated drug delivery to tumor neovasculature. Biomaterials 2014;35(4):1215–26. Available from: https://doi.org/10.1016/j.biomaterials.2013.08.091.

[45] Veldhoen S, Laufer SD, Restle T. Recent developments in peptide-based nucleic acid delivery. Int J Mol Sci 2008;9(7):1276–320. Available from: https://doi.org/10.3390/ijms9071276.

[46] Borrelli A, Tornesello AL, Tornesello ML, Buonaguro FM. Cell penetrating peptides as molecular carriers for anti-cancer agents. Molecules 2018;23(2). Available from: https://doi.org/10.3390/molecules23020295.

[47] Margus H, Padari K, Pooga M. Cell-penetrating peptides as versatile vehicles for oligonucleotide delivery. Mol Ther 2009;20(3):525–33. Available from: https://doi.org/10.1038/mt.2011.284.

[48] Trefilov A, Imendoerffer M, Sekot G, Strobl F, Jungbauer A, Hahn R. A microscale method of protein extraction from bacteria: interaction of *Escherichia coli* with cationic microparticles. J Biotechnol 2015;207:21–9. Available from: https://doi.org/10.1016/j.jbiotec.2015.04.023.

[49] Bais MV, Kumar S, Tiwari AK, Kataria RS, Nagaleekar VK, Shrivastava S, et al. Novel Rath peptide for intracellular delivery of protein and nucleic acids. Biochem Biophys Res Commun 2008;370:27–32. Available from: https://doi.org/10.1016/j.bbrc.2008.03.023.

[50] Park J, Maltzahn G, Von, Jue M, Fogal V, Ramana V, et al. Cooperative nanomaterial system to sensitize, target, and treat tumors. Proc Natl Acad Sci USA 2010;107(3). Available from: https://doi.org/10.1073/pnas.0909565107.

[51] Lee T, Lin C, Kuo S. Peptide-mediated targeting to tumor blood vessels of lung cancer for drug delivery. Cancer Res 2007;67(22):10958–65. https://doi.org/10.1158/0008-5472.CAN-07-2233.

[52] Rapoport TA. Protein translocation across the eukaryotic endoplasmic reticulum and bacterial plasma membranes. Nature 2007;450:663–9. https://doi.org/10.1038/nature06384.

[53] Martin RB. Free energies and equilibria of peptide bond hydrolysis and formation. Biopolymers 1998;45(5):351–353.

[54] Oehlke J, Scheller A, Wiesner B, Krause E, Beyermann M, Klauschenz E, et al. Cellular uptake of an alpha-helical amphipathic model peptide with the potential to deliver polar compounds into the cell interior non-endocytically. Biochim Biophys Acta 1998;1414:127–39.

[55] Zhou XH, Po ALW. Peptide and protein drugs: I. Therapeutic applications, absorption and parenteral administration. Int J Pharm 1991;75:97–115.

[56] Lundberg P, Langel Ü. A brief introduction to cell-penetrating peptides. J Mol Recognit 2003;16(5):227–33. Available from: https://doi.org/10.1002/jmr.630.

[57] Tanaka G, Kawaguchi Y, Kuwata K, Takeuchi T, Nakase I, Futaki S. Photoaffinity labeling methods to explore internalization mechanisms of arginine-rich cell-penetrating peptides. In: Hatanaka Y, Hashimoto M, editors. Photoaffinity labeling for structural probing

within protein. Japan: Springer; 2017. p. 225–40. Available from: https://doi.org/10.1007/978-4-431-56569-7_11.
[58] Deshayes S, Morris MC, Divita G, Heitz F. Interactions of amphipathic CPPs with model membranes. Biochim Biophys Acta, Biomembr 2006;1758(3):328–35. Available from: https://doi.org/10.1016/j.bbamem.2005.10.004.
[59] Duchardt F, Fotin-Mleczek M, Schwarz H, Fischer R, Brock R. A comprehensive model for the cellular uptake of cationic cell-penetrating peptides. Traffic 2007;8(7):848–66. Available from: https://doi.org/10.1111/j.1600-0854.2007.00572.x.
[60] Kosuge M, Takeuchi T, Nakase I, Jones AT, Futaki S. Cellular internalization and distribution of arginine-rich peptides as a function of extracellular peptide concentration, serum, and plasma membrane associated proteoglycans. Bioconjugate Chem 2008;19(3):656–64. Available from: https://doi.org/10.1021/bc700289w.
[61] Thorén PEG, Persson D, Esbjörner EK, Goksör M, Lincoln P, Nordén B. Membrane binding and translocation of cell-penetrating peptides. Biochemistry 2004;43(12):3471–89. Available from: https://doi.org/10.1021/bi0360049.
[62] Tünnemann G, Martin RM, Haupt S, Patsch C, Edenhofer F, Cardoso MC. Cargo-dependent mode of uptake and bioavailability of TAT-containing proteins and peptides in living cells. FASEB J 2006;20(11):1775–84. Available from: https://doi.org/10.1096/fj.05-5523com.
[63] Elmquist A, Hansen M, Langel Ü. Structure–activity relationship study of the cell-penetrating peptide pVEC. Biochim Biophys Acta, Biomembr 2006;1758(6):721–9. Available from: https://doi.org/10.1016/j.bbamem.2006.05.013.
[64] Guterstam P, Madani F, Hirose H, Takeuchi T, Futaki S, Andaloussi SEL, et al. Elucidating cell-penetrating peptide mechanisms of action for membrane interaction, cellular uptake, and translocation utilizing the hydrophobic counter-anion pyrenebutyrate. Biochim Biophys Acta, Biomembr 2009;1788(12):2509–17. Available from: https://doi.org/10.1016/j.bbamem.2009.09.014.
[65] Kaplan IM, Wadia JS, Dowdy SF. Cationic TAT peptide transduction domain enters cells by macropinocytosis. J Controlled Release 2005;102(1):247–53. Available from: https://doi.org/10.1016/j.jconrel.2004.10.018.
[66] E. J, A. Aguilar AO-ZJ, D. P. Antimicrobial peptides: diversity and perspectives for their biomedical application. Biomedical engineering, trends, research and technologies. InTech; 2011. Available from: https://doi.org/10.5772/13058.
[67] Pouny Y, Rapaport D, Mor A, Nicolas P, Shai Y. Interaction of antimicrobial dermaseptin and its fluorescently labeled analogs with phospholipid membranes. Biochemistry 1992;31(49):12416–23. Available from: https://doi.org/10.1021/bi00164a017.
[68] Lee M-T, Hung W-C, Chen F-Y, Huang HW. Many-body effect of antimicrobial peptides: on the correlation between lipid's spontaneous curvature and pore formation. Biophysical J 2005;89(6):4006–16. Available from: https://doi.org/10.1529/biophysj.105.068080.
[69] Guo Z, Peng H, Kang J, Sun D. Cell-penetrating peptides: possible transduction mechanisms and therapeutic applications. Biomed Rep 2016;4(5):528–34. Available from: https://doi.org/10.3892/br.2016.639.
[70] Gordon S. Phagocytosis: an immunobiologic process. Immunity 2016;44(3):463–75. Available from: https://doi.org/10.1016/j.immuni.2016.02.026.
[71] Ambrose CT. The Osler slide, a demonstration of phagocytosis from 1876. Reports of phagocytosis before Metchnikoff's 1880 paper. Cell Immunol 2006;240(1):1–4. Available from: https://doi.org/10.1016/j.cellimm.2006.05.008.

[72] Mayor S, Parton RG, Donaldson JG. Clathrin-independent pathways of endocytosis. Cold Spring Harb Perspect Biol 2014;6(6). Available from: https://doi.org/10.1101/cshperspect.a016758.

[73] Koren E, Torchilin VP. Cell-penetrating peptides: breaking through to the other side. Trends Mol Med 2012;18(7):385–93. Available from: https://doi.org/10.1016/j.molmed.2012.04.012.

[74] Freeman SA, Grinstein S. Phagocytosis: mechanosensing, traction forces, and a molecular clutch. Curr Biol 2020. Available from: https://doi.org/10.1016/j.cub.2019.11.047.

[75] Richards DM, Endres RG. The mechanism of phagocytosis: two stages of engulfment. Biophys J 2014;107(7):1542–53. Available from: https://doi.org/10.1016/j.bpj.2014.07.070.

[76] Thomsen ARB, Plouffe B, Cahill TJ, Shukla AK, Tarrasch JT, Dosey AM, et al. GPCR-G protein-β-arrestin super-complex mediates sustained G protein signaling. Cell 2016;166(4):907–19. Available from: https://doi.org/10.1016/j.cell.2016.07.004.

[77] Derakhshankhah H, Jafari S. Cell penetrating peptides: a concise review with emphasis on biomedical applications. Biomed Pharmacother 2018;108:1090–6. Available from: https://doi.org/10.1016/j.biopha.2018.09.097.

[78] Nabi IR, Le PU. Caveolae/raft-dependent endocytosis. J Cell Biol 2003;161(4):673–7. Available from: https://doi.org/10.1083/jcb.200302028.

[79] Davis SS. Biomedical applications of nanotechnology—implications for drug targeting and gene therapy. Trends Biotechnol 1997;15(6):217–24. Available from: https://doi.org/10.1016/S0167-7799(97)01036-6.

[80] Prabha S, Zhou WZ, Panyam J, Labhasetwar V. Size-dependency of nanoparticle-mediated gene transfection: studies with fractionated nanoparticles. Int J Pharmaceutics 2002;244(1–2):105–15. Available from: https://doi.org/10.1016/S0378-5173(02)00315-0.

[81] Desai MP, Labhasetwar V, Walter E, Levy RJ, Amidon GL. The mechanism of uptake of biodegradable microparticles in Caco-2 cells is size dependent. Pharm Res 1997;14(11):1568–73. Available from: https://doi.org/10.1023/A:1012126301290.

[82] Matsuoka H, Harada K, Mashima K, Inoue M. Muscarinic receptor stimulation induces TASK1 channel endocytosis through a PKC-Pyk2-Src pathway in PC12 cells. Cell Signal 2020;65:109434. Available from: https://doi.org/10.1016/j.cellsig.2019.109434.

[83] Luisoni S, Greber UF. Biology of adenovirus cell entry: receptors, pathways, mechanisms. Adenoviral vectors for gene therapy: second edition. Elsevier Inc.; 2016. p. 27–58. Available from: https://doi.org/10.1016/B978-0-12-800276-6.00002-4.

[84] Garrett WS, Mellman I. Studies of endocytosis. Dendritic cells. Elsevier; 2001. p. 213-cp1. Available from: https://doi.org/10.1016/b978-012455851-9/50055-9.

[85] Jones AT. Macropinocytosis: searching for an endocytic identity and role in the uptake of cell penetrating peptides. J Cell Mol Med 2007;11(4):670–84. Available from: https://doi.org/10.1111/j.1582-4934.2007.00062.x.

[86] Francis CL, Ryan TA, Jones BD, Smith SJ, Falkow S. Ruffles induced by Salmonella and other stimuli direct macropinocytosis of bacteria. Nature 1993;364(6438):639–42. Available from: https://doi.org/10.1038/364639a0.

[87] Alpuche-Aranda CM, Racoosin EL, Swanson JA, Miller SI. Salmonella stimulate macrophage macropinocytosis and persist within spacious phagosomes. J Exp Med 1994;179(2):601–8. Available from: https://doi.org/10.1084/jem.179.2.601.

[88] Watarai M, Makino SI, Fujii Y, Okamoto K, Shirahata T. Modulation of Brucella-induced macropinocytosis by lipid rafts mediates intracellular replication. Cell Microbiology 2002;4(6):341–55. Available from: https://doi.org/10.1046/j.1462-5822.2002.00195.x.

[89] Weiner A, Mellouk N, Lopez-Montero N, Chang Y-Y, Souque C, Schmitt C, et al. Macropinosomes are key players in early shigella invasion and vacuolar escape in epithelial cells. PLoS Pathog 2016;12(5):e1005602. Available from: https://doi.org/10.1371/journal.ppat.1005602.

[90] de Carvalho TMU, Barrias ES, de Souza W. Macropinocytosis: a pathway to protozoan infection. Front Physiol 2015;6. Available from: https://doi.org/10.3389/fphys.2015.00106.

[91] Fosgerau K, Hoffmann T. Peptide therapeutics: current status and future directions. Drug Discovery Today 2015;20(1):122–8. Available from: https://doi.org/10.1016/j.drudis.2014.10.003.

[92] Lee ACL, Harris JL, Khanna KK, Hong JH. A comprehensive review on current advances in peptide drug development and design. Int J Mol Sci 2019;20(10). Available from: https://doi.org/10.3390/ijms20102383 MDPI AG.

[93] Sheridan C. Proof of concept for next-generation nanoparticle drugs in humans InNat Biotechnol 2012;30(6):471–3. Available from: https://doi.org/10.1038/nbt0612-471.

[94] Schulte I, Tammen H, Selle H, Schulz-Knappe P. Peptides in body fluids and tissues as markers of disease. Expert Rev Mol Diagnostics 2005;5(2):145–57. Available from: https://doi.org/10.1586/14737159.5.2.145.

[95] Camacho PM, Petak SM, Binkley N, Clarke BL, Harris ST, Hurley DL, et al. American Association of Clinical Endocrinologists and American College of Endocrinology clinical practice guidelines for the diagnosis and treatment of postmenopausal osteoporosis-2016-executive summary. Endocr Pract 2016;22(9):1111–17. Available from: https://doi.org/10.4158/EP161435.ESGL.

[96] Fujita T, Ohue M, Nakajima M, Fujii Y, Miyauchi A, Takagi Y. Comparison of the effects of elcatonin and risedronate on back and knee pain by electroalgometry using fall of skin impedance and quality of life assessment using SF-36. J Bone Miner Metab 2011. Available from: https://doi.org/10.1007/s00774-011-0259-7.

[97] Russell FA, King R, Smillie SJ, Kodji X, Brain SD. Calcitonin gene-related peptide: physiology and pathophysiology. Physiol Rev 2014. https://doi.org/10.1152/physrev.00034.2013

[98] Nebbioso M, Evangelista M, Librando A, Plateroti AM, Pescosolido N. Iatrogenic dry eye disease: an eledoisin/carnitine and osmolyte drops study. Biomedicine Pharmacotherapy 2013. Available from: https://doi.org/10.1016/j.biopha.2013.07.001.

[99] Hata T, Ogihara T, Mikami H, Nakamaru M, Mandai T, Kumahara Y. Effects of two angiotensin II analogues on blood pressure, plasma aldosterone concentration, plasma renin activity and creatinine clearance in normal subjects on different sodium intakes. Eur J Clin Pharmacol 1980;18(4):295–9. Available from: https://doi.org/10.1007/BF00561385.

[100] Warkentin TE, Greinacher A, Koster A. Bivalirudin. Thromb Haemost 2008;99(05):830–9. https://doi.org/10.1160/TH07-10-0644.

[101] Jordan MA. Suppression of microtubule dynamics by binding of cemadotin to tubulin: possible mechanism for its antitumor action. Biochemistry 1998. Available from: https://doi.org/10.1021/bi9817414.

[102] Barnett A. Exenatide. Expert Opin Pharmacother 2007. Available from: https://doi.org/10.1517/14656566.8.15.2593.

[103] Lamb HM, Wiseman LR. Pexiganan acetate. Drugs 1998;56:1047–52. https://doi.org/10.2165/00003495-199856060-00011

[104] McGivern JG. Ziconotide: a review of its pharmacology and use in the treatment of pain. Neuropsychiatric Dis Treat 2007. Available from: https://doi.org/10.2147/nedt.2007.3.1.69.

[105] Maertens C, Wei L, Tytgat J, Droogmans G, Nilius B. Chlorotoxin does not inhibit volume-regulated, calcium-activated and cyclic AMP-activated chloride channels. Br J Pharmacology 2000. Available from: https://doi.org/10.1038/sj.bjp.0703102.

[106] Tarcha EJ, Olsen CM, Probst P, Peckham D, Muñoz-Elías EJ, Kruger JG, et al. Safety and pharmacodynamics of dalazatide, a Kv1.3 channel inhibitor, in the treatment of plaque psoriasis: a randomized phase 1b trial. PLoS One 2017. Available from: https://doi.org/10.1371/journal.pone.0180762.

[107] Fu S, Hirte H, Welch S, Ilenchuk TT, Lutes T, Rice C, et al. First-in-human phase I study of SOR-C13, a TRPV6 calcium channel inhibitor, in patients with advanced solid tumors. Investig N Drugs 2017. Available from: https://doi.org/10.1007/s10637-017-0438-z.

[108] Aguirre TAS, Teijeiro-Osorio D, Rosa M, Coulter IS, Alonso MJ, Brayden DJ. Current status of selected oral peptide technologies in advanced preclinical development and in clinical trials. Adv Drug Deliv Rev 2016;106:223–41. Available from: https://doi.org/10.1016/j.addr.2016.02.004.

[109] Ritschel WA. Microemulsion technology in the reformulation of cyclosporine: the reason behind the pharmacokinetic properties of Neoral. Clin Transplant 1996;10(4):364–73.

[110] Hauss DJ. Oral lipid-based formulations. Adv Drug Deliv Rev 2007;59(7):667–76. Available from: https://doi.org/10.1016/j.addr.2007.05.006.

[111] Mannucci PM. Desmopressin (DDAVP) in the treatment of bleeding disorders: the first 20 years. Blood 1997;90(7):2515–21.

[112] Weinberg RS, Grecco MO, Ferro GS, Seigelshifer DJ, Perroni NV, Terrier FJ, et al. A phase II dose-escalation trial of perioperative desmopressin (1-desamino-8-d-arginine vasopressin) in breast cancer patients. SpringerPlus 2015;4(1). Available from: https://doi.org/10.1186/s40064-015-1217-y.

[113] Agersø H, Larsen LS, Riis A, Lövgren U, Karlsson MO, Senderovitz T. Pharmacokinetics and renal excretion of desmopressin after intravenous administration to healthy subjects and renally impaired patients. Br J Clin Pharmacol 2004. Available from: https://doi.org/10.1111/j.1365-2125.2004.02175.x.

[114] Vande Walle J, Stockner M, Raes A, Norgaard J. Desmopressin 30 years in clinical use: a safety review. Curr Drug Saf 2008;2(3):232–8. Available from: https://doi.org/10.2174/157488607781668891.

[115] Fukuchi I, Asahi T, Kawashima K, Kawashima Y, Yamamura M, Matsuoka Y, et al. Effects of taltirelin hydrate (TA-0910), a novel thyrotropin-releasing hormone analog, on in vivo dopamine release and turnover in rat brain. Arzneimittel-Forschung 1998;48 (4):353–9.

[116] Zheng C, Chen G, Tan Y, Zeng W, Peng Q, Wang J, et al. TRH analog, taltirelin improves motor function of hemi-pd rats without inducing dyskinesia via sustained dopamine stimulating effect. Front Cell Neurosci 2018;12. Available from: https://doi.org/10.3389/fncel.2018.00417.

[117] Puga L, Alcántara-Alonso V, Coffeen U, Jaimes O, de Gortari P. TRH injected into the nucleus accumbens shell releases dopamine and reduces feeding motivation in rats. Behav Brain Res 2016;306:128–36. Available from: https://doi.org/10.1016/j.bbr.2016.03.031.

[118] Crespi F, Keane PE, Morre M. In vivo evaluation by differential pulse voltammetry of the effect of thyrotropin-releasing hormone (TRH) on dopaminergic and serotoninergic synaptic activity in the striatum and nucleus accumbens of the rat. Exp Brain Res 1986;62(2):329–34. Available from: https://doi.org/10.1007/BF00238852.

[119] Kalivas PW, Stanley D, Prange AJ. Interaction between thyrotropin-releasing hormone and the mesolimbic dopamine system. Neuropharmacology 1987;26(1):33–8. Available from: https://doi.org/10.1016/0028-3908(87)90041-4.

[120] Ogata A, Nagashima K, Yasui K, Matsuura T, Tashiro K. Sustained release dosage of thyrotropin-releasing hormone improves experimental Japanese encephalitis virus-induced parkinsonism in rats. J Neurol Sci 1998;159(2):135–9. Available from: https://doi.org/10.1016/s0022-510x(98)00150-6.

[121] Tanabe M, Tokuda Y, Takasu K, Ono K, Honda M, Ono H. The synthetic TRH analogue taltirelin exerts modality-specific antinociceptive effects via distinct descending monoaminergic systems. Br J Pharmacol 2007;150(4):403–14. Available from: https://doi.org/10.1038/sj.bjp.0707125.

[122] Urayama A, Yamada S, Kimura R, Zhang J, Watanabe Y. Neuroprotective effect and brain receptor binding of taltirelin, a novel thyrotropin-releasing hormone (TRH) analogue, in transient forebrain ischemia of C57BL/6J mice. Life Sci 2002. Available from: https://doi.org/10.1016/S0024-3205(02)02268-3.

[123] Nakamura T, Honda M, Kimura S, Tanabe M, Oda SI, Ono H. Taltirelin improves motor ataxia independently of monoamine levels in rolling mouse Nagoya, a model of spinocerebellar atrophy. Biol Pharm Bull 2005;28(12):2244–7. Available from: https://doi.org/10.1248/bpb.28.2244.

[124] Cappiello A, Famiglini G, Palma P, Pierini E, Termopoli V, Trufelli H. Overcoming matrix effects in liquid chromatography–mass spectrometry. Anal Chem 2008;80(23):9343–8. Available from: https://doi.org/10.1021/ac8018312.

[125] Kinoshita K, Yamamura M, Sugihara J, Suzuki M, Matsuoka Y. Taltirelin hydrate (TA-0910): an orally active thyrotropin-releasing hormone mimetic agent with multiple actions. CNS Drug Rev 1998;4(1):25–41. Available from: https://doi.org/10.1111/j.1527-3458.1998.tb00039.x.

[126] Busby RW, Bryant AP, Bartolini WP, Cordero EA, Hannig G, Kessler MM, et al. Linaclotide, through activation of guanylate cyclase C, acts locally in the gastrointestinal tract to elicit enhanced intestinal secretion and transit. Eur J Pharmacol 2010;649(1–3):328–35. Available from: https://doi.org/10.1016/j.ejphar.2010.09.019.

[127] Deresinski S. Vancomycin: does it still have a role as an antistaphylococcal agent? Expert Rev Anti-Infective Ther 2007;5(3):393–401. Available from: https://doi.org/10.1586/14787210.5.3.393.

[128] Rule AM. American society of health-system pharmacists' pain management network. J Pain Palliat Care Pharmacother 2004;18(3):59–62.

[129] Dubos RJ. Studies on a bactericidal agent extracted from a soil bacillus: I. Preparation of the agent. Its activity in vitro. J Exp Med 1939;70(1):1–10. Available from: https://doi.org/10.1084/jem.70.1.1.

[130] Davey P, Wilcox MH, Irving W, Thwaites G. Antimicrobial chemotherapy. Oxford University Press; 2015. Available from: https://doi.org/10.1093/med/9780199689774.001.0001.

[131] Hattori M, Taylor TD. The human intestinal microbiome: a new frontier of human biology. DNA Res 2009;16(1):1–12. Available from: https://doi.org/10.1093/dnares/dsn033.

[132] Pocai A. Unraveling oxyntomodulin, GLP1's enigmatic brother. J Endocrinol 2012;215(3):335–46. Available from: https://doi.org/10.1530/JOE-12-0368.

[133] Gault VA, Bhat VK, Irwin N, Flatt PR. A novel glucagon-like peptide-1 (GLP-1)/Glucagon hybrid peptide with triple-acting agonist activity at glucose-dependent insulinotropic polypeptide, GLP-1, and glucagon receptors and therapeutic potential in high

fat-fed Mice. J Biol Chem 2013;288(49):35581–91. Available from: https://doi.org/10.1074/jbc.M113.512046.

[134] Mattes JA, Pettinati HM, Nilsen SM, Robin SE, Willis KW. Vassopressin for ECT-induced memory impairment: a placebo-controlled comparison. Psychopharmacol Bull 1989.

[135] Carter CS. Oxytocin pathways and the evolution of human behavior. Annu Rev Psychol 2014. Available from: https://doi.org/10.1146/annurev-psych-010213-115110.

[136] Rudich L, Butcher FR. Effect of substance P and eledoisin on K+ efflux, amylase release and cyclic nucleotide levels in slices of rat parotid gland. BBA - Gen Subj 1976. Available from: https://doi.org/10.1016/0304-4165(76)90317-2.

[137] Cecanho R, De Luca LA, Ranali J. Cardiovascular effects of felypressin. Anesthesia Prog 2006; https://doi.org/10.2344/0003-3006(2006)53[119:CEOF]2.0.CO;2.

[138] Chesnut CH, Azria M, Silverman S, Engelhardt M, Olson M, Mindeholm L. Salmon calcitonin: a review of current and future therapeutic indications. Osteoporos Int 2008. Available from: https://doi.org/10.1007/s00198-007-0490-1.

[139] Lenz K, Hortnagl H, Druml W, Grimm G, Laggner A, Schneeweisz B, et al. Beneficial effect of 8-ornithin vasopressin on renal dysfunction in decompensated cirrhosis. Gut 1989. Available from: https://doi.org/10.1136/gut.30.1.90.

[140] Han J, Jung JH, Bakker CJ, Ebell MH, Dahm P. Desmopressin for treating nocturia in men. Cochrane Database Syst Rev 2017. Available from: https://doi.org/10.1002/14651858.CD012059.pub2

[141] Bernadich C, Bandi JC, Melin P, Bosch J. Effects of F-180, a new selective vasoconstrictor peptide, compared with terlipressin and vasopressin on systemic and splanchnic hemodynamics in a rat model of portal hypertension. Hepatology 1998. Available from: https://doi.org/10.1002/hep.510270206.

[142] Lux AL, Edwards SW, Hancock E, Johnson AL, Kennedy CR, Newton RW, et al. The United Kingdom infantile spasms study comparing vigabatrin with prednisolone or tetracosactide at 14 days: a multicentre, randomised controlled trial. Lancet 2004. Available from: https://doi.org/10.1016/S0140-6736(04)17400-X.

[143] Brogden RN, Buckley MMT, Ward A. Buserelin: a review of its pharmacodynamic and pharmacokinetic properties, and clinical profile. Drugs 1990. Available from: https://doi.org/10.2165/00003495-199039030-00007.

[144] McKelvie RS, Yusuf S, Pericak D. Comparison of candesartan, enalapril, and their combination in congestive heart failure: randomized evaluation of strategies for left ventricular dysfunction (RESOLVD) pilot study. Congestive Heart Fail 1999.

[145] Morel S, Ugazio E, Cavalli R, Gasco MR. Thymopentin in solid lipid nanoparticles. Int J Pharm 1996. Available from: https://doi.org/10.1016/0378-5173(95)04388-8.

[146] Romero E, Velez De Mendizabal N, Cendrós JM, Peraire C, Bascompta E, Obach R, et al. Pharmacokinetic/pharmacodynamic model of the testosterone effects of triptorelin administered in sustained release formulations in patients with prostate cancer. J Pharmacol Exp Ther 2012. Available from: https://doi.org/10.1124/jpet.112.195560.

[147] Brain SD, Williams TJ, Tippins JR, Morris HR, MacIntyre I. Calcitonin gene-related peptide is a potent vasodilator. Nature 1985. Available from: https://doi.org/10.1038/313054a0.

[148] Picard-Hagen N, Lhermie G, Florentin S, Merle D, Frein P, Gayrard V. Effect of gonadorelin, lecirelin, and buserelin on LH surge, ovulation, and progesterone in cattle. Theriogenology 2015. Available from: https://doi.org/10.1016/j.theriogenology.2015.03.004.

[149] Holst JJ. The physiology of glucagon-like peptide 1. Physiol Rev 2007. Available from: https://doi.org/10.1152/physrev.00034.2006

[150] Huff DS, Snyder HM, Rusnack SL, Zderic SA, Carr MC, Canning DA. Hormonal therapy for the subfertility of cryptorchidism. Hormone Res 2001. Available from: https://doi.org/10.1159/000049962.

[151] Wang J, Yadav V, Smart AL, Tajiri S, Basit AW. Stability of peptide drugs in the colon. Eur J Pharm Sci 2015. Available from: https://doi.org/10.1016/j.ejps.2015.06.018.

[152] Comi G, Martinelli V, Rodegher M, Moiola L, Bajenaru O, Carra A, et al. Effect of glatiramer acetate on conversion to clinically definite multiple sclerosis in patients with clinically isolated syndrome (PreCISe study): a randomised, double-blind, placebo-controlled trial. Lancet 2009. Available from: https://doi.org/10.1016/S0140-6736(09)61259-9.

[153] Bar-Hava I, Zahavi Z, Brzezinski A, Orvieto R, Dicker D, Levy T, et al. P-15. Menopause 1997. Available from: https://doi.org/10.1097/00042192-199704040-00064.

[154] Su L-L, Chong Y-S, Samuel M. Carbetocin for preventing postpartum haemorrhage. Cochrane Database Syst Rev 2012. Available from: https://doi.org/10.1002/14651858.cd005457.pub4.

[155] Schirger J, Chen H, Cataliotti A, Boerrigter G, Costello L, Martin F, et al. Renal resistance to BNP in severe experimental heart failure is overcome with high dose Nesirtide. J Card Fail 2004. Available from: https://doi.org/10.1016/j.cardfail.2004.06.097.

[156] Jiang Y, Zhao JJ, Mitlak BH, Wang O, Genant HK, Eriksen EF. Recombinant human parathyroid hormone (1-34) [teriparatide] improves both cortical and cancellous bone structure. J Bone Miner Res 2003. Available from: https://doi.org/10.1359/jbmr.2003.18.11.1932.

[157] Lalezari JP, Henry K, O'Hearn M, Montaner JSG, Piliero PJ, Trottier B, et al. Enfuvirtide, an HIV-1 fusion inhibitor, for drug-resistant HIV infection in North and South America. N Engl J Med 2003. Available from: https://doi.org/10.1056/NEJMoa035026.

[158] Ravussin E, Smith SR, Mitchell JA, Shringarpure R, Shan K, Maier H, et al. Enhanced weight loss with pramlintide/metreleptin: an integrated neurohormonal approach to obesity pharmacotherapy. Obesity 2009. Available from: https://doi.org/10.1038/oby.2009.184.

[159] Klotz L, Boccon-Gibod L, Shore ND, Andreou C, Persson BE, Cantor P, et al. The efficacy and safety of degarelix: A 12-month, comparative, randomized, open-label, parallel-group phase III study in patients with prostate cancer. BJU Int 2008. Available from: https://doi.org/10.1111/j.1464-410X.2008.08183.x.

[160] Kuter DJ, Bussel JB, Lyons RM, Pullarkat V, Gernsheimer TB, Senecal FM, et al. Efficacy of romiplostim in patients with chronic immune thrombocytopenic purpura: a double-blind randomised controlled trial. Lancet 2008. Available from: https://doi.org/10.1016/S0140-6736(08)60203-2.

[161] Deeks ED. Icatibant. Drugs 2010;70:73–81. Available from: https://doi.org/10.2165/11204500-000000000-00000

[162] Frampton JE. Mifamurtide. Pediatric drugs 2010. Available from: https://doi.org/10.2165/11204910-000000000-00000.

[163] Allen PJ, Goñen M, Brennan MF, Bucknor AA, Robinson LM, Pappas MM, et al. Pasireotide for postoperative pancreatic fistula. N Engl J Med 2014. Available from: https://doi.org/10.1056/NEJMoa1313688.

[164] Kortuem KM, Stewart AK. Carfilzomib. Blood 2013;121(6). Available from: https://doi.org/10.1182/blood-2012-10-459883

[165] Corsetti M, Tack J. Linaclotide: a new drug for the treatment of chronic constipation and irritable bowel syndrome with constipation. U Eur Gastroenterol J 2013. Available from: https://doi.org/10.1177/2050640612474446.

[166] Buchman AL, Katz S, Fang JC, Bernstein CN, Abou-Assi SG. Teduglutide, a novel mucosally active analog of glucagon-like peptide-2 (GLP-2) for the treatment of moderate to severe Crohn's disease. Inflamm Bowel Dis 2010. Available from: https://doi.org/10.1002/ibd.21117.

[167] Langendonk JG, Balwani M, Anderson KE, Bonkovsky HL, Anstey AV, Bissell DM, et al. Afamelanotide for erythropoietic protoporphyria. N Engl J Med 2015. Available from: https://doi.org/10.1056/NEJMoa1411481.

[168] Hamano N, Komaba H, Fukagawa M. Etelcalcetide for the treatment of secondary hyperparathyroidism. Expert Opin Pharmacother 2017. Available from: https://doi.org/10.1080/14656566.2017.1303482.

[169] Leder BZ, O'Dea LSL, Zanchetta JR, Kumar P, Banks K, McKay K, et al. Effects of abaloparatide, a human parathyroid hormone-related peptide analog, on bone mineral density in postmenopausal women with osteoporosis. J Clin Endocrinol Metab 2015. Available from: https://doi.org/10.1210/jc.2014-3718.

[170] Shailubhai K, Comiskey S, Foss JA, Feng R, Barrow L, Comer GM, et al. Plecanatide, an oral guanylate cyclase C agonist acting locally in the gastrointestinal tract, is safe and well-tolerated in single doses. Digestive Dis Sci 2013. Available from: https://doi.org/10.1007/s10620-013-2684-z.

[171] Garoff H, Simons K. Location of the spike glycoproteins in the Semliki Forest virus membrane. Proc Natl Acad Sci USA 1974;71(10):3988–92. Available from: https://doi.org/10.1073/pnas.71.10.3988.

Chapter 28

Therapeutic peptides: market and manufacturing

Bryan J. Xie[1] and Vandon T. Duong[2]
[1]*Department of Medicine, Stanford University School of Medicine, Stanford, CA, United States,*
[2]*Department of Bioengineering, Stanford University School of Medicine, Stanford, CA, United States*

28.1 Historical perspectives on the therapeutic peptide market

For over a century, peptides have been developed as treatments for a wide variety of human diseases, from ameliorating hormone dysregulation to delivering cytotoxic payloads. As early as the 1920s, peptides like insulin and adrenocorticotropic hormone (ACTH) were isolated from animals as medicines [1]. Insulin was used to manage blood glucose levels for diabetic patients, and ACTH was used to drive cortisol production and release for patients with endocrine disorders. By the 1960s, advances in peptide chemistries allowed for peptides to be synthesized for clinical use. Examples of these chemically synthesized peptides are oxytocin and vasopressin—oxytocin was used for inducing labor and supporting delivery in women with slow births, and vasopressin was used for modulating tonicity for managing antidiuretic hormone deficiency. Collectively, these early applications represented how exogenous peptides, either from animals or synthetic origins, could treat human diseases.

Technological advances in molecular biology and bioengineering have enabled the development of more innovative therapeutic peptides. One such breakthrough was the ability to incorporate synthetic genes into bacteria for expressing human proteins. The founders of Genentech, Bob Swanson and Herb Boyer, had the grand idea to utilize recombinant DNA technology for drug development and commercialization, beginning with human versions of biologics [2]. Animal peptides, like insulin extracted from bovine or porcine pancreas, were prone to cause immunogenicity, thereby limiting therapeutic efficacy [3]. If instead human peptides were produced and administered for therapy, then the prevalence of allergic reactions would decrease considerably.

Peptide and Peptidomimetic Therapeutics. DOI: https://doi.org/10.1016/B978-0-12-820141-1.00030-3

After demonstrating proof-of-concept with somatostatin, which is a small single-chain peptide with multiple neuroendocrine inhibitory effects, Genentech was able to chemically synthesize the genes for insulin and express it within *Escherichia coli* [4]. Genentech went on to form a partnership with Eli Lilly to fund efforts like establishing facilities, optimizing plasmid designs, scaling up production, and conducting preclinical and clinical trials. In 1982, the FDA had approved Humulin as the first therapeutic peptide from recombinant DNA technology. Since then, Genentech and other biotechnology companies have continued to discover and develop innovative therapeutic peptides.

28.2 Current perspectives on the therapeutic peptide market

Today, there are around 80 peptide therapeutics on the global market spanning a wide breadth of therapeutic areas. Oncology, urology, infertility/obstetrics-gynecology, and metabolic diseases comprise over half of the approved indications, with oncology and metabolic diseases accounting for the largest revenue (Fig. 28.1). In 2019, peptide therapeutics accounted for approximately 5% of the global pharmaceutical market, equating to global sales of US $50 billion [5]. The market is projected to grow at a 9% compound annual growth rate, with near-term growth primarily driven by cancer and metabolic disease therapeutic areas.

The projected growth of therapeutic peptides can be attributed to the recognition that peptides occupy a therapeutic niche in between large biologics and small molecules. Similar to monoclonal antibodies, peptide therapeutics have high specificity toward their target, but benefit from having lower production complexity and thus lower cost. In addition, their smaller size can

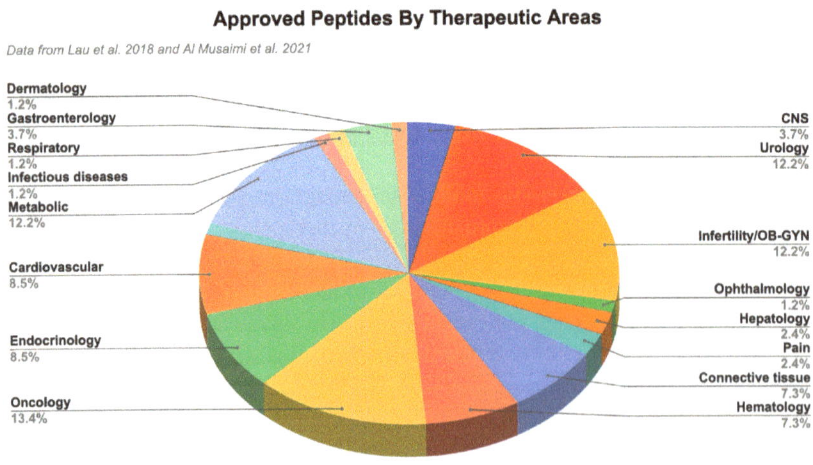

FIGURE 28.1 Approved peptides by therapeutic areas.

potentially allow certain peptides to cross the intestinal tract so that these therapeutics can easily be administered via ingestion [6]. They have excellent efficacy profiles without a number of off-target effects that are typically experienced with small molecules. Ideal targets for peptide therapeutics are extracellular receptors with multiple functional subtypes [7]. In particular, the ion channel and G-protein coupled receptor (GPCR) class of targets are where peptide therapeutics have excelled; in the past decade, almost half of the peptide therapeutics entering clinical development target these receptors [1]. They remain an attractive therapeutic modality for the ion channel and GPCR class of targets because they resemble their endogenous ligands of peptidic hormones and cytokines [8]. In addition, large biologics and small molecules have inherent limitations that prevent them from effectively targeting these receptors. Monoclonal antibodies, a popular class of large biologics, rely on recognizing defined epitopes through their complementarity-determining regions (CDRs). Ion channels and GPCRs, however, lack defined epitopes that are compatible with antibody CDRs, and generating robust monoclonal antibodies to bind to these proteins have been unfruitful [8]. For small molecules, studies have typically shown a lack of potency and poor selectivity with off-target binding [8]. This is due to a lack of binding pockets on ion channels and GPCRs and also a physical requirement to cover a large surface area for specific recognition [5]. The physical features of peptides allow them to overcome the inherent limitations of large biologics and small molecules in targeting ion channels and GPCRs; thus, peptide drugs satisfy an important therapeutic niche and their market will continue to grow.

At present, it is estimated that there are more than 150 peptides in clinical development with another 400–600 peptides in preclinical studies [5]. These developments are primarily concentrated in metabolic diseases, oncology, and cardiovascular diseases. Notably, out of the 17 peptide therapeutics approved in the past 5 years, seven of them were indicated for oncology [9]. Interestingly, all seven were peptide conjugates: four radionuclide conjugates and three monoclonal antibody conjugates. This is indicative of a growing interest in using conjugation to enhance the properties of drug candidates, and the proportion and variety of peptide conjugates in development have increased over time. In fact, 30% of peptide therapies entering the clinic in the past decade were conjugated, compared to 24% in the decade prior and 7% from 1980 to 1999 [1]. This approach expands the potential targets that could be drugged, and thus broadens the indications for which therapeutic peptides can be developed. Therefore, peptide conjugation will likely serve as a significant driver of the growth of the therapeutic peptide market.

While there are many advantages to therapeutic peptides, there are risks inherent to this class of therapeutics that temper market sentiment. Potential risks include chemical and physical instability, the tendency for aggregation, short half-life, and immunogenicity [10]. With risks, though, comes innovation to minimize or eliminate those risks such as optimizing peptide sequences, designing novel spatial configurations, and developing stable formulations. This will help bolster sentiment of the therapeutic peptide market.

28.3 Assessing the market value of therapeutic peptides

The value of a given peptide drug is tied to its approved indication(s); the approved indications define a group of patients that makes up the annual sales for a peptide drug. Herein are several examples of peptide drugs and their market potential:

- Leuprolide acetate (Lupron Depot, AbbVie), a gonadotropin-releasing hormone agonist, was approved in 1989 and is still on the market today. Lupron recently achieved global sales of US $181 million in 2020 for the palliative treatment of advanced prostate cancer.
- [^{177}Lu]-DOTA-TATE (Lutathera, Novartis), a radionuclide-conjugated peptide drug approved in 2018 for the treatment of somatostatin receptor-positive gastroenteropancreatic neuroendocrine tumors, was developed by Advanced Accelerator Applications and bought by Pfizer for US $3.9 billion. Lutathera sales reached US $441 million in 2019, with peak sales expected to reach US $2 billion in the next 3–5 years as clinical trials for label expansion conclude.
- Liraglutide (Victoza/Saxenda, Novo Nordisk), a glucagon-like peptide-1 (GLP-1) receptor agonist indicated for glycemic control in adults with type 2 diabetes mellitus and approved by the FDA in 2010, achieves US $4 billion sales annually, but this figure is expected to decline as semaglutide (Ozempic, Novo Nordisk), another GLP-1 receptor agonist for the same indication, was recently approved in 2017 by the FDA. The key difference between Victoza and Ozempic is how often they are used; Victoza is injected once per day, whereas Ozempic is injected once per week. A less-frequent dosing schedule is preferred by patients and their physicians, and it is expected that many patients currently on Victoza will switch to Ozempic. Ozempic reached US $2.8 billion in sales in 2020, and is projected to reach US $6 billion in sales by 2024.
- Enfortumab vedotin-efjv (Padcev, Astellas Pharma and Seattle Genetics), a peptide-monoclonal antibody conjugate, was approved in December 2019 for the treatment of adult patients with urothelial cancers who have previously received checkpoint inhibitors and chemotherapy. In its first full year on the market, Padcev recorded US $222 million in sales. Astellas Pharma and Seattle Genetics are currently applying for approval in treatment-naive patients, representing a market opportunity upwards of US $1 billion.
- Etelcalcetide (Parsabiv, Amgen), a calcium-sensing receptor agonist indicated for secondary hyperparathyroidism adults with chronic kidney disease on hemodialysis and approved by the FDA in 2017, logged US $716 million in sales in 2020.

Owing to the commercial success of therapeutic peptides, there is ample interest by investors in this class of biotechnology. This is exemplified by Stuart Therapeutics, Bicycle Therapeutics, and Protagonist Therapeutics. Stuart Therapeutics, a clinical-stage company focused on developing unique therapeutic peptides for ophthalmic diseases with its platform PolyCol, completed a US $11 million series A in 1Q 2021. These proceeds are aimed to fund its lead asset ST-100, a treatment for keratitis sicca-related ocular damages, out to phase 2 clinical trials. Bicycle Therapeutics completed an initial public offering of US $60 million in the first half of 2019, and as of 2Q 2021, grew to a valuation of approximately US $650 million. Its lead asset BT1718 targets Membrane Type 1 Matrix Metalloproteinase, an enzyme highly expressed in multiple solid tumors, and is currently in phase 2 trial for solid tumors. In addition, Bicycle Therapeutics' value is also in its platform's capability of generating novel bicyclic peptide therapeutics for both oncology and nononcology indications. As of 2Q 2021, Protagonist Therapeutics has a valuation of approximately US $1.6 billion with its lead asset in phase 2 trial for hematologic disorders such as polycythemia vera and hereditary hemochromatosis. Its lead asset is rusfertide (PTG-300), a synthetic mimetic of hepcidin but with improved potency and physicochemical properties.

For peptide therapeutics companies and investors, determining the potential value of peptide drugs factors into whether investors will provide capital to fund the company's development of novel therapeutic peptides. This is crucial for bringing to market new and improved medicines for helping patients. There are multiple ways to value a company, and using a discounted cash flow (DCF) model is a common and useful method for evaluating therapeutic companies. DCF modeling utilizes future free cash flow projections and discounts them to determine the present value of an asset and/or company. The discount is a rate that captures the opportunity cost and time value of not investing in other assets (i.e., money today is worth more than money in the future). It takes on average 9–12 years for therapeutic innovations to reach the market and start generating revenue from the time of discovery. Biotechnology companies have little to no earnings until then and are thus valued for their future revenues. Because the DCF model uses future free cash flows, as opposed to other valuation approaches based on present earnings, it is well-suited for evaluating the market value of therapeutic innovations. It is important to note that a major limitation of the DCF model is that it is only as accurate as its inputs and assumptions. Typically, multiple DCF models are iterated with different variables (e.g., revenue, expenses, indication population, therapy penetration rate, therapy cost, time to market, risks) to enable sensitivity analysis and ascertain an approximate range of the present value of a therapeutics company. A DCF model supplied with well-supported inputs and assumptions can provide a reasonable market perspective on emerging varieties of small biotech and large pharma therapeutic innovations.

There is no strict rule in determining a discount rate to use, and deciding on an appropriate discount rate to use will be based on the modeler's own experiences. However, to provide a general framework from our experiences: if the company's lead asset is in preclinical development to early phase 1, a discount rate of 25% is appropriate; if the company's lead asset is in late phase 1 to late phase 2, a discount rate of 20% is appropriate; and if a company's lead asset is in phase 3, a discount rate of 15% is appropriate. These provided discount rates are independent of risk, but it is not an uncommon practice to also factor risk into the discount rates. When the risk is factored in, these discount rates are higher in commensuration with the degree of risk.

In assessing the market value of a therapeutics company, it is good practice to diligently examine the risks and use that information to inform valuation. The value of a therapeutics company is tied to its drug pipeline. If the drug does not make it to market, then the value of that company drops. Thus, the higher risk of a therapeutic asset failing to reach the market, the lower the valuation of the company. Risks come in the forms of clinical, scientific, commercial, regulatory, financial, and management. These six risk categories reflect the different factors that ultimately influence whether a therapeutic innovation will make it to market. Here, we provide a simplified risk framework for determining the market value of a therapeutics company and a briefcase study on incorporating risk into a DCF valuation for a fictional peptide drug company (Fig. 28.2).

28.4 Approaches to manufacturing therapeutic peptides

The length and production scale of the therapeutic peptide will largely dictate the manufacturing process used. Modern methods for peptide

Case study: Company X is publicly-traded biotechnology company developing a peptide drug for patients with non-small cell lung cancer (NSCLC). Its lead asset is in phase 2 clinical trials. Preclinical studies were published in reputable journals, and phase 1 results indicate strong safety profile and response rates. Due diligence (DD) on Company X showed a proven management team with experience bringing a drug to market, but the company will have to raise funds to cover their phase 3 expenses. DCF modeling placed the value of Company X at $1 billion.

Risk	Key Question	Metric	Effect on DCF
Scientific	Is the rationale and mechanism of action for the drug robust?	Probability of observing a clinical effect	90% (based on phase 1 results and publications)
Clinical	Will the drug be safe and efficacious in humans?	Probability of successful clinical development	18% (based on historical overall success rates from phase 2)
Regulatory	Will regulators approve the drug?	Probability of successful NDA	85% (based on historical FDA approval rates)
Financial	Is or will there be sufficient funds for the company to develop the drug?	Dilution risk	90% (based on predicted fundraising requirements)
Commercial	Will anyone pay for and use the drug given current and future competition?	Peak market penetration	25% (DD revealed 3 other drugs in use for target indication)*
Management	Will the company be able to develop, market, and scale the drug?	Years to peak market penetration	4 (estimated faster than average based on strong management)*

Risk-incorporated valuation: $1B(DCF)*0.9(scientific)*0.18(clinical)*0.85(regulatory)*0.9(financial) = **$124M**
*Commercial and management input is included in the DCF model

FIGURE 28.2 Risk framework and case study for market valuation.

manufacturing can be generally described through a dichotomy between chemical synthesis or biological production [11,12]. Chemical synthesis is capable of synthesizing peptides up to 100 amino acids in length and can be done in solution-phase, solid-phase, or the combination of the two as hybrid synthesis. Solution-phase synthesis is the coupling of individual amino acids, then short fragments, on soluble supports in solution. While this approach is cost-efficient and scalable, purification is complex and laborious as it requires the growing peptide to be transferred between apparatuses or isolated via a phase transfer. With solid-phase synthesis, the first amino acid is attached to a resin, then subsequent amino acids are added to the anchored, nascent peptide. In comparison to the multitude of purification steps required for solution-phase synthesis, the nonreacted derivatives and coupling reagents can be simply filtered and washed off the resin. Therefore, purification is quicker and simplified in solid-phase synthesis, but is more costly and less efficient than solution-phase synthesis as excess reagents are needed to overcome physical limitations imposed by solid-phase chemistries to drive peptide synthesis [13]. Hybrid synthesis first involves synthesizing short fragments of five to eight amino acids in solid-phase and then constructing the fragments into the full-length peptide in solution-phase. This enables the synthesis of longer peptides than either solution-phase or solid-phase synthesis alone. Hybrid synthesis requires fewer reactions and offers intermediate fragments at higher yields and purity. Compared to biological production, chemical synthesis can have more flexibility and be less expensive for manufacturing scaled up to multiple hundreds of kilograms.

Biological production is generally preferred when specific residue blocking or modification strategies are not needed when the peptide is primarily composed of natural amino acids, or when the peptide is greater than 100 amino acids in length [8]. Biological production can be done through recombinant expression or with cell-free systems. Recombinant expression is especially suited for peptide-conjugates with another protein domain such as human serum albumin or the Fc region of an antibody. Bacteria, yeast, insect, and mammalian cells all have the capability of being used as expression systems [8]. Furthermore, noncanonical amino acids can be added site-specifically, allowing for extremely precise modification of peptides. This strategy is accomplished by modifying the host cell with an aminoacyl-tRNA synthetase/tRNA orthogonal pair recognizing the amber stop (UAG) codon, thereby allowing a noncanonical amino acid to be incorporated wherever the amber stop codon is designed on the DNA template [14]. Typically, the noncanonical amino acid sports a bioorthogonal chemical reporter such as an azide, and corresponding bioorthogonal chemistries are used to append moieties of interest to modify the peptide.

For greater control over peptide modifications and folding, using cell-free systems is advantageous as the in vitro reactions contain only the required components for peptide synthesis. Stochastic posttranslational modifications are avoided, and the addition of peptide modifying factors such as chaperones, isomerases, and/or posttranslational modifying enzymes is controlled. Cell-free systems can be derived from bacterial, plant, and mammalian cells [14]. While the past 5 years has seen an exponential increase in publications detailing the use of cell-free systems to produce a myriad of proteins, challenges exist with scalability that prevents cell-free systems from being used to manufacture in a GMP setting [15]. The ability to precisely control peptide modifications, however, makes it an attractive and promising manufacturing approach.

28.5 Innovations in therapeutic peptide discovery and manufacturing

The field of peptide drug discovery and manufacturing is continually evolving, particularly around the speed and flexibility of production platforms. This is exemplified by two biotechnology startups, Tierra Biosciences and AbSci, and the PEPSTAR approach developed through a collaboration between Eli Lilly, the University of KwaZulu-Natal, and the Livingston group at Imperial College London.

Tierra is developing an automated workflow platform for cell-free synthesis and testing of designed peptides. Cloning and in vivo transformation of circular plasmids is a laborious and slow process, but linear DNA could be quickly assembled and used for rapid prototyping. Such an ex vivo approach was demonstrated with crude lysates from *E. coli*, provided that the linear DNA is protected from exonuclease degradation [16]. Together with other enabling technologies, like on-demand synthetic DNA synthesis from Twist Biosciences or acoustic liquid handling from Labcyte, the cycle times and costs for producing and testing peptides at scale could be dramatically reduced.

AbSci is developing a platform for screening and manufacturing biologics in proprietary *E. coli* cell lines. Generally, bacterial cells have lower production costs relative to the mammalian cells that are typically deployed. AbSci has engineered bacterial strains to have a semioxidizing cytoplasm and a dual-inducible promoter system to tune the rate of protein biosynthesis. This facilitates the robust expression of soluble proteins and the screening of millions of variants in high-throughput assays. Moreover, AbSci can conduct drug optimization in the same cell lines for downstream production, thereby accelerating the timeline between discovery and development. Both Tierra and AbSci are building innovative technologies for fast and adaptable peptide manufacturing and testing.

The PEPSTAR method is a "one-pot" solution-phase approach where synthesis and purification happen in one vessel without the need for material and phase transfers [13]. Briefly, this is performed by growing the nascent peptide on soluble support and incorporating a sieve into the reaction vessel where only the reaction byproducts can pass and be filtered out with organic solvent. This eliminates the complex and numerous purification steps in traditional solution-phase synthesis without compromising the advantages that it has over solid-phase synthesis. The PEPSTAR approach, however, has only been tested for peptides up to 10 amino acids long, and further modifications are needed to allow for the synthesis of longer peptides and for industrial applications.

28.6 Conclusions

Peptides are an impressive therapeutic modality with a diverse range of medicinal applications. Historically, peptide drugs were extracted from animals. As techniques in peptide chemistry and recombinant biology were discovered, more intriguing therapeutic peptides emerged that were compatible with the clinic and feasible to manufacture. These included peptides with conjugated moieties or human sequences, which could be useful for delivering molecular cargo or modulating physiological signaling without immunogenicity. Such peptide drugs are highly valuable to both patients and investors. There continue to be considerable efforts in peptide drug development and commercialization. Numerous small biotechnology startups and large pharmaceutical companies are advancing technology platforms to accelerate peptide drug innovation.

References

[1] Lau JL, Dunn MK. Therapeutic peptides: historical perspectives, current development trends, and future directions. Bioorg Med Chem 2018;26(10):2700–7.

[2] Hughes SS. Genentech: the beginnings of biotech. University of Chicago Press; 2011.

[3] Fineberg SE, et al. Effects of species of origin, purification levels, and formulation on insulin immunogenicity. Diabetes 1983;32(7):592–9.

[4] Riggs AD. Making, cloning and expression of human insulin genes in bacteria: the path to humulin. Endocr Rev 2020.

[5] Muttenthaler M, et al. Trends in peptide drug discovery. Nat Rev Drug Discov 2021;20 (4):309–25.

[6] Fasano A. Innovative strategies for the oral delivery of drugs and peptides. Trends Biotechnol 1998;16(4):152–7.

[7] Davenport AP, et al. Advances in therapeutic peptides targeting G protein-coupled receptors. Nat Rev Drug Discov 2020;19(6):389–413.

[8] Furman JL, et al. Early engineering approaches to improve peptide developability and manufacturability. Am Assoc Pharm Sci 2015;17(1):11–120.

[9] Al Musaimi O, et al. 2020 FDA TIDES (peptides and oligonucleotides) harvest. Pharmaceuticals 2021;14(2):145.

[10] Fosgerau K, Hoffmann T. Peptide therapeutics: current status and future directions. Drug Discov Today 2015;20(1):122–8.

[11] Zompra AA, et al. Manufacturing peptides as active pharmaceutical ingredients. Future Med Chem 2009;1(2):361–77.

[12] Chandrudu S, Simerska P, Toth I. Chemical methods for peptide and protein production. Molecules 2013;18(4):4373–88.
[13] Yeo J, et al. Liquid phase peptide synthesis via one-pot nanostar sieving (PEPSTAR). Angew Chem 2021;60(14):7786–95.
[14] Dondapati SK, et al. Cell-free protein synthesis: a promising option for future drug development. BioDrugs 2020;34(3):327–48.
[15] Voyvodic PL, Bonnet J. Cell-free biosensors for biomedical applications. Curr OpBiomed Eng 2020;13:9–15.
[16] Sun ZZ, et al. Linear DNA for rapid prototyping of synthetic biological circuits in an Escherichia coli based TX-TL cell-free system. ACS Synth Biol 2014;3(6):387–97.

Further reading

Dutton G. Bacterial platforms can rival mammalian platforms: AbSci says that its *Escherichia coli* platforms are engineered to defy known cell line limitations and enable high-titer scale-up of challenging biologics. Genet Eng Biotechnol N 2020;40(6):10–11.
Itakura K, et al. Expression in *Escherichia coli* of a chemically synthesized gene for the hormone somatostatin. Science 1977;198(4321):1056–63.
Sperry JB, et al. Thermal stability assessment of peptide coupling reagents commonly used in pharmaceutical manufacturing. Org Process Res Dev 2018;22(9):1262–75.

Chapter 29

Future perspectives on peptide therapeutics

Özge Ünsal[1,*], Jamie Liu[1,*], Kathleen M. Sicinski[1] and Krishna Kumar[1,2]

[1]*Department of Chemistry, Tufts University, Medford, MA, United States,* [2]*Department of Biomedical Engineering, Tufts University, Medford, MA, United States*

29.1 Introduction

Peptides are an integral part of modern medicine. Peptides made their debut as a therapeutic when insulin was first isolated from the bovine pancreas in the 1920s and used as a life-saving medicine for the management of type 1 diabetes mellitus (T1D) [1]. Although this approach was soon met with its own set of challenges in organ availability, cost of production, and undesired immunogenic responses to animal insulin [2], this nonetheless invigorated the idea that endogenous peptides could be used to treat human disease [3]. Recombinant technology further paved the road for mass production of peptides inexpensively, leading to the regulatory approval of recombinant insulin as the standard of care for both T1D and type 2 diabetes mellitus (T2D) [4,5]. Meanwhile, breakthroughs in chemical synthesis and purification of peptides enabled thousands of peptidic ligands to be synthesized, including synthetic oxytocin and vasopressin, two mainstays in the arsenal of clinically used compounds today [6] (Fig. 29.1).

Most peptides of interest are derived from natural sources and function as signaling molecules, growth factors, ion channel ligands, and antibiotics [6]. Having gone through rounds of evolution, peptides are highly specific and potent, with minimal off-target effects or toxicity, making them superior pharmaceutical agents compared to traditional small molecules. An example of a blockbuster peptide drug is liraglutide (Victoza), a glucagon-like peptide 1 (GLP-1) analog that is used for the management of T2D and glycemic control [7]. To date, over 60 therapeutic peptides have been approved in the United States, Europe, and Japan, with more than a hundred in various clinical and pre-clinical stages of investigation [6].

* These authors contributed equally to this work.

Peptide and Peptidomimetic Therapeutics. DOI: https://doi.org/10.1016/B978-0-12-820141-1.00003-0

FIGURE 29.1 Peptide therapeutics have desirable properties, such as high potency and selectivity coupled with low toxicity. On the other hand, they suffer from short plasma half-lives and renal clearance. Peptide delivery remains a challenge because of low oral bioavailability. The balance between these advantages and disadvantages of peptides are tilted to the former to highlight their promise as therapeutics.

Native peptides, despite their favorable properties, are still burdened with inherent problems that limit therapeutic use [7]. Unmodified peptides have short in vivo half-lives—usually a few minutes—and poor membrane permeability that prevents oral delivery and necessitates the need for injection, compromising patient compliance. Poor membrane penetration limits access to intracellular targets, restricting broader utility. In effect, the peptide's intrinsic properties—absorption, distribution, metabolism, and excretion—must be optimized before it can be deployed as a therapeutic. In this review, we outline the fundamental challenges of peptides as therapeutics (plasma instability and poor membrane penetration) followed by general approaches to address each obstacle and highlight recent efforts to enhance oral bioavailability (BA) and blood–brain barrier (BBB) penetration (Figs. 29.2 and 29.3), with selected clinical examples (Table 29.1).

29.2 Proteolytic stability

The inherently short plasma half-life of most peptides is largely attributed to degradation by proteases. Among the approximately 550 known human proteases, many are ubiquitously dispersed, such as dipeptidyl peptidase-4 (DPP4), while others are localized, for example, in the lumen of the gastrointestinal (GI) tract (pepsins, trypsin, and chymotrypsin) and at the brush border membrane (endopeptidases and amino/carboxy peptidases). Because proteases serve important physiological functions, such as digestion, activation or inactivation of signaling molecules, and regulation of protein turnover, broad inhibition is typically not a viable option. Significant research efforts have therefore been directed to the design of protease-resistant peptides that simultaneously maintain native efficacy and potency [34].

29.2.1 Rational design

Medicinal chemistry approaches to optimize a potential peptide drug candidate includes finding the shortest active sequence by identifying key residues required for bioactivity. The elucidated structure-activity relationship (SAR)

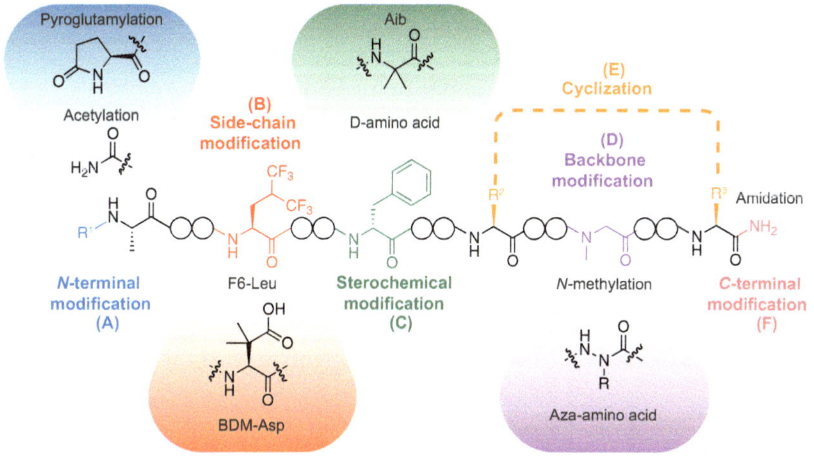

FIGURE 29.2 Selected modifications of peptides that enhance proteolytic stability and oral bioavailability. (A) *N*-terminal modifications such as acetylation [8] and pyroglutamylation [9] afford protection against aminopeptidases and some endopeptidases by altering electronic and steric properties. (B) Sidechain modifications with unnatural residues like hexafluoroleucine (F6-Leu) [10] and β,β-dimethyl-Asp (BDM-Asp) [11] render peptides resistant to protease degradation. (C) Stereochemical modifications such as incorporation of D-amino acids [12,13] or α-aminoisobutyric acid (Aib) [14] change the secondary structure of the peptide and result in enhanced metabolic stability. (D) Backbone modifications: *N*-methylation [15] confers protease resistance and enhanced BA by increasing conformational rigidity and lipophilicity of peptides. Aza peptides [16] provide proteolytic stability by inducing secondary structural changes to the peptide backbone. (E) Cyclization [15] restricts conformational freedom of peptides while reducing the polar exposed area, leading to improved metabolic stability and oral BA. (F) C-terminal amidation [8] sometimes affords protection against carboxypeptidases.

enables optimization of "drug-like" properties, namely potency, selectivity, and pharmacokinetic (PKs) properties [7]. Critical residues are traditionally identified through alanine (or other amino acid) scanning—a method that replaces each amino acid individually and assesses whether the change causes any loss in activity. These efforts have yielded peptide mimics that exhibit exceptional metabolic stability without compromising activity.

29.2.2 Identification of cleavage site(s)

To generate proteolytic resistant analogs, it is crucial to identify cleavage sites within the sequence and tease out putative proteases in action. Although the latter may not always be feasible, labile peptide bonds are identified via incubation of peptides with biological matrices containing proteolytic enzymes (i.e., blood serum, liver, or stomach hydrolysates) followed by liquid chromatography-mass spectrometry analysis to determine peptide fragments [34]. Sites of cleavage often point to a particular protease and in some

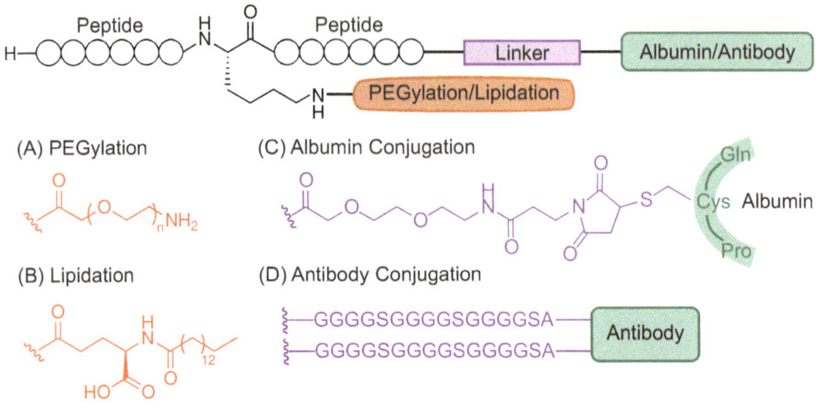

FIGURE 29.3 The pharmacokinetic properties of therapeutic peptides are improved through the selected modifications depicted above. These modifications contribute to a longer half-life upon common changes, such as increasing the molecular weight and adding steric bulk around the peptide to block immune and protease recognition. (A) PEGylation [17] provides higher solubility and increased apparent size due to the extended hydrodynamic radius. (B) Lipidation [18,19] promotes "micelle-like" oligomeric structures and lipid groups can bind to albumin, which contributes to circulation half-life. (A) and (B) are often utilized as sidechain modifications. (C) Albumin [20,21] and (D) antibody conjugation [22] are similar strategies where the peptide is bound to long-lived plasma proteins and half-life is improved through the neonatal Fc receptor (FcRn) salvage pathway. (C) and (D) are usually conjugated *via* a linker (shown in purple) to the C-terminus of the peptide since these groups (shown in green) are extremely large.

cases, the hydrolytic enzyme may remain unknown. Alternatively, one can utilize insight from initial SAR studies regarding critical residues to introduce known protective motifs at the labile site—including the incorporation of unnatural amino acids or other side chain and backbone chemical modifications to disrupt protease-substrate recognition and/or foil the catalytic machinery, conferring stability. Modulation of secondary structure by conformational constraints offers another avenue for stabilization. Using the knowledge of the cleavage site and library scanning, medicinal chemists can design peptides with superior selectivity, potency, and enzymatic stability [7].

29.2.2.1 *Modification of amino and carboxy termini*

Stability toward exopeptidases can be frequently achieved by modifying the peptide's amino or carboxy terminus. *N*-acetylation and *C*-amidation are commonly utilized (Fig. 29.1A,F) [34]. Enfuvirtide (Fuzeon, Trimeris), a 36-residue peptide with both such modifications, is a clinically used antiretroviral drug that inhibits human immunodeficiency virus (HIV) entry by preventing fusion between the viral and host-cell membranes (Fig. 29.1) [23]. Other *N*-terminal modifications include *N*-pyroglutamylation [9], lipidation [35], glycosylation [36], and PEGylation [37] that have found use for increasing stability of peptides. Tesamorelin, a 44-residue human growth

TABLE 29.1 Selected therapeutic peptides in the clinic with their respective modifications and influence on drug properties.

Modification(s)	Effect(s)	Clinical Examples
N-acetylation and *C*-amidation	Stabilization of secondary structural motifs	Enfuvirtide [23]
N-hexenoyl acetylation	Protection against DPP4 & enhanced potency	Tesamorelin [24]
N-terminal deamination & D-Arg	Protection against exopeptidases and trypsin-like proteases	Desmopressin [12,13]
D-Phe and D-Nle	Proteolytic stability & enhanced potency	Afamelanotide [25–27]
N-terminal deamination, O-methylation [Tyr(OMe)], and thioether cyclization	Protection against disulphidase, aminopeptidase and oxidoreductase	Carbetocin [28]
N-methylation & cyclization	Proteolytic stability and enhanced bioavailability	Cyclosporin [15]
Cα-methylation & lipidation	Metabolic stability against DPP4 w/enhanced plasma half-life	Semaglutide [14]
Aza dipeptide	Proteolytic stability & enhanced bioactivity	Atazanavir [29,30]
D-Ser(tBu) and aza-Gly	Proteolytic stability	Goserelin [31]
Cyclization	Proteolytic stability & enhanced oral bioavailability	Linaclotide [32,33]
PEGylation	Increased molecular weight and hydrodynamic radius	PEG-adrenomedullin [17]
Lipidation	Increasing molecular weight and "micelle-like" oligomeric structures and albumin binding	Liraglutide [18,19]
Albumin conjugation	Increased molecular weight and FcRn recycling	CJC-1134-PC [20,21]
Antibody conjugation	Increased molecular weight and FcRn recycling	Dulaglutide [22]

hormone-releasing hormone analog adorned with a *N*-hexenoyl terminus, is an FDA-approved treatment for excess abdominal fat in HIV patients (Table 29.1). The drug was shown to resist DPP4 cleavage with an extended half-life compared to the parent peptide (1–6.8 h) [24]. While terminal modifications have found tremendous utility in the clinic, peptides are also degraded by endopeptidases and require additional routes for protection.

29.2.2.2 D-amino acid incorporation

Peptides composed of D-amino acids are resistant to proteolysis and are a great alternative to enhance plasma half-life (Fig. 29.1C). In a comparative study investigating proteolytic stability of common backbone modifications, D-amino acid substitution was shown to be the most effective in preventing chymotrypsin degradation, while affording remote protection for up to four residues away from the cleavage site [38]. Many pharmaceuticals containing D-amino acids exhibit significant increases in half-lives compared to their L counterparts. Desmopressin (1-desamino-8-d-Arg-vasopressin), a D-Arg analog of antidiuretic hormone (vasopressin), has a 10-fold longer half-life while maintaining picomolar potency and is bioavailable by both oral and intranasal administration (Table 29.1) [12,13]. Another example is the 13-residue analog of human α-melanocyte-stimulating hormone, afamelanotide (Scenesse), which contains a D-Phe and L-Nle substitution and is approved for the treatment of sunlight intolerance. The modifications resulted in a four- to fivefold increase in plasma stability [39] as well as potency in a cyclic adenosine monophosphate (cAMP) assay (Table 29.1) [25].

However, not all D-amino acid substitutions lead to improved potency and stability. For instance, dermorphin, a heptapeptide and powerful analgesic isolated from the skin of frogs harbors a D-Ala in position two but is susceptible to cleavage at other positions [40]. Dermorphin analogs with either D-Tyr^1 or D-Phe^3 substitution resulted in even more rapid degradation, decreasing the half-life by 50% compared to native dermorphin [40]. This shows that other factors such as the peptide's secondary structure as well as the specific interactions within the proteolytic enzyme active site (electrostatic/H-bonding) need to be considered in designing stable peptides. Nonetheless, incorporation of D-amino acids has shown to be a facile way to obtain potent and stable peptide analogs with clinical relevance.

29.2.2.3 Amino acid modification

Aside from D-amino acids, other noncanonical amino acids can be introduced at the labile site to improve PK of a peptide. Unnatural amino acids are comprised of residues that interfere with protease recognition by introducing side chains that differ in sterics, polarity, or electrostatics. For example, substitution of GLP-1 with either β,β-dimethyl-Asp (BDM-Asp) or *tert*-Leu at position 3 produced DPP4 resistant analogs that were completely intact upon in vitro incubation with enzyme at 24 h compared to 0.5 h for native GLP-1 (Fig. 29.1B) [11]. Potency of each analog, measured bycAMP stimulation, deviated little from native GLP-1, but *tert*-Leu substitution resulted in the 20-fold loss of binding affinity. Alternatively, GLP-1 substitution with hexafluoroleucine (F6-Leu) at position 2 conferred complete DPP4 resistance while maintaining efficacy, but suffered from moderate loss of potency (cAMP production) [10]. In all cases, incorporation of unnatural side chain with altered sterics and hydrophobicity have shown to be a versatile way to enhance peptide stability (Fig. 29.1B).

"Bioisosterism" [41] is frequently used in the design of unnatural residues that enhance the drug-like properties of the peptide. In this approach, each amino acid can be mimicked with a similar isosteric amino acid (canonical or noncanonical) with comparable size, hydrophobicity, electronic properties, number of hydrogen bond donors or acceptors, and reactivity. Amino acids that bear structural and electronic resemblance to Arg (Lys, ornithine, homoarginine, citrulline, or *N*-isopropylornithine) have been used to produce potent and selective agonists of vasopressin receptors [42]. Aromatic residues (Phe, Tyr, or Trp) can be mimicked by a number of heterocycles [43] or methylated [44], giving rise to a diverse set of isosteres. One clinically relevant example is carbetocin, a potent oxytocin analog approved for postpartum hemorrhage (Table 29.1). Carbetocin contains two isosteric substitutions: *O*-methylated Tyr [Tyr(OMe)], and thioether cyclization [28]. Another peptide, icatibant [45], illustrates versatility by utilizing multiple unnatural residues to generate a selective antagonist of bradykinin B2 receptors for the treatment of acute angioedema. Icatibant contains an *N*-terminal D-Arg, a proline and histidine isostere, 4-hydroxy-proline and 2'-thienylalanine, and bicyclic analogs, tetrahydro-isoquinoline-3-carboxylic acid (Tic) and octahydroindole-2-carboxylic acid (Oic), that add further stability by imposing conformational restraints in the linear peptide [45]. Thus, the incorporation of unnatural amino acids is an effective way to generate stable peptide therapeutics with enhanced selectivity and potency.

29.2.3 Peptidomimetics

In addition to altering hydrolytic enzyme recognition sites, incorporation of noncanonical amino acids can increase BA by enhancing secondary structure [46]. Conformational constraints, induced by cyclization or enforcing structural elements such as an α–helix [47] or β–sheet, help achieve metabolic stability as proteases have evolved to bind substrates in an extended conformation. The biological activity of peptides can be enhanced if "locked" in a favorable conformation to bind the target with fewer degrees of freedom and lower entropy cost than linear peptides.

Molecules that bear resemblance to a peptide and maintain comparable bioactivity have been coined peptidomimetics [48]. Linear versions are obtained by the incorporation of unnatural amino acids capable of inducing structural changes [46]—such as D-amino acids, residues alkylated on the alpha carbon or amine, and peptides with an extended or modified backbone.

This review will focus on modifications that gave rise to clinically relevant peptides either in the late or final stages of drug development.

29.2.3.1 N-methylation

N-methylation of backbone amides has proven to be an effective way to make peptides proteolytically stable, while simultaneously creating potent antagonists or agonists of biological targets (Fig. 29.1D) [49,50]. This simple modification constraints conformational motifs and adds steric bulk to the peptide backbone, abrogating protease binding and recognition. In particular, amide alkylation lowers the energy barrier between the cis and trans conformations, promoting small molecules [51] and short peptides [52] to adopt the cis-orientation in solution based on nuclear magnetic resonance studies. *N*-methylated peptides lack the ability to act as H-bond donors, and this characteristic proved beneficial in the case of the *N*-methylated analog of the luteinizing hormone-releasing hormone (LHRH). The methylated amide in LHRH disrupted an H-bond in the chymotrypsin active site but was 20 times more potent than the parent peptide [53].

N-methylation not only confers metabolic stability but also enhances cell permeability, leading to analogs with higher oral BA. The rationale is twofold: (1) increase in lipophilicity facilitates molecules to partition across the cell membrane and (2) methylation serves to stabilize intramolecular hydrogen bonding essential for membrane permeation [54]. In fact, *N*-methylated and cyclic peptides are commonly seen in nature and often exhibit superior stability and membrane permeability. Cyclosporin, an immunosuppressant isolated from the fungus *Trichoderma polysporum* contains seven *N*-methylated amino acids and is orally available. Cyclosporin is approved for use during organ transplantation (Table 29.1) [15]. Drawing inspiration from nature, the strategy to improve oral BA using *N*-methylation at multiple sites was applied to a cyclic somatostatin analog cyclo(-PFwKTF-) [55]. Also known as Verber-Hirschmann peptide, a tri-*N*-methylated analog exhibited a five-fold increase in half-life and a 10% increase in oral BA.

N-methylation has been shown to improve pharmacological properties such as enhanced metabolic stability and oral BA. However, care must be exercised to strategically place *N*-methyl residues, as the introduced modification can frequently disrupt the secondary structure of peptides and compromise bioactivity [50].

29.2.3.2 α-Methylation (Aib)

Methylation at the α-carbon offers an alternative avenue to make peptides refractory to proteases by conformational restriction and addition of steric bulk at the *now* quaternary α-carbon. One residue, α-aminoisobutyric acid (Aib) [56], promotes α-helicity and is most commonly deployed (Fig. 29.1C). According to crystal structures of Aib containing compounds, shorter Aib sequences adopt a 3_{10} helical structure, while longer peptides favor α-helices. The change in secondary structure renders the peptide unrecognizable by proteases, thereby escaping degradation. This phenomenon is further supported by Ramachandran analyses of backbone conformations of typical chymotrypsin substrates in comparison to Aib containing peptides, which revealed that the latter falls under the unfavored region due to restricted torsional angles of Aib residues [38,57].

The unnatural amino acid Aib was discovered in a naturally occurring peptide antibiotic, peptailbol [56]. The primary sequences of peptaibols include Aib residues and a *C*-terminal amino alcohol that give rise to their interesting conformational properties. One example is alamethicin, a helical peptailbol with 8 Aib residues, which exhibits antimicrobial activities by aggregating and forming cation-selective channels across lipid bilayers—disrupting the ionic gradient and membrane integrity essential to bacterial survival [58]. Despite its promising antimicrobial activity, alamethicin never progressed to the clinic due to associated cytotoxicity [59].

Aib has been utilized in the development of numerous therapeutic peptides. Notably, GLP-1, a secretin hormone that regulates blood glucose homeostasis, has been an attractive remedy for T2D and obesity [60]. However, GLP-1 is rapidly inactivated by endogenous proteases, such as DPP4—and has an in vivo half-life of ≤ 2 min [61]. DPP4 hydrolyzes the two *N*-terminal residues (His1-Ala2) of GLP-1, rendering the cleavage product impractical for clinical use. To preserve the therapeutic efficacy of GLP-1, significant research efforts have been spent in the design of GLP-1 analogs that resist proteolysis while maintaining potency at its cognate receptor, GLP-1R [62,63]. This metabolic instability is solved by a Ala2Aib substitution, giving rise to a long-acting GLP-1 analog, semaglutide (Ozempic, Novo Nordisk), which has a half-life of 165 h and can be subcutaneously (s.c.) administered weekly [14]. Though the dramatic improvement in plasma half-live is largely influenced by the tighter albumin binding via a C18 diacid tail at Lys26, Aib substitution confers proteolytic protection when the peptide is free from albumin.

29.2.3.3 Aza modifications

Another backbone modification involves isosteric replacement of the C_α of an amino acid of a peptide with nitrogen, giving rise to "aza" peptides that confer enhanced proteolytic stability and retention of bioactivity (Fig. 29.1D) [16]. This change causes loss of chirality at the C_α and gain of rigidity about the C_α-C(O) bond, which favors peptides to adopt β-turn conformations, observed in crystal structures [64] and computational models [65] of short aza constructs. Owing to the favorable PK properties of aza peptides, there has been extensive use in drug development, particularly in the design of protease inhibitors that target viruses [66,67]. Atazanavir (Reyataz; Bristol-

Myers Squibb) is an antiretroviral protease inhibitor approved as a prophylactic and treatment for HIV-1 infection and acquired immunodeficiency syndrome [29,30]. Atazanavir, containing an aza-dipeptide isostere and bis (tert-leucine), mimics the transition state in the substrate-enzyme complex of HIV-1 protease and exhibits potent antiviral activity and oral BA. Additionally, aza amino acids can be exploited to generate a long lasting "super agonist," exemplified by goserelin (Zoladex, AstraZeneca), a gonadotropin-releasing hormone (GnRH) analog. Goserelin (D-Ser(t-Bu)^6, azaGly^{10}) has a potency 100 times greater than GnRH and is used in the clinic for the treatment of prostate and breast cancer [31].

29.2.3.4 Cyclization

Another way to constrain secondary structures of peptides is *via* cyclization. Peptide cyclization often leads to improved binding and selectivity, as well as proteolytic stability due to shielded backbone amides. Cyclized peptides are generally not recognized by proteases, because most proteases recognize substrates in an extended conformation and a lack of charged termini. Many clinically relevant cyclic peptides are derived from natural products, making them an important class of antibiotics and antimicrobial agents [68]. For example, vancomycin is a glycopeptide antibiotic that inhibits cell wall synthesis of Gram-positive and certain Gram-negative bacteria and is the first line of defense against methicillin-resistant *Staphylococcus aureus* [69].

One key benefit of cyclic peptides is their propensity to be membrane permeable—promoted by increasing numbers of internal hydrogen bonds and decreasing polar exposed area [70]. This facilitates membrane insertion by lowering the free energy cost of desolvating peptides into the lipid bilayer, which is essential for gut–blood barrier penetration and oral delivery. As mentioned earlier, the cyclic and *N*-methylated peptide cyclosporin was one of the first orally available peptide drugs for the treatment of autoimmune conditions and inflammation related to organ transplant [15]. Another example that highlights how cyclic structures help with oral BA and stability is linaclotide, a 14-mer with three disulfide linkages that exhibits exceptional stability against simulated gastric fluid and in vitro incubation with pepsin. Linaclotide is FDA-approved for irritable bowel syndrome with constipation by acting as a potent agonist of intestinal epithelial guanylate cyclase C (GC-C) [32].

Cyclic peptides can be linked in various fashions (head to tail, head/tail to side chain, or side chain to side chain) and through different chemical means—lactamization, lactonization, and disulfide bond formation [46]. In addition to providing metabolic stability, cyclization often enhances bioactivity and sometimes alters the selectivity profile of peptides. For instance, a synthetic opioid analog of enkephalin, [D-Pen^2-D-Pen^5]-Enk (DPDPE), exhibits excellent selectivity for the δ-opioid receptor [71] over two other subtypes (μ and κ) while maintaining proteolytic resistance [72]. The selectivity and protective effect likely stem from the presence of a disulfide linkage formed between the D-penicillamine (D-Pen) residues at the 2nd and 5th positions, conformationally constraining the overall structure [73]. In other cases, cyclization may be absolutely essential for bioactivity, as demonstrated by the development of a potent anticoagulant, eptifibatide, approved for acute coronary syndrome [74]. Eptifibatide, a cyclic heptamer composed of one mercaptopropionyl residue and a single disulfide bond, is derived from a disulfide-rich snake-venom disintegrin barbourin and prevents platelet aggregation by binding to a cell-surface receptor glycoprotein. Eptifibatide exhibits potency comparable to the native ligand, likely due to the presence of a disulfide linkage that holds the peptide in the correct conformation to enable receptor binding [75].

In addition to traditional cyclization, stapled peptides have emerged as useful tools for disrupting interfaces between proteins. Protein–protein interactions (PPIs) may now be therapeutically targeted that may not otherwise be accessible by small molecules [76]. The term "stapled peptides" generally refers to cyclic peptides that are linked in a way that facilitates helix formation, by forming a covalent bond between the side chains of residues on the same face of a helix. Among other bridge forming chemistries (lactam, thioether, triazole, etc.), hydrocarbon staples have been most widely used and are achieved by ring-closing metathesis of olefin-bearing amino acids [46]. The utility of hydrocarbon stapling was demonstrated by the enhanced bioactivity and serum stability of BH3, a pro-apoptotic protein with activity against a leukemia cell line, as a result of staple incorporation [77].

In summary, protease protection can be achieved by site-specific modification using noncanonical residues or via a more global approach to induce secondary structural changes to the peptide backbone. Most successful lead compounds with optimized PK likely contain a combination of strategies that have been described thus far and in the upcoming sections to best preserve bioactivities. Often, medicinal chemists have to grapple with finding a fine balance between the added benefits of the modifications and loss in potency and efficacy, along with toxicity. The good news is this is achievable with modern technologies, evidenced by the number of peptide drugs in clinical usage today, with extended metabolic half-life.

29.3 Renal clearance

Providing stability toward proteases in various ways is an important step for enhancing the half-life of peptide therapeutics. However, this often does not resolve the problem entirely. Kidneys contribute to the short plasma half-life of peptides since they rapidly remove them from the bloodstream by glomerular filtration. The glomerular capillary wall is a barrier that filters many compounds depending on their size, shape, and charge [78,79]. The cutoff size is approximately 60 kDa; however, the clearance rate is rapid for proteins with molecular weight less than 30 kDa, and peptides (<5 kDa) fall within this range [78]. The first mechanism is the reabsorption of macromolecules into endocytic vesicles into the proximal tubules upon glomerular filtration and hydrolysis into smaller metabolites such as amino acids and truncated peptides [78]. Another mechanism involves glomerular filtration followed by macromolecule decomposition by exopeptidases situated in the proximal tubule membrane whereafter the smaller fragments are reabsorbed into the plasma [78]. In addition to size, charge is another important criterion for renal clearance. The filtration barrier accommodates negatively charged molecules that include heparan sulfate proteoglycans in the glomerular basement membrane which block the anionic molecules from crossing, leading to a low filtration [78,79]. On the other hand, cationic and neutral compounds are able to pass more readily through the barrier that is coated with negative charges and this elevates the clearance rate for nonanionic molecules from circulation. To overcome this complication, modifications have been made to change the PK of therapeutic peptides, which include but are not limited to PEGylation, lipidation, albumin binding, fusion with the immunoglobulin G (IgG) Fc region, and glycosylation. Conjugation of these groups with peptides increases the overall molecular weight, leading to a reduced renal clearance rate. On the other hand, the addition of these large structures onto peptides is often a challenging step that requires optimization to find the ideal site for modification without sacrificing the binding and potency of the peptide.

29.3.1 PEGylation

One of the most common strategies to overcome renal clearance is a conjugation of the peptide to a large molecular weight polyethylene glycol (PEG) polymer. PEG is comprised of repeating units of ethylene oxide $(-CH_2-CH_2-O-)_n$ and participates in hydrogen bonding with water, therefore increasing the effective molecular weight and hydrodynamic radius of the peptide [80,81]. Since kidneys filter substances by size, the addition of PEG slows renal clearance, which eventually leads to a prolonged plasma half-life. It is also important to note that conjugation to PEG can improve the PK properties in additional ways, such as blocking immune and protease recognition. The large hydrodynamic radius, however, may interfere with target binding, requiring optimization of sites for modification.

Adrenomedullin is a 52-amino acid endogenous peptide known for its properties of vasodilation and induction of cell growth [17,82]. Despite these positive effects, adrenomedullin suffers from a short half-life of 22 mins [83]. Additionally, native adrenomedullin is a fast-acting potent compound that leads to an elevated heart rate. In an effort to both increase circulation and develop a long-acting analog, Kubo and coworkers explored PEGylation of adrenomedullin. The *N*-terminal tyrosine of human adrenomedullin peptide was conjugated to PEG5000 (5 kDa) and injected into rats intravenously (i.v.), which resulted in an increase of plasma half-life by sevenfold compared to native human adrenomedullin [17]. PEG-Adrenomedullin with a 40-kDa PEG unit (BAY1902607, Bayer) is now in Phase I clinical trials for lung disease [80].

Although PEGylation prolongs the half-life of therapeutic peptides, the long-term safety profile of these unnatural polymers is not fully understood. PEG is considered to be biologically inert; however, there are some concerns including hypersensitivity and immunogenic responses caused by PEG conjugation [84]. A synthetic peptide containing a 20-kDa PEG moiety, peginesatide (Omontys, 2012 Affymax/Takeda), used for the treatment of chronic kidney disease was recalled for hypersensitivity [85]. Despite concerns regarding PEG conjugation, this is still widely used in medicinal chemistry as a way to increase the circulating half-life of peptides.

29.3.2 Lipidation

Another strategy for prolonging the plasma half-life of peptides is lipidation. While both PEGlyation and lipid conjugation utilize the same concept of increased molecular weight, lipidation also increases hydrophobicity, which can promote "micelle-like" oligomeric structures [86]. An important characteristic of lipidated peptides is their ability to bind to an abundant plasma protein, serum albumin, which can bind up to seven fatty acids simultaneously [87,88]. Albumin is the most abundant serum protein in humans with a half-life of approximately 19 days, which can assist with decreasing renal clearance and shielding peptides from nearby proteases. The binding of the peptide to albumin is beneficial because it can dramatically increase the molecular weight by 65 kDa to slow renal clearance. Albumin prefers to bind fatty acids of 10–18 carbons in length and recognizes a polar carboxylate headgroup. Lipid chains are accommodated differently in each binding site followed by similar mechanisms, such as hydrophobic interactions with methylene tails and electrostatic interactions with basic and polar side chains for carboxylate moieties. Interactions between the lipid and albumin are reversible, and the lipid-peptide complex serves as a depot for the free peptide in circulation.

There are many FDA-approved drugs that increase peptide half-life by albumin binding. Insulin detemir (Levemir, 2004, Novo Nordisk) was the first clinically approved lipidated insulin used for the treatment of T1D. The analog lacks threonine at the B30 position (desB30) and has a fourteen carbon lipid chain myristoyl moieties appended to the lysine at B29 [Lys^{B29}(*N*-tetradecanoyl)des(B39)human insulin], which generates oligomers to help absorption and prolong the action of the native hormone [89]. C14 fatty acid chains aid self-association to form hexamers. Human serum albumin (HSA) is able to bind both forms of insulin detemir, and more than 98% of circulating analog is albumin-bound [89,90]. Insulin detemir has a prolonged half-life of 4–6 h after s.c. injection, whereas the native hormone is absorbed within an hour [91–93].

Native GLP-1 is another endogenous peptide that has an extremely short half-life (~90 s) mediated by proteolysis and rapid renal clearance [61]. Many GLP-1 analogs have been investigated to determine if they display increased circulation time. Liraglutide (Victoza, Novo Nordisk) is a GLP-1R agonist that has an improved half-life due to addition of a palmitic acid on Lys26 with an L-γ-glutamic acid linker [18]. Various GLP-1 derivatives were assessed to find the optimum fatty acid chain length and an appropriate site for lipid addition. The half-life of liraglutide is extended to 11–15 h through albumin binding and oligomer formation. This allows for liraglutide to be administered once a day by s.c. injection for T2D management [18,94]. An improved version is Semaglutide (Ozempic, Novo Nordisk) requires once weekly ($t_{1/2}$ ~ 165 h) s.c. injection for T2D [95,96]. Unlike liraglutide, this analog contains a modified Lys26 with two oligoethylene glycol (OEG) spacers (8-amino-3,6-dioxaoctanoic acid), a L-γ-glutamic acid, and an octadecanoic diacid lipid. These linkers aid flexibility and solubility of the peptide, whereas the free carboxylate group prolongs the plasma half-life by tighter binding to albumin [95].

29.3.3 Plasma protein fusion to extend circulation times

29.3.3.1 Albumin conjugation

Modifying peptides with a moiety that has a high affinity to albumin is a useful strategy to increase half-life. This can also be achieved by direct conjugation to albumin. Albiglutide (Tanzeum, GlaxoSmithKline), another GLP-1 analog, is such a construct covalently bound to recombinant HSA, extending the half-life to approximately 5 days [97]. Although the half-life of albiglutide is enhanced to make it available for once-weekly administration, the potency for GLP-1R is impaired because of the sheer bulk of the appendage. In vitro experiments showed that albiglutide potency decreased 100-fold compared to a potent GLP-1R agonist, exendin-4 [60]. Albiglutide was able to make it to the clinic; however, the drug was withdrawn four years after its approval by the FDA for economic reasons. Another GLP-1 analog (CJC-1134-PC, ConjuChem) is being investigated in Phase II clinical trials . This is a modified exendin-4 analog where a reactive maleimidopropionic acid group installed at Lys40 allows albumin conjugation (Ex-4, 1–39 aa), leading to longer circulation times (~8 days) [20,21]. Recombinant HSA is conjugated to exendin-4 ex vivo by a reaction between a maleimide group on the peptide and the Cys34 residue on albumin. In sum, albumin conjugation can yield functional constructs with improved half-lives, but sometimes this comes with diminished potency.

29.3.3.2 Conjugation to immunoglobulin G

Attachment of peptides to long-lived plasma proteins can be expanded to other large biologics such as antibodies, forming "peptibodies" [98]. These constructs combine the bioactivity of a peptide with the longer plasma time afforded by the antibody moiety. Peptibodies of target peptides and nonvariable fragment crystallizable (Fc) domain of an antibody is engineered in *Escherichia coli* through recombinant technologies or phage libraries [98,99]. The Fc domain of IgG slows renal clearance via two mechanisms: (1) increasing the molecular weight of the peptide by approximately 70 kDa and (2) improving the half-life through the neonatal Fc receptor (FcRn) salvage pathway [98]. The IgG-Fc domain has an affinity for FcRn, which can lead to internalization of the peptibody. Once the peptibody is endocytosed, it is delivered to the plasma membrane or transcytosed to the extracellular milieu [98].

Romiplostim (Nplate, Amgen) is a recombinant therapeutic developed as a thrombopoietin (TPO) receptor-ligand for chronic immune thrombocytopenia treatment. The drug contains two TPO receptor binding peptides linked to two disulfide-bonded Fc fragments of human IgG1 via polyglycine [100,101]. This construct had equipotent biological activity with multiple peptide units activating the TPO receptor and a half-life prolonged by antibody recycling, enabling once weekly s.c. injection [102]. Dulaglutide (Trulicity, Eli Lilly) is another therapeutic that has two DPP-IV resistant GLP-1 mimetic peptides connected to two Fc fragments of IgG4 [22]. Insulinotropic effects of the endogenous peptide were retained, and the plasma half-life was improved to 90 h [103]. Dulaglutide incorporates the Fc fragment of IgG4 instead of IgG1 for the following reasons: (1) in vitro assays suggested that the biological activity of the peptide is dramatically reduced with IgG1 fragment, (2) IgG1 has higher affinity for Fc receptors, and (3) cytotoxicity levels are elevated when IgG1 is used potentially through complement-dependent cytotoxicity and/or antibody-dependent cell-mediated cytotoxicity (ADCC) [22].

The fusion of Fc fragments to peptides has led to successful therapeutics. However clinical outcomes are not always predictable. AMG 819, a nerve growth factor antagonist developed for pain management, is one such example [99]. The peptibody has an IgG1-Fc domain fused to a 20 amino acid peptide generated by screening a phage-display library. Unfortunately, AMG 819 trials were terminated at Phase I due to forming strong immunogenic responses [99]. In sum, generating peptibodies by appending antibody scaffolds has advantages, but also comes with liabilities, such as the potential for immune reactions.

29.4 Oral bioavailability

In addition to in vivo lability, peptides also suffer from low membrane permeability and have a low oral BA (~1%). Only 4% of currently approved peptide drugs (over 60) in the United States, Europe, and Japan are delivered orally, and 90% are injectables [6,104]. This disparity comes from physical and biochemical barriers of the GI tract that prevent oral absorption of peptides [105]. The gut–blood barrier is lined with epithelial cells connected via tight junctions that limit the passage of molecules. Small peptide fragments can passively diffuse through tight junctions (paracellularly) or epithelial cells (transcellularly) via active transport. Either route depends on physicochemical properties of the peptide with emphasis on size, charge, lipophilicity, and solubility at the pH range of the intestine.

The biochemical barriers to oral delivery include vast numbers of proteolytic enzymes present in the GI tract that degrade proteins and peptides [105]. Some of the most prominent proteases are pepsin, an aspartate protease-activated in the low pH environment of the stomach, and pancreatic proteases secreted in the small intestine, including serine proteases (trypsin, α-chymotrypsin, and elastase) and exopeptidases. To overcome biochemical barriers of peptide oral delivery, strategies to afford metabolically stable peptides have been discussed in-depth in the previous section (see Section 29.3), some of which are also applicable to enhancing membrane permeability.

Common scaffolds that facilitate membrane permeation are cyclization and *N*-methylation, by increasing lipophilicity and reducing conformational freedom, along with replacing external H-bonds with internal ones [34]. This is exemplified by some orally available peptide drugs that contain the above epitopes: lipophilic cyclosporin (cyclic and heavily *N*-methylated) affords 30%–40% oral BA, and potent desmopressin (cyclic and containing D-amino acids) affords 0.17% oral BA [106].

29.4.1 Technologies for oral delivery of peptides

29.4.1.1 Permeation enhancers

One technology involves the use of amphiphilic molecules that can be coadministered with peptide drug candidates to increase hydrophobicity and facilitate passive diffusion across the lipid membrane. Surfactants, bile salts, chelating ligands, medium-chain fatty acids, and chitosan are effective in this respect and are collectively termed intestinal permeation enhancers (PEs) [105].

One of the leading carriers for impermeable peptides across the GI tract is sodium N-[8-(2-hydroxybenzoyl) aminocaprylate] (SNAC, or salcaprozate sodium), an *N*-acetylated derivative of salicylic acid. SNAC was developed by Emisphere as part of a large library screen for *N*-acylated α-amino acids that have high PE potential [107]. The most recent success story of SNAC comes from the 2019 FDA approval of Rybelsus oral tablets (Novo Nordisk), dosed once daily, coformulated with semaglutide for T2D [108]. The approval stems from successful Phase II and III clinical trials with once daily oral administration of the tablet showing a 1.4% reduction in glycated hemoglobin (HbA1c) and weight loss—results comparable to weekly s.c. delivery of semaglutide without SNAC [108]. However, the mechanism of SNAC remains largely undefined in molecular terms.

Until recently, absorption of small molecules and peptide-based pharmaceuticals with or without enhancers were thought to take place in the intestine due to its absorptive nature and large surface area [109]. A study in 2018 of oral administration of semaglutide and SNAC in human volunteers and canine models revealed that absorption of the carrier-drug mixture takes place in the stomach [110]. Human volunteers were orally dosed with a semaglutide (10 mg)/SNAC (300 mg) tablets containing [111] In-labeled ion exchange resin, and complete tablet erosion (CTE) was observed over 57 min using scintigraphic imaging. To ascertain the site of absorption, semaglutide was intragastrically administered to canine models with and without ligation of the canine pyloric ring to prevent intestinal absorption. The concentration of semaglutide in plasma was quantified, and results revealed no significant difference in detected semaglutide in the plasma of the ligated and nonligated dogs. These results demonstrated that absorption of the semaglutide/SNAC tablet indeed occurs in the stomach [110]. Moreover, SNAC affects the membrane fluidity by partitioning across the lipid bilayer to enable transcellular absorption of semaglutide, as evidenced by the decrease in membrane melting temperature (Tm) and the intracellular increase of semaglutide accumulation upon SNAC exposure to monolayers of NCI-N87 human gastric epithelial cells. Contrary to previously reported mechanism [112], SNAC promotes monomerization of semaglutide by changing surrounding polarity and weakening hydrophobic interactions necessary for self-assembly instead of forming direct and noncovalent complexes with the peptide to aid in absorption. In addition, the inherent buffering capacity of SNAC (pK_a values 4.5 and 8.6) further enhances oral BA by (1) neutralizing the harsh acidic environment of the stomach, (2) preventing activation of pepsin, and (3) promoting solubility when in the deprotonated and charged state. However, there exists an intricate interplay between SNAC in charged and neutral states; the former results in higher effective concentration due to increased solubility, and the latter enables membrane penetration. There is also a spatial dependence between semaglutide and SNAC that requires the two components to be coformulated instead of coadministered. Coadministration of SNAC and lisinopril, a model nonpermeable small molecule, resulted in no absorption [111]. The absorption mechanism of SNAC seems to depend heavily on the nature of the peptide drug; liraglutide, a close analog of semaglutide, was not absorbed when coformulated with SNAC [110]. Therefore, detailed experiments and analysis may be necessary to determine the absorption mechanism and PK of individual peptide drugs with SNAC. Nonetheless, oral semaglutide (Rybelsus) is the first orally available GLP-1 analog approved for management of T2D, but requires dosing in the fasting state, as food may interfere with stomach absorption of the drug.

29.4.1.2 Microneedle approaches

One of the most ground-breaking technologies developed for oral delivery of peptides is a small ingestible capsule, named self-orienting millimeter-scale applicator (SOMA), which houses an injection system—capable of delivering therapeutic agents through the stomach wall in a time and orientation-dependent manner [113]. The self-orienting property of the device comes from its unique shape of a leopard tortoiseshell (highly curved top and flat bottom) and low center of mass that enables the applicator to quickly adjust itself to the right position while refraining from tipping over in the upright orientation. The upper hollow part of the device contains a biodegradable millipost that stores the drug, along with a stainless-steel spring held together by a dissolvable sugar coating that quickly disintegrates once in contact with stomach gastric acid, after which the drug is injected into the GI wall. After drug delivery, the capsule passes through the GI tract and safely exits the body. In swine models, oral delivery of insulin using SOMA showed glycemic control comparable to s.c. injection of insulin, without any significant adverse effects [114]. SOMA may also be used to deliver other biomacromolecules, such as protein and nucleic acid therapeutics.

Similar technologies have been developed for intestinal delivery of biomolecules including peptides, proteins, and antibodies [106,115]. The "robotic pill" from Rani Therapeutics is a capsule containing hollow and sugar-based microneedles loaded with drugs and compartments separating citric acid and sodium bicarbonate [106]. The capsule and its barriers are stable until it reaches the intestine at a pH of 6.5–7, in which it disintegrates, and a reaction is initiated between citric acid and sodium bicarbonate to produce CO_2. The chemical reaction and generated gas ultimately lead to injection and drug delivery across the intestinal wall. Over 50% BA with the robotic pill has been reported for insulin and adalimumab, an immunosuppressive antibody used to treat autoimmune diseases.

In conclusion, the inherent shortcomings of peptides with low oral BA can be salvaged by peptidomimetic modifications that alter the secondary structure of peptides, along with recent technologies such as PE and microinjection modalities, which are likely to advance peptide therapeutics to new frontiers in the future.

29.5 Peptide therapeutics targeting the central nervous system

Peptides have shown promise for the treatment of neurological diseases including brain cancer, anxiety, depression, neurodegeneration, and pain.

Unfortunately, peptides targeting the central nervous system (CNS) encounter even more daunting challenges. To reach targets in the CNS, peptides need to cross a highly selective BBB. Penetration of peptides is challenging because brain homeostasis is maintained by the BBB shielding foreign substances and toxins [116]. Strategies to achieve passage across the BBB include chemical modification of peptides to increase BA, decrease elimination rates, limit degradation, and/or enhance absorption. These chemical modifications include lipidation, halogenation, methylation, polymer conjugation, structural change, and glycosylation that have been previously reviewed [116,117]. Another approach is to take advantage of the working mechanisms and transporters of the BBB (i.e., carriers for vitamins, amino acids, and glucose) [116]. Some preliminary work with nanoparticles enabling passage is also encouraging [117,118].

ANG1005 is a promising peptide therapeutic that is in Phase II clinical trials for the treatment of brain tumors . The drug candidate is angiopep-2 conjugated to three chemotherapeutic paclitaxel groups [119]. Angiopep-2 is built from 19 amino acids and targets a receptor in the brain called lipoprotein receptor–related protein (LRP-1). The native function of angiopep-2 supports its function as a BBB shuttle that induces transcytosis of conjugated molecules across the BBB [120]. In situ mouse brain perfusion experiments showed that ANG1005 is able to cross the BBB 4–5 times more than a single paclitaxel molecule, resulting in up to 15-fold increased paclitaxel transported into the brain [119].

Oxytocin, a neuropeptide in Phase II clinical trials for autism treatment, improves cognitive function by enhancing social stimuli and emotional registration [121]. It is not fully understood whether or not it is necessary for oxytocin to cross the BBB, since there is conflicting evidence regarding bypass of the barrier and effects on the brain through an alternative mode like the vagus nerve [122–124]. Earlier studies have shown that peripheral administration of oxytocin in rats yielded only 0.002% of the dose in the CNS at 10 min after injection [123]. Targeting the CNS presents many challenges during drug development, of which drug route of administration is particularly notable. Some routes involve invasive administration, such as convection-enhanced delivery, intracerebroventricular (i.c.v.) injection, and intracerebral administration [117]. On the other hand, intranasal administration offers noninvasive drug delivery for neuropeptides [125]. Oxytocin is in clinical trials for nasal administration, which is more convenient than targeting the BBB with invasive administration routes. Although there is much unknown about the mechanism of action for oxytocin, it shows promise as a therapy for autism [126].

29.6 Future perspectives

Peptides represent an important reservoir of agents in the therapeutic arsenal available to treat and manage diseases. Selectivity and potency epitomize the promise of peptide drugs, while the liabilities include short half-lives and the need for injection as the most common route of administration. Screening technologies have provided a powerful avenue to quickly identify sequences that can be further modified to afford viable candidates for further drug development.

In the case of G protein–couple receptors, which are involved in the mechanism of action for several current blockbuster peptide drugs, binding of certain ligands can selectively engage particular downstream effectors over others, resulting in "biased" signaling favoring one channel of activity over another [127]. This has led to the idea that selective modulation of certain channels may result in therapeutic ligands with fewer side effects. This would lead to additional peptides that can function as drugs with more precision and likely greater patient compliance. This is an active area of research that will also require a fundamental understanding of receptor dynamics with regard to how downstream signaling pathways are dictated. In addition, this would also support "personalized medicine," as pathways that are tolerated or beneficial to one individual may not be in another.

Multiagonism is a powerful tool that enables a single peptide to activate multiple receptors to varying degrees, where the combined biological benefits of two to three peptide hormones are elicited in one molecule. The main goal is to maximize efficacy and limit the adverse effects by tuning up desirable effects and tuning down unwanted outcomes. Having to administer one compound over three different drugs is also more favorable in terms of safety and patient compliance. For example, a single hybrid peptide, containing a combination of GLP-1 and glucagon, can induce agonism at either receptor to affect the feeling of satiety (benefits of GLP-1) and increased energy expenditure (benefits of glucagon) and aid weight loss [128]. This dual agonist provides greater antidiabetic effects compared to GLP-1R agonists alone, demonstrating the promising potential of multiagonism in the management of metabolic diseases [129]. Similarly, a single hybrid molecule can be engineered to incorporate the sequences of GIP (insulinotropic effects), glucagon (greater energy expenditure), and GLP-1 (glycemic control), generating a multifunctional triagonist with superior pleiotropic effects through stimulating three different receptors individually [130]. Finally, the tenability of hybrid peptides for different receptors can be applied to the treatment of a variety of illnesses, ranging from metabolic disorders to cancers, highlighting the vast potential of this technology.

To expand the accessibility of peptide drugs to the brain, BBB penetration is an important hurdle to overcome. One way to achieve this is by utilizing nanoparticles—not only making peptides useful as therapeutics but also as diagnostics in monitoring disease states [131]. Nano-scale biomaterials are widely used and can carry several peptide units per carrier, which allows multifunctional properties such as simultaneous diagnosis and treatment of neurological diseases [132]. Peptidic drugs can be incorporated into such nanoparticles to enable stability and delivery to the CNS. To bypass obligate crossing of the BBB but still exert effects on the CNS, one may also turn to the gut microbiome in the future, as this has been linked to some neurological disorders such as anxiety and depression [133,134].

Peptides remain indispensable in the therapeutic toolkit with excellent selectivity and good safety profiles but have limited application due to short in vivo half-lives and poor membrane penetration. This review focused on peptidomimetic modalities that ameliorate metabolic instability and low barrier permeability, including the use of unnatural amino acids, side chain and backbone modifications, cyclization, as well as conjugation strategies with polymers and biologics. Leading technologies that boost oral BA (such as PEs and microneedle interventions) or BBB penetration can propel peptide therapeutics to the next level with broadened utility and accessibility in medical use. In combination with computational analysis and artificial intelligence, we have the potential to expand beyond benchtop experiments creating a new subset of lead peptide therapeutics that could target orphan diseases and/or advance the promise of personalized medicine.

Only a small fraction of the total chemical and structural space has been explored by the variety of modifications employed to date. The enormous diversity that can be achieved by juxtaposition of sequence diversity and further developing the number of available decorations to the backbone, side chains and termini remains the goal to tailor peptides to be suitable for any therapeutic challenge at hand.

References

[1] Banting FG, Best CH, Collip JB, Campbell WR, Fletcher AA. Pancreatic extracts in the treatment of diabetes mellitus. Can Med Assoc J 1922;12:141–6.

[2] Leader B, Baca QJ, Golan DE. Protein therapeutics: a summary and pharmacological classification. Nat Rev Drug Discovery 2008;7:21–39.

[3] Elkinton JR, Hunt AD, Godfrey L, McCrory WW, Rogerson AG, Stokes J. Effects of pituitary adrenocorticotropic hormone (ACTH) therapy. J Am Med Assoc 1949;141:1273–9.

[4] Goeddel DV, Kleid DG, Bolivar F, Heyneker HL, Yansura DG, Crea R, et al. Expression in *Escherichia coli* of chemically synthesized genes for human insulin. Proc Natl Acad Sci 1979;76:106.

[5] Clark AL, Knight G, Wiles P, Keen H, Ward J, Cauldwell J, et al. Biosynthetic human insulin in the treatment of diabetes: a double-blind crossover trial in established diabetic patients. Lancet 1982;320:354–7.

[6] Lau JL, Dunn MK. Therapeutic peptides: historical perspectives, current development trends, and future directions. Bioorg Med Chem 2018;26:2700–7.

[7] Fosgerau K, Hoffmann T. Peptide therapeutics: current status and future directions. Drug Discov Today 2015;20:122–8.

[8] Brinckerhoff LH, Kalashnikov VV, Thompson LW, Yamshchikov GV, Pierce RA, Galavotti HS, et al. Terminal modifications inhibit proteolytic degradation of an immunogenic mart-127–35 peptide: implications for peptide vaccines. Int J Cancer 1999;83:326–34.

[9] Green BD, Mooney MH, Gault VA, Irwin N, Bailey CJ, Harriott P, et al. N-terminal His (7)-modification of glucagon-like peptide-1(7−36) amide generates dipeptidyl peptidase IV-stable analogues with potent antihyperglycaemic activity. J Endocrinol 2004;180:379−88.

[10] Meng H, Krishnaji ST, Beinborn M, Kumar K. Influence of selective fluorination on the biological activity and proteolytic stability of glucagon-like peptide-1. J Medhem 2008;51:7303−7.

[11] Heard KR, Wu W, Li Y, Zhao P, Woznica I, Lai JH, et al. A general method for making peptide therapeutics resistant to serine protease degradation: application to dipeptidyl peptidase IV substrates. J Med Chem 2013;56:8339−51.

[12] Gruber CW, Koehbach J, Muttenthaler M. Exploring bioactive peptides from natural sources for oxytocin and vasopressin drug discovery. Future Med Chem 2012;4:1791−8.

[13] Vilhardt H, Lundin S, Falch J. Plasma kinetics of DDAVP in Man. Acta Pharmacologica et Toxicologica 1986;58:379−81.

[14] Lau J, Bloch P, Schäffer L, Pettersson I, Spetzler J, Kofoed J, et al. Discovery of the once-weekly glucagon-like peptide-1 (GLP-1) analogue semaglutide. J Med Chem 2015;58:7370−80.

[15] Rüegger A, Kuhn M, Lichti H, Loosli H-R, Huguenin R, Quiquerez C, et al. Ein immunsuppressiv wirksamer peptidmetabolit aus *Trichoderma polysporum* (LINK ex PERS.) Rifai. Helvetica Chim Acta 1976;59:1075−92.

[16] Boeglin D, Lubell WD. Aza-amino acid scanning of secondary structure suited for solid-phase peptide synthesis with fmoc chemistry and aza-amino acids with heteroatomic side chains. J Combinat Chem 2005;7:864−78.

[17] Kubo K, Tokashiki M, Kuwasako K, Tamura M, Tsuda S, Kubo S, et al. Biological properties of adrenomedullin conjugated with polyethylene glycol. Peptides 2014;57:118−21.

[18] Knudsen LB, Nielsen PF, Huusfeldt PO, Johansen NL, Madsen K, Pedersen FZ, et al. Potent derivatives of glucagon-like peptide-1 with pharmacokinetic properties suitable for once daily administration. J Med Chem 2000;43:1664−9.

[19] Knudsen LB, Lau J. The discovery and development of liraglutide and semaglutide. Front Endocrinol (Lausanne) 2019;10:155.

[20] Baggio LL, Huang Q, Cao X, Drucker DJ. An albumin-exendin-4 conjugate engages central and peripheral circuits regulating murine energy and glucose homeostasis. Gastroenterology 2008;134:1137−47.

[21] Kim I, Kim TH, Ma K, Lee ES, Kim D, Oh KT, et al. Synthesis and evaluation of human serum albumin-modified exendin-4 conjugate via heterobifunctional polyethylene glycol linkage with protracted hypoglycemic efficacy. Bioconjugate Chem 2010;21:1513−19.

[22] Glaesner W, Vick AM, Millican R, Ellis B, Tschang SH, Tian Y, et al. Engineering and characterization of the long-acting glucagon-like peptide-1 analogue LY2189265, an Fc fusion protein. Diabetes Metab Res Rev 2010;26:287−96.

[23] Kazmierski WM, Kenakin TP, Gudmundsson KS. Peptide, peptidomimetic and small-molecule drug discovery targeting HIV-1 host-cell attachment and entry through gp120, gp41, CCR5 and CXCR4†. Chem Biol Drug Des 2006;67:13−26.

[24] Ferdinandi ES, Brazeau P, High K, Procter B, Fennell S, Dubreuil P. Non-clinical pharmacology and safety evaluation of TH9507, a human growth hormone-releasing factor analogue. Basic Clin Pharmacol Toxicol 2007;100:49−58.

[25] Sawyer TK, Sanfilippo PJ, Hruby VJ, Engel MH, Heward CB, Burnett JB, et al. 4-Norleucine, 7-D-phenylalanine-alpha-melanocyte-stimulating hormone: a highly potent alpha-melanotropin with ultralong biological activity. Proc Natl Acad Sci 1980;77:5754.

[26] Sawyer TK, Hruby VJ, Wilkes BC;, Draelos MT, Hadley ME, Bergsneider M. Comparative biological activities of highly potent active-site analogs of alpha.-melanotropin. J Med Chem 1982;25:1022–7.

[27] Ugwu SO, Blanchard J, Dorr RT, Levine N, Brooks C, Hadley ME, et al. Skin Pigmentation and pharmacokinetics of melanotan-I in humans. Biopharm Drug Dispos 1997;18:259–69.

[28] Hunter DJS, Schulz P, Wassenaar W. Effect of carbetocin, a long-acting oxytocin analog on the postpartum uterus. Clin Pharmacol Therap 1992;52:60–7.

[29] Bold G, Fässler A, Capraro H-G, Cozens R, Klimkait T, Lazdins J, et al. New aza-dipeptide analogues as potent and orally absorbed HIV-1 protease inhibitors: candidates for clinical development. J Med Chem 1998;41:3387–401.

[30] Raja A, Lebbos J, Kirkpatrick P. Atazanavir sulphate. Nat Rev Drug Discov 2003;2:857–8.

[31] Bolla M, Collette L, Blank L, Warde P, Dubois JB, Mirimanoff R-O, et al. Long-term results with immediate androgen suppression and external irradiation in patients with locally advanced prostate cancer (an EORTC study): a phase III randomised trial. Lancet 2002;360:103–8.

[32] Busby RW, Kessler MM, Bartolini WP, Bryant AP, Hannig G, Higgins CS, et al. Pharmacologic properties, metabolism, and disposition of linaclotide, a novel therapeutic peptide approved for the treatment of irritable bowel syndrome with constipation and chronic idiopathic constipation. J Pharmacol Exp Therap 2013;344:196.

[33] Busby RW, Bryant AP, Bartolini WP, Cordero EA, Hannig G, Kessler MM, et al. Linaclotide, through activation of guanylate cyclase C, acts locally in the gastrointestinal tract to elicit enhanced intestinal secretion and transit. Eur J Pharmacol 2010;649:328–35.

[34] Di L. Strategic approaches to optimizing peptide ADME properties. AAPS J 2015;17:134–43.

[35] Toth I, Flinn N, Hillery A, Gibbons WA, Artursson P. Lipidic conjugates of luteinizing hormone releasing hormone (LHRH) + and thyrotropin releasing hormone (TRH) + that release and protect the native hormones in homogenates of human intestinal epithelial (Caco-2) cells. Int J Pharmaceut 1994;105:241–7.

[36] Varamini P, Mansfeld FM, Blanchfield JT, Wyse BD, Smith MT, Toth I. Synthesis and biological evaluation of an orally active glycosylated endomorphin-1. J Med Chem 2012;55:5859–67.

[37] Zhang G, Han B, Lin X, Wu X, Yan H. Modification of antimicrobial peptide with low molar mass poly(ethylene glycol). J Biochem 2008;144:781–8.

[38] Werner HM, Cabalteja CC, Horne WS. Peptide backbone composition and protease susceptibility: impact of modification type, position, and tandem substitution. ChemBioChem 2016;17:712–18.

[39] Kim ES, Garnock-Jones KP. Afamelanotide: a review in erythropoietic protoporphyria. Am J Clin Dermatol 2016;17:179–85.

[40] Darlak K, Benovitz DE, Spatola AF, Grzonka Z. Dermorphin analogs: resistance to in vitro enzymatic degradation is not always increased by additional D-amino acid substitutions. Biochem Biophys Res Commun 1988;156:125–30.

[41] Thornber CW. Isosterism and molecular modification in drug design. Chem Soc Rev 1979;8:563–80.

[42] Wiśniewski K, Galyean R, Tariga H, Alagarsamy S, Croston G, Heitzmann J, et al. New, potent, selective, and short-acting peptidic V1a receptor agonists. J Med Chem 2011;54:4388–98.

[43] Frey V, Viaud J, Subra G, Cauquil N, Guichou J-F, Casara P, et al. Structure–activity relationships of Bak derived peptides: affinity and specificity modulations by amino acid replacement. Eur J Med Chem 2008;43:966–72.

[44] Haskell-Luevano C, Toth K, Boteju L, Job C, Castrucci AMd L, Hadley ME;, et al. β-Methylation of the Phe7 and Trp9 melanotropin side chain pharmacophores affects ligand – receptor interactions and prolonged biological activity. J Med Chem 1997;40:2740–9.

[45] Bork K, Yasothan U, Kirkpatrick P. Icatibant. Nat Rev Drug Discov 2008;7:801–2.

[46] Grauer A, König B. Peptidomimetics—a versatile route to biologically active compounds. Eur J Org Chem 2009;2009:5099–111.

[47] Shepherd NE, Hoang HN, Abbenante G, Fairlie DP. Single turn peptide alpha helices with exceptional stability in water. J Am Chem Soc 2005;127:2974–83.

[48] Giannis A, Kolter T. Peptidomimetics for receptor ligands—discovery, development, and medical perspectives. Angew Chem Int Ed Engl 1993;32:1244–67.

[49] Cody WL, He JX, Reily MD, Haleen SJ, Walker DM, Reyner EL, et al. Design of a potent combined pseudopeptide endothelin-A/endothelin-B receptor antagonist, Ac-dBhg16-Leu-Asp-Ile-[NMe]Ile-Trp21 (PD 156252): examination of its pharmacokinetic and spectral properties. J Med Chem 1997;40:2228–40.

[50] Linde Y, Ovadia O, Safrai E, Xiang Z, Portillo FP, Shalev DE, et al. Structure-activity relationship and metabolic stability studies of backbone cyclization and N-methylation of melanocortin peptides. Peptide Sci 2008;90:671–82.

[51] Kessler H. Detection of hindered rotation and inversion by NMR spectroscopy. Angew Chem Int Ed Engl 1970;9:219–35.

[52] Vitoux B, Aubry A, Cung MT, Boussard GUY, Marraud M. N-methyl peptides: III. solution conformational study and crystal structure of N-pivaloyl-L-prolyl-N-methyl-N'-isopropyl-L-alaninamide. Int J Peptide Protein Res 1981;17:469–79.

[53] Haviv F, Fitzpatrick TD, Swenson RE, Nichols CJ, Mort NA, Bush EN, et al. Effect of N-methyl substitution of the peptide bonds in luteinizing hormone-releasing hormone agonists. J Med Chem 1993;36:363–9.

[54] White TR, Renzelman CM, Rand AC, Rezai T, McEwen CM, Gelev VM, et al. On-resin N-methylation of cyclic peptides for discovery of orally bioavailable scaffolds. Nat Chem Biol 2011;7:810–17.

[55] Veber DF, Freidinger RM, Perlow DS, Paleveda WJ, Holly FW, Strachan RG, et al. A potent cyclic hexapeptide analogue of somatostatin. Nature 1981;292:55–8.

[56] Tanaka M. Design and synthesis of chiral alpha,alpha-disubstituted amino acids and conformational study of their oligopeptides. Chem Pharm Bull 2007;55:349–58.

[57] Schweitzer-Stenner R, Gonzales W, Bourne GT, Feng JA, Marshall GR. Conformational manifold of α-aminoisobutyric acid (Aib) containing alanine-based tripeptides in aqueous solution explored by vibrational spectroscopy, electronic circular dichroism spectroscopy, and molecular dynamics simulations. J Am Chem Soc 2007;129:13095–109.

[58] Laver DR. The barrel-stave model as applied to alamethicin and its analogs reevaluated. Biophys J 1994;66:355–9.

[59] Bechinger B. Structure and functions of channel-forming peptides: magainins, cecropins, melittin and alamethicin. J Membr Biol 1997;156:197–211.

[60] Baggio LL, Huang Q, Brown TJ, Drucker DJ. A recombinant human glucagon-like peptide (GLP)-1-albumin protein (albugon) mimics peptidergic activation of GLP-1 receptor-dependent pathways coupled with satiety, gastrointestinal motility, and glucose homeostasis. Diabetes 2004;53:2492–500.

[61] Meier JJ, Nauck MA, Kranz D, Holst JJ, Deacon CF, Gaeckler D, et al. Secretion, degradation, and elimination of glucagon-like peptide 1 and gastric inhibitory polypeptide in patients with chronic renal insufficiency and healthy control subjects. Diabetes 2004;53:654.

[62] Manandhar B, Ahn J-M. Glucagon-like peptide-1 (GLP-1) analogs: recent advances, new possibilities, and therapeutic implications. J Med Chem 2015;58:1020–37.

[63] Graaf Cd, Donnelly D, Wootten D, Lau J, Sexton PM, Miller LJ, et al. Glucagon-like peptide-1 and its class b g protein-coupled receptors: a long march to therapeutic successes. Pharmacol Rev 2016;68:954–1013.

[64] Abbas C, Pickaert G, Didierjean C, Grégoire BJ, Vanderesse R. Original and efficient synthesis of 2:1-[α/aza]-oligomer precursors. Tetrahedron Lett 2009;50:4158–60.

[65] Lee H-J, Song J-W, Choi Y-S, Park H-M, Lee K-B, Theoretical A. Study of conformational properties of N-methyl azapeptide derivatives. J Am Chem Soc 2002;124:11881–93.

[66] Epinette C, Croix C, Jaquillard L, Marchand-Adam S, Kellenberger C, Lalmanach G, et al. A selective reversible azapeptide inhibitor of human neutrophil proteinase 3 derived from a high affinity FRET substrate. Biochem Pharmacol 2012;83:788–96.

[67] Huang Y, Malcolm BA, Vederas JC. Synthesis and testing of azaglutamine derivatives as inhibitors of hepatitis A Virus (HAV) 3C proteinase. Bioorg Med Chem 1999;7:607–19.

[68] Abdalla AM, McGaw JL. Natural cyclic peptides as an attractive modality for therapeutics: a mini review. Molecules 2018;23.

[69] Liu J, Volk KJ, Lee MS, Pucci M, Handwerger S. Binding studies of vancomycin to the cytoplasmic peptidoglycan precursors by affinity capillary electrophoresis. Anal Chem 1994;66:2412–16.

[70] Rezai T, Bock JE, Zhou MV, Kalyanaraman C, Lokey RS, Jacobson MP. Conformational flexibility, internal hydrogen bonding, and passive membrane permeability: successful in silico prediction of the relative permeabilities of cyclic peptides. J Am Chem Soc 2006;128:14073–80.

[71] Mosberg HI, Hurst R, Hruby VJ, Gee K, Yamamura HI, Galligan JJ, et al. Bis-penicillamine enkephalins possess highly improved specificity toward delta opioid receptors. Proc Natl Acad Sci 1983;80:5871.

[72] Kramer TH, Toth G, Haaseth RC, Matsunaga TO, Davis P, Hruby VJ, et al. Influence of peptidase inhibitors on the apparent agonist potency of delta selective opioid peptides in vitro. Life Sci 1991;48:881–6.

[73] Haaseth RC, Sobczyk-Kojiro K, Medzihradsky F, Smith CB, Mosberg HI. Single residue modifications of the delta opioid receptor selective peptide, [d-Pen2,d-Pen5]-enkephalin (DPDPE). Int J Peptide Protein Res 1990;36:139–46.

[74] Scarborough RM. Development of eptifibatide. Am Heart J 1999;138:1093–104.

[75] Scarborough RM, Naughton MA, Teng W, Rose JW, Phillips DR, Nannizzi L, et al. Design of potent and specific integrin antagonists. Peptide antagonists with high specificity for glycoprotein IIb-IIIa. J Biol Chem 1993;268:1066–73.

[76] Cromm PM, Spiegel J, Grossmann TN. Hydrocarbon stapled peptides as modulators of biological function. ACS Chem Biol 2015;10:1362–75.

[77] Walensky LD, Kung AL, Escher I, Malia TJ, Barbuto S, Wright RD, et al. Activation of apoptosis in vivo by a hydrocarbon-stapled BH3 helix. Science 2004;305:1466.

[78] Meibohm B, Zhou H. Characterizing the impact of renal impairment on the clinical pharmacology of biologics. J Clin Pharmacol 2012;52:54S–62S.

[79] Kriz W, Elger M. CHAPTER 1 - Renal anatomy. In: Floege J;, Johnson RJ;, Feehally J, editors. Comprehensive clinical nephrology. Fourth Edition Philadelphia: Mosby; 2010. p. 3−14.
[80] Swierczewska M, Lee KC, Lee S. What is the future of PEGylated therapies? Expert Opin Emerg Drugs 2015;20:531−6.
[81] Harris JM, Martin NE, Modi M. Pegylation. Clin Pharmacokinet 2001;40:539−51.
[82] Schönauer R, Els-Heindl S, Beck-Sickinger AG. Adrenomedullin—new perspectives of a potent peptide hormone. J Peptide Sci 2017;23:472−85.
[83] Meeran K, O'Shea D, Upton PD, Small CJ, Ghatei MA, Byfield PH, et al. Circulating adrenomedullin does not regulate systemic blood pressure but increases plasma prolactin after intravenous infusion in humans: a pharmacokinetic study. J Clin Endocrinol Metab 1997;82:95−100.
[84] Zhang P, Sun F, Liu S, Jiang S. Anti-PEG antibodies in the clinic: current issues and beyond PEGylation. J Control Rel 2016;244:184−93.
[85] Bennett CL, Jacob S, Hymes J, Usvyat LA, Maddux FW. Anaphylaxis and hypotension after administration of peginesatide. N Engl J Med 2014;370:2055−6.
[86] Wang Y, Lomakin A, Kanai S, Alex R, Benedek GB. Transformation of oligomers of lipidated peptide induced by change in pH. Mol Pharmaceut 2015;12:411−19.
[87] Zorzi A, Linciano S, Angelini A. Non-covalent albumin-binding ligands for extending the circulating half-life of small biotherapeutics. Medchemcomm 2019;10:1068−81.
[88] Bhattacharya AA, Grune T, Curry S. Crystallographic analysis reveals common modes of binding of medium and long-chain fatty acids to human serum albumin. J Mol Biol 2000;303:721−32.
[89] Havelund S, Plum A, Ribel U, Jonassen I, Vølund A, Markussen J, et al. The mechanism of protraction of insulin detemir, a long-acting, acylated analog of human insulin. Pharm Res 2004;21:1498−504.
[90] Whittingham JL, Havelund S, Jonassen I. Crystal structure of a prolonged-acting insulin with albumin-binding properties. Biochemistry 1997;36:2826−31.
[91] Williams RF, Gleason RE, Soeldner JS. The half-life of endogenous serum immunoreactive insulin in man. Metabolism 1968;17:1025−9.
[92] Kurtzhals P. Engineering predictability and protraction in a basal insulin analogue: the pharmacology of insulin detemir. Int J Obes Relat Metab Disord 2004;28(Suppl 2): S23−8.
[93] Brunner GA, Sendhofer G, Wutte A, Ellmerer M, Sogaard B, Siebenhofer A, et al. Pharmacokinetic and pharmacodynamic properties of long-acting insulin analogue NN304 in comparison to NPH insulin in humans. Exp Clin Endocrinol Diabetes 2000;108:100−5.
[94] Elbrond B, Jakobsen G, Larsen S, Agerso H, Jensen LB, Rolan P, et al. Pharmacokinetics, pharmacodynamics, safety, and tolerability of a single-dose of NN2211, a long-acting glucagon-like peptide 1 derivative, in healthy male subjects. Diabetes Care 2002;25:1398−404.
[95] Lau J, Bloch P, Schaffer L, Pettersson I, Spetzler J, Kofoed J, et al. Discovery of the once-weekly glucagon-like peptide-1 (GLP-1) analogue semaglutide. J Med Chem 2015;58:7370−80.
[96] Marbury TC, Flint A, Jacobsen JB, Derving Karsbol J, Lasseter K. Pharmacokinetics and tolerability of a single dose of semaglutide, a human glucagon-like peptide-1 analog, in subjects with and without renal impairment. Clin Pharmacokinet 2017;56:1381−90.
[97] Rosenstock J, Reusch J, Bush M, Yang F, Stewart M. Potential of albiglutide, a long-acting GLP-1 receptor agonist, in type 2 diabetes: a randomized controlled trial exploring weekly, biweekly, and monthly dosing. Diabetes Care 2009;32:1880−6.

[98] Cavaco M, Castanho M, Neves V. Peptibodies: an elegant solution for a long-standing problem. Biopolymers 2017;.

[99] Shimamoto G, Gegg C, Boone T, Queva C. Peptibodies: a flexible alternative format to antibodies. MAbs 2012;4:586–91.

[100] Wang B, Nichol JL, Sullivan JT. Pharmacodynamics and pharmacokinetics of AMG 531, a novel thrombopoietin receptor ligand. Clin Pharmacol Ther 2004;76:628–38.

[101] Stasi R, Evangelista ML, Stipa E, Buccisano F, Venditti A, Amadori S. Idiopathic thrombocytopenic purpura: current concepts in pathophysiology and management. Thromb Haemost 2008;99:4–13.

[102] Perreault S, Burzynski J. Romiplostim: a novel thrombopoiesis-stimulating agent. Am J Health Syst Pharm 2009;66:817–24.

[103] Jimenez-Solem E, Rasmussen MH, Christensen M, Knop FK. Dulaglutide, a long-acting GLP-1 analog fused with an Fc antibody fragment for the potential treatment of type 2 diabetes. Curr Opin Mol Ther 2010;12:790–7.

[104] Usmani SS, Bedi G, Samuel JS, Singh S, Kalra S, Kumar P, et al. THPdb: Database of FDA-approved peptide and protein therapeutics. PLoS One 2017;12:e0181748.

[105] Karsdal MA, Riis BJ, Mehta N, Stern W, Arbit E, Christiansen C, et al. Lessons learned from the clinical development of oral peptides. Br J Clin Pharmacol 2015;79:720–32.

[106] Aguirre TAS, Teijeiro-Osorio D, Rosa M, Coulter IS, Alonso MJ, Brayden DJ. Current status of selected oral peptide technologies in advanced preclinical development and in clinical trials. Adv Drug Deliv Rev 2016;106:223–41.

[107] Leone-Bay A, Santiago N, Achan D, Chaudhary K, DeMorin F, Falzarano L, et al. N-Acylated alpha.-amino acids as novel oral delivery agents for proteins. J Med Chem 1995;38:4263–9.

[108] Davies M, Pieber TR, Hartoft-Nielsen M-L, Hansen OKH, Jabbour S, Rosenstock J. Effect of oral semaglutide compared with placebo and subcutaneous semaglutide on glycemic control in patients with type 2 diabetes: a randomized clinical trial. JAMA 2017;318:1460–70.

[109] Lennernäs H. Regional intestinal drug permeation: biopharmaceutics and drug development. Eur J Pharm Sci 2014;57:333–41.

[110] Buckley ST, Bækdal TA, Vegge A, Maarbjerg SJ, Pyke C, Ahnfelt-Rønne J, et al. Transcellular stomach absorption of a derivatized glucagon-like peptide-1 receptor agonist. Sci Transl Med 2018;10:eaar7047.

[111] McIntyre C, Schmidt J, Castelli MC, Bittner B. Study on the impact of SNAC (sodium N-[8-(2-hydroxybenzoyl) amino] caprylate) on the bioavailability of ibandronate (IBN) in postmenopausal women. J Drug Deliv Sci Technol 2011;21:521–5.

[112] Leone-Bay A, Paton DR, Variano B, Leipold H, Rivera T, Miura-Fraboni J, et al. Acylated non-α-amino acids as novel agents for the oral delivery of heparin sodium, USP. J Controlled Rel 1998;50:41–9.

[113] Abramson A, Caffarel-Salvador E, Khang M, Dellal D, Silverstein D, Gao Y, et al. An ingestible self-orienting system for oral delivery of macromolecules. Science 2019;363:611.

[114] Abramson A, Caffarel-Salvador E, Soares V, Minahan D, Tian RY, Lu X, et al. A luminal unfolding microneedle injector for oral delivery of macromolecules. Nat Med 2019;25:1512–18.

[115] Traverso G, Schoellhammer CM, Schroeder A, Maa R, Lauwers GY, Polat BE, et al. Microneedles for drug delivery via the gastrointestinal tract. J Pharm Sci 2015;104:362–7.

[116] Witt KA, Gillespie TJ, Huber JD, Egleton RD, Davis TP. Peptide drug modifications to enhance bioavailability and blood-brain barrier permeability. Peptides 2001;22:2329–43.

[117] Furtado D, Bjornmalm M, Ayton S, Bush AI, Kempe K, Caruso F. Overcoming the blood-brain barrier: the role of nanomaterials in treating neurological diseases. Adv Mater 2018;30:e1801362.

[118] Lalatsa A, Schatzlein AG, Uchegbu IF. Strategies to deliver peptide drugs to the brain. Mol Pharm 2014;11:1081–93.

[119] Régina A, Demeule M, Ché C, Lavallée I, Poirier J, Gabathuler R, et al. Antitumour activity of ANG1005, a conjugate between paclitaxel and the new brain delivery vector Angiopep-2. Br J Pharmacol 2008;155:185–97.

[120] Thomas FC, Taskar K, Rudraraju V, Goda S, Thorsheim HR, Gaasch JA, et al. Uptake of ANG1005, a novel paclitaxel derivative, through the blood-brain barrier into brain and experimental brain metastases of breast cancer. Pharm Res 2009;26:2486–94.

[121] Guastella AJ, Einfeld SL, Gray KM, Rinehart NJ, Tonge BJ, Lambert TJ, et al. Intranasal oxytocin improves emotion recognition for youth with autism spectrum disorders. Biol Psychiatry 2010;67:692–4.

[122] Reichel CM. Unraveling oxytocin's peripheral vs. central mechanisms. Neuropsychopharmacology 2020;.

[123] Mens WB, Witter A, van Wimersma Greidanus TB. Penetration of neurohypophyseal hormones from plasma into cerebrospinal fluid (CSF): half-times of disappearance of these neuropeptides from CSF. Brain Res 1983;262:143–9.

[124] Leng G, Ludwig M. Intranasal oxytocin: myths and delusions. Biol Psychiatry 2016;79:243–50.

[125] Born J, Lange T, Kern W, McGregor GP, Bickel U, Fehm HL. Sniffing neuropeptides: a transnasal approach to the human brain. Nat Neurosci 2002;5:514–16.

[126] Preti A, Melis M, Siddi S, Vellante M, Doneddu G, Fadda R. Oxytocin and autism: a systematic review of randomized controlled trials. J Child Adolesc Psychopharmacol 2014;24:54–68.

[127] Gurevich VV, Gurevich EV. Biased GPCR signaling: possible mechanisms and inherent limitations. Pharmacol Therapeut 2020;211:107540.

[128] Day JW, Ottaway N, Patterson JT, Gelfanov V, Smiley D, Gidda J, et al. A new glucagon and GLP-1 co-agonist eliminates obesity in rodents. Nat Chem Biol 2009;5:749–57.

[129] Finan B, Ma T, Ottaway N, Müller TD, Habegger KM, Heppner KM, et al. Unimolecular dual incretins maximize metabolic benefits in rodents, monkeys, and humans. Sci Transl Med 2013;5:209ra151.

[130] Finan B, Yang B, Ottaway N, Smiley DL, Ma T, Clemmensen C, et al. A rationally designed monomeric peptide triagonist corrects obesity and diabetes in rodents. Nat Med 2015;21:27–36.

[131] Sharma M, Dube T, Chibh S, Kour A, Mishra J, Panda JJ. Nanotheranostics, a future remedy of neurological disorders. Expert Opin Drug Deliv 2019;16:113–28.

[132] Li M, Guan Y, Zhao A, Ren J, Qu X. Using multifunctional peptide conjugated Au nanorods for monitoring β-amyloid aggregation and chemo-photothermal treatment of Alzheimer's disease. Theranostics 2017;7:2996–3006.

[133] Zheng X, Zhang X, Kang A, Ran C, Wang G, Hao H. Thinking outside the brain for cognitive improvement: is peripheral immunomodulation on the way? Neuropharmacology 2015;96:94–104.

[134] Lach G, Schellekens H, Dinan TG, Cryan JF. Anxiety, depression, and the microbiome: a role for gut peptides. Neurotherapeutics 2018;15:36–59.

Index

Note: Page numbers followed by "*f*" and "*t*" refer to figures and tables, respectively.

C

D

E

H

I

J

K

L

N

O

Q

R

T

X

Y

Z

Kolibri 360 GmbH
Ferdinand-Jühlke-Straße 7
99095 Erfurt, Deutschland
produktsicherheit@kolibri360.de